T0296644

# Urban Climates

*Urban Climates* is the first full synthesis of modern scientific and applied research on urban climates. The book begins with an outline of what constitutes an urban ecosystem. It develops a comprehensive terminology for the subject using scale and surface classification as key constructs. It explains the physical principles governing the creation of distinct urban climates, such as airflow around buildings, the heat island, precipitation modification, and air pollution, and then illustrates how this knowledge can be applied to moderate the undesirable consequences of urban development and help create more sustainable and resilient cities. With urban climate science now a full-fledged field, this timely book fulfills the need to bring together the disparate parts of climate research on cities into a coherent framework. It is an ideal resource for students and researchers in fields such as climatology, urban hydrology, air quality, environmental engineering, and urban design.

**Timothy R. Oke** is Professor Emeritus and former Head of Geography at the University of British Columbia. His research interests focus on urban climate and the energy and water balances of cities. Dr. Oke has published more than 200 scientific publications, and is author of the widely used text *Boundary Layer Climates* (1987). He founded and was the first president of the International Association for Urban Climate. He has won several medals, including the Meteorological Service of Canada's Patterson Medal in 2002, for distinguished service to meteorology, and the Royal Canadian Geographical Society's Massey Medal in 2005, for substantial contributions to understanding the microclimates of cities.

**Gerald Mills** is Associate Professor in the School of Geography at University College Dublin, where he is also Head of School. He teaches in the areas of climatology, sustainable urbanism and spatial analysis. His research interests lie mainly in the climates of cities, especially at the human scale, and the potential for managing these through climate-sensitive urban design. He is a former president of the Geographical Society of Ireland and a former president of the International Association for Urban Climate.

**Andreas Christen** is Associate Professor in the Department of Geography and in the Atmospheric Science Program at the University of British Columbia, where he teaches fundamentals in weather, climate, climate change and urban system sciences. His research focuses on interactions between complex land surfaces – including cities – and the atmosphere. He develops new methods to quantify, attribute, and model the exchange of energy, water and trace-gases between complex land surfaces and the atmosphere, in order to benefit applications such as climate-sensitive urban design, air pollution and emission management, urban weather forecasting and climate modeling.

**James A. Voogt** is Associate Professor in the Department of Geography at Western University in London, Ontario. His broad area of research interest is urban climatology, with a specialization in the measurement and modeling of urban surface temperatures. He was the co-principal investigator of the Environmental Prediction for Canadian Cities Network, a multi-institutional research network intended to develop and evaluate an urban-atmosphere model for the Canadian weather forecasting system. Dr. Voogt is currently President of the International Association for Urban Climate and former chair of the American Meteorological Society's Board of the Urban Environment.

"This is a very important book for anyone interested in understanding the climates of cities – their characteristics, controls, causes and implications. Comprehensive, clearly written, richly illustrated, and with broad-ranging examples and well-documented sources of data, this is a book that should be read by researchers, students and practitioners interested in the urban environment, urban planning and design, and smart cities. Its presentation makes it accessible, valuable and insightful, to those new to the field as well as established experts. This book will be a classic reference that will stand up to being read many times."

– Sue Grimmond,
*Department of Meteorology, University of Reading*

"As an architect and urban planner, I find this scientific text by Oke *et al.* to be comprehensive, insightful and useful for my next project. It makes my job designing healthy and resilient cities much easier. It is a must for those who care about the future of our cities."

– Edward Ng,
*School of Architecture, The Chinese University of Hong Kong*

"Because the proportion of the world's population living in urban areas is expected to approach 65–70% by 2050, it is urgent and timely to gain a comprehensive understanding of "...physical mechanisms underlying the workings of urban atmospheres..." (p.xx). The four authors, all world-renowned physical geographers and urban climatologists, have created the ultimate book on urban climate for gaining this understanding, meant primarily for upper undergraduate and beginning graduate levels and for those in other related fields, such as urban hydrology, air quality, environmental engineering, and urban design. The authors suggest that the book represents "...the first full synthesis of modern scientific and applied research on urban climates..." (p.i). Of this there is no doubt, and thus it is a book of monumental importance for researchers, educators, and students of urban climate science and urban environments."

– Anthony J. Brazel,
*School of Geographical Sciences and Urban Planning, Arizona State University*

"Monumental summary of urban climate, which could not be written by any other team of researchers than this, headed by the international face of the field for the last 50 years, Tim Oke. From its 19th century beginnings, to latest research results, and then to its critical applications in weather forecasting, air quality, health, climate change, and planning, it will be indispensable for anyone interested in the subject, from students to researchers, and most importantly, planners."

– Robert Bornstein,
*Department of Meteorology and Climate Science, San Jose State University*

"This book is without peer in its field. Written by four of the most eminent scientists in urban climate, this excellent book is destined to become a classic and a fundamental reference for students, teachers and researchers alike."

– Nigel Tapper,
*School of Earth, Atmosphere and Environment, Monash University*

"A signature textbook has been lacking for some time now within the field of Urban Climate. We now have it. The content, quality and scope of *Urban Climates* are just what you would expect from some of the most respected urban climatologists in the world. I look forward to using it for my Applied Climatology in the Urban Environment class at the University of Georgia."

– Marshall Shepherd,
*University of Georgia and former President of the American Meteorological Society*

"*Urban Climates* is a must-read for students and scientists. From climatology to urban planning, it is very clear and complete, from concepts and processes to practical implementation and adaptation of cities to climate. It is illustrated with explicative diagrams of exceptional quality and many examples. Beyond the clear and rigorous overview of the physics of the urban atmosphere, *Urban Climates* offers a fantastic travel through the history of climate in cities around the world, from preindustrial cities and before, to modern high-rise megacities."

– Valéry Masson,
*Météo-France and Centre National de la Recherche Scientifique*

# Urban Climates

**Timothy R. Oke**
University of British Columbia

**Gerald Mills**
University College Dublin

**Andreas Christen**
University of British Columbia

**James A. Voogt**
University of Western Ontario

CAMBRIDGE
UNIVERSITY PRESS

# CAMBRIDGE
## UNIVERSITY PRESS

University Printing House, Cambridge CB2 8BS, United Kingdom

One Liberty Plaza, 20th Floor, New York, NY 10006, USA

477 Williamstown Road, Port Melbourne, VIC 3207, Australia

4843/24, 2nd Floor, Ansari Road, Daryaganj, Delhi–110002, India

79 Anson Road, #06–04/06, Singapore 079906

Cambridge University Press is part of the University of Cambridge.

It furthers the University's mission by disseminating knowledge in the pursuit of education, learning, and research at the highest international levels of excellence.

www.cambridge.org
Information on this title: www.cambridge.org/9780521849500
DOI: 10.1017/9781139016476

First published 2017

*A catalogue record for this publication is available from the British Library.*

*Library of Congress Cataloging-in-Publication Data*
Names: Oke, T. R.
Title: Urban climates / Timothy R. Oke, University of British Columbia [and three others].
Description: Cambridge : Cambridge University Press, 2017. | Includes bibliographical references and index.
Identifiers: LCCN 2016050848| ISBN 9780521849500 (hardback : alk. paper) |
ISBN 9781107429536 (pbk. : alk. paper)
Subjects: LCSH: Urban climatology. | Urban ecology (Biology) | Architecture and climate. | Urban hydrology. |
Urban heat island. | City planning.
Classification: LCC QC981.7.U7 U7327 2017 | DDC 551.6/6091732–dc23 LC record available at
https://lccn.loc.gov/2016050848

ISBN 978-0-521-84950-0 Hardback
ISBN 978-1-107-42953-6 Paperback

# Contents

| | | | |
|---|---|---|---|
| Symbols and Units | | *page* xii |
| | Symbols | xii |
| | Units | xvii |
| | Scientific Notation | xvii |
| Preface | | xix |
| Acknowledgements | | xxi |

**1 Introduction**     1

| 1.1 | Urban Ecology | 2 |
|---|---|---|
| | 1.1.1 Urban Ecosystems | 2 |
| | 1.1.2 Urban Metabolism | 3 |
| | 1.1.3 Urban vs Rural | 4 |
| 1.2 | Environmental Impacts of Urban Development | 5 |
| | 1.2.1 Urban Pedosphere and Lithosphere | 5 |
| | 1.2.2 Urban Hydrosphere | 5 |
| | 1.2.3 Urban Biosphere | 5 |
| | 1.2.4 Urban Atmosphere | 6 |
| 1.3 | Urbanization and Urban Form | 6 |
| | 1.3.1 First Phase of Urbanization | 7 |
| | 1.3.2 Second Phase of Urbanization | 9 |
| | 1.3.3 Future Urbanization Pathways | 10 |
| 1.4 | Planning for More Sustainable Cities | 10 |
| | 1.4.1 Renewable and Nonrenewable Resources | 10 |
| | 1.4.2 The Ecological Footprint of Cities | 10 |
| | 1.4.3 Planning for More Sustainable and Resilient Cities | 12 |
| Summary | | 13 |

**2 Concepts**     14

| 2.1 | The Urban 'Surface' | 15 |
|---|---|---|
| | 2.1.1 Defining the Land-Atmosphere Interface in Cities | 15 |
| | 2.1.2 The Hierarchy of Urban Units | 18 |
| | 2.1.3 Description of Urban Surface Properties | 20 |
| | 2.1.4 Classification of the Urban Surface | 25 |
| 2.2 | The Urban Atmosphere | 29 |
| | 2.2.1 Scales of Urban Climate Phenomena | 29 |
| | 2.2.2 The Vertical Structure of the Urban Atmosphere | 30 |
| | 2.2.3 Linking the Vertical Structure to Horizontal Scales | 34 |
| 2.3 | Defining an Urban Climate | 35 |
| | 2.3.1 The Superposition of Urban and Non-Urban Influences on Climate | 35 |
| | 2.3.2 Dealing with the Complexity of the Urban System | 36 |
| | 2.3.3 Isolating Urban Effects | 38 |
| | 2.3.4 Identifying Causes of Urban Effects | 40 |
| | 2.3.5 Transferability of Results and Answers | 42 |
| Summary | | 42 |

**3   Methods**                                                                   44
  3.1   Field Observations                                                45
    3.1.1   Instruments and Their Exposure                          46
    3.1.2   Measurements at Fixed Stations                          52
    3.1.3   Mobile Measurements                                     55
    3.1.4   Flow-Following Techniques                               58
    3.1.5   Remote-Sensing Techniques                               59
  3.2   Physical Modelling                                                60
    3.2.1   Scaling and Similitude                                  60
    3.2.2   Laboratory Models                                       61
    3.2.3   Outdoor Scale Models                                    63
  3.3   Numerical Modelling                                               66
    3.3.1   Governing Equations                                     66
    3.3.2   Numerical Experimental Design                           69
    3.3.3   Micro- and Local-Scale Urban Climate Models             71
    3.3.4   Mesoscale Urban Models                                  72
  3.4   Empirical Models                                                  74
  Summary                                                                 75

**4   Airflow**                                                                   77
  4.1   Basics of Wind and Turbulence                                     78
    4.1.1   Mean and Turbulent Parts                                78
    4.1.2   Production of Turbulence                                80
    4.1.3   Dissipation of Turbulence                               81
  4.2   Flow in the Roughness Sublayer                                    82
    4.2.1   Isolated Buildings                                      82
    4.2.2   Uniform Building Arrays                                 86
    4.2.3   Streets and Intersections                               89
    4.2.4   Arrays with Uneven Heights and Tall Buildings           93
    4.2.5   Spatially Averaged Flow and Turbulence Statistics       95
  4.3   Flow in the Intertial Sublayer                                    98
    4.3.1   The Profile of Mean Wind                                99
    4.3.2   Urban Surface Roughness                                102
    4.3.3   Turbulence and Turbulent Exchange                      105
    4.3.4   Local-Scale Advection                                  107
  4.4   Flow in the Mixed Layer                                          109
    4.4.1   Roughness Influences                                   109
    4.4.2   Thermal Influences                                     114
    4.4.3   Combined Roughness and Thermal Influences              118
  Summary                                                                120

**5   Radiation**                                                                122
  5.1   Basics of Radiation Exchanges and Budgets                        123
    5.1.1   Basic Radiation Principles and Laws                    123
    5.1.2   Radiation-Mass Interactions                            124
    5.1.3   The Surface Radiation Budget                           126
  5.2   Radiation in the Urban Canopy Layer                              128
    5.2.1   Radiation Properties of the Urban Canopy               128
    5.2.2   Isolated Buildings                                     132

|  |  | 5.2.3 | Urban Canyon | 134 |
|  |  | 5.2.4 | Albedo and Emissivity of Urban Systems | 140 |
|  | 5.3 | Radiation in the Urban Boundary Layer | | 145 |
|  |  | 5.3.1 | Urban Aerosol | 145 |
|  |  | 5.3.2 | Shortwave Radiation | 146 |
|  |  | 5.3.3 | Longwave Radiation | 150 |
|  | 5.4 | Surface Net Allwave Radiation Budget | | 151 |
|  |  | 5.4.1 | Urban–Rural Differences of Net Radiation | 151 |
|  | Summary | | | 154 |

**6  Energy Balance**  156

| | 6.1 | Basics of Energy Transfer and Balance | | 157 |
| | | 6.1.1 | Energy Balance of a Flat Surface | 157 |
| | | 6.1.2 | Energy Balance of Urban Systems and Elements | 158 |
| | | 6.1.3 | Case Studies | 159 |
| | 6.2 | Anthropogenic Heat Flux | | 160 |
| | | 6.2.1 | Estimating the Anthropogenic Heat Flux | 162 |
| | | 6.2.2 | Controls on Anthropogenic Heat Flux | 164 |
| | | 6.2.3 | Significance of Anthropogenic Heat Flux | 166 |
| | 6.3 | Heat Storage Change | | 168 |
| | | 6.3.1 | Controls on Heat Storage Change | 168 |
| | | 6.3.2 | Estimating and Modelling Urban Heat Storage Change | 171 |
| | | 6.3.3 | Typical Values of Urban Heat Storage Change | 174 |
| | 6.4 | Turbulent Heat Fluxes | | 175 |
| | | 6.4.1 | Modelling Turbulent Heat Transfer | 176 |
| | | 6.4.2 | Measurement of Turbulent Heat Fluxes | 178 |
| | | 6.4.3 | Controls on Turbulent Heat Fluxes | 180 |
| | 6.5 | Example Energy Balances in Cities | | 184 |
| | | 6.5.1 | Energy Balance of Individual Facets | 184 |
| | | 6.5.2 | Energy Balance of Buildings | 187 |
| | | 6.5.3 | Energy Balance of Urban Canyons | 187 |
| | | 6.5.4 | Energy Balance of an Urban Canopy | 190 |
| | | 6.5.5 | Urban–Rural Energy Balance Differences | 193 |
| | Summary | | | 195 |

**7  Urban Heat Island**  197

| | 7.1 | Urban Temperatures and Heat Island Types | | 198 |
| | | 7.1.1 | Surface Temperatures | 200 |
| | | 7.1.2 | Air Temperatures | 202 |
| | | 7.1.3 | Subsurface Temperatures | 205 |
| | | 7.1.4 | Heat Island Magnitude | 205 |
| | 7.2 | Surface Heat Island | | 206 |
| | | 7.2.1 | Observation | 206 |
| | | 7.2.2 | Spatial and Temporal Variability | 206 |
| | | 7.2.3 | Genesis of Surface Heat Islands | 210 |
| | 7.3 | Canopy Layer Heat Island | | 213 |
| | | 7.3.1 | Observation | 213 |
| | | 7.3.2 | Spatial Morphology | 213 |
| | | 7.3.3 | Maximum Canopy Layer Heat Island | 215 |
| | | 7.3.4 | Diurnal Variations | 216 |

|       | 7.3.5 | Effects of Weather and Surface State | 218 |
|       | 7.3.6 | Seasonal Variations | 222 |
|       | 7.3.7 | Genesis of Canopy Heat Islands | 223 |
|       | 7.3.8 | Final Remarks | 226 |
| 7.4 | | Boundary Layer Heat Island | 226 |
|       | 7.4.1 | Spatial Structure | 228 |
|       | 7.4.2 | Variability | 232 |
|       | 7.4.3 | Genesis of Boundary Layer Heat Islands | 233 |
| 7.5 | | Subsurface Heat Island | 234 |
|       | 7.5.1 | Genesis of Subsurface Heat Islands | 236 |
|       | Summary | | 236 |

**8    Water**    238

| 8.1 | | Basics of Surface Hydrology and Water Balances | 239 |
|       | 8.1.1 | Hydrologic Cycle | 239 |
|       | 8.1.2 | The Surface Water Balance | 239 |
| 8.2 | | Water Balance of Urban Hydrologic Units | 241 |
|       | 8.2.1 | Urban Development of River Catchments | 241 |
|       | 8.2.2 | The Water Balance of an Urban Catchment | 241 |
|       | 8.2.3 | Water Balance of an Urban Neighbourhood | 242 |
|       | 8.2.4 | Water Balance of an Entire City | 242 |
| 8.3 | | Urban Effects on Water Balance Components | 243 |
|       | 8.3.1 | Precipitation | 243 |
|       | 8.3.2 | Piped Water Supply | 244 |
|       | 8.3.3 | Anthropogenic Water Vapour | 247 |
|       | 8.3.4 | Evaporation | 248 |
|       | 8.3.5 | Runoff | 249 |
|       | 8.3.6 | Change in Storage | 251 |
|       | 8.3.7 | Groundwater | 252 |
|       | 8.3.8 | Advection | 252 |
|       | Summary | | 253 |

**9    Atmospheric Moisture**    254

| 9.1 | | Basics of Atmospheric Moisture | 255 |
|       | 9.1.1 | Humidity | 255 |
|       | 9.1.2 | Condensation | 256 |
| 9.2 | | Urban Effects on Humidity | 257 |
|       | 9.2.1 | Urban Canopy Layer | 257 |
|       | 9.2.2 | Urban Boundary Layer | 262 |
|       | 9.2.3 | Genesis of Effects | 264 |
| 9.3 | | Urban Effects on Condensation | 264 |
|       | 9.3.1 | Dew | 264 |
|       | 9.3.2 | Fog | 265 |
|       | Summary | | 268 |

**10    Clouds and Precipitation**    270

| 10.1 | | Basics of Cloud and Precipitation Formation | 271 |
|        | 10.1.1 | Warm Cloud Processes | 271 |
|        | 10.1.2 | Cold Cloud Processes | 272 |
|        | 10.1.3 | Thunderstorms | 272 |

10.2  The Methodological Challenge                                          273
      10.2.1  The Discrete Nature of Cloud and Precipitation               273
      10.2.2  Observing Cloud and Precipitation                            273
10.3  Urban Observations                                                   275
      10.3.1  Clouds                                                       275
      10.3.2  Precipitation                                                276
      10.3.3  Snowfall and Freezing Rain                                   279
10.4  Hypotheses of Urban Effects                                          281
      10.4.1  Modification of Moisture and Thermodynamic Processes         282
      10.4.2  Modification of Dynamical Processes                          286
      10.4.3  Modification of Microphysical Processes                      288
Summary                                                                    292

**11  Air Pollution**                                                      **294**
11.1  Basics of Air Pollution                                             295
      11.1.1  Emissions                                                    298
      11.1.2  Dispersion and Transport                                     299
      11.1.3  Removal and Transformation                                   301
      11.1.4  Scales of Air Pollution                                     302
11.2  Micro- and Local-Scale Air Pollution in Cities                      303
      11.2.1  Indoor Air Pollution                                         304
      11.2.2  Outdoor Air Pollution in the Urban Canopy Layer             307
      11.2.3  Air Pollution from Elevated Point Sources                    312
11.3  Urban-Scale Air Pollution                                           315
      11.3.1  Meteorological Controls on Air Quality in the Urban Boundary Layer  315
      11.3.2  Smog                                                         316
      11.3.3  Modelling Urban Air Pollution                                322
11.4  Regional and Global Effects of Urban Air Pollution                  325
      11.4.1  Urban Plumes                                                 325
      11.4.2  Effects of Urban Pollutants on Ecosystems Downwind           328
Summary                                                                    330

**12  Geographical Controls**                                             **332**
12.1  The Macroclimatic Context of Cities                                 333
      12.1.1  Urban Population and Macroclimates                           333
      12.1.2  Urban Effects in Different Macroclimates                     339
      12.1.3  The Urban Surface Energy Balance in Different Macroclimates  339
      12.1.4  The Urban Heat Island in Different Macroclimates             341
12.2  Topography                                                          342
      12.2.1  Cities Affected by the Mechanical Influences of Orography on Airflow  344
      12.2.2  Cities Affected by the Thermal Influences of Orography on Airflow     347
      12.2.3  Cities Affected by Coastal Wind Systems                      350
      12.2.4  City Airflow in Diverse Topography                           353
12.3  Synoptic Controls                                                   353
      12.3.1  Storm Systems                                                355
      12.3.2  Floods                                                       357
      12.3.3  Heatwaves                                                    358
Summary                                                                    359

**13   Cities and Global Climate Change**                                              360
   13.1   Urban Impacts on the Global Climate System                                   361
          13.1.1  Land-Cover Change                                                    361
          13.1.2  Greenhouse Gases                                                     362
   13.2   Greenhouse Gas Emissions From Cities                                         365
          13.2.1  Tracking Carbon in Cities                                            365
          13.2.2  Directly Measuring Greenhouse Gas Emissions from Cities              370
          13.2.3  Urban Land Cover and Greenhouse Gas Exchange                         374
   13.3   Global Climate Change in Urban Environments                                  376
          13.3.1  Monitoring Climate Change in Urban Environments                      377
          13.3.2  Projecting Future Climates in Cities                                 378
          13.3.3  Impacts of Global Climate Change on Cities                           379
          Summary                                                                      383

**14   Climates of Humans**                                                            385
   14.1   Basics                                                                       386
          14.1.1  Managing Heat                                                        386
          14.1.2  Maintaining Postural Balance                                         386
   14.2   The Human Energy Balance                                                     387
          14.2.1  Internal Energy Exchanges                                            387
          14.2.2  External Energy Exchanges                                            388
          14.2.3  The Radiation Budget                                                 389
          14.2.4  Sensible Heat Flux                                                   391
          14.2.5  Latent Heat Flux                                                     392
          14.2.6  Clothing                                                             393
   14.3   Thermal Stress and Body Strain                                               394
   14.4   Thermal Comfort and Its Assessment                                           395
          14.4.1  Indoors                                                              396
          14.4.2  Outdoors                                                             397
          14.4.3  Thermal Indices                                                      400
   14.5   Wind and Comfort                                                             401
   14.6   The Urban Effect on Human Climates                                           402
          14.6.1  Microclimates                                                        402
          14.6.2  Neighbourhood                                                        404
          14.6.3  Urban Scale                                                          406
   14.7   Indoor Climates                                                              406
          Summary                                                                      407

**15   Climate-Sensitive Design**                                                      408
   15.1   Basics of Climate-Sensitive Planning and Design                             409
          15.1.1  Political Context and Policy Mechanisms                             409
          15.1.2  Climate Assessments                                                  410
          15.1.3  Guiding Principles                                                   411
   15.2   Design Interventions at Different Scales                                     418
          15.2.1  Cities                                                               419
          15.2.2  Neighbourhoods                                                       423
          15.2.3  Facets                                                               424
          15.2.4  Buildings                                                            427
          15.2.5  Streets and Urban Blocks                                             431

15.2.6  Trees                                                    433
15.2.7  Gardens and Parks                                        438
15.2.8  Water as a Design Feature                                439
15.3  The Well-Planned and Designed City                         441
15.3.1  Resource Use                                             444
15.3.2  Comfort                                                  444
15.3.3  Air Quality                                              446
15.3.4  Weather Extremes                                         448
Summary                                                          451

## Epilogue                                                      453

A1    History of Urban Climatology                               454
      A1.1  Pioneer Urban Climatographies (Prior to 1930)        454
      A1.2  Advances in Micro- and Local Climatology (1930 to 1965)   455
      A1.3  Towards a Physical Climatology of Cities (1965 to 2000)   457
      A1.4  Consolidation and Prediction (Since 2000)            459
A2    Site Codes and Data Sources                                460
A3    Glossary and Acronyms                                      469

References                                                       486
Subject Index                                                    510
Geographical Index                                               520

# Symbols and Units

## Symbols

Some symbols have more than one entry; the different meaning is assumed to be self-evident from the context.

### Roman Capital Letters

| Symbol | Name | Unit |
|---|---|---|
| $A$ | Surface area | (m$^2$) |
| $A$ | Available energy flux density ($A = Q^* + Q_F - \Delta Q_S$) | (W m$^{-2}$) |
| $A$ | Area of advective influence (e.g. urban) | (m$^2$) |
| $A_b$ | Plan area of buildings | (m$^2$) |
| $A_{body}$ | Surface area of the human body | (m$^2$) |
| $A_c$ | Complete surface area | (m$^2$) |
| $A_f$ | Frontal area | (m$^2$) |
| $A_s$ | Area of a shadow | (m$^2$) |
| $A_T$ | Plan area of total surface | (m$^2$) |
| $C$ | Heat capacity of a substance | (J m$^{-3}$ K$^{-1}$) |
| $C$ | Carbon dioxide flux density due to combustion (emissions) | (kg C m$^{-2}$ y$^{-1}$; μmol m$^{-2}$ s$^{-1}$) |
| $C_D$ | Drag coefficient | (-) |
| $D$ | Diffuse irradiance | (W m$^{-2}$) |
| $D$ | Distance between building centroids | (m) |
| $D$ | Rate of removal of air pollutants | (kg s$^{-1}$ or kg m$^{-2}$ s$^{-1}$) |
| $E$ | Evaporation or evapotranspiration | (mm per time, kg m$^{-2}$ s$^{-1}$) |
| $E$ | Emittance or total energy flux density | (W m$^{-2}$) |
| $E$ | Rate of emissions of air pollutants from a source | (kg s$^{-1}$) |
| $F$ | Anthropogenic water release due to combustion | (mm per time, kg m$^{-2}$ s$^{-1}$) |
| $F_B$ | Force exerted by the wind on a person | (kg m s$^{-2}$) |
| $\vec{F}_{Co}$ | Coriolis force | (kg m s$^{-2}$) |
| $F_{CH_4}$ | Methane (mass or molar) flux density | (kg C m$^{-2}$ y$^{-1}$; μmol m$^{-2}$ s$^{-1}$) |
| $F_{CO_2}$ | Carbon dioxide (mass or molar) flux density | (kg C m$^{-2}$ y$^{-1}$; μmol m$^{-2}$ s$^{-1}$) |
| $\vec{F}_f$ | Frictional force | (kg m s$^{-2}$) |
| $F_P$ | Air pollutant (mass or molar) flux density | (μg m$^{-2}$ s$^{-1}$, μmol m$^{-2}$ s$^{-1}$) |
| $\vec{F}_{pn}$ | Pressure gradient force | (kg m s$^{-2}$) |
| $Fr$ | Froude number | (-) |
| $G$ | Rate of infiltration into groundwater | (mm per time, kg m$^{-2}$ s$^{-1}$) |
| $H$ | Height (of a building or those adjacent to an urban canyon, of a chimney stack, of a hill or mountain, of a body) | (m) |
| $I$ | Piped water supply per unit horizontal area | (mm per time, kg m$^{-2}$ s$^{-1}$) |
| $I_{cl}$ | Insulation due to clothing | (K m$^2$ W$^{-1}$) |
| $K^*$ | Net shortwave radiation flux density | (W m$^{-2}$) |
| $K_\downarrow$ | Shortwave irradiance | (W m$^{-2}$) |
| $K_\uparrow$ | Shortwave reflectance | (W m$^{-2}$) |
| $K_H$ | Eddy conductivity for heat | (m$^2$ s$^{-1}$) |
| $K_M$ | Eddy viscosity | (m$^2$ s$^{-1}$) |
| $K_V$ | Eddy diffusivity for water vapour | (m$^2$ s$^{-1}$) |
| $L$ | Length (of a building, of an urban canyon) | (m) |
| $L$ | Obukhov stability length | (m) |
| $L^*$ | Net longwave radiation flux density | (W m$^{-2}$) |
| $L_\downarrow$ | Incoming longwave radiation flux density | (W m$^{-2}$) |
| $L_\uparrow$ | Outgoing longwave radiation flux density | (W m$^{-2}$) |

| Symbol | Name | Unit |
|--------|------|------|
| $\mathcal{L}_f$ | Latent heat of fusion | $(\text{J kg}^{-1})$ |
| $\mathcal{L}_s$ | Latent heat of sublimation | $(\text{J kg}^{-1})$ |
| $\mathcal{L}_v$ | Latent heat of vaporization | $(\text{J kg}^{-1})$ |
| $\mathcal{M}$ | Molar mass | $(\text{g mol}^{-1})$ |
| $\mathcal{M}_a$ | Molar mass of dry air | $(28.97\,\text{g mol}^{-1})$ |
| $N$ | Natural (pristine) area | $(\text{m}^2)$ |
| $N$ | Number of data points | $(-)$ |
| $N_{BV}$ | Brunt-Väisälä frequency | $(\text{rad s}^{-1})$ |
| $P$ | Precipitation | (mm per time) |
| $P$ | Rate of photosynthesis (carbon dioxide flux density due to photosynthesis) | $(\text{kg C m}^{-2}\,\text{y}^{-1};\,\mu\text{mol m}^{-2}\,\text{s}^{-1})$ |
| $P$ | Population of a settlement | |
| $Q$ | Heat flux | $(\text{W})$ |
| $Q^*$ | Net allwave radiation flux density | $(\text{W m}^{-2})$ |
| $Q_\downarrow$ | Total incoming short- and longwave radiation flux density | $(\text{W m}^{-2})$ |
| $Q_\uparrow$ | Total outgoing short- and longwave radiation flux density | $(\text{W m}^{-2})$ |
| $Q_E$ | Turbulent latent heat flux density | $(\text{W m}^{-2})$ |
| $Q_F$ | Anthropogenic heat flux density | $(\text{W m}^{-2})$ |
| $Q_G$ | Substrate heat flux density | $(\text{W m}^{-2})$ |
| $Q_H$ | Turbulent sensible heat flux density | $(\text{W m}^{-2})$ |
| $Q_M$ | Energy flux density due to snow and ice melt/ freezing | $(\text{W m}^{-2})$ |
| $Q_M$ | Energy flux due to human metabolism | $(\text{W})$ |
| $R$ | Run-off | (mm per time, $\text{kg m}^{-2}\,\text{s}^{-1}$) |
| $R$ | Rate of respiration (carbon dioxide flux density due to photosynthesis) | $(\text{kg C m}^{-2}\,\text{y}^{-1};\,\mu\text{mol m}^{-2}\,\text{s}^{-1})$ |
| $R$ | Rural area | $(\text{m}^2)$ |

| Symbol | Name | Unit |
|--------|------|------|
| $R_c$ | Absorptive efficiency of an urban canyon to trap radiation | $(-)$ |
| $\mathcal{R}$ | Universal gas constant | $(8.314\,\text{J K}^{-1}\,\text{mol}^{-1})$ |
| $\mathcal{R}_a$ | Specific gas constant for dry air | $(287.04\,\text{J K}^{-1}\,\text{kg}^{-1})$ |
| $Re$ | Reynolds Number | $(-)$ |
| $RH$ | Relative humidity | $(\%)$ |
| $Ri$ | Richardson's Number | $(-)$ |
| $S$ | Direct-beam irradiance | $(\text{W m}^{-2})$ |
| $S$ | Volumetric soil moisture content | $(\%)$ |
| $T$ | Temperature | $(\text{K or }^\circ\text{C})$ |
| $T_a$ | Air temperature | $(\text{K or }^\circ\text{C})$ |
| $T_d$ | Dewpoint (temperature) | $(\text{K or }^\circ\text{C})$ |
| $T_g$ | Deep soil temperature | $(\text{K or }^\circ\text{C})$ |
| $T_{\text{MRT}}$ | Mean radiant temperature | $(\text{K or }^\circ\text{C})$ |
| $T_s$ | Soil temperature | $(\text{K or }^\circ\text{C})$ |
| $T_0$ | Surface temperature, equilibrium surface temperature | $(\text{K or }^\circ\text{C})$ |
| $T_{0,B}$ | Brightness temperature | $(\text{K or }^\circ\text{C})$ |
| $U$ | Urban area | $(\text{m}^2)$ |
| $V$ | Volume | $(\text{m}^3)$ |
| $V$ | Measured value in Lowry (1977) analysis of urban effects | $(-)$ |
| $V_b$ | Normalized building volume $(V_b = \lambda_b z_H)$ | $(\text{m}^3)$ |
| $V_f$ | Ventilation factor | $(\text{m}^2\,\text{s}^{-1})$ |
| $W$ | Width (of a building, of an urban canyon) | $(\text{m})$ |
| $X$ | Rate of air pollutant venting | $(\text{m}^3\,\text{s}^{-1})$ |
| $Z$ | Solar zenith angle | $(^\circ)$ |

## Roman Small Letters

| Symbol | Name | Unit |
|--------|------|------|
| $a$ | Wind attenuation coefficient | $(\text{m}^{-1})$ |
| $\bar{c}$ | Mean cloud droplet concentration | $(\text{cm}^{-3})$ |

| Symbol | Name | Unit |
|---|---|---|
| $c$ | Specific heat of a substance | $(\text{J kg}^{-1}\text{ K}^{-1})$ |
| $c_p$ | Specific heat of air at constant pressure | $(\text{J kg}^{-1}\text{ K}^{-1})$ |
| $\bar{d}$ | Mean cloud droplet diameter | $(\mu\text{m})$ |
| $e$ | Vapour pressure | $(\text{Pa})$ |
| $e^*$ | Saturation vapour pressure | $(\text{Pa})$ |
| $f_{cl}$ | Clothing area factor | $(-)$ |
| $f_p$ | Peak factor (gusts) | $(-)$ |
| $g$ | Acceleration due to gravity (also as a vector $\vec{g}$) | $(\text{m s}^{-2})$ |
| $g$ | Conductance | $(\text{m s}^{-1})$ |
| $h_c$ | Convective heat transfer coefficient | $(\text{W m}^{-2}\text{ K}^{-1})$ |
| $h_r$ | Radiative heat transfer coefficient | $(\text{W m}^{-2}\text{ K}^{-1})$ |
| $h_v$ | Latent heat transfer coefficient | $(\text{W m}^{-2}\text{ K}^{-1})$ |
| $\hbar$ | Planck constant | $(\text{J s})$ |
| $j$ | Photolytic rate | $(\text{s}^{-1})$ |
| $k$ | Thermal conductivity | $(\text{W m}^{-1}\text{ K}^{-1})$ |
| $k$ | Cloud type factor | $(-)$ |
| $k$ | Kinetic rate | $(\text{s}^{-1})$ |
| $k_p$ | Molecular diffusivity for an air pollutant | $(\text{m}^2\text{ s}^{-1})$ |
| $k$ | von Kármán's constant | $(0.4)$ |
| $m$ | Mass | $(\text{kg})$ |
| $n$ | Cloud amount | $(\text{tenths})$ |
| $p$ | Pressure | $(\text{Pa})$ |
| $q$ | Specific humidity | $(\text{kg kg}^{-1})$ |
| $r$ | Correlation coefficient | $(-)$ |
| $r$ | Radius of droplet or aerosol | $(\mu\text{m})$ |
| $r$ | Molar mixing ratio | $(\text{mol mol}^{-1};$ $\mu\text{mol mol}^{-1}=$ $\text{ppm})$ |
| $r_a$ | Aerodynamic resistance | $(\text{s m}^{-1})$ |
| $r_b$ | Excess resistance | $(\text{s m}^{-1})$ |
| $r_c$ | Canopy resistance | $(\text{s m}^{-1})$ |
| $r_{cl}$ | Thermal resistance of clothing | $(\text{s m}^{-1})$ |
| $r_l$ | Laminar boundary layer resistance | $(\text{s m}^{-1})$ |
| $r_s$ | Surface resistance | $(\text{s m}^{-1})$ |
| $s$ | Slope of the saturation vapour versus temperature curve | $(\text{Pa K}^{-1},$ $\text{kg m}^{-3}\text{ K}^{-1})$ |
| $t$ | Time | $(\text{s})$ |
| $u$ | Longitudinal ($x$-axis) wind component | $(\text{m s}^{-1})$ |
| $\vec{u}$ | Three dimensional wind vector $\vec{u}=(u,v,w)$ | $(\text{m s}^{-1})$ |
| $\hat{u}$ | Typical gust wind speed | $(\text{m s}^{-1})$ |
| $\vec{u}_g$ | Gradient wind (vector) | $(\text{m s}^{-1})$ |
| $u_h$ | Horizontal wind speed $(u_h = \sqrt{u^2 + v^2})$ | $(\text{m s}^{-1})$ |
| $u_{\text{GEM}}$ | Gust equivalent wind speed | $(\text{m s}^{-1})$ |
| $u_*$ | Friction velocity | $(\text{m s}^{-1})$ |
| $u_0$ | Upstream wind speed | $(\text{m s}^{-1})$ |
| $v$ | Lateral ($y$-axis) wind component | $(\text{m s}^{-1})$ |
| $v_d$ | Deposition velocity | $(\text{m s}^{-1})$ |
| $v_s$ | Terminal settling velocity | $(\text{m s}^{-1})$ |
| $vdd$ | Vapour density deficit | $(\text{kg m}^{-3})$ |
| $vpd$ | Vapour pressure deficit | $(\text{Pa})$ |
| $w$ | Vertical ($z$-axis) wind component | $(\text{m s}^{-1})$ |
| $w_*$ | Convective velocity scale | $(\text{m s}^{-1})$ |
| $x$ | Horizontal (along-wind) distance | $(\text{m})$ |
| $y$ | Lateral horizontal (across-wind) distance | $(\text{m})$ |
| $\hat{y}$ | Predicted value of variable $y$ | |
| $z$ | Vertical distance, height above ground | $(\text{m})$ |
| $z_0$ | Aerodynamic roughness length (usually momentum) | $(\text{m})$ |
| $z_{0m}$ | Aerodynamic roughness length for momentum | $(\text{m})$ |
| $z_{0H}$ | Aerodynamic roughness length for sensible heat | $(\text{m})$ |
| $z_{0V}$ | Aerodynamic roughness length for water vapour | $(\text{m})$ |
| $z_d$ | Zero-plane displacement length | $(\text{m})$ |

| Symbol | Name | Unit |
|---|---|---|
| $z_e$ | Height of inflection point in the wind profile | (m) |
| $z_H$ | Mean height of buildings in an area | (m) |
| $z_i$ | Depth of the mixed layer (base of inversion) | (m) |
| $z_m$ | Height of measurement | (m) |
| $z_r$ | Blending height (height of the roughness sublayer) | (m) |

## Greek Capital Letters

| Symbol | Name | Unit |
|---|---|---|
| $\Delta$ | finite difference approximation (i.e. difference or net change in a quantity over time or space) | |
| $\Delta A$ | Advection of water vapour per unit volume or per unit horizontal area | (mm per time, kg m$^{-2}$ s$^{-1}$; kg m$^{-3}$ s$^{-1}$) |
| $\Delta G$ | Net ground water import or export due to change in water table | (mm per time, kg m$^{-2}$ s$^{-1}$) |
| $\Delta Q_A$ | Net energy (sensible and latent) advection; rate per unit volume or per unit horizontal area | (J; W m$^{-3}$; W m$^{-2}$) |
| $\Delta Q_S$ | Net heat storage; rate per unit volume or per unit horizontal area | (J; W m$^{-3}$; W m$^{-2}$) |
| $\Delta S$ | Net change of mass (water, air pollutant, etc.) rate per unit volume or per unit horizontal area. | (mm per time; kg m$^{-2}$ s$^{-1}$; kg m$^{-3}$ s$^{-1}$) |
| $\Delta S_0$ | Difference between stress fractions of sweeps and ejections | (-) |

| Symbol | Name | Unit |
|---|---|---|
| $\Delta R$ | Net runoff | (mm) |
| $\hat{\Theta}$ | Angle between direct solar beam and normal to the surface | (°) |
| $\Phi_H$ | Dimensionless stability function for sensible heat | (-) |
| $\Phi_m$ | Dimensionless stability function for momentum | (-) |
| $\Phi_V$ | Dimensionless stability function for water vapour | (-) |
| $\Omega$ | Solar azimuth angle | (°) |
| $\Omega$ | McNaughton-Jarvis coupling factor | (-) |
| $\hat{\Omega}$ | Surface aspect angle | (°) |
| $\vec{\Omega}$ | Spin of the Earth | (s$^{-1}$) |

## Greek Small Letters

| Symbol | Name | Unit |
|---|---|---|
| $\alpha$ | Surface albedo | (-) |
| $\alpha$ | Power law exponent | (-) |
| $\beta$ | Bowen ratio ($\beta = Q_H/Q_E$) | (-) |
| $\beta$ | (Solar) altitude angle | (°) |
| $\hat{\beta}$ | Surface slope angle | (°) |
| $\gamma$ | Psychrometric constant | (Pa K$^{-1}$; kg m$^{-3}$ K$^{-1}$) |
| $\varepsilon$ | Emissivity | (-) |
| $\zeta$ | Stability parameter ($\zeta = z/L$) | (-) |
| $\theta$ | Potential temperature | (K) |
| $\theta_*$ | Friction temperature | (K) |
| $\kappa$ | Thermal diffusivity of a substance ($\kappa = k/C$) | (m$^2$ s$^{-1}$) |
| $\lambda$ | Wavelength of radiation | (μm) |
| $\lambda_b$ | Building plan area fraction | (-) |
| $\lambda_c$ | Complete, or three-dimensional aspect ratio ($\lambda_c = A_c/A_T$) | (-) |
| $\lambda_f$ | Frontal area aspect ratio ($\lambda_f = A_f/A_T$) | (-) |
| $\lambda_{\text{floor}}$ | Floor space ratio | (-) |

xvi    Symbols and Units

| Symbol | Name | Unit |
|---|---|---|
| $\lambda_i$ | Impervious surface plan area fraction | (-) |
| $\lambda_s$ | Canyon aspect ratio ($\lambda_s = H/W$) | (-) |
| $\lambda_v$ | Vegetation plan area fraction | (-) |
| $\mu$ | Thermal admittance | (J m$^{-2}$ s$^{-1/2}$ K$^{-1}$) |
| $\nu$ | Frequency | (s$^{-1}$) |
| $\nu$ | Kinematic viscosity | (m$^2$ s$^{-1}$) |
| $\rho_a$ | Density of a dry air | (kg m$^{-3}$) |
| $\rho_w$ | Density of liquid water | (kg m$^{-3}$) |
| $\rho_i^*$ | Saturation vapour pressure over an ice surface | (kg m$^{-3}$) |
| $\rho_v$ | Absolute humidity (vapour density) | (kg m$^{-3}$) |
| $\rho_v^*$ | Saturation vapour density over a liquid water surface | (kg m$^{-3}$) |
| $\sigma$ | Stefan-Boltzmann constant | (W m$^{-2}$ K$^{-4}$) |
| $\sigma$ | Standard deviation of a variable | |
| $\hat{\sigma}_y$ | Lateral standard deviation of a concentration field | (m) |
| $\hat{\sigma}_z$ | Vertical standard deviation of a concentration field | (m) |
| $\tau$ | Reynolds stress (turbulent momentum flux per unit surface area), also as a vector $\vec{\tau}$ | (Pa) |
| $\tau_\lambda$ | Transmissivity of radiation for wavelength $\lambda$ | (-) |
| $\phi$ | Latitude | (°) |
| $\varphi$ | Wind direction (direction of origin from geogr. N) | (°) |
| $\varphi_\lambda$ | Absorptivity of radiation for wavelength $\lambda$ | (-) |
| $\varphi_c$ | Orientation of street (canyon) axis relative to wind | (°) |
| $\chi$ | Mass (or molar) concentration of an air pollutant / trace gas in air | (µg m$^{-3}$; µmol m$^{-3}$) |
| $\psi$ | View factor | (-) |
| $\omega_\lambda$ | Reflectivity of radiation for wavelength $\lambda$ | (-) |

## Common Subscripts

| | |
|---|---|
| $\cdots_a$ | Air, outdoor atmosphere, aerodynamic |
| $\cdots_B$ | Background effects in Lowry (1977) scheme, brightness |
| $\cdots_b$ | Building, laminar boundary layer, blood |
| $\cdots_{bot}$ | Bottom |
| $\cdots_c$ | Complete, canopy, canyon |
| $\cdots_{cl}$ | Clothing |
| $\cdots_{env}$ | Environment |
| $\cdots_f$ | Frontal |
| $\cdots_{floor}$ | Floor (of a canyon) |
| $\cdots_g$ | Ground |
| $\cdots_H$ | Human-caused effects in Lowry (1977) scheme |
| $\cdots_i$ | Inversion base impervious input; indoor |
| $\cdots_L$ | Landscape effects in Lowry (1977) scheme |
| $\cdots_m$ | Measured, measurement height |
| $\cdots_{max}$ | Maximum |
| $\cdots_{min}$ | Minimum |
| $\cdots_o$ | Output |
| $\cdots_p$ | Pollutant, plan area |
| $\cdots_R$ | Rural |
| $\cdots_{ref}$ | Reference |
| $\cdots_{Res}$ | Residual |
| $\cdots_{roof}$ | Roof |
| $\cdots_S$ | Suburban |
| $\cdots_s$ | Soil or substrate value, or inter-element spacing |
| $\cdots_{sky}$ | Sky |
| $\cdots_{Sub}$ | Subsurface |
| $\cdots_{Surf}$ | Surface |
| $\cdots_t$ | Trees |
| $\cdots_{top}$ | Top (of a canyon) |
| $\cdots_U$ | Urban |
| $\cdots_{U-R}$ | Urban–rural difference |
| $\cdots_u$ | Longitudinal wind |
| $\cdots_v$ | Vegetated, lateral wind |
| $\cdots_w$ | Vertical wind |
| $\cdots_x$ | Along-wind (longitudinal) direction |
| $\cdots_y$ | Across-wind (lateral) direction |
| $\cdots_z$ | Vertical direction |
| $\cdots_0$ | Surface value |

## Common Superscripts

| | |
|---|---|
| $\overrightarrow{\cdots}$ | Vector |
| $\overline{\cdots}$ | Time-averaging operator |

$\cdots'$  Instantaneous deviation from a time-averaged value

## Units

| Symbol | Unit (meaning) |
| --- | --- |
| Bq | Becquerel (radioactivity) |
| °C | Degree Celsius (temperature) |
| cap | Capita |
| d | Day (time) |
| h | Hour (time) |
| ha | Hectare |
| J | Joule (energy, $J = N\,m = kg\,m^2\,s^{-2}$) |
| K | Kelvin (temperature) |
| kg | Kilogram (mass) |
| $\ell$ | Litre (volume) |
| m | Metre (distance) |
| mol | Mole (amount of a substance) |
| mo | Month (time) |
| N | Newton (force, $N = kg\,m\,s^{-2}$) |
| Pa | Pascal (pressure, force per area, $Pa = N\,m^{-2} = kg\,m^{-1}\,s^{-2}$) |
| ppm | Parts per million ($\mu mol\,mol^{-1}$) |
| ppb | Parts per billion ($nmol\,mol^{-1}$) |
| s | Second (time) |
| W | Watt (power, $W = J\,s^{-1} = kg\,m^2\,s^{-3}$) |
| wk | Week (time) |
| y | Year (time) |
| ° | Degree (angle) |

## Scientific Notation

| Symbol | Prefix | Scientific notation | Decimal notation |
| --- | --- | --- | --- |
| P | peta- | $10^{15}$ | 1,000,000,000,000,000 |
| T | tera- | $10^{12}$ | 1,000,000,000,000 |
| G | giga- | $10^{9}$ | 1,000,000,000 |
| M | mega- | $10^{6}$ | 1,000,000 |
| k | kilo- | $10^{3}$ | 1,000 |
| c | centi- | $10^{-2}$ | 0.01 |
| m | milli- | $10^{-3}$ | 0.001 |
| μ | micro- | $10^{-6}$ | 0.000,001 |
| n | nano- | $10^{-9}$ | 0.000,000,001 |

# Preface

Urban climatology is concerned with interactions between a city and the overlying atmosphere. While interactions are two-way, the prime focus of this book is the impact of the city on the atmosphere. Urban development so fundamentally transforms the preexisting biophysical landscape that a city creates its own climate. To a lesser extent, the book considers the effects of weather and climate on the city.

As an object of study, a city initially presents a climatologist with a gloriously elaborate set of knotty challenges. They include questions of how to handle a dauntingly wide array of surface elements of very different sizes and compositions, along with the fact that the vast majority of them are alien to the natural landscape and include pulses of energy, water, gases and particles controlled by people rather than geophysical activity.

Given these challenges and the desire of the rapidly growing world to live in cities, the book begins with an outline of the idea of urban ecosystems and suggests ways to approach the study of urban climates. Chapter 2 sets out a central theme of the book: to understand and effectively communicate about urban climate systems, a set of common terms, symbols, units, and descriptions of the urban surface is required. Here we adopt the Oke (1984) classification of urban climate systems that is built on scales of surface organization set by the roughness elements (mainly built structures) and scales of atmospheric motion and vertical stratification to systematize discussion. Chapter 3 is an overview of techniques used to obtain valid field observations and model results.

The main exchange processes governing the budgets of momentum and radiation and the balances of heat, water, and carbon in cities are outlined. This permits description and analysis of the spatial distribution and dynamics of airflow, temperature, humidity, greenhouse gases, and air pollutants in urban areas in Chapters 4 through 11. These and other cloud processes are relevant to the potential effects of cities on cloud development, precipitation, and severe weather. Urban air pollution has been a bane of urban living for centuries, but the mix of emissions keeps changing over history, as does the urban atmosphere into which

it must be dispersed. Again, it is useful to view things through the prism of scale.

The text to this point deals with cases where micro- and local effects are the prime controls on climate. On the other hand, Chapter 12 considers the role of orographic and coastal controls on urban climate, and the significance of the synoptic and macroclimatic context of a city. In Chapter 13, the scale expands further to consider the increasing impacts of cities on global climate and how the altered state of that system in turn imposes impacts on city life.

Chapter 14 introduces the fundamental climatic requirements of humans, our need for shelter and a comfortable environment to live and work, and how they set the context for the construction of appropriate buildings and urban infrastructure. In Chapter 15 we appeal to the principles outlined in the rest of the book to discuss ideas about intelligent and effective use of design elements such as construction materials, shade, shelter, water, and vegetation to create or modify urban climates at all scales.

*Urban Climates* is thought to be a 'first' because it is a text designed to elucidate the general principles of the subject. There was an early attempt to do this in Chinese (Shuzhen and Chao, 1985), but this may be the first in English. Of course there are several reviews of research in the field (e.g., Kratzer, 1937, 1956; Daigo and Nagao, 1972; Landsberg, 1981; Yoshino and Yamashita, 1998; Kanda, 2012), but they do not develop a synthesis of the subject suitable for teaching. There are others that describe and analyze aspects of the climate of a particular city, for example, the first book in the field: *The Climate of London* (Howard, 1833). Similarly outstanding are *The Climate of Uppsala* (Sundborg, 1950), *The Climate of London* (Chandler, 1965), *Das Klima von Berlin* (Hupfer and Chmielewski, 1990), *The Climate of Mexico City* (Jáuregui, 2000), the compilation *Urbanization and the Atmospheric Environment in the Low Tropics* (Sham, 1987) and *Das Klima von Essen* (Kuttler et al., 2015). There are also excellent monographs dealing with individual urban weather elements and those written for particular professional groups (e.g.

Changnon, 1981; Givoni, 1998; Akan and Hought-alen, 2003; ASCE, 2011; and Erell et al., 2011).

*Urban Climates* is a unified synthesis that brings together explanations of the ways cities interact with their atmospheres over scales that extend from walls and roofs up to whole cities. It aims to be an introduction accessible to students at the upper undergraduate and beginning graduate levels and professionals in cognate fields with interests in urban environments. It provides a coherent system to describe, study, and understand the essentials of urban climates. It is based on recognition of the climatic scales at play in a city and the resulting structure of urban atmospheres. Scales and layers are recurrent themes to organize the structure of the book. It resists giving the latest research findings if they have yet to be replicated, preferring to rely on material with a degree of established support. Examples are no more complex than is necessary to illustrate modern understanding.

*Urban Climates* deals primarily with the physical principles of urban climates. It is not comprehensive and tries to avoid overlap with topics that are the domains of wind engineering, urban hydrology, or urban planning. Further, since knowledge has developed unevenly, some topics are well documented while others are in a more rudimentary state or contain significant gaps. Urban systems and environments are complex and inhomogeneous such that they can seem daunting, even for those accustomed to work in other surface boundary layers. The challenges are in part due to unnaturally sharp changes of roughness, wetness and other properties. Unlike most other climate systems, human actions strongly affect the operation of urban climates. People continually change the form of the built environment through their daily living patterns, the use of indoor climate control, the emission of heat, water, gases, and particulates, all of which add forcings not connected directly to natural cycles. These are some of the reasons that explain why it has taken almost two hundred years for the field to become a recognized subfield of meteorology and climatology and only now make it possible for a textbook to be written.

This text assumes an introductory background in meteorology or climatology, ideally including some knowledge of small-scale climates and the boundary layer. Its explanatory approach dovetails with *Boundary Layer Climates* (Oke, 1987). To bridge the gap for readers lacking a background in small-scale climates, most chapters start with a short section outlining the basic physics in the topic; advanced readers may choose to skip these. The key concepts and terminology used in microclimate and micrometeorology can also be found in texts such as Arya (2001); Campbell and Norman (1998); and Monteith and Unsworth (2008).

*Urban Climates* began in the 1970s as Dr. Oke's notes for a course for physical geographers that established most of the present concepts and overall structure. The authors of the present text are physical geographers experienced in university teaching and the conduct of urban climate research in Europe and North America. Each has expertise in field observation, and both scale and numerical modeling. We share a pedagogic view that emphasizes *understanding* climates, not just *describing* them (i.e., physical climatology). *Urban Climates* seeks to assist in the teaching and application of this emergent predictive science based on understanding the physical mechanisms underlying the workings of urban atmospheres.

The first time a new term is introduced in any chapter, it is **highlighted** and defined in the Glossary. Mathematical equations are used only when necessary; often, an equation's meaning is also spelled out in words. There is a complete list of all symbols (used either as shorthand text or in equations). Système International (SI) units are used exclusively. We have received many valuable comments, criticisms, and insights on early drafts, from students and colleagues around the world. Any errors or omissions remain the sole responsibility of the authors.

We express deep gratitude to Midge, Maeve, Tanja and Ileana, our life partners, for their unfailing support, inspiration, and love. Also to Nick, Kate, Valentin, Noémi and Anouk for the joy they bring to our lives and the unspoken spur they give us to improve all urban futures. We are indebted to them for the time they have afforded us.

# Acknowledgements

A project such as the writing of this book benefits greatly from the insight, expertise and support provided by many colleagues, students and professionals. The authors are particularly indebted to Eric Leinberger, the University of British Columbia Department of Geography cartographer who drafted all the figures. It was a joy to work with such a skilled and empathetic professional.

We are also deeply grateful to our many colleagues who reviewed drafts of individual chapters and provided wise counsel and valuable criticism. In particular: Robert Bornstein (San Jose State University), Tony Brazel (Arizona State University), Omduth Coceal (University of Reading), Johan Feddema (University of Victoria), Krzysztof Fortuniak (University of Lodz), Julie Futcher (Urban Generation), Sue Grimmond (University of Reading), Björn Holmer (University of Gothenburg), Manabu Kanda (Tokyo Institute Technology), Scott Krayenhoff (Arizona State University), Frederik Lindberg (University of Gothenburg), Ian McKendry (University of British Columbia), Fred Meier (Technical University Berlin), Dev Niyogi (Purdue University), Eberhard Parlow (University of Basel), Mathias Rotach (University of Innsbruck), Matthias Roth (National University Singapore), David Sailor (Arizona State University), Hans-Peter Schmid (Karlsruhe Institute of Technology), Marshall Shepherd (University of Georgia), Chris Smart (Western University), Iain Stewart (University of Toronto), Douw Steyn (University of British Columbia), Jennifer Vanos (University of California, San Diego), Roland Vogt (University of Basel) and Shuji Yamashita (Gakugei University).

Permission to reproduce figures and photographs has been granted by the originators, owners, or publishers, and if necessary, a fee has been paid. Special assistance with the sourcing of figures and photographs came from Alain Bertaud (New York University), Eric Christensen (McGill University), David Hall (Envirobods), Bernd Leitl (University of Hamburg), Helmut Mayer (University of Freiburg), Erich Plate (University of Karlsruhe), Alan Robins (University of Surrey), Herbert S. Saffir (Saffir Associates), Rachel Spronken-Smith (University of Otago), Michael Steven (Nottingham University), Welsh School of Architecture and Elvin Wyly (University of British Columbia); processing satellite or Lidar imagery was provided by Corinne Frey (DLR), European Space Agency, USGS/NASA Landsat Program; arranging for custom output from computer models/data analysis was provided by Chris Adderly (University of British Columbia), Omduth Coceal (University of Reading), Meghan Hannon (Western University), Manuela Hayn (University of British Columbia), Eugénie Paul-Limoges (ETH Zürich), Paul Schmid (Purdue University) and Kevin Strawbridge (Environment Canada).

# 1 | Introduction

**Figure 1.1** Part of Mexico City showing the completeness of landscape change that accompanies intensive urbanization. The character of the underlying topography is evident in the parallel ridges, but the surface cover has been utterly replaced. Further, what is not visible is the infrastructure (water, electricity, transport, communication systems and so on) that sustains the population (Credit: P. Lopez Luz).

For most of us, the urban environment is the norm. Over the past 200 years, the global population has increased sevenfold, from 1 billion in 1800 to more than 7 billion by 2015; during the same period the fraction of people living in urban areas increased from 3% to more than 50% (UN, 2015). The intensity of landscape change that can accompany urbanization is exemplified in Figure 1.1. It shows part of greater Mexico City, which has a current population of more than 20 million and covers an area greater than 2,500 km$^2$. Its impact on environmental systems, including the atmosphere, is

profound and this influence is transmitted to regional and global systems far downwind. To place this landscape change in context, Figure 1.2 shows the Valley of Mexico in 1875 when its natural lake setting was substantially intact and the original settlement had been replaced by a planned settlement with a population of less than 200,000. Even at this moderate size, urban effects on environmental systems were evident as land was 'developed' and the city grew in extent and population. One of the most obvious impacts was the gradual draining of the Texcoco lake system as water was

**Figure 1.2** The Valley of Mexico from the Santa Isabel Mountain Range by José María Velasco (1875). This landscape painting taken from an elevated vantage point shows the Ciudad de México in the middle background. The extensive lakes on the valley floor are still prominent in the left middle and background, but urban development of the valley has begun as transportation networks extend from the nascent megacity (Source: Google Arts and Culture, Creative Commons).

diverted to human uses. Urban effects on the environment are present to a greater or lesser extent in every city. In *Urban Climates*, we identify, describe and quantify the impact of urban growth on atmospheric processes and the consequent development of distinct urban climates.

Urban climates are a prime example of inadvertent climate modification – the unintended impact of human activities on the atmosphere. Cities contribute to changes of climate and atmospheric composition at local, regional and even global scales. In turn, the atmosphere has impacts on the infrastructure, health and safety in cities as they struggle to cope with extreme events such as storms, floods, droughts, etc. Proper understanding, description and modelling of these interactions is needed to intelligently minimize unwanted and maximize beneficial, aspects. Ultimately, this provides the essential scientific data necessary to design, manage and operate safer, healthier, more sustainable and more resilient settlements.

This chapter sets the context for the subject by exploring the history and environmental implications of global urbanization.

## 1.1 Urban Ecology

Earth's population continues to grow and people continue to be drawn to settlements. Such development directly and indirectly leads to atmospheric changes that are the focus of this book, but it also affects the biological and physical components of the ecosystems that existed in the area (i.e., the vegetation, animals, soil, landforms and water) prior to urban development.

### 1.1.1 Urban Ecosystems

Urban ecology is the science of the relationships among living organisms, their communities and their abiotic environment in cities (Sukopp, 1998). It is a central tenet of ecology that living and abiotic components are inextricably linked in an ecosystem. **Urban ecosystems** are formed by the biological population of organisms (vegetation, animals, people) and the abiotic environments of cities (Figure 1.3). The presence of people means the environments are both cultural and biophysical. The cultural environment is entirely due to people's activities and its various social, political, economic and other attributes are studied by social scientists. Its most obvious external signs are the physical cultural artifacts of a built system such as buildings, industries, roads, etc. The biophysical environment can be divided into subsystems following the classical ecological 'spheres': the urban atmosphere, biosphere, hydrosphere, pedosphere (soils) and lithosphere (geology).

In practice, the cultural and biophysical environments overlap in urban ecosystems. For example, a road is clearly part of the cultural landscape, but its construction and presence also disrupts the biophysical realm through removal of vegetation, destruction

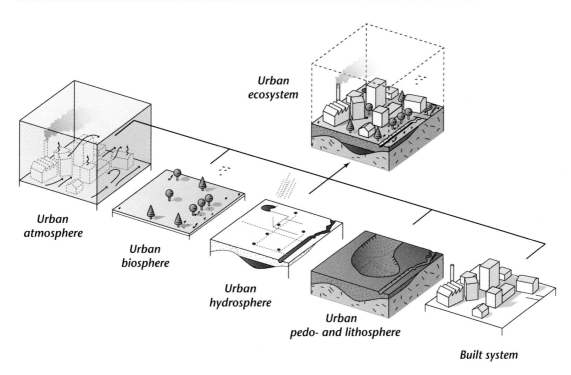

**Figure 1.3** The biophysical components that comprise an urban ecosystem. They include all aspects of the preurban natural environment subsequently modified by the introduction of built infrastructure.

of animal habitat, terrain smoothing, soil contamination, surface waterproofing and changes to the surface climate. In fact, the road itself is subject to natural environmental processes and, if left untended, becomes weathered, breaks down, is recolonized by plants and animals, decomposes and eventually turns into a rubble soil.

Urban ecosystems are the home (habitat) of the majority of humanity. Urban ecosystems are created and reshaped in the process of urban development and urban transformation. From an environmental perspective the process of urban development is not total, instantaneous or uniform. This produces the great diversity typical of urban ecosystems each associated with different land uses and degrees of human management or disturbance. The ecological spectrum in an urban area extends from remnant ecosystems (e.g. undisturbed ponds, lakes, ravines, escarpments, forests and other parkland) through managed ecosystems (e.g. fields, tended parks, gardens, cemeteries, golf courses and other open ground) to totally **anthropogenic** systems (i.e., those dominated by the built system, such as roads, buildings parking lots, industrial tips and

ponds). Each urban ecosystem provides a distinct set of habitats for plants, wildlife and humans, embedded in a unique arrangement of abiotic elements.

### 1.1.2 Urban Metabolism

The city is considered 'an integrated open system of living things interacting with their physical environment' (Douglas, 1983). It is 'open' to the import and export of both energy and mass. That means the built-up area of a city cannot exist without support from outside the boundaries of the system. In analogy to the metabolism of a living organism that 'transforms substances into tissue with an attendant release of energy and waste', **urban metabolism** describes the flow and transformation of materials and energy in a city (Figure 1.4). But biological analogies (including viewing a city as a monster, parasite, or cancer) are of limited value: a city is not imbued with a life of its own. Yet studying the urban metabolism can be of great benefit because it enables us to quantify limits and dependencies, assess the impact of new technologies or other changes and predict future needs.

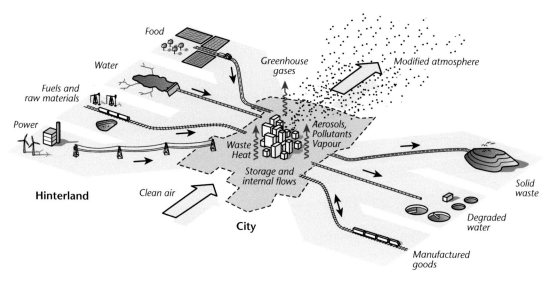

**Figure 1.4** Representation of inputs to, and outputs from, an urban ecosystem (Modified after Christen 2014; © Elsevier, used with permission).

Urban metabolism can be studied through material and energy flow analysis. The physical inputs to the system are power, fuels and raw materials, water, food and air, which are needed to feed people, operate a city and fabricate infrastructure, manufacture goods and generate services. The main physical outputs from the system are manufactured goods and the wastes excreted as sewage, solid waste, **air pollutants**, waste heat and **greenhouse gases**. Injection into the urban atmosphere through release of waste by-products, **combustion**, manufacturing, irrigation, construction, etc. varies according to human activity cycles with time of day, day of week, seasonal cycles, cultural practice and changes in technology.

The concept of urban metabolism shown in Figure 1.4 relates to the energy and materials of direct importance to the operation of the built-up system. More specific budgets of urban metabolism can be written for chemical elements (e.g. carbon or nitrogen), chemical compounds (e.g. water), or specific forms of energy being cycled through the biophysical subsystems. The sum of the inputs does not necessarily equal that of the outputs over a period of inventory. However, since the principle of conservation (of mass and energy) must apply, inequality between input and output signals will change net storage by the system. An excess of input over output indicates net storage of mass and/or energy, i.e., growth of the system. This includes physical increases in size of the built

environment, the inventory of goods, population increase, embodied or stored energy, etc. Conversely, if output exceeds input, the system is in a state of 'thinning down' or perhaps of decay.

In this book we discuss the budgets of **momentum** (e.g. Chapter 4), energy (e.g. Chapters 5–7), water (Chapters 8 and 9), air pollutants (Chapter 11) and carbon (Chapter 13) that exist at the interface between cities and atmosphere. We are also concerned with any differences between their present and preurban state, and the ways in which future urban development leads to alteration of climatic processes and elements. The aim is to demonstrate cause and effect in urban climate; in systems terminology, this is referred to as a process-response system.

### 1.1.3 Urban vs Rural

Few if any cities were established on a pristine or **natural** landscape devoid of human impacts. Most emerged in places that had already been settled for a variety of purposes, including agriculture, extraction of natural resources, trade, defense or transport. As the settlement emerged, the adjacent landscape was altered to provide resources (food, water, raw materials, fuels and power). In this book, we refer to the countryside surrounding an urban area as being **rural**. To most people, especially in mid-latitudes, this means a landscape consisting of mixed farming and woodland with fields, woods, scrub, streams, hedges, open land and a few buildings. Therefore, it is better classed as a 'managed', not a natural, system. That means it has been substantially altered by human activity through deforestation, drainage, cultivation, grazing, fertilization, land enclosure (by fences, walls or hedges) and even the construction of roads and scattered houses or villages. However, we should not overlook the fact that a few cities, especially at high and low latitudes, are surrounded by undisturbed ecosystems such as deserts, bushland, jungle, tundra or mountains where the extent of human modification is much less. Nevertheless, in this book all are referred to as rural.

## 1.2 Environmental Impacts of Urban Development

### 1.2.1 Urban Pedosphere and Lithosphere

Urban development is capable of modifying the surface form (geomorphology) of the landscape and certainly of altering the soils of the original site. Bulldozers and similar earth-moving equipment are emotive symbols of urban development, and justly so because their effects are fundamental and lasting. In preparing the ground to construct roads, bridges, houses, etc., it is considered necessary or at least convenient, to flatten hummocks, fill depressions, excavate large openings, fill or straighten water courses, remove, rearrange and compact topsoil and so on. Such activity is capable of totally destroying small ecosystems (e.g. a pond, marsh or stream can be obliterated by the dumping of rubble and refuse); whole valleys can be lost by flooding behind a dam, and completely new islands or coastlines can be created by land reclamation. Alterations may be less obviously disruptive, such as when mining or melting permafrost leads to land subsidence, or removal of vegetation cover produces nutrient leaching and soil erosion. The disruption of the upper layer of soil destroys its natural structure. These changes are compounded when imported materials, such as sand, gravel, the detritus accompanying construction, topsoils and mulches, are added to the original soil materials and perhaps become exported by subsequent wind and water erosion. These processes radically change the composition and layering of urban soils resulting in significant changes to the fertility and hydrology of the ground in an urban ecosystem.

### 1.2.2 Urban Hydrosphere

The hydrologic impacts of land disturbance can also be profound. The land is often drained and wetlands, ponds and lakes are in-filled. The water régime is drastically interrupted when impervious materials like asphalt and concrete roads, parking lots and buildings are constructed, thereby partially sealing the surface and greatly reducing the **infiltration** of water into the ground. Even if the soil is not completely sealed, disturbance to the terrain upsets the patterns and rates of both overland and subsurface drainage. This decreases percolation to deeper layers and increases the speed and amount of surface **runoff**. The lack of drainage to deeper layers desiccates the underlying ground and reduces **groundwater** storage. Enhanced runoff leads to the possibility of damaging flash floods. Perhaps surprisingly, some districts or whole cities may actually receive an increased availability of water compared with their preurban state because they access irrigation water from deep wells or pipes. Unfortunately, all these changes to the amount and routing of water are accompanied by degradation of water quality. Many characteristics are impacted, including increased turbidity, greater chemical and biological pollutant loads and increased water temperature.

### 1.2.3 Urban Biosphere

Urban development accelerates or even completes the loss of vegetation started by land clearance for agriculture or other activities prior to settlement on a site. Except for remnant patches (usually unsuitable for construction, or purposely preserved for parkland) original native plants are eliminated or replaced by exotic species of cultivars and weeds. The process is a combination of removal and replanting with more favored urban species (grass, flowering plants and shrubs, vegetables, ornamental trees, fruit trees, shade and shelter trees, etc.), and the gradual loss of native vegetation, sometimes accelerated by degraded environmental quality (damage from pollutants, disease and desiccation). All of these impacts on the biophysical environment of cities affect the ecology of their nonhuman inhabitants – wild and domesticated mammals, birds, reptiles and insects. In general, wildlife is threatened by the loss of natural habitat and food sources (especially vegetation) and by pollution and changes in the predatory system.

Of course some animal species are able to adapt to urban life and some even thrive as a result of finding new habitat and food supplements (cockroaches, gulls, rats, foxes).

### 1.2.4 Urban Atmosphere

In broad perspective, there are two sets of urban features that modify the atmosphere: those related to changes in surface properties and those due to anthropogenic **emissions**. The former are associated with aspects of **urban form**, the latter with **urban function**. In addition to providing a useful classification of the actors generating urban climates, this duality is a core principle in architecture and urban planning; therefore, it aids our discussion of the role of climate in urban design and vice versa.

### How Urban Form Affects the Atmosphere

The overall dimensions of a city such as its area, diameter, shape (i.e., circular, radial, linear or cellular), its skyline and whether its core is central, ex-central or multiple, can all play a role in the spatial form of urban climates. At finer scales, the form of any urban area affects the atmosphere as follows:

- **Fabric**: The natural and construction materials that form urban elements such as buildings, roads and vegetation. Fabric determines the radiative, thermal and moisture properties of a surface and therefore its abilities to absorb, reflect and emit **radiation**, and to accept, transfer and retain heat and water.
- **Surface cover**: The fractions of the surface interface occupied by various patches made up of different fabrics: built-up, paved, vegetated, bare soil, water. Cover fractions are especially relevant to the partitioning of heat, for example, a dry patch becomes very warm on a sunny day because it cannot dispose of solar energy through **evaporation**; vegetated patches stay relatively cool, because plants and soils use solar energy to evaporate water.
- **Urban structure**: the 3-D configuration of urban elements: dimensions of buildings and the spaces between them, street widths and street spacing. Structure is important at two scales: that of the complete city where it helps determine its **albedo** and aerodynamic roughness, and that of individual buildings and streets where it controls patterns of radiative exchange and airflow.

### How Urban Function Affects the Atmosphere

Metabolic cycles operate over different time scales: the most obvious being the rhythm of the work day and work week that modulates the pulses of traffic, domestic water use, space heating and cooling, industrial activity and so on. In many cities, there are also characteristic weekday–weekend cycles and seasonal changes in human activity. These rhythms create fluctuations in anthropogenic output that must be added to those of the natural solar cycle. Anthropogenic emissions include the release of water vapor, heat and liquid, gaseous or particulate air pollutants: all can be considered waste by-products of urban metabolism. Over much longer time periods there are surges of building and other infrastructure construction, decay, destruction and reconstruction. These dynamics are due to human occupation, activity and decision-making in settlements. The impacts of anthropogenic emissions are either direct or indirect. Direct impacts are heat losses from houses or automobiles that warm the air nearby, or vapor injections from cooling towers that condense into cloud. Indirect impacts include processes where air pollutants interfere with radiative transfer in the atmosphere, or form condensation nuclei around which cloud droplets grow, or greenhouse gases that modify Earth's radiation budget. Anthropogenic emissions mean people are actors in the climate system at all scales. Human decisions on an ongoing basis affect the spatial and temporal nature of these impacts. Examples include decisions regarding the scheduling of the work day and week, whether to commute or work at home, irrigate the garden, use an air conditioner, adjust a thermostat, use a car or public transit and so on.

## 1.3 Urbanization and Urban Form

This profound transition in the living patterns of humans has come about very quickly (Figure 1.5). The word 'urbanization' refers to the socioeconomic processes that lead to the concentration of people in urban areas where work is typically associated with manufacturing and service industries rather than agriculture and fishing. Humans seem to have adapted reasonably well to the urban environments they have largely created. That is not to say people are immune from negative impacts attributable to the city; nevertheless, judging the popularity of cities by their

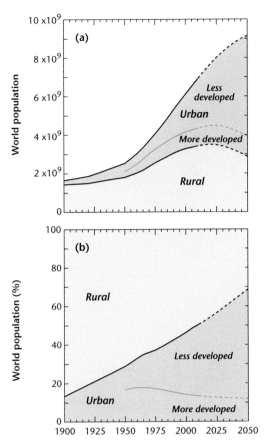

**Figure 1.5** Population growth since 1900, divided into those living in urban and rural areas. **(a)** Total population of the world divided. **(b)** The percentage of the population living in urban areas. Urban population is divided into less and more economically developed countries since 1950. Based on data from UN (2015).

drawing power (measured by their demographic statistics), they are a great success.

Although the earliest cities can be traced back 5,000 years to the development of an interconnected urban system in the Fertile Crescent (centred on the Euphrates and Tigris rivers), there was by 1000 A.D. only perhaps 25 cities with populations over 100,000 and just 1 city (Baghdad) with a population exceeding 1 million. The rate of urbanization and the growth of large cities increased rapidly during the modern period (1700 onwards): by 1800 there were four cities of over 1 million (Beijing, Guangzhou and Tokyo in Asia; London in Europe); by 1900 there were 16 very

large cities (mostly located in Europe and eastern North America) and by 2000 there were nearly 300 cities with populations greater than 1 million, some even include > 10 million residents (termed megacities) (Modelski, 2003). Most urban dwellers live in more modest settlements of less than half a million. The global distribution of the current population and several of the largest cities is shown in Figure 1.6a.

This concentration of Earth's population into urban areas has occurred in two phases, one of which is substantially complete and the other of which is still happening. While we know something of the urban functions and forms that resulted from the first phase, our knowledge about the cities emerging during the second phase is rudimentary at best.

### 1.3.1 First Phase of Urbanization

The first urbanization phase was initiated by an Industrial Revolution that transformed agricultural production, manufacturing activity and transportation through the use of technology and by employing fossil fuels on a grand scale. Initially, in places such as Manchester, United Kingdom, historic centres became the nucleus of a growing, densely occupied settlement. These places grew rapidly as rural migrants moved into towns and cities, often enduring overcrowded and unhealthy environmental conditions in order to secure employment. Limited means of transport meant that most of them lived within walking distance of work. Later, investment in urban infrastructure (water supply, sewage treatment, urban public transport) improved conditions and the cities grew in physical size and the population density fell. **Suburban** residential areas developed at the edge of the city along arterial routes. At this stage of the process, the urban population grew primarily as a result of a natural increase as rural migration slowed.

The arrival of the automobile in the early twentieth century provided greatly increased freedom of movement and reduced the need for close proximity of factories, offices and residences. In many places, these tendencies were actively supported by urban planning policies that zoned land-use for single-use purposes. The net result was a distinctive urban form with increased building density towards the urban centre where buildings were taller and, relatively speaking, more closely spaced. By 1950, the first phase of urbanization had peaked in more

## (a) Population density

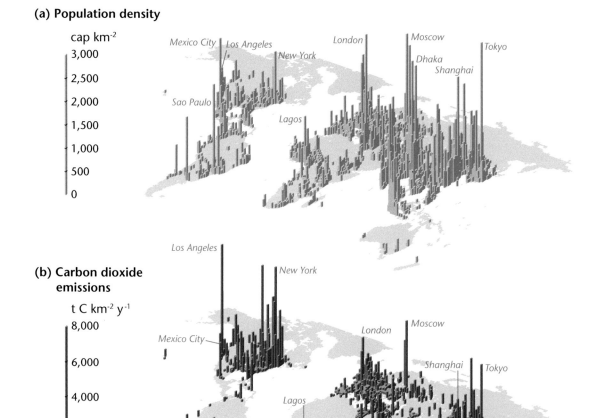

**Figure 1.6 (a)** Population density for 2010 at a resolution of $1 \times 1°$. Bars are only shown for grid cells with more than 50 cap. $km^{-2}$ (Data source: Center for International Earth Science Information Network Columbia University, 2015). **(b)** Carbon dioxide emissions from fossil-fuel burning for 2010 at a resolution of $1 \times 1°$. Bars are only shown for grid cells with more than 100 t C $m^{-2}$ $year^{-1}$ (Data source: Andres et al., 2015).

economically developed regions (e.g. North America, Europe and Japan) where 53% of people lived in urban areas. The expansion of those cities slowed, but their continued attractiveness meant that by 2010, over 85% of the population in these regions lived in urban areas.

For many cities, the urban form established by the 1950s has remained. Whereas tall, closely spaced buildings occupy the core, the urban edge extends into surrounding nonurban areas at low building densities, causing 'urban sprawl'. This is an

oversimplification, and there are examples of both central city decay and nodes of high density in outlying areas, but generally it holds for cities created in the first phase of urbanization in more economically developed regions. This is reflected here by use of the following terms:

• **Central city** refers to the core district of densely packed, sometimes multistory buildings that house government, institutional, major commercial and similar operations.

**Table 1.1** The ten most populous cities over time (Sources: Chandler and Fox, 1974; United Nations, 2008).

| Rank | 1750 | | 1850 | | 1900 | | 1950 | |
|------|------|------|------|------|------|------|------|------|
| 1 | Beijing | 900 | London | 2,320 | London | 6,480 | New York | 12,300 |
| 2 | London | 676 | Beijing | 1,648 | New York | 4,242 | London | 8,860 |
| 3 | Constantinople | 666 | Paris | 1,314 | Paris | 3,330 | Tokyo | 7,547 |
| 4 | Paris | 560 | Canton | 800 | Berlin | 2,424 | Paris | 5,900 |
| 5 | Yedo | 509 | Constantinople | 785 | Chicago | 1,717 | Shanghai | 5,406 |
| 6 | Canton | 500 | Hangchow | 700 | Vienna | 1,662 | Moscow | 5,100 |
| 7 | Osaka | 403 | New York | 682 | Tokyo | 1,497 | Chicago | 4,906 |
| 8 | Kyoto | 362 | Bombay | 575 | St. Petersburg | 1,439 | Ruhr | 4,900 |
| 9 | Hangchow | 350 | Yedo | 567 | Philadelphia | 1,418 | Calcutta | 4,800 |
| 10 | Nanking | 325 | Soochow | 550 | Manchester | 1,255 | Buenos Aires | 4,600 |

| Rank | 1975 | | 2000 | | 2015 | |
|------|------|------|------|------|------|------|
| 1 | Tokyo | 19,771 | Tokyo | 26,444 | Tokyo | 27,190 |
| 2 | New York | 15,880 | Mexico City | 18,066 | Dhaka | 22,766 |
| 3 | Shanghai | 11,443 | Sao Paulo | 17,962 | Mumbai | 22,577 |
| 4 | Mexico City | 10,691 | New York | 16,732 | Sao Paulo | 21,229 |
| 5 | Sao Paulo | 10,333 | Mumbai | 16,086 | Delhi | 20,884 |
| 6 | Osaka | 9,844 | Los Angeles | 13,213 | Mexico City | 20,434 |
| 7 | Buenos Aires | 9,144 | Calcutta | 13,058 | New York | 17,944 |
| 8 | Los Angeles | 8,926 | Shanghai | 12,887 | Jakarta | 17,268 |
| 9 | Paris | 8,885 | Dhaka | 12,519 | Calcutta | 16,747 |
| 10 | Beijing | 8,545 | Delhi | 12,441 | Karachi | 16,197 |

- **Suburban areas** are the predominantly residential rings around central cities consisting mainly of single and multifamily homes with scattered shopping, school and hospital services and some light industry.
- The terms **urban area** or **city** here refer to the whole settlement and to the quasi-continuous built-up area seen in satellite and aircraft imagery that have sufficient density to require organization into recognizable block-type patterns by the road system.

## 1.3.2 Second Phase of Urbanization

The second phase is evident in the changing ranks of the world's largest cities (Table 1.1). This demographic transition is focused in economically less-developed regions and is unprecedented in both magnitude and rapidity (Figure 1.5). As a result, nearly one billion people occupy slum settlements on the outskirts of formal cities and few of them have basic amenities or protections associated with urban living. While first-phase cities are largely built and will remain in place for the foreseeable future, those associated with the second phase are in process of creation and transformation. The new urban forms often do not correspond to the model associated with the first phase. For example, the urban edge is often difficult to detect – urban functions and their infrastructure seemingly merge into nonurban landscapes. In China, semirural places that consist of intense mixtures of urban, industrial, agricultural, transport and other uses with scattered nodes of denser settlement occur in the region surrounding large cities – this pattern is called 'desakota' (McGee, 1991). Land-use control as practiced in Europe, and to a lesser extent in North America, is hardly evident. This more diversified pattern, which is variously described as decentralized, dispersed and polynucleated, renders Western land classes like urban, suburban and rural as moot. This also makes it more difficult to identify what is 'urban' about their climate.

### 1.3.3 Future Urbanization Pathways

It remains to be seen what form urban settlements will take in response to changes in modern and future economic forces, technological advances and urban policies. It may be that a counter urbanization process occurs in highly urbanized 'postindustrial' societies where a desire to live outside cities can be facilitated by electronic communications and rapid transit systems. The economic, cultural and political processes in economically less-developed countries may produce very different outcomes. Although some known patterns are being repeated, albeit at a much faster pace, other patterns including the smearing of urban–rural contrasts and in-fill development are producing semicontinuous urban regions where the concept of 'rural' has little merit. Nevertheless, it is clear that urban environments will be home to the majority of people in the future, so the rationale for seeking better understanding of urban-environment interactions should be self-evident. Only through improved knowledge and its intelligent application will it be possible to both lessen the impacts of humans on nature and preserve, or possibly improve, the state of the everyday environment experienced by most people on the planet.

### 1.4 Planning for More Sustainable Cities

The term 'sustainable' is used in so many different contexts that it has taken on many different meanings, some of which may be incompatible. There are three broad components to sustainability when viewed from a human perspective: ecological, economic and sociopolitical. While urban planning must consider all of these components and their interrelationships, our focus here is mainly on the ecological.

### 1.4.1 Renewable and Nonrenewable Resources

There is ample evidence that the current human use of the planet's resources is ecologically not sustainable

(e.g. Rogers and Feiss, 1998; Simmons, 1995). At the global scale, ecological sustainability would mean that human systems only draw on Earth's resources at a rate that allows them to recover or be replaced naturally. For example, energy resources held in long-term storage (e.g. fossil fuels) are nonrenewable, their use depletes the resource and is, by definition, not sustainable. On the other hand, those energy sources renewed directly by the Sun (e.g. solar, wind) or replenished through annual growth (e.g. biomass) can be sustainable. In such terms, modern urban systems are unsustainable because they require intense use of fossil fuels and fossil water to supply energy, food and water, to produce goods and to manage waste.

In this sense, a truly sustainable city is one able to feed itself, has sufficient water for drinking and industrial uses, enough power for all purposes and raw materials to build its infrastructure and manufacture goods and that all of this is achieved without depleting resources in its surrounding landscape. Historically, most cities were reliant on their immediate surroundings (hinterland) for resources; buildings were constructed from local materials, water was drawn from nearby rivers, food and fuel acquired from the surrounding land and so on. However, the development of networks that allow resources to be transferred across great distances has allowed cities to become physically divorced from their immediate geographical locations (Rees, 1997).

### 1.4.2 The Ecological Footprint of Cities

A widely used approach to examine the resource base associated with human activities at all scales is the **ecological footprint**, which expresses the cumulative effect of human consumption and waste practices in terms of the ecologically productive area needed to sustain these activities (Rees and Wackernagel, 1996). While this approach is open to debate, it has the advantage of linking a wide variety of activities to a simple measure of ecological sustainability. By this definition, a city is not, and cannot be, sustainable because it must gather resources from areas outside its boundaries and often deposits its waste there as well. For illustrative purposes, consider the resources required by a major metropolitan area and the area needed to gather them sustainably according to this criterion (Table 1.2). The resource consumption of Greater London, United Kingdom, is equivalent to ecological productivity generated by over 95% of the land area of Britain. It may be more useful, then, to consider cities as the foci of human activities through which global resource use is modulated. As such, cities occupy a critical scale for implementing strategies to attain sustainability at a global, rather than city, scale.

**Table 1.2** London's ecological footprint (Source: Girardet, 1999).

| Resource | Area (km²) |
| --- | --- |
| London's built-up area | 1,578 |
| Farmland area (food) | 84,984 |
| Forest area (wood) | 7,689 |
| Carbon sequestration | 105,218 |
| Total footprint | 197,891 |
| Britain's productive land area | 210,437 |

In planning usage, 'sustainability' has a much broader meaning that includes economic, social and political concerns. In essence, sustainable planning attempts to plot a course for human systems that allows economic development and ensures that benefits are shared equitably while simultaneously protecting environmental resources and hence reducing the ecological footprint. This approach has developed alongside the growing evidence of the significant impacts humans have on Earth and its 'natural' systems and the growing recognition that economy and society fundamentally depend on the continued health of the environment. Cities represent a particularly relevant scale in which to apply these ideas because, for the most part, the global economic system *is* an urban one. Figure 1.6 shows global maps of the distributions of population density along with anthropogenic emissions of the greenhouse gas **carbon**

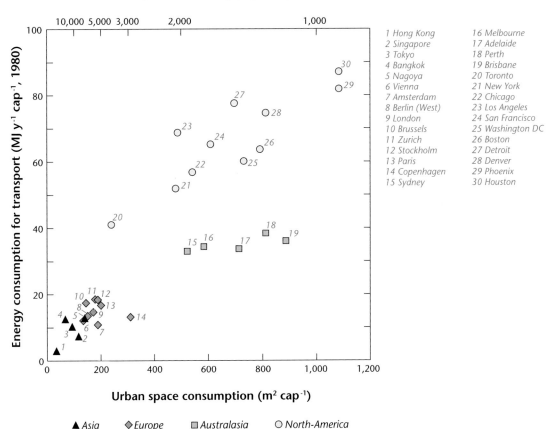

**Figure 1.7** The relationship between per-capita urban space consumption (the inverse of population density) and per-capita energy use for transportation (Modified after: Newman and Kenworthy 1989).

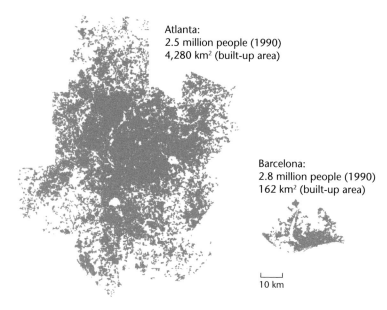

Atlanta:
2.5 million people (1990)
4,280 km² (built-up area)

**Figure 1.8** Built-up area of Atlanta, United States, and Barcelona, Spain: these are cities with an equivalent population but built at very different densities (Source: Bertaud, 2003; © Elsevier, used with permission).

Barcelona:
2.8 million people (1990)
162 km² (built-up area)

10 km

dioxide ($CO_2$). Juxtaposition of the two clearly shows a good correlation between them in the cities of Europe, North America and Japan. By contrast, much of South America, Africa and India are characterized by low levels of $CO_2$ emissions, despite the large population of cities such as Sao Paulo, Lagos and Dhaka.

### 1.4.3 Planning for More Sustainable and Resilient Cities

A well-planned and designed city from a climatic perspective would be efficient in its use of resources (sustainable) and designed to protect people and infrastructure from extreme weather events (resilient). Sustainable urban planning focuses on technology, connectivity, services and zoning as keys to managing an urban area. One of its tenets is that compact, high density cities composed of mixed land-uses and walkable communities are inherently more efficient. Rather than allowing cities to expand into nonurban ('green') sites that require new services (roads, sewers, etc.), existing underused or derelict urban land (so called 'brown' sites) is redeveloped. There is some evidence to support this view; information from a variety of world cities in more developed economies indicates that increased population density reduces per-capita

energy use associated with travel (Kenworthy, 2006). Figure 1.7, which is based on the classic study of this relationship captures many complex interactions between urban form and function that are not immediately apparent. For example, low-density, widely dispersed populations usually must rely on private transport requiring residents to own one or more vehicles per family, whereas densely occupied urban areas are often conveniently served by a mass transit system. This generally carries additional benefits of lower air pollutant emissions.

The data in Figure 1.7 are clearly clustered according to region. The North American results appear to follow a relation on their own, where the amount of energy used per person is approximately double that for other parts of the world, and their use of urban land per person is also much greater. The remarkable space and energy efficiency of the group of European and Asian cities in the bottom left-hand corner is also evident in Figure 1.7. It shows that population alone is not the reason for different $CO_2$ emissions by cities. The spatial dimensions of this are dramatically illustrated using the urban land footprint of two cities of similar population from each of the two regions (Figure 1.8). Clearly, addressing the urban form and functions of cities is central to limiting the magnitude of their ecological footprints and regulating their aggregate impact on the planet.

**Summary**

The current global trend of population growth and urbanization causes substantial environmental impact on the atmosphere, biosphere, lithosphere, pedosphere and hydrosphere; taken together, they form **urban ecosystems,** wherein living organisms, including humans, interact with their built environment. We can study the environmental impacts of cities through energy and material flow analysis, using the concept of the **urban metabolism,** i.e., cities are supplied with power, fuels, water and food that they convert to goods and use to sustain organisms but as by-products they generate wastes including solids, degraded water and polluted air.

- Urban systems can be broadly characterized by their **urban form** (fabric, land cover, geometric structure) and **urban function** (land uses, industrial and other processes, transport, etc.). Both affect the atmosphere through either physical (e.g. roughness) or chemical influences (e.g. air pollutant emissions).
- Global urbanization can be separated into two phases. The **first phase of urbanization** was initiated by the **Industrial Revolution** when technology employing fossil fuels on a grand scale transformed the way of living, creating central cities and later with the arrival of the automobile suburban sprawl. The **second phase** of urbanization in economically less-developed regions is **ongoing** and unprecedented in magnitude. It results in uncontrolled, rapid development in scattered nodes of denser settlements surrounding large cities.
- Modern **cities are not sustainable** because they depend largely on **fossil resources** and a global economic network. Transforming the form and function of cities to make them more resilient and to allow for more efficient use of local resources are among the most challenging tasks in the twenty-first century. Urban climatology can help to make informed decisions during this **transformation**, for example, to **efficiently use energy and water** in cities, mitigate and adapt to **global climate change** and combat **air pollution**.

The organization of *Urban Climates* follows the logic of these interests. It starts with the concepts and methods needed to study urban climates (Chapters 2 and 3), then describes and explains urban climatic effects on atmospheric variables (Chapters 4 to 12) and finishes with discussion of how this knowledge can be used to mitigate and adapt to global climate change (Chapter 13) and ideally to design more sustainable and resilient cities with respect to human comfort, health and energy use (Chapters 14 and 15).

# 2 | Concepts

**Figure 2.1** The diversity of an urban neighbourhood in Sao Paulo, Brazil (Credit: N. de Camaret).

The study of the physical, chemical and biological processes operating to produce or change the state of the urban atmosphere is urban meteorology and the study of the statistically preferred states of urban weather is **urban climatology**. Climatology includes the quantitative description of the climates themselves (climatography), the use of meteorological knowledge to explain climatic differences and phenomena (physical climatology) and the use of climatic data to solve practical problems (applied climatology).

All of these topics are covered in *Urban Climates*, but special emphasis is on the physical climatology of cities. That requires integration of the observed climate with the characteristics of the place and an appreciation of the governing processes. This is at the interface between climatography and meteorology: seeking to explain the former in terms of the

latter. This creates the cause-and-effect (or process-response) framework necessary to understand how climate systems work which, in turn, creates a basis for prediction and intelligent decision-making in the design and management of cities.

Although the study of **urban climates** started in the early nineteenth century (for a historical overview, see Appendix A1), much of its scientific underpinnings were only established in the last few decades. This is partly owing to the scope and complexity of **urban form** and **urban function** and their interactions with the overlying atmosphere. Figure 2.1, which shows a neighbourhood of Sao Paulo, Brazil, is an example of the extraordinary spatial heterogeneity present in urban landscapes. Looking at that image it becomes clear there is great diversity in urban form and function – reflected in the numerous types of **fabric**, patterns of **surface cover**, complex 3-D **urban structure** and **urban metabolism** associated with **emissions** of heat, water and **air pollutants** into the atmosphere. This presents challenges: for example, how to identify a site where representative measurements can be obtained, or how best to simplify and model an urban environment.

Successful study of urban climates relies on some essential concepts, definitions and approaches that are the subject of this chapter. We start with the question of describing an urban surface and how that affects the scales of urban climates (Section 2.1). It cannot be overemphasized that full appreciation of the role of scale is probably the single most important key to understanding urban climates. Scale is an organizing theme throughout *Urban Climates*.

## 2.1 The Urban 'Surface'

### 2.1.1 Defining the Land-Atmosphere Interface in Cities

Definition of the **active surface** is an essential first step to study the **boundary layer** climate of any environment. The exact nature of Earth's surface at a place controls much of the boundary layer characteristics. It does so via its role in **absorption**, reflection and **emission** of **radiation**; the transformation of energy and mass (e.g. **radiant energy** to thermal energy, liquid water to water vapour); the **interception** of **precipitation** and air pollutants; and the deflection of airflow and the slowing of the wind. As a result, the land surface controls the exchanges of energy, mass and **momentum** and it usually experiences the most extreme climate (i.e. where it is hottest or coldest, driest or most moist, where flow is brought to rest). Near the surface is also where the greatest variability in **microclimates** exists (i.e. the greatest range of values from place to place and day to night).

### Definitions of the Urban Land Surface

Definition of the relevant 'surface' is not always simple and this is especially the case in an urban environment. Identification and quantification of the urban surface poses methodological difficulties. Figure 2.2 illustrates six potential ways to define 'the surface'; the convoluted interface between an urban system, and the atmosphere:

(a) The **ideal** or **complete three-dimensional (3-D) surface** where every detail of the interface is considered. In

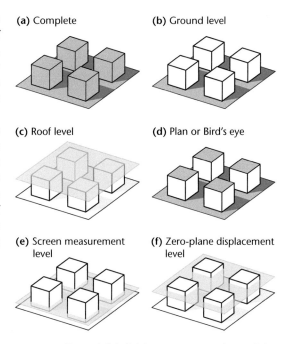

**(a)** Complete    **(b)** Ground level

**(c)** Roof level    **(d)** Plan or Bird's eye

**(e)** Screen measurement level    **(f)** Zero-plane displacement level

**Figure 2.2** Potential definitions, or perspectives, of the 'surface' (in blue) of a highly simplified representation of an urban system.

practice it is usually impossible to measure or model the properties of this surface (especially in a city); therefore simplification is necessary.

(b) The perspective of an observer who considers the surface to be synonymous with that of the **ground**. This relegates the influence of upstanding elements to an insignificant role and is an increasingly incomplete definition as the density of elements increases and/or if the relevant properties of the elements are very different from those of the ground.

(c) The **roof-top view** that essentially ignores the spatial complexity of the interface itself and treats the system below as a 'black box'. Processes and properties at smaller scales than the box are lumped together and only the net result at the top of the box is considered.

(d) The **plan,** or **bird's-eye view** is that 'seen' by a remote **sensor** looking down vertically upon the system. From this perspective, vertical surfaces like walls cannot be seen and other surfaces are obscured from view (e.g. under trees).

(e) The **screen-measurement level view**, refers to the height of a weather screen (typically 1.5 to 2 m above ground) where standard (sometimes loosely

*Facet*    *Building*    *Street Canyon*

**Figure 2.3** Urban units (Table 2.1) and their typical dimensions in four cities from different geographical/cultural regions.

called surface) climate and meteorological observations are made.

(f) The **zero-plane view** is commonly adopted in micrometeorology when dealing with exchange

over crops, and forests that have a canopy layer. The location of the zero-plane is based on the vertical **profile** of atmospheric entities like wind above the canopy. When we extrapolate the profile

## Block    ## LCZ or Neighbourhood    ## City

**Figure 2.3** (*cont.*)

downwards to its apparent origin, this height is displaced from the ground (e.g. the **zero-plane displacement** of the wind profile; see Chapter 4 and Equation 4.9 therein). This 'surface' is defined mathematically (or statistically) as the source of exchanges for the entity of concern (e.g. heat, moisture, momentum) above the canopy. Unfortunately this effective 'surface' cannot be sensed directly and its height for a given system may be different for different entities.

One reason why we need to define the location of the surface might be to measure or model a particular characteristic of the surface (e.g. surface temperature, **albedo** or wetness). However, definitions (b) to (f) are necessary compromises to the 'ideal' case in (a). A reasonable choice of the definition of an urban surface depends on the entity involved, the sensing system available, the **accuracy** required and the scale of interest. The scale difficulty becomes apparent if we consider say the challenges involved to measure or calculate the temperature of each of the 'surfaces' in Figure 2.3. For example, what if it turns out that the surface depicted in the figure is the microscopic surface of an individual roof tile, or alternatively a whole urban neighbourhood?

### Properties of the Urban Surface

An urban system is composed of an almost limitless number of climatically-active surfaces. They consist of a range of fabrics each of which has different climatic properties including the following:

- **Radiative** – geometry, **absorptivity, reflectivity, transmissivity, emissivity.**
- **Thermal** – **specific heat, heat capacity, thermal conductivity, thermal diffusivity, thermal admittance.**
- **Moisture** – interception and storage capacity, permeability, **stomatal** characteristics, chemical nature.
- **Aerodynamic** – roughness, zero-plane displacement, porosity.

For typical values of these properties, see Chapters 4–6. These surfaces are also arranged in distinctive structural configurations: sizes, shapes and relationships relative to their surrounding surfaces. Knowledge of these active surfaces is fundamental to understanding urban climates, so it is necessary to map the distribution of those that possess climatically-significant physical characteristics. There are examples throughout this book including the distributions of surface albedo, surface roughness, **anthropogenic heat** release, thermal characteristics and surface water availability.

These mosaics of active surfaces generate a similarly large array of microclimates. For example, there is a remarkably wide variety of atmospheric conditions co-existing in the environs of a single house lot during a fine summer day. It is not unusual to find sunny portions of a roof with surface temperatures of more than 50°C whilst a nearby shaded wall or plants

in the garden are 30 degrees cooler; over an open lawn it may be breezy yet just around the corner of the house the air is almost still. Further, whilst most of the lawn is moist, under the overhanging roof and beneath tree canopies the soil is arid. What may be the most favoured outdoor spot to sit for breakfast is unlikely to be the same as that for an after-lunch nap or to relax in the evening air. Similarly, every street, block and neighbourhood within an urban area has its own unique mix of active surfaces which contribute a distinctive climate at that scale. As a result, urban climates at the pedestrian level are often described as a 'collection of microclimates'.

### 2.1.2 The Hierarchy of Urban Units

The array of active surfaces and associated microclimates found in cities is initially bewildering in its complexity and the fact that spatial variability extends over distances from a few metres (a wall or courtyard) to tens of kilometres (whole cities). Hence, the question of scale is central to the intelligent design of measurement, modelling and applied schemes in urban climatology.

To a first approximation the integrated climate of a city is the sum of repeating combinations of materials and the arrangement of active surfaces into what we might call **urban units**. Table 2.1 and Figure 2.3 show a hierarchy of such urban units, using the names of common features, which can be recognized in most cities of the modern world.

### Urban Facets and Elements

In this book an **urban element** (also called 'roughness element' in the context of airflow) is regarded as the primary 3D unit, the repetition of these across a city produces urban climate effects. Urban elements can be buildings, trees and building lots. They are made up from smaller units which we refer to as **facets** (e.g. roofs, walls, lawns and paths) often distinguished by their directional aspect (compass direction, vertical, sloped or horizontal) The detailed treatment of the climatic effects of building architecture is part of the related field of building climatology. The geometric placement of an element, and its roughness, radiative, thermal and moisture properties are all of interest because each creates its own microclimate and influence upon the surroundings by creating **turbulent vortices** and **wakes, thermals** or pollutant plumes, and spatial patterns of temperature and rainfall on the

**Table 2.1** Classification of urban morphological units, built and green, and their urban climate phenomena, based on typical horizontal length scales (modified after Oke, 1984, 1989). Dimensions of units 4 to 6 are based on a city with about 1M inhabitants. See also Figure 2.3.

| Urban units | Built features | Green and water features | Urban climate phenomena | Typical horizontal length scales | Climate scale[1] |
|---|---|---|---|---|---|
| **Facet** | Roof, wall, road | Leaf, lawn, pond | Shadows, storage heat flux, dew and frost patterns | $10 \times 10$ m | Micro |
| **Element** | Residential building, high-rise, warehouse | Tree | Wake, stack plume | $10 \times 10$ m | Micro |
| **Canyon** | Street, canyon | Line of street trees or gardens, river, canal | Cross-street shading, canyon vortex, pedestrian bioclimate, courtyard climate | $30 \times 200$ m | Micro |
| **Block** | City block (bounded by canyons with interior courtyards), factory | Park, wood, storage pond | Climate of park, factory cumulus | $0.5 \times 0.5$ km | Local |
| **Neighbourhood** or **Local Climate Zone** | City centre, residential (quarter), industrial zone | Greenbelt, forest, lake, swamp | Local neighbourhood climates, local breezes, air pollution district | $2 \times 2$ km | Local |
| **City** | Built-up area | Complete urban forest | Urban heat island, smog dome, patterns of urban effects on humidity, wind | $25 \times 25$ km | Meso |
| **Urban region** | City plus surrounding countryside | | Urban 'plume', cloud and precipitation anomalies | $100 \times 100$ km | Meso |

[1] Scales used in this book.

ground due to shade and wind effects. Buildings, being essentially hollow, enclose a climate space designed to maintain human comfort in the face of a range of exterior climate conditions. Buildings are a major focus of applied research into **thermal comfort**, energy conservation, distributed renewable energy systems (e.g. photovoltaic and wind) and indoor **air pollution**. Trees also create their own microclimates and contribute significantly to the aesthetic, acoustic, hydrologic and ecological environment of cities. In parts of the world where trees are part of the natural vegetation, as one moves from the periphery to the city core generally the density of buildings increases, and of trees decreases. In the core, the morphometry is dominated by buildings and street pattern whereas, in lower density urban districts streets buildings and trees contribute more equally.

## Urban Canyon

The **urban canyon** or street canyon unit is the structure formed by the common arrangement of a street and its flanking buildings. It is most clearly defined in the densely built central areas of cities but its basic form is usually echoed across the urban landscape. Its geometric form is its most characteristic feature. Canyons are described by their two-dimensional cross-section, referred to by the dimensionless ratio $H/W$ where $H$ is the height of the buildings adjacent to the street and $W$ is its width. $H/W$ is known as the **canyon aspect ratio** ($\lambda_s$), it is important because it is relevant to many features of urban climates including radiation access, shade and trapping, wind effects, thermal comfort and the **dispersion** of vehicle pollutants. If roads are not a major feature of the settlement this unit describes the height and spacing of the dominant elements.

## Urban Block

The **urban block** unit is formed by the road network, which is usually comprised of a number of adjacent street canyons that have similar structure. A grid road pattern produces square or rectangular blocks with buildings around its perimeter and gardens, courtyards, etc. in its core; in the city centre a block may be a single massive building. Other road patterns may result in triangular and crescent shaped blocks. Blocks can also be irregular and block-scale phenomena are produced by special building groups like shopping centres, clusters of apartment towers, institutional buildings and factories. The block unit structure can also be manifest by the absence of development such as in parks, cemeteries, lakes, etc. Such units produce a patchwork of climates and are of sufficient size to be able to generate localized breeze systems (e.g. parks) or clouds (e.g. factory complexes). The block unit may not be easily identifiable in all cities.

### Neighbourhood

The **neighbourhood** unit includes the land use zones used in urban planning such as industrial, residential, commercial, major parkland and undeveloped land. These zones are often subdivided based on some criterion, e.g., heavy and light industrial, or low- and high-density residential areas. In fact description according to land use is not very meaningful in climatic terms (see the Local Climate Zone scheme, Section 2.1.4).

### City

The city itself forms a set of structures and materials which contrast to those of its **rural** environs; this gives rise to distinct climatic phenomena such as the '**urban heat island**' and the 'urban pollutant plume', both of which crystallize the idea of an urban climate. Cities, which sprawl, may have diffuse edges that make it difficult to know where the urban area ends and the rural area starts. Indeed, the intensity of management for agriculture and other purposes may mean no rural area exists. Urban climate impacts extend beyond a city's physical borders, into downstream areas, due to transport of atmospheric properties (temperature, humidity, air pollutants) by the wind.

### 2.1.3 Description of Urban Surface Properties

The classification of urban units in Table 2.1 gives a semblance of order to what at first appears to be an impossibly varied and jumbled set of surface scales and properties. Those scales are conferred by the

dimensions and climatic properties of the features that make an urban landscape. There are four classes of urban surface properties: fabric (materials), surface cover, urban structure, and urban metabolism. To properly study and model urban climates, and to compare climates at different sites and cities, we need to describe and quantify these properties.

### Fabric

The materials used to construct cities are extremely varied (e.g. concrete, asphalt, stone, brick, wood, metal, glass, tile) plus natural materials (e.g. soils, vegetation, water) are still present. Each material has its own distinct mix of radiative, roughness, thermal and moisture properties, so the climatic behaviour of an urban surface, which comprises a variety of materials in different proportions, is unique. At the microscale, these differences must be understood and taken into account, but at the neighbourhood and city scale it is acceptable, indeed usually necessary, to average or use bulk properties.

### Surface Cover

A simple way to describe the main components of an urban system at coarser scales than facets is to express the plan view area occupied by a cover or element type ($A_x$) within a total ground surface area ($A_T$) that is large enough to be representative of the area of interest (i.e. it contains a representative sample population of the $x$-th type, for example, an urban block, neighbourhood or city). This is a **plan area fraction**,

$$\lambda_x = \frac{A_x}{A_T} \qquad\qquad \textbf{Equation 2.1}$$

where the subscript $x$ in Equation 2.1 stands for the cover type that can be $\lambda_b$, $\lambda_v$, $\lambda_i$, the plan area fractions of buildings ($b$), vegetation ($v$) and impervious ground ($i$) (roads, parking lots, etc.) (Figure 2.4a–c).

Figure 2.5 shows the fractions of the three cover types at several urban sites where urban climates have been studied extensively in the literature. It provides a convenient way to compare study sites and assess where they sit in the universe of possible urban cover fractions. As one moves to the right and upward, sites are more heavily developed, i.e. there

**Urban cover**          **Length scales**          **Urban structure**

(a) $\lambda_b = A_b/A_T$    (d) Building dimensions    (g) $\lambda_{\text{floor}} = A_{\text{floor}}/A_T$

(b) $\lambda_v = A_v/A_T$    (e) Building spacing       (h) $\lambda_c = A_c/A_T$

(c) $\lambda_i = A_i/A_T$    (f) $\lambda_s = H/W$       (i) $\lambda_f = A_f/A_T$

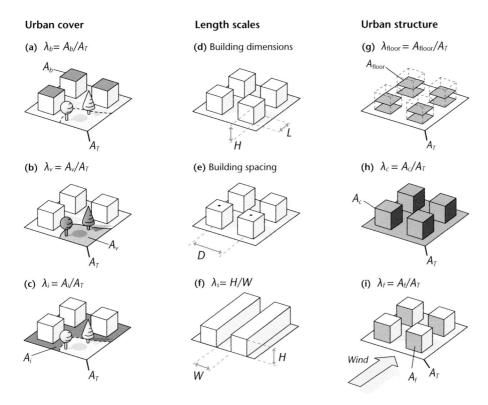

**Figure 2.4** Parameters used to describe urban cover, length scales and urban structure.

is an increase in the plan area occupied by built infrastructure (buildings, impervious ground).

## Dimensions and Structure

The dimensions of urban elements (buildings and trees) and their spatial arrangement are significant to airflow and radiation exchange in the **urban canopy**. Most of the relevant measures used herein are illustrated using the case of buildings in Figure 2.4d–f. The simplest are the element dimensions: the height ($H$), width and length ($L$), their spacing, i.e. the distance between centroids ($D$). For example, the canyon aspect ratio discussed above is a universal way to express built density as the ratio of height of the canyon walls ($H$) to the width of the canyon street ($W$).

In this book the height of buildings appears with one of two symbols: $H$ is used for the height of an individual urban element (a building, tree or canyon) and $z_H$ is the mean height of all elements (buildings, trees, etc.) in a larger area. The former relates to the scale of an individual element whereas the latter applies to a population of elements at the scale of a neighbourhood or city.

Figure 2.4g–i also depicts universal measures of structure like the floor space ratio ($\lambda_{\text{floor}}$) that gives the total area of all floors per unit plan area, as commonly used in urban planning and design and is also relevant in energy consumption.

The **complete aspect ratio** ($\lambda_c$) relates the total three-dimensional external surface area of all elements including the ground surrounding them, to the total plan area they occupy. The value of $\lambda_c$ for a flat surface is unity. As the number and size of elements increases so does $\lambda_c$, which expresses the important fact that the absolute size of the active surface area available for exchange increases. Hence, other things being equal, a convoluted urban area has more potential to absorb, reflect, emit, intercept, retard, store and release meteorological entities like energy, water and air pollutants than an equivalent flat plan area.

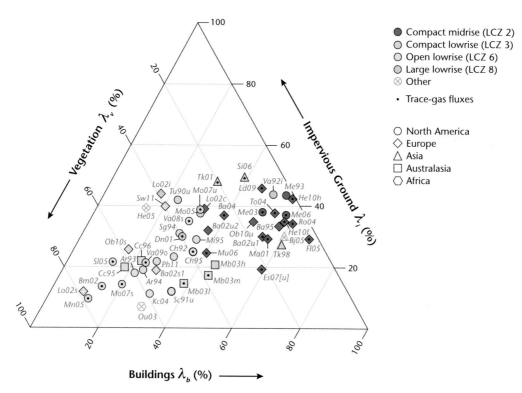

**Figure 2.5** Urban land cover domain seen as the plan area fractions of buildings, vegetation and impervious (other than building) surfaces. The symbols are urban research sites that have been studied extensively in the literature. The color identifies the Local Climate Zone (LCZ) representative of a site (see Section 2.1.4, Figure 2.9 and Table 1.1). Letters and numbers are a code that identifies the location, year of observation and a lower case letter identifies a site within the city. Observational results from these sites appear in later figures and tables in this text. A full listing is given in Appendix A2, Table A2.1 (Source: Grimmond and Christen, 2012).

The **frontal aspect ratio** ($\lambda_f$) is a parameter relevant to airflow. It indicates the fractional barrier presented by the buildings as 'seen' by the oncoming flow, it is similar to an inverse porosity. It is the ratio of the sum of the windward area of all elements to the total plan area they occupy. $\lambda_f$ varies with wind direction, the shape of the elements and their configuration as an array, because together they determine the area of the windward facets that is exposed.

Unlike the measures of structure discussed up to this point that quantify a larger urban area (block, neighbourhood), the **sky view factor** ($\psi_{sky}$) is a three-dimensional measure for a single point on a surface (Figure 2.6). It is defined as the fraction of the radiative **flux** leaving the surface at this point that reaches the atmosphere above the urban canopy, i.e. the 'sky' (Johnson and Watson, 1984). Its value depends on the position and orientation of a surface relative to the

amount of sky obstruction overhead. For example, on top of a roof with no **horizon screening** by other buildings or hills $\psi_{sky}$ is unity. But in the middle of a street canyon floor $\psi_{sky}$ depends on the depth and width of the canyon (i.e. it is loosely related to $H/W$; see Section 5.2.3 and Figure 5.12). In the bottom corners of the canyon it is even lower than in the middle. The sky view factor is significant for radiation calculations such as solar access and the nocturnal cooling of street canyons.

A fish-eye lens view upward from the point gives an excellent feel for this measure (Figure 2.7). Open areas such as the parking lot in (a) experience high $\psi_{sky}$, whereas highly obstructed configurations (d,e,f) have their overlying hemisphere nearly completely screened out by the walls of adjacent buildings and trees. Depending on the nature and amount of view occupied by vegetation, such as in (b) and (c), seasonal

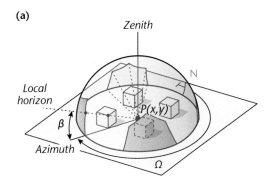

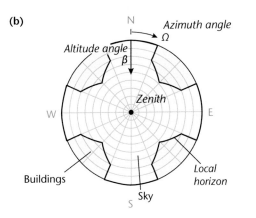

**Figure 2.6** **(a)** The sky view factor ($\psi_{\text{sky}}$) conceptualized for a point $P$ on the ground at an urban site. Dashed lines are sight lines from $P$ towards the sky hemisphere, the dots are where the line intersects the edge of an obstruction (building, tree, hill). The sum of all edge intersections around the compass defines the local horizon from $P$. **(b)** A polar plot illustrating the angles in (a) projected onto a flat plane. The outer circle is the theoretical horizon for an unobstructed plane. The local horizon joins the edges of the obstructions (in yellow). The sky view factor is the fraction of the radiation that leaves the surface at point $P$ that reaches the light blue area, i.e. the sky, whilst the remainder is intercepted by the buildings.

changes in leaf cover cause changes in $\psi_{\text{sky}}$. Calculating $\psi_{\text{sky}}$ requires consideration of spherical geometry, i.e. solid not planar angles and the orientation of the surface (cosine law, see Johnson and Watson, 1984). Projecting the horizon from a given location onto a polar plot conveys a good sense of $\psi_{\text{sky}}$, but it should be appreciated that the value is not simply the ratio of a circle on the ground to the area of sky seen in Figure 2.6.

Finally, the spatial organization of urban elements can also be important (Figure 2.8). Common

simplified patterns of element arrays include: aligned (elements located with approximately equal spacing and aligned both front-to-back and side-to-side with clear channels through), staggered (elements offset front-to-back and/or side-to-side) and random (elements scattered with no organized pattern across the array). Such patterns can exert control upon aspects of urban climates, for example airflow paths and the **turbulence** generated as air passes through and over the different array types. Figure 2.8a–c emphasize differences due to alignment, whereas Figure 2.8d,e show additional degrees of freedom created by varying the height and orientation of the elements. Real cities present an almost limitless set of element properties (Figure 2.8f).

Measures necessary to calculate or estimate the dimensions and structure of urban features can be extracted from architectural drawings, maps, street-level and aerial photographs, high-resolution satellite images and aerial Light Detection and Ranging data (**lidar**). Ground surveys and photograph are useful in determining the nature of surfaces beneath overhanging objects like trees that can be obscured when viewed from space- or airborne platforms. Lidar, **radar** and microwave sensors can help assess overall surface roughness at coarser scales. Further, many city governments maintain digital databases with detailed information about location and dimensions of features like buildings, roads, water pipes, sewers and trees.

### Urban Metabolism

The **urban ecosystem** runs on resources like fuels, water and materials mostly imported from elsewhere (Figure 1.4). These are incorporated into myriad human activities many of which end up expelling, often degraded, versions of these resources back into the environment, including the atmosphere, where they may play a role in urban climate. The three major types of emissions into the atmosphere are:

- Large amounts of **sensible heat** are released in the **combustion** of fuels in transport, industrial processes and in the engineered cooling, heating and air conditioning of the internal space of buildings (see Section 6.2).
- Additional **water** is released in vapour or liquid form during combustion, in industrial processing, due to air conditioning and due to garden irrigation (see Section 8.3.3).

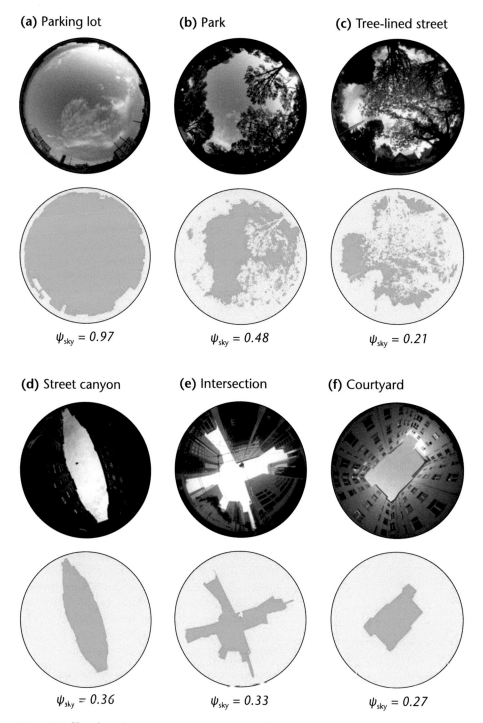

**(a)** Parking lot

$\psi_{sky} = 0.97$

**(b)** Park

$\psi_{sky} = 0.48$

**(c)** Tree-lined street

$\psi_{sky} = 0.21$

**(d)** Street canyon

$\psi_{sky} = 0.36$

**(e)** Intersection

$\psi_{sky} = 0.33$

**(f)** Courtyard

$\psi_{sky} = 0.27$

**Figure 2.7** Sky view photographs obtained using a fish-eye lens at a range of urban sites. The sky view factors ($\psi_{sky}$) are calculated for points at the ground. Note the values of $\psi_{sky}$ must consider the angular distortion of the lens and the cosine response of the surface. They are not simply the area covered by sky (in blue) projected onto a flat circle (Credit: (a)–(e) I. Stewart, (f) F. Meier, with permission).

**(a)** Aligned       **(b)** Staggered       **(c)** Random distribution

**(d)** Random height       **(e)** Random orientation       **(f)** Realistic 3D-structure

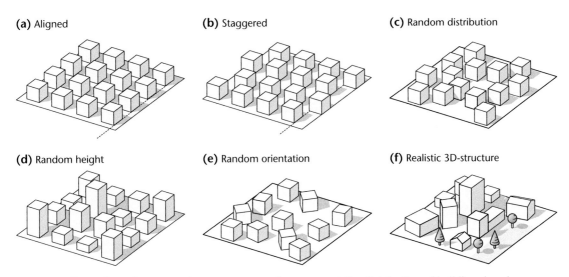

**Figure 2.8** Illustration of common element arrays used to represent the distribution of buildings in urban areas.

- Particulate and gaseous **air pollutants**, are emitted into the air from vehicles, buildings, waste management and industrial processes (see Section 11.1.1).

The timing and magnitude of all three sets of emissions operate on daily and weekly cycles that are largely dictated by human activities: the commute to and from work, work schedules and shifts, cultural practices, vacations and sleep, so they are called **anthropogenic** (i.e. human generated).

### 2.1.4 Classification of the Urban Surface

The four climatically relevant controls on urban climates (fabric, land cover, structure and metabolism) tend to cluster together in a city. For example, the core of many cities has relatively tall buildings packed together densely, the ground is mostly covered with buildings or paved with materials that are impervious, usually dry and good heat stores and the concentration of human activity gives large emissions of heat, perhaps moisture and air pollutant releases from furnaces, air conditioners and vehicles. Toward the other end of the spectrum of urban development are districts with low-density housing, one- or two-storey buildings of relatively light construction, surrounded by considerable garden, agricultural and open space with relatively low emission of heat but perhaps large injections of water by irrigation. The spatial correspondence between fabric, land-cover, structure and metabolism means that distinct climate modifications can be linked to urban landscape types.

### The Local Climate Zone Scheme

Such clustering underlies the notion of **Local Climate Zones** (LCZ, Figure 2.9 and Table 2.2) (Stewart and Oke, 2012). The criteria on which the classification is based are known to exert control on aspects of micro- and **local** climates (wind, temperature and moisture). The LCZ are clustered by their approximate ability to modify local surface climates due to their typical fabric, land cover, structure and metabolism. In Table 2.2, these controls are expressed through common properties that describe impermeability, roughness, thermal behaviour and the use of energy and water. The classification covers both built and natural ecosystems. Most cities around the world incorporate several of the built LCZ, or variants of them, depending on their particular cultural and economic character. The classification can be implemented using objective criteria that are easy to obtain from aerial photographs, **remote sensing** techniques, and maps including direct measurements of structure such as the canyon aspect ratio ($H/W$) known to be a critical control on flow régimes, solar shading and **heat island magnitude**, and a measure of surface permeability which is closely related to surface moisture availability. This scheme is more suited to urban climatic studies and projects than are land use classes (e.g. commercial, industrial, residential) which are designed to express urban function rather than the physical properties of the site that control its climatic responses. For example, the 'residential' land use class could include a wide range

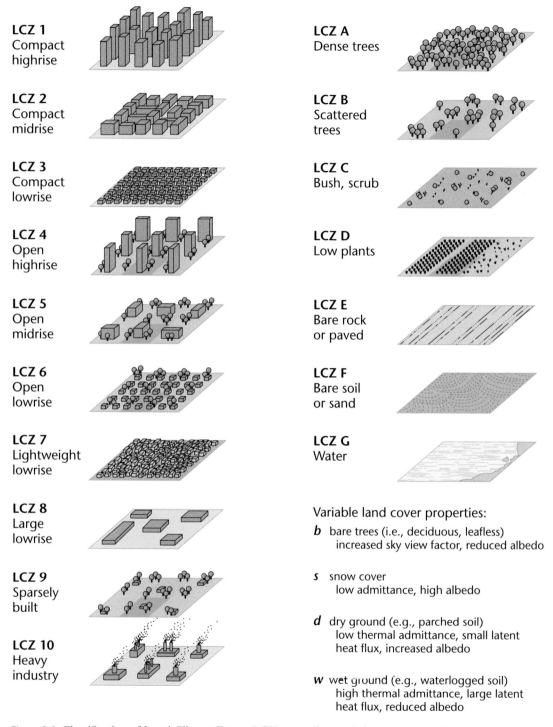

**LCZ 1**
Compact
highrise

**LCZ 2**
Compact
midrise

**LCZ 3**
Compact
lowrise

**LCZ 4**
Open
highrise

**LCZ 5**
Open
midrise

**LCZ 6**
Open
lowrise

**LCZ 7**
Lightweight
lowrise

**LCZ 8**
Large
lowrise

**LCZ 9**
Sparsely
built

**LCZ 10**
Heavy
industry

**LCZ A**
Dense trees

**LCZ B**
Scattered
trees

**LCZ C**
Bush, scrub

**LCZ D**
Low plants

**LCZ E**
Bare rock
or paved

**LCZ F**
Bare soil
or sand

**LCZ G**
Water

Variable land cover properties:

*b*  bare trees (i.e., deciduous, leafless)
      increased sky view factor, reduced albedo

*s*  snow cover
      low admittance, high albedo

*d*  dry ground (e.g., parched soil)
      low thermal admittance, small latent
      heat flux, increased albedo

*w*  wet ground (e.g., waterlogged soil)
      high thermal admittance, large latent
      heat flux, reduced albedo

**Figure 2.9** Classification of Local Climate Zones (LCZ) according to their perceived ability to modify local climate (Source: Stewart and Oke, 2012; © American Meteorological Society, used with permission). For quantitative measures of urban zone properties (first ten zones), see Table 2.2.

**Table 2.2** Typical properties found in the Built Zone series of the Local Climate Zone (LCZ) classes illustrated in Figure 2.9. A detailed tabulation of all LCZ classes including additional properties can be found in Stewart and Oke (2012).

| Local Climate Zone | Building plan fraction[1], $\lambda_b$ (%) | Impervious plan fraction[2], $\lambda_i$ (%) | Canyon aspect ratio[3], $\lambda_s = H/W$ | Sky view factor, $\psi_{sky}$ | Mean height of roughness elements, $z_H$ (m) | Thermal admittance[4] of system, $\mu$ (J m$^{-2}$ s$^{-\frac{1}{2}}$ K$^{-1}$) | Anthropogenic heat flux density[5], $Q_F$ (W m$^{-2}$) |
|---|---|---|---|---|---|---|---|
| **LCZ 1** Compact high-rise | 40–60 | 40–60 | > 2 | 0.2–0.4 | > 25 | 1,500–1,800 | 50–300 |
| **LCZ 2** Compact midrise | 40–70 | 30–50 | 0.75–2 | 0.3–0.6 | 10–25 | 1,500–2,200 | < 75 |
| **LCZ 3** Compact lowrise | 40–70 | 20–50 | 0.75–1.5 | 0.2–0.6 | 3–10 | 1,200–1,800 | < 75 |
| **LCZ 4** Open high-rise | 20–40 | 30–40 | 0.75–1.25 | 0.5–0.7 | > 25 | 1,400–1,800 | < 50 |
| **LCZ 5** Open midrise | 20–40 | 30–50 | 0.3–0.75 | 0.5–0.8 | 10–25 | 1,400–2,000 | < 25 |
| **LCZ 6** Open lowrise | 20–40 | 20–50 | 0.3–0.75 | 0.6–0.9 | 3–10 | 1,200–1,800 | < 25 |
| **LCZ 7** Lightweight lowrise | 60–90 | < 20 | 1–2 | 0.2–0.5 | 2–4 | 800–1,500 | < 35 |
| **LCZ 8** Large lowrise | 30–50 | 40–50 | 0.1–0.3 | > 0.7 | 3–10 | 1,200–1,800 | < 50 |
| **LCZ 9** Sparsely built | 10–20 | < 20 | 0.1–0.25 | > 0.8 | 3–10 | 1,000–1,800 | < 10 |
| **LCZ 10** Heavy industry | 20–30 | 20–40 | 0.2–0.5 | 0.6–0.9 | 5–15 | 1,000–2,500 | > 300 |

[1] Plan area fraction of ground covered by buildings.
[2] Plan area fraction of ground covered by impervious surfaces.
[3] Ratio of mean height of buildings to mean street width (LCZ 1–7) or distance between houses and trees (LCZ 8–10).
[4] Thermal property governing ease with which a body accepts or releases heat at its surface. Values are typical range for surfaces in each LCZ (e.g., buildings, roads, soils). Varies with soil wetness and density of materials (see Section 6.3).
[5] Heat released per area as a result of human activities, e.g. due to combustion of fuels. Mean annual values at local, not building, scale. Varies with heating/cooling degree days and season (see Section 6.2).

of urban forms: inner city housing with dense packing (perhaps row housing) and small or no gardens, or tall apartment towers or low-density housing with isolated buildings on relatively large vegetated lots. Clearly the climatic effects of these three exemplary 'residential' areas are likely to be very different given their contrasting fabric, land cover, structure and metabolism. LCZs are likely to be more meaningful ways to classify urban districts at the local (neighbourhood) scale for urban climate purposes. The scheme has been tested against thermal climate results from observations in real cities and as simulated by **numerical models** (Stewart et al., 2014).

### Climatopes

There is an alternate method to classify and map urban (and rural) microclimates based on

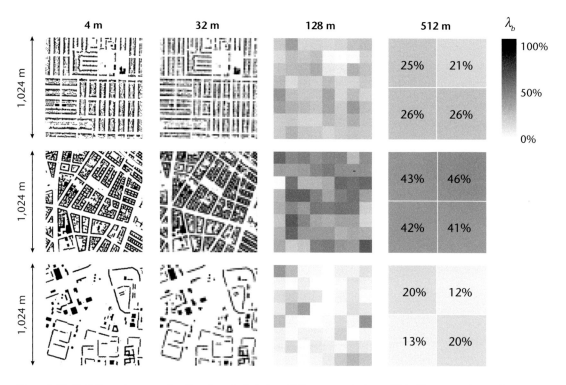

**Figure 2.10** Building plan area fraction $\lambda_b$ calculated at different resolutions. This illustrates the scale dependence of surface homogeneity and heterogeneity in three real urban environments with $\lambda_b$ that vary from an open lowrise (top), to compact midrise (centre), and open high-rise (bottom) LCZ. At the 4 m scale the urban surface is highly heterogeneous and patchy, but at > 250 m it blends into a more homogeneous raster of relatively uniform $\lambda_b$ in all three cases (Based on an idea of H.-P. Schmid).

geographically coherent areas that exhibit roughly the same microclimatic characteristics. Its spatial units are called **climatopes**, which describe areas with similar near-surface atmospheric properties. Unlike LCZs, they do not explicitly describe the surface that characterizes them. Climatopes are defined by aspects of the near-surface climate that are relevant to planning at a place and usually correspond to distinct geographic features such as valley systems that generate cold air drainage flows. Their purpose is to allow urban planners to design cities that make best use of the atmospheric ecosystem services and to identify areas that may worsen near-surface climates by inhibiting airflow paths that provide natural ventilation. The input information for a climatope map comes in the form of 'layers' of spatial data concerning atmospheric properties (airflow, temperatures), urban form (structure, cover) and place-specific geography (**topography**, population, land use). Maps are

created by experts who are informed by measured and modelled climatic elements; hence, they are unique to a particular place.

## Scale and Surface Homogeneity

In this book we consider an urban area to be 'homogeneous' in respect of a particular surface property, if the spatial variation of that characteristic lies within some statistical criterion. Homogeneity depends on the scale at which the property is determined and on the size of the **domain** over which it is being examined and applied. If the criterion for homogeneity is not met the area or scale is said to be heterogeneous. Commonly the smallest urban unit at which homogeneity can be found is the urban block and the largest unit where homogeneity is possible is probably a neighbourhood.

Spatially something homogeneous at a one length scale might be heterogeneous at another. Figure 2.10

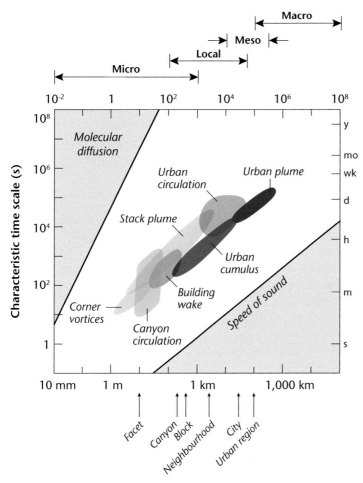

**Figure 2.11** Time and horizontal space scales of selected urban climate dynamics and wind phenomena. The common names of these scales are shown above and the corresponding urban scales (Table 2.1) below.

**Characteristic horizontal distance scale (m)**

shows that at small horizontal scales (< about 50 m) urban surfaces are highly heterogeneous but at larger scales (> about 250 m) you see that averaging renders the same surface to be almost homogeneous. Hence the 'neighbourhood' scale is typically relatively homogeneous, which conceptually justifies our treatment of the urban surface at this scale as one-dimensional. This is acceptable because changes in surface and atmospheric properties in the vertical are stronger than those in any horizontal direction, i.e. there is horizontal surface homogeneity at the study scale. On the other hand, when studying climates at the resolution of a single building, an urban block or an entire city, it is necessary to treat the problem as three-dimensional, because changes in properties in all directions are relevant, i.e. there is surface heterogeneity at the study scale.

## 2.2 The Urban Atmosphere

### 2.2.1 Scales of Urban Climate Phenomena

The scales of some urban climate phenomena related to atmospheric motion are given in Figure 2.11 according to their characteristic space and time dimensions. The lower horizontal space limit is set by the size of a building or the size of a smoke plume as it exits a chimney, and the time limit is set by the lifetime of a vortex behind a building, or an across-canyon circulation. The upper limits are the dimensions and lifespan of the pollutant plume of a whole city. Therefore, the majority of urban climate phenomena lie in the microscale, local scale and **mesoscale** domains. Climatic features cannot remain discrete in a diffusive medium like the atmosphere; they are part of a continuum. Each climate feature combines to form larger ones up to the scale of the whole urban boundary layer. The primary process that merges scales is mixing due to atmospheric turbulence.

Urban phenomena are just part of the complete spectrum of atmospheric scales (e.g. see Figure 1.1 in Oke, 1987). At greater scales than those of urban effects (i.e. moving to the top right in Figure 2.11) lie the motions of **synoptic** weather systems including **anticyclones**, mid-latitude **cyclones** and tropical storms. Urban climates are always under the control of synoptic weather events that on longer time scales create the background climate of a place. Weak synoptic flow combined with cloud free skies supports the development of thermal effects on climate. Stronger winds, on the other hand, dilute thermal effects at the expense of mechanical flow effects. Even stronger winds and cloud further dampen heating and stir up the atmosphere so urban effects are almost destroyed, except for those due to the flow deflecting and roughness effects of cities. At smaller scales than urban effects (moving to the bottom left corner of Figure 2.11), lie the motions of increasingly smaller turbulent eddies.

### 2.2.2 The Vertical Structure of the Urban Atmosphere

The lowest part of Earth's atmosphere that is in direct contact with Earth's surface is called the **atmospheric boundary layer** (ABL). The ABL is between 100 and 3,000 m deep and controlled by the roughness, thermal mixing and the injections of moisture and air pollutants from Earth's surface. The ABL can be subdivided into an outer and inner region. In the outer region thermal effects of Earth's surface dominate. In the inner portion, roughly the lowest 10% of the ABL and more commonly called the **surface layer** (SL), flow is dominated by friction with Earth's surface.

During daytime, surface heating usually creates large, buoyant thermals which effectively carry surface influences upward until they reach the top of the ABL where further lifting is halted by a **capping inversion**. This inversion at height $z_i$ is the base of the overlying **free atmosphere** (FA), where influences of Earth's surface are negligible. Just below the FA is the **entrainment zone** (EZ) where buoyant thermals 'bombard' the underside of the inversion; some overshoot into the FA through their own inertia and when

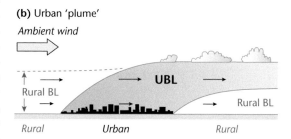

**Figure 2.12** Typical overall form of urban boundary layers at the mesoscale: **(a)** urban 'dome' when regional flow is nearly calm, and **(b)** urban internal boundary layer and downwind 'plume' in moderate regional airflow.

settling back carry cleaner, warmer and drier air down into the ABL. This daytime situation in the outer layer is often termed **mixed layer** (ML). The ML refers to the top 90% of the ABL excluding the SL. As its name implies, the ML homogenizes atmospheric properties so that vertical profiles of **potential temperature**, water vapour, wind speed and direction are almost uniform with height.

At night the ABL shrinks as cooling at Earth's surface usually creates a stagnant layer near the ground about 200 to 400 m deep which inhibits vertical mixing – this is the **nocturnal boundary layer** (NBL). Above the NBL, extending roughly up to the height of the daytime ABL, is a layer with properties preserved from the previous afternoon. This **residual layer** (RL) is capped by the inversion carried over from the daytime entrainment zone. This layer is mixed but little active mixing is going on.

### The Urban Boundary Layer

The ABL over a large city has its own structure. Theoretically when a regional wind is absent the climatic influence of a city is restricted to a self-contained **urban dome** (Figure 2.12a). More commonly, there is a regional wind and an **internal boundary layer** grows upward with distance, starting at the

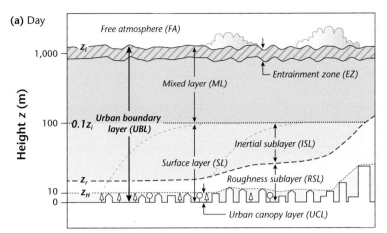

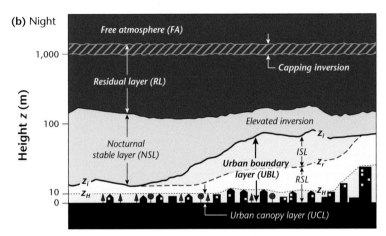

**Figure 2.13** Schematic of typical layering of the atmosphere over a city **(a)** by day, and **(b)** at night. Note the height scale is logarithmic, except near the surface.

upwind rural-urban border, until it fills the whole ABL – this is the **urban boundary layer** (UBL, Figure 2.12b), that part of the ABL influenced by the presence of a city.

At the downwind urban–rural border a new rural boundary layer forms; so the layer of urban-modified air becomes isolated aloft, forming what Clarke (1969) dubbed the **urban plume** (since it seems to spew forth as from a giant chimney, Figure 2.12b). The plume contains the thermal, moisture and kinematic effects of the city for tens of kilometres, and it carries air pollutants for hundreds of kilometres downwind. If the background wind happens to align plumes of several cities together they can combine to form a megalopolitan plume. These include the so called 'brown clouds' originating from the eastern seaboard of North America, south Asia and east Asia, which are detectable thousands of kilometres downstream.

A summary of the internal structure of the UBL by day and by night and its scaling variables is given in Figure 2.13a,b, respectively and Table 2.3.

In daytime, heating at the urban surface is stronger in most cities compared to the rural surrounding. This creates more vigorous mixing in the ML and **entrainment** at the top of the UBL which makes it deeper than the equivalent rural ABL; hence, the height of the capping inversion $z_i$ is elevated over cities (see Section 4.4). The strong mixing in the urban ML is responsible for the commonly uniform murkiness of the polluted atmosphere over cities during daytime.

Figure 2.14 helps to make the shape, dimensions and even the internal structure of a daytime boundary layer visible. The image comes from a downward-facing lidar system mounted in an aircraft as it flies across the coast and the city of Vancouver, Canada, early in the afternoon of a fine day in August. This lidar emits a laser beam towards the surface and it is

**Table 2.3** Classification of atmospheric layers comprising the urban climate system based on typical vertical length scales (after Oke, 1984). For illustration of the layering see Figure 2.13. Symbols of common scaling parameters are as follows: $z$ – height above ground; $z_H$ – mean building / tree height; $W$ – width of street canyons; $z_0$ – roughness length, $z_d$ – height of zero-plane displacement; $z_i$ – depth of mixed layer; $u_*$ – friction velocity; $w_*$ convective velocity scale; $\theta_*$ – friction temperature, $\theta_*^{ML}$ – mixed layer temperature scale (see Chapter 4 for details on those scaling parameters).

| Name of layer | Definition | Typical vertical dimension[3] | Common scaling parameters | Scale[3] |
|---|---|---|---|---|
| Urban canopy layer (UCL) [1,2] | From ground to the mean height of buildings/trees. It consists of exterior (outdoors) and interior (inside buildings) atmosphere. | Tens of metres | $z_H$, $W$ | Micro |
| Roughness sublayer (RSL) [1,2] | From ground up to two to five times the height of buildings/trees including the UCL. In the RSL flow is affected by individual elements. | Tens of metres | $z$, $z_H$, $W$ | Micro |
| Inertial sublayer (ISL) [1,2] | Above the RSL, where shear-dominated turbulence creates a logarithmic velocity profile and variation of turbulent fluxes with height is small ($< 5\%$). | ~25–250 metres | $(z - z_d)$, $z_0$, $u_*$, $\theta_*$ | Local |
| Mixed layer (ML) [2] | Above the ISL, where atmospheric properties are uniformly mixed by thermal turbulence and usually capped by an inversion. | ~250–2,500 metres | $z_i$, $w_*$, $\theta_*^{ML}$ | Meso |

[1] The entire layer from ground to the top of the ISL is called the 'surface layer' (SL), i.e. the SL includes UCL, RSL and ISL.
[2] The entire layer from ground to the top of the ML that is influenced by an urban surface is called the 'urban boundary layer' (UBL), i.e. includes UCL, RSL, ISL and ML.
[3] Scales used in this book.

so sensitive that it records the light scattered back from suspended **particulate matter (aerosols)** in the atmosphere as well as the ground. The lighter the tone in Figure 2.14 the greater is the particle backscatter, i.e. greater the number of particles in the air. Airflow is from the left across sea, where there is a marine layer extending up to about 230 m that contains mainly wavy horizontal layers. The layers dip down as they approach the warmer land, probably because the flow speeds up as it crosses the gradient at the coast, towards the lower pressure over the warmer land. At about 7 km from the left edge of the image a thermal internal boundary layer (IBL) develops near the coast as the air over the sea advects across sand flats, over the cliffs and encounters the warmer, drier and rougher surface of the land and city (i.e. it becomes a UBL). The internal structure of the UBL is more vertically oriented than over the sea. The light vertical streaks are thermals, as air 'bubbles' up from the hotter surface and punches into the overlying inversion, which seems to induce horizontal waves aloft. The streaks are most pronounced over the more developed parts

of the city and an industrial zone. The top of the UBL reaches up to a height of about 700 m.

At night the NBL found in rural areas is greatly modified in cities. Mixing, caused by heating of the surface due to the nocturnal urban heat island (Section 7.4) and the greater urban roughness (Section 4.4.1), cause the atmosphere over the city to be better mixed at night. Although the cooling might restrict exchange in less built-up areas all the way to the ground, in densely developed areas the near-surface atmosphere may experience often weak mixing and patchy elevated inversions at 50–300 m. The depth of the nocturnal UBL is much less than in the daytime; perhaps only a quarter of it, depending on the strength of the heat island and roughness effects.

### The Surface Layer

The surface layer (SL) occupies the lowest 10% of the ABL. Here, surface influences such as heating, cooling and roughness dominate even more, and frictional forces create most mixing. The SL over an urban area has two parts (Figure 2.13a):

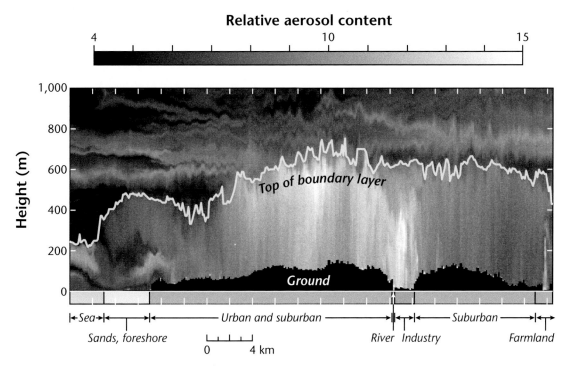

**Figure 2.14** The atmospheric boundary layer over Vancouver, Canada on August 15th, 2003, at 1337–1343 PST. The image comes from a downfacing lidar system mounted in an aircraft flying across the city. Airflow is from left to right and the yellow line is the approximate top of the mixed layer ($z_i$) (Source: Dr Kevin Strawbridge, Science and Technology Branch, Environment Canada).

- **Inertial sublayer** (ISL) is the upper part of the SL where the atmosphere responds to the integral effects of urban neighbourhoods. Over homogeneous land cover (e.g. within a LCZ), horizontal variations are small and the ISL can be treated in a one-dimensional way (i.e. only changes in the vertical matter). Well-known simplifications include the logarithmic wind profile (Equation 4.9). The characteristics and models applicable in the ISL are discussed in detail in Section 4.3. Also, we will see in Section 4.3 that variation of **turbulent fluxes** with height is small ($< 5\%$), so the ISL is sometimes also termed a 'constant flux layer'.
- **Roughness sublayer** (RSL) is the layer below, where the flow responds to the nature of the individual roughness elements themselves. It is a zone of substantial flow deflection, zones of up- and downflows, overturning, vortices and plumes arising from facets that are warmer/cooler, moister/drier, cleaner/more polluted than average. The situation in the RSL needs to be approached in a fully 3-D context. The characteristics of the flow in the RSL are discussed in detail in Section 4.2. Common relationships such

as the logarithmic wind profile are not applicable to the RSL which typically extends from the ground up to 1.5 to 3 times the height of urban elements (buildings and trees). The upper boundary of the RSL is termed the **blending height** $z_r$ (Figure 2.13).

Over horizontally extensive surfaces with low roughness, the SL is mostly an ISL. Over much rougher surfaces as one approaches the ground the flow is increasingly affected by the presence of individual roughness elements and the RSL becomes the dominant part of the SL. This is illustrated in Figure 2.13a by the increasing building height and the ascent of the blending height $z_r$, as one moves from left to right.

In city centres, it may not even be certain that a true ISL exists (e.g. at the right edge of Figure 2.13). There are two competing criteria for its existence. Firstly, the ISL must be above the effects of individual roughness elements, i.e. above the blending height $z_r$ where individual urban elements together create a uniform blended structure, and secondly, its properties must be in equilibrium with a homogeneous surface.

Since buildings and trees are typically 5 to 10 m tall the first requirement means the bottom of the ISL starts at least at $z_r = 10$ to 20 m above the ground, and since some elements can be 100 m or more in height, that suggests the blending height $z_r$ may be at altitudes of hundreds of metres in the core of large cities. An ISL is unlikely to form over clusters of such tall buildings and development of rather **chaotic flow** in a deep layer tends to 'squeeze out' any pre-existing ISL as its upper boundary is defined as the lowest 10% of the total UBL height from ground.

The second requirement for the existence of an ISL is the equilibrium with the local surface. This requires that an internal boundary layer (IBL) forms having been initiated at the border separating two LCZ surfaces. In Fig 2.12a the dashed lines curving upward from transitions in the building size, i.e. different LCZ, are IBLs. Such layers grow relatively slowly (say 1 m in depth for every 100 to 300 m of horizontal travel). Downstream from a change in surface properties to a new LCZ it takes a minimum of 1 to 3 km for an equilibrium layer to develop up to the top of the RSL (Figure 2.13a), and it could take 30 km or more. Summing up the two requirements, it emerges that whether an equilibrium ISL exists or not, depends on the height and spacing of the elements and on the distance in the upwind direction, to where the first different LCZ boundary is encountered. Hence, given the patchiness and changing physical structure of real cities, it cannot be assumed that an ISL exists at a given site, which is unfortunate because most of the common micrometeorological theory and simplifications applies only to the ISL but not to the complex flow in the RSL (see Section 4.2).

### The Urban Canopy Layer

In a city the lower part of the RSL, that is the layer below $z_H$, i.e. the height of the main urban elements, is termed the **urban canopy layer** (UCL) (Oke, 1976). The UCL is the site of intense human activity and exchange and transformation of energy, momentum and water. The top of the UCL is defined as the height of the urban elements – buildings and/or trees (Figure 2.13). Defining the height of the urban elements is not an easy task, as evidence suggests that a few unusually tall elements (isolated high buildings, tall trees) play a particularly significant role in the generation of roughness and turbulence (Section 4.2.4).

Distinctly different processes and phenomena occur near roof-level compared with those within the canopy. At roof-level there is intense **wind shear** and mixing.

Deeper down within the canyon-like streets, conditions are quite different; the canopy may provide shelter from the main impact of the wind speed. Further, the radiation exchange is disrupted by the increasingly restricted view of the Sun and sky. The climate at any point in the UCL is controlled by the unique mix of surface properties within a radius of only a few hundred metres. Their influence decays quickly with distance away from the point.

The volume of air within the UCL consists of an exterior and an interior part (Mills, 1997a). The exterior part is the air outdoors in pedestrian and managed spaces (courtyards, parks), and the interior part is the living space within buildings, the climates of which are often controlled by space heating and/or cooling. The coupling between the exterior and interior parts depends on the transmissivity of the windows and the porosity of the building membranes (walls and roofs).

### The Subsurface Layer

An urban surface modifies not only the atmospheric climate but also the subsurface thermal and moisture régimes, from the micro- to the urban scale (Sections 7.5 and 8.3.6). The subsurface beneath the city that includes urban soils and the below ground portions of buildings, tunnels and pipes, collectively exchange heat, water and air pollutants with the UCL above. The subsurface is coupled through the surface interface with the atmosphere by heat **conduction** and **diffusion**, hence modelling urban climates requires consideration of subsurface processes.

### 2.2.3 Linking the Vertical Structure to Horizontal Scales

To gain appreciation of the different layers of the urban atmosphere it is helpful to view it from different vantage points and scales. The microscale UCL perspective is largely that of a pedestrian walking between the elements in the street or looking out from a ground floor window (these are the facets, buildings and canyons seen in Figure 2.3). The climate at this scale is spatially highly variable, so exact location in relation to the surrounding urban elements is very important. The local-scale ISL perspective, on the other hand, can be compared to looking down on a neighbourhood from the window or roof of a tall building (the block and neighbourhood scale in Figure 2.3). The mesoscale perspective of the ML and the complete UBL is what one sees looking out of an aircraft as it flies over a city (satellite images in Figure 2.3). The organization of this book emphasizes the significance of the division of the urban atmosphere into these various layers (Table 2.3) and the corresponding horizontal scales of the urban surface (Table 2.1).

## 2.3 Defining an Urban Climate

### 2.3.1 The Superposition of Urban and Non-Urban Influences on Climate

This text focuses on atmospheric phenomena generated, or processes modified by, cities. Hence we are interested in phenomena and processes acting at the scales of urban systems (facets, buildings, urban blocks, neighbourhoods, cities). But remember that atmospheric phenomena and processes are themselves subject to modulation, even domination, by larger atmospheric scales, especially synoptic-scale weather and regional wind systems. So we need to separate urban from non-urban influences when studying urban climates.

### Lowry's Framework

Here we formally define an urban climate and 'urban effects on climate' using a slightly modified version of the conceptual framework of Lowry (1977). The measured value of a weather variable $V_M$ (such as air temperature, humidity, wind speed, etc.) at a station is conceptually assumed to consist of the linear sum of three contributions:

$$V_M = V_B + V_L + V_H \qquad \text{Equation 2.2}$$

where $V_B$ is the background 'flat-plane' value of the variable due to the **macroclimate** of the region, $V_L$ is the departure from $V_B$ due to landscape or local climate effects such as relief or water bodies in the vicinity and $V_H$ is the departure from $V_B$ due to the effects of human activities, including urban effects. The contribution from each term depends on: (i) the location of the station which is fixed; (ii) the suite of synoptic weather types that define a climate over a period of time; and (iii) the variable and changing character of human influences.

Consider the situation portrayed in Figure 2.15 where the location of a weather station regulates its source area and the contribution of $V_H$ to its observations. If it is located in:

- an **urban area** ($U$) the human impacts are obvious ($V_H$ is likely a major term);

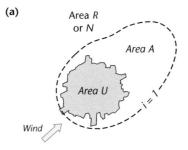

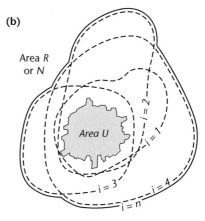

**Figure 2.15** Hypothetical form of the areas $U$ (urban), $A$ (urban-affected) and $R$ (rural) or $N$ (natural) – see text for details. In **(a)** the zone bounded by the dashed line is an example of an area resulting from a single weather type with a dominant wind direction from the south-west. In **(b)** the area inside the solid bounding line is the average urban-affected zone resulting from many weather types each with a dominant wind direction.

- a **rural area** ($R$) where effects of human activity are present due to land management (e.g. agriculture, forestry) human impacts are detectable ($V_H$ is likely relevant);
- a **natural area** ($N$) where human changes to ecosystems are not evident ($V_H = 0$);
- a rural or natural area ($A$) that is **open to transport of urban effects** from settled areas upstream. Here, the contribution of urban effects to $V_M$ varies with weather patterns. Over time, as the settlement grows, the areas $U$ and $A$ increase, $R$ may increase or decrease, and $N$ generally decreases.

Using this framework we can define the urban climate as the ensemble of $V_M$ values for the station located in $U$, but it also includes an amalgam of background climate, topographic and human influences. In this location the urban effect is the dominant human influence and is represented by $V_H$ which must be separated from $V_B$ and $V_L$. Similarly, for a station located in $R$, $V_H$ represents the impacts of human landscape changes but if it lies within the area $A$, an urban influence is also present. Only when a station is located in $(N)$ and outside the influence of the urban area $(A)$ can we state that $V_H$ is absent, that is $V_M = V_B + V_L$. However, definition is a lot easier than evaluation, due to several knotty methodological questions, with which we must now grapple. Moreover, this scheme does not account for the human contribution to global climate change that affects $V_B$; this is a topic taken up in Chapter 13.

### Conceptual Challenges

There is little, if any, methodology that belongs exclusively to urban climatology. This state of affairs is because the diverse nature of urban systems pose several challenging issues to the use of standard climatological approaches and methods. This section identifies some of the recurring difficulties of conceptualization and analysis encountered in the study of urban climates, and some commonly used experimental approaches to avoid or minimize them. Methodological aspects of the measurement and modelling of urban climates are dealt with in Chapter 3.

### Controls on Urban Climates

The physical processes and phenomena found in the urban atmosphere are subject to an almost bewildering list of **forcings**, influences and controls. In addition to the usual ones concerning the general climate of a place there are others due to the special nature of cities. The former are given, or *extrinsic* controls on a city's climate, whereas the latter are a function of its particular makeup, they are *intrinsic* factors. Intrinsic features are created by the builders and inhabitants of a city over its history and they retain the power to alter them through planning, design and new construction.

Some extrinsic controls such as latitude, altitude, proximity to a water body, relief and the biophysical character of the surrounding area are relatively, although not always completely, static, and hence give a relatively fixed contribution to the background climate of the city. Over shorter time scales further extrinsic controls govern the internal dynamics of weather: the time of day and season in conjunction with the solar geometry of the place and its prevailing synoptic weather situation. In particular the air mass dynamics and the accompanying wind, cloud and **static stability** conditions modulate the urban weather.

A city's intrinsic controls are its particular mix of urban surface properties, i.e. its fabric, land cover, structure, and metabolism (Section 2.1.1). To those we must add the size of the city (e.g. its diameter) which sets the distance over which the urban surface can interact with the atmosphere and indirectly affects the depth of the UBL (Figure 2.12, Figure 2.13). Later chapters will show how surface properties impact the processes that transport momentum, heat, water vapour and air pollutants between the surface and the air over a city and hence how distinctive urban effects are created in the fields of wind, temperature, humidity, air pollutants, clouds and precipitation.

## 2.3.2 Dealing with the Complexity of the Urban System

This Section grapples with the methodological challenges posed by having a multiplicity of controls acting on urban climates at several time and space scales and some strategies available to try to establish experimental 'control' and to isolate the influence of individual controls on urban climate processes and phenomena.

### The Multi-Scale Nature of Urban Climates

Section 2.1.2 listed the range of scales contributing to urban climates and possible urban units that form larger aggregates. In the course of urban climate research it is crucial to retain a sense of the scale appropriate to the process(es) and controls that govern the phenomena associated with it. This is true for both measurement and modelling. For example, if interest is in the effect of the city's roughness on the airflow in the ISL or ML, it is inappropriate to use wind measurements observed *within* the UCL. Nor should we expect a numerical model resolving boundary layer processes to be able to predict the wind field around a single building in the canopy layer with any accuracy. These are examples where the relevant scales (here UBL and UCL) have been mixed up. This problem may seem self-evident, but it is surprising how often it occurs in practice.

### The Role of Advection

Further, it is important to identify the active surface relevant to the scale of the enquiry, even though the spatial complexity of the system does not make this

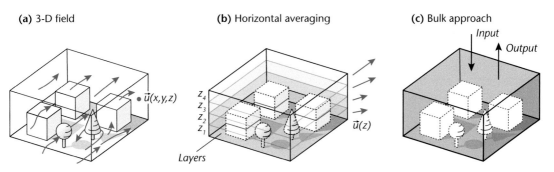

**Figure 2.16** Three approaches to reducing complexity in urban climate analysis of the UCL. In **(a)** the atmospheric state at a point is the result of forces and fluxes in three spatial dimensions and time. Any variable is described in a fully 3-D field. In **(b)** the atmospheric state is averaged horizontally to give a set of vertical layers that is described with an average value. In **(c)** a bulk approach assigns a bulk average to the transfer of the component through the top of a 'black box'; internal details of the system are not considered.

easy (Section 2.1.1). For example, working in the SL over many non-urban surfaces (bare ground, low vegetation, water, ice and snow) the system can be treated as one-dimensional (vertical) for the exchanges of heat, mass and momentum, as long as the surface is horizontally extensive. This is not appropriate for a study in the urban RSL where the structure of the active surface means that three-dimensionality is at play. Therefore, horizontal exchanges must be included in the analysis, because air flowing from one active surface to a different one carries with it the characteristics it acquired upwind (a process generally called **advection**). If the first surface is warmer than the second there is a positive net heat transfer from the first to the second, thereby adding to the energy balance of the latter. If the first surface is cooler, the net transport is negative and if the two surfaces are at similar temperature there is no *net* advection of heat. Analogous net transports of mass (e.g. water vapour, air pollutants) and momentum are also carried out by this process. Given the immense diversity of the urban units outlined in Section 2.1.2, it follows that advection is the norm, and it occurs at all of the scales in Table 2.3. It is most prevalent in the UCL because diversity is greatest there (e.g. from one facet to another), and it is also responsible for the transport of urban influences to downwind rural areas in the urban plume (Figure 2.12b).

### The Role of Anthropogenic Emissions

The urban climate system is further complicated by anthropogenic emissions of heat, vapour and air pollutants into the atmosphere (see Section 2.1.1). Emissions may be concentrated or diffuse, single or

multiple, intermittent or continuous, surface-based or elevated. In short, emissions are difficult to characterize and all but impossible to measure. Nevertheless, they represent a very real part of the climate system and must not be ignored – to do so would be akin to the over-simplification in early studies of **evaporation** that neglected the active role of plants in the process of **transpiration**.

### Simplifying the System by Spatial Averaging

The most obvious complexity of the UCL and RSL is its 3-D, multi-faceted and multi-layered structure. Everything appears to be specific to a unique location; the particular overwhelms the general, almost defying analysis. Within the UCL active surfaces occur at several different levels. For example, water vapour may be released from lawns at ground level, tree canopies at an intermediate height, water puddles on flat roof tops and even from chimneys above that.

This complexity dissuaded most meteorologists from working in urban environments until it was shown in the 1980s that simplification is possible with relatively little loss of rigour. The apparent chaos of the RSL can be simplified by two main approaches as illustrated in Figure 2.16. At any point, like the **vector** $\vec{u}(x, y, z)$ in Figure 2.16a, the 3-D properties of the wind field, and other variables, are the result of it being subject to the influences of all obstacles (buildings, trees, etc.) and surfaces in its path. However, it would take a multitude of such points to try to characterize fully the domain illustrated. To measure the complete wind field in that way is impossible with presently available sensors, even using the most advanced and detailed numerical models it is a major undertaking to simulate it. In many cases such detail is neither needed nor useful.

Conceptual simplification, however, is possible in measurement (e.g. some instruments perform path-averaging) and in numerical models (Chapter 3). Horizontal averaging is a simplification of the urban system to just one effective dimension – the vertical ($z$) direction (Figure 2.16b). That means the property of interest is averaged over 'layers'. We are no longer interested in differences in the horizontal ($x$ and $y$) directions because we assume atmospheric variables like wind, temperature and humidity reach horizontal homogeneity within a relatively uniform urban unit (e.g within a neighbourhood, Section 2.1.4). For this to be true requires that the domain be large enough. This then returns one value (the effective average) for the entire layer. If enough layer averages are used a vertical profile of the variable emerges (Figure 2.16b). Horizontal averaging commonly considers only outdoor air, not that inside buildings. The 'indoor' airspace is typically considered to be decoupled, often mechanically and thermally, from that outdoors, by the walls.

The bulk transfer approach is by far the simplest (Figure 2.16c). In systems terminology that means the domain is essentially a 'black box' – none of the interior dynamics or patterns of the property under study are resolved. We just describe input and output through the box. A box volume, whose depth spans the height of the UCL or RSL, is a useful choice, because it effectively eliminates concerns about the distribution of sources and sinks (of heat, mass and momentum) and limits exchanges to those through the top of the box. For surface properties (e.g. albedo) an equivalent horizontal active surface can be defined at the top of the box by using an appropriate weighting of wall and floor surface areas. It is probably not surprising that the complexity of model output decreases from left to right in Figure 2.16, and the computing time to run such models decreases in the same direction.

All is not solved by the approaches in Figure 2.16. Simplifying the system by using spatially averaged layers, active surfaces and volumes has its own set of conundrums. For example, it becomes more difficult to assign values to 'surface' properties, such as the aerodynamic roughness, thermal admittance or moisture availability. There are few methods to provide appropriate spatial averaging or weighting of such properties, and sometimes it is doubtful if they retain

their same connotation. Equally, it is difficult to know how to interpret the output of models that use spatially averaged surface values as input. This can make validation of such models ambiguous.

### Establishing Experimental Control

Scientific research which attempts to experiment, using field data, faces the problem of the variability of natural conditions, i.e. one is unable to 'control' many of the factors influencing the phenomenon under study. Careful experimental design can reduce the problem and thereby greatly enhance the value of the investigation. Unfortunately, many urban climate studies have been conducted without adequate attention to provision of control (e.g. poor choice of sites, inadequate record length, lack of stratification of data by weather conditions, and so on). Lack of control leaves an unstructured and incomplete data set with little chance of obtaining meaningful or transferable results. The intent of this discussion is to raise awareness of potential complications or pitfalls lurking in observational and modelling studies or in the interpretation of the work of others, it is not to 'freeze' the reader into inaction.

Attention to experimental control is essential if one is to provide scientifically sound and transferable answers to the following basic questions in urban climatology:

- Do urban climate effects exist, and if so what is their nature? (Section 2.3.3)
- What are the causes of such effects? (Section 2.3.4)
- Are the answers to these questions transferable to other times or in other cities? (Section 2.3.5)

### 2.3.3 Isolating Urban Effects

Much of the history of urban climatology is tied up with trying to answer the question: do urban climate effects exist and what is their nature? To employ Lowry's framework and identify urban effects, requires us to isolate $V_H$ for an urban weather station (located in $U$ in Figure 2.15).

### Pre-urban vs. After Urbanization

This is possible if we have continuous measurements ($V_M$) for a station that was originally located in a natural setting (i.e. located in area $N$ where $V_H = 0$) that subsequently experienced urban development. As long as we control for the ensemble of weather

**Table 2.4** Methods to attempt estimation of the 'urban effect' using observations.

| Method | Calculation | Assumptions | Issues |
|---|---|---|---|
| **Urban–pre-urban difference** | $V_H = V_{M(U)} - V_{M(N)}$ Difference between observations at the same station as its environment changes from natural (N) to urban (U) due to urbanization. | Contributions of the macroclimate ($V_B$) and topographic context ($V_L$) do not vary over period since urbanization began. | Stations that capture the transformation of a landscape from natural to urban are very rare. Background climate may change due to long term global or regional climate change. |
| **Urban–rural difference** | $V_H \approx \left( V_{M(U)} - V_{M(R)} \right)$ Difference between observations at two stations located in adjacent rural (R) and urban (U) areas. | Contributions of $V_B$ and $V_L$ are the same at each site for weather conditions and period examined. | Measurements in a rural area (R) are not equivalent to pre-urban values; the character of the rural area is not static so the contribution to $V_H$ changes; station may open to advection from urban area, hence urban-affected (A). |
| **Upwind–downwind difference or ratio** | $V_H \approx \left( V_{M(R)\ up} - V_{M(A)\ down} \right)$ $\left( V_{M(U)\ up} / V_{M(A)\ down} \right)$ Difference (or ratio) between observations at two stations located in a rural area upwind ((R) up), and one downwind, of an urban area (urban-affected (A) down). | Contributions of $V_B$ and $V_L$ are the same at each site for weather conditions (especially wind direction) and period examined. | Difficult to find stations that meet the requirements because urban-affected area (A) is unknown at start of the study and its shape and extent oscillates with weather, especially wind direction. |
| **Weekday–weekend difference** | $V_H \approx \left( V_{M(U)\ w'day} - V_{M(U)\ w'end} \right)$ Difference between observations at the same station, subdivided into those on weekdays (w'day) and weekends or holidays (w'end). | Contributions of $V_B$ and $V_L$ are the same for weekday and weekend datasets. Also magnitude and pattern of human activities have not changed significantly over period examined. | Weekend or holiday observations are not the equivalent of pre-urban values because human activity is not absent; effects of urban form (fabric, cover and structure) are present in both sets of values. |

types that govern the contributions of background and topographic contributions then the difference between $V_M$ before and after urban development represents $V_H$.

Implementing this procedure requires urban and pre-urban values of $V_M$ that are stratified by variables such as weather type, season, phenology, etc. In practice, pre-urban values are rarely available and even if they are, it is unlikely that they were observed at a natural site completely unaffected by human activity. Climate stations are usually established because of human interest in the location, such as an existing settlement, agricultural or forest uses, transport infrastructure, mining or other projected development. Even prehistoric populations may have changed the surface cover for example by burning or logging so finding a truly natural site is rare indeed. In the absence of appropriate observations Lowry suggests three means to try to generate surrogate pre-urban values: (i) run a site-specific numerical or **physical model** without the urban area included; (ii) use

observations from non-urban stations whose $V_B$ and $V_L$ characteristics are expected to be a close analogue of the urban site; or (iii) use statistics to extrapolate backwards in time combined with appropriate historical information about urbanization since pre-urban time. None of the possibilities is completely satisfactory.

The requirements to implement this ideal scheme are demanding in practice. Recent developments in numerical modelling make Lowry's means (i) the most promising approach, especially if the model has been tested against field observations. However, a variety of methods have been employed in attempts to isolate the urban effect (Table 2.4).

### Urban–Rural Differences

The most common approach has been to take the difference between present-day measured values of a weather element at one (or more) stations located in 'typical' urban (U) and rural (R) areas (Figure 2.15). If the assumption that the selected stations all have the same background climate and topographic influences, and that the rural station is outside the advective influence of the city (area $A$ in Figure 2.15) are fulfilled, then the difference between $V_{M(U)}$ and $V_{M(R)}$ could be taken as a measure of the urban effect ($V_H$).

In practice, it is nearly impossible to satisfy all the assumptions, although they can be approximated. Lowry notes that ensuring the rural station is truly in area $R$, not $A$, creates a circular argument because that decision depends on knowing whether urban effects on the weather element are present at the non-urban site or not. Also note that this approach assumes the present-day rural area is a reasonable surrogate for pre-urban conditions. Depending on the degree of agricultural or other management of the non-urban site this assumption may or may not be a good one; it may be better to use a station in a natural area (area $N$ in Figure 2.15). To retain an ideal case we also need to know if the present-day natural and the pre-urban sites on which the settlement has been built are different. It can be concluded that urban–rural differences can be used as second-order estimates of urban effects, but only if great care is exercised in the choice of stations, especially the non-urban one.

### Upwind–Downwind Differences

Similar thinking underlies this approach, but it only relates to urban influences on weather elements subject to advection downstream from the city. Thus, for a given weather type, when the urban-effect areas $U$ and $A$ take on a pattern like that in Figure 2.15a, it is assumed that upwind observations represent areas $R$ (or $N$), those in the city represents area $U$ and those downwind represent area $A$. Again, if we may assume $V_M$ at each site has the same values for $V_B$ and $V_L$, any resulting differences between $V_M$ at upwind and downwind stations are estimates of the urban signal. In practice such estimates are subject to similar flaws to those afflicting urban–rural differences, unless differences of regional climate and **topography** can be objectively ignored, and comparisons are restricted to conditions with similar wind directions (not just similar weather types).

If the study is longer-term, more climatological in scope, the conceptual definition of the urban-affected area takes on a more amorphous or quasi-circular shape (Figure 2.15b). It is the amalgam of the shapes of many weather types, each of which has its own characteristics (air masses, **fronts**, convective patterns, wind speeds and wind directions).

### Weekday–Weekend Differences

These are used on the premise that their existence in climatic records can only be due to impacts of the weekly human activity cycle on the atmosphere. For a given station, $V_M$ is divided into two subsets representing workdays, and weekend days (or holidays). Since there is no reason for $V_B$ or $V_L$ to vary between weekdays and weekends at the same station then the difference is taken to represent the urban effect ($V_H$). Of course weekend days are not the same as pre-urban days, so this measure cannot reliably estimate the magnitude of the total urban effect of a city on the weather element and it only signals the possible existence of effects, not their cause. Moreover, even if evidence of human activity differences on the two sets of days is found it is only circumstantial support, not a physical indication as to their cause.

### Other Indicators

There are other estimators of urban effects on weather and climate, including the urban–rural ratio, which uses the notion that if the ratio of $V_M$ values for an urban and a downwind station exceeds unity, for an element like precipitation, it might be due to downwind enhancement by urban influences. Another uses the idea that a trend in the time series of a weather element (or its urban–rural difference or ratio) is the result of an urban effect. Neither approach is free of complicating factors. Ratios are sensitive to spatial variability of $V_B$ and $V_L$, and time trends are sensitive to temporal trends at other scales.

### 2.3.4 Identifying Causes of Urban Effects

Lowry provides us with a useful framework within which to design experiments on urban effects and to help ensure that they maintain maximum control. His scheme helps to identify the main pitfalls to avoid, and provides a basis for assessing the quality of results. However, because the requirements are very hard to meet, it is not a scheme that can be readily implemented.

There is need for similar care in all investigations regarding the causes of urban effects. Indeed, if the two are to be linked in process-response models there is need to employ methods which isolate both cause and effect. Isolating causes is complicated by the fact that a change in one characteristic of the urban system often leads to modification of more than one process. An example of this correlation is the growth of a city's central core. An increase in the height-to-width ratio of the canyons alters not only the receipt and loss of solar radiation but also **longwave** exchanges and the state of airflow. Further, in practice it is likely that such changes in structure are accompanied by different building materials (e.g. to those necessary to support taller structures), an increase of space heating/cooling demand and a decrease of the area covered by vegetation on the lot. In turn, these will cause changes in **heat storage**, **evapotranspiration** and heat released in the burning of fuels. As a consequence of such changes to the central city it is quite probable that the mean temperature is increased – a phenomenon widely known as the **urban heat island**. However, if we ask 'what caused the warming?' the answer is 'everything'; all the modifications mentioned here are probably involved in a complex set of interwoven relations.

This should caution us to avoid trying to attribute effects to single causes. There is no set methodology to isolate individual or group agents in an urban climate system. With informed insight into operation of the climate system it may be possible to devise measurement and analysis strategies that permit the influence of certain controls to be examined. Four such examples are:

### Conditional Sampling by Source Area

Firstly, one can filter raw observational data to eliminate unwanted influences or isolate desired ones. Let's say we want to measure the wind properties (e.g. its **roughness length** $z_0$) representative of an LCZ that has a particular urban structure. But even if the roughness assessment is upset because upstream of the instruments there is a cluster of much taller buildings, or an open patch that is much smoother, the measurements can still be useful. As long as wind direction is gathered at the same time as the variable of interest, the information can be binned into different wind sectors and the analysis can be conducted, excluding times when flow is from problematic directions. **Source area** analysis is a more sophisticated approach to identify areas that influence a sensor (see Section 3.1.1). It allows identification of times when the signal from an instrument is being 'contaminated' by the effects of anomalous surfaces, so those data can be excluded from further analysis. Alternatively, one can exploit the contrasting effect of different surface properties as the source area of a sensor changes, to isolate surface controls on processes in the urban atmosphere (Crawford and Christen, 2014).

### Conditional Sampling by Weather Situation

An even simpler filter can be used to extract the effect of a weather variable. Under clear skies and light airflow, micro- and local phenomena and the effects of surface properties are best expressed. Subtle changes of properties like slope angle and aspect, albedo and emissivity, thermal properties, surface moisture and roughness spawn a myriad of atmospheric phenomena, and their statistical average establishes different urban microclimates. On the other hand, if the synoptic pattern brings cloudy, wet and windy conditions there is much less chance for small-scale differentiation of urban climates. This is because: heating and cooling is muted by the effects of cloud, thermodynamic responses are damped by surface wetness, and temperature and humidity differences are homogenized by mixing when it is windy.

The influence of weather on urban climate phenomena is visible when studying the magnitude of the urban heat island (UHI). A simple graphical plot of UHI versus wind speed shows increasing UHI magnitude with decreasing wind (see Section 7.3.5) however, often this relationship is associated with considerable scatter in the data. This is partly due to other influences on the UHI, such as cloud amount and cloud type that confound the relation. If there are sufficient auxiliary observations, a simple way to elucidate the relation better is to use only the data from runs with cloudless skies. When freed from the confounding effects of cloud, a more clearly defined relation often emerges.

### Economic, Technological and Societal Changes in Urban Systems

Opportunities to observe experimental controls arise spontaneously due to special events or when a particular control is terminated or altered. For example, a labour strike can allow the impact of industrial pollutants on climate to be better gauged because factory emissions cease. Traffic restrictions applied during large events can provide control to study urban air pollution. For example, Song and Wang (2012) show that measured air pollutant emissions from Beijing, China, decreased during the 2008 Olympic Games, due to traffic restrictions and factory closures. Other examples include the ability to assess the effects of space heating by comparing observations before and after a new building is occupied, or the effects of removing vegetation to pave an area may show up as differences in climate variables gathered before and after, or a municipal ban on external water use may allow the effect of garden irrigation on evaporation to be examined. In extreme cases, born of devastating circumstances, observations before and after the destruction of a city district by fire or aerial bombing have been used to examine the existence of urban effects (e.g. Imamura, 1949). These are occasions when an identifiable physical control is changed, and provides a degree of experimental control.

### Hardware and Numerical Modelling

Development of urban climate models (hardware or numerical) that simulate field conditions (Sections 3.2 and 3.3) is the most active area of present research seeking to identify cause-effect relations. If such models have been successfully tested against observations, they hold great potential. For example, they can allow the impact of proposed changes to the urban landscape to be assessed before construction. Such models depend upon full and accurate knowledge of how the climate system operates, before they can be expected to simulate conditions correctly. An interesting approach is to use an out-of-doors scale model of a building or city. Then, if observations of both the scale and full-scale cases are conducted simultaneously, the synoptic weather conditions are common to both.

### 2.3.5  Transferability of Results and Answers

There is need for methods to establish the comparability and transferability of urban climate knowledge and results from one city to another. One should not transplant results about urban effects on climate from one location to another (perhaps in a different macroclimate), without an adequate basis for generalization having been established. Some support may be provided by empirical evidence found in closely analogous cities, or from the output of a verifiable urban climate model, whose input characteristics can be specified in detail.

Munn (1973) recommends the use of **dimensional analysis** (Section 3.2.1) to help overcome some of the difficulties inherent in making results more transferable. For example, in this book we attempt to provide inter-city comparison of flux estimates by normalizing data from different cities and expressing all results as non-dimensional ratios. Considering energy fluxes ($Q_H$, $Q_E$ and $\Delta Q_S$) we use the radiant energy forcing the fluxes (e.g. $Q^*$). Thereby comparison of individual fluxes is standardized to a common level of energetic input.

If this final section appears daunting, that is not the aim. Rather, it tries to realistically point out that cities are very different from most rural environments, and therefore require their own methods of study. Perfectly designed urban climate studies are rare, but that should not deter work in cities. This section has been successful if it induces a note of caution when reading the work of others and in the design of new studies. It directs the reader to where others may have made errors and where most of the common pitfalls lie when thinking about undertaking a new study. There is plenty of scope to improve the practice of urban climate work.

### Summary

This chapter has laid out much of the conceptual framework and lexicon for what follows in this book. The overriding message is the paramount significance of **scale** in studying urban climates; whereas the physical processes are immutable their relative magnitude varies with scale and as a consequence so does the atmospheric response.

- The smallest unit of the urban landscape is a **facet** like a wall or roof, which has consistent properties, including material, slope and aspect. Facets may be combined to create an **urban element** such as a building or a tree. The combination of elements creates urban features such as **canyons** and **blocks**. The diversity of these elements makes the urban landscape spatially heterogeneous at a microscale, however, large areas ($> 1$ km$^2$) of cities tend to have similar mix of such diversity; these are **neighbourhoods**, which represent a type of homogeneity at a larger scale. The **city** as a whole is comprised of neighbourhoods of varying extent and makeup that reflects the historical development of the settlement.

- The **urban boundary layer (UBL)** describes the envelope of air that is modified by the presence of an urban area; it can be 1–2 km deep by midday but shrinks to 100s of metres at night. The lowest 10% of the UBL is the **surface layer (SL)** where the effects of the city are most profound, its properties are gradually diluted by mixing into the overlying **mixed layer (ML)**. The SL is subdivided into an **inertial sublayer (ISL)** and a **roughness sublayer (RSL)**. The RSL is shallow and highly turbulent, it exhibits the microscale effects of individual facets and elements; it extends from 1.5 to 3 times the height of buildings and incorporates the **urban canopy layer (UCL)** that exists below mean roof-level. The ISL separates the RSL from the ML; here the effects of turbulence in the RSL blend the contributions of the facets and elements that comprise neighbourhoods.
- To isolate an **urban effect** requires that other, **background climatic effects** are removed; these effects include weather and climate at the global and regional scales and their **interactions with topographic variations** (e.g. valleys, coasts, etc.). This is not a simple task and often requires methodological ingenuity.

The concepts introduced in this chapter provide the framework to organize the knowledge contained in much of the remainder of this book by scale.

# 3 | Methods

**Figure 3.1** Physical scale model of central area of Oklahoma City, United States, in a wind tunnel illustrating that the geometric similarity of building shapes is preserved (Credit: B. Leitl; with permission).

**Urban climate** studies are characterized by a great variety of methodological approaches. Figure 3.1 shows a scaled **physical model** of the downtown area of a city (Oklahoma City, United States). The model is located in a **wind tunnel** that can simulate the ambient airflow and how the cluster of buildings interferes with the wind and its ability to disperse **air pollutants**. Note the exact scaling of the buildings in the study area and obstacles placed upstream of the model city, which are used to generate **turbulence** and create an appropriate wind **profile**. Our understanding of urban climate has been, and continues to be, developed through scientific exploration that employs different but complementary

methods, each of which offers its own advantages. This exploration should be guided by the concepts outlined in Chapter 2, which provide a context for the application of methods and the interpretation of results. A successful study of urban climates is founded on clearly stated scientific and/or applied objectives that stipulate the

- **properties** and **processes** of interest (such as air temperature or **turbulent fluxes**),
- **physical extent** (**domain**) of the system under consideration, and
- strategy for capturing the **horizontal, vertical and temporal variation** within the system.

For many urban climate studies, the intention is to isolate and understand the effect that can be attributed to the processes that occur almost exclusively in urban areas (see Section 2.3). To accomplish this, the study must be designed to control, to the extent possible, extraneous effects that are not under study. The urban atmosphere is fluid and its properties adjust spatially and temporally to changing conditions at both its boundaries (the edges of the domain) and its internal processes (within the domain). Characterizing this continuous field and establishing its urbanized properties requires careful sampling both in space and time.

Here we introduce the methods used by urban climate scientists to study climate effects. It is not an exhaustive examination of these techniques and it does not try to instruct you how to conduct studies. Rather it allows the reader to understand the rationale employed by scientists in grappling with the great variation of effects that occur at different time- and space-scales in a city and the difficulty in extracting the distinctly urban contribution. This chapter employs conceptual diagrams to illustrate principles and uses examples from a range of studies that have studied the urban modification of atmospheric processes or phenomena.

The chapter is organized into four sections dealing with the main methods used in **urban climatology**: field observation (Section 3.1) that relies on measurements of surface and atmospheric properties using **sensors**; physical modelling (Section 3.2) in which observations are made using a scaled representation of the urban surface such as that shown in Figure 3.1; **numerical models** (Section 3.3) that represent the urban surface and simulate atmospheric processes using mathematical equations; empirical generalization and synthesis (Section 3.4), which establishes statistical relationships between parameters of the urban climate system using any, or a combination of, the previous approaches. The chapter concludes with a summary assessment of the relative strengths and limitations of each method (Section 3.5).

## 3.1 Field Observations

The careful recording of observations and their analysis is at the heart of any scientific enterprise. Luke Howard, working in the early nineteenth century, utilized instruments installed on his properties to measure atmospheric elements in the vicinity of a city.

His data, gathered on a daily basis over 25 years, allowed him to conclude that the climate of London, United Kingdom was distinct from that in the surrounding area (see Appendix A1.1). This observational approach has provided a large proportion of knowledge that underpins the study of urban climates.

Observations continue to play a central role in the evaluation of urban climate effects and in the development of understanding of the processes responsible. Given the growth in the number and size of cities described in Chapter 1, and the relative paucity of urban meteorological stations, the need for regular observations of the atmosphere in cities is probably greater now than ever.

Observations can be as simple as written records of events and phenomena seen. For example, the date of spring flowering of a given tree species across a city enables examination of temperature trends over history and spatial patterns of phenology; the behaviour of chimney plumes to study wind; and the type and abundance of sensitive lichens to assess the presence and **concentrations** of certain air pollutants. Some routine meteorological observations such as cloud amount and type are still gathered by visual inspection. Observations also include measurements made by physical devices (sensors) that respond to changes in the environment to which they are exposed. If sensors are calibrated so their response to change can be recorded on an agreed scale, then there is a basis for comparison and scientific communication of results. A modern sensor can generate an electrical output proportional to an environmental signal that is captured at regular intervals and stored on a computer designed for the purpose (a data-logger). An instrument consists of one or more sensors and incorporates additional signal processing of the sensor output. An instrument platform is used to mount and position sensors and instruments in the desired location; they include fixed towers and ground-, air- and space-borne vehicles. The ability to measure and record has advanced greatly in recent decades, but the question remains of how best to observe in a complex and heterogeneous setting like a city.

Two main types of observations are deployed. The first is an instrumented station situated to provide long-term but routine urban meteorological information (e.g. hourly readings of air temperature, pressure, wind and pollutants). Such stations are used by national weather service agencies or municipal authorities, for operational and regulatory purposes.

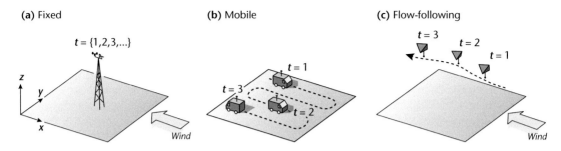

**Figure 3.2** Measurement over time and space using **(a)** fixed, **(b)** mobile and **(c)** flow-following approaches. A fixed measurement approach observes properties at regular time intervals as the atmosphere passes by a fixed point in space. A mobile approach observes properties through the atmosphere with a path determined by the observing platform, here represented by a vehicle. The flow-following approach is illustrated here by a neutrally buoyant balloon that follows airflow at a constant air density level.

The second is a field campaign established to evaluate some aspect of the urban climate, often to answer specific questions. These campaigns have a shorter life but their particular *foci* may require instruments that are uncommon, or are placed in dense configurations, or record at unusually high frequency, etc. Such campaigns use fixed observing sites (Figure 3.2a) similar to routine meteorological monitoring stations as well as mobile (Figure 3.2b) or flow-following approaches (Figure 3.2c). Whatever the method, it is essential to know the particular attributes of the immediate urban environment influencing the observations, as outlined in Chapter 2, as well as the methods employed to gather data.

### 3.1.1 Instruments and Their Exposure

The choice of an instrument depends on the variable(s) of interest, but none are unique to the study of climates in cities. Given the variety of spatial and temporal scales of urban climates it is essential to understand the properties of sensors and how to deploy them. Key concepts are instrument **exposure** and the related concepts of **fetch** and **source area**.

#### Instruments and Their Sensors

Instrument sensors are either immersed (*in situ*) in the medium of interest or they record it from afar (remote sensors). An *in situ* sensor is intended to come into equilibrium with ambient conditions, so that its response can be taken as a measure of the behaviour of the medium itself. This class of sensor includes conventional meteorological instruments to measure air temperature (thermometer), humidity (**hygrometer**)

and wind speed (**anemometer**) and direction (e.g. vane). They respond to the varying properties of the atmosphere as it passes by the sensor. Their response can be designed to follow fluctuations slowly or rapidly. Some sensors can respond to very small fluctuations that occur over short time ($\leq 0.1$ s) and space ($\leq 0.1$ m) scales; they are associated with **turbulent** motions. Instruments built with such sensors are used to measure the variables necessary to calculate turbulent exchanges of mass, **momentum** and energy.

The observational challenge is to expose instruments appropriately, so the recorded data provide useful information. In urban areas this means deciding where to situate instruments within the **urban boundary layer** (UBL), taking account of its spatial and vertical character (Section 2.2.2).

**Remote-sensing** instruments have sensors that are sensitive to energy associated with **radiation** (or sound) that emanates from a surface or volume. These sensors filter the signals linked to particular exchange processes and capture both the direction and quantum of the energy transfer. Many media absorb, scatter (reflect) and transmit radiation selectively based upon its wavelength. Remote sensors exploit this feature by tuning to different wavelengths so as to discriminate between different surface types and to examine atmospheric responses to turbulence, air pollutants, etc. Some instruments are active, meaning they both generate a wave signal (or beam) in a given direction and their sensors record the depleted beam that is transmitted through, or scattered back from, the object or medium of interest. These instruments can regulate the strength and direction of the signal they emit so that observations can be spatially attributed. Others

**Table 3.1** Remote-sensing instrument systems used in urban meteorology and climatology. With the exception of the radiometer these are active sensors that emit a signal and record its passage through a layer of atmosphere. The distortion of that signal is used to obtain information about the properties of the intervening atmosphere or surface. The choice of instrument depends on the property under examination.

| Instrument system | Principle of operation | Purpose |
|---|---|---|
| **Radiometer** | Solar or infrared radiation | Passively measures radiant fluxes. Facing upwards instrument 'sees' down-welling fluxes. Facing downwards it 'sees' radiation emitted and/or reflected from the surface. |
| **Radar** | Radiowaves | Locates and tracks the movement of suspended materials in the atmosphere. Can be used to acquire information on airflow, aerosols, clouds and precipitation. |
| **Sodar** | Soundwaves | Measures the wind field in the urban atmosphere. Can be used to estimate the vertical wind profile and the turbulent state of the atmosphere up to a few hundred metres. |
| **Microwave wind profiler** | Microwaves | Measures the wind field (can include all three components) in the atmosphere and the overlying troposphere. Can be used to estimate the turbulent state of the atmosphere. |
| **RASS** | Microwaves / Soundwaves | A combination of sodar and microwave wind profiler. Measures simultaneously air temperature and wind profile in the atmosphere up to at least 1.5 km. |
| **Lidar** | Light (Monochromatic laser) | Facing upwards, locates aerosols and clouds droplets. Can be used to determine mixing depth, cloud base and thermal inversions. By tracking movements of aerosols some systems can also infer wind and turbulence. Facing downwards, it can acquire a digital surface model of an urban environment (e.g. buildings, trees). |
| **Scintillometer** | Light (Monochromatic laser) | Measures the refractive index of air which changes with turbulent air temperature and humidity fluctuations. Oriented horizontally it can be used to obtain area-averaged convective fluxes over an urban area. |

are passive, meaning they simply record the absorption of 'natural' **radiant energy**, such as incoming solar radiation or its reflection from urban surfaces. Remote-sensing instruments are also characterized by their **field of view** (FOV, i.e. the solid angle from which the signal is received; see Figure 3.6). The FOV of instruments range from hemispheric, which describes a sensor that can 'see' from horizon to horizon in all directions, to narrowly focussed ones that are used on airborne or satellite platforms to observe features on Earth's surface (Table 3.1).

### Exposure

All sensors record the environment to which they are exposed; near the surface the challenge is to ensure they are positioned to obtain their signals from appropriate surfaces/areas of interest. To record the influence of a particular surface on its adjacent atmosphere requires careful consideration of the scales of the surfaces of interest and of the **boundary layer(s)** they generate.

To illustrate, consider a simple study based on observations of air temperature made over an extensive, flat surface of short grass (Figure 3.3a). The thermometer is housed in a ventilated shelter so its temperature is in equilibrium with the adjacent air. The thermometer is located away from the edges of the grassed area at a height of 2 m, so it is located within the boundary layer established by the homogenous grass surface. The thermometer responds to the passage of a great number of turbulent eddies of various size, many of which have interacted with the underlying grass surface and have been warmed/cooled as a result.

In steady wind conditions these eddies form a statistical ensemble that can be used to infer the position and size of the surface area that probably generated the signal being monitored. This patch is called the turbulent source area (or sometimes turbulent **footprint**); in this case it corresponds to a patch of grass upwind. Cases 1, 2 and 3 illustrate different atmospheric situations in which the source area varies in

**(a)**

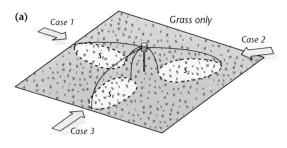

**(b)**

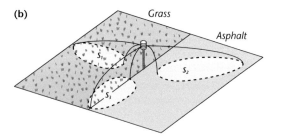

**Figure 3.3** Illustration of the concepts of the exposure and turbulent source area of a thermometer located: **(a)** in the middle of an extensive flat surface of short grass, and **(b)** at the border between extensive surfaces of grass and asphalt. See text for explanation of the three cases.

response to wind direction (location of the source area) and **dynamic stability** (size of the source area). Thus, sensor exposure may be translated into a turbulent source area: that patch upwind to which the sensor responds in those conditions. In these cases, because the thermometer is located within the same **internal boundary layer** (IBL), the signal is always derived from the grass surface, even though the location and size of the source area changes.

Now consider the challenge of obtaining equivalent, representative air temperature readings over a more complex landscape (Figure 3.3b). Imagine the thermometer is positioned at 2 m above the boundary between the moist grass and an asphalt surface. Both individual surfaces are extensive, homogeneous and flat, but each has a distinct surface climate that creates a unique boundary layer. On a cloudless sunny day the asphalt will usually be warmer by day and cooler at night, compared to the grass. Depending on wind direction the sensor is situated within the IBL of either (or both) surfaces, and the thermometer monitors eddies affected by the different surface types. If the source area is asphalt (grass), the observed air temperature is relatively warm (cool) during the day, and

cool (warm) at night. On the other hand, if airflow meanders across the grass-asphalt boundary, the air temperature fluctuates as the thermometer records a mixture of signals. In other words, if the source area straddles the boundary it is not simple to link the recorded temperature to either surface type. Poor site selection has introduced the complication of **advection** (i.e. the horizontal transfer of atmospheric properties, see Section 2.3.2). Careful planning is needed to reduce such complications, especially where the **surface cover** is heterogeneous in type and in size.

### Turbulent Source Areas

Suppose we are interested in the effect of a **neighbourhood (local scale)** on the overlying atmosphere. As indicated in Chapter 2, the urban landscape is comprised of a mosaic of climatically-active **facets** that affects the adjacent atmosphere and may be advected downstream. This heterogeneity is also scale dependent, such that common assemblages of facets describe **Local Climate Zones** (LCZ, Section 2.1.4, Table 2.2 and Figure 2.9) that may be considered relatively homogeneous surfaces at a local scale. Thus, the critical decision for the correct exposure of a sensor is based on the scale of climate effect being observed. In this case, the instrument should be located within the IBL developed by the neighbourhood of interest at a height where the microscale effects of facets have been thoroughly mixed. This distinct IBL develops at the windward edge of the LCZ, known as the **leading edge**, and grows in depth downwind (typically at a slope of 1:100 to 1:300). The distance of the sensors from this upwind edge in the windward direction is termed the **fetch** (Figure 3.4a). In the case of an urban neighbourhood, the microscale effects of individual surface facets extend through the **roughness sublayer** (RSL), which extends upwards to approximately 1.5 to 3 times the height of the urban elements (trees and buildings). So, to capture the effect of the target neighbourhood (LCZ), the sensors must be located both sufficiently far from its windward edge and at a height where the effects of individual facets have been blended.

If the sensors are correctly positioned, the turbulent source area of the signal at the measurement level can be assessed: it will be located upwind of the sensor, but its size and shape vary with height, wind speed and stability. In Figure 3.4a, the fetch of the observation site is sufficient that it is within its IBL and the sensor is positioned well above the roughness elements. In

**(a)** Side view

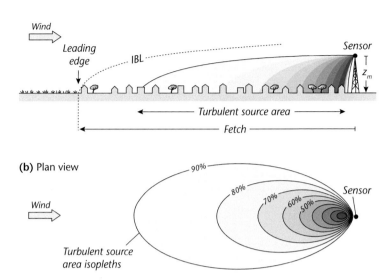

**(b)** Plan view

**Figure 3.4** Cross-section **(a)** and plan **(b)** views of an envelope enclosing the ensemble of eddies recorded by a sensor at height $z_m$. The plan view shows the turbulent source area of these eddies at the surface. Isopleths represent a given percentage of the entire source area, darker shading indicates a higher probability of influencing the sensor. The source area for the sensor is fully contained within the internal boundary layer (IBL) that develops with distance downwind (fetch) from the leading edge of a new LCZ. The sensor is assumed to be at a height $z_m$ sufficient for it to be within the inertial sublayer (ISL). The vertical scale in the figure is exaggerated relative to the horizontal scale.

the case illustrated, the source area is elliptical with its long axis aligned with the wind, however other shapes are possible. The isopleths indicate the likely contribution of that portion of the source area to the measured signal. Note the surface area closest to the sensor is where a significant proportion of the signal originates and that the contribution diminishes with distance upwind. Although this case is more complex than the simple grass-asphalt case, the logic remains the same. If the sensor were located at the edge of two distinct LCZ, then it can become immersed in one (or both) IBLs, depending on the wind direction. As long as the instruments are appropriately placed within the IBL of a single LCZ class, then although the source areas vary their position, size and shape they all sample from the same surface type.

Turbulent source areas cannot be defined with confidence for observations made *within* the RSL because of the variety of microclimatic environments to which the sensor is exposed. Within the **urban canopy layer** (UCL) the nature of the immediate environment is by far the most important influence and this includes the channelling or blocking of airflow and the presence of objects or surfaces that exert strong or anomalous climatic influence. As the influence of individual surface facets become more diluted with height above the UCL (i.e. as the local-scale signal begins to dominate

the microscale signals), the turbulent source area becomes more defined. This area will vary in size depending on the height of measurement (larger at greater heights), surface roughness, stability (increasing size from **unstable** to **stable**) and whether the measured quantity is a **flux density** or a **scalar** (larger for a scalar). As an example, a sensor that is mounted at 30 m above a LCZ of medium height and density will acquire 50% of its signal from a source area that extends to about 0.25 km in the upwind direction under typical daytime (unstable) conditions. However at night (stable conditions) that distance may increase to at least 0.75 km.

### Radiation Source Areas

The source areas for radiation sensors are somewhat simpler. In a conventional observational system a remote-sensing instrument (a **radiometer**) with a hemispheric FOV that points towards a flat surface acquires its information from a circular area. Technically such an instrument 'sees' to the horizon but the majority of its signal is obtained from a source area directly under the sensor. The response of many radiation sensors has a cosine-shape, which influences how they 'see' the surface below. Figure 3.5 illustrates the differences between: (a) the plan view of a regular array of cubic elements to crudely represent

**(a)** Plan view

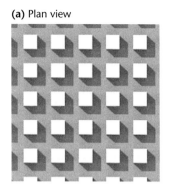

**(b)** Oblique view

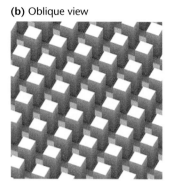

**(c)** Hemispherical view

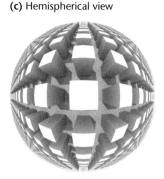

**Figure 3.5** Views of the same surface covered by cuboid elements: **(a)** plan; **(b)** oblique; **(c)** the view seen by a downward-facing radiation sensor with a hemispheric field of view (FOV). Solar azimuth is approximately 315° as shown in panel (a) (Source: Adderley et al., 2015; CC 3.0).

'buildings', (b) an oblique view of them such as one might observe from a tall building or a low flying aircraft, and (c) the view from a hemispheric radiometer mounted above the same array. The plan view sees only horizontal surfaces (here 'roofs' and 'roads'). The oblique view more explicitly shows the 3-D structure of the buildings but sees only select parts of the elements, for example, only two of the four walls can be seen and the view of one of those walls is partially obscured. The radiometer source area shows that the instrument receives some of its signal from the sides of the elements (building 'walls') in addition to horizontal surfaces. It also shows that the radiometer signal is dominated by surfaces located closest to the sensor. A useful rule-of-thumb for a downward directed sensor over a plane surface is that one half of the radiant influence of the surface originates from an area with radius equal to the height of the sensor.

A remote sensor with a narrower FOV that is directed obliquely at the urban surface, 'sees' a variety of different facets of the UCL, including roofs, walls and roads. Each facet contributes to the total signal intercepted by the sensor based on its radiative output and its area projected in the direction of the sensor. For narrow FOV sensors used to observe the surface, the information gathered depends on the viewing angle of the device relative to the surface. As the distance between the sensor and surface, and/or the FOV, increases the radiation source area usually increases to include multiple facets (Figure 3.6). Further, in urban areas, buildings and vegetation can partially block the sensor's view of the ground so the shape of the radiation source area becomes more complex. The representation of surfaces viewed within the source area may be biased (e.g. shaded surfaces may be under sampled or over sampled) depending on the sensor FOV, position and viewing geometry relative to that of the **solar azimuth** and **solar zenith angle**. This bias means the resultant measurement is directionally dependent, or **anisotropic**.

In many urban field studies, both radiation and turbulence sensors are used. However, their source areas may not fully overlap (Figure 3.7). If the source areas of sensors draw from different surfaces it is difficult to combine their observations into a coherent description of the energy balance. For example, in (Figure 3.7) the turbulence-based sensors have a source area that is mostly influenced by a vegetated patch but the radiation-based sensors have a source area firmly within a built-up area. To avoid this mismatch, measurement sites should be selected to have sufficiently extensive and homogeneous surface properties. Where such sites are not available, turbulent observations must be filtered by wind direction to avoid non-representative surfaces and the measurement height of the radiation and turbulent instruments may have to be different.

## Study Design

Planning an observational study must necessarily be clear on its purpose so that decisions can be made on: the site selection(s); the meteorological parameters to be measured and; the required spatial (horizontal and vertical) and temporal resolution. Clarity on these issues allows the researcher to select appropriate instruments and deploy them efficiently.

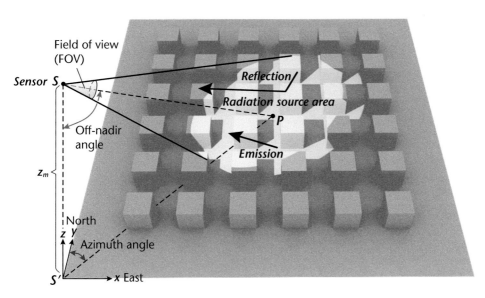

**Figure 3.6** Radiation source area of a remote sensor that receives radiation emitted by and reflected from a highly simplified and regular urban surface. The surfaces viewed within the source area depend on the sensor field of view (FOV), location $S$, height $z_m$ and, viewing geometry defined by the off-nadir and azimuth angles. The sensor line of sight is the distance between points $S$ and $P$, where $P$ is any position in the FOV. The off-nadir angle is the angle between the vertical ($S - S'$) and the sensor line of sight ($S - P$), typically defined for a point $P$ in the centre of the FOV. The azimuth angle of the sensor is the angle between true North and the line of sight projected on the horizontal plane ($S' - P$) measured in clockwise (geographic) convention. Not shown is the interference due to the radiative properties of the atmosphere (particles and gases that absorb, scatter or emit the type of radiation involved).

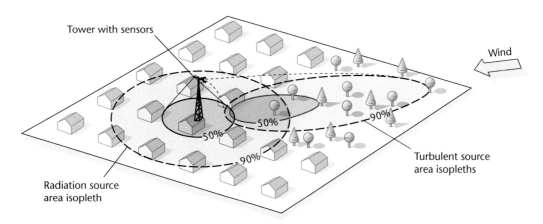

**Figure 3.7** Conceptual diagram of source areas showing disagreement between circular radiation (red) and elliptical turbulent source area (blue) isopleths, for sensors within the ISL, above an urban surface. In this example, the radiometer samples the signal from the mostly built-up surface, whereas the turbulent source area encompasses mostly the adjacent vegetated patch.

The instrumented mast shown in Figure 3.8, illustrates elements of instrument siting and sensor exposure. The mast is located within the fetch of an extensive LCZ and has been placed in a typical street canyon. Note the positioning of sensors at different heights to capture the climate effects generated at different urban scales. The lowest sensors are situated within the RSL, some distance from the roof, wall and

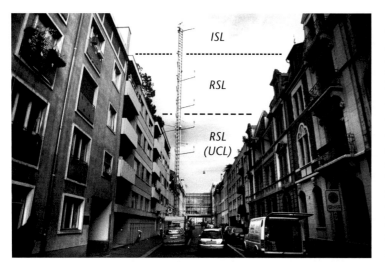

**Figure 3.8** Sensor levels on a tall tower located in Basel, Switzerland installed as part of the Basel Urban Boundary Layer Experiment (BUBBLE; site 'Ba02u1', see Table A2.1). The tower extends from ground level through the urban canopy layer (UCL) and roughness sublayer (RSL) into the inertial sublayer (ISL). Instruments located within the UCL and RSL yield measurements that vary spatially whereas those from within the ISL should not since they are representative of the neighbourhood scale (Credit: A. Christen).

floor facets of an **urban canyon** unit. These sensors record the microscale climate effects of individual facets and of the canyon itself. The sensors at the highest level are situated within the **inertial sublayer** (ISL), where the turbulent flux densities are nearly constant with height and represent the integrated effect of the underlying local scale, neighbourhood surface type.

Figure 3.9 illustrates a variety of platforms used to study climate variables, processes and phenomena at different urban scales. In the following we discuss four strategies to measure the urban effect: fixed stations (Section 3.1.2); mobile measurements (Section 3.1.3); flow-following approaches (Section 3.1.4); and remote-sensing techniques (Section 3.1.5). None are unique to urban climate studies.

### 3.1.2 Measurements at Fixed Stations

The simplest way to sample an urban climate over time is to use a conventional weather station, designed for long-term use (decades) to observe climate properties (**precipitation**, wind, temperature, etc.) comparable to those gathered at other sites. Many of the concepts used to site fixed stations apply to more specialized measurement programmes (Figure 3.9, Labels 1 to 3)

If it is only possible to have a single station, the LCZ selected might be the largest in the urban area, or that occupied by the highest population, or that located where urban effects are expected to be

most marked. The minimum size of a suitable LCZ is > 2 km$^2$ so the station can be located away from any leading edge of the zone, and hence be in the IBL of the LCZ. The exact placement of the instruments (horizontally and vertically) sets their exposure and the scale of climate effects that can be detected. Depending on the instrument array deployed such a site can be used to sample microscale environments within the UCL, or the local-scale climate above the RSL, or both.

### Observing Microclimates at Fixed Stations

Stations within the UCL (Figure 3.9, Label 1) are probably affected by the environment within a radius of between 500–1,000 m, although this distance is difficult to quantify and varies considerably with the nature of the UCL and the exact location of the instruments. For example, in a narrow street with tall buildings on either side (LCZ 1), exposure of the instruments is relatively symmetrical, aligned with the street axis and includes mostly similar materials. But in an open-set low-rise LCZ 6, characteristic of a medium-density residential area, the source area may incorporate a wide range of structures and surface cover types. The site should be located within a reasonably homogeneous part of the LCZ and avoid places with anomalous structure, cover, materials or other properties (like a relatively moist patch in an otherwise relatively dry area, or one near a concentrated heat source such as a heating plant) within the source area. If the site is *within* a street, ideally the

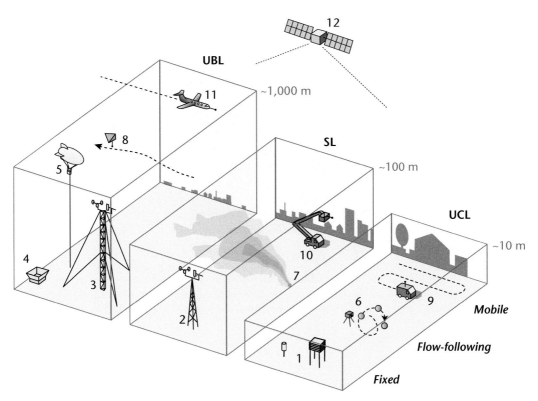

**Figure 3.9** Conceptual diagram of ground-based, aerial and remote-sensing observational platforms, sorted by their suitability to sample the entire urban boundary layer (UBL, left), the surface layer (SL, centre) or the urban canopy layer (UCL, right), and their sampling approach (up right side). Fixed platforms (along front) include a Stevenson Screen (1), a meteorological tower at least $1.5z_H$ high, (2), a tall tower (3), and ground-based remote-sensing platforms (e.g. sodar, 4) or a tethered balloon with instruments that are winched up and down (5). Flow-following approaches (along middle) include small-scale balloons tracked by cameras (6), a tracer release experiment (7), or tetroon balloons (8). Mobile / traversing approaches (along back) include: vehicles (9), mobile crane platforms (10), helicopters, airplanes or drones (11). Satellite remote-sensing can characterize the surface or atmosphere across a range of scales (12).

dimensions (height, width and length) of that canyon should be representative of the surrounding neighbourhood. If the site meets these criteria, the sensors can be expected to record the UCL **microclimate** that is 'typical' of that zone. Figure 3.10a shows sensors exposed to represent the climate experienced by a pedestrian within the UCL in order to assess their **thermal comfort**. This setup is designed for a research project rather than to provide long-term measurements representative of a given LCZ. Table 3.2 summarizes the general guidelines for measurements of commonly observed parameters at the microscale.

A network of stations is required if the spatial character of urban climate within the UCL is to be resolved. The location of each station should follow a similar protocol to that described here but there is no

universally-applicable design for an urban network. The best network may be specific to the city and elements(s) to be measured.

### Observing Local Climates at Fixed Stations

Above roof-level, measurements begin to represent the influence of the full LCZ. To get an integrated view sensors must be located fully in the ISL (Figure 3.9, Label 2). The minimum height for sensors is $> 1.5z_H$ but the horizontal and vertical position should reflect the growth of the IBL, which depends on surface roughness and atmospheric stability (Figure 3.4). For example, in a neutral atmosphere, the IBL over buildings of medium height (7–10 m) and density ($z_0 > 1$ m) grows at a ratio of about 1:12 but it is the lower 10% that comprises the ISL;

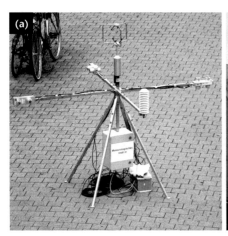

**Figure 3.10** Fixed measurement sites in urban environments. **(a)** A temporary site used to assess pedestrian thermal comfort in urban canyons within the UCL. Instruments for air temperature and humidity (within the white, louvred shield), 3-D airflow (3 axis instrument on top of the tripod) and radiation in the vertical and two horizontal planes (instruments on the three other arms) are visible (Credit: H. Mayer; with permission). **(b)** Access to the ISL above a compact mid-rise district requires a tall tower mounted on top of a six storey building (Credit: J. Voogt).

a platform at 15 m lies within the ISL of that surface if the fetch is > 350 m. It follows that districts with tall roughness elements require (i) use of a very tall tower, and (ii) a large LCZ. Sometimes the tower can be mounted on top of a building (Figure 3.10b), but only as long as it extends above the flow induced by the building itself (see Chapter 4) and into the ISL if one exists. Guidelines for making appropriate measurements at the local scale are available (Oke, 2008), as briefly summarized in Table 3.2.

The region between the top of the UCL and the top of the RSL poses difficulties to measurements of airflow. Sensors placed near roofs are unlikely to provide representative readings, despite their elevation with regard to the ground. Roof facets are unusually dry, warm by day, cool at night and generate complex flows. Even when the platform is located above the UCL but within the RSL, turbulence is unlikely to have sufficiently mixed the contributions from individual facets to be fully representative of the LCZ. As a consequence, instrument signals vary greatly, simply by moving the sensors horizontally or vertically, because that exposes them to a very different mix of microscale influences. Rooftops can however be an ideal site to measure incoming radiation, if there are no obstructions above the horizon.

Very tall towers can reach into the lower part of the **mixed layer** (ML) of the UBL (Figure 3.9, Label 3). To access this layer, the tower must have a height of at least 0.1 the depth of the ABL. Over, or downwind of, extensively developed urban surfaces this can require heights well in excess of 100 m. Suitable towers are rarely available; where they exist the effects of their substantial structure on the recorded signal may need to be carefully considered.

### Metadata for Fixed Stations

Observation sites must be accompanied by **metadata** to fully describe the location and the instrument system (including the sensors, their exposure and recording schedule). If such data are absent the record is incomplete and it is difficult to attribute any variations over time to changes in climate alone. It is likely to be necessary to update the metadata regularly to account for any changes affecting the site. Metadata describe both the local (~1 km) and microscale (~100 m) conditions around the station, including the surface cover, structure, **fabric** and **urban metabolism**. At the local scale, the topographic setting of the site, its relation to the larger urban area and the distance to major changes in surface character should be recorded. At the microscale, it is important to record the properties of the instrumentation and observing

**Table 3.2** General guidelines for making meteorological / climatological measurements in the urban environment. Choice of the height of the measurement platform ($z_m$) relative to the urban surface ($z_H$, height of buildings) is critical (see Oke, 2008).

| Parameter | Microscale | Local scale | General |
|---|---|---|---|
| **Air temperature & humidity** | Screen-level ($1.5 < z_m < 5$ m). Avoid roofs that are anomalous climatic environments. | Ideally $z_m \approx 1.5 \, z_H$ to ensure adequate mixing of surface components. | Avoid anomalous local influences such as dry/wet patches and significant sources of anthropogenic heat (e.g. HVAC systems). |
| **Wind speed and direction** | Measurements are site specific. Position of sensor with respect to surroundings and roughness elements is critical. Avoid roof sites. | Must be in the ISL (i.e. above RSL) and below IBL of the LCZ of interest. Use tall mast, typically $z_m > 1.5 \, z_H$. | Sensors must be 2–3 tower diameters' distance from mounting tower. Open tower construction is preferred. Avoid wake zones of tall buildings. |
| **Precipitation** | Measurement is site specific and sensitive to wind conditions. Avoid roof-level measurements due to anomalous wind patterns. | Same as for wind; $z_m > 1.5 \, z_H$ (to avoid building effects). | Locate gauges at open urban sites that meet standard exposure criteria. Co-locate with wind instruments on tall towers and correct for wind effects. Snow depth requires extensive spatial sampling to account for drifting. |
| **Surface temperature (via infrared radiometry)** | Position of sensor relative to surface should match source area with objectives. | Representative values may require sampling from airborne or satellite platforms. | Surface type, geometry and incident radiation required to correct for surface emissivity effects. |
| **Radiation (incoming)** | Site specific measurements. Sky view factor determines shading and contribution from terrain. | Representative values require an unobstructed view of the sky. | Measurements usually made on a plane, unobstructed surface except for specialized measurements. |
| **Radiation (outgoing and net)** | Site specific measurements. Adjust height to give a source area for the surface of interest. | For LCZ scale representative values $z_m > 2 \, z_H$ | Ensure surface directly below radiometer is not anomalous. Radiative divergence may need to be considered for large $z_m$. |
| **Turbulent fluxes** | Measurements are site specific. Avoid roof-level measurements due to anomalous wind patterns. | For representative values at LCZ scale: must be above RSL and within IBL of the LCZ of interest ($z_m > 1.5$–2 $z_H$). | Same as for wind measurements. |

platform, including sensor heights, locations, tower characteristics, surfaces underlying individual instruments and information about changes in instrument location or observing practice.

### 3.1.3 Mobile Measurements

Another way to sample urban climates over space is to move the sensors *through* the urban atmosphere (Figure 3.2b and Figure 3.9, Labels 9–11). Ideally, such traverses are conducted during near steady-state conditions so variations during the sampling period can be attributed to the character of the place rather than to changes in weather at larger scales (e.g. **synoptic**). Traverse observations are usually conducted while the measurement platform is in motion, hence in the case of the UCL, the sensors pass through a succession of microclimates. This is ideal for recording scalars like pressure, temperature, humidity and air pollutant concentrations. Traverses are not suited for wind observations unless there is a way to account for the effects of the sensor's movement relative to the flow. Flow-following approaches (Section 3.1.4) are available to assess airflow in all layers of the UBL.

### Traverses in the Urban Canopy Layer

It is common in urban climatology to make observations using instruments mounted on vehicles such as cars and bicycles (Figure 3.11). Traverses can identify spatial patterns in air temperature, humidity, surface temperature, incoming short- and longwave radiation and air pollutants. They are especially suited to examine parts of the urban landscape that are rarely explored or where spatial gradients in some property (such as air temperature) are expected to be large. Adequately monitoring these situations with fixed stations is expensive and perhaps not possible. Traverses can also give an overview of the diversity of microclimates in an urban area, to supplement fixed station networks and to aid in the placement of new fixed stations.

It is possible to conduct a 'stop-and-go' strategy, whereby instruments are mounted on a mobile platform that stops at fixed points to gather data, before moving to another location. Naturally, to interpret such results it is necessary to know the location of the platform at the time of measurement, including its elevation with respect to a benchmark, so corrections for changes of relief can be made. Inexpensive Global Positioning System (GPS) units can be included as part of the instrument system. This has the advantage of recording its location, and if this is done at regular intervals the velocity of the platform can also be derived.

This measurement strategy, while providing spatial information, introduces difficulties that must be addressed before urban effects can be assessed. For example, studies of the near-surface air temperature, are often conducted at night under clear skies with little airflow. In these conditions, thermal differences between surface types are apparent in air temperature so that a traverse exposes the thermometer to a sequence of microclimates. However, to explain the recorded temperature differences in terms of urban surface types requires that other confounding effects be removed first. This involves:

- **Removing the effect of elevation changes** along the route. A first order estimate can be made by applying an average environmental **lapse rate** (e.g. 0.64 K per 100 m) to points along the route, but this does

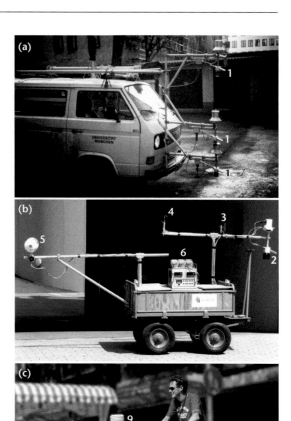

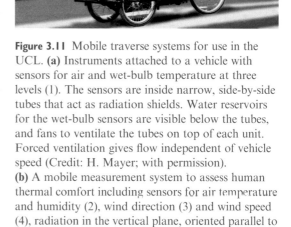

**Figure 3.11** Mobile traverse systems for use in the UCL. **(a)** Instruments attached to a vehicle with sensors for air and wet-bulb temperature at three levels (1). The sensors are inside narrow, side-by-side tubes that act as radiation shields. Water reservoirs for the wet-bulb sensors are visible below the tubes, and fans to ventilate the tubes on top of each unit. Forced ventilation gives flow independent of vehicle speed (Credit: H. Mayer; with permission).
**(b)** A mobile measurement system to assess human thermal comfort including sensors for air temperature and humidity (2), wind direction (3) and wind speed (4), radiation in the vertical plane, oriented parallel to the canyon axis, (5) all recorded by a central datalogger (6) (Credit: H. Mayer; with permission).
**(c)** A traverse system using a cargo-bike as platform, with radiometers (7), wind (8) and temperature/humidity sensors (9) (Credit: B. van Hove; with permission).

**Figure 3.12** A specially modified aircraft outfitted with meteorological sensors for observing conditions in the ABL. The instrumentation suite includes sensors to measure aerosols (1), static pressure and airspeed (2), wind and turbulence (3), fast-response temperature (4) and fast-response humidity (5) (Credit: D. Lenschow; with permission).

not address the effects of microscale advection, such as cold air drainage (see Section 12.2.2).

* **Correcting for cooling or warming** that occurred during the traverse. An average cooling rate can be estimated by having the vehicle return to the starting point, or by 'calibrating' measurements against observations at fixed stations on the route especially if they have a continuous record. One must also consider that cooling rates in the different LCZs traversed may be different.
* Removing occasions when the **vehicle is stationary or moves slowly** through traffic. At such times extraneous heat sources, including the vehicle's own exhaust, may contaminate the signal.

Detailed information about the exposure of the sensor to these microscale environments over the route are also needed. As a first step, the traverse route could be designed using LCZs as the sampling frame, that can provide a general guide for subsequent analysis. A limitation is that automobile traverses are restricted to the road network, therefore they give a biased sample of many LCZs, e.g. under-sampling of large parks or areas without access for vehicles. Of course it is possible to supplement with bicycle or foot surveys to fill-in some of the non-road temperature field. Careful design of UCL traverse routing prior to running the survey is invaluable.

### Traverses in the Urban Boundary Layer

Most of the UBL is not accessible by routine *in situ* observations. Airborne traverses therefore offer an attractive, but often expensive, way to sample the UBL air volume. Aircraft (including unmanned vehicles such as drones) can conduct horizontal and vertical traverses through the atmosphere, especially the ML (Figure 3.12). They have been used to measure radiant exchanges, turbulent flux densities, thermodynamic structure, air pollutant concentrations, cloud microphysical parameters, and so on, in addition to conventional meteorological elements.

Ideally these data can be used to build a three-dimensional picture of conditions and processes in the UBL. In practice, cost and regulation mean aircraft operations are often limited to line traverses at a few levels, or vertical profiles over a few locations, perhaps organized to form a cross-section or similar. Building a fuller 3D picture requires the combination of many systems: aircraft, balloons, and both upward- and downward-facing remote-sensing systems (Section 3.1.5, Figure 3.9).

### Vertical Soundings

Observations above the RSL can also be made by attaching sensors to buoyant balloons that freely float up through the UBL, or are tethered at a location (Figure 3.13). The simplest approach is to assume a free-flying balloon rises at a constant rate, and to track its path using a theodolite and record its position relative to the observation point. These data can be used to construct a wind profile, showing how winds change direction and speed with height. Large balloons can carry an instrument package that includes sensors for temperature, humidity and pressure. These are radiosondes, which transmit their monitoring to a ground receiving station (Figure 3.13a). On-board GPS sensors can give the positional information necessary to derive information about winds.

**Figure 3.13** Vertical soundings of the UBL. **(a)** Free-flying balloon and radiosonde instrument package (1) about to be launched from a roof (Credit: M. Roth; with permission). **(b)** Tethered balloon launch from an urban cemetery within a residential neighbourhood. The lightweight instrument package (1) is suspended beneath the helium-filled balloon. The balloon ascent and descent is controlled by a winch in the box (2) (Credit: A. Christen).

Radiosondes provide a comprehensive vertical profile of the state of the UBL throughout its depth. It is difficult to obtain information near the ground with good height resolution; for that purpose there are smaller lightweight minisonde balloons. Alternatively, a tethered balloon (Figure 3.13b), with an instrument package suspended below, can give detailed information at fixed altitudes in the lower part of the UBL. This provides several advantages over free-flying balloons: increased vertical resolution of observations, control over the altitude and sampling time of the sensors, and the ability to deploy more expensive instruments which can be recovered at the end of the flight. On the other hand tethered balloons are limited to use in relatively fine weather and are subject to aviation flight rules. Oscillation of the balloon in the wind can affect the measurements and requires correction.

### 3.1.4 Flow-Following Techniques

Flow-following methods are a special form of mobile measurement used to track the movement of selected air parcel(s) (Figure 3.2c and Figure 3.9 Labels 6–8). A very simple method is to release **neutrally** buoyant visual markers (e.g. coloured gas, bubbles, smoke or balloons) that drift with the flow. These can visualize the flow in city streets, around buildings or even across cities. If the movement of the marker can be tracked by photography, film, video, radar or GPS useful quantitative data on flows can be obtained (Figure 3.14). Tetroon balloons can be used to study circulation systems and 3D winds (speed and direction) with

sufficient **precision** to estimate the flux of momentum and the movement of pollutant plumes over urban areas (Chapter 4, e.g. Figure 4.28 and Figure 4.30).

### Tracers

**Tracer** methods are ideally suited to study the movement and dilution of air pollutants. A tracer is released to simulate **emissions** from a point (e.g. chimney) or line (e.g. road) source. Such tracers are usually an artificial, non-toxic, non-reactive (inert) and non-depositing compound that can be detected easily. The height of the tracer release (UCL, RSL, ISL or ML) must be carefully noted because this greatly affects the movement and dilution of the plume through the urban atmosphere. Figure 3.15 shows the results of a field study where the tracer is released within the urban canopy of Salt Lake City, United States. It depicts **dispersion** of the tracer both at the **mesoscale** (Figure 3.15a) and in the immediate vicinity (local scale) of the release (Figure 3.15b). At the larger scale, sampling points are positioned to form a series of arcs downwind of the release site; they capture the dispersion of the tracer as it drifts downwind as a plume. At the smaller scale, a different sampling strategy is used: a grid of locations surround the release site to account for the complex flows within the urban canopy that mix the plume. Once the spatial concentrations of the tracer have been measured they can be mapped (e.g. Figure 3.15) and a series of maps at different times, during and after the release, can show the temporal evolution of the tracer plume.

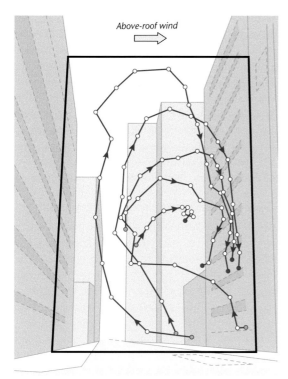

**Figure 3.14** Circulation within a street canyon in Chicago, United States. The above canyon wind is perpendicular to the axis of the street and drives a vortex across the canyon cross-section. The flow pattern is revealed by the motion of neutrally buoyant balloons, whose positions were photographed at regular intervals (Source: DePaul and Sheih, 1986; © Elsevier, used with permission).

### 3.1.5 Remote-Sensing Techniques

Remote-sensing instruments (Table 3.1) are deployed near-ground level to 'look' upwards to explore the overlying atmosphere, especially the UBL (Figure 3.9, Label 4), or they are pointed down from an elevated viewing position, to observe properties of the urban surface (Figure 3.6 and Figure 3.7). Their innate ability to integrate over volumes and surfaces and their ability to access the urban atmosphere from a single location, makes these systems especially valuable in urban climate studies.

Remote-sensing of surface characteristics is often able to provide **boundary conditions** (surface temperature, wetness, state of cover) to run numerical models. Their observations of surface conditions can also be

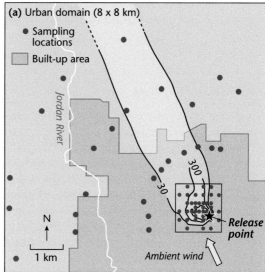

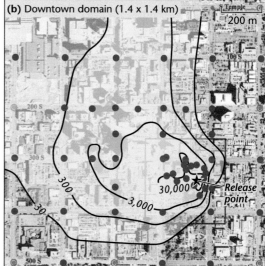

**Figure 3.15** Isopleths of half-hourly average sulfur hexafluoride ($SF_6$) concentration in parts per trillion at the end of a one hour tracer release, conducted during a field observation campaign in Salt Lake City, United States. Results are from 0100 MST 26 Oct 2000 and show: **(a)** a portion of the urban domain and the associated arcs of sampling locations, **(b)** the downtown domain (Modified after: Allwine et al., 2002; © American Meteorological Society; used with permission).

important in the evaluation of the output from numerical models. Successful deployment of remote-sensing must match carefully the scale of the source area to the application. It must also consider whether there is any potential sampling bias due to anisotropy (Section 3.1.1) or the limitations imposed by the FOV of the sensor to 'see', and therefore to sample, all facets of the 3-D urban surface.

## 3.2 Physical Modelling

A physical model, also called a hardware or scale model, is a simplified surrogate of a real-world system. Physical modelling permits quasi-controlled experiments, to simulate the climatic effects of one or more variables on the system. The urban elements (e.g. buildings, trees) are scaled and/or greatly simplified (e.g. using cubes as buildings). Models are constructed at full or reduced scale and its climatic effects result from exposure to selected conditions, sometimes described as **forcings**.

If the model is located outdoors, forcing depends upon the prevailing weather, but in a laboratory facility (e.g. wind tunnel, water tank) conditions can be fully controlled. For example, flow can be maintained in a steady-state, surface conditions can be modified at will (e.g. 'buildings' can be rearranged) and observations can be made wherever wanted (e.g. well above the model in its boundary layer). True experiments are possible because it is possible to examine the effect of changing one part of the system, while holding others constant. Practically these models are more efficient than field observations because the latter require a fully appropriate site and waiting for the required weather conditions to occur. These advantages of laboratory modelling have been exploited in many urban studies including turbulent flow, dispersion and transport, wind energy, mechanical forces on bridges and buildings, snow and sand drifting, human comfort and safety, solar access and shadows, the **surface radiation budget** and **energy balance**. The main limitation is physical modelling requires special facilities and care to establish similitude (next section).

### 3.2.1 Scaling and Similitude

Physical modelling requires similarity (or **similitude**) between the model and the real-world system (or prototype) it seeks to mimic. Similitude is achieved by ensuring a consistent match between critical ratios of physical and atmospheric properties in both the scale model and the prototype. This is informed by a **dimensional analysis** in which fundamental quantities such as mass, length, time and temperature are

examined with a view to preserving key relationships between the prototype and model, albeit at a different scale. Complete similitude requires the fluxes of mass, momentum and energy are matched between the model and the prototype. Generally this is not feasible, because some aspects of the prototype cannot be scaled but, since the purpose of a model is to simplify a complex system for controlled study, it is enough to achieve similarity of key aspects. Physical models usually seek similitude in one or more of the following: surface geometry, fluid flow and heat transfer.

### Geometric Similarity

Geometric similarity is the most basic form of similitude, it simply requires the length dimensions and angles of the prototype to be scaled consistently in the model (Figure 3.1). This preserves areas and volumes, the slope and aspect of facets, the shapes of elements and their geometric relationships to each other. For urban surfaces this means that non-dimensional descriptors such as the **canyon aspect ratio** $(H/W)$ , **plan area fractions** $(\lambda_b, \lambda_v$ etc.) and **complete aspect ratio** $(\lambda_c$, see 2.1.3 for definitions) are preserved. Geometric similarity is needed in studies of radiation exchange including analysis of shadow patterns, reflection and exchange of radiation between urban facets. If the study only deals with **direct-beam irradiance**, geometric similarity may suffice. However, if other processes (e.g. reflection, transmission, emission) are to be considered, further model design, including the radiative properties of the materials is needed.

### Dynamic Similarity

Achieving dynamic similarity is much more difficult because fluid flow and momentum transfer are scale dependent. One measure of the relationship between the model and the prototype is to compare the respective Reynolds Numbers (*Re*), which is a non-dimensional measure of the ratio between the inertial and viscous forces in fluids:

$$Re = \frac{uL}{v} \qquad \text{Equation 3.1}$$

where $u$ and $L$ represent a characteristic velocity (in m s$^{-1}$) and length scale (in m) and $v$ is the kinematic viscosity (m$^2$ s$^{-1}$) of the fluid. *Re* is used to classify flow regimes as being laminar or turbulent in nature; the threshold at which this occurs is $< 10^5$, however the transition from laminar to turbulent flow occurs at lower $Re$ over rough surfaces. The value of $v$ for air is very small ($2 \times 10^{-5}$ m$^2$ s$^{-1}$) so laminar flow is generally restricted to the thin ($< 1$ cm) layer of air immediately next to a surface. In the **surface layer** (SL) of an ABL with a depth ($L$) of 100 m and $u$ of 5 m s$^{-1}$, $Re$ is $> 10^7$ which means its flow is turbulent. A physical model designed to examine flow in the UBL therefore needs to produce a flow field with comparable values if it is to achieve dynamic similarity. Alternatively, if the model can be constructed at closer to 1:1 scale, this is more easily assured.

### Thermal Similarity

Thermal similarity requires that differences of temperature (air, surface, subsurface) in the model match those in the prototype. This is difficult to achieve in scaled urban models intended to examine the temperature of building facets, or indoor-outdoor exchange. For example, heat transfer into a building interior depends on the exterior and interior facet temperatures and the thermal mass of the building envelope. The latter is a function of both the volume and thermal properties of its materials. To preserve similitude a model building has to have the thermal mass of its smaller volume increased to match the thermal response of the prototype. Practically, this can be done by adding water containers to the interior of the scaled model thereby increasing its thermal mass of the model bringing it closer to the prototype.

### Model Domain and Boundary Conditions

The model domain is the physical extent (horizontal and vertical dimensions) of the prototype under study, often this is simply the volume enclosing the model. At the faces of the volume, boundary conditions set the values that 'force' the model and allow the investigator to examine urban effects. The lower boundary is set by the model itself and the domain extends to include where processes relevant to the objectives of the study exist. For example, if the focus is the UBL that develops over and downstream from a city then the domain must include not only that over the scale model but also the upstream zone, where the ambient airflow was established and downstream in the plume. The nature of model domains and their boundaries becomes clearer by considering specific examples.

### Modelling Strategy

Physical models can be exposed to weather outdoors or conditions generated in a laboratory. There is also a distinction between models to solve applied problems and those designed to answer research questions. The former includes studies located in a particular urban context where the physical model attempts to duplicate that place (e.g. Figure 3.1). These can provide information specific to that location and allow examination of the potential effects of say introducing a new building, or group of buildings or an entire neighbourhood into that context. Research models, on the other hand, tend to represent the urban surface using generic forms, such as cubes, to represent buildings. These are used to gather data about the climatic effects of different geometric arrangements of buildings and the impacts of altering the design by changing the size of the elements or re-arranging them. Such studies are ideally suited to extract general relationships that are not specific to a place.

### 3.2.2 Laboratory Models

An advantage of laboratory models is the ability to change the climatic environment at will. Here we review models to simulate radiation exchange, thermal effects, airflow and dispersion.

### Laboratory Models of Radiation Exchange

A heliodon is a laboratory model to study solar radiation in a building or urban context (Figure 3.16). The light source that simulates the solar beam is sufficiently small that parallel radiation is generated. This source can be moved to any position in the sky hemisphere to capture variations in Earth-Sun geometry over the course of a day or year. Alternatively, the source can be fixed and the physical model rotated to achieve the same effect. The **diffuse irradiance** originating from the sky dome can be regulated to match the anisotropic distribution of natural light from the sky. Heliodons are used to study shadow patterns cast by obstructing elements (buildings, trees), access to solar radiation in streets and on building facets, illumination of building interior by daylight, solar heat gain and optimal siting of solar panels in complex geometries. Hence, the model at the centre must preserve geometric similarity and sensors within the model to record the environment must be small enough so as not to disturb the relationship to the prototype.

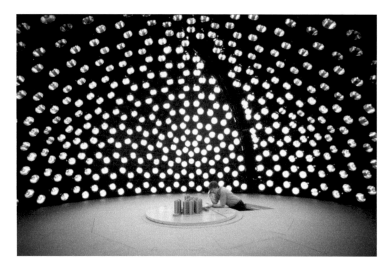

**Figure 3.16** A physical scale model of an urban neighbourhood exposed in a heliodon (Sun machine). The lights of the surrounding dome are dimmable to provide control over the skylight, i.e. the diffuse solar radiation arriving at the surface from the sky in directions other than directly from the Sun. The dark vertical track carries a single light source that represents the Sun (Credit: Welsh School of Architecture, Cardiff University; with permission).

Photographs can be taken to examine natural lighting indoors or shadow patterns outdoors.

### Laboratory Models for Temperature

Figure 3.17 shows a model designed to study nocturnal cooling in and around an urban grass park. The model is heated to a uniform temperature and then placed in an isothermal chamber to simulate cooling after sunset on a clear sky and calm conditions. In these circumstances the relative rates of surface cooling are controlled largely by the **sky view factor**, which scales geometrically. Only geometric similarity is preserved: the base is plywood; cube-shaped wood blocks represent buildings; the trees that border the park are made of foam rubber and thermometers are attached to the floor to record surface temperature. The entire apparatus is enclosed in a polyethylene tent to minimize heat exchange by **convection**. The focus is to study the role of urban and park structure in night-time cooling. In another configuration the park material was changed to simulate the effect of different 'fabric' (**thermal admittance** of the ground).

### Laboratory Models to Study Airflow

Laboratory physical models can simulate atmospheric flows and dynamics using a wind tunnel, water flume or water tank. Flow through a wind tunnel is regulated by an intake or exhaust fan at one end. The tunnel may be open (air passes through the system once) or closed (air is circulated), water flumes similarly circulate the fluid. In a wind tunnel or fluid flume

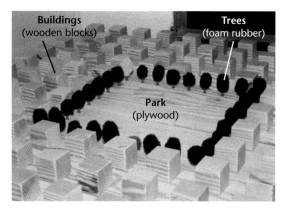

**Figure 3.17** An indoor physical scale model to represent thermal conditions of surfaces in urban neighbourhoods with parks. The model is within a cold chamber that duplicates nighttime conditions and the surface temperatures of the model are measured (Credit: R. Spronken-Smith).

the model is placed in a suitable test area and its effect on flow is explored using visual and chemical markers and by measuring turbulence and velocities at selected positions. A few elaborate systems also simulate temperature stratification (**static stability**) through the use of heating or cooling sources to create lapse (unstable) or **inversion** (stable) conditions. The effect of airflow on the pressure distribution across the facets of a building can also be studied by a special model of a building with tiny pressure ports built into its facets.

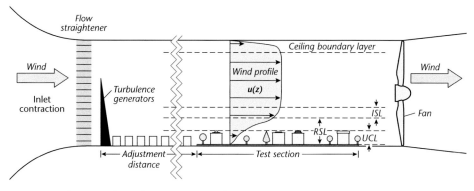

**Figure 3.18** Cross-section of an open path wind tunnel. Adjustment distance is large relative to test section length and is truncated as shown by vertical jagged lines (Modified after: Plate, 1999).

A static fluid tank can be used to examine flow in nearly calm atmospheric conditions. Tanks have the advantage that they can be stratified relatively easily to simulate the static stability using fluid layers of differing density (e.g. saltiness).

In an open path wind tunnel a physical model of a city, or part of it, is mounted in the central test section of a tunnel, sufficiently far from the lateral and top sides of the channel to avoid wall effects. Upwind of the urban model spires are installed to generate an appropriate profile of turbulent properties. These are followed by a section with many roughness elements, to create a vertical profile of horizontal wind speed to match that of the prototype (Figure 3.1, Figure 3.18, Figure 3.19). The shape of the wind profile changes with stability so for non-neutral conditions ideally the tunnel should be equipped to produce temperature stratification.

There are limitations to the types of flows that can be studied using wind tunnels. For example, it is not possible to simulate large scale weather events (such as atmospheric fronts) or wind systems that are affected by the **Coriolis force** due to Earth's rotation. At small scales, these processes are of less concern. To study the advection and dispersion of atmospheric scalars, such as air pollutants, it is necessary to simulate both the airflow and stability characteristics of the prototype, which affect horizontal and vertical spread. Moreover, the nature of emission sources should be modelled including their configuration (point, area, line), whether the release point is at the ground or elevated and if the release is continuous or intermittent (Figure 3.19 illustrates the point). It shows a model of a real industrial district used to measure the

dispersion of a pollutant depending on the position of the release point. If the pollutant is introduced in the ISL, the pattern takes on a standard plume form (Figure 3.19b) that expands in the downwind direction. However, when emissions enter the atmosphere from within the UCL (Figure 3.19c), the pollutant spreads much more laterally due to deflection by the branches of the street network.

Investigations using simple shapes such as cubes and canyons to represent buildings and streets have generated insight into flows in the UCL and RSL (see Figure 4.3, Figure 4.4 and Figure 4.6). This understanding would not have been possible using only field observations. Scale model studies often employ visualization techniques to record properties of the atmosphere, including smoke (Figure 3.19), oil films and laser light sheets that illuminate particles as they move with the flow using high-speed cameras. Such techniques can provide quantitative data as well as demonstrate the complex nature of urban flows that are often transient. Moreover, generic representations of the urban surface have allowed the derivation of simple empirical relationships between **urban structure** (e.g. $H/W$) and surface roughness, they have found wide applicability in urban meteorology and urban design.

### 3.2.3 Outdoor Scale Models

Outdoor physical models use the weather to force the investigation, so only the lower boundary conditions (i.e. the urban surface) can be controlled. Although real-world fields such as radiation, wind and temperature are provided, this also limits options, including

**Figure 3.19** Physical model of an industrial plant in Ludwigshafen, Germany to study the effect of source height on dispersion in an urban area. **(a)** The wind tunnel, with geometrically scaled buildings (foreground), general urban roughness elements (middle-ground) and turbulence generators (spires in background). **(b)** Dispersion of a tracer from source released above canopy height (in the ISL), and **(c)** below the canopy top (in the UCL) (Credit: University of Karlsruhe; see Theurer et al., 1996 for example).

**Figure 3.20** Photograph of a small-scale (approximately one-thirtieth) outdoor physical model to assess radiative characteristics within and above an urban area. Two radiometers aligned with the 'street canyon' orientations are mounted on a vertical traversing system to move them up and down and measure vertical profiles of radiation. The profiles extend from within the 'UCL' to well above the 'canopy'. Radiative source areas of the instruments change with the height of the instruments (Credit: J. Voogt).

the processes that can be examined. For example, dynamic similarity can only be achieved for prototypes at near 1:1 scale, which effectively restricts their use to airflow studies in the RSL. Nevertheless, outdoor models have been used to study radiation and energy exchanges at facets, the mean and turbulent properties of airflow and other near-surface properties of air.

### Outdoor Models of Radiation Exchange

Physical models have been used to examine the effects of surface geometry on radiation exchange. Figure 3.20 shows the setup of a study where urban structure is represented using concrete blocks to simulate cube-shaped buildings, so that the fabric and surface cover can be kept constant. The construction itself is placed on a rooftop with an unobstructed view of the sky and radiometers are positioned above the model's 'canopy' to record unobstructed upward and downward radiation fluxes. The large array is needed to ensure that the source area for radiation, which integrates the signal

**(a)** Plan view

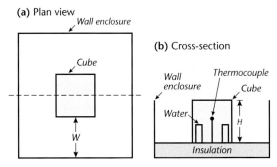

**(b)** Cross-section

**Figure 3.21** A physical scale model consisting of a cube, representing a building, surrounded by a wall enclosure, representing the neighbouring buildings. The wall approximates the effect of surrounding buildings on radiation exchanges at the facets of the central cube. See text for details (Source: Mills 1997b; reprinted with permission from *Physical Geography*, 1997, 18, 3, pp. 197. ©Bellwether Publishing, Ltd. All rights reserved).

**Figure 3.22** A one-fifth scale outdoor physical model, built as part of the Comprehensive Outdoor Scale Model Experiment (COSMO) on the campus of the Nippon Institute of Technology, Saitama Prefecture, Japan. The model (both base and cubes) is built from concrete coated with dark gray paint to provide identical radiative and thermal properties. The 512 cubes are approximately 1.5 m high, have a wall thickness of 0.1 m and have a plan area fraction $\lambda_b = 0.25$. An 11 m high measurement tower is located in the centre of the array (Credit: M. Kanda and A. Inagaki; with permission). See Kanda and Moriizumi, (2009) for example results.

from the underlying street, walls and roofs, falls entirely within the urban area. The design allows the blocks to be rearranged, to modify the urban structure. A disadvantage is that because the model is outdoors it is subject to wetting by precipitation that changes its surface properties. Hence, observations are restricted to dry conditions.

## Outdoor Models for Temperature

Figure 3.21 shows the design of a model to examine the effects of urban structure on the internal temperature of 'buildings' made of plywood. Each is represented by a hollow box, within which the air temperature is measured using a thermocouple. The rest of the urban environment (i.e. the neighbouring buildings) is represented by a wall, also made of plywood that surrounds the box. The model is placed on a roof and sits on a plywood base insulated from the roof. Four such models were created, each with a different size of area enclosed by the wall, this simulates different urban densities. To address the thermal similitude issue, each box included water containers to enhance its thermal mass – while this did not achieve similitude it did allow the 'buildings' to respond more realistically to the diurnal energy gains and losses. Data from calm and sunny days (to limit the effects of advection) were analyzed to explore the relationship between urban structure and air temperature inside the building.

## Outdoor Models of Airflow

Outdoor models capable of examining airflow and turbulent exchange in the urban atmosphere are uncommon. To meet the need for a homogeneous upwind fetch (so that the turbulent source area of the instruments falls within the model) and to address the issue of dynamic similitude, the physical model must be large. Figure 3.22 shows an array of cube-shaped 'buildings', not dissimilar to those in Figure 3.20, but at a much larger scale (one-fifth versus one-thirtieth). An instrument mast is mounted above the array. Results are restricted to times when instruments are within the IBL of the urban model, analysis is restricted to occasions when the sensor's turbulent source area lies within the model area. This model is used to examine the effect of urban structure on airflow, turbulence, dispersion and radiation exchanges. Further, it has been used to examine the effect of wind on precipitation receipt within the **urban canopy**, which would be difficult to achieve in a laboratory.

## 3.3 Numerical Modelling

Numerical models simulate real-world phenomena using a set of equations that link properties (e.g. air temperature) to processes (e.g. **sensible heat flux density**). Similar to physical models they can be used to conduct quasi-controlled experiments. A variety of models of varying sophistication have been employed to understand exchanges of energy, mass and momentum within the UBL or at the urban surface. Consequently, either the state of the atmosphere is a boundary condition for a model of the urban surface climate or the properties of the urban surface provide the boundary conditions for a mesoscale atmospheric model. Increasingly, these two types of model are coupled in so called **urban climate models** (UCMs), where the climates of the urban atmosphere and surface evolve together in response to exchanges to/from the surface-atmosphere domain. In this section, the general properties of the two types of model are introduced separately, before discussing experimental design and types of coupled UCMs.

### 3.3.1 Governing Equations

A numerical model of the atmosphere has a core set of governing equations that ensures the conservation of momentum, mass and energy. These describe relationships between the pressure, energy and density changes experienced by an atmospheric parcel as it passes through a flow field, in response to various forces that cause it to vary its speed and/or direction, its temperature and humidity and its density.

#### Conservation of Momentum, Mass and Energy in the Atmosphere

The principle of conservation of momentum is encapsulated in the equation of motion:

$$D\vec{u}/Dt = -2\,\vec{\Omega} \times \vec{u} - 1/\rho\nabla p + \vec{g} + \vec{\tau} \qquad (\mathrm{m\ s^{-2}})$$

**Equation 3.2**

where $\vec{u}$ is wind velocity[1] (m s$^{-1}$), $\vec{\Omega}$ is the spin of the Earth (s$^{-1}$), $p$ is air pressure (Pa = kg m$^{-1}$ s$^{-2}$) and $\rho$ is the density of air (kg m$^{-3}$). This states that the acceleration ($D\vec{u}/Dt$) of a mass of air is a result of the sum

of forces acting upon it, which include the Coriolis force ($\vec{F}_{Co} = 2\,\vec{\Omega} \times \vec{u}$), the **pressure gradient force** ($\vec{F}_{pn} = 1/\rho\nabla p$), the gravitational force ($\vec{g}$) and frictional ($\vec{\tau}$) forces[2]. These forces act on a small volume of air with uniform properties known as a 'parcel'. In steady-state conditions, the sum of these forces is zero and the parcel maintains its **trajectory**, travelling at the same speed in the same direction. This description of flow is complemented by the statement of the conservation of mass:

$$D\rho/Dt = -\rho\nabla\cdot\vec{u} + E - D \qquad (\mathrm{kg\ m^{-3}s^{-1}})$$

**Equation 3.3**

where $E$ is the rate of emission (creation), and $D$ the rate of decay or destruction of atmospheric constituents through chemical reactions per unit volume. If the fluid is treated as incompressible ($D\rho/Dt = 0$) and, if the chemical reactions that result in ($E - D$) are ignored, then the remaining term in Equation 3.3 can be used to link the vertical and horizontal flow fields. This approach is frequently employed because vertical velocities are relatively small and difficult to obtain compared to horizontal velocities.

The conservation of energy applied to an air mass is:

$$DI/Dt = -p(D\rho^{-1}/Dt) + Q \qquad (\mathrm{J\ kg^{-1}s^{-1}})$$

**Equation 3.4**

This states that the change in the internal energy content ($DI/Dt$) of a unit mass is the result of changes in density that result from compression/expansion of that mass (adiabatic processes) and the addition/subtraction of heat $Q$ (diabatic processes). The latter concept refers to the net **convergence** or **divergence** of the radiative and convective fluxes that cause warming and cooling, respectively.

These conservation laws (Equation 3.2, Equation 3.3 and Equation 3.4) are linked through a shared set of state variables, namely the pressure ($p$), density ($\rho$) and temperature ($T$), which are related via the Ideal Gas Law:

$$p = \rho\mathcal{R}_a T \qquad (\mathrm{kg\ m^{-1}\ s^{-2}}) \qquad \textbf{Equation 3.5}$$

$\mathcal{R}_a$ is the specific gas constant for dry air (287.04 J kg$^{-1}$ K$^{-1}$, defined as $\mathcal{R}_a = \mathcal{R}/\mathcal{M}_a$, with $\mathcal{R}$ the universal gas

---

[1] Symbols with an arrow denote three-dimensional vectors, i.e. they have both a magnitude and direction. For example the velocity $\vec{u}$ is a three-dimensional vector that can be decomposed into orthogonal components in the $x(u)$, $y(v)$ and $z(w)$ directions: $\vec{u} = (u, v, w)$ (see Section 4.1).

[2] $\frac{D}{Dt}$ is the total or material derivative that represents the change with respect to time anywhere along the parcel's trajectory and $\nabla = \frac{\partial}{\partial x} + \frac{\partial}{\partial y} + \frac{\partial}{\partial z}$

constant and $\mathcal{M}_a$ the molar mass of dry air). This, combined with the laws of motion, allows prediction of the thermal properties of the atmosphere.

## Surface Boundary Conditions

The UBL acquires its distinctive properties through interaction with the urban surface. So the aerodynamic, radiative and thermal properties of this interface are the critical boundary conditions for the development of the UBL at all scales. The roughness of this surface acts as a **drag** on the overlying atmosphere and causes removal of momentum from airflow (these effects are discussed in Chapter 4). This action is transferred through the depth of the UBL via turbulent air motions. These motions also respond to exchanges of energy at the urban surface, which act to transfer surface properties through the UBL. Not surprisingly then, a considerable amount of urban climate research has focussed on surface-air exchange of energy at the urban surface.

The statement of the conservation of energy at the surface (see Chapters 5, 6) states that **net allwave radiation** ($Q^*$) is partitioned into a turbulent flux of **sensible** ($Q_H$) and **latent** ($Q_E$) heat with the atmosphere and the conductive exchange of sensible heat ($Q_G$) with the substrate:

$$Q^* = Q_H + Q_E + Q_G \quad (\mathrm{W\,m^{-2}}) \qquad \textbf{Equation 3.6}$$

Similarly the **surface water balance** (see Chapter 8) is a statement of the conservation of mass of water, at an urban-atmosphere interface.

## Obtaining Solutions

The essence of numerical modelling, as a scientific exercise, is to make judicious use of Equation 3.2 to Equation 3.6 in different settings. Depending on the type of model (and its application) some of the terms in individual equations (or entire equations) may be simplified or even omitted. This may be done to focus attention on specific aspects of the system or to make a solution possible, easier or quicker. For example, a model concerned with the properties of flow only, and does not consider the effect of heating, may just apply rules for the conservation of momentum and mass. If such a model is applied to a small area ($< 50$ km) and examines low velocities, then the Coriolis force can be ignored as being too small to have a significant effect. Furthermore, if the focus of attention is on steady-state conditions then $D\vec{u}/Dt = 0$. Thus, an operational model might have relatively few terms, but despite being limited in scope, its focus on particular processes and interactions offers great advantages. We

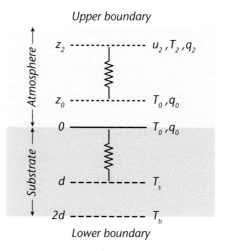

**Figure 3.23** Schematic structure of the Myrup (1969) urban surface energy balance (SEB) model. The 'surface' is located at a height $z_0$ and the layer between the ground (0 m) and $z_0$ represents an isothermal urban canopy. Transfers between the surface and overlying air at a height $z$ well above the urban canopy and between the surface and substrate are shown by a resistance symbol, i.e. the flux is proportional to the difference in properties across a distance.

refer to the set of equations that describe the urban system under examination, as its governing equations.

Numerical models solve the governing equations of a system by seeking the unique value(s) that ensure conservation of momentum, mass and energy as needed. The following example, using a simple surface energy balance (SEB) model may help to understand how solutions are obtained. Myrup (1969) introduced a model to study the urban energy balance (Equation 3.6) that represented the city as an extensive homogenous surface. As such, the model examined the vertical exchange of energy at the interface between the atmosphere and the substrate (Figure 3.23). The challenge was to couple these two systems, which have different response times, to energy exchanges: whereas, turbulent sensible ($Q_H$) and latent ($Q_E$) heat are rapidly transferred into the overlying atmosphere that responds quickly, the **conduction** of sensible ($Q_G$) heat into the substrate is slow and so is its response.

In Myrup's model $Q^*$ was computed as a function of Earth-Sun geometry and the turbulent fluxes $Q_H$ and $Q_E$ were obtained as functions ($f$) of surface-air differences in air temperature ($\Delta T_a$) and **specific humidity** ($\Delta q$), respectively, across an air layer of depth ($\Delta z$). Similarly, the conductive flux $Q_G$ was obtained from the difference in temperature ($\Delta T_s$) across a layer of soil ($\Delta z$). In summary:

| Term | Boundary conditions | Unknown | Equation |
|---|---|---|---|
| $L_\uparrow = f(T_0)$ | | $T_0$ | Equation 3.7 |
| $Q_H = f\dfrac{\Delta T_{air}}{\Delta z} = f\dfrac{T_2 - T_0}{z_2 - z_0} T_0$   $T_{a(z)}$ | | $T_0$ | Equation 3.8 |
| $Q_E = f\dfrac{\Delta q}{\Delta z} = f\dfrac{q_2 - q_0}{z_2 - z_0}$   $q_{(z)}$ | | $T_0$ | Equation 3.9 |
| $Q_G = f\dfrac{\Delta T_{soil}}{\Delta d} = f\dfrac{T_s - T_0}{d}$   $T_{s(b)}$ | | $T_0$ | Equation 3.10 |

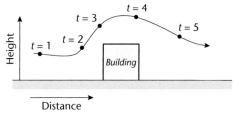

(a) Lagrangian

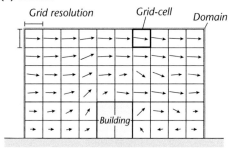

(b) Eulerian

$t = \{1,2,3,...\}$

**Figure 3.24** Conceptual diagram illustrating the difference between **(a)** Lagrangian and **(b)** Eulerian modelling perspectives. The Lagrangian approach follows a parcel as it moves through the flow field, experiencing accelerations and decelerations along its path ($du/dt$). The position of this parcel at regular time intervals ($t = 1$, $t = 2$, etc.) is shown. An Eulerian approach establishes a bounded space (domain) that is sub-divided into cells for which the time varying flow at the centre point is obtained ($\partial u/\partial t$).

Myrup expressed specific humidity ($q$) as a function of **relative humidity** and related this to surface temperature so that each of the terms in Equation 3.7 to Equation 3.10 contains a common variable, known as the equilibrium surface temperature ($T_0$). As illustrated in Figure 3.23 the model is 'forced' by upper boundary conditions that specify the air temperature ($T_2$) and humidity ($q_2$) at a height ($z_2$). At a lower boundary (below depth $2d$) the temperature of the soil is fixed ($T_{s(b)}$) and $T_s$ is allowed to respond to changes in the surface temperature, which generates $Q_G$. The function terms ($f$) are either established values for material conductivity (in the case of $Q_G$) or are obtained from **parameterizations**, which link unknown properties (such as turbulent exchange) with observed properties (such as wind speed) as is the case for $Q_H$ and $Q_E$. A solution is achieved by finding the value of $T_0$ that balances Equation 3.6 using numerical methods.

Obtaining solutions for complex models is more difficult, but follows the same rationale. The equation of motion, for example can be decomposed into components ($u$, $v$ and $w$) along three orthogonal axes ($x$, $y$ and $z$) and the forces that act in each direction:

$$\frac{D\vec{u}}{Dt} = \frac{du}{dt} + \frac{dv}{dt} + \frac{dw}{dt} \quad (\text{m s}^{-2}) \qquad \textbf{Equation 3.11}$$

In this form the equations are described as **Lagrangian** because they describe the path of an individual air parcel over time (Figure 3.24a). This path is analogous

to the flow-following approach shown in Figure 3.2c. Acceleration along each axis represents an instantaneous response to the sum of forces that it encounters at a particular point in the flow field, identified by its $x$, $y$ and $z$ coordinates. These equations can be written differently, so that the perspective is **Eulerian**, that of a budget at fixed locations in space ($x$, $y$ and $z$) and describes the flow as it passes by (Figure 3.24b). The terms in these Eulerian equations of motion are now expressed as partial derivatives (changes in that property, while all other changes are held constant) and a solution is sought for the acceleration at a fixed location. For example, the acceleration of air along the $x$-axis (Equation 3.11) can be rewritten as,

$$\frac{du}{dt} = \frac{\partial u}{\partial t} + \left[ u\frac{\partial u}{\partial x} + v\frac{\partial u}{\partial y} + w\frac{\partial u}{\partial z} \right] \quad (\text{m s}^{-2})$$

**Equation 3.12**

The new term in brackets represents the transfer of momentum in the $x$ direction by the $u$, $v$ and $w$ components of airflow. Equivalent expansions can be done for $dv/dt$ and $dw/dt$. Even so, these Eulerian equations are written in terms of instantaneous rates of change, evaluated along the surfaces of a spatially continuous velocity field.

Practically, a solution would generate an endlessly varying velocity field that would require considerable computer resources. Instead, more commonly, we seek the average flow that emerges over a given period of time associated with a given set of conditions, such as the airflow over a building array. This is accomplished by decomposing the terms in the equations of motion (e.g. Equation 3.12) into mean and fluctuating (turbulent) components, and averaging these over a suitable time interval such that the net force is zero (i.e. no acceleration). Consequently the steady-state flow along each axis results from the transfer of momentum to that location by the average and turbulent flows.

### Numerical Discretization

Numerical modelling on digital computers requires that differential equations, which describe atmospheric fields that are continuous in space and time, be converted to a difference form $(\partial t \rightarrow \Delta t; \partial x \rightarrow \Delta x; \partial y \rightarrow \Delta y; \partial z \rightarrow \Delta z)$. To implement a model of airflow, the domain is partitioned into discrete cells (with dimensions $\Delta x$, $\Delta y$ and $\Delta z$) at which computations are performed. Those determine the spatial resolution or cell size of the model. The model can only resolve atmospheric phenomena that are at least twice the size of the cell. At the boundaries of the domain appropriate conditions must be established. Typically, flow at the upper boundary is fixed for a simulation run. At the inflow boundary a wind profile describes the approaching air and, at the outflow boundary, the modified air leaves. At the lower boundary (the surface) the airflow velocity is zero.

While the distance between cells is often invariant in the horizontal directions ($\Delta x$, $\Delta y$), that in the vertical direction ($\Delta z$) usually increases with height. This variation is in response to the nature of vertical exchanges of heat, mass and momentum that vary most in the layer close to Earth's surface. This principle, that the cell resolution should reflect the atmospheric processes under examination, holds generally in modelling. In many urban airflow models, where buildings are represented explicitly by cells that are **impermeable**, the horizontal cell spacing is closest adjacent to building surfaces and farther apart where the effects of individual buildings have been mixed and diluted. The calculations are performed at the centre point of each cell.

Like space, time must be decomposed into discrete intervals at which calculations are performed. The choice of the temporal resolution or time step ($\Delta t$) depends on the nature of the processes and the responsiveness of the system under study. The model can only resolve processes and phenomena that last longer than twice the time step. Those that simulate airflow must select a sufficiently small time scale so that the effects of changes at one cell must affect neighbouring cells first. In other words, the choice of $\Delta t$ depends on the decisions made with regard to cell size $\Delta x$, $\Delta y$ and $\Delta z$. If changes are advected to distant neighbour cells during one time increment, the model calculations will be unstable because effects will have been communicated to downstream cells without affecting intervening grid cells.

Unlike Myrup's SEB example, solving the equations of motion (and their associated scalar fields, like air temperature) is much more difficult because each cell interacts with its neighbouring cells, and all cells in the domain must simultaneously at least satisfy the equations of motion and the conservation of mass. Several computational 'sweeps' through the domain are needed, with repeated calculations that adjust grid cell values until there is little variation from sweep to sweep. When the variation has diminished sufficiently, the modelled system is said to have come into equilibrium with the conditions established at the domain boundaries.

### 3.3.2 Numerical Experimental Design

Numerical modelling is one of the fundamental tools (together with observations and physical modelling) to understand urban climates. As noted, unlike observational studies, numerical models can be used to conduct laboratory-like experiments that allow the complexity of the real world to be simplified and to isolate and examine the effects of modifying one part of the system. Like physical models, there is a spectrum of model designs that range from the highly abstract that are only used to guide research, to those used to predict real-world outcomes or even simulate the climate under future urban development scenarios. At the outset we can distinguish between urban numerical models by their scale and dimensions, the processes included and their description of the urban surface. Decisions on each of these are related.

## Scales and Dimensions

If the purpose is to examine the effect of an entire city on the overlying atmosphere then the model domain must encompass the horizontal and vertical extents of that influence, which go far beyond the dimensions of the urban area. On the other hand, the domain for a model designed to examine the climate of a single city street will be smaller. In any case, it is necessary to impose conditions at the boundaries of the domain that represent the relationship between the domain space and the external environment. The scale of the model will inform decisions on cell size and time step. For many studies, it is desirable (or even necessary) to limit the dimensions of the domain to focus on exchanges in one or two directions. Not surprisingly, given the importance of surface-air exchanges, the vertical dimension is nearly always retained. Thus, a one-dimensional model only considers exchanges along the $z$-axis and is applied to a surface that is horizontally extensive and homogeneous so that horizontal transfers can be ignored with justification. Such models are sometimes termed 'column' models. A two-dimensional model describes a domain that represents a cross-section ($x$ and $z$ axes) or a plane ($x$ and $y$ axes) at a fixed level. In a model of airflow along a cross-section, the $x$-axis is oriented parallel to the flow field and exchanges along this axis are treated explicitly. This is acceptable when forces along that axis (such as pressure gradient) dominate over the forces that are perpendicular to the flow field. Three-dimensional models are necessary when forces in all three directions are relevant, such as in calculations of the specific flow around an individual building.

## Processes

Numerical models include only those processes relevant to the properties under study. In the case of weather prediction or climate models that are used for several different purposes, the range of processes has to be extensive. On the other hand, for air quality models, the treatment of transport by airflow and the atmospheric chemistry are the over-riding considerations. For urban models, the processes associated with surface-air interactions are the primary focus. These may be restricted still further if the intent is to examine particular features, such as the formation of the urban heat island. The complexity of a model is related to both the range of processes it includes and

their interaction. For example, it has been common practice for urban models to focus on either airflow or surface energy fluxes. Although both sets of processes are thoroughly interdependent in reality, most models parameterize either airflow properties (in the case of energy balance models) or surface-air energy exchanges (in the case of airflow models) rather than fully model their interaction. Rather than being a disadvantage, this ability to 'switch off' certain processes and to observe the effect of others in isolation is one of the defining attributes of numerical models.

## Surface Description

The chief distinguishing feature of urban models is the treatment of the lower boundary or urban surface (Section 2.1). About the simplest representation is that of a two-dimensional 'slab' that has specified aerodynamic, radiative and thermal properties. This may be appropriate for mesoscale models that cover a large domain with coarse grid cells that do not have the vertical resolution or computational resources needed to resolve processes within the UCL. However, the resulting simulations of surface properties, such as temperature, are difficult to validate, because they do not correspond with any location in the real-world. On the other hand, the smaller domains of sub-city scale models allow for more realistic treatment of the urban surface including horizontal and vertical heterogeneity. For example, many models represent the UCL as a street canyon (see Section 2.1.2) in the urban structure that is defined by its dimensions (wall height, street width and length) and orientation. Decisions on all of these matters are based on the purpose of the modelling experiment, our understanding of the urban climate systems and the capacity of the computing resources. Obviously there is an inverse relationship between model simplicity and computational speed, however, it does not follow that the most complex model is best for every investigation. It may be the case that a well-designed, simple model that requires few resources is the most appropriate.

The results of modelling experiments must be validated if they are to provide useful insights into and prediction of surface-atmosphere processes. Thus, modelling and observational techniques are complementary – just as modelling can provide direction for observational studies, the latter can support the modelling decisions and generate the data necessary to evaluate model simulations. It is common practice to compare the results of models with the published results of field campaigns or laboratory experiments.

However, genuine validation would require that the model simulates the exact circumstances associated with a field campaign and that variables common to both are selected for comparison.

### Types of Urban Climate Models

There are a great many different UCMs that could be categorized, based on the properties discussed above, many of which are functions of the size of the modelling domain and the degree of simplification that this entails. Here we choose a simple twofold division into micro-/ local-scale UCMs (with an emphasis on the RSL and the UCL) and mesoscale UCMs (with an emphasis on the UBL above the RSL). This division is one of convenience only because these models can be coupled, so the results of one can provide boundary conditions for the other. As computing resources have increased, the number of surface-atmosphere processes included in models has increased, the description of the urban surface has become more detailed, the domain extent has increased and cell size has diminished. At the same time, our understanding of the urban-atmosphere has also developed so we have become more attuned to the processes that are significant at different scales and where modelling attention should be concentrated.

### 3.3.3 Micro- and Local-Scale Urban Climate Models

The horizontal and vertical dimensions of micro- and local-scale modelling domains are approximately equal and their focus is to capture the climatic effects of the properties of the urban surface; its cover, structure and fabric. Within this category of models we can distinguish between those that focus on the fluxes of momentum (airflow), or energy (the SEB) or mass (water balance, air quality). The more sophisticated models couple these exchanges, so the properties of the surface and atmosphere evolve together. Models may also incorporate a representation of some aspects of urban metabolism, for example the **anthropogenic heat flux** $Q_F$ (see Section 6.2), water applied for irrigation or air pollutant emissions.

### Computational Fluid Dynamics Applied to Cities

Models able to simulate airflow around obstacles are usually classed as **computational fluid dynamics** (CFD) models. These solve the equations of motion numerically but the complexity of the UCL means that these computations are usually performed on generic urban surfaces, such as building arrays, or around individual buildings where considerable attention is paid to the description of the building surface and form. Their focus on the airflow disturbance by the urban surface is often reflected in the selection of a cell size that is smallest adjacent to obstacles. There are several types of CFD models, based on their approach to implementing the equations of motion and the resources required to do so. They include: RANS (Reynolds Averaged Navier-Stokes), LES (**Large Eddy Simulation**) and DNS (Direct Numerical Simulation) models.

Not surprisingly, CFD models can consume an extraordinary amount of computer resources if the situation under examination is complex and simulations proceed by very small increments towards a solution. Initially, such models were limited to studies in the architectural engineering field to establish the effect of individual large buildings on airflow and *vice versa*. Until recently, computer resource limitations prevented their application to urban meteorological problems, such as the dispersal of hazardous airborne materials within the UCL. However, this is changing. As an example, Figure 4.10 shows the instantaneous turbulent flow calculated by a CFD model for a simple cube-shaped building that is part of a regularly-spaced array of identical buildings. Similar flows are observed in wind tunnel models showing highly complex patterns in building **wakes** as eddies are shed downstream.

### Energy Balance Models

The number and range of SEB models, reflects the history of urban climate research (Figure 3.25). Whereas first generation models considered the effects of changes in fabric associated with urban areas (e.g. Myrup, 1969; Figure 3.25 1a), the second address the effects of urban structure. Faced with the diversity of urban features, model designers adopted the 2D canyon form to represent city streets as the most common repetitive microscale **urban form** after single buildings. This 2D structure forms the basis for many urban SEB models. For example, the original **Town Energy Balance** (TEB) model (Masson, 2000) is a single layer UCL model that represents the 2D structure of an urban canyon (Figure 3.25 1b; Figure 6.9). It simulates the SEB for a representative UCL microscale setting characterized using a set of geometric, radiative and thermal properties that describe the basic canyon structure and fabric. Sources of **anthropogenic** heating may be prescribed by the user. These

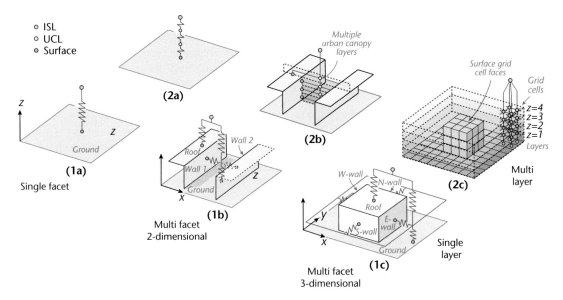

**Figure 3.25** Selected modelling strategies for urban canopy models with increasing complexity. From left to right (**a to c**) complexity of the surface increases from (**a**) a point, to (**b**) a 2D canyon cross-section, to (**c**) representing ground, wall and roof of a simplified cubic building with four facets; the increase from front to back (**1 to 2**) shows the addition of additional vertical layers in the urban canopy layer (UCL). For simplicity, the most complex case (**2c**) does not show all points of the grid, just a few representative ones in the lower right corner of the model. The computational requirements increase dramatically from left to right (**a to c**) and from front to back (**1 to 2**) depending on the number of layers added.

models represent the atmospheric properties in the UCL at single point for each grid cell, along with surface conditions for the facets represented. To incorporate urban vegetation cover the model may be combined with a separate vegetation model in a tile scheme (see Figure 3.27).

For canyons the SEB is solved using the equilibrium surface temperature ($T_0$) approach (Equations 3.7–3.10) for each canyon facet (wall or street). Since the SEB for each facet depends on that of the other canyon facets, iterative techniques are used to obtain $T_0$ values for each facet that are consistent with both the established boundary conditions and the $T_0$ at other facets: only then can the SEB for the entire canyon unit be obtained. As these models have evolved, the horizontal extent of the domain has expanded to include both buildings and the intervening street canyons so as to better represent the 3D structure of the urban surface (Figure 3.25 1c).

The addition of multiple layers in the UCL and RSL provide greater resolution to the UCL (Figure 3.25 2a–c) and can output specific values for screen-level, mid-canyon, roof-level, etc. Urban surface properties may be coupled to the multiple layers; for example the model of Martilli et al. (2002) provides for a distribution of building heights rather than a single mean building height. Multi-layer numerical models may represent the urban surface structure in 2D or 3D form (Figure 3.25 2b,c). Other model characteristics may be used to further categorize SEB models (e.g. Grimmond et al., 2010). These can include: which SEB terms are included in the model, for example, some models represent only dry impervious surfaces and thus do not represent $Q_E$, the method by which individual fluxes such as $\Delta Q_S$ and $Q_F$ are calculated, the ability of the model to represent the orientation of street canyons and the manner in which multiple reflections of radiation are calculated.

### 3.3.4 Mesoscale Urban Models

These models focus on the development of the UBL as a result of interactions with the underlying urban surface. Their domain must encompass the urban effect, which extends vertically into the ABL and horizontally beyond the city's edge. As a consequence, the horizontal extent of the domain is much larger (perhaps two orders of magnitude) than its vertical extent and the grid cells generally have a coarse resolution (often far greater than 1 km in the horizontal direction). For most, the UCL is too 'thin' to be included explicitly as a vertical layer(s) and its effect must be parameterized. There have been few numerical models designed explicitly to study urban climates at this scale. Instead, mesoscale models that have been designed for general use are modified for urban applications by introducing urban surface properties at the lower boundary to account for the distinctive roughness and thermal properties and the emission of air pollutants.

## Coupled Urban Climate Models

As numerical boundary layer models have evolved, greater consideration has been given to exchanges within the SL, including process within both vegetation and urban canopies. Modern numerical models with large regional-scale domains ($10^5$ to $10^8$ m in the horizontal) can have cell dimensions that are small enough ($< 10^3$ m) to include many of the horizontal variations found in urban areas. Nevertheless, dealing with the complexity of processes within the RSL (including the UCL) is still prohibitive in this context.

A common approach is to couple a mesoscale model (that incorporates the city) with a suitable local scale model and apply the concept of horizontal averaging to the RSL and UCL (see Figure 2.16b). Coupled UCMs are configured so that a mesoscale model provides the upper boundary conditions necessary for the smaller model, which in turn provides the lower boundary conditions for the larger model (Figure 3.27). Such models are sometimes referred to as being 'nested' and in some instances there may be multiple levels of nesting. The local-scale model may also itself consist of multiple parts. For example, some UCMs treat vegetation completely separately from the built portion of the urban surface. In this land-surface scheme, two models are required, referred to as 'tiles', one representing the vegetation and one the built portion. The outputs of the individual tiles are weighted according to the overall vegetated and built fraction in the grid cell they represent and then combined. Tiles may also be defined for other major surface types such as bodies of water. The combined results may then be used in further calculations in both the local and mesoscale models. In contrast, some UCM integrate vegetation directly into the urban structure. This adds significant complexity to the model but has the potential to provide more accurate results than the tile approach.

The coupling between models can also be configured to be 'one way'; in this mode, typically the larger scale model output is interpolated to the grid of the local scale model and used, along with surface conditions, to force the local scale model. These results are then saved for output without being communicated back to the larger scale model. This approach is computationally less expensive than a full two-way coupling but may be less accurate.

Validation of these models needs consideration of the spatial and temporal representativeness of both the model simulations and the field observations. In many cases these models use spatially averaged surface characteristics that are reflected in their simulations. Hence, field observations that are used for validation purposes must recognize the source areas of the measurements and the need for proper instrument exposure, as described in Section 3.1.

Appropriately validated coupled models have many important applications, including weather and air pollution forecasting, assessment of climate change impacts on cities and **adaptation** measures, scheduling of water and energy resources, and testing urban design.

## A Modelling Experiment

To illustrate the power of numerical models, especially when confronted with a challenging urban problem, we present the results of a pioneering experiment conducted more than three decades ago using the Urban Meteorology (URBMET) model (Bornstein, 1975). URBMET applied the two-dimensional form of the equations of motion to a simulated domain that extended 100 km in the along-wind direction (of which the urban area occupied 12.5 km) and 1.65 km in the vertical. This domain was divided into just $16 \times 16$ cells (owing to limited computer resources) with uniform spacing in the horizontal direction and variable spacing in the vertical. Results of three experiments are shown in Figure 3.26. In each simulation the initializing flow is for a neutral atmosphere that is in equilibrium with a non-urban surface. In the first experiment a 'rough' urban surface (equivalent to about six-story buildings) is introduced (Figure 3.26a). As the airflow encounters the rougher city, flow near the ground is retarded and convergence that generates uplift occurs. Similarly, at the lee edge of the city, rapid acceleration over the smoother **rural** area causes divergence and **subsidence**. After six hours

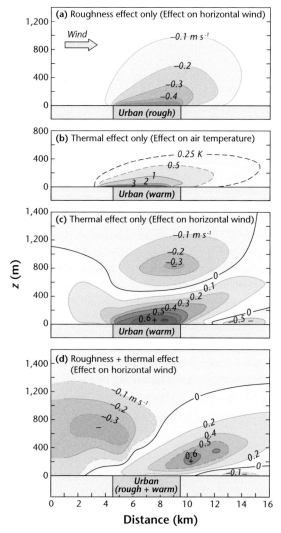

**Figure 3.26** Three experiments using URBMET (see text) are shown. The first **(a)** incorporates the urban surface as a rough boundary and examines the influence on horizontal wind. The second **(b)** treats the urban surface as warm and shows its effect on air temperature in the UBL and **(c)** horizontal wind. The third **(d)** treats the urban surface as both warm and rough and shows its impact on horizontal wind (Source: Bornstein, 1975; ©American Meteorological Society, used with permission).

of simulated time, the effects of the city are seen to affect conditions up to a height of several hundred metres and extend several kilometres downwind of the city.

In the second experiment, the urban surface is not distinguished by its roughness but is prescribed to cool more slowly than surrounding rural areas after sunset. After 12 hours of simulated time the city surface is 6 K warmer than the rural surface. This warmth generates a heat island of 3 K in atmosphere that spreads vertically and horizontally as a plume of warm air that drifts downwind (Figure 3.26b). This warming results in a weak, thermally-driven circulation above the city that draws near-surface air from the rural area into the city. Superimposed on the general airflow pattern, this circulation enhances horizontal flow through the city, which extends downwind at higher elevation. Flow above, and near the surface downwind of the city is retarded, because the thermally-driven circulation opposes the regional airflow (Figure 3.26c).

In the final experiment, the effects of both city roughness and warming are examined. The urban airflow effect is shown as an elevated zone of weaker winds, above and upwind of the city and a zone of enhanced winds near the urban surface and downwind of the city. The highest velocities are located at an elevation of 200 m, 3.5 km downwind of the city centre (Figure 3.26d). This relatively simple model demonstrates the potential of numerical models generally to investigate problems that would be virtually impossible to address using field observations or physical models.

## 3.4 Empirical Models

Models based on statistical relationships between system variables, obtained from observations or from more sophisticated models, are described as empirically-based. They suffer physical rigour because they are fundamentally restricted by the information upon which they are based. As a result they offer little diagnostic insight, but they can be of descriptive value and of great practical value by providing rapid simulations using only simple computers (such as those needed by emergency response teams). Many of these models often take the form of a regression equation, which predicts the value of one (dependent) variable ($y_i$) as a function of another (independent) variable ($x_i$), e.g.:

$$\hat{y}_i = a + b(x_i) \qquad \textbf{Equation 3.13}$$

where $\hat{y}_i$ is the predicted value of $y_i$ for a given value of $x_i$, $a$ is the intercept (the value of $\hat{y}_i$ at $x_i = 0$) and $b$ is the slope of the relation (the change in $y_i$ for a change in $x_i$). Any parameterized function can be used to describe the result as a function of several input parameters.

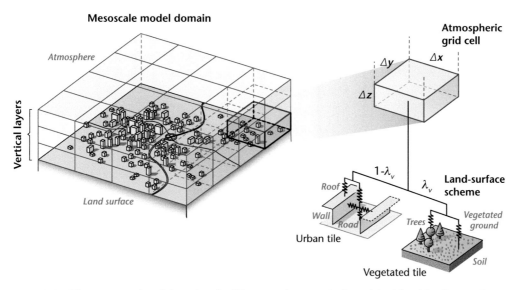

**Figure 3.27** The computational domain of a 3D mesocale numerical model with grid cells superimposed on an urban landscape. The right-hand side shows an enlargement of a single grid cell, with the associated tiles for urban and vegetated areas of the land-surface scheme (Modified after: Krayenhoff et al., 2014).

A simple example that illustrates both the power and limits of such models is the convective heat transfer coefficient ($h_c$) which is used to estimate the sensible heat flux density ($Q_H$) at building facets:

$$Q_H = h_c(T_0 - T_a) \qquad (\text{W m}^{-2}) \qquad \textbf{Equation 3.14}$$

where $T_0$ is the surface temperature and $T_a$ is the adjacent air temperature. Values for $h_c$ have been estimated in many studies using different construction materials (e.g. glass, wood and brick) under both laboratory and field conditions. Observational data of $Q_H$, $T_0$, $T_a$ and air velocity ($u$) are needed to derive a statistical relationship between $Q_H$ and the adjacent air velocity for a given surface-air temperature difference. One of the best known expressions is based on

observations of heat loss made at the external vertical surface of a rectangular, glazed and tall ($> 45$ m) building. Airflow was measured 1 m away from the facet and these were statistically linked to observations of horizontal wind ($u_{10}$) at a nearby weather station. Two expressions were derived for the windward ($h_c = 0.2 + 1.8u_{10}$) and leeward ($h_c = 1.7 + 0.2u_{10}$) faces of a building (Sharples, 1984). Such results are of great practical use, because they permit estimates of heat loss at the outer surfaces of buildings without the need to undertake detailed measurements. However, the model is limited: it contains no explanatory power and its application is, strictly speaking, restricted to circumstances similar to those under which it was derived.

## Summary

This chapter has attempted to draw together the variety of **techniques used to explore the urban effect**, and there are a great many. A number of texts either discuss the methods above in considerable detail, or else demonstrate the method in practice. The essential points relate to each of the major methodologies discussed here:

- Our knowledge of the urban effect is derived from **field observations** of the urban surface and the overlying air. The data that are derived must be interpreted within the context of the measurement environment, which is heterogeneous in character. Consequently, one must be aware of the instrumentation and its exposure to the urban environment under study, which for some instruments will vary with the state of the atmosphere (and wind direction, in particular). The important decisions are the **positioning** of the instrument platform **horizontally** and **vertically** within the urban landscape. A well situated instrument will record data that can be reasonably attributed to the character of the underlying surface through the instrument's source area.

- Examining the urban climate effect through observations alone is limited by the ability of technology to record the effect and the impossibility of observing the effects of all combinations of urban elements. Practical restrictions on accessing parts of the urban atmosphere also limit observations. **Numerical modelling** allows us to examine and test our understanding of how cities affect climate and *vice versa*. There are a large number of such models of varying sophistication that have developed, which can be fruitfully employed to simulate the urban effect and to conduct experiments by **isolating certain processes of interest**. For studying many urban effects, the subtle nature of the changes and the complex interactions between the urban drivers means that numerical models are the only viable tools. Successfully evaluated numerical models are an important urban planning and design tool (Chapter 15).
- For a great many studies, **scaled physical models** can provide a useful alternative to either field observations or numerical modelling because they allow the researcher to regulate both the nature of the urban surface and the properties of the overlying atmosphere. Moreover, they allow instruments to be positioned precisely with regard to the effect under examination.

It is important to understand that these methods are **complementary**. Numerical models require input parameters and meteorological drivers and data to evaluate their simulations. Decisions about where and what to measure are informed by the insights provided by modelling efforts. Each method has both advantages and disadvantages (Table 3.3); usually no one method is sufficient to investigate or to fully understand a phenomenon. Coherent understanding of the urban-atmosphere system is most likely to be gained when the insights acquired by using different **approaches are combined and synthesized**.

**Table 3.3** Relative strengths and limitations of methods used in urban meteorology and climatology.

| Method | Advantages | Disadvantages |
| --- | --- | --- |
| **Field observation** | Records 'real-world' urban conditions including all scales of influence. Provides data to test models. | Lack of experimental control. Vagaries of weather may limit measurement period or otherwise constrain planned study. Measurement errors always present. Equipment can be costly. |
| **Physical modelling** | Provides experimental control and allows detailed observation of urban effects. | Requires careful design to ensure similitude. Requires access to specialized facilities, e.g. flume, wind tunnel. Expensive. Requires testing against field observations or numerical results. |
| **Numerical modelling** | Gives complete experimental control and can account for all scales of climate. Can give predictions that possess practical utility. | Assumptions can be restrictive, unrealistic or too theoretical. Requires testing against field observations to establish confidence. Output can be voluminous. |
| **Empirical generalization** | Useful summary of results that are often quite complex. Simple to apply. | Little diagnostic value, portrays descriptive relations. Limited applicability, little transferability to other locations. Requires input data from another method. |

# 4 | Airflow

**Figure 4.1** Pedestrians at street level struggle with wind during a storm in Tokyo, Japan (Credit: Y. Tsuno / AFP / Getty Images; with permission).

When it is windy a pedestrian in the street finds strong gusts in some places with winds seemingly coming from almost any direction, whilst it is almost calm in other places. This apparent chaos, although complex is in fact an amalgam of flows responding to different forces acting across the range of urban scales discussed in Chapter 2. Here we concentrate on unravelling the contributing flows. The large-scale driving force is usually the **pressure gradient force** responding to the pressure patterns seen on a weather map, but the resulting wind is modified by the **atmospheric boundary layer** (ABL) of the region. That in turn is modified by the roughness and thermal effects of the city itself, including responses to the structural form of **neighbourhood**, the street network and the microscale effects of buildings, trees and even the effect of moving vehicles. Figure 4.1 illustrates the capricious nature of wind at street level during a wind storm in Tokyo, Japan. In the foreground a pedestrian is bent into the wind, lowering the body's centre of gravity and taking short tentative steps to make progress. In the background, another person sheltered by the building, occupies a calm spot, just metres away.

In this chapter large-scale (regional and **synoptic**) **forcing** is taken as given, the focus is unpacking the smaller-scale urban wind patterns to show their nature and causes. We start at the scale of a building (microscale) and progress up to the scale of an entire urban region (**mesoscale**). The study of airflow is central to understanding the formation of distinct climates close to the ground because it is the **turbulent** exchange between the surface and the atmosphere that defines the ABL. The structure of this chapter is based on the layering of the **urban boundary layer** (UBL) described in Chapter 2. We work our way upwards through the UBL, starting with the **roughness sublayer** (RSL, Section 4.2), followed by the **inertial sublayer** (ISL, Section 4.3) above, where the effects of individual buildings and trees have been thoroughly blended. We then explore the overlying **mixed layer** (ML, Section 4.4) where contributions from different structural zones within a city are mixed together to form the outer envelope of the UBL.

Before describing urban winds Section 4.1 reviews basic principles governing airflow to refresh understanding and establish nomenclature for what follows. More experienced readers can move directly to Section 4.2.

## 4.1 Basics of Wind and Turbulence

### 4.1.1 Mean and Turbulent Parts

Wind – the flow of air in the atmosphere – is described as a 3-D **vector**, $\vec{u} = (u, v, w)$, that is it includes the horizontal velocity components $u$, $v$, and vertical velocity component $w$ (all in m s$^{-1}$). In most cases, the wind blows roughly parallel to the terrain because flow is constrained by the presence of the ground. Hence, over flat terrain, the horizontal components $u$ and $v$ are typically much larger than the vertical component $w$. This is why weather stations can reasonably simplify wind to a two dimensional vector, using a (horizontal) wind speed $u_h = \sqrt{u^2 + v^2}$, and a horizontal direction of origin $\varphi$ (in degrees from geographic north, clockwise). However, when the wind is forced to flow around and over sharp-edged buildings and trees it can undergo substantial uplift and sinking and the average vertical component $w$ is not necessarily zero. To illustrate, revisit the experiment of DePaul and Sheih (1986) who released flow-following balloons

in an **urban canyon** in downtown Chicago, United States (Figure 3.14). Their balloons rise on the left (leeward) side of the canyon, then move across the top to the windward wall and subsequently sink down the right (windward) side. All balloons, although released at different times, consistently follow that **trajectory**. Hence, wind in the RSL is characterized by flow with non-negligible vertical motions – there are distinct regions of uplift and downdraft. This is best described by a 3-D field of vectors $\vec{u}$ $(x, y, z)$.

Further from the surface, in the ISL and ML, the influence of individual buildings decreases and the assumption that the average vertical wind component $w$ is much smaller than the horizontal wind components $u$ and $v$, is again fulfilled. The quasi-planar flow in the ISL is a major reason to conceptually separate the ISL from the complex 3-D flow in the RSL below (Chapter 2). However, as we will also see, the average vertical wind component in the ISL and the ML is not exactly zero in all cases, because we might experience large-scale uplift and **subsidence** over cities, which is sufficient to affect the regional flow and for example cloud formation (see Chapter 10).

At Earth's surface, airflow is brought to a halt and the kinetic energy of the wind is transformed into thermal energy (heat). The presence of the surface extends upward as a frictional force into the atmosphere, the depth of which depends on the roughness of the surface and the dynamics of the atmosphere. At some height, this force becomes negligible and flow is governed by the remaining forces, especially **Coriolis force** ($\vec{F}_{Co}$) and the synoptic-scale pressure gradient force ($\vec{F}_{pn}$) – this is the **free atmosphere** (FA). In steady-state conditions, the resulting flow is the **gradient wind** ($\vec{u_g}$). The effects of cities on airflow are only relevant in the ABL, where the effects of surface friction and heating are significant.

### Reynolds Decomposition

If one were to measure wind at a given point over a period of time of any length, the record would show considerable fluctuation over short intervals. This apparently haphazard sequence of gusts and lulls is produced by **turbulence**, which describes short-term motions embedded within the airflow.

Reynolds decomposition refers to a scheme that decomposes the wind velocity components into a mean and a deviation from the mean, which describe the mean wind and the turbulent fluctuation, respectively, hence,

$$u = \bar{u} + u'$$
$$v = \bar{v} + v' \qquad (\text{m s}^{-1}) \qquad \textbf{Equation 4.1}$$
$$w = \bar{w} + w'$$

the overbar (e.g. $\bar{u}$) indicates an average obtained over a relatively long time period (~30 min), whilst the prime (e.g $u'$), indicates the instantaneous deviation from the average. Wind components with an overbar describe the average 3D wind field, whilst the deviations represent short-term turbulent fluctuations superimposed on this field. Turbulence generally refers to **perturbations** operating at time scales smaller than 30 min. Conceptually, we conceive of these quasi-random fluctuations, as caused by **eddies**, parcels of the flow that rotate at higher and lower velocities relative to that of the mean flow. One can visualize eddies as pockets of air that are moving faster or slower than the average. If a slower moving pocket passes a measurement point, it is felt as a lull, if a faster moving eddy passes by, it is felt as a gust. The three graphs in Figure 4.2 show measured traces of $u$, $v$, and $w$. The time traces are separated into a mean (thick line) and a fluctuating part (grey shaded areas). Eddies that cause the gustiness of wind are visible as events that fluctuate around the mean. Note that by

definition the sum of the deviations from the mean is zero, that is, $\overline{u'} = 0, \overline{v'} = 0$ and $\overline{w'} = 0$.

The turbulence embedded in a flow can be described in several ways. One approach is to average the squared deviations from the mean and taking the square root thereof (i.e. the standard deviation $\sigma$).

$$\sigma_u = \sqrt{\overline{u'^2}}, \;\; \sigma_v = \sqrt{\overline{v'^2}}, \text{and } \sigma_w = \sqrt{\overline{w'^2}} \qquad (\text{m s}^{-1})$$
$$\textbf{Equation 4.2}$$

The **turbulence intensities** of the wind components are retrieved by dividing $\sigma_u$, $\sigma_v$ or $\sigma_w$ by the magnitude of the mean wind vector $|\vec{u}| = \sqrt{\bar{u}^2 + \bar{v}^2 + \bar{w}^2}$. It is useful to present the distinction between the mean and turbulent components by partitioning the total kinetic energy of the flow into its parts. The mean kinetic energy per unit mass (MKE/$m$) is the kinetic energy of the mean wind components:

$$\frac{\text{MKE}}{m} = \frac{\bar{u}^2 + \bar{v}^2 + \bar{w}^2}{2} \qquad (\text{m}^2 \text{ s}^{-2}) \qquad \textbf{Equation 4.3}$$

The **turbulent kinetic energy** (TKE) per unit mass of a flow is determined by the sum of the contributions by each component,

$$\frac{\text{TKE}}{m} = \frac{\overline{u'^2} + \overline{v'^2} + \overline{w'^2}}{2} \qquad (\text{m}^2 \text{ s}^{-2}) \qquad \textbf{Equation 4.4}$$

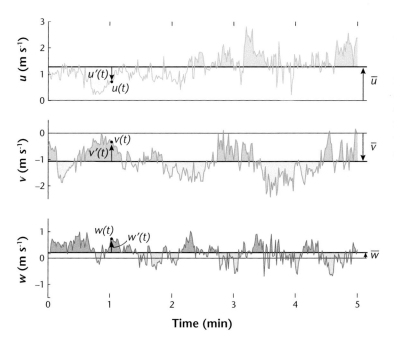

**Figure 4.2** Traces of the three components of the wind vector $\vec{u} = (u, v, w)$ recorded inside an urban canyon over a period of 5 minutes. Reynolds decomposition separates the components $(u, v, w)$ at any time $t$ into a mean wind $(\bar{u}, \bar{v}, \bar{w})$ and a turbulent deviation from it $(u', v', w')$.

Time (min)

## 4.1.2 Production of Turbulence

Turbulence is generated by both mechanical and thermal processes, which together ensure the ABL is nearly always in a turbulent state. The relative contribution of either process depends on the structural properties of the surface (its roughness or smoothness) and the strength of the **sensible heat flux density** ($Q_H$) at the surface (the result of the **surface energy balance**).

### Production of Mechanical Turbulence

When air flows in different directions and/or with different speeds, the resulting velocity gradients or **wind shears** create internal forces in the atmosphere. If these forces are sufficiently large, they become dynamically unstable and initiate the transfer eddies, parcels of air that are displaced from their current position into slower/faster moving flow resulting in gusts/lulls. The development of these eddies causes disruption in the flow and converts MKE to TKE.

These velocity gradients are greatest close to Earth's surface due to the frictional force imposed by the ground and its roughness elements (e.g. buildings and trees) on the flow, called drag. Over a flat, smooth surface, **skin drag** dominates because of the 'no-slip' condition that states air movement must come to a standstill very close ($\ll 1$ cm) to a surface. At this scale, the kinetic energy of the atmosphere must be transferred to a laminar layer where molecular viscosity takes over. Over rough surfaces, it is **form drag** that dominates as airflow interacts with roughness elements protruding into the atmosphere. The pressure perturbations induced around these elements by the wind create chaotic flow patterns with lower velocity, including **wakes**, that become detached from the elements, thus creating strong velocity gradients. Figure 4.3 visualizes the nature of mechanical turbulence using smoke in a **wind tunnel**. The turbulence in (a) is primarily caused by skin drag, whilst in (b) there is also turbulence caused by form drag of the building, in particular behind the building. If the drag imposed by the surface on the airflow (in N) is expressed per horizontal surface area (in m²), this is a stress (in N m$^{-2}$ = Pa).

In a turbulent atmosphere, this stress causes air in layers close to the surface to transfer horizontal **momentum** vertically from the higher layers, where wind is typically faster, to layers closer to the surface, where wind is slower. This **momentum flux** in the ABL

**Figure 4.3** Smoke visualization of dispersion from a point source release located at ground level with roughness elements upstream. **(a)** Flow over a flat plane and **(b)** same but with a cubic 'building' model added (Source: Hall et al., 1997; with permission).

is primarily made possible by displacement of turbulent eddies, that randomly move up and down and mix their properties with the surrounding air. This is a turbulent momentum flux also called **Reynolds stress** ($\tau$). The turbulent eddies carry momentum surplus (or deficit) downward (or upward) to a new level where they mix with the flow, and add (or subtract) momentum to (or from) that level. At a given height downdrafts are on average sensed as relative gusts (called **sweeps**), which are faster eddies that originate from higher levels where momentum is greater. Updrafts are, in the majority of cases, sensed as relative lulls in speed (called **ejections**) and come from lower levels where momentum is lower. This turbulent mixing of momentum eventually reduces the wind gradient.

A momentum transport can also be achieved if momentum surplus or deficit is transported vertically by a non-zero mean wind (called momentum **advection**), or even on a molecular scale due to the viscosity of the atmosphere (called viscous stress). Whilst momentum advection is relevant in the urban atmosphere at a range of scales from flow around individual buildings (Section 4.2.1) to flow across entire cities (Section 4.3.1), viscous stress is only relevant in the laminar region of the atmosphere, directly adjacent to any surface (< 1 cm).

### Production of Thermal Turbulence

Thermally generated turbulence results typically from surface heating. Parcels heated more effectively by the surface impose a **buoyancy** force relative to colder and denser air surrounding them. The buoyancy force (working against gravity) accelerates the less dense, warmer parcels called **thermals** or plumes. Thermals are made visible when we see a cloud rising from a chimney. Thermals create friction and mixing between themselves and the surrounding air as they rise. TKE from thermal turbulence is greatest when buoyancy is strong, which results in a strong $Q_H$ that transports heat away from the surface. Vertical motion can be also suppressed by density stratification of the atmosphere. If the surface is cooler than the atmosphere above, the coldest (and densest) air stays closer to the surface and vertical motion will be dampened, i.e. TKE will be consumed.

### Scales of Turbulent Motion

Mechanically generated eddies are relatively small and scale with the size of the roughness elements. On the other hand, thermally generated eddies scale with height above the surface and are only constrained by the presence of the ground and the depth of the ABL. As a consequence, the two sources of turbulence result in eddies with different scales. When flow is dominated by mechanical eddies, we only observe small fluctuations in the recorded wind velocity that reflect the size of the obstacles (roughness elements) embedded in the flow. Over an extensive grass surface, most mechanically produced eddies are so small that most of them take as little as a fraction of a second to pass, but over an urban surface, most eddies are larger and may take tens of seconds to pass.

When thermal production is dominant, flow is dominated by even larger thermal eddies with a wide range of sizes, the largest of which can take several minutes to transit and are clearly felt as gusts with intermittent calm periods. Typically, both production processes operate so that eddies of varying size coexist in the atmosphere. Whilst the influence of smaller mechanically generated eddies tends to decrease with distance from the surface, thermally produced eddies scale with height above ground and become more efficient in mixing properties in the ML. What is typical about the urban atmosphere is that both, mechanical and thermal production are exceptionally large. Over a wide range of space and time scales, they cause the strong, incessant fluctuations that characterize the traces of wind in Figure 4.2.

### Dynamic Stability

The sum of thermal and mechanical production determines the total production of TKE, and hence at a given wind velocity the turbulence intensity. The ratio between thermal production (or suppression) to mechanical production is called **dynamic stability** and determines the nature of the turbulence. This includes forces acting on the air and size and shape of eddies. Table 4.1 summarizes different dynamic stability classes and the corresponding values of $Q_H$ and Reynolds stress ($\tau$). The closer to **neutral**, the more the flow is characterized by mechanical turbulence production. The more **unstable** a flow is, the more dominated it is by thermal production. In a **stable** situation part of TKE is consumed by the density stratification of the flow, although mechanical production in cities is typically so strong part of the turbulence can be maintained.

### 4.1.3 Dissipation of Turbulence

Eddies are by definition transient features with short lifetimes. Larger-scale eddies are asymmetrically shaped; for mechanical turbulence, the kinetic energy of the vertical wind component $\overline{w'^2}$ is typically much less than the horizontal components ($\overline{u'^2} = \overline{v'^2}$). Again this is because vertical motions are constrained by the ground but horizontal motions remain relatively unfettered. These larger eddies break up into smaller ones, which causes a continuous transfer of TKE to progressively smaller eddies with shorter lifespans. Smaller eddies show no preference for spatial elongation and flow at this scale is therefore said to be statistically **isotropic** (i.e. $\overline{u'^2} = \overline{v'^2} = \overline{w'^2}$). Eventually the smallest eddies reach a level whereupon the TKE is opposed by the air's viscosity, this turns their kinetic energy into **sensible heat**. This energy transfer from kinetic energy to heat is called **dissipation** of

**Table 4.1** Turbulence régimes in the atmosphere.

| Dynamical stability | Processes | Typical situation | Sensible heat flux $(Q_H)$ and Reynolds stress $(\tau)$ | Stability parameter[1] $(\zeta)$ |
|---|---|---|---|---|
| **Very unstable** (free convection) | Buoyant plumes produce exclusively thermal turbulence. Mechanical production is absent. | Calm daytime, no cloud, substantial irradiance | $Q_H > 0$ <br> $\tau \approx 0$ | $\zeta \ll -1$ |
| **Unstable** | Mechanical and thermal turbulence production both important. | Daytime, moderate wind, moderate irradiance | $Q_H > 0$ <br> $\tau > 0$ | $-1 < \zeta < 0$ |
| **Neutral** | Only mechanical turbulence produced. Buoyancy neither enhances nor suppresses turbulence. | Daytime or nighttime with strong wind, overcast, no or little irradiance | $Q_H > 0$ <br> $\tau > 0$ | $\zeta \sim 0$ |
| **Stable** | Turbulence produced mechanically, but buoyancy suppresses vertical fluctuations and consumes turbulence. | Nighttime, moderate wind | $Q_H < 0$ <br> $\tau > 0$ | $0 < \zeta < 1$ |
| **Very stable** | No turbulence created. Buoyancy suppresses vertical motions. | Nighttime, calm, cloudless | $Q_H \ll 0$ <br> $\tau \approx 0$ | $1 < \zeta$ |

[1] The stability parameter $\zeta = (z - z_d)/L$ is defined by the Monin-Obukhov Similarity Theory (see Section 4.3.3) and only valid in the ISL.

turbulence. The heat released by the process of dissipation is so small it is not considered in the heat balance (Chapter 6).

## 4.2 Flow in the Roughness Sublayer

### 4.2.1 Isolated Buildings

The primary determinants of urban surface roughness are the buildings. Unlike vegetative elements buildings are **impermeable**, inflexible and usually sharp-edged. Their intrusion into the mean wind causes form drag, and forces strong perturbations in their vicinity and downstream, spawning wakes characterized by decreased MKE and increased TKE. Buildings also often possess distinct radiative, thermal and moisture properties so they become the source of dry thermal plumes causing thermal turbulence. Together these mechanical and thermal attributes dictate the length scales of flow and turbulence in the surrounding atmosphere, especially in the **urban canopy layer** (UCL).

Gaining full understanding of the three-dimensional wind field around a real building from observations is a daunting task, because the spatial complexity requires deployment of multiple wind sensors and the temporal variability requires long sampling periods. As a result most of our knowledge comes from **physical** or **numerical models** where experimental control is more readily achieved (see Chapter 3). Buildings can then be miniaturized or are virtual, and details of the surrounding structures and properties of the approach flow (speed, direction, turbulence, and static stability) can be simulated. Firstly, we look at the simple case of an isolated cubical building where the mean approach flow is perpendicular to its windward face (Figure 4.4). In reality the wind is turbulent so its direction continually varies and the instantaneous angle of incidence is not fixed. The effect on the building is to generate pressure perturbations on its **facets** that pulsate with the variations in the approach flow. Flow in the vicinity of a building is highly dynamic for two further reasons: (i) the air must deflect around and over the building, (ii) flow over an extensive surface can stay attached to it (the no-slip condition), but when it encounters sharp edges it must detach, and **flow separation** takes place. Immediately behind the corner of a building the lower surface pressure causes flow to recirculate against the approach flow creating

(a)

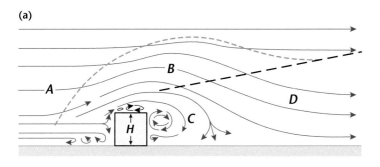

(b)

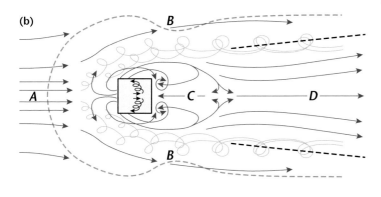

(c)

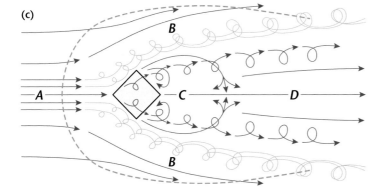

**Figure 4.4** Typical patterns of airflow around an isolated cubic 'building'. (a) Side view with unobstructed flow A from the left at normal incidence (0°), showing the displacement zone B, the cavity C and the wake D. (b) Plan view near ground level of same flow as in (a). (c) Plan view of flow with the building oriented at 45° to the approach flow (Modified after: Meroney, 1982 using Yakhot et al., 2006).

a **vortex**. Although the flow field in Figure 4.4, and other diagrams here appear fairly complex, most emphasize the mean flow rather than its turbulent nature, which is difficult to depict. This should be borne in mind when following the discussion below. For a more realistic appreciation of flow around buildings, it is instructive to watch continuous visualizations of wind tunnel experiments or animations of computer simulations.

As oncoming flow (A) approaches a building the mean flow streamlines are displaced, due to the build-up of a positive pressure perturbation on the windward face (see displacement zone B, Figure 4.4). Air diverges vertically and laterally to flow over and around the building. The maximum positive pressure perturbation is about two thirds of the way up the centre of the windward face of the building, the stagnation point (Figure 4.5). There all kinetic energy of the flow is transformed to a pressure head and the air is brought to rest. The flow diverges in all directions from this point: up over the roof, around the sides and down the front. There are regions of

**(a)** Wind facing side

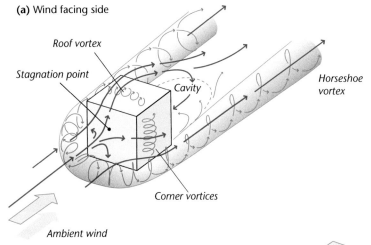

**Figure 4.5** Features of the flow around an isolated cubical building with approach flow normal to one face. Viewed from the **(a)** windward, and **(b)** leeward side (Modified after: Hunt et al., 1978; Martinuzzi and Tropea, 1993 and Meinders et al., 1999).

**(b)** Lee side

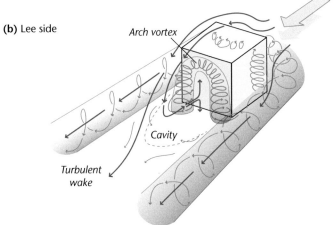

negative pressure perturbation on the roof, side walls and leeward face of the building which create suction and induce air to circulate in the reverse direction to the main flow. If the building length in the direction of the flow ($L$) is greater than its height ($H$, i.e. $L/H > 1$) the flow re-attaches to the roof and side walls.

Some of the air deflects down the face of the building to the ground. There it streams away from the face and back towards the approach flow, under-cutting it and causing it to separate from the ground upwind and forms a standing eddy vortex near the base. This roll-like structure often includes more than one vortex line oriented with its axis of rotation normal to the approach flow, i.e. parallel to the front wall. The ends of these vortices stretch around the

sides of the building and are carried downstream to form a characteristic horseshoe shape (Figure 4.4, Figure 4.5). The 'arms' of the horseshoe are a pair of counter-rotating vortices which define the lateral edges of the building wake downstream near the surface.

Behind the building the flow structure is complex; flow separation dominates, as it streams past the rear edges of the side walls and roof leaving a suction zone in the lee of the building. This is the **cavity zone** C, which is part of the overall wake zone D, of the building (Figure 4.4). The cavity is a three-dimensional recirculation zone formed by both an along-wind roll due to flow over the roof, plus two horizontally-oriented rolls due to lower pressure behind each side wall. The dimensions of the cavity

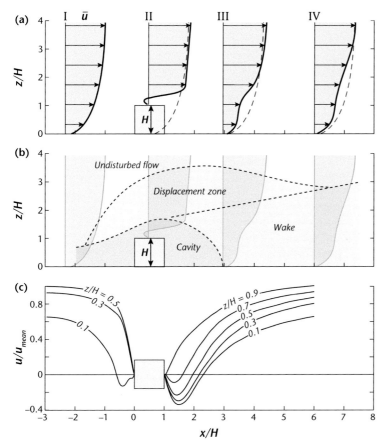

**Figure 4.6** Side view of profiles of **(a)** velocity upwind (I), over (II) and downwind (III and IV) of an isolated cubical building with approach flow from left and normal to front face. **(b)** Zones of the flow (Modified after: Meroney, 1982). **(c)** Time-averaged $u$-velocity at different heights along the centre-line of the building, with $u_{mean}$ the mean undisturbed wind speed at given height (Modified after: Yakhot et al., 2006).

are approximately those of the building in the vertical and lateral directions (i.e. typically 1 to $1.5H$ or its width, $L$), but in the along-wind direction the cavity length extends about 2 to $3H$ from the lee face for cubic obstacles, growing to about $12H$ for those where $H/L$ is small (Meroney, 1982). Behind the rear corners of the building vertically-oriented separation zones create strong spiral vortices rising like pillars (Figure 4.4b, Figure 4.5). These entrain some fluid from the horseshoe vortex and whirl it upward meeting to form an arch across the rear face. The top of the pillars feed fluid with high turbulent kinetic energy into the flow coming over the roof and a pair of elevated counter-rotating vortices trails downstream (Figure 4.5). The cavity circulation is not completely self-contained but it restricts exchange between air inside and outside the cavity. It causes **downwash**, which occurs when

**air pollutants** are drawn down into a recirculating roll behind an obstacle (smokestack, building). Material can be transported into it via the main streamline over the roof and can exit vertically via the corner spiral vortices as they trail downstream. The overall wake zone includes the effects of the building in extracting MKE from the flow (i.e. producing shelter) and the generation of TKE. The height of the wake typically grows to about 3 or $4H$ at $10H$ downwind, and wake effects on the flow persist from 5 to $30H$ downwind. The pair of elevated vortex streams persists even further downstream.

The effects of a single isolated building on the vertical **profile** of the horizontal mean wind ($\bar{u}$) are shown in Figure 4.6a. Upwind of the building (I), the profile is in equilibrium with the underlying surface. At location (II) the profile responds by accelerating,

due to the confluence of the flow in the displacement zone, as it 'squeezes' over the top of the building. The flow in the displacement zone is made visible through dynamically formed clouds above the buildings in Figure 4.7.

Closer to the roof surface, in the cavity zone, there is a sharp decrease of speed in the zone of lower pressure above the roof, and counter-flow, especially immediately adjacent to its windward edge. In very strong winds this may produce lift forces that can raise loose roof materials or even cause the front part of the roof to flap or be ripped off. Behind the building (III), the wind profile also shows return flow and a marked velocity deficit (shelter) in the cavity. At location (IV), flow starts to restore the wind profile to its undisturbed state, however the velocity deficit persists for some distance ($x$), decaying at a rate of about $x^{-1.5}$ and fading out by about $x = 15H$.

The mean velocity profile responds to turbulence generated by the building. Figure 4.8 shows isolines of the TKE for the three wind components. The main zones containing TKE are the roll vortices at the base of the front wall (which initiate the horseshoe), and in the counter-rotating vortex on the roof. The along-wind turbulence ($\overline{u'^2}$) is highest above the roof and decays downstream, the $v$ and $w$ components have a similar maximum above the roof and a secondary one at about $x = 5H$ behind the building due to enhanced mixing there.

If the angle-of-attack of the approach flow is say 45° rather than normal to the windward face, the flow and pressure fields are different (Figure 4.4c). The principal change is the development of two counter-rotating vortices, one from each leading roof edge. As they trail downstream they entrain faster moving air down into the wake. This reduces the cavity height, slows the rate of wake growth, and increases velocities in the wake compared with Figure 4.4a,b. Because these stack vortices may persist to $80H$ downstream (i.e. beyond where shelter is normally found), their ability to draw momentum downwards can give velocity *excesses* in the far wake.

Differences in building geometry, rounding corners, roof design, exterior protrusions (balconies, chimneys, etc.) produce special features, but the essential flow around an isolated bluff object remain. What is very much more difficult to unravel is the interaction between overlapping building wakes that take place in an urban setting.

### 4.2.2 Uniform Building Arrays

For most buildings in urban areas the approach flow has already been disturbed by upwind structures, so the character of the wind is complex. There may be some organization at larger scales, due to the regularity of the spacing between buildings and/or the width

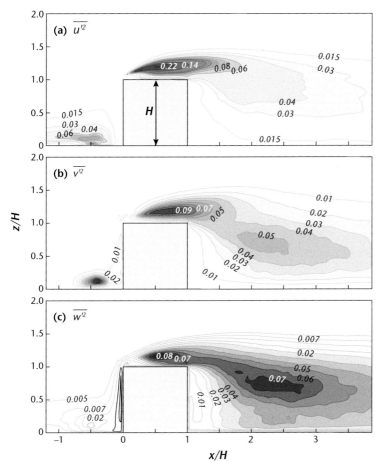

**Figure 4.8** Isolines of components of the turbulent kinetic energy (velocity variances) on a plane through the centre of an isolated cubical building. Intensity of turbulent motions (variance) in the **(a)** along-wind, **(b)** across-wind, and **(c)** vertical directions (*Source:* Yakhot et al., 2006; © Elsevier, used with permission).

and orientation of streets. In areas of reasonably uniform building height the building spacing (e.g. aspect ratio, $H/W$) sets three canonical flow régimes in the RSL (Figure 4.9):

- **Isolated roughness flow** – When buildings are widely spaced ($H/W < 0.35$) individual wakes form, much like those of isolated buildings (Figure 4.9a), except that in an array the approach flow to each element is disturbed by residual effects of elements upwind. Depending on the across-wind spacing, and whether the array is regular, staggered or random (see Figure 2.8), there will be lateral wake interactions, but inter-element interactions are small. In an area characterized by such widely spaced houses the concept of an urban canyon and of a canopy is of limited use.
- **Wake interference flow** – At greater densities ($0.35 < H/W < 0.65$) the distance separating the roughness elements in the along-wind direction begins to

match the horizontal extent of the building cavity (e.g. $W \leq 2H$). Now the vortex in the cavity behind the upwind building is reinforced by the flow down the windward face of the next building (Figure 4.9b). Note that such vortices depict the mean flow, but at shorter (turbulent) time scales they are often not steady, they fluctuate in velocity, form and re-form.
- **Skimming flow** – With even closer spacing ($H/W > 0.65$) the above-roof flow 'skips' across the tops of the buildings with less tendency to enter into the street canyons. Here the **urban canopy** concept finds its fullest expression: the flow above the roof has become partially de-coupled from that in the canyons (Figure 4.9c). The mean flow above provides a weak tangential force maintaining a vortex circulation flow within the street canyon. In weak wind conditions even that influence may inject momentum to the base of the canyons where near

**(a)** Isolated roughness flow

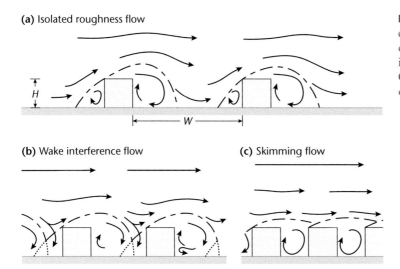

**Figure 4.9** Effect of packing density ($H/W$) on flow régimes over urban-like 'building' arrays in a wind tunnel (Modified after: Oke, 1988 constructed using data of Hussain and Lee, 1980).

**(b)** Wake interference flow          **(c)** Skimming flow

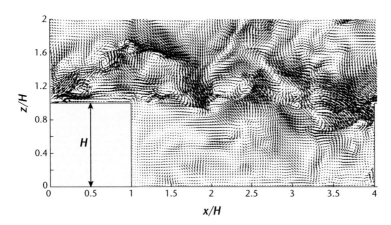

**Figure 4.10** Detailed numerical simulation of the instantaneous turbulent flow behind a cubical building. Flow from left to right showing the fluctuating wind components $u$, $w$ (i.e. along-wind and vertical motion) in the streamwise-vertical ($x$, $z$) plane through the middle of a building (in yellow at bottom left). This building is located in a staggered array with streamwise spacing of $3H$ and $\lambda_f = 0.33$ (i.e. isolated roughness flow). The next downwind element is at the bottom right edge (Source: Coceal et al., 2006; © Springer Science+Business Media B.V., used with permission of Springer).

stagnation prevails. If the packing is greater in one direction (to form elongated canyons rather than cubes) the threshold aspect ratio for the transition of flow type is slightly lower.

Figure 4.10 provides a detailed snapshot of the instantaneous flow field in an isolated roughness flow in a staggered array of cubical buildings. It views the simplified patterns in Figure 4.9 from another perspective. In fact what appears in Figure 4.9a as simple descending flow into the inter-element space is actually intermittent gusting. The sources of the intermittency, in addition to the inherent variability of turbulence, are 'flapping' of the **shear zone** (zone of rapid change in wind speed with height) that is shed from the upwind roof, or larger eddies descending from higher in the flow. Much of the exchange of momentum between the UCL and the overlying air is thought to be associated with Reynolds stress, caused by these strong downward gusts (sweeps). They bring higher velocity air into the UCL, which initiates less vigorous upward motions of lower velocity air, out of the canyon (ejections).

It is important to bear in mind that these patterns are abstractions that provide a framework to describe general effects of **urban form** on airflow. They are only intuitive 'models' helping us to consider practical implications of street geometry on wind and pedestrian shelter, drifting of snow and sand, **dispersion** of air pollutants and transport of heat water vapour and momentum. Spacing is also important to lateral dispersion from a point source in the canopy.

### 4.2.3 Streets and Intersections

Understanding and controlling flow and dispersion in urban canyons, is important for **air pollution** management (Chapter 11). Highest exceedances of air pollutants are typically found in canyons with a high $H/W$ and high traffic volume for two obvious reasons: reduced mixing of the air inside the canyon with the atmosphere above, and the high rate of **emissions** of air pollutants by vehicles. **Concentrations** of air pollutants are highly variable, depending on the

location within a canyon relative to direction and magnitude of the approaching flow.

### Flow Patterns in Urban Canyons

Flow in urban canyons is reasonably well understood, at least for the relatively simple case of intermediate aspect ratios, neutral flow (strong winds and/or weak heating/cooling) and long streets with flanking buildings of similar height and flat roofs. Again, flow is driven by the above-roof wind, especially its horizontal direction relative to the axis of the canyon (angle-of-attack, $\varphi_c$):

- **Cross-canyon vortex** – When the approach flow is perpendicular to the canyon axis (i.e. $\varphi_c = 90°$) the observed mean flow creates a recirculation vortex of the cavity of the upwind building, reinforced by deflection down the wall of the next building downwind (Figure 4.11a). This cross-canyon vortex is strongest with strong winds and in areas with built structure that favours wake interference flow. If the buildings flanking the canyon have pitched (or slanted), rather than flat, roofs the effects are significant. Kastner-Klein et al. (2004) show that pitched roofs inhibit the formation of a cross-canyon vortex which lowers the mean velocity in the canyon and decreases exchange between the atmosphere above roof and the UCL.

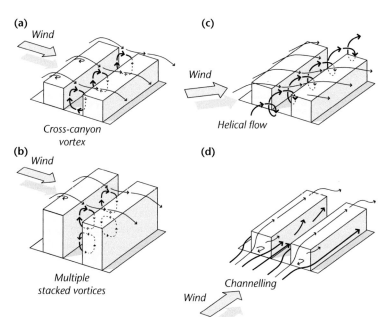

**(a)** Wind
Cross-canyon vortex

**(b)** Wind
Multiple stacked vortices

**(c)** Wind
Helical flow

**(d)** Wind
Channelling

**Figure 4.11** Typical flow patterns in urban canyons: **(a)** cross-canyon vortex, **(b)** multiple stacked vortices in a deep canyon, **(c)** helical flow along a canyon, and **(d)** along channelling and jetting along a canyon (Modifed after: Oke, 1997; Belcher, 2005).

- **Stacked vortices** – If $H/W$ is very large and $\varphi_c = 90°$, physical and numerical models show the main vortex slows and develops into one or more secondary cells towards the floor. They rotate counter to each other and closer to the floor they rotate more slowly (Figure 4.11b). Sometimes small secondary cells appear in one or both corners of the canyon near the floor. These cells tend to recirculate polluted air internally rather than mix it with the cleaner air above, hence dispersion is poor.
- **Helical vortex** – When the external wind is at an intermediate angle to the canyon axis ($\varphi_c \sim 45°$) transport is carried along as the vector sum of the cross-canyon vortex and the channelling flow. The result is a helical path, spiralling down the canyon (Figure 4.11c), akin to a stretched canyon vortex.
- **Channelling** – The in-canyon vortex may disappear when $\varphi_c < 30°$ and the flow is channelled along the canyon instead. This angle is dependent on street canyon length and the height to width ratio. If the flow has just entered the end of the canyon, the constriction causes it to accelerate, this is **jetting** flow (Figure 4.11d).

The specific dimensions of buildings, roofs, possibly trees, thermal effects and even traffic, affect flow and turbulence, and hence air pollutant concentrations in a canyon. Flow structures, such as those in Figure 4.11, depend on the mean direction of the approach flow, but not on short-term changes (of the order of seconds); they are simply mixed into the generally turbulent buffeting of canyon winds.

Figure 4.12 shows wind directions measured at different heights on a tower inside and above an urban canyon. The measurements illustrate for a given background wind above the RSL, how wind direction changes with decreasing height down into the UCL. Any wind is directed into the canyon's axis (shaded bands) causing a turning wind direction with height. The exception being cases with wind perpendicular to the axis, where the horizontal wind direction at the floor is opposite to the wind direction above the RSL due to the cross-canyon vortex (e.g. leftmost profile in Figure 4.12).

Flow patterns in canyons can be enhanced or diminished by thermal effects, especially if the ambient airflow is weak. A wall strongly heated by solar **irradiance** warms the adjacent air and a narrow sheet of buoyant air streams up the wall. The impact of this on the canyon vortex depends on the wind direction.

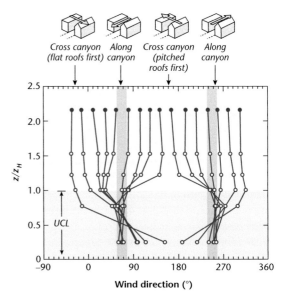

**Figure 4.12** Turning and channelling of wind measured at multiple heights on a tower in and above an urban canyon in Basel, Switzerland (site 'Ba02u1', Table A2.1 and Photo in Figure 3.7). For a given background wind direction (closed circles, measured at the top of the tower), the lines connect simultaneously measured mean wind direction at the five lower heights (open circles). Profiles are the average from several hundred measurements in each wind sector (Source: Christen, 2005; with permission).

If it is the leeward wall (i.e. on the upward limb of the circulatory cell) the additional uplift enhances the vortex rotation, but if it is the windward wall (with the downward limb) it opposes the mechanical flow and retards the circulation, or can create two vortices (Sini et al., 1996). A different type of thermally-induced canyon exchange can occur at night. In canyons with large aspect ratios, wall, floor and air temperatures do not cool as rapidly as on the roof, due to differences of the **radiation** and **heat storage** of canyons and roofs (see Chapters 6 and 7); hence canyon air is unstable with respect to that above-roof level. At intervals, especially if triggered by cold air drainage off the roofs, plumes of buoyant canyon air 'bubble' upward. This thermal canyon venting is a mechanism that helps reduce canyon pollutant concentrations and maintains an upward flux of sensible heat, even through the night.

In canyons with a high traffic volume, moving vehicles affect its flow by producing additional mechanical turbulence, called traffic produced turbulence (TPT). If the external flow is weak, TPT can become the primary mechanism for dispersion of emitted air pollutants within an urban canyon. In streets with asymmetric traffic (e.g. one-way streets) the flow is accelerated by moving vehicles in the direction of the traffic.

### Intersections

Airflow and dispersion at street intersections are much less studied. The practical need is obvious, given that intersections are nodes where traffic converges, slows down and idles, leading to high vehicle emissions in a relatively confined space. Moreover, they are located where the along-canyon transports of air pollutants (helical and channelled) meet. Take the case of a common four-way street intersection bounded by buildings. It is the focus of competition between the relative energies of the contributing street flows. Given the 3D complexity of the resulting flow patterns, field observations (even at several levels and locations), give an incomplete and often confusing picture. Physical and numerical models give a better understanding of the situation.

Figure 4.13 shows flow simulations for a case with four streets of equal aspect ratio, named for their cardinal directions (N, E, S and W), that join at a symmetric intersection bounded by buildings of equal height. The figure illustrates the effect of a change in the angle-of-attack $\varphi_c$ of the above-roof approach flow relative to the street grid. With flow parallel to one of the street axes ($\varphi_c = 0°$, Figure 4.13a) the horizontal flow field at $0.5H$, shows the expected channelling along one street (here the W-E street), but a small amount of the flow is drawn into the side streets (N and S) where it develops corner vortices and mixes for a distance of about one street width. As a consequence of the vortices and the turbulence they create, the air and some pollutants from the W-E street are mixed into the side streets. At $\varphi_c = 15°$ (Figure 4.13b) more of the flow from the W-street is side-tracked into the northern street, and a small part of the weak helical flow in the southern street is entrained into the main channelling street (E) towards the east. At $\varphi_c = 30°$ (Figure 4.13c) the flow from W into side street N is even more pronounced, because the angle-of-attack now drives a west-to-east helical flow in the W-E street and a weaker helical flow from south to north in the S-N street. Flow in the W-E street tends to hug the northern side preferentially and the small corner vortices are stronger. As might be anticipated at $\varphi_c = 45°$ (Figure 4.13d) flow is split symmetrically. There is a helical flow in both canyons and air is bled away from the W-E street into

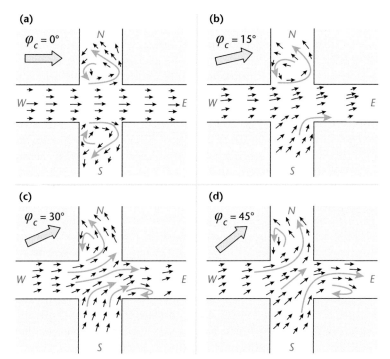

**Figure 4.13** Typical flow patterns at street intersections. Numerical simulations of mean horizontal velocity at half canyon height ($z = 0.5z_H$) in a 4-way street intersection. The different cases are for above-roof wind directions ($\varphi_c$, indicated by the block arrow) at **(a)** $0°$, i.e. parallel to the east-west street, and at **(b)** $\varphi_c = 15°$, **(c)** $\varphi_c = 30°$ and **(d)** $\varphi_c = 45°$. Results for $60°$, $75°$ and $90°$ (not shown) are simply rotated and reflected versions of those for their complementary angles ($90°-\varphi_c$) (Modified after: Soulhac et al., 2009).

**(a)**

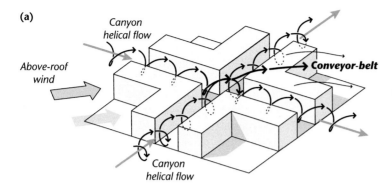

Canyon
helical flow

Above-roof
wind

Conveyor-belt

Canyon
helical flow

**Figure 4.14** Schematic depiction of 3D flow at **(a)** an intersection on an orthogonal street grid with above-roof flow diagonal to the grid, and **(b)** similar, but at a T-junction. Light blue arrows show the mean flow direction whilst the black arrows depict the actual helical motion (Drawn based on observations reported in Robins et al., 2004).

**(b)**

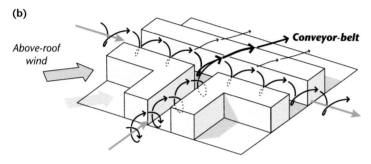

Above-roof
wind

Conveyor-belt

the northern one, but it is augmented by some energy that is tranferred from the flow in the N-S street into the eastern street.

Figure 4.14a gives an idea of the 3D flow structures present at an idealized four-way intersection. It includes flows that the slice at $0.5H$ in Figure 4.13 is unable to depict, but are seen in wind tunnel studies. For example, it has been observed that with an above-roof flow at $45°$ to the street grid, when the helical flows join at the intersection it creates a 'conveyor belt' tilted upwards that helps vent air pollutants to above roof-level. In the case of a T-junction with the same approach flow relative to the street grid (Figure 4.14b) the helical flow patterns from the two streets mostly converge into a single receiving street. This causes an increase in wind velocities in the receiving canyon (right) and even stronger uplift than in the four-way case. Again, the conveyor belt across the building row facilitates venting of air pollutants upwards to roof level.

### Effects of the Street Grid

In real cities there are many possible element configurations, but take the simple case of flow parallel to a street through a grid array of cubic elements. If the elements are wide apart (say $H/W \approx 0.4$) the wakes

of buildings on either side of a street are not wide enough to interact effectively with each other (Figure 4.15a). Similarly if $H/W \approx 4$ they are so close that the wakes are confined, so there is little lateral motion and much of the flow is deflected above the elements (Figure 4.15c). The greatest potential for lateral spread of air pollutants (or other entities) exists at intermediate densities $(0.7 < H/W < 1.0)$ (Figure 4.15b). The key to the spread is that there is sufficient lateral motion to carry the plume contents into side streets where it is deflected off the front walls of buildings. The plume widens at each street branch in a zigzag pattern that is tied to details of the **urban structure**, especially the street pattern.

If the above-roof flow is at an angle-of-attack to a street grid, the lateral spread of a plume released in the UCL is even greater than for an area with equivalent roughness but without the steering conduits of the urban street network. Figure 4.16 shows concentration visibly decreases with distance of travel through the grid. Note also that there is a build-up in concentration at the first intersection, where flow is slowed and smoke becomes visible above roof-level, especially above the corner most in line with the above-roof wind direction and in accord with Figure 4.14a.

**(a)** $\lambda_b = 8\%$, $H/W = 0.4$    **(b)** $\lambda_b = 25\%$, $H/W = 1$    **(c)** $\lambda_b = 64\%$, $H/W = 4$

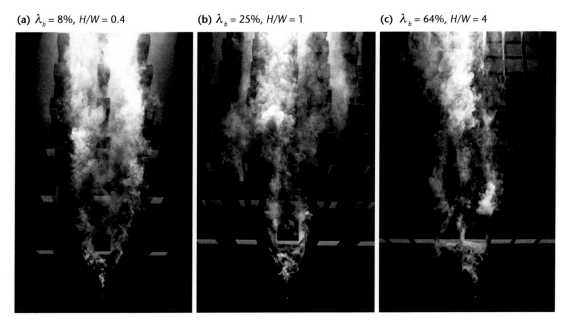

**Figure 4.15** Visualization of dispersion of smoke released from a ground-level source for aligned arrays of different packing densities (Source: Hall et al., 1997; with permission).

In the real world other variables come into play: wind directions are not steady; street directions are not orthogonal; **canyon aspect ratios** are neither symmetric nor continuous; street widths vary; some streets have trees, awnings; balconies and traffic whilst others do not. Whilst there is much to learn, nevertheless, there is hope that application of simple fluid dynamic principles and models that resolve buildings can give a reasonable degree of understanding.

### 4.2.4 Arrays with Uneven Heights and Tall Buildings

Real cities are usually patchily developed with different elements (buildings and trees) of uneven height often with winding roads and rivers and located in rolling terrain. Measures of the variability of element height and patchiness are recognized to be important contributors to the drag and turbulence of urban arrays. For example, if we consider two arrays each with the identical mean element height, but one built of elements with different heights (Figure 2.8d) the other with elements of the same height (Figure 2.8a), the form drag of the former can be almost double the latter.

The flow around an anomalously tall building has been studied for several reasons. Firstly, such a large obstacle juts up into higher level flow where momentum is greater, this means there are large wind loads, both positive and negative, which need to be addressed in the structural engineering of the building. Secondly, the blocking effect of the tall building deflects high velocity air down into the UCL where it can be a hazard to lightweight structures and pedestrians. It also generates vigorous swirling of materials like leaves and litter and sometimes leads to problematic deposition of materials like snow and sand. Thirdly, the cavity and stack vortices of the building affect the transport of air pollutants, both positively and negatively.

The flow pattern around an isolated tall building is an exaggerated version of that around a solitary cubical building (Figure 4.4 and Figure 4.5). Considering the height of most tall buildings that means the stagnation point is commonly 70 to 200 m above the ground; at 200 m the wind speed is at least double that at 30 m. This means that the magnitude of the perturbation to the pressure and velocity fields can be greater and the momentum injected downward is potentially larger. Such a case arises if the building is also wide and thereby blocks much of the flow. Then some air escapes around the sides but most rushes down to ground level and forms one or more roll

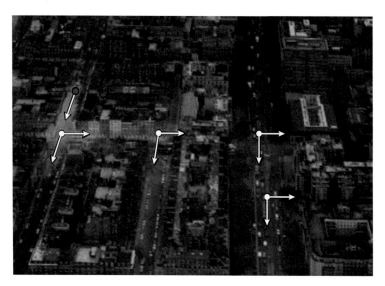

**Figure 4.16** Example of dispersion in the urban canopy layer. Results from a wind tunnel simulation using coloured smoke released from within a canyon (at circle in top left) in a model of a neighbourhood in London, UK. The model is a faithful scale representation of the urban structure, the aerial photograph is an overlay. Above-roof wind direction is from top left to bottom right, hence at intersections part of the smoke travels further along the same street towards you, and part bleeds into a street on the right of the image (left in flow direction) (Credit: A. Robins, Univ. of Surrey; with permission).

vortices that elongate towards lower pressure at the sides where they wrap around as corner streams in the classic horseshoe shape. These can be a danger to pedestrians, not only due to their strength but their very localized nature. A pedestrian can suddenly encounter a blast of strong wind, the force of which can be sufficient to knock them over (see Section 14.5).

Many cities have adopted the requirement that such buildings must be subject to wind tunnel or other testing in the planning stage. It is important also to model the surrounding structures in detail, not just the building, and include a range of wind directions. If problems are foreseen, given the wind climate of the location, remedial measures can be imposed such as redesign of the structure (change its height, or shape, place it on a podium such as a lower wider building, add shelter over walkways such as fixed canopies or closely spaced mature trees). Even so pressure perturbations may mean either it becomes difficult to open doors, or conversely they are sucked open by themselves. An opening under the building to allow pedestrian traffic through should also be avoided because it is a point of lower pressure where air rushing down the face will be sucked into as a strong jet. Behind the building the cavity zone extends from the ground to the building top. Whilst the mean horizontal wind might be reduced inside the cavity at ground level, especially compared with that in the corner streams, the cavity is gusty and a focus for swirling loose leaves and litter.

If a tall building is placed in a neighbourhood of lower mean height, some of the force of the downward displaced flow is intercepted at roof level of the lower buildings. This somewhat lessens difficulties at pedestrian level. Smoke visualizations of the flow show remarkably effective vertical transport in the rising limb of the lee eddy, up the back face and within the stack vortices shed off the back corners. The visualization in Figure 4.17 clearly shows mixing in the turbulent wake persists downstream. In fact simulations show this mechanism can make a positive contribution to neighbourhood pollutant concentrations, by injecting cleaner air down into the UCL and expelling dirtier air upwards. Adding a few strategically-placed tall buildings in an otherwise relatively low-rise district therefore can decrease overall mean air pollutant concentrations near the ground.

Tall buildings have certain climatic benefits for their occupants. In hot regions, especially hot, humid ones the extra air motion is very welcome, as is the likelihood of cooler air and perhaps cleaner air and less noise. Most negative side-effects accrue to surrounding areas, including loss of sunshine by shading (Chapter 5) and the dangerous wind effects noted above. Particularly worrisome, at both the **local scale** and the mesoscale, is the placement of tall and/or wide buildings that block existing flow for districts in their lee. Many cities depend on the comfort and cleanliness provided by different types of breezes from their surroundings, especially cold air drainage down slopes / valleys and sea / **lake breezes** (Chapter 12). Poor

**Figure 4.17** Smoke visualization of dispersion from a release located at ground level in front of a staggered array of cubes (plan area density $\lambda_b = 16\%$) with a 'tall' building present. Compare with Figure 4.3b (Source: Hall et al., 1997; with permission).

placement of a massive building can deprive large areas of those benefits. The situation becomes critical when clusters of buildings combine their shelter and deny the benefits of clean, cool breezes to downstream districts. The development of such 'walls' of high-rise apartments and hotels along coasts and lakeshores has become the bane of many coastal cities (Hong Kong, China; Miami, United States; Tel Aviv, Israel; Tokyo, Japan, e.g. Figure 4.7). These 'walls' block the invigorating and healthy incursion of cooler, cleaner **sea breezes** into a large urban region because of lack of foresight or failure to incorporate climatic considerations into legislation to limit development.

Flow around tall buildings is unique to the particular structures and their arrangement, even when clustered in a modern office or apartment district their response differs with wind direction. The flow around such clusters does not fit into the types in Figure 4.9. They are better classed as **chaotic flow** regimes, where flow is so specific to those particular buildings that analysis requires detailed modelling (Grimmond and Oke, 1999 a; Davenport et al., 2000).

### 4.2.5 Spatially Averaged Flow and Turbulence Statistics

In many applications the exact 3-D structure of buildings and trees in the UCL is not of relevance, or cannot be modelled due to lack in computing power. In such cases we can conceptually simplify the RSL from 3D $(x, y, z)$ to 1D $(z)$. This is achieved by averaging over horizontal layers ('slices'), which gives spatially-averaged variables (for detailed discussion and illustration of the procedure see Section 2.3.2 and Figure 2.16b). Spatially-averaged variables are indicated by angle brackets ($\langle \ldots \rangle$) and are only a function of height $z$. In typical applications, we aim to describe the horizontally-averaged mean flow and the characteristics of turbulence simply as a function of height $z$ (relative to mean building height $z_H$).

For simplicity, we redefine the coordinate system that underlies the wind components in the spatially-averaged frame. We rotate the coordinate system around the $z$-axis into the spatially-averaged mean wind direction in the RSL. Doing so, the length of the spatially-averaged horizontal wind vector becomes simply $\langle \overline{u} \rangle$ and due to mass conservation, we can also assume **divergence** is small, so $\langle \overline{v} \rangle \approx 0$ and $\langle \overline{w} \rangle \approx 0$. This convention reduces the complexity of the spatially-averaged mean wind field from three to one component $\langle \overline{u} \rangle$.

### Momentum Flux

We have seen that the RSL is characterized by complex but decipherable flow patterns. In dense arrays, deep in the UCL the mean flow is for the most part sheltered, except for channelling along some canyons, so spatially-averaged horizontal wind speeds $\langle \overline{u} \rangle$ at this height are relatively weak, but near roof-level wind speeds rapidly increase up as the merge into the general flow above the roofs. Hence the strongest velocity gradient $\partial \langle \overline{u} \rangle / \partial z$ occurs around roof level

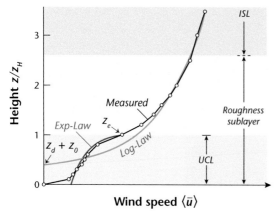

Figure 4.18 Wind profile in the RSL and ISL measured in a wind tunnel over an array of cubes ($\lambda_f = 0.16$). Values are horizontal spatial averages (see Figure 2.15b) and include an inflection point at height $z_e$. Also shows the extrapolation of the exponential law ('Exp-Law', see Equation 4.5) down through the UCL, and the logarithmic profile ('Log-Law', see Equation 4.9) down into the UCL to its apparent momentum sink at ($z_d + z_0$) (Modified after: Macdonald, 2000).

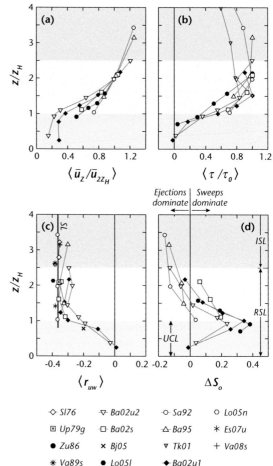

| ◇ Sl76 | ▽ Ba02u2 | ○ Sa92 | ○ Lo05n |
| ▣ Up79g | □ Ba02s | △ Ba95 | ✳ Es07u |
| ● Zu86 | ✕ Bj05 | ▼ Tk01 | + Va08s |
| ✳ Va89s | ● Lo05l | ◆ Ba02u1 | |

Figure 4.19 'Family portrait' of spatially-averaged mean wind and momentum exchange in the RSL as a function of non-dimensional height ($z/z_H$, here $z$ is the height of measurement). **(a)** Mean horizontal wind $\langle \bar{u} \rangle$ normalized by mean horizontal wind at twice the average height of the buildings. **(b)** Reynolds stress $\langle \tau \rangle$ at a given height in the RSL relative to the maximum Reynolds stress $\tau_0$ under neutral conditions. **(c)** The efficiency of the turbulence to exchange momentum at a given height in the RSL expressed as $\langle r_{uw} \rangle$ (see text) in neutral conditions. **(d)** Relative contributions of ejections and sweeps ($\Delta S_0$) to the total momentum exchange, under neutral conditions. Data compiled from 15 different field sites; codes, references and site properties are given in Table A2.1.

(Figure 4.18, Figure 4.19a). We call this region of the RSL in dense arrays the shear zone, which is characterized by an inflection point (at height $z_e$) in the average wind profile, typically located around $z_H$ ($z_e$ can be higher or lower than $z_H$, depending on variability of building heights and packing density). At $z_e$ curvature of the wind profile changes from concave to convex.

Below the inflexion point, each layer of the canopy imposes some drag on the flow i.e. it is a momentum sink. Momentum is absorbed first by the tallest roofs, then progressively by walls and in less densely packed districts, also by the canyon floor. Therefore, the average momentum flux decays with distance into the urban canopy, i.e. each layer removes a certain fraction of the momentum flux injected from the level above. Figure 4.19b shows measured momentum fluxes (precisely measured Reynold stress $\langle \tau \rangle$) from different field studies. $\langle \tau \rangle$ at any height is normalized by the maximum measured value of $\tau$ above the canopy. All the sites included in Figure 4.19b are relatively densely packed ($\lambda_b > 0.25$) and below about $0.5z_H$ no significant $\langle \tau \rangle$ is registered. Also note that $\langle \tau \rangle$ starts to decrease *above* $z_H$ probably because $z_H$ is the *average* height of buildings. Some structures are taller

than $z_H$, and they are disproportionally more significant in terms of the drag they impose on the flow (Rotach, 2001).

Figure 4.19c shows how effectively momentum is transferred by turbulence at different heights in the

RSL. The efficiency of the transport is expressed using the correlation coefficient $r_{uw} = \overline{u'w'}/(\sigma_u \sigma_w)$; the more negative is the value of $r_{uw}$ the more organized is turbulence, and momentum is transported by turbulent eddies in an orderly fashion. On the other hand, values closer to zero mean there is little Reynolds stress and eddies are completely disorganized. Figure 4.19c underlines that above the urban canopy, momentum is efficiently transferred, and $r_{uw}$ approaches the value of –0.32 which is a typical number found in **surface layer** (SL) scaling over much smoother surfaces. As we move down into the canopy, $\tau$ is reduced, which decreases the efficiency of the flow in transporting momentum. In the lowest part of the canopy, eddies are essentially random and do not contribute to turbulent momentum exchange.

Figure 4.19d tells us about the turbulent actors – the events that exchange momentum in the RSL. The parameter $\Delta S_0$ describes the relative contributions of sweeps and ejections (see Section 4.1.2) to the total Reynolds stress at a given height. Although the turbulent exchange of momentum is the result of both, sweeps and ejections, their contribution is not necessarily equal ($\Delta S_0 = 0$). Sweeps – the events that bring eddies with higher velocity downwards – dominate in most of the RSL, particularly around $z_H$. Ejections – the lulls that bring lower momentum upwards – are dominant in the upper part of the RSL and ISL. The reason for their uneven contribution with height is that most mechanical turbulence is created in the shear zone. Part of the turbulence from this highly energetic zone is ejected upwards, part of it makes it down into the deeper UCL as sweeps.

Figure 4.19b–d implies the momentum flux in the spatially-averaged case is transferred solely by turbulence (Reynolds stress), however, there is also the possibility that it is transferred by smaller-scale features of the mean flow around obstacles. This form-induced advection of momentum at a scale below that of spatial averaging is commonly referred to as the 'dispersive stress'. Dispersive stresses are relevant if there is a correlation between mean horizontal flow and mean vertical flow around roughness elements in the RSL. Momentum can be transferred by regions that experience downdrafts on average and faster horizontal winds and also by regions that experience updrafts on average and have statistically slower horizontal winds. Numerical simulations by Martilli and Santiago (2007) show that in regular arrays the contribution of dispersive stresses to the overall momentum flux in the UCL can be substantial.

### The Exponential Wind Law

The progressive absorption of momentum with depth in the UCL can be used to approximate the decay for wind in the UCL as an exponential function, assuming that momentum reduction and the typical length scale of turbulent eddies are both constant with depth. The 'exponential law' developed for wind within vegetation canopies (Cionco, 1965), was used by Macdonald (2000) to describe the spatially-averaged wind speed $\langle \bar{u} \rangle$ with depth in the UCL ($z < z_e$) of simple arrays in wind tunnels:

$$\langle \bar{u} \rangle_z = \langle \bar{u} \rangle_{z_e} \exp \left[ a \left( \frac{z}{z_e} - 1 \right) \right] \quad (\mathrm{m\ s^{-1}})$$

**Equation 4.5**

where $a$ is an empirical wind attenuation coefficient. Macdonald (2000) proposed that $a = 9.6$ for wind tunnel experiments using in-line and staggered arrays of cubes of uniform height. Naturally Equation 4.5 only applies in the middle and upper part of the UCL, closer to the floor (say $z < 0.15\ z_H$) skin drag of the floor produces a more logarithmic decay until the flow is brought to rest at the floor. Nevertheless, for most of the UCL an exponential decay fits reasonably well with wind tunnel and field measurements (Figure 4.19a).

### Horizontally Averaged Turbulent Kinetic Energy

Some means to describe turbulence in the RSL is essential to be able predict the dispersion of heat, water and air pollutants. However predicting turbulence at any location around a single building is a huge challenge and to do it for an entire city is at the limit of current computational power. Hence, most dispersion models rely on spatially-averaged statistics of turbulent exchange and parameterize the average effect of urban structure on turbulence in a highly simplified way.

Figure 4.20a shows measured ratios of turbulent kinetic energy (TKE) to mean kinetic energy (MKE) as a function of height, compiled from many different field studies that represent locations in canyons, above roofs, on towers, and others from various cities. Despite the large number of surfaces, all sites show a similar pattern: TKE increases relative to the MKE as one descends towards the urban canopy and the mean wind speed decreases. Generally turbulence

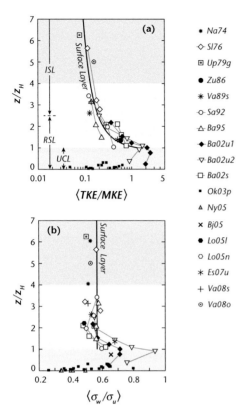

Figure legend (symbols):
* Na74
◇ Sl76
▣ Up79g
● Zu86
✳ Va89s
⊖ Sa92
△ Ba95
◆ Ba02u1
▽ Ba02u2
▢ Ba02s
▪ Ok03p
▲ Ny05
✕ Bj05
● Lo05l
○ Lo05n
✳ Es07u
+ Va08s
⊙ Va08o

**Figure 4.20** Typical vertical profiles of turbulence statistics in the UCL, RSL and ISL in neutral atmospheric conditions. Measured ratios of **(a)** turbulent kinetic energy (TKE) to mean kinetic energy (MKE) and **(b)** vertical to horizontal turbulent motions. Data from 18 field studies with surface properties and references listed in Table A2.1.

intensity peaks in the shear zone at around the height $z_e$. Here the ratios of TKE to MKE are highest. Further down into the UCL turbulence intensities again seem to decay due to lack of production (other than TPT), although channelling and standing vortices may keep the mean flow (and hence MKE) relevant.

The shear zone is the dominant source region of mechanical TKE within the RSL, however, not all of its TKE dissipates there. Part is injected down into the UCL and part is transferred into higher layers of the RSL and ISL by advection, turbulent transport and probably by pressure transport processes. The shear zone is the primary export source of TKE in the RSL (more is produced than is dissipated locally). The UCL below and higher regions of the RSL are import

zones, where more TKE is dissipated than is produced locally (Christen et al., 2009).

Common scaling relationships used in traditional air pollution dispersion modelling rely on local closure of the TKE budget (i.e. the amount of TKE produced at a location is also dissipated locally) and assumes a roughly constant flux of momentum with height. Clearly neither prerequisites is fulfilled in the RSL. This renders common scalings, including the **Monin-Obukhov Similarity** (Section 4.3.3), unable to predict turbulence in that layer. Local scaling has been used with some success in the RSL; that uses empirical **parameterization** relations to describe the changing profile of $\tau$ and $Q_H$. That eliminates the need for fluxes to be constant with height, however, it does not capture all processes properly and this approach only works adequately above the shear zone.

To give an example of the fundamentally different nature of turbulence in the shear zone, we look at the distribution of TKE broken into its different wind components. In Figure 4.20b measured turbulent fluctuations in the vertical direction (standard deviation $\sigma_w$) are compared to those in the along-wind direction ($\sigma_u$, longitudinal). Well above the urban canopy, the ratios measured over different cities are relatively constant and match well the SL prediction for smoother surfaces, i.e. $\sigma_w/\sigma_u = 0.57$. There, more TKE is found in horizontal than vertical wind fluctuations. Approaching the shear zone, however, flow features around obstacles and flapping of the turbulent wakes behind them, greatly enhance vertical motions and turbulence is not only greater overall, but also more random in space, in some cases closer to a state of isotropy, i.e. $\sigma_w/\sigma_u \sim 1$. These differences make an integral treatment of turbulence using SL predictions, impossible. Unfortunately, a universal scaling framework for turbulence in the RSL remains elusive.

## 4.3 Flow in the Intertial Sublayer

Above the RSL lies the urban inertial sublayer (ISL, Figure 2.13). It differs from the layers beneath by the fact that its mean characteristics are essentially homogeneous in the horizontal and that there is no significant mean vertical wind. Being above the **blending height** ($z_r$) that separates RSL from ISL, the spatial variability of atmospheric properties around the roughness elements of the RSL are homogenized. Assuming the surface is reasonably uniform and spatially extensive, the effects of turbulent mixing are so effective that mean properties such as wind, temperature and humidity in the ISL are independent of horizontal position, only depending on height above ground and time.

Realistically given the physical structure and patchiness of real world cities an ISL is not always present. The best chance to have a well-developed ISL is found over horizontally extensive parts of urban areas with a relatively homogeneous and dense urban structure. It is never likely to exist over localized groups of tall buildings, and it can be entirely consumed by the RSL when elements over an extensive zone are so tall they reach up to more than 10% of the UBL (Rotach, 1999 and Fig 2.12, right).

Conceptually it is straightforward to define the blending height, but setting an actual level for $z_r$ is less simple because the transition from RSL to ISL is gradual. From a practical perspective this challenge becomes clear when installing sensors. If an instrument to measure atmospheric quantities is placed below $z_r$ it registers **microclimate** anomalies such as wakes of buildings or plumes of heat, but if it is above $z_r$ it mostly 'sees' a blended, spatially-averaged signal representative of the local scale. Rule-of-thumb estimates and measurements indicate that $z_r$ can be as low as 1.5 $z_H$ at densely built-up (skimming flow régime) and very homogeneous sites. However, in extreme cases $z_r$ may be greater than 4 $z_H$ such as in sparsely built areas with an isolated roughness flow régime. As a first rule-of-thumb $z_r = 1.5$–$3z_H$ for medium to high density areas with relatively homogeneous roughness often proves to be reasonable.

Because horizontal wind velocity and direction do not change with horizontal position in the ISL and the mean vertical wind is zero, we simplify description of airflow from managing all three components of the wind vector $\vec{u} = (u, v, w)$, to a single wind component only, similar to Section 4.2.5. Rotating the coordinate system around the z-axis into the mean wind, we simplify since the mean horizontal wind $\bar{u} = \left| \vec{u} \right|$ and $\bar{v} = 0$ (and as discussed above $\bar{w} = 0$). This convention is used in the following sections.

## 4.3.1  The Profile of Mean Wind

Because over a given period, the mean horizontal wind ($\bar{u}$) in the ISL is independent of horizontal position, we only need to know how $\bar{u}$ changes with height to be able to predict wind at any location in the ISL. Hence our primary goal is to describe $\bar{u}$ as a

function of height $z$, for the current atmospheric state and surface structure. Sufficiently above the roughness elements, an urban surface can be described by its integral roughness, rather than the specific structure of all its roughness elements (buildings, trees). This means we envision the urban surface as a uniform rough plane at an effective surface height $z_d$ (called **zero-plane displacement**) and neglect its three-dimensionality. This has the advantage that we can disregard details of the complex forces in the flow around the 3-D roughness elements. This explains why we dropped the spatial averaging operator introduced in 4.2.5 for flow in the ISL.

### The Vertical Flux of Momentum

Because all of the momentum in the ISL is exchanged by turbulence (Reynolds stress) we investigate the role of turbulent eddies in more detail. The amount of momentum added or subtracted by a single eddy is $mu'$, where $m$ is the mass of the eddy that causes the exchange and $u'$ its horizontal velocity surplus (or deficit) relative to the mean velocity at the given level (Figure 4.21). If we express the momentum exchanged per square metre and second (i.e. kg m s$^{-1}$ m$^{-2}$ s$^{-1}$, i.e. the units of pressure, Pa) we can write:

$$\tau = -\rho\, w'u' \quad \text{(Pa)} \qquad \textbf{Equation 4.6}$$

where $w'$ is the instantaneous vertical wind fluctuation of the eddy (m s$^{-1}$), and the product $\rho\, w'$ determines the mass exchanged per second and square metre, with $\rho$ being air density (kg m$^{-3}$). The net motions of many eddies, over a longer period, create a mean **turbulent flux** of momentum. Using an overbar to show we average all realizations of Equation 4.6 over a longer period (e.g. 30 min):

$$\tau = -\overline{\rho w'u'} = -\rho\, \overline{u'w'} \quad \text{(Pa)} \qquad \textbf{Equation 4.7}$$

Note that since $\rho$ does not vary significantly during a 30 min period it can be regarded as a constant, i.e. independent of time, and can be taken out of the averaging operator (as done on the r.h.s.). The remaining term $\overline{u'w'}$ is the covariance of horizontal and vertical wind fluctuations. Because sweeps carry a momentum surplus ($u' > 0$) down ($w' < 0$) and ejections carry momentum deficits ($u' < 0$) up ($w' > 0$), the averaged product, $\overline{u'w'}$, is negative in the ISL. The definition of Reynolds stress $\tau$, adopted is that it is positive when directed downward, which explains the minus sign in Equation 4.6 and 4.7. There is the possibility of a vertical flux of v-momentum, due to

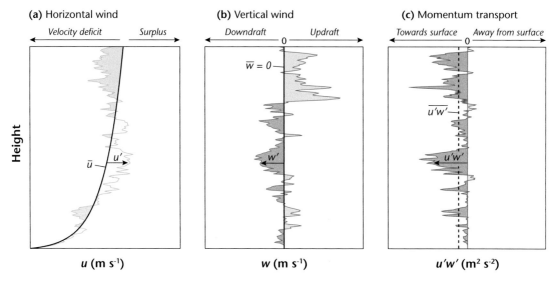

**(a)** Horizontal wind   **(b)** Vertical wind   **(c)** Momentum transport

**Figure 4.21** Schematic profiles of instantaneous (very variable, thin curves in color) and time mean (black line) wind velocities. The difference between instantaneous velocity and time-averaged velocities is the turbulent departure $u'$, which is in color. **(a)** Horizontal (along-wind) velocity $u$, **(b)** vertical velocity $w$, and **(c)** the related kinematic downward flux of horizontal momentum as the product of $u'$ and $w'$ (Modified after: Oke, 1997).

the mean covariance $\overline{v'w'}$, but in practice it is rarely significant because there is no reason to anticipate a preferential association between up- and downdrafts and horizontal motion in any one direction, unless the wind turns significantly with height.

The magnitude of $\tau$ decreases with height through the ABL until it approaches zero in the FA. In the ISL, $\tau$ can be assumed to vary only weakly with height and is close to the value that characterizes the urban surface ($\tau_0$). The total momentum flux in the ISL that represents the integral effect of the urban canopy is often expressed in the form of a scaling velocity, called the **friction velocity** $u_*$:

$$u_* = \sqrt{\tau_0/\rho} \quad (\text{m s}^{-1}) \qquad \textbf{Equation 4.8}$$

The $u_*$ is widely used to express the momentum flux in the ISL. It is considered a global scaling parameter valid for the entire ISL. The friction velocity $u_*$ changes with time, but once determined at one location in a situation, it is valid everywhere else in the ISL and facilitates the calculation of not only wind but also turbulence statistics (see Section 4.3.3).

### Controls on the Momentum Flux

Other influences being equal, the flux of momentum increases with surface roughness and wind speed. It is

further modified by the sensible heat flux density – unstable conditions enhance the exchange stable conditions suppress it.

The rate of mechanical turbulence production is controlled by surface roughness, therefore, everything else being constant, a rough surface creates more TKE than a smooth one. If the atmosphere is in a highly turbulent state, such as above a rough surface, then momentum is easily mixed between layers, $\tau$ is greater (or $u_*$ is faster). This causes the difference in mean wind between layers to be lessened. Indeed, the wind over a rough surface changes least with height (Figure 4.22a). A smooth surface in contrast causes a flow that is less turbulent and therefore creates smaller and less energetic eddies. With less turbulence, momentum is not as easily mixed between layers, $\tau$ is small (or $u_*$ is slow), which allows larger differences between layers, and a steeper wind gradient (Figure 4.22b). Thus urban areas, being particularly rough, are expected to have larger $\tau$ (and therefore $\tau_0$) than their surrounding countryside. Under neutral conditions, at 50 m over a typical urban surface, drag reduces the wind speed to about 50% of the gradient wind in the FA above. Over a crop the wind at 50 m would be about 70% of the gradient wind, and over the ocean about 80% of the gradient wind (Figure 4.22a).

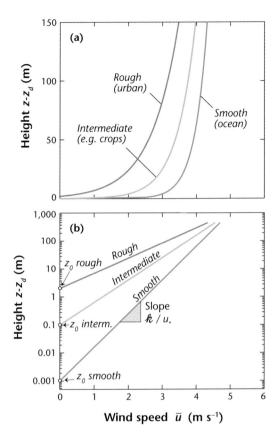

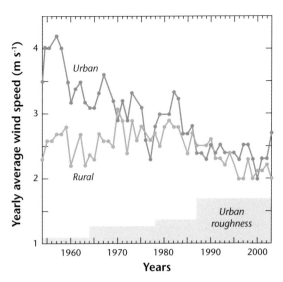

**Figure 4.23** Observations of wind speed at 15 m height at an urban site in Moscow, Russia (Moscow University) and at a rural site outside Moscow (Podmoskovnaya) (Source: Lokoshchenko, 2005; with permission).

**Figure 4.22** General form of the logarithmic wind profile in the lower atmosphere in neutral stability. **(a)** Profiles of mean horizontal wind speed $\bar{u}$ over three types of surface roughness under the same large-scale forcing: rough, intermediate roughness and smooth. **(b)** Same as (a) but plotted with logarithmic height scale.

This effect is seen in climatological records made at urban observation sites where new construction in the vicinity increases form drag. Figure 4.23 shows observed wind speed at a site above the canopy in the centre of Moscow, Russia, where the recorded average wind speed has fallen by about 30% since 1954, in response to increases in roughness of the surrounding urban terrain. Observations at a **rural** site outside Moscow, Russia, that has experienced little change over the same period, shows the background wind climate has not changed.

The production of mechanical turbulence also increases with wind speed. Over an urban surface, the ambient wind speed is reduced compared to a smoother rural surface, so it might be expected that the production of mechanical turbulence is dampened.

However, the relatively few urban–rural observational studies and numerical models of the UBL show that $\tau$ is in most cases greater in an urban atmosphere compared to a rural surrounding, which suggests that the increased surface roughness more than compensates for the lowered ambient wind speed.

Dynamic stability describes how the exchange of momentum in the urban atmosphere is enhanced or suppressed. Enhancement is greatest in mid-afternoon at the time of greatest $Q_H$. $\tau$ is also greater in the city at night, probably due to weak **convection** caused by the urban heat island (UHI, Chapter 7), whereas exchange is suppressed in the more stable rural atmosphere, creating even steeper gradients of wind with height.

## Logarithmic Wind Profile

In the ISL the variation of horizontal wind speed ($\bar{u}$) with height under neutral conditions is described using the logarithmic wind profile equation (e.g. Oke, 1987; Stull, 1988):

$$\bar{u}_z = \frac{u_*}{k} \ln\left(\frac{z - z_d}{z_0}\right) \qquad (\text{m s}^{-1}) \qquad \textbf{Equation 4.9}$$

where, $\bar{u}_z$ is the mean wind at height $z$, $z_0$ is the aerodynamic **roughness length** (m) and $z_d$ is the zero-plane displacement length (m), $u_*$ is the friction velocity, and $k$ is **von Kármán's constant** (0.4). Except for inclusion of $z_d$, this is the same form as the equation that applies to smoother surfaces (such as low plant covers, bare soil, water and ice) where the roughness elements are small. The zero-plane displacement length $z_d$ is included to account for the fact that momentum is not absorbed throughout the entire canopy, but only by the upper canopy (Figure 4.19b). The displacement length is the distance by which the height scale has to be adjusted upward to effectively become the 'ground surface' 'felt' by the flow (Figure 4.18). The $z_d$ parameter is not urban-specific; it is also used for other land surfaces if their elements are tall and close enough together to form a distinct canopy (e.g. tall stands of plant and forests).

When horizontal wind speed ($\bar{u}$) is graphed against logarithmic height above $z_d$, $\ln(z - z_d)$ in a neutral ISL the values form a straight line (Figure 4.22b). If we extrapolate from these measurements to where $\bar{u}$ equals zero, we have a measure of the roughness of the canopy, known as the roughness length ($z_0$).

The logarithmic wind profile is a valuable model that can link sites where measurements are made to others that lack observations. For example, suitable wind observations are often not available at several heights above an urban surface of interest. So if there is need to know the wind at other than the measurement heights, as long as they are within the ISL, extrapolation is achieved by applying Equation 4.9 to two heights:

$$\frac{\bar{u}_z}{\bar{u}_{z_{ref}}} = \frac{\ln\left(\dfrac{z - z_d}{z_0}\right)}{\ln\left(\dfrac{z_{ref} - z_d}{z_0}\right)} \quad (\text{m s}^{-1}) \qquad \textbf{Equation 4.10}$$

where $z_{ref}$ is a reference height at which data are available, $z$ is the height of interest and $z_0$ is the roughness length. The reference height $z_{ref}$ should be well above the displacement height $z_d$.

It is also possible to calculate the wind at the reference height using wind observations at another nearby station such as an airport by using the logarithmic transformation model of Wieringa (1986):

$$\bar{u}_{zA} = \bar{u}_{zB} \left[ \frac{\ln\left(\dfrac{z_{ref}}{z_{0B}}\right) \ln\left(\dfrac{z_A}{z_{0A}}\right)}{\ln\left(\dfrac{z_B}{z_{0B}}\right) \ln\left(\dfrac{z_{ref}}{z_{0A}}\right)} \right] \quad (\text{m s}^{-1})$$

$$\textbf{Equation 4.11}$$

where the subscript A refers to the site of interest and B a site where standard wind measurements are available and the logarithmic law is applicable. $z_{ref}$ is any

height above the blending height, which could be taken to be between 1.5 to 3 $z_H$ – the method is not sensitive to this term. The logarithmic wind profile (Equation 4.9) and hence Equation 4.10 and 4.11 only apply in neutral conditions, that means in absence of any effects of thermal turbulence. The law can be extended to unstable and stable ISLs (Section 4.3.3).

### Power Law Wind Profile

The logarithmic law is consistent with fluid theory and is preferred by meteorologists, however, in practice the parameters $z_0$ and $z_d$, and the friction velocity $u_*$, are not easily obtained. An alternative empirical form, favoured by engineers, is the power law, which is used where it is desirable to obtain the wind at one height based on observations made at another height:

$$\bar{u}_{z1} = \bar{u}_{z2} \left[ \frac{z_1 - z_d}{z_2 - z_d} \right]^{\alpha} \quad (\text{m s}^{-1}) \qquad \textbf{Equation 4.12}$$

The subscripts 1 and 2 represent two levels in the atmosphere and $\bar{u}_{z2}$ represents the observed wind speed at a suitable reference height ($z_2$). The exponent $\alpha$ depends on the surface roughness and atmospheric stability. Typical values of $\alpha$ are listed in Table 4.2. For neutral conditions Counihan (1975) related $\alpha$ to the roughness length $z_0$:

$$\alpha = 0.096 \log_{10} z_0 + 0.016 (\log_{10} z_0)^2 + 0.24 \quad (\text{unitless})$$

$$\textbf{Equation 4.13}$$

Unfortunately, the power law does not approach satisfying asymptotic limits near the surface and so its use should be restricted to heights substantially greater than $z_d$.

Neither the logarithmic nor the power laws account for the effects of the Coriolis force, the influence of which increases, as that of the friction force diminishes with height. The result is that both wind speed and direction change through the ABL.

### 4.3.2 Urban Surface Roughness

The key to applying either the logarithmic wind profile (Equation 4.9) or finding the correct coefficient $\alpha$ in the power law (Equation 4.12) is knowledge of the aerodynamic roughness of the underlying surface. Cities are about the roughest surfaces to be found, and further the magnitude of their roughness varies greatly because the urban structure varies greatly from one land-cover zone to another, or from one city to another. Two approaches are taken to estimate the

**Table 4.2** Typical values of properties related to surface roughness including: mean height of roughness elements ($z_H$), roughness length ($z_0$), zero-plane displacement ($z_d$), power law exponent ($\alpha$) for neutral conditions and normalized friction velocity ($u_*/u_{\mathrm{ref}}$) for natural and urban terrain. The categories are consistent with the Davenport classification scheme that is used in wind engineering (Data sources: Oke, 1987; Wieringa, 1993; Grimmond and Oke, 1999a and Davenport et al., 2000).

| Surface or terrain | Mean height of roughness elements $z_H$ (m) | Roughness length $z_0$ (m) | Zero-plane displacement $z_d$ (m) | Power law exponent $\alpha$ | Normalized friction velocity $u_*/u_{\mathrm{ref}}$ |
|---|---|---|---|---|---|
| **Rural** | | | | | $u_{\mathrm{ref}}$ at 10 m |
| Mud flats, ice, tarmac | | 0.001–0.01 mm | | 0.08 | 0.03 |
| Snow, water (average state)[1] | | 0.1–1 mm | | 0.09 | 0.03 |
| Desert sand | | 0.3–0.5 mm | | 0.10 | 0.04 |
| Bare soil, cut grass[1] | 0.02–0.05 | 0.01–0.02 m | | 0.10–0.11 | 0.04–0.06 |
| Grass[1], stubble field | 0.2–0.5 | 0.03–0.06 m | 0.1–0.3 | 0.11–0.13 | 0.06–0.07 |
| Farmland, crops[1] | 0.4–1 | 0.05–0.15 m | 0.2–0.7 | 0.14–0.18 | 0.07–0.10 |
| Bushland[1], orchards[1], savannah | 2–4 | 0.4–1 m | 1.3–2.5 | 0.18–0.24 | 0.1–0.17 |
| Forest[1]– range from temperate to tropical | 12–30 | 0.8–2 m | 9–24 | 0.23–> 0.27 | > 0.16 |
| **Urban** | | | | | $u_{\mathrm{ref}}$ at 30 m |
| *Low height and density — houses, gardens, trees; warehouses* | 5–8 | 0.3–0.8 m | 2–4 | 0.2–0.25 | 0.09–0.12 |
| *Medium height and density — row and close houses, town centres* | 7–14 | 0.7–1.5 m | 3.5–8 | 0.23–0.27 | 0.11–0.14 |
| *Tall and high density — less than six floors, row and block buildings* | 11–20 | 0.8–2 m | 7–15 | 0.26–0.29 | 0.13–0.16 |
| *High-rise — office and apartment tower clusters[2]* | > 20 | > 2 m | > 12 | 0.29–0.35 | > 0.16 |

[1] Values depend on wind speed as vegetation is flexible.
[2] Estimates have little field support.

parameters needed to describe the roughness of a given urban surface. Micrometeorological (or anemometric) approaches use field measurements of wind or turbulence to solve for roughness length ($z_0$) and zero-plane displacement length ($z_d$) using Equation 4.9. Alternatively morphometric approaches use algorithms that relate the dimensions of the elements and their packing density to $z_0$ and $z_d$. Several algorithms have been derived from measurements of scale models or field experiments over idealized arrays of elements (ones simple enough to be described by measures such as $\lambda_b$ or $\lambda_f$; see Figure 2.4).

## Micrometeorological Approaches

The logarithmic wind profile is the basis of several methods to obtain the parameters $z_0$ and $z_d$ at selected sites. By measuring $\bar{u}$ at multiple heights, or alternatively measuring simultaneously $\bar{u}$ and the Reynolds stress $\tau$ at any height in the ISL, one can solve for $z_0$ and/or for $z_d$. For extrapolation and curve fitting to be successful it requires an extensive and relatively uniform site, instruments free of measurement errors (such as cup **anemometer** overspeeding or flow distortion by the tower), multiple levels of measurement to minimize statistical errors and a long enough observation record to incorporate all required wind directions under strictly neutral conditions. Using multiple levels of mean wind measurements can create errors because they require several instruments each subject to error, and long **fetch** to achieve flow equilibrium over a deep layer. Methods using instruments that simultaneously measure $\bar{u}$ and $\tau$, such as those resolving fast-response fluctuations of $u$, $v$ and $w$ (i.e. ultrasonic anemometers to determine $\overline{u'w'}$), have the advantage of requiring only one instrument located at a single height in the ISL. Micrometeorological methods are good because observations capture the local response of the atmosphere to the specific urban structure of the site, so there is no necessity to measure all the roughness elements in detail, which can be laborious. On the other hand, an extensive site with a tower and expensive equipment are needed.

## Morphometric Approaches

The morphometric approach relies on functional relations between roughness parameters and easily accessible measures of urban structure (height, density, array configuration, etc.). In practice only the main elements, buildings, trees and shrubs, are considered. Obstacles such as utility poles, fences and traffic signs are neglected.

For low roughness surfaces the rule-of-thumb is that $z_0$ and $z_d$ depend mainly on the element height: to a first approximation $z_0 \approx 0.1 z_H$ and $z_d \approx 0.65 z_H$. However, using only $z_H$ ignores the significance of the packing density of the elements. The same elements in a sparse array have a very different roughness effect when crowded together.

The relation between the element density and the roughness parameters $z_0$ and $z_d$ is illustrated in Figure 4.24. The range of real city densities shows that in most cases they experience wake interference or skimming flow. As the packing density of the buildings $\lambda_b$ increases from zero the height $z_d$, which describes the effective height of the momentum sink for the flow in the ISL, moves higher inside the UCL until theoretically the packing becomes total and the canopy top becomes a new surface plane.

Over the most typical range of real city densities $z_d$ has values between about 0.6 and 0.8 of the mean element height $z_H$. In fact if no detailed estimate is possible a value of $0.7 z_H$ is a reasonable first value to use for cities. The form of the relation for $z_0$ shows a peak roughness effect at intermediate density. Initially

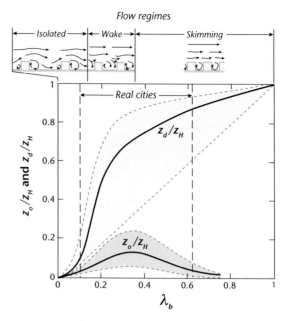

**Figure 4.24** Schematic representation of relation between roughness element density (packing) and wind profile roughness parameters $z_0$ and $z_d$ (both normalized by average element height $z_H$). The areas in color include most estimates from scale models and real cities (Source: Grimmond and Oke, 1999a; © American Meteorological Society, used with permission).

as elements are added to the underlying surface they add to the drag and increase the roughness. However, this effect reaches its maximum in the wake interference régime, adding elements beyond that point restricts flow into the inter-element spaces and stifles their roughness effect, because some of the flow skips over the spaces and skimming is initiated. Again, a first idea of roughness is given by the $z_0 \approx 0.1 z_H$ relation, but clearly it also does not account for the influence of density.

Morphometric methods need no equipment and once the urban structure of the site is described calculations can be done for any direction of interest and spatial roughness patterns can be produced. However, a weakness is that most datasets used to derive the algorithms were collected in wind tunnels with idealized arrays (e.g. regular, staggered and random arrays often with a uniform building height, Figure 2.8). They do not include the true variability of field sites and hence the measures of surface structure may oversimplify the real world case.

## Typical Values

Table 4.2 lists typical values of $z_0$ and $z_d$ for urban canopies, taken from both anemometric and morphometric analyses. This summary places the magnitude of the parameters in context with their non-urban counterparts. The urban roughness categories of Table 4.2 were suggested by Grimmond and Oke (1999a) and are illustrated in Figure 4.25. The magnitude of these rural and urban $z_0$ values confirms that cities are absolutely rough. Urban–rural $z_0$ differences are often one to two orders of magnitude. The exception is a city surrounded by forest where urban-rural differences are likely to be smaller.

Within a city, trees also contribute to drag. In areas with low building heights tall trees generate significant extra roughness, on the other hand if streets are lined with trees they can help to 'cushion' the space between buildings by filling in the canopy and thereby limiting building wakes.

Measured urban roughness parameters show large variability for two reasons. The first is simply due to limitations of the methods. The second is a result of the variation of roughness elements with compass direction around a site due to the street pattern or other structural control on flow which varies with wind direction (i.e. angle-of-attack, Figure 4.12, Figure 4.13). Seasonal change of deciduous foliage also alters the aerodynamic drag of trees. Error in estimating $z_d$ is less critical than in $z_0$. This is because the former is only a correction to the overall datum $(z - z_d)$, and when using Equation 4.9 interest is usually at heights much larger than $z_d$ so the impact of $z_d$ is small.

### 4.3.3 Turbulence and Turbulent Exchange

The turbulent exchange of horizontal momentum $(\tau)$ and also the turbulent fluxes of sensible heat $(Q_H)$, latent heat $(Q_E)$, or air pollutants change little with height in the ISL. Hence the ISL is also known as the 'constant' flux layer. This consistency of fluxes with height in the ISL is a very helpful attribute. For example it follows that determination of the flux of any entity (sensible heat, water, air pollutants) at any level in the ISL, also determines the integral flux from the underlying surface. Further, it aids development of relationships and numerical models by allowing certain simplifying assumptions to be fulfilled.

## Monin-Obukhov Similarity Theory

The most widely accepted framework of relationships in the ISL is the Monin-Obukhov Similarity Theory (MOST). MOST has been applied very successfully to the SL over smooth surfaces (see Kaimal and Finnigan, 1994). Observations confirm that the MOST framework also applies in the ISL above cities (Roth, 2000). This finding is one of the most significant results from decades of **urban climate** research. However, it is important to note that MOST does not apply to the flow in most of the RSL or the ML.

Based on **dimensional analysis** MOST provides a unified means of relating turbulent fluxes to the mean gradient of an associated property, and of making predictions about turbulence intensities and other characteristics of the turbulent flow. MOST recognizes four key parameters that govern the flow in the ISL under all conditions, the:

1. integral **momentum flux** of the urban interface on the overlying flow $\tau_0$ (in Pa)
2. integral **sensible heat flux density** of the urban interface $Q_H$ (in W m$^{-2}$)
3. effective **height above ground** $z - z_d$ (in m)
4. **buoyancy ratio** $g/\overline{\theta}$ (in m s$^{-2}$ K$^{-1}$)

Together, these four parameters completely determine the gradients of the mean properties and the characteristics of turbulence in the ISL. These can be combined to form the following velocity, temperature and length scales:

$$u_* = \sqrt{\tau_0/\rho} \qquad (\text{m s}^{-1}) \qquad \textbf{Equation 4.14}$$

$$\theta_* = -\frac{Q_H}{C_a u_*} \qquad (\text{K}) \qquad \textbf{Equation 4.15}$$

$$L = \frac{\overline{\theta} u_*^2}{kg\theta_*} \qquad (\text{m}) \qquad \textbf{Equation 4.16}$$

$$z - z_d \qquad (\text{m}) \qquad \textbf{Equation 4.17}$$

where $u_*$ is the friction velocity (m s$^{-1}$) that has been already introduced in Equation 4.8, $\theta_*$ is the friction temperature (K), and $L$ is the Obukhov length (m). $L$ is proportional to the (negative) ratio of mechanical and thermal forces producing turbulence (see Section 4.1.2).

Out of these scales, one dimensionless group can be formed:

$$\zeta = (z - z_d)/L \qquad (\text{unitless}) \qquad \textbf{Equation 4.18}$$

which is a measure of the dynamic stability of the ISL. $\zeta$ is 0 in neutral, positive in stable, and negative in

Low height and density

Medium height and density

Tall and high density

High-rise

**Figure 4.25** Photographs of the four urban and suburban roughness classes in Table 4.2, arranged in order downward of increasing roughness (Source: Grimmond and Oke, 1999a; © American Meteorological Society, used with permission).

dynamically unstable atmospheric conditions (Table 4.1). The larger the absolute value of $\zeta$ the greater the role of thermal production relative to mechanical production. This dimensionless group is extremely handy, because any other parameter describing the flow in the ISL such as the mean gradient of velocity or temperature, or the statistics describing their fluctuations, is a universal function of $\zeta$ when non-dimensionalized by an appropriate scale.

### Effects of Turbulence on Mean Wind

MOST can be used to extend the logarithmic wind profile (Equation 4.9) to non-neutral conditions. Dynamic stability alters the shape and size of eddies and it changes the curvature of the wind profile. In unstable conditions buoyancy aids vertical transport of momentum and the gradient is lessened. In stable conditions vertical transport is hindered and the gradient steepens. The form of the logarithmic law still applies if a stability function for momentum is included to account for these effects on curvature of the profile:

$$\frac{\partial \overline{u}}{\partial z} = \frac{u_*}{k(z - z_d)} \Phi_m \quad \left(\text{s}^{-1}\right) \qquad \textbf{Equation 4.19}$$

Moriwaki and Kanda (2005) show for the ISL above an urban surface that $\Phi_m$ follows the conventional semi-empirical relationships as a function of $\zeta$ that were derived for flat surfaces (e.g. Dyer and Hicks et al., 1970; Högström, 1988). However, Rotach (1993) and Moriwaki and Kanda (2005) find that relationships diverge from the prediction when approaching the urban canopy from above (i.e. entering the RSL). In the RSL $\Phi_m$ is typically lower than predicted with MOST. In other words, the gradient of mean wind becomes less steep compared to the MOST prediction, likely because of non-local turbulence enhancing the size of eddies when approaching the shear zone.

### Effects of Dynamic Stability on Turbulence Intensities

Turbulence intensities of longitudinal ($\sigma_u/\overline{u}$), lateral ($\sigma_v/\overline{u}$) and vertical wind ($\sigma_w/\overline{u}$) relate the magnitude of turbulent fluctuations to the mean flow. Along with the wind profile, turbulence intensities allow determination of how rapidly entities such as heat, water or air pollutants are mixed and dispersed in the urban atmosphere. Revisiting Figure 4.20 shows the vertical variation of the turbulence intensities from ground level into the ISL. The magnitude of the intensities is greatest near the shear zone at the top of the roughness elements and then steadily decays with height.

Observations suggest a large city increases turbulence intensities by as much as 50%, but care must be taken in making urban–rural comparisons. Notice that the definition of intensity includes the mean wind speed. Since the greater drag of cities reduces wind speeds by about 20–30% (Section 4.4.1), this reduces the turbulence excess to say 10–20%. In the central city where drag is greatest the excess could still approach 50%.

Within MOST the appropriate scaling parameter for turbulent fluctuations is $u_*$ and empirical relationships suggest:

$$\sigma_i/u_* = \begin{cases} a(1 - b\zeta)^c & \text{for } \zeta \leq 0 \\ a & \text{for } \zeta \geq 0 \end{cases} \quad \text{(unitless)}$$

$$\textbf{Equation 4.20}$$

where $i = u$, $v$, or $w$. Results from rural, suburban and urban studies show remarkable agreement in the values of $a$, $b$ and $c$ (Table 4.3), which gives little reason to discriminate between terrain types and suggests they apply as long we are in the ISL.

### 4.3.4 Local-Scale Advection

The discussion so far has assumed that the ISL is formed above an urban area of extensive, uniform urban structure, such as those described in Table 4.2. It follows then that the airflow in the ISL can be linked unambiguously to the roughness and thermal properties of the underlying extensive surface. However, urban surfaces are often heterogeneous at the local (neighbourhood) scale and consist of distinct zones that differ significantly in their properties, yet are of limited extent. As a result we may expect the properties of one zone can be transferred to another zone through airflow (i.e. advection). Here we briefly consider the effects of roughness and of thermal patchiness on airflow in the ISL.

### Internal Boundary Layers

In Chapter 2, we considered a city of sufficient extent that it develops a unique UBL that grows with distance from the upwind urban edge until, eventually, it occupies the depth of the ABL (Figure 2.12b). At the city's downwind edge, the development of a new rural boundary layer begins to replace the UBL from below, so that the remnants of the UBL exist as an

**Table 4.3** Semi-empirical constants $a$, $b$ and $c$ used to scale turbulence in the ISL based on Equation 4.20. Data compiled as average of 11 field studies in the ISL and from Roth (2000).

| Wind component | Constant | Typical values in urban ISL[1] | Typical surface layer values[2] |
|---|---|---|---|
| **Longitudinal turbulence** $(\sigma_u/u_*)$ | $a$ | 2.29 (2.21–2.39) | 2.39 |
| | $b$ | 0.15 | 3.0 |
| | $c$ | 0.94 | 0.33 |
| **Lateral turbulence** $(\sigma_v/u_*)$ | $a$ | 1.85 (1.71–1.96) | 1.92 |
| | $b$ | 3.34 | 3.0 |
| | $c$ | 0.31 | 0.33 |
| **Vertical turbulence** $(\sigma_w/u_*)$ | $a$ | 1.24 (1.19–1.30) | 1.25 |
| | $b$ | 2.09 | 3.0 |
| | $c$ | 0.33 | 0.33 |

[1] Urban ISL values are compiled from Roth (2000) plus data from more recent studies.
[2] Surface Layer values compiled from Panosky et al. (1971), Panofsky and Dutton (1984) and Wyngaard et al. (1971).

elevated plume that becomes progressively shallower with distance from the city. At smaller scales, this process is repeated, as for example, when air passes from a low-rise and low density suburban zone into a zone dominated by tall and closely spaced apartment blocks. The adjustment of airflow to the change in roughness causes an increased momentum flux in the second zone, which is transferred upward. The effects on the overlying atmosphere as airflow crosses a change in surface properties is dealt with in several micrometeorology textbooks (Oke, 1987; Stull, 1988; Garratt, 1992; Arya, 2001).

If a zone within a city is sufficiently extensive in the along-wind direction then we expect an **internal boundary layer** (IBL) to develop from the surface that adjusts to the structure of the underlying zone. The IBL deepens with the distance travelled across that zone and is eroded from below when the air crosses the leeward edge and enters a new urban zone, with different structure, which develops its own IBL. It follows that, within the urban atmosphere, there are a myriad of IBLs that form over zones of different structure. The contributions of these zones are eventually blended in the ML of the UBL, just as the ISL blends contributions from the urban elements (buildings, trees) that comprise a distinct urban zone.

In neutral conditions, the growth in depth of an IBL is approximately logarithmic to the distance of fetch from the upwind border. The actual rate of growth depends on the roughness of the new surface and

stability conditions. Example of relations can be found in Garatt (1992) or Savelyev and Taylor (2005). It is faster (slower) than the neutral case over rougher (smoother) surfaces and faster (slower) when unstable (stable). We can expect that IBLs fluctuate in depth according to the stability of the atmosphere ever varying wind direction as it changes the path across and the zone boundaries.

### Implications for Measurements

If we wish to obtain measurements of wind or turbulent fluxes that represent a particular urban structural zone, two conditions have to be met: Firstly our observation platform has to be within the IBL of that zone. This means that the measurement location should be sufficiently downwind of a **leading edge**, such as a border between two structural zones. Secondly, it should be located above approximately 1.5–3 times the height of the roughness elements that characterize that zone in order to avoid microscale effects within the RSL. This highlights the difficulty in finding locations suitable for the conduct of wind and flux measurements. In cities with structural complexity the challenges are great. In some instances, these can be addressed through careful choice of measurement location that accounts for the extent of structural zones in relation to typical wind direction and the stability of the atmosphere. Nevertheless, there may be many urban zones in any city that are simply too small in extent to be examined using the theories applicable in the ISL. For these patches, a more complex experimental design is required that must account for both horizontal (e.g. advective) and vertical exchanges.

### Intra-Urban Circulations

During periods of calm weather, under clear skies, the different characteristics of urban land-covers result in different surface and air temperatures. Large areas of green-space, such as urban parks (grassed, forested or mixed vegetation), playing fields, open ground or water bodies such as lakes and large rivers have different thermal characteristics compared to built-up areas (see Table 6.4). By day and by night they are usually cooler than their surroundings, however the differences are usually greatest at night. The colder, denser near-surface air in the park causes a baroclinic gradient between the park and adjacent warmer built-up areas. If the pressure differences become sufficiently large, they initiate an outward flow of air that **advects** the thermal properties of the cooler park air into it built-up periphery. Observations of the night-time canopy layer temperatures readily detect the presence of parks (and lakes) as areas of lower temperature reaching into adjoining streets (Chapter 7). These near-surface 'park breezes' are part of a 3-D local circulation system. The outflow causes divergence and subsidence over the park and **convergence** with mild uplift over peripheral areas of the park. Only the near-surface arm of this circulation has been observed. Rivers passing through cities are also local sources of cool air and give relatively unobstructed, low roughness routes that aid dispersal of air pollutants, insects and cool air.

## 4.4  Flow in the Mixed Layer

SL fluxes and climate effects blend upward into the ML portion of the UBL (Figure 2.13). The effects of surface roughness decrease with height, and the turbulent momentum flux (Reynolds stress) $\tau$ becomes negligible at the top of the ML, in the FA. In the FA, wind speed and direction correspond to the geostrophic flow and are not affected by surface friction.

In the ML, surface-generated turbulence is still in evidence but the dominance of mechanically produced turbulence (roughness) declines relative to that of thermally produced turbulence (buoyancy). The largest thermal eddies that can efficiently mix properties within the ML scale with $z_i$, the depth of the **inversion-**

capped convective ABL by day. As a result in the ML, the importance of $z$ as a scale length diminishes relative to $z_i$. The ML responds to urban scale phenomena such as the growth and dynamics of the UBL, advection, local or mesoscale convection and circulations. There is a general lack of direct observations of urban effects on ML dynamics. This is partly due to the difficulty of gaining access to the atmosphere above the reach of tall towers. Our information relies on insight gained from **remote sensing**, balloon ascents, aircraft surveys and both scale and numerical models.

Airflow in the urban ML is subject to modification by two main characteristics: surface roughness (mechanical effects), and the urban heat island (thermal effects) (e.g. see Figure 3.26). The greater roughness of the city compared to its rural surroundings creates larger drag on the flow and an increase in mechanically produced turbulence. The excess warmth of the urban heat island (UHI) can change spatial patterns of air pressure both at the surface and aloft, thereby altering the balance of forces governing motion. Air temperature distributions in the UBL are covered in Chapter 7; here it is sufficient to say that the UHI forms a node of relatively lower air pressure over the city, in the lower layers of the ML. The spatial gradient of temperature, and therefore of air pressure, is usually greatest around the urban–rural perimeter. The central area of a UHI is not a smooth horizontal temperature field, so a degree of patchiness is to be expected in the pressure field. The UHI also creates enhanced production of thermal turbulence, which progresses up into the ML.

A city is subject to mechanical and thermal effects on its wind field at the same time. To aid understanding this section is artificially subdivided into the effects of extra roughness and of the UHI. The separation can be approximated by the strength of the regional flow. Firstly, with strong regional flow the influence of roughness tends to dominate, because the associated mixing and advection dampen or eliminate the UHI. Secondly, with weak flow thermal differences can develop and persist and the UHI is best displayed, especially if skies are cloud free. Thirdly, we consider the more general, intermediate case when both roughness and thermal effects are at play, and compete with each other.

### 4.4.1  Roughness Influences

When airflow crosses from countryside to the city and back again in a traverse of the city, we expect

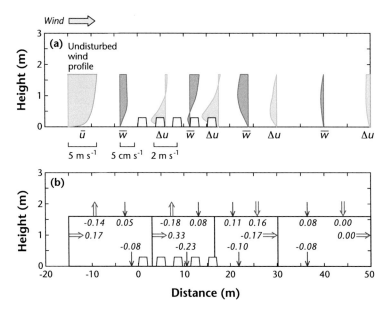

**Figure 4.26** Scale model experiment using upturned baskets on lake ice to simulate roughness effect of a whole city. **(a)** Velocity profiles: at extreme left is the undisturbed upwind horizontal velocity distribution; profiles at the right of each set are horizontal velocity changes ($\Delta \bar{u}$) from the undisturbed case; and the computed vertical velocity ($\bar{w}$) at positions over and in the lee of the roughness array. **(b)** Local budgets of horizontal momentum (all units Pa) — single arrows (in purple) are vertical turbulent fluxes of $u$-momentum; double arrows (in red) are $u$-momentum transport by mean flow. Note: fivefold vertical scale exaggeration (Modified after: Lettau, 1963).

profound changes in surface roughness that will cause adjustments to the forces that govern ABL winds, and hence change in both wind speed and direction.

### Changes in Horizontal and Vertical Wind Speed

A classic micrometeorological study helps to illustrate these adjustments. A rough surface was created by placing 210 upturned bushel baskets out on an otherwise unobstructed ice-covered lake. The baskets were arranged in a regular grid to represent a 'city' ($z_H = 0.3$ m, $z_0$ ~10 mm, $\lambda_f = 0.06$) and the ice represents the surrounding rural landscape ($z_0$ ~1 mm). Wind was measured upwind, within and downwind of the obstacle array and Figure 4.26 shows the evolution of the horizontal wind profile and of the associated momentum budget.

In Figure 4.26a the vertical profiles of mean vertical wind speed ($\bar{w}$) and the difference in mean horizontal wind speed ($\Delta \bar{u}$) are shown. Note that the background airflow is in equilibrium with the smooth ice surface and there is no mean vertical wind. As the flow encounters the obstacles, an updraft is generated ($\bar{w} > 0$) and horizontal motion is slowed ($\Delta \bar{u}$ is negative up to $3z_H$). Deeper into the array a downdraft

($\bar{w} < 0$) occupies the lower part but remains positive above. The profile of horizontal wind now shows a marked deficit up to $4z_H$, as horizontal flow decelerates. At the leeward side of the array, the transition to the smoother ice surface results in acceleration of horizontal motion ($\Delta \bar{u}$ is less negative) as downward flow ($\bar{w} < 0$) becomes negative through the entire layer. The upwind horizontal profile is slowly re-established downwind of the array.

This varying flow along the traverse is a response to the flux of momentum, which occurs horizontally and vertically by mean flow and turbulence as the surface roughness changes. Figure 4.26b shows the momentum budget (in units of force per unit area, or Pa) for a series of four boxes (1.65 m deep) that divide the traverse into discrete sections. Note that the significant transfers accounted for are advective fluxes caused by the mean flow ($\bar{u}$ and $\bar{w}$, double arrows) through the sides and top and those by turbulent exchange at the top and bottom (single arrows). The effect of changing surface roughness is seen clearly as the oncoming air becomes squeezed above the obstacles and exerts a force (+0.33 Pa) at the upwind boundary of the second box. Here the enhanced surface roughness is apparent in enhanced surface drag (–0.23 Pa) that decelerates the overlying air, exports this effect upward via $\bar{w}$ (–0.18 Pa), where there is a compensating downward transfer by turbulence (+0.08 Pa) that achieves balance. Downwind of the obstacles (third box) the reverse occurs. The return to a smoother surface is evident in reduced surface drag (–0.10 Pa), this causes: an acceleration that generates a suction force (–0.17 Pa) at the far boundary and a compensating downward transfer of momentum by advection and turbulent exchange (+0.11 Pa and +0.16 Pa, respectively). In the final box, the net flux of momentum is greatly diminished: transport of horizontal momentum by the mean flow is negligible; momentum sink at the surface has returned to its upwind value and; wind profile has almost recovered its upwind shape. Although this scale model greatly simplifies the urban effect on airflow it mimics the broad features that characterize the mechanical influences of cities as illustrated by the following field results.

Although Figure 4.27 looks dauntingly complex this is what the ML looks like in the air volume occupied up to about 2 km above an area about $80 \times 80$ km encompassing a large city. The example is St. Louis, United States, on a summer afternoon with strong southwesterly flow (8–10 m s$^{-1}$). Horizontal convergence/divergence fields have been computed from winds observed using pilot balloons released and tracked by theodolites at 11 stations along two cross-sections that span the built-up area and the Mississippi and Missouri rivers. The small inset map shows the location of the cross-sections: the upper panel is approximately normal to the regional flow (i.e. the wind is flowing from behind you into the page) and the lower panel is almost parallel to it (i.e. you see the flow going from left to right across the city). The horizontal component of the arrows is not simply wind speed but the difference between the horizontal wind speed at any point and the network mean at the same height above ground. Arrows to the right show regions with enhanced horizontal wind, arrows to the left are regions with decreased horizontal wind. The vertical component of the arrows reveals local areas of updraft and downdraft. The spatial variation of mean horizontal winds record deceleration over, and downwind of the city. As a result the urban area is dominated by convergence (negative divergence) in the layer below 1 km. Such areas (in orange in Figure 4.27) are clearly associated with the underlying built-up area. The convergence was fairly symmetric in the across-

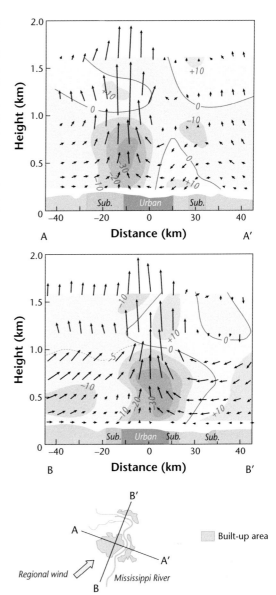

**Figure 4.27** Flow field in the boundary layer over St. Louis, United States under strong regional winds. Shown is the distribution of areas of divergence ($\times 10^{-5}$ s$^{-1}$) at noon on August 8, 1973 during strong southwesterly flow at 8–10 m s$^{-1}$. Inset map shows cross-section **(a)** is approximately normal, and **(b)** parallel, to the regional wind direction, and they intersect near the centre of the urban area. Areas in red indicate zones of horizontal convergence; areas in blue show zones of divergence. Arrows show the deviation of the horizontal wind at a given height from the network mean (to the left: slower, to the right: faster), the vertical component is calculated from the divergence. Note 40-fold vertical scale exaggeration (Source: Ackerman, 1978; with permission).

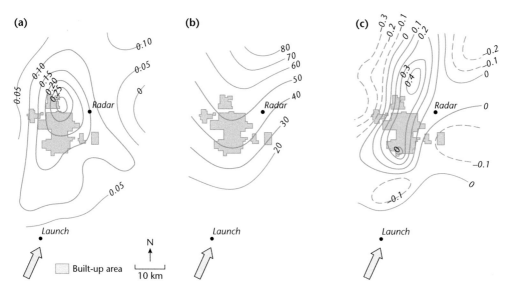

**Figure 4.28** Measured dynamics of boundary layer flow across Oklahoma City, United States. Isolines are smoothed distributions of **(a)** Reynolds stress $\tau$ (Pa) based on tetroon oscillations of 1–30 min period, **(b)** percentage of the Reynolds stress due to oscillations of 1–10 min period, and **(c)** distribution of derived air parcel vertical velocity $\overline{w}$ (m s$^{-1}$, positive values indicate upward motion). Tetroons were released in daytime at an average height of 400 m into a mean southerly flow of 13 m s$^{-1}$ (Source: Angell, et al., 1973; © American Meteorological Society, used with permission).

wind plane. In the along-wind direction it was displaced downwind, reaching maximum values of $3 \times 10^{-4}$ s$^{-1}$. Horizontal convergence is accompanied by vertical stretching which produced maximum uplift of 0.35 m s$^{-1}$, more typical would be 0.02–0.05 m s$^{-1}$. The most substantial uplift is located above the city. Conversely, downwind from the city and aloft, horizontal divergence of about $1 \times 10^{-4}$ s$^{-1}$ prevails (blue areas) producing vertical shrinking and hence an area of subsidence. Even small urban uplift combined with rural subsidence produces a tendency for the ML to 'dome' up over the city. Net differences of the order of 0.01–0.02 m s$^{-1}$ can produce doming upward of several hundred metres over the course of a day.

Oklahoma City, United States, shows similar effects of increased roughness, drag and stress, extending up into the ML and downwind from the city. Flow over the city is observed using tetroon (tetrahedral-shaped neutral buoyancy balloons, see Section 3.1.4) flights launched upwind of the city and tracked by **radar** as they cross the city at a mean height of 400 m in strong (~13 m s$^{-1}$) southerly daytime flow (Figure 4.28). A total of 32 flights are analyzed and compared against measurements from a

460 m tower near the downwind edge of the city. The Reynolds stress $\tau$ is computed from short-term velocity changes of the tetroons. These in turn are based on radar fixes of balloon positions every 1 s, averaged to give 30 s values. Their deviations ($u'$, $w'$) from a 5 min average are used to calculate $\tau$ at mean flight level. Flight paths cover a range of directions so it is possible to construct a spatially-smoothed field of $\tau$ in the vicinity of the city (Figure 4.28a). Clearly there is a maximum $\tau$ over and slightly downwind of the city. Values increase from about 0.05 to 0.1 Pa in rural areas to a maximum on the northern edge of the city of about 0.3 Pa. The increase is attributed to the urban structure (building height and density) and the maximum value of $\tau$ is located about 10 km downwind from the city centre. Results also show the city increases the fraction of $\tau$ attributable to smaller-scale eddies (Figure 4.28b), indicates the structure of turbulence is also changed for tens of kilometers downstream.

### Changes in Wind Direction

An increase in $\tau$ leads to deceleration of airflow at lower heights in the SL (Figure 4.22a). Changes of

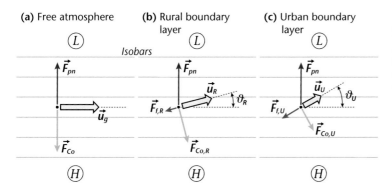

**Figure 4.29** Schematic of vector forces producing **(a)** the geostrophic wind $(\vec{u}_g)$ in the FA, **(b)** resultant winds in the ABL over rural terrain $(\vec{u}_R)$, **(c)** and in the UBL $(\vec{u}_U)$. Note that $|\vec{u}_g| > |\vec{u}_R| > |\vec{u}_U|$ and $\vartheta_R < \vartheta_U$.

wind speed across an urban area produce two important changes in the trajectory of air parcels: cyclonic/anticyclonic turning; uplift/subsidence. Direction changes occur because a new balance is struck by the forces of motion. Above the ABL, the balance is as Figure 4.29a, with vectors due to the horizontal pressure gradient force $(\vec{F}_{pn})$ and Coriolis forces $(\vec{F}_{Co})$. The resultant geostrophic wind $(\vec{u}_g)$ flows parallel to the isobars. Within the ABL, the friction force $(\vec{F}_f)$ must be included in the balance, it increases with surface roughness and proximity to the surface and this decreases the magnitude of $\vec{F}_{Co}$. The new balance, creates a resultant wind that is weaker, and flows at an angle $\vartheta$ to the isobars (Figure 4.28b) compared to $\vec{u}_g$, that is the wind turns cyclonically (towards lower pressure). This happens continuously with height through the friction layer towards the surface, creating the Ekman wind spiral. Similarly the balance of forces, and therefore the speed and direction, changes when air flows at the same height from over rural (relatively smooth) to urban (rough) terrain. The wind slows and turns cyclonically (compare the resultant wind vectors $\vec{u}_U$ and $\vec{u}_R$ and cross-isobar angles $\vartheta_U$ and $\vartheta_R$ between Figure 4.28b and 4.28c). The reverse changes in speed and direction happen when the wind blows from city to country, i.e. it accelerates and turns anticyclonically (towards higher pressure).

These principles are illustrated by the trajectory of the tetroons flowing across Oklahoma City in strong southerly flow (Figure 4.30). The average of the 32 flights indicate about 5° cyclonic turning as flights enter the built-up area and a similar anticyclonic recovery in the lee. Given the dimensions of the city (i.e. the time turning was in effect) this results in a 2–3 km offset in the path downwind of the city. The effect shows a diurnal pattern with a 10° turning in the morning, 4° in the afternoon and negligible effect in

the evening. This appears to be related to the depth of the ABL. Since all tetroon flight levels flew at about 400 m, that means in the morning the tetroons are within the developing ML with a strong Reynolds stress. By afternoon the ML is much deeper (about 1 km) so frictional influences are mixed and diluted, and in the evening the balloons are above the UBL so almost no frictional effects are seen. There is a suggestion that flow is around the city on both sides, i.e. the city acts like a 'barrier' to the flow.

The Oklahoma City results suggest competing influences in the lee of the city. With strong daytime flow the anticipated subsidence downwind of the city is not evident at the 400 m level where any sinking is overcome by residual upward motion acquired over the city. There is a 'plume' of ascending air that extends at least 30 km downwind (Figure 4.28c) in which maximum mean upward motion is about 0.4 m s$^{-1}$. There is sinking on the lateral outskirts of the city that resemble roll vortices shed from the side edges of buildings (Section 4.2.1). Vertical influences are greatest in the afternoon (mean maximum uplift of 0.7 m s$^{-1}$) when the ML is well-developed. In the evening, however, the weak ML of the UHI is below tetroon flight level. The surface-generated uplift is therefore constrained vertically by the stable air above. This may be partially relieved by air flowing around the city as the corresponding trajectory diagram shows (Figure 4.30c).

### Cities as Physical Barriers

The physical mass of a large, heavily built-up city causes a **barrier effect**, i.e. the city is a significant barrier to flow, especially in districts where high-rise buildings are closely clustered (e.g. Manhattan in New York, United States, Hass et al., 1967, see also Section 10.4.2). It might also be evident at the windward

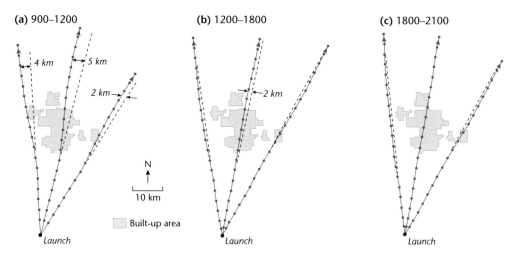

**(a)** 900–1200

**(b)** 1200–1800

**(c)** 1800–2100

4 km    5 km

2 km

2 km

N

10 km

Built-up area

Launch    Launch    Launch

**Figure 4.30** Mean trajectories of tetroons (solid) as they drift across Oklahoma City, United States, at three different times of day with strong southerly flow. The dashed lines are linear extrapolations of the flow direction before it encountered the city (i.e. the expected path had there been no urban effect) (Source: Angell et al., 1973; © American Meteorological Society, used with permission).

urban/rural boundary if there is an abrupt change of height of the effective surface ($z_H - z_d$). The combined roughness and barrier effects of a city produce an urban 'wake', within which speeds are decreased and turbulence is increased for distances of 50 km or more, based on numerical model simulations.

### 4.4.2 Thermal Influences

When viewed at the mesoscale, the UHI appears as a localized, thermal anomaly approximately coincident with the built-up area, extends up into the UBL and can form a plume downwind (see Chapter 7). Temperature differences and gradients, in both the vertical and especially in horizontal, are sufficient to generate atmospheric motion. The UHI modifies vertical **lapse rates** and **static stability**. The horizontal pressure gradient force ($\overrightarrow{F_{pn}}$) near the surface enters the balance of forces as a vector directed towards lower pressure in the UHI core, especially near the surface. This produces acceleration of flow into the city. The lapse rate is also relevant because inversions dampens vertical mixing, so surface drag is restricted to a shallow layer.

### The Urban Heat Island Circulation

When background regional winds are calm, or very weak (say $< 1.5 \text{ m s}^{-1}$), and skies are sufficiently cloud free to favour solar input (by day) and **longwave** radiative cooling (especially at night), conditions are suitable for a UHI-induced thermal circulation, similar to that of land and sea breezes. Indeed the 'island' analogue is particularly apt; the breezes are similar to the thermal circulation of a heated island surrounded by cooler sea. Figure 4.31a is an idealized distribution of air pressure in the vicinity of a heat island showing that isobars over a city follow a baroclinic stratification (i.e. isothermic and isobaric surfaces are not parallel to each other). The warmth of the urban air column causes it to expand and diverge aloft. The isobaric surfaces tilt outward, a horizontal pressure gradient force and outflow is created. Outflow causes the surface pressure to drop in the city, whereas addition of air above the surrounding countryside causes an increase in surface pressure there (Figure 4.31a). So a country-to-city pressure gradient exists at low-level and there is a confluence of near-surface 'country breezes' into the city. This centripetal drift of breezes causes convergence of cooler rural air into the city where there is uplift, from both dynamic and thermal effects (Figure 4.31b). Over a wide city the distance of travel is great enough to make the acceleration subject to Coriolis turning, giving a cyclonic spiral into the centre near the surface, and an anticyclonic outward spiral aloft, which is not shown in 4.31b.

This **urban heat island circulation** (UHIC) was simulated over Paris, France on a summer afternoon using

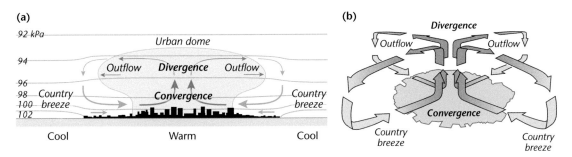

**Figure 4.31** Schematic of the urban heat island circulation (UHIC). **(a)** Idealized 2D air pressure distribution (thin horizontal arrows represent horizontal pressure gradient forces), and dotted lines are isobars (lines of equal atmospheric pressure in kPa). The thick lines are the resulting circulation. **(b)** Highly simplified view of the 3D circulation pattern (neglecting the Coriolis force).

the **Town Energy Balance** (TEB) model (Chapter 3) coupled to a 3-D mesoscale model of the ABL for a 2-day period. A realistic simulation including the city was run and then repeated excluding the urban area properties; as if Paris had never been built. The difference between the with- and without-city runs pinpoints the effects of the city on the ABL. In mid-afternoon, when synoptic flow in the region is disorganized and light ($< 5$ m s$^{-1}$), the realistic simulation shows a UHI of about 2 K (Figure 4.32a) near the surface, when the city is removed the temperature field is relatively uniform (Figure 4.32b). At the same time the sensible heat flux to the air in the city is about double that in the countryside. The mean vertical wind speed above the city shows uplift of about 1 m s$^{-1}$ and the height of the ML is domed upward by an additional 0.7 km over the city (**urban dome**, see Figure 2.12a). The difference of horizontal wind between the two cases, both near the surface and towards the top of the ML (Figure 4.33), show the city induces confluence of winds with cyclonic curvature flowing into the lower pressure of the city at low-level and diffluent winds with anticyclonic curvature aloft. The surface outline of the UHI itself is roughly coincident with the built-up area but the UHIC affects an area about 2 to 3 times larger. This is a UHIC that broadly resembles Figure 4.31. The same study showed that when the flow at 10 m is $> 7$ m s$^{-1}$ the UHIC does not form and the urban dome is replaced by an **urban plume** (e.g. Figure 2.12).

It is important to keep the form of the complete system in mind, not just the perspective of a ground-based observer. This is illustrated in a study by Shreffler (1979) in St. Louis, United States. He investigated periods of weak flow (average speed across a 25 station

network $< 1.5$ m s$^{-1}$) and divided the cases into 'weak' or 'strong' UHI classes. He calculated wind vectors at each station as the difference between the local and network mean to identify any local wind perturbations. These difference vectors show propensity both to accelerate and to converge on the centre of the UHI, for both UHI classes. What is initially surprising however is that the strength of the converging low-level breezes is greater for the weaker heat islands. Average hourly convergence was $2.2 \times 10^{-4}$ s$^{-1}$ for the weak UHI class, and $0.9 \times 10^{-4}$ s$^{-1}$ for the strong class, with peak (1 min) values about four times greater. This seems to run counter to the idea that the circulation should be proportional to the strength of the driving force (i.e. the thermal, horizontal pressure difference). The key to this apparent oddity is that the two UHI classes are associated with very different stability régimes. The weak UHI are from daytime, in convectively unstable conditions, whereas the strong UHI are nocturnal cases with much lower mixing depths over the city and strong rural stability. This shows the strength of the urban centripetal circulation depends as much on the vertical structure of the ABL, as it does on the **screen-level** horizontal urban–rural temperature difference. Facilitation of vertical motion by instability favours the weak UHI daytime circulation over the nocturnal one, even with a stronger UHI (see also Chapter 12). Similarly over the city of Tokyo, Japan, convergence and cyclonic **vorticity** are found to be strongest over the city by day, when the UHI is weakest. Even in cloudy, rainy conditions with weak heat islands the circulation is better developed than on clear nights, with strong UHI (Fujibe and Asai, 1980).

Haeger-Eugensson and Holmer (1999) suggest that in the evening after a nocturnal UHIC has been

**(a)** Realistic – air temperature and wind

**(b)** City removed – air temperature and wind

**(a)** z = 200 m

**(b)** z = 2,000 m

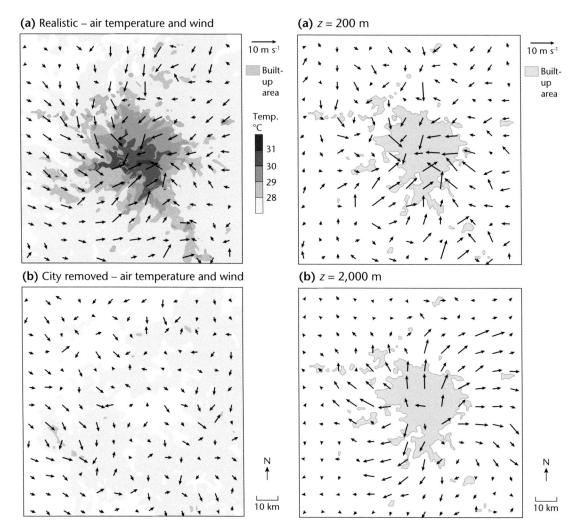

**Figure 4.32** Simulations of the ML dynamics in the vicinity of Paris, France on July 12, 1994 at 1500 h local time. The simulated fields of air temperature at 200 m above ground **(a)** with the city included (realistic) and **(b)** with the city removed. The vector arrows are the corresponding simulated wind fields (Source: Lemonsu and Masson, 2002; © 2002 Kluwer Academic Publishers, used with permission of Springer).

**Figure 4.33** The *difference* in horizontal wind speed between the realistic and a non-city simulations of the boundary layer over Paris, France, at **(a)** 200 m and **(b)** 2,000 m, at the same time as in Figure 4.32 (Source: Lemonsu and Masson, 2002; © 2002 Kluwer Academic Publishers, used with permission of Springer).

initiated, rural cooling rate decreases due to subsidence of warmer urban air. Later, advective transport of cooler rural air into the city, near the surface, lowers the urban cooling rate, thereby equalizing rates in the two environments. The authors consider this to be a self-regulating effect on UHI magnitude.

The vertical structure of the divergence and wind fields (difference from mean at given height) are

shown in Figure 4.34 for the daytime circulation in St. Louis. This is a case with very weak (1–2 m s$^{-1}$) westerly synoptic flow. Both the nearly parallel, and normal cross-sections show strong convergence (in orange) up to about 1 km, centred over the most heavily built-up area. The core area of uplift vectors is found in the same area, with uplift up to at least 2 km. Inflow comes from all directions near the surface. The main areas of subsidence occur over the

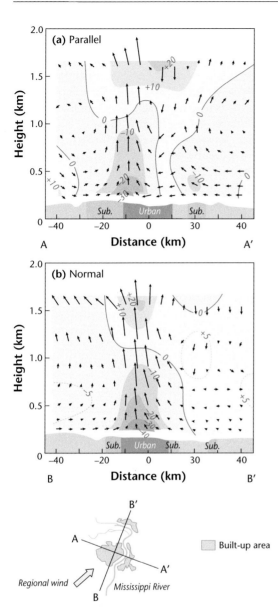

**Figure 4.34** Flow field in the boundary layer over St. Louis, United States with very weak regional winds. Vertical cross-sections of divergence ($\times\ 10^{-5}$ s$^{-1}$) on the afternoon of August 4, 1973 with regional winds of 1–2 m s$^{-1}$ from the west. Inset map: shows cross-section **(a)** is approximately parallel, and **(b)** normal, to the primary wind direction. Areas in red indicate convergence, areas in blue are divergence. Arrows are the deviation of the horizontal wind at a given height from the network mean (to the left: slower, to the right: faster) and the vertical component calculated from the divergence. Note 40-fold vertical scale exaggeration (Source: After Ackerman, 1978; with permission).

cooler Mississippi valley, which because of the bend in the river is located downwind of the main city in both cross-sections (Figure 4.34).

### The Variability of the Urban Heat Island Circulation

Several laboratory scale models of the UHIC have been run using thermally stratified air in a chamber, or liquid in a tank. The control such models provide gives valuable theoretical insight into the role played by factors like diameter of the UHI, temperature lapse rate and strength of the heat flux and their combination in non-dimensional parameters such as the Froude Number (*Fr*) stability criterion (the ratio of the inertial force to the force of gravity, see also Section 12.2.1) in setting the dimensions of the UHIC (its height, width of updraft, width of complete system). However, as the real city examples above demonstrate the centripetal circulation is neither as symmetrical, nor as simple, as Figure 4.31 depicts. Weak thermal wind systems are notoriously variable in both space and time and the urban circulation is no exception. In the city heat sources and the horizontal distributions of surface and air temperature are complex (e.g. Figure 7.7 and 7.12) and ever changing the array of obstacles and momentum sinks is intricate and the effects sensitive to shifts in wind direction. The periphery of the UHI often exhibits the steepest temperature gradients whereas the bulk of suburbia may have relatively slack gradients. Localized warm or cool spots (e.g. a commercial centre or a park, respectively) can generate their own mini-circulations within the city (see Section 4.3.4), and individual industries may generate thermal plumes from their operations. The influx of country air is far from steady because the flow has to overcome the inertia of the surface drag and barrier effects. The forces appear to build up to a critical threshold before a pulse of cooler country air is able to invade the urban area. The picture is one of an intermittent, agitated, multicellular system. The situation is vividly described by Munn (1970) as 'the motions of an amorphous, slowly-pulsating jellyfish'.

The UHIC is important to the air pollution meteorology of a city. Clearly the dome (Figure 2.12 and Figure 4.31) is a semi-closed system in which air pollutants largely recirculate, rather than are transported away. Therefore, over time the concentration increases until the circulation is breached or transforms into a plume. Of course if the city is located in a valley or basin, topographic constraints and other local breeze systems interact or override the UHIC (see Chapter 12).

### 4.4.3 Combined Roughness and Thermal Influences

The previous two sections sought to isolate urban roughness and thermal influences. The more normal state of the ABL sees both controls co-existing and interacting. Four controls are likely to come into play: friction due to roughness; the physical barrier effect of the city as an obstacle; the isotherm-created pressure field; and the enhancing or suppressing effect of atmospheric stability on vertical motion. Winds in a city at any given time can thus be anticipated to be a complex mix of forces even on a flat plain, but in a given city the peculiarities of the local topographic relief and the influence of water bodies must also be added. Figure 4.35 is a simplified summary of understanding of these competing controls and gives a set of first order 'rules' governing flow over cities in flat terrain.

With respect to roughness effects in strong synoptic flow the UHI can be assumed to be small. Winds entering a rough city will decelerate and curve cyclonically, when they revert back to smoother rural terrain on the downwind side, the flow accelerates and the curvature is reversed (Figure 4.35a). In the Northern Hemisphere, on the left side of the city (in flow direction), the slower flow over the city is next to faster flow over the countryside, which causes confluence and convergence. On the right side of the city (in flow direction), the flow discrepancy is the reverse, giving flow diffluence and divergence. If the physical bulk of a city's buildings is significant (large increase in $z_d$ and/or large clusters of tall buildings) the barrier effect may add further changes in flow direction, such as flow around the obstacle. These effects are more likely to appear if the ML is shallow.

Several studies suggest a critical threshold value of the regional wind speed: above that threshold, roughness effects and deceleration dominate (e.g. Figure 4.35a), below it thermal effects and acceleration are significant or even dominant. In the large

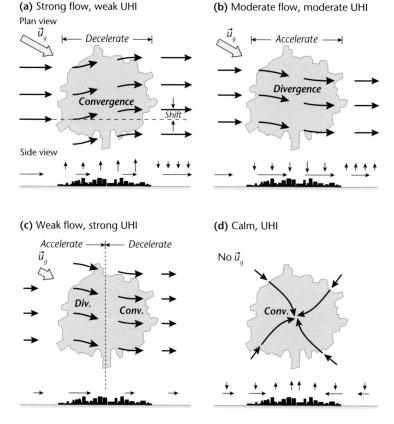

**(a)** Strong flow, weak UHI
Plan view
$\vec{u}_g$ |← Decelerate →|
Convergence
Shift
Side view

**(b)** Moderate flow, moderate UHI
$\vec{u}_g$ |← Accelerate →|
Divergence

**(c)** Weak flow, strong UHI
Accelerate →|← Decelerate
$\vec{u}_g$
Div. Conv.

**(d)** Calm, UHI
No $\vec{u}_g$
Conv.

**Figure 4.35** Schematic summary of flow adjustment of boundary layer flow in response to the greater surface roughness and the UHI of cities. In **(a)** strong flow advection and mixing weaken the UHI, so the main effects on wind speed and vertical motion are mechanical. In **(d)** with almost calm air the UHI is often large and thermal effects dominate urban flow patterns, initiating an UHIC. For the intermediate cases **(b)** and **(c)** see text. These 'rules' apply to cities on flat terrain in the Northern Hemisphere.

cities, of London, UK, New York, United States, and Paris, France the critical speed is reported to lie between about 3.5 and 5 m s$^{-1}$, using 10 m level speeds (observed at a rural site). It seems likely that the value is related to physical properties of the urban surface, such as its **fabric**, **surface cover**, structure and **metabolism** but that has yet to be shown.

With regional flow just below the threshold wind speeds may be increased in the city by two mechanisms (Figure 4.35b). Firstly, in such flow and fine weather the UHI is still capable of exerting some influence in a large city. This adds a vector component to the flow in the upwind half of the city, causing it to accelerate and perhaps to adopt anticyclonic curvature, depending on competition between the strength of the Coriolis and horizontal pressure gradient forces. However, downwind of the UHI centre the pressure gradient subtracts a vector quantity from the balance of forces, which may reduce or eliminate speed and direction changes.

Secondly, it is possible that the increase of thermal and mechanical turbulence production over the city induces greater downward transfer of momentum from faster moving upper air layers into the SL, sometimes called momentum 'down-folding'. In an example from Pietermaritzburg, South Africa, a **katabatic** mountain wind that is cool and stable with a well-defined maximum speed at about 75 m, flows out of a valley and across the warm rough city (Figure 4.36). Extra mixing over the city results in speed-up in the wind profile up to about 40 m which causes divergence. Conversely, in the layer between 40 and 90 m (from which momentum was extracted), there is deceleration and convergence. As it travels further across the city the profile form readjusts and

the effect diminishes. **Entrainment** of momentum into the SL at night has also been observed over Sapporo, Japan, from balloon-borne profiles of turbulence. The fluxes of momentum and heat were close to zero above the nocturnal UHI but sharply increased in the **entrainment zone** just below (Uno et al., 1988). More evidence of momentum down-folding comes from a numerical modelling study over Washington DC, United States, by Draxler (1986). A 3D numerical model, using geostrophic flow of 6 m s$^{-1}$ (giving a surface wind of about 4 m s$^{-1}$) and a nocturnal UHI of 5 K, increased the downward mixing of momentum and altered flow across the city. The speed at 50 m was accelerated by 17%, sufficient to produce anticyclonic turning of 1° km$^{-1}$ over the city and readjustment when it exited the urban area.

With weaker regional flow, acceleration over the upwind half the UHI may exceed the frictional retardation, but over the downwind half the UHI combines with the roughness to decelerate flow again (Figure 4.35c). The changes are unlikely to result in significant vertical motion at the city scale.

The final case (Figure 4.35d) is when regional flow is calm and the UHI is strong. There is lower pressure in the city, so flow entering from the countryside accelerates down the pressure gradient (if the force is large enough to overcome the extra friction) towards the city centre, and curves anticyclonically. Since the isotherm pattern is essentially concentric this happens from all directions, so country breezes curl and converge into the UHI core leading to uplift. If the lapse rate does not suppress vertical motion this might initiate an UHIC (i.e. similar to that in Figure 4.31). The impact of the lapse rate within the city is fairly uniform because the UHI moves conditions towards

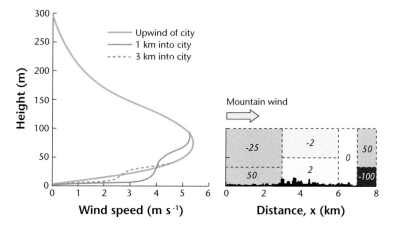

**Figure 4.36** Redistribution of momentum as katabatic airflow (a mountain wind) crosses the warm rough city of Pietermaritzburg, South Africa. **(a)** The observed wind profile is progressively modified to greater heights as the air moves further across the city, and **(b)** the divergence field calculated from the profiles (all values $\times 10^{-5}$ s$^{-1}$) in the lowest 100 m (Modified after: Tyson et al., 1973).

neutrality (Chapter 7), therefore any differential effect on the flow depends mostly on the urban–rural contrast in stability.

Observed wind fields are unfortunately often very complex. Trying to tease out the relevant controls from field observations, given the limited sampling possible, usually leaves an incomplete picture. Analysis shows airflow is disturbed by the city on most occasions, almost irrespective of weather type, cloud cover, wind speed or direction. In pre-rain, and rainy conditions patterns are further complicated by local circulations accompanying shower and storm clouds, which may mask any urban effects. The ability of cities to modify external weather phenomena like fronts, storms, sea breezes or valley winds is discussed in Chapter 12.

## Summary

The apparently chaotic urban winds described at the beginning of this chapter are the response of the ambient wind to the roughness and thermal influences of buildings, neighbourhoods and cities, each of which affect airflow at their specific scale: the roughness sublayer that responds to building and block-scale features; the inertial sublayer represents the blended adjustment of a neighbourhood, and the mixed layer reacts to changes at the scale of an entire city. These simplifications provide a framework to apply laws and scalings that are common in other boundary layers (see also Table 2.3).

- The buildings that comprise the urban surface are extraordinarily diverse in terms of their dimensions and placement. The net result is a deep **roughness sublayer** (RSL) where **mechanical turbulence dominates** usually up to 2–3 times the mean height of the buildings. In this layer mean airflow is reduced but turbulence is increased. Classical laws and scaling such as the logarithmic wind profile equation and Monin-Obukhov Similarity do not apply here. The main controls affecting mean flow and turbulence in the RSL are the typical **height** ($H$) and **spacing** ($W$) of the roughness elements and the pattern of the street network. Within the urban canopy layer (UCL) **microscale circulations** can exist. In dense urban areas, street canyons may be aerodynamically separated from flow above (skimming flow régime), and **vertical exchange may be limited**.
- Over extensive neighbourhoods the airflow effects of individual buildings become blended together and flow becomes horizontally homogeneous in the **inertial sublayer** (ISL). Identifying the position of the ISL is critical if we aim to measure the **integral effects of a neighbourhood**. Here, atmospheric variables (wind, temperature, concentrations) are only a function of height and their fluxes (momentum, sensible heat, water vapour, other gases and particulates) are roughly constant with height. The **logarithmic wind profile** and **Monin-Obukhov Similarity** apply in the ISL. Measurements at this level confirm that cities are absolutely rough with roughness lengths ($z_0$) usually one to two orders of magnitude higher than for surrounding rural areas.
- Above the ISL, the contributions of different urban neighbourhoods become blended into the **mixed layer** (ML), which has received comparatively little attention. This is unfortunate because some urban effects, such as those on convection and precipitation, are intrinsically linked to airflow in the ML as it **responds to the roughness and thermal effects of an entire city**. Flow patterns over cities most commonly lead to convergence and uplift in the ML. The reason for convergence may, however, be quite different: it could be slowing due to roughness effects, or a low pressure zone over the warmer city that initiates an **urban heat island circulation** (UHIC).

Many of the concepts discussed in this chapter are relevant in the management of the urban atmosphere from an applied perspective, for example to improve the **comfort and safety** of pedestrians (Chapter 14), to calculate **wind loads on buildings**, reduce heat loss from buildings, control snow drifting and **predict dispersion** of gases and aerosols (Chapter 11). Further, the concepts and theories in this chapter inform simplifications and parameterizations for use in urban climate models. They are especially relevant to the prediction of the vertical exchange of entities like heat, water vapour, pollutants and momentum between their sources and sinks in the UCL and the overlying boundary layer (Chapters 6–12). Wind is also a key variable in the design of more sustainable urban neighbourhoods and cities in concert with the climate of the location (Chapter 15).

# 5 | Radiation

**Figure 5.1** An aerial image of old Marrakech, Morocco (Credit: Y. Arthus-Bertrand; © Y. Arthus-Bertrand/ Corbis; used with permission).

**Radiation** from the Sun is the most important driver of climates near the ground. In cities, the dynamic pattern of sunlight and shadow in streets, as the solar beam is blocked by obstructions like buildings and trees, is a distinctive feature. Pedestrians often choose to walk on the shaded, rather than the sunny side of the street if the climate is hot or *vice versa*. In many places, homeowners jealously guard their access to sunshine, which may be encoded in local laws. In another environment, the mutual shade provided by buildings is a useful urban feature. For example, Figure 5.1 is an aerial image of old Marrakech, Morocco, which is situated in a hot and arid climate.

Note how the **urban structure** limits solar access below roof-level and creates shaded spaces for pedestrians and outdoor markets. This design minimizes the exposure of building walls and streets to direct sunshine and potentially excessive heat gain. Further, reflection of sunlight by the urban 'surface' from this aerial perspective is dominated by the light-coloured rooftops.

Less obvious, indeed not visible but equally important, are infrared radiation exchanges among the urban **facets** and between the urban surface and the atmosphere. At the surface and at points within the urban canopy layer (UCL), **sky view factor,**

($\psi_{sky}$, Section 2.1.3, Figure 2.6) is reduced, this impacts the relative amounts of diffuse solar and thermal infrared radiation received from the sky and surrounding buildings that affect the radiation budget for the point. These examples illustrate that surface structure, orientation to the Sun and sky and the radiative properties of surfaces are important to understanding urban radiation exchanges. Not surprisingly, managing radiation exchange through building and landscape design is one of the most important tools available to architects, urban designers and landscape planners to control **microclimates** (see Chapter 15).

This Chapter focuses on radiation exchanges and the resulting **surface radiation budget**, which is one component of the **surface energy balance** (SEB, Chapter 6). In cities these exchanges assume special significance because urban development alters them, owing to changes to both the atmosphere and surface. The radiative properties of the atmosphere are altered by an increase in the **concentration** of **air pollutants** and changes to air temperature, humidity and cloud. Cloud affects transmission and **emission** of incoming radiation and increases interception of outgoing radiation. Consequently the magnitude, directionality and spectral makeup of radiation arriving at the bottom of the **urban boundary layer** (UBL) are likely to be very different from that of a nearby **rural** ABL. In addition, the urban surface is comprised of a wide mix of facets, made of natural (lawns, trees, etc.) and manufactured (buildings, roads, etc.) **fabric** arranged into a 3-D structure. The result is a multi-faceted urban 'surface' that emits and reflects to itself and generates myriad distinct radiation budgets.

Here we build understanding of the system by moving vertically through the UBL, first looking at micro- and **local-scale** effects of the surface on radiation budgets and exchanges and their combined impacts at the city-scale (Section 5.2). Next, the effects of air pollutants and other modified properties of the UBL on radiation transmission are considered (Section 5.3). Finally surface and atmospheric controls are combined in an integrated discussion of urban effects on radiation (Section 5.4). We begin with general background about the nature of radiation and how it interacts with surface materials and atmospheric constituents (Section 5.1). Readers familiar with radiation principles can skip to Section 5.2.

## 5.1 Basics of Radiation Exchanges and Budgets

### 5.1.1 Basic Radiation Principles and Laws

Radiation is emitted by all objects with a temperature ($T$) greater than absolute zero (0 K). It may be described as a series of electromagnetic waves of differing wavelengths ($\lambda$) that emanate from the radiating object. The energy emitted by a body at a given wavelength (W m$^{-2}$ μm$^{-1}$) and temperature is described by Planck's Law. When plotted for all wavelengths a characteristic Planck curve emerges, it has a single peak wavelength of maximum emission and is positively skewed. These curves describe emission of radiation by a body that is a perfect emitter (**blackbody**). Curves for bodies at the temperatures of the Sun and Earth-Atmosphere System (EAS) are given in Figure 5.2a. The photosphere of the Sun has a temperature of ~5,780 K and emits most of its radiation in the range 0.1 to 3 μm, with a peak at 0.47 μm (Figure 5.2a). The EAS has a mean temperature of 288 K and emits in the range 3 to 100 μm (Figure 5.2a, right), with its peak at 10 μm. In other words, these sources emit radiation in distinctly different spectral ranges. Hence, for ease we refer to radiation from the Sun as **shortwave** ($K$), also called 'solar' radiation. Radiation emitted within and from the EAS is **longwave** radiation ($L$), also called 'terrestrial' (or 'thermal infrared') radiation.

The area under these emission curves is the total energy **flux density** ($E$), also called **emittance**, given by the Stefan-Boltzmann Law,

$$E = \varepsilon \sigma T^4 \qquad (\text{W m}^{-2}) \qquad \textbf{Equation 5.1}$$

where $\sigma$ is the Stefan-Boltzmann constant (5.67 × 10$^{-8}$ W m$^{-2}$ K$^{-4}$) and $\varepsilon$ is the average **emissivity** of the emitting body, which indicates its radiative efficiency. If $\varepsilon$ were equal to 1.0 then the object would be a perfect emitter (blackbody). Objects are however imperfect emitters and $\varepsilon$ values are spectrally dependent, but for our purposes it is sufficient to employ average $\varepsilon$ values over a suitable spectral range (such as the short- and longwave radiation bands).

Kirchhoff's law states that the emissivity $\varepsilon_\lambda$ of an object at a given wavelength $\lambda$ equals its ability to absorb radiation (**absorptivity** $\varphi_\lambda$) at the same wavelength:

$$\varepsilon_\lambda = \varphi_\lambda \qquad (\text{unitless}) \qquad \textbf{Equation 5.2}$$

Similar to emissivity, absorptivity $\varphi_\lambda$ is a number between 0 and 1 that describes the fraction of

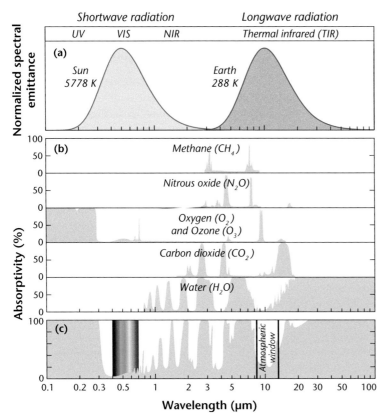

**Figure 5.2** **(a)** The radiation curves of perfect (blackbody) emitters at the temperatures of the Sun and Earth-Atmosphere System (EAS) respectively. **(b)** the absorptivity and emissivity of individual atmospheric gases, and **(c)** the total atmospheric absorptivity and emissivity.

radiation absorbed in the given wavelength. A thick cloud layer for example is an efficient emitter and consequently an excellent absorber in the longwave part of the radiation spectrum (Figure 5.2c).

Again, when the respective temperatures of the Sun and EAS are entered into Equation 5.1, the former is seen to emit over $10^5$ times more energy per square metre than the latter. However as the Sun is, on average, $1.5 \times 10^8$ km distant only a very small fraction of the output is intercepted by the EAS. At the top of the Atmosphere, on a surface perpendicular to the solar beam this energy flux amounts to an annual average of $1,361$ W m$^{-2}$, known as the solar constant. Of course, the amount received at Earth's surface is reduced owing to depletion of the beam as it passes through the atmosphere, and it depends on the angle of incidence, the angle at which the solar beam strikes the surface.

## 5.1.2 Radiation-Mass Interactions

In the absence of any medium (e.g. in space), the propagation of radiation is relatively straightforward,

as radiation travels in straight lines and at the speed of light. However, in the EAS radiation interacts with mass in the atmosphere and at the surface, and those interactions are a fundamental driver of the climate system.

### Conservation of Energy

When radiation of a given wavelength encounters a medium it experiences one of three fates: **absorption** which in most cases results in heating of the medium or sometimes enables a chemical reaction; reflection which redirects the path of the radiation backwards and; transmission which allows the radiation to pass through the medium. We can therefore write a statement of **radiant energy** conservation:

$$1 = \varphi_\lambda + \omega_\lambda + \tau_\lambda \quad \text{(unitless)} \qquad \textbf{Equation 5.3}$$

where $\varphi_\lambda$ is the absorptivity, $\omega_\lambda$ the **reflectivity** and $\tau_\lambda$ the **transmissivity**, all with values between 0 and 1, at the same wavelength. The fraction of total radiant energy partitioned into absorption, reflection or transmission depends strongly on wavelength (Figure 5.2b):

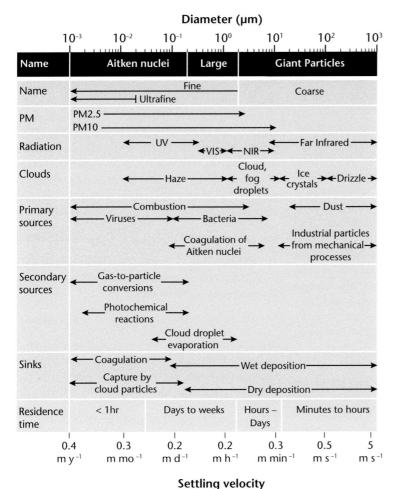

**Figure 5.3** Size classifications, sources, sinks and characteristics of aerosols in the troposphere along with relative sizes of radiation wavelengths and cloud droplets in the atmosphere (Modified after: Wallace & Hobbs, 2006 and Turco, 2002).

Earth's atmosphere for example, transmits a great deal in the shortwave, but very little in the longwave part of the radiation spectrum (Figure 5.2c). However, for simplicity we often express these fates as averaged values over the short- or longwave part of the spectrum. If radiation strikes a surface that is solid and opaque there is no transmission, hence the input is either reflected back from the surface or absorbed, so the reflectivity is:

$$\omega_\lambda = 1 - \varphi_\lambda \quad \text{(unitless)} \qquad \textbf{Equation 5.4}$$

### Radiation Interaction with Aerosols

**Aerosols** are small solid or liquid particles suspended in the atmosphere. They play an important role in how radiation interacts with the atmosphere, and are responsible for urban-specific effects on radiation transfer (Section 5.3), for the formation of **fog** (Chapter 9), cloud and **precipitation** (Chapter 10) and for aspects of atmospheric chemistry and **air pollution** (Chapter 11).

Aerosols are categorized according to size, represented in Figure 5.3 by diameter, and for simplicity typically are assumed to be spherical in shape. Aerosol size, which varies over several orders of magnitude, is an important attribute that influences **residence time**, settling velocity, interactions with radiation and impacts precipitation formation. Several size classifications are used in different applications, each with its own nomenclature. One includes 'Aitken nuclei' with radius $r < 0.1$ μm: 'large aerosols' with $0.1 \leq r \leq 1.0$ μm and 'giant particles' with $r > 1.0$ μm (Figure 5.3). A second uses 'ultrafine', 'fine' and 'coarse particle' categories. Air pollution often refers to **particulate matter** (PM) which is usually divided into two main size classes relevant to human health: particles with diameter < 10 μm (**PM10**) and < 2.5 μm (**PM2.5**) (see Section 11.1). Finally, visibility assessments often use a measure of the total suspended particulate of all sizes expressed as a density or concentration (μg m$^{-3}$).

Aerosols are derived from a variety of natural and **anthropogenic** sources, including primary sources that provide aerosols directly to the atmosphere, and secondary sources, wherein aerosols are formed from gas-to-particle conversions, photochemical reactions and **evaporation** of cloud droplets. Aerosol concentrations are greatest near the ground and decrease with height. That is firstly, because that is where most sources are located and secondly, because gravitational settling is the main removal process together with **rainout** and **washout** (see also Section 11.1.3). The rate of settling increases with size (Figure 5.3) and the residence time is influenced by both settling velocity and other processes that act as aerosol sinks.

Aerosols have direct and indirect effects on the transfer of radiation. Direct effects include **scattering**, i.e. redirection of the beam from its original path, and absorption. Backscattering refers to a redirection in the hemisphere towards the source and is equivalent to reflection. The most efficient particle diameter for scattering is approximately $0.2\lambda$. Indirect effects include the growth of **haze** and cloud droplets which, in turn, affect the radiative properties of the atmosphere. Soluble aerosols are particularly important to the formation of precipitation because they enable the growth of cloud droplets (Chapter 10).

### 5.1.3 The Surface Radiation Budget

The radiation budget at any surface may be stated:

$$Q^* = K^* + L^* = K_\downarrow - K_\uparrow + L_\downarrow - L_\uparrow \quad (\text{W m}^{-2})$$

**Equation 5.5**

where $Q^*$, $K^*$ and $L^*$ are the budgets of **net allwave**, net shortwave and net longwave radiation flux density, respectively. Arrows indicate incoming and outgoing fluxes at the surface. For urban areas the 'surface' to which Equation 5.5 applies, extends from a leaf or pane of glass, through urban facets, like a wall, roof, lawn and combined units like a building or an **urban canyon**, all the way up to the fully integrated urban surface referred to in Chapter 2.

## Shortwave Irradiance

Incident shortwave flux density is called shortwave **irradiance** ($K_\downarrow$). It depends on Sun-Earth geometric relations (latitude, longitude, time of year and time of day), the ability of the atmosphere to transmit and absorb shortwave, and the ability of clouds, sky and surrounding surfaces to scatter and reflect radiation.

Shortwave irradiance consists of two streams: that coming directly through the atmosphere as a parallel beam from the direction of the Sun, the **direct-beam irradiance** ($S$), and that arriving from all parts of the sky hemisphere and surrounding objects due to scattering and reflection, the **diffuse irradiance** ($D$), so

$$K_\downarrow = S + D \quad (\text{W m}^{-2})$$

**Equation 5.6**

The magnitude of $S$ depends on the geometric relationship between the surface and the solar beam, so in absence of any obstructions

$$S = S_b \cos \hat{\Theta} \quad (\text{W m}^{-2})$$

**Equation 5.7**

where $S_b$ is the direct-beam irradiance received on a surface perpendicular to the beam and $\hat{\Theta}$ is the angle between the beam and an axis perpendicular to the surface which is calculated (Figure 5.4):

$$\cos \hat{\Theta} = \cos \hat{\beta} \sin \beta + \sin \hat{\beta} \cos \beta \cos (\Omega - \hat{\Omega}) \quad (\text{unitless})$$

**Equation 5.8**

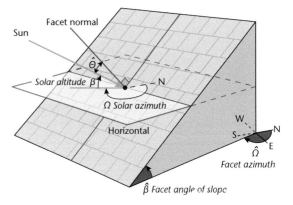

**Figure 5.4** Radiation geometry for direct-beam irradiance ($S$) on an inclined urban facet, such as the roughly south-facing sloping roof illustrated. Whilst for any particular surface, the facet angle of slope ($\hat{\beta}$), facet azimuth ($\hat{\Omega}$), and the location of the normal to the facet are fixed (red angles), solar altitude ($\beta$) and solar azimuth ($\Omega$) change constantly over the course of a day and with seasons (yellow angles).

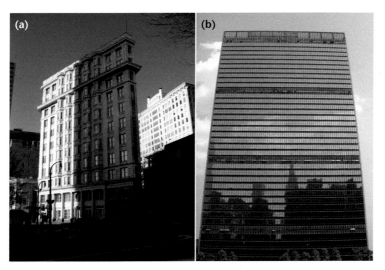

**Figure 5.5 (a)** Distribution of light and shadow on a building in Atlanta, United States, and **(b)** specular reflection of buildings and clouds on the glass face of a building in New York, United States (Credit: G. Mills).

where $\beta$ is the current **solar altitude**; $\Omega$ is the current **solar azimuth**; $\hat{\beta}$ is the surface slope; and $\hat{\Omega}$ is the surface aspect. The first two measures change over time but can be obtained from standard solar geometry calculations (e.g. Stull, 2000), the second two terms are fixed and available from analysis of surface structure. If the surface is horizontal, $\hat{\Theta}$ is simply the **solar zenith angle** $Z$.

If a facet is flat and has an unobstructed view of the sky, then $D$ is shortwave irradiance scattered downward by the atmosphere. From the perspective of the facet, this radiation is received from an overlying hemisphere occupied by sky. Within the sky hemisphere, there are variations in the diffuse shortwave **radiance** that make up $D$ (for clear sky conditions it is much greater from around the position of the Sun and under overcast skies the radiance is larger at the **zenith** compared to the horizon; Monteith and Unsworth, 2008. However, for climate purposes $D$ is often approximated as an **isotropic** source of diffuse shortwave radiation (same radiance from any direction in the hemisphere). The addition of air pollutants to the UBL makes the distribution of diffuse shortwave radiance across the sky hemisphere closer to being isotropic (Section 5.3.2).

### Shortwave Reflectance

Upwelling shortwave flux ($K_\uparrow$) is dominantly due to surface **reflectance**. Emittance of $K_\uparrow$ is essentially negligible, even in cities where there is considerable fugitive light from artificial sources (a.k.a. light pollution). Therefore $K_\uparrow$ is primarily a function of $K_\downarrow$ and the average surface reflectivity for the shortwave band known as the surface **albedo** ($\alpha$), i.e.:

$$K_\uparrow = \alpha\, K_\downarrow \quad \left(\text{W m}^{-2}\right) \qquad \textbf{Equation 5.9}$$

The spectral nature of natural sunlight varies in response to seasonal, atmospheric and cloud conditions, hence a single albedo value for a surface is not possible. Shortwave reflectance, on the other hand, can be controlled in a laboratory setting. In this book the reported albedos are measured in the field.

The nature of the reflection from a surface may be specular (beam-like) or diffuse. The former occurs from smooth surfaces where the beam is reflected at right angles to the angle of incidence; glazed and polished building facets can act in a specular manner (e.g. Figure 5.5b); even rough surfaces seem to act as if they are smooth when the Sun is near the horizon. In most circumstances, rough surfaces reflect diffusely (perfectly diffuse reflectors are described as 'Lambertian'). Finally, if there is no significant transmission through the surface plane:

$$K^* = (1 - \alpha)\, K_\downarrow \quad \left(\text{W m}^{-2}\right) \qquad \textbf{Equation 5.10}$$

### Longwave Radiative Exchanges

Longwave radiation exchanges are also diffuse. Incoming longwave ($L_\downarrow$) on a surface depends on the ability of the overlying atmosphere and surrounding surfaces to emit, and to a smaller extent their ability to reflect, $L_\downarrow$ towards the surface. Outgoing longwave ($L_\uparrow$) from a surface or medium depends on its ability to both emit and reflect longwave radiation. Emission depends on the temperature ($T$) and emissivity ($\varepsilon$) of the emitting surface or medium (Equation 5.1), and reflection depends on its longwave reflectivity ($\omega_\lambda$). Because Kirchhoff's law (Equation 5.2) states that the emissivity of a surface is equal to its absorptivity at the same wavelength, and since most solid surfaces are virtually opaque to longwave radiation (i.e. they allow almost no transmission) it follows from Equation 5.2 and Equation 5.4 that at the same wavelength:

$$\omega_\lambda = (1 - \varepsilon_\lambda) \quad \text{(unitless)} \qquad \textbf{Equation 5.11}$$

Hence for a simple surface part of the outgoing radiation is emitted (based on Equation 5.1) and part is reflected incident longwave ($L_\downarrow$):

$$L_\uparrow = \varepsilon\sigma T_0^4 + (1 - \varepsilon)L_\downarrow \quad \left(\text{W m}^{-2}\right) \qquad \textbf{Equation 5.12}$$

where $T_0$ is the absolute temperature of the surface. Since the magnitude of $\varepsilon$ for natural materials is commonly greater than 0.9, reflection of longwave (the second term on the RHS of Equation 5.12) is usually minor, but not inconsequential. When $L_\uparrow$ from Equation 5.12 is converted to a temperature under the assumption the body is a perfect emitter (blackbody), the result is referred to as a **brightness temperature** ($T_{0,B}$); these temperatures appear anomalously cold when low-emissivity surfaces are viewed under clear sky conditions (e.g. Figure 7.4).

Longwave radiation exchanges within the atmosphere are more complex than those at a surface. Whilst aerosols and cloud water droplets largely behave as if they are tiny 'solid' surfaces, the gases are selective absorbers – they have an affinity to absorb certain wavelengths but permit others to be transmitted (Figure 5.2c). From Kirchhoff's law (Equation 5.2) this means they are also selective emitters at the same wavelengths. In a cloud-free atmosphere the spectrum of atmospheric emission (sensed as $L_\downarrow$ at the ground) is deficient in the wavelengths where the atmosphere is 'open', including the **atmospheric window** between about 8 and 13 µm, and richer in wavelengths where gases such as

water vapour, **carbon dioxide** ($CO_2$), ozone ($O_3$), methane ($CH_4$), nitrous oxide ($N_2O$) and others are good absorbers and therefore emitters (Figure 5.2b). The greater the concentration of these gases and aerosols the higher is the atmospheric emissivity and hence at a given temperature the greater is $L_\downarrow$. Cloud is able to virtually close the atmospheric window because the emissivity of liquid water and ice are very high and thereby greatly increase $L_\downarrow$. The longwave emission by clouds depends on the cloud base temperature which is a function of cloud base height and cloud type.

Under clear conditions, longwave radiation from the sky hemisphere is not isotropic; directional variation is mainly due to path length differences through the atmosphere. Smallest radiance comes from the zenith, where path length is shortest and increases to a maximum at the horizon. Under a fully overcast layer of homogeneous cloud the radiance distribution is close to isotropic. Partly cloudy skies are characterized by a complex longwave radiance distribution.

## 5.2 Radiation in the Urban Canopy Layer

There are three attributes of the urban surface at greater than the **neighbourhood** scale that distinguish it from a simple flat and homogenous surface. Firstly, is the great diversity of its constituent materials with their distinctive reflectivities and emissivities; secondly, its 3D structure has innumerable facets, each of which has a unique slope and aspect, and thirdly, the possibility that facets can emit or reflect to each other (Section 5.1.1) or block such exchanges. Here we discuss these characteristics in relation to the radiation budget of individual buildings (Section 5.2.2), urban canyons (Section 5.2.3), and whole urban areas (Section 5.2.4).

### 5.2.1 Radiation Properties of the Urban Canopy

#### Albedo and Emissivity of the Urban Fabric

The radiative properties of materials used in roads and buildings cover a wide range (Table 5.1 and Table 5.2). Commonly road and roof materials are relatively dark ($\alpha$ typically 0.05 to 0.15) and walls are lighter (0.15 to 0.30). The great majority of surfaces

**Table 5.1** Albedos ($\alpha$) of individual urban and natural materials and of urban areas. Values given as a range or average (Sources: tables of values (see source for exact reference) – Arnfield, 1982; Bailey et al., 1997; Bretz et al., 1998; Gubareff et al., 1960; Maykut, 1985; Oke 1987, 1988).

| Surface | Albedo $\alpha$ |
|---|---|
| **Natural surfaces** | |
| **Bare ground**[1] | |
| Soil (dark colour, wet) | 0.05–0.10 |
| Soil (dark colour, dry) | 0.10–0.13 |
| Soil (light colour, wet) | 0.12–0.18 |
| Soil (light colour, dry) | 0.18–0.30 |
| Desert sands | 0.20–0.45 |
| **Low vegetation** | |
| Grass (long → short) | 0.16–0.26 |
| Crops | 0.18–0.25 |
| Wetlands | 0.07–0.19 |
| Tundra | 0.08–0.19 |
| **Forests** | |
| Deciduous (bare → leaf) | 0.13–0.20 |
| Orchards | 0.07–0.15 |
| Coniferous | 0.11–0.13 |
| **Water** [2] | |
| $\beta > 60°$ | 0.03–0.10 |
| $10° < \beta < 60°$ | 0.10–0.50 |
| Overcast | 0.05–0.10 |
| **Snow and ice**[3] | |
| Fresh, cold, clean snow | 0.80–0.90 |
| Wet, clean snow | 0.50–0.75 |
| Old, porous, dirty snow | 0.40–0.50 |
| Sea ice, multi-year | 0.55–0.75 |
| Sea ice, first year | 0.30–0.60 |
| Glacier ice | 0.20–0.40 |
| **Urban surface materials** | |
| **Roads** | |
| Asphalt (fresh → weathered) | 0.05 0.27 |
| Concrete[4] | 0.10–0.35 |
| **Walls** | |
| Concrete | 0.10–0.35 |
| Brick (colour, red → white) | 0.20–0.60 |
| Grey and red stone | 0.20–0.45 |
| Limestone | 0.40–0.64 |
| Wood | 0.22 |
| **Roofs**[5] | |
| Tile (clay, old → fresh) | 0.10–0.35 |
| Shingles (dark → light) | 0.05–0.25 |
| Tar and gravel, bitumen | 0.08–0.18 |
| Slate | 0.10–0.14 |
| Thatch | 0.15–0.20 |

| | |
|---|---|
| Corrugated iron | 0.10–0.16 |
| Galvanized steel (weathered → new) | 0.37–0.45 |
| **Windows** | |
| Clear glass | 0.08 |
| $\beta > 50°$ | 0.09–0.52 |
| $10° < \beta < 50°$ | 0.40–0.80 |
| **Paints** | |
| White, whitewash | 0.50–0.90 |
| Red, brown, green | 0.20–0.35 |
| Black | 0.02–0.15 |
| **Metals** | |
| Polished metals | 0.50–0.90 |
| **Urban areas** | |
| Urban (snow-free) | **0.14** |
| | 0.09–0.23 |
| Urban (with snow) | 0.14–0.41 |
| Suburban (snow-free) | **0.15** |
| | 0.11–0.24 |
| **Urban–rural differences** | $\Delta\alpha_{U-R}$ |
| Urban – rural | **-0.05** |
| | -0.09–+0.03 |
| Urban – rural (with snow)[6] | -0.55–-0.11 |

[1]Lower if rough, tilled.
[2]Lower if surface agitated.
[3]Depends on crystal structure, density, soiling.
[4]Range due to aggregate and soiling.
[5]High reflectance roof materials raise values into the range 0.4 to 0.8.
[6]Snow increases values but by less than in the countryside.

have emissivities greater than 0.90, but there are a few with $\varepsilon$ values that are significantly lower. Hence facets made of metals, especially polished ones, and treated window glass appear as objects of lower emittance ('cold') in thermal infrared imagery (e.g. skylights, metal roofs, vehicles; Sobrino et al., 2012).

Materials or coatings with relatively 'anomalous' radiative properties (reflectivity, absorptivity and transmissivity) are used to control radiatively-driven aspects of building climates and exterior environments (Chapter 15). Exterior paints and roof coverings can be chosen to minimize heat absorption, control glare and so on. Similarly window glass can be selected to help control interior daylighting or minimize interior solar heat gain or reflect longwave radiation. It can also be engineered to filter certain wavebands whilst allowing others to pass

**Table 5.2** Broadband longwave emissivities ($\varepsilon$) of individual urban and natural materials and for urban areas. Values given as a range or average. Spectral emissivity databases are also available (e.g. Kotthaus et al., 2014) (Sources: tables of values (see source for exact reference) – Arnfield, 1982; ASHRAE, 2009; Oke 1987, 1988; Campbell & Norman, 1998; Wittich, 1997).

| Surface | Emissivity $\varepsilon$ |
|---|---|
| **Natural surfaces** | |
| **Bare ground** | |
| Soil (light, dry → dark, wet) | 0.89–0.98 |
| Sand or rock | 0.84–0.92 |
| **Low vegetation** | |
| Grass (short → long) | 0.90–0.98 |
| Crops, tundra | 0.90–0.99 |
| **Forests** | |
| Deciduous (bare → leaf) | 0.90–0.99 |
| Coniferous | 0.97–0.99 |
| **Water** [1] | 0.92–0.97 |
| **Snow and ice** | |
| Old, soiled snow | 0.82–0.89 |
| Fresh snow | 0.90–0.99 |
| Sea ice [2] | 0.92–0.97 |
| Glacier ice [3] (melting → frozen) | 0.97–0.98 |
| **Human skin** | 0.98 |
| **Urban surface materials** | |
| **Roads** | |
| Asphalt | 0.89–0.96 |
| Concrete | 0.85–0.97 |
| **Walls** | |
| Concrete | 0.85–0.97 |
| Rough plaster | 0.89 |
| Brick (colour, red – white) | 0.90–0.92 |
| Stone | 0.85–0.95 |
| Wood | 0.90 |
| **Roofs** [1] | |
| Tile | 0.90 |
| Shingles | 0.90 |
| Tar and gravel, bitumen | 0.92 |
| Slate | 0.90 |
| Corrugated iron | 0.13–0.28 |
| Galvanized steel | 0.25 |
| **Windows** | |
| Clear glass, uncoated | 0.87–0.95 |
| Tinted, film-covered and coated (low-emissivity) glass | 0.05–0.30 |
| **Paints** | |
| White, whitewash | 0.85–0.95 |
| Red, brown, green | 0.85–0.95 |
| Black | 0.90–0.98 |
| **Metals** | |
| Polished metals | 0.02–0.06 |
| Aluminum foil | 0.03 |
| **Urban areas** [4] | |
| Urban and suburban (snow-free) | 0.94–0.96 |
| Urban and suburban (with snow) | 0.97–0.98 |

[1] Depends on surface roughness.
[2] Depends on bubble content and age.
[3] Pure state, in practice depends on age and surface and internal soiling.
[4] Results of canyon radiation model (Arnfield, 1982).

through. Whole facades of high-rise buildings can be covered with low-emissivity glass. Kirchhoff's law (Equation 5.2) states that a surface with low emissivity has an equally low longwave absorptivity, and since canyon facets transmit almost no longwave radiation they are very good reflectors of longwave arriving from facets lying in their view. This affects the heat loading of nearby objects. Albedo and emissivity are powerful tools for architects and urban designers (Sections 15.2.3–15.2.5). The exterior appearance of cities in warm and hot climates is evidence of the efficacy of simple choices such as painting walls with whitewash or roofing with light-coloured tiles. Such choices also affect the street, neighbourhood or entire urban radiation budget.

## Urban Structure

Viewed looking directly downward from above the **urban canopy layer** (UCL), the urban 'surface' appears to be located at mean roof-level and comprises roof facets and the inter-element spaces (Figure 5.1). The latter are complex radiative environments that incorporate the effects of shadowing and multiple exchanges between facets, below roof-level. The ratio of roof-level (or plan) area to complete surface area (see $\lambda_c$ in Figure 2.4) is an indicator of the 'enlargement' of the apparent surface area due to the convoluted form of the built elements. In **suburban** areas this increase is 30–80%, whilst in dense urban areas it ranges from 50–100% and in high-rise districts it can exceed 100%. This matters because it alters the area over which the surface

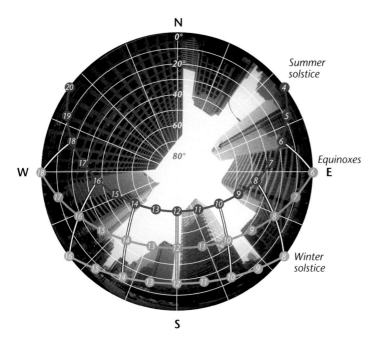

**Figure 5.6** A vertical fish-eye photograph that depicts the viewing hemisphere from a point on the road at the centre of a city street. Much of the sky dome is obscured by buildings. Superimposed on the image are two graphs: the first is a spherical co-ordinate system that indicates azimuth and altitude and the second is the path of the solar orb across the sky at the times of the Equinoxes and Solstices. The position of the orb at a given time is indicated by the labels (Credit: T. Oke).

absorbs, reflects and emits radiation. The impact of this is evident in the albedo values for urban areas (Table 5.1), that are lower than one might expect from the albedo of individual urban surface materials (Section 5.2.1).

Diffuse shortwave ($D$) and longwave ($L_\downarrow$) irradiances at a facet are the result of the sum of contributions from all **source areas** in its field of view (FOV). Conceptually, we view these as originating from the inner surface of an imaginary hemisphere centred above the facet of interest (Figure 2.6) and the resulting irradiance is the integral of this hemispheric flux. It is common to distinguish among the sources of diffuse radiation and calculate their contributions to overall (short- or longwave) irradiance based in part on the proportion of the hemisphere (**view factor**) that that source area represents. At its simplest, the 'view' of a flat, unobstructed facet is occupied by sky only and its view factor ($\psi_{sky}$) equals 1. More likely, a portion of the view is occupied by the surroundings (buildings, trees, ground, etc.) such that

$$\psi_{sky} + \psi_{env} = 1 \quad \text{(unitless)} \qquad \textbf{Equation 5.13}$$

Figure 5.6 gives a sense of the complexity of the exchanges from within an **urban canopy**. It shows the

path of the Sun across the sky during the year, plotted onto a photograph representing a view of the sky hemisphere from the centre of a city street at 49°N. The effect of the built environment is to reduce $\psi_{sky}$ (and increase $\psi_{env}$) and partially block access to direct shortwave irradiance. Note this surface is in shade until 0900, and after 1500 h, throughout the year. At noon, the solar beam is accessible through a narrow gap between buildings. The diffuse radiation environment is even more complex owing to the diversity of surface types, each of which has a unique radiation budget and geometric relationship with the receiving point. One can see that $\psi_{env}$ is a gross simplification of the realities of radiation exchange in the UCL.

Although these processes are difficult to measure, they are well suited to **numerical modelling** that describes the urban surface and simulates radiative transfers from the Sun and among various urban facets (walls, roofs, ground etc.). In following sections we use such a model, to simulate energy exchanges in simple urban configurations, to illustrate the course of fluxes on a building and within urban canyons. We begin with the radiation exchanges at the facets of a cubic building, then interactions between buildings, and finally the integral response of the urban 'surface' that characterizes the UCL.

## 5.2.2 Isolated Buildings

The shadows around even simple forms, such as a cube, are revealing. Obtaining the shadow mask for any time is an exercise in geometry. In essence, the shadow is cast in the opposite direction to the solar beam and has a length equal to the height of the obstruction ($H$), divided by the tangent of the solar altitude angle (tan $\beta$). In Figure 5.7, the loss of direct sunshine over the course of the day is mapped for a cubic building located at 42°N. At this time of year, the altitude of the Sun at noon is just 23.5° and the shadow cast by an object extends to more than twice its height. Note the shadows over the course of a day create a distinct pattern of solar loss. The length of shadow is longest in the early morning and evening hours (shown as elongated shaded areas extending east and west) and is shortest closest to noon. The area just to the north of the building experiences the greatest shadowing effect where a small patch adjacent to the north-facing wall receives no direct sunshine over the course of the day.

A corollary is that the loss of sunshine in the shadowed area is a measure of the direct shortwave irradiance ($S$) intercepted by the building. For an isolated cube, the roof has unobstructed access to the Sun, so the shadow area represents the energy intercepted by the walls. In Figure 5.7, the path of the Sun is close to the southern horizon because it is the winter solstice, so the focus of solar interception is on the south-facing wall (Equation 5.7 and Equation 5.8). At noon, $\beta$ is 23.5° so a flat surface (like the roof) receives just 40% of the radiation intensity compared with a surface placed perpendicular to the solar beam. On the other hand, a vertical south-facing wall (slope angle 90°) has a local solar altitude of 66.5° at the same time and acquires 92% of the intensity of the beam. Clearly, ensuring good solar access for the equator-facing wall of a building at a high latitude location with a cold climate, is an important design principle.

In the following, to explore radiation exchanges ($K_\downarrow$, $K_\uparrow$, $L_\downarrow$ and $L_\uparrow$) at the facets of a cube-shaped

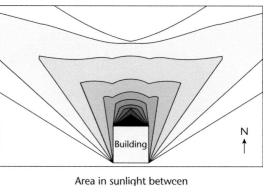

Area in sunlight between

| 9-8h | 8-7h | 7-6h | 6-5h | 5-4h | 4-3h | 3-2h | 2-1h | 1-0h | none |

**Figure 5.7** Shadow curves during the winter solstice (22 December) for a cubic building (white box) located at 42°N. Shaded areas represent the potential hours of available sunshine. Thus, if a similar building were to be placed to the north of this cube, it must be outside the 5–6 hour area to ensure access to that many hours of sunlight (Source: Atkinson, 1912).

building, we use a numerical model to simulate the fluxes at 49°N on August 15th. The walls are oriented towards the cardinal directions – east (E), west (W), north (N) and south (S). The roof (R) has an unobstructed view of the cloudless sky ($\psi_{sky} = 1$), but a wall facet 'sees' both the sky and the surrounding ground ($\psi_{sky} < 1$). The distribution of both shortwave and longwave radiation received from the sky are assumed isotropic. Inputs to the model include: location and time of year, the prevailing weather and the geometric, radiative and thermal properties of the building surfaces (Table 5.3). To obtain all terms of the radiation budget the model must simulate the full surface energy balance (of which the radiation budget is just a part) because longwave radiation is a function of surface temperature, which depends on conductive and convective exchanges (Chapter 6).

### Shortwave Radiation Exchange on Walls and Roofs

Simulations allow us to 'unpack' the diurnal pattern of irradiance on each facet that was alluded to in the discussion of Figure 5.7. A flat roof has an unimpeded view of the sky but receives no radiation from

**Table 5.3** Materials, dimensions and radiative and thermal properties of the building and canyons simulated by TUF-3D.

| Facet | Material | Depth $d$ (m) | Albedo $\alpha$ | Emissivity $\varepsilon$ | Thermal conductivity $k$ (W m$^{-1}$ K$^{-1}$) | Heat capacity $C$ (MJ m$^{-3}$ K$^{-1}$) |
|---|---|---|---|---|---|---|
| **Walls** | Concrete | 0.2 | 0.25 | 0.90 | 0.80 | 1.33 |
| **Roof** | Gravel, insulation over concrete | 0.1 | 0.12 | 0.92 | 0.06 | 1.00 |
| **Floor** | Asphalt over concrete | 0.3 | 0.15 | 0.94 | 1.21 | 1.95 |

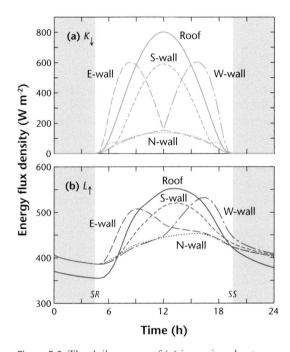

**Figure 5.8** The daily course of **(a)** incoming shortwave and **(b)** outgoing longwave radiation fluxes for the walls and roof of an isolated cube-shaped building at 49°N on August 15$^{th}$. Results calculated using TUF-3D.

the ground around the building (Figure 5.8a). Its shortwave irradiance is therefore composed of direct and diffuse shortwave radiation from the sky (i.e. $K_\downarrow = S + D_{sky}$). For the wall facets, there are two sources of diffuse shortwave radiation: the sky and reflection from the ground ($K_\downarrow = S + D_{sky} + D_{env}$). Unlike the roof, some vertical wall facets are oriented such that the solar beam is obstructed by the building itself.

For the E and W walls, solar receipt is concentrated in the morning and afternoon, respectively. The E wall faces the rising Sun and is one of the first to be directly irradiated. Moreover, this vertical facet has almost ideal exposure to the beam that arrives from low in the sky and strikes at a nearly perpendicular local angle. It is not uncommon for sunlit walls to receive $> 500$ W m$^{-2}$ soon after sunrise. As morning progresses the Sun's orb moves higher in the sky and towards the south (at this northern hemisphere location). By mid-morning the angle of receipt starts to become less favourable and solar receipt by the facet declines. At noon the beam reaches the grazing angle for the E wall (i.e. cos $\hat{\Theta}$ approaches zero, Figure 5.4) and it goes into shade, receiving only diffuse shortwave, the same as the N wall. The daily pattern of receipt for the W wall is the mirror image of that for the E wall. The S wall displays a pattern that is symmetric about solar noon. It does not become directly irradiated until the Sun is south of E in the morning, and goes into shade when it turns north of W in the evening. The N wall of the building receives a short input burst just after sunrise when the Sun is N of E, then it enters shade for most of the day, receiving only diffuse shortwave. There is a similar short period of beam input just before sunset when the Sun is N of W – in other words it experiences two local 'sunrises' and 'sunsets'. When a wall is sunlit, the contribution by $D_{env}$, which includes reflection off the adjacent illuminated ground, is increased. Similarly, for much of the period when a wall is in shade, so is the adjacent ground (see Figure 5.7). The daily course of reflected shortwave at each facet (not shown) is simply a reduced mirror image of the input, the fraction depending on the surface albedo.

These patterns change according to the latitude of the building. For example, in tropical latitudes the high altitude angle of the Sun ($\beta$) at noon generally means the bulk of intercepted shortwave radiation is concentrated on the E and W walls and especially the roof. On the other hand, at higher latitudes in winter, low solar altitude means S walls often receive the greatest shortwave irradiance. This point only relates to instantaneous values because, due to the greater time of exposure on a S wall, the input integrated over a complete summer day is often greater than for other walls, but it is not necessarily greater than the roof or ground.

### Longwave Radiation Exchange on Walls and Roofs

Longwave input ($L_\downarrow$) to a roof is dominated by irradiance from the sky, which varies little through a day. However, on the ground near walls, $L_\downarrow$ has two distinct sources; the sky and adjacent wall (i.e., $L_\downarrow = L_{\downarrow sky} + L_{\downarrow env}$). The contribution of wall facets varies over the day, as their temperature changes. Further away from the building, $\psi_{env}$ diminishes until $L_\downarrow \approx L_{\downarrow sky}$. Input to the walls also has two sources with the nearby ground contributing most to $L_{\downarrow env}$ but its temperature also varies considerably over a day. However, the difference between facets is seen most clearly in the emitted longwave radiation (Figure 5.8b). Notice that $L_\uparrow$ for the S wall and roof facets are nearly symmetrical around noon, whilst the patterns of the E and W wall facets have peaks in the morning and afternoon, respectively. These curves correspond to the patterns of shortwave irradiance ($K_\downarrow$) at each facet, the absorption of which is mainly responsible for their surface temperatures and the magnitude of emittance. The effect of the longwave input ($L_\downarrow$) can be seen in the higher $L_\uparrow$ flux at the W wall, which faces ground that has been warmed during the day so it receives a greater contribution of $L_{\downarrow env}$ than the E wall. The effect of ground warming is also evident in the asymmetry of $L_\downarrow$ irradiance on the N wall, even though the surface itself is in shade most of the day.

The net longwave budget ($L^*$) is negative for all facets throughout the day (not shown). By day the magnitude of $L^*$ is greatest for the roof, which is both hot and exposed to the cool sky. In contrast the cooler N wall, which views a shaded surface, has a net loss that is one-third that of the roof. At night greatest losses are from the ground because it is *relatively* warm, due to its larger **heat storage** and sky view. The roof is also exposed, it has a lower temperature and so emits less. The smallest loss is from the relatively cool N wall.

### Net Allwave Radiation of Walls and Roofs

Net allwave radiation ($Q^*$) largely follows the shortwave radiation pattern during daytime (Figure 5.8). The roof has the greatest $Q^*$ surplus because its shortwave absorption, due to its lower albedo, is sufficient to overcome its greater longwave loss, which differentiates it from the ground. At night $Q^*$ is simply equal to $L^*$ (discussed above).

## 5.2.3 Urban Canyon

One of the defining attributes of **urban climates** is the extent to which buildings interact with each other to create distinct microclimates *within* the UCL. To cope with the great variety of **urban forms** and yet still extract general relations, we consider radiation exchange in a simple street canyon and further we assume that all reflection and emission events are diffuse, that way exchanges are simply a function of the view factors linking the exchanging surfaces. Finally, we subdivide facets (such as walls) into smaller surface units so that spatial variation of radiation budgets across them can be considered. The integral effect on radiation exchange is examined at an imaginary plane across the top of canyons as a function of their aspect ratio ($H/W$) and orientation.

### Principles of Shortwave Radiation Exchange in a Canyon

Figure 5.9 shows the effects of **urban structure** on a quantity of radiation within a two-dimensional canyon. Initially, a facet receives direct-beam irradiance (Figure 5.9a), a portion of which is reflected. This radiation spreads in a diffuse manner: some exits the canyon, and some is intercepted by the opposite wall and the canyon floor (Figure 5.9b). These recipient surfaces also absorb and reflect in turn. After the second reflection (Figure 5.9c), each unit has exchanged radiation with every other unit within the canyon. This process of multiple reflection is repeated *ad infinitum* with progressively less energy being exchanged, because each unit absorbs a portion of the radiation they receive.

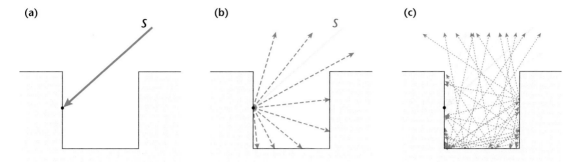

**Figure 5.9** Impact of the geometric configuration of an urban canyon on radiation exchanges. The case is simplified by limiting consideration to a single incoming ray with two reflections and to 2D; in reality the form is 3D (i.e. also into and out of the page). **(a)** The receipt of direct-beam irradiance through the canyon top, **(b)** first reflection from the canyon facets, **(c)** second reflection. Input and reflection of diffuse sky radiation is not depicted.

Formally, we can write the shortwave radiation receipt on a facet unit ($i$) as

$$K_{\downarrow i} = S_i + D_{\mathrm{sky}_i} + D_{\mathrm{env}_i} \qquad (\mathrm{W\,m^{-2}}). \quad \textbf{Equation 5.14}$$

If the unit is in shade then ($S_i = 0$) and $K_{\downarrow i}$ consists of only diffuse terms, which originate from the sky ($D_{\mathrm{sky}_i}$) and reflection from the surrounding urban environment ($D_{\mathrm{env}_i}$). For an isotropic sky the diffuse irradiance from the sky on the facet is simply

$$D_{\mathrm{sky}_i} = D_{\mathrm{sky}_0} \psi_{i\to\mathrm{sky}} \qquad (\mathrm{W\,m^{-2}}) \quad \textbf{Equation 5.15}$$

where $\psi_{i\to\mathrm{sky}}$ is the sky view factor for that facet unit. Obviously $D_{\mathrm{sky}_i}$ is smaller within a canyon compared with that received on a horizontal facet with an unobstructed view of the sky ($D_{\mathrm{sky}_0}$); this reduction is due to **horizon screening**. Figure 5.10 depicts the variation of $\psi_{i\to\mathrm{sky}}$ as a function of position within the canyon; it is smallest at the base of the canyon and near the sides.

The urban contribution to diffuse shortwave irradiance ($D_{\mathrm{env}_i}$) is the integral of all reflection events that impinge on the unit. At its simplest,

$$D_{\mathrm{env}_i} = K_{\uparrow\mathrm{env}} \psi_{i\to\mathrm{env}} \qquad (\mathrm{W\,m^{-2}}) \quad \textbf{Equation 5.16}$$

i.e. intercepted diffuse irradiance equals the shortwave radiation reflected from surfaces in the surrounding environment multiplied by the corresponding environment's view factor for that facet unit ($\psi_{i\to\mathrm{env}}$). The complex geometry within the UCL leads to an extraordinarily heterogeneous distribution of diffuse radiation. Hence urban facets are commonly divided into smaller units or patches

($j$) to determine their individual contributions. The total $D_{\mathrm{env}}$ received at patch ($k$) must include the reflection contributions from all other patches for which $\psi_{k\to j} > 0$,

$$D_{\mathrm{env}_k} = \sum_{j=1}^{N-1} K_{\uparrow j} \psi_{k\to j} \qquad (\mathrm{W\,m^{-2}}).$$

**Equation 5.17**

This calculation must be performed for each of the $N$ patches that comprise the canyon facets. However, it is apparent that $D_{\mathrm{env}_k}$ depends on $K_\uparrow$ at all other units within the canyon, each of which also include $D_{\mathrm{env}}$ (Equation 5.17). Hence these calculations must be performed repeatedly to account for multiple reflection events, each of which contributes a progressively smaller magnitude than the event preceding it (Figure 5.9). Numerical simulation studies conclude that in practice it is sufficient to limit consideration to two events (Harman et al., 2004). If all view factors are known, an exact solution for the diffuse radiation received from multiple reflections gives greater **accuracy** and is recommended where the surface albedo is greater than 0.2 (Harman et al., 2004).

Multiple reflection makes a canyon a more efficient absorber of shortwave radiation than the albedo of the individual canyon facets would suggest, i.e. the albedo of a canyon is lower than that of its constituent facets. To illustrate, in Figure 5.11 a canyon has been 'flattened' so that its three facets are shown as segments of a horizontal surface the overall albedo (for a canyon of unit length) of which is

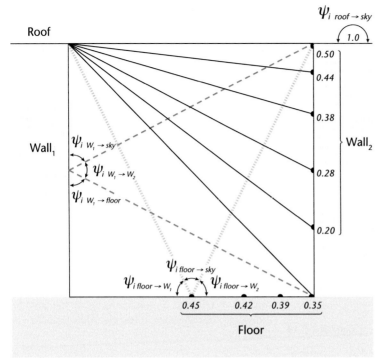

**Figure 5.10** View factors in a canyon with $H/W = 1.0$. The view factors ($\psi$) of Wall$_1$, and the Floor, for the other facets and sky are shown. The fractions on the Floor and Wall$_2$ are the sky view factors of the points for the case of an infinitely long canyon. The roof has the sky view factor of a horizontal surface with an unobstructed horizon.

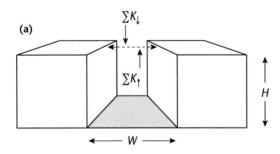

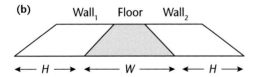

**Figure 5.11** The effect of 2D urban structure on canyon system albedo, evaluated by comparing measured solar fluxes: **(a)** in and out through the top of a canyon of unit length (the spatial summation $\Sigma$ is conducted across the canyon top $W$), versus **(b)** the sum of the fluxes in and out of the same surface facets when laid flat.

$$\alpha_{\text{Surf}} = \frac{W\alpha_{\text{floor}} + H(\alpha_{\text{wall}_1} + \alpha_{\text{wall}_2})}{2H + W} \quad \text{(unitless)}$$

**Equation 5.18**

The absorptive 'efficiency' of the canyon ($R_c$) can be evaluated by comparing the absorptive ability of the surface facets to that of the canyon as a whole, as measured at the canyon top ($\alpha_{\text{top}}$),

$$R_c = \frac{1 - \alpha_{\text{top}}}{1 - \alpha_{\text{Surf}}} \quad \text{(unitless)} \qquad \textbf{Equation 5.19}$$

where $\alpha_{\text{top}}$ is obtained as the ratio of $K_\downarrow$ entering the canyon, to that exiting, $K_\uparrow$ (see Figure 5.9).

This effect is greater in deeper and narrower canyons (i.e. higher $H/W$ value) where the $\psi_{i\to\text{env}}$ increases but $\psi_{i\to\text{sky}}$ decreases (Figure 5.10). **Canyon trapping** means that, other things being equal, the albedo of a canyon is inversely related to $H/W$.

## Principles of Longwave Radiation Exchange in a Canyon

Incoming longwave radiation within the UCL is broadly analogous to the shortwave case,

$$L_{\downarrow i} = L_{\downarrow sky_0} \psi_{i \to sky} + \sum_{j=1}^{N} \left( L_{\uparrow j} \psi_{i \to j} \right) \quad (\mathrm{W \ m^{-2}})$$

**Equation 5.20**

i.e. $L_{\downarrow i}$ is comprised of input from the portion of the sky hemisphere 'seen' ($\psi_{i \to sky}$), which, similar to $D_{sky_0}$, is reduced due to horizon screening, and from the surrounding environment ($L_{\uparrow j}$), including emission (and some reflection, see Equation 5.12) from neighbouring facets. At solid surfaces almost all intercepted $L_{\downarrow}$ is absorbed, only a small proportion is reflected $(1 - \varepsilon_j)$. The portion absorbed by a unit causes an increment of surface warming, that affects longwave emission ($L_{\uparrow j}$) and hence $L_{\downarrow i}$ at all points in its FOV, i.e. the behaviour of all facets is mutually dependent. The augmented longwave flux received is a form of canyon trapping.

At facet units in the canyon with small $\psi_{sky}$, the proportion of $L_{\uparrow i}$ escaping the canyon, rather than being intercepted and absorbed by other facets, is small. Reciprocally the fraction of the radiation received that comes from the 'cold' sky is proportionately small, most input derives from relatively warmer facets. In a canyon, the points with the lowest $\psi_{i \to sky}$ values are near the base of the walls; largest values are in the middle of the floor and near the top of the walls (Figure 5.10). Hence, other things being equal, cooling is least at the bottom of a canyon near the junction of the walls and floor; cooling is less in deep, narrow canyons than in more open ones, and greatest on rooftops and flat rural areas open to the sky.

Figure 5.12 gives an overall appreciation of view factors in canyons showing how they vary with $H/W$. It shows facet-averaged $\psi$ values, not only for the sky from the canyon floor, but also for the sky from either wall, the floor from one of the walls and the canyon walls from each other. Notice that at $H/W = 1$ the value of $\psi_{floor \to sky}$ equals 0.4 (as does $\psi_{wall \to wall}$) and $\psi_{floor \to wall}$ equals 0.3 (and *vice versa*), so the view factor for the floor is chiefly occupied by canyon walls. As a consequence, 60% of the radiation emitted by the floor is intercepted by the canyon itself, most of which is absorbed and subsequently re-emitted to other canyon surfaces.

## Patterns of Radiation Exchange in a Canyon

The numerical model and **boundary conditions** used in Section 5.2.2 are used here to simulate exchanges in a long, dry symmetric canyon (base case $H/W = 1$). Calculations are performed at a number of points to

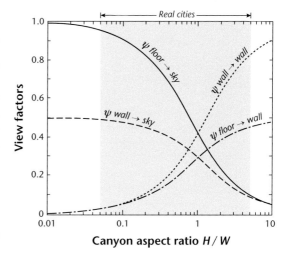

**Figure 5.12** Facet-averaged view factors of infinitely long and across-canyon symmetric, urban canyons over a range of aspect ratios (Modified after: Harman et al., 2004; © 2003 Kluwer Academic Publishers with permission of Springer).

sample each facet, but results are presented as averages for the entire facet. Experiments are conducted for streets oriented north-south (N-S), east-west (E-W) and for several $H/W$ ratios.

The pattern of shortwave radiation receipt on the canyon facets (Figure 5.13a and Figure 5.14a) resemble those on the corresponding facets of an isolated building (Figure 5.8a), differences are mainly due to shadowing and canyon trapping. In the N-S canyon (Figure 5.13a) a solar day for the E wall starts at about the same time as that on the isolated building and the favourable angle of incidence produces a similarly rapid increase in receipt. However, the average receipt is less because the lower part of the wall is shaded by the building opposite, hence it receives only diffuse shortwave radiation. Towards solar noon the azimuth of the Sun aligns with the axis of the street and the E wall becomes increasingly less favourable to receipt. After noon the E wall falls into complete shade and thereafter receives only diffuse shortwave irradiance from the sky or that reflected from other canyon facets. There is a small 'hump' in the afternoon due to reflection from the W wall that is largely sunlit at that time. Indeed the daily pattern for the W wall is the mirror image of that for the E wall, and the 'hump' in diffuse radiation on the E wall corresponds to the time of maximum illumination of the

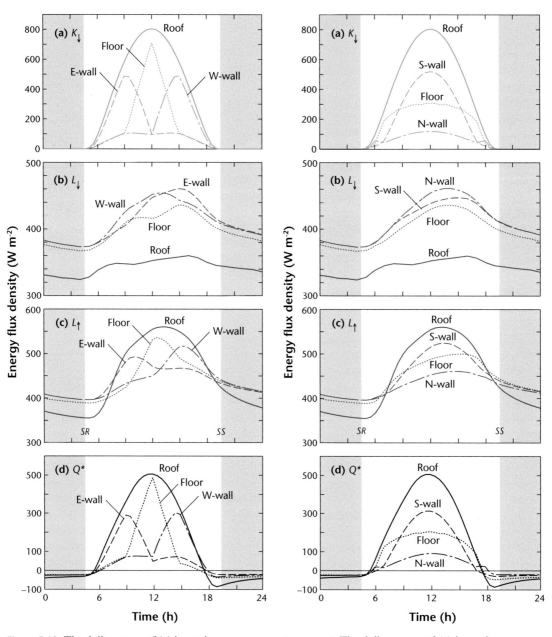

**Figure 5.13** The daily course of **(a)** incoming shortwave, **(b)** incoming longwave, **(c)** outgoing longwave and **(d)** net radiation fluxes in a N-S oriented urban canyon. The walls are identified by the direction in which the facet is facing. Results are facet averages from the TUF-3D model, see text for input conditions.

**Figure 5.14** The daily course of **(a)** incoming shortwave, **(b)** incoming longwave, **(c)** outgoing longwave and **(d)** net radiation fluxes in an E-W oriented urban canyon. The walls are identified by the direction in which the facet is facing. Results are facet averages from the TUF-3D model.

W wall. The floor of a N-S canyon has a short-lived daily solar climate: with peaks at solar noon, and 'shoulders' of diffuse radiation in the morning and afternoon (Figure 5.13a).

Solar irradiance on facets of the E-W street display a simple symmetric pattern centred on solar noon (Figure 5.14a). Receipt for the S and N walls look much like that of their counterparts on the isolated building, differences are due to canyon shading which is greatest near sunrise and sunset but continues through the day for lower parts of the S wall and over most of the N wall. The exact pattern is set by the $H/W$ ratio and the latitude. The solar beam briefly illuminates the floor of the E-W canyon in the early morning and late evening but only just grazes the walls. Thereafter, whether the beam reaches the floor during the day depends on the angle of the beam and the $H/W$ ratio.

In real cities these exchanges are complicated by the inherent variability of building height, street orientation, surface albedo, the presence of trees, seemingly minor features like roof eaves, balconies and so on, but solar geometry remains the underlying control. Shade is a powerful, and often under-appreciated, regulator of surface climates. Obstruction of the solar beam produces a local 'sunset' that suddenly decreases input several fold (e.g. the E, W and floor facets in Figure 5.13a).

Shadows cast by buildings that are significantly taller than others in a neighbourhood can rob affected areas of valuable daylight and heat. Numerical and **physical models** (e.g. the heliodon, Chapter 3) are useful to calculate the location, size and dynamics of building shadows. Concerns are greatest at higher latitudes and in winter when shadows become most extensive and long-lasting and the need for light and warmth is greatest.

There are strong differences between the longwave radiation emitted from the sky and from canyon facets due to differences in the temperature and emissivity of the two environments (Equation 5.1). Under clear skies, the **anisotropic** distribution of longwave sky radiance means that whilst $\psi_{i \to sky}$ is important to surface cooling at any point, the specific part of the sky in its FOV is also relevant.

Throughout a day roof facets receive least $L_\downarrow$, because they do not 'see' other warm facets, only much colder sky (Figure 5.13b and Figure 5.14b). Of the canyon facets the floor receives least $L_\downarrow$, because its sky view is largest. During daytime the variation of $L_\downarrow$ is greater inside the canyon than on the roof, due to the greater range of facet temperatures compared to the sky. Notice also that $L_\downarrow$ on the W and E walls in the N-S canyon reverse their relative positions between morning and afternoon, in response to the change in $T_0$ of the wall opposite caused by the temporal shift of irradiation and shadow patterns. At night $L_\downarrow$ for each facet is mainly inversely related to its $\psi_{i \to sky}$ value.

Under cloudy skies differences of temperature and longwave exchange between the surface and sky become muted, therefore cooling is slowed. However, the impact of cloud on exchanges within an urban canyon, is less than in rural areas at the same time. That is because, viewed from within a canyon, only part of $L_\downarrow$ comes from the sky, the rest is from other canyon facets that are warmer than the cloud base, whereas at most rural sites $\psi_{sky}$ is greater.

By day, it is the absorption of shortwave radiation by canyon facets that is the dominant control on $T_0$ and $L_\uparrow$ (Figure 5.13c and Figure 5.14c). The facet most directly irradiated by the Sun generates greatest $L_\uparrow$ and receives the least amount back from other facets, hence it experiences the greatest net longwave loss ($L^*$).

The daily course of $Q^*$ on canyon wall and floor facets versus that on the roof (Figure 5.13d and Figure 5.14d) shows the effect of canyon geometry. The roof experiences the greatest and range over the course of the day because it is the most exposed to the direct controls of solar geometry and sky view. Canyon facets on the other hand, experience the additional mediating effects of beam interception, shading and inter-facet reflection in the shortwave and absorption-reemission and sky view in the longwave.

To simulate the effects of canyon structure four configurations of the TUF-3D model are run corresponding to canyons with $H/W = 0.5$, $1.0$, $2.0$ and $4.0$ and the complete system (canyon plus adjacent flat roof) (Table 5.4). Since $K_\downarrow$ and $L_\downarrow$ at the top of the UCL are fixed for a given case, any $Q^*$ differences are a result of the canyon materials (Table 5.1) and/or structure ($H/W$), which control $K_\uparrow$ and $L_\uparrow$. During daytime, the effect of geometry is to lower albedo ($K^*$ is increased, more positive) due to multiple reflection and larger $\lambda_c$, and also to lower the mean canyon surface temperature ($L^*$ is increased, less negative at the canyon top), due to shading. $Q^*$ at the canyon top is therefore more positive during the day due to canyon trapping (Table 5.4) and the effect increases with $H/W$ (up to 28% at $H/W = 4$). Whilst canyon orientation has a large influence on radiation budget terms for individual facets, when terms are aggregated and expressed as a flux at the canyon top within-canyon interactions compensate thereby diminishing directional bias.

**Table 5.4** Effect of aspect ratio ($H/W$) on net radiation ($Q^*$, $L^*$) in the UCL, as calculated using TUF-3D numerical model. Energy totals in MJ m$^{-2}$ d$^{-1}$, percentages are difference from a flat roof made of similar material to the walls and floor. Note: nocturnal values are small and approach model capability hence percentages are approximate.

| System | Urban form | Daytime | | Night | | Daily | |
|---|---|---|---|---|---|---|---|
| | | $Q^*$ canyon top | $L^*$ canyon top | $L^*$ canyon floor | $L^*$ system | $Q^*$ canyon top | $Q^*$ system |
| Flat surface | ———————— | 13.3 | -1.6 | -1.6 | -1.6 | 11.7 | |
| Canyon, $H/W = 0.5$, $\lambda_b = 0.33$ | | 14.8 (+11%) | -2.5 (+56%) | -1.4 (-13%) | -2.0 (+25%) | 12.3 (+5%) | 12.0 (+3%) |
| Canyon, $H/W = 1$, $\lambda_b = 0.5$ | | 15.9 (+20%) | -2.6 (+63%) | -1.0 (-38%) | -2.1 (+31%) | 13.2 (+13%) | 12.5 (+7%) |
| Canyon, $H/W = 2$, $\lambda_b = 0.66$ | | 16.5 (+24%) | -2.7 (+69%) | -0.5 (-69%) | -2.1 (+31%) | 13.8 (+18%) | 12.8 (+9%) |
| Canyon, $H/W = 4$, $\lambda_b = 0.8$ | | 17.0 (+28%) | -2.8 (+75%) | -0.2 (-88%) | -2.2 (+38%) | 14.2 (+21%) | 13.0 (+11%) |

At night, horizon screening (smaller $\psi_{sky}$) within the canyon increases $L_\downarrow$ and reduces the loss of longwave radiation from canyon facets ($L^*$ on the canyon floor is less negative). Decreased nocturnal $L^*$ retards surface cooling; hence canyon facet temperatures remain warmer. The difference between $L^*$ at the canyon top (a loss) and that for the roof is even greater, due to the relative warmth of the canyon system, although the effect of $H/W$ is relatively small. However, the difference between $L^*_{floor}$ and $L^*_{roof}$ is the reverse: the loss at the floor is considerably less due to the reduced sky view. Consequently, cooling rates for the floor near sunset (when they are greatest), are 2.6, 2.0, 1.4 and 0.9 K h$^{-1}$ for $H/W = 0.5$, 1, 2 and 4, respectively.

Viewed from above the canopy the planar area at the top of the UCL is comprised of roofs and canyon top areas. Whilst increasing $H/W$ alters $Q^*$ at the canyon top, it also reduces the proportion of the UCL surface occupied by canyon top, and increases that occupied by roofs. In other words $Q^*$ at the top of the UCL becomes dominated by the roof signal, owing to its greater surface area. This is especially true in districts with large $H/W$ (e.g. Figure 5.1).

### 5.2.4 Albedo and Emissivity of Urban Systems

#### Effect of Urban Structure on Albedo

The corrugate form of the urban surface favours multiple reflection of shortwave radiation so, all other things being equal, absorption by the urban surface is increased in comparison with a flat, horizontal surface made of the same material. But further, because the same energy input is spread over a greater surface area, compared to a horizontal surface (i.e. $\lambda_c$ is larger), the average flux density received by facets is smaller than on a flat surface of the same material.

To isolate these effects of urban structure on albedo, Aida (1982) constructed an outdoor physical model of basic urban geometry with concrete blocks arranged on a flat concrete base to simulate a range of $H/W$ and $\lambda_c$ values. The albedo of the blocks and the base were identical (0.4). Incident and reflected shortwave radiation were measured above the model to obtain the integrated 'effective' albedo of the 'urban' surface. Results show that as $H/W$ increases from 0 (a flat plane) to 2, the albedo of the system decreases from 0.40 to 0.23, an increase in absorption of 27% achieved through the effects of geometry alone (Table 5.5).

**Table 5.5** The effect of urban form upon the surface albedo. Values obtained using an outdoor scale model constructed from concrete blocks placed on a base of the same material and observed over a full one year period by Aida (1982). 'Change in absorption' is calculated compared with flat case.

| Urban form | $H/W$ | $\lambda_c$ | Albedo $\alpha$ | Change in absorption |
|---|---|---|---|---|
| | 0 | 1 | 0.40 | |
| | 0.5 | 1.5 | 0.32 | +17% |
| | 1 | 2 | 0.27 | +21% |
| | 2 | 3 | 0.23 | +27% |

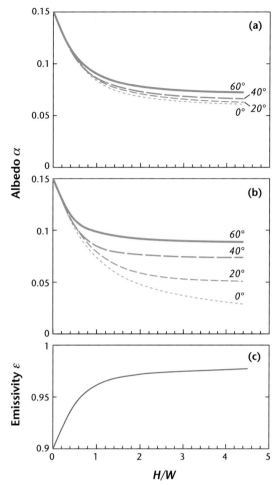

**Figure 5.15** Relation between effective albedo, averaged over a complete day, and $H/W$ for a 2D canyon (albedo of walls 0.25 and floor 0.15) the axis of which is oriented **(a)** N-S, and **(b)** E-W, plotted for latitudes 0°, 20°, 40° and 60° at the time of an Equinox. **(c)** The equivalent effective emissivity $\varepsilon$ (**Equation 5.23**) of a 2D isothermal canyon at 295 K, with wall and floor emissivities of 0.9 and incoming $L_\downarrow$ of 275 W m$^{-2}$. Calculated by Scott Krayenhoff using the TUF-2D model (Krayenhoff and Voogt, 2007).

Figure 5.15 gives results from a numerical model of a 2D canyon for a range of $H/W$ at different latitudes, summed over a complete day period, at the time of an Equinox. The effective albedo values are lower than the average of the three facet values, both due to canyon trapping and the fact that significant fractions of the facets are in shade. The shape of the relation shows the impact of geometry on albedo becomes much smaller beyond an aspect ratio of unity. It also illustrates that the effect of latitude is most apparent for an E-W oriented canyon and is relatively minor for a N-S case (Figure 5.14b and a, respectively). The altitude angle ($\beta$) is critical because it determines the extent to which $S$ penetrates into the canyon. Increasing the absolute albedo of the walls leads to a greater geometric effect as well as an overall increase in the effective canyon albedo. Streets oriented between the orthogonal directions have an intermediate geometric effect. The diurnal variations of canyon effective albedo are influenced by both the albedo of individual components of the canyon and by $H/W$ (Fortuniak, 2008).

### Effect of Urban Fabric on Albedo

The range of albedo values of surface materials found in cities is greater than for rural areas; it includes more with high albedos (Table 5.1) due to the diversity of natural and manufactured materials used in urban construction. Some facets are engineered specifically to have unusual radiative properties, such as albedos that are either particularly low (dark paint, roof shingle, tinted glass) or high (light paint, bright metals, glass films and coatings). Roof materials have relatively low albedos, especially in temperate and high latitudes. This can result in high surface temperatures of roofs in the middle of a summer day; values $> 50°C$ are not unusual and up to $80°C$ is possible on blackened surfaces (Gaffin et al., 2012). Modifying the albedo of urban facets through intelligent selection of materials or coatings, can yield several benefits including greater indoor comfort, less use of energy for air-conditioning and a reduction of the UHI (Section 15.2.3).

### Typical Urban Albedo Values

Measured albedos at the city and local scales combine the reflection from many roofs, trees, canyons, open spaces and so on, but notwithstanding the diversity of materials, structure and geographical setting of cities, there is remarkable consistency in the measured values (Table 5.1). The slightly lower mean albedo of urban (0.14) vs suburban (0.15) areas in snow-free conditions may be due to the greater $H/W$ typical of more densely built-up sites but the mean difference is small.

The measured albedo of cities show a dependence on solar altitude and zenith angle (Figure 5.16). The relationship is not strong, but it broadly agrees with the known pattern for vegetated surfaces and other natural surfaces where albedo increases rapidly at higher solar zenith angles. The greater curvature at high zenith angles (i.e. when the Sun is lower in the sky) is probably due to increased specular reflection from horizontal roofs relative to the interior of canyons, which again is related to $H/W$.

Overall, in snow-free conditions urban and suburban albedo values are slightly ($< 0.10$) lower than in the surrounding countryside. Hence urban–rural albedo differences ($\Delta\alpha_{U-R}$) in snow-free conditions are usually negative (Table 5.1). However, $\Delta\alpha_{U-R}$ depends on *two* environments and the rural one is often the more variable. Although intraurban albedo differences are fairly small the range of possible rural terrain values is large (Table 5.1). So it is quite possible that cities surrounded by tropical or coniferous forest may not be better absorbers (i.e. $\Delta\alpha_{U-R}$ could be positive), similarly a desert city may well show greater $\Delta\alpha_{U-R}$ differences than those from

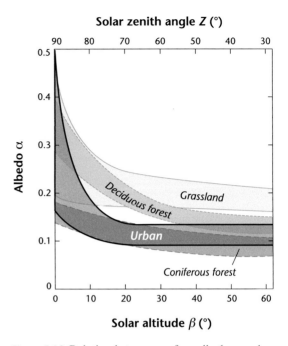

**Figure 5.16** Relation between surface albedo $\alpha$ and solar altitude angle $\beta$ (bottom axis) or zenith angle $Z$ (top axis) for snow-free conditions. Envelopes are a summary of the curves, obtained from many field sites. The vegetation results include the data of forests and grassland sites, those for cities include Hamilton, Canada; St. Louis, United States; Vancouver, Canada; and Basel, Switzerland (Data sources: Rouse & Bello, 1979; White et al., 1978; Grace, 1983; Christen & Vogt, 2004 and unpublished data).

temperate cities (Table 5.1). To date most large negative values of $\Delta\alpha_{U-R}$ have been observed in warmer, drier cities where light-coloured fabric is favoured.

Satellite data provide a spatial distribution of albedo in urban regions, however, such results may differ from those using surface-based tower platforms (Chapter 3) because of differences in the spectral sensitivity of the **sensors**, the effects of scattering and absorption of the atmosphere between the surface and the sensor and differences in source areas that comprise the FOV (Figure 3.6). Nevertheless, satellite data can give an idea of geographical and seasonal variations in urban albedo. For example, Figure 5.17 shows albedo patterns in the Greater Cairo, Egypt, area which clearly identify differences between urban areas and those of the surrounding agricultural (largely irrigated) and desert environs.

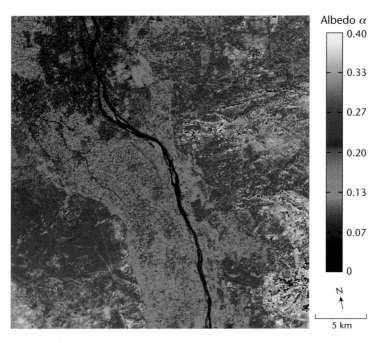

Albedo $\alpha$

0.40

0.33

0.27

0.20

0.13

0.07

0

N

5 km

**Figure 5.17** A satellite image of albedo $\alpha$ in the Greater Cairo area and the surrounding landscape. The dark blue line extending south-north is the River Nile. Variations in albedo are associated with topography, crop type and growth stage. Blue (lowest albedo) is urbanized; green is irrigated farmland and; reds and yellows are desert or bare rock. Dark lineaments on the eastern side of the image are artefacts caused by shadows in narrow valleys. Calculated from ASTER satellite data taken at approximately 1100 h (LST) on 24 December, 2007 using data from Frey and Parlow (2012); (Credit: C. Frey, 2010; with permission).

### The Effect of Snow Cover on Albedo

The arrival of fresh snow changes the magnitude of $\Delta a_{U-R}$ to $\geq -0.50$ (Table 5.1, Figure 5.18). With full snow cover urban albedos typically increase to about $0.20 - 0.35$, depending on the urban structure and the age of the snow cover. At the same time, if the countryside is flat farmland, its albedo may initially be $> 0.70$ (Figure 5.28). If the city is in a forested area $\Delta a_{U-R}$ is smaller.

There are five main reasons why urban areas are better absorbers in snowy conditions:

• **Walls remain virtually snow-free** and these are important facets for solar receipt in winter, when solar altitude is low.
• Many winter cities **physically remove snow** from roads, sidewalks and parking areas and spread salt to melt the snow and ice (Figure 5.18a).
• The **snow surface is soiled** by the sedimentation of aerosols, vehicle tire tracks and the sand or gravel spread to improve vehicle traction. These practices both lower the albedo and hasten the melting of snow cover.
• **Urban warmth assists snow melt**. This is caused by heat loss from buildings, especially their roofs, and the **canopy layer urban heat island** (Chapter 7) that keeps the canyons warm.

• It is possible that the **boudary layer urban heat island reduces snowfall in cities**. When air temperatures are near the freezing point, snowflakes may melt as they fall through the warmer UBL so they are deposited as sleet or rain rather than snow (Chapter 10).

It follows that after snowfall cities stand out in aerial photographs or satellite images; the darker skeleton of roads and dirty snow contrast with the countryside (Figure 5.18d).

### Emissivity of Urban Systems

The configuration of canyons also favours greater absorption of $L_\downarrow$. Although the average reflectivity of $L_\downarrow$ at facets is very small, multiple reflection between facets does lead to a minor net increase in absorption. We define an effective canyon emissivity following Harman et al. (2004):

$$\varepsilon_{\text{top}} = \frac{L^*_{\text{top}}}{L^*_{\text{top (blackbody)}}} \qquad \text{(unitless)} \qquad \textbf{Equation 5.21}$$

i.e. the ratio of $L^*$ at the canyon top to the same quantity if the system is assumed to be an isothermal blackbody. Hence, the emissivity of a canyon system is slightly greater than that of the area-averaged sum of its individual facets (Figure 5.15).

**Figure 5.18** Impacts of urban structure and human activity on the darkness (related to the albedo) of Montreal, Canada, when covered in snow at different scales: **(a)** canyon (Credit: E. Christensen), **(b)** canyon/urban block (Credit: J. Voogt), **(c)** neighbourhood (Credit: J. Voogt), and **(d)** city (Credit: USGS/NASA Landsat Program). Impacts and surfaces: 1 – snow clearance, 2 – soiling/tracking by cars and snow piles, 3 – snow-free walls, 4 – snow removal on main roads, 5 – snow-free pitched roof, 6 – snow pack on flat roof, 7 – undisturbed snow in park, 8 – residential area, 9 – city, 10 – countryside, 11 – snow on river ice, 12 – ice-free river.

There are few estimates of emissivity for whole urban areas. Satellite or aircraft-based **remote sensing** techniques can provide estimates of emissivities at neighbourhood and urban scales; but they are often are challenged by the existence of low-emissivity urban surfaces such as metal roofs (Sobrino et al., 2012) and the biased sampling of horizontal surfaces from aerial platforms. Urban-scale emissivities calculated using numerical models (e.g. Arnfield, 1982) are relatively large ($\varepsilon \geq 0.94$).

The addition of snow increases urban emissivity values slightly (Table 5.2). $\Delta\varepsilon_{U-R}$ in snow-free conditions is usually small ($< 0.05 - < 0.03$). Knowledge of $\Delta\varepsilon_{U-R}$ is of utility in converting remotely-sensed brightness values into temperatures, e.g. to estimate the surface urban heat island (UHI$_{Surf}$; Chapter 7).

## Net Radiation of an Urban Canopy

Despite all the potential complications introduced by the many different aspects, orientations and radiative properties offered by a canyon system, the resultant flux at canyon top ($Q_{top}^*$) follows a smooth diurnal regime (Figure 5.13, Figure 5.14). Notwithstanding the radically different radiation budgets of each facet, the course of the net result for the urban system is still driven by: $K_\downarrow$ across the canyon top in the daytime, and the relatively steady drain of $L^*$ due to the sky-canyon temperature difference. The net allwave radiation budget at the top of the UCL is simply the sum: ($Q_{top}^* + Q_{roof}^*$), after each is weighted by the fraction of the interface area each occupies (Table 5.4). Again the daily pattern of the budget looks similar in form to that of simple natural surfaces. This observation is important because the integrated radiation budget of cities (Section 5.4) drives their energy balance (Chapter 6).

## 5.3  Radiation in the Urban Boundary Layer

The radiative properties of the UBL are very different to those of the ABL in non-rural environments. Urban pollution changes the aerosol and gaseous composition of the UBL. This results in changes to the absorption, transmission and reflection of radiation within the UBL that in turn impacts visibility, the spectral energy content of radiation received by the urban surface and the temperature structure of the UBL. We first consider urban aerosol and then short- and longwave radiation in the UBL.

### 5.3.1  Urban Aerosol

The UBL abounds with aerosols emitted from **combustion** and industrial processing sources, and disturbed dust due to vehicles and construction. There is also secondary production of aerosol through gas-to-particle conversion, i.e. gases chemically and physically transformed into tiny particles (Figure 5.3). Urban aerosols are added to pre-existing aerosol concentrations injected by natural sources including

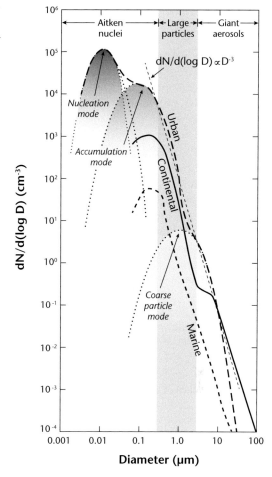

**Figure 5.19** Aerosol number distributions for urban, continental rural and maritime air (Modified after: Hobbs, 2000 and Whitby, 1978).

wind-blown dust, salt from sea spray, forest fires, and biological emissions, including pollen and particles such as viruses and bacteria.

Figure 5.19 shows the number distribution (number per volume of air) of aerosols typically present in three different types of air mass: urban, rural and marine. Clearly the total aerosol loading is much higher in urban areas than in the other types and the largest component of the modified urban distribution is in the ultrafine, and Aitken size ranges. In turn, the aerosol load of rural air is an order of magnitude greater than that of marine air. In all cases, particle concentration drops off rapidly as particle size increases.

The number distribution is composed of three modes. The nucleation mode comes from both primary combustion products and gas-to-particle conversion (Figure 5.3). The number distribution is dominated by this mode, but it represents relatively little surface area or volume. The accumulation mode combines **condensation** of gases onto existing aerosols, notably the oxidation of sulfur dioxide ($SO_2$) emissions, coagulation of nuclei and particles that remain following evaporation of cloud droplets. This mode tends to become more prominent as the aerosol load ages. The coarse particle mode includes aerosols generated by industrial processes and road and construction dust.

Within urban areas, the size distribution of aerosols shows strong spatial variability in both the UCL and UBL. In the UCL, proximity to major sources, such as roads or point source emissions can lead to an order of magnitude increase in aerosols (Seinfeld and Pandis, 2006; Chapter 11).

In the UBL, airborne traverses during **METROMEX** in St. Louis, United States (see Section 10.3.1), reveal spatial variations in the concentration of Aitken nuclei associated with plumes from point sources such as power plants, refineries and along major highways In Figure 5.20 these plumes are superimposed on background concentrations of just under $10 \times 10^3$ cm$^{-3}$. Further downwind of the city the plumes merge to form a polluted **UBL**.

The chemical composition of urban aerosol is diverse because of the multiple sources, but generally they are rich in sulphates, benzene-soluble organic matter, nitrates and a long list of trace metals, including iron, lead, zinc, manganese, barium and carbon (soot). Sodium chloride (from sea salt) is a significant component in coastal cities.

Cities are also the source of many radiatively-active gases such as carbon dioxide ($CO_2$), methane ($CH_4$), nitrous oxide ($N_2O$), water vapour, halocarbons (CFCs) and others (see Section 13.1). Further, in warm, sunny conditions photochemical reactions can transform **primary pollutants** to form $O_3$ in the **urban plume** (Section 11.3). These radiatively-active gases selectively absorb shortwave radiation, and both absorb and emit in certain longwave bands (Figure 5.2b).

### 5.3.2 Shortwave Radiation

#### Processes

Urban changes to aerosol, gases, humidity, cloud and the temperature structure create an atmospheric layer

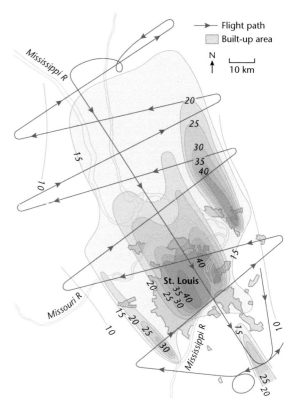

**Figure 5.20** Aitken nuclei concentrations in the St. Louis, United States area determined from airborne measurements at 450 m above mean sea level between 1400 and 1800 CDT 9 August, 1976 with winds 5 to 6 m s$^{-1}$ from 140 to 170° (Modified after: Komp and Auer 1978; © American Meteorological Society, used with permission).

that affects scattering, reflection and absorption of shortwave and the reflection, absorption and emission of longwave radiation. Figure 5.21 conceptualizes these effects for the case of a homogeneously polluted UBL and simplifies what is actually a continuum into three subsystems: the surface, the polluted layer and the clean atmosphere.

Incoming shortwave radiation at the top of the polluted layer (Flux 1) encounters enhanced concentrations of aerosols and gases that increase scattering and absorption. As a result a greater fraction of $K_\downarrow$ is scattered: both backwards (Flux 2) and forwards towards the urban surface. The flux transmitted to the surface (Flux 3), is attenuated in total energy and includes an increased fraction of diffuse ($D$). The

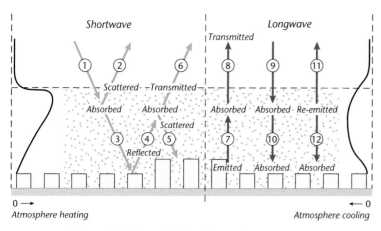

**Figure 5.21** Schematic of the radiative effects of a polluted UBL on short- (left) and longwave (right) radiation processes. Note in practice not all of these radiation streams can be measured separately, and although drawn as a single stream the transfer is often diffuse. Similarly, whilst processes of scattering, absorption and emission are drawn at a specific point, in reality they are continuous through the layer. The profiles on each side anticipate the sign and shape of the net impact of the radiation exchange in changing air temperature. See text for explanation of numbered exchanges (Modified after: Atwater, 1971; © American Meteorological Society, used with permission).

shortwave irradiance is reflected by the urban surface (Flux 4), it travels back up through the same layer and undergoes similar scattering and absorption again. This results in a flux of backscattered radiation at the surface (Flux 5), the remainder is transmitted out of the top (Flux 6). In Figure 5.21 processes are necessarily truncated but in theory Flux 5 is re-reflected *ad infinitum*. The net result is likely to cause slight warming of the polluted layer, especially near its top where the strength of the incoming stream is greatest (see the simplified **profile** at left).

In the right-hand longwave panel of Figure 5.21 each of the three subsystems has a different ability to emit and absorb radiation, depending on its emissivity and temperature. The surface emits upward (Flux 7) some part of which is absorbed, adding to the temperature of the layer, whilst the rest is transmitted (Flux 8). Flux 9 is the longwave input from the sky which similarly is partially absorbed and partially transmitted through to the surface (Flux 10). The polluted layer itself emits both up- and downward (Fluxes 11 and 12, respectively). The combination of the absorbed short- and longwave radiation in the layer may result in net cooling, due to longwave emission (profile at right).

The processes captured in Figure 5.21 have several implications for measurement of radiation fluxes in the UBL. For example, it is probable that a measured

albedo depends on the height of the sensors (Chapter 3). At greater heights absorption and scattering in both the layer beneath, and in the rest of the polluted layer above, skew the ratio relative to that at the surface. Estimates of 'urban' albedo are often made using up- and down-facing sensors mounted on aircraft or satellites flying well above the polluted layer. The observed $K_\uparrow$ therefore includes both surface reflection and backscattering emanating from below. Such albedo estimates are likely to be lower than ones measured on towers because the former combine upwelling radiation (Fluxes 2 and 6), that has travelled both down and up through the polluted layer, with the initial stream (Flux 1) that has experienced no UBL **attenuation**. If the sensors are at two different heights (e.g. one on a building for $K_\downarrow$ and one on an aircraft or a satellite for $K_\uparrow$) the validity of the ratio is compromised even more.

### Visibility

The reduction of **visibility**, due to polluted air is, regrettably, almost a defining characteristic of cities. Visibility is judged by the ability of an observer to perceive objects in the distance. It might be thought that decreased visibility is caused by absorption, but in fact the main process responsible is the scattering of shortwave radiation by aerosols in the visible part of the spectrum (0.40–0.70 μm). Polluted air between the viewed object and the observer diffuses sunlight and thereby it gains luminance. The scattered light is added to that coming from the object and reduces the contrast with its surroundings. It is this drop in contrast that makes it difficult to distinguish a distant object, causing it to appear hazy. The effect increases with distance to the object (explaining the tendency for more distant features to appear lighter) and is even greater if the Sun is in front of the observer (i.e. in the direction of the target object).

Scattering depends on the size, shape and transmissivity of the aerosols present and the wavelength of the radiation. Rayleigh scattering applies when the gas molecule or particle is less than one-fifth the wavelength of the light; it diffuses light in a fairly symmetric pattern around the object and the intensity of scattering varies inversely in proportion to the fourth power of wavelength. In a clear atmosphere, made of air molecules and small particles, blue light is scattered most effectively.

When the sizes of aerosols approach or exceed the wavelengths of light, Mie scattering dominates. Urban atmospheres have high loadings of aerosols in this size range and these particles are much more effective at scattering light than are air molecules. Mie scattering is not strongly wavelength dependent so the scattered sunlight is whitish and the sky becomes a lighter blue, eventually becoming milky and hazy in appearance with more aerosols. The shape of the scattered light around the object is asymmetric with a preference for 'forward' scattering (i.e. in the original direction of the ray). This increases with the size of the object, but at sizes greater than the wavelength of the light the 'forward' portion becomes much larger than the 'backwards' part. The net result is that in dirty atmospheres overall scattering increases, more of the scattered energy arrives at the surface and visibility decreases. The enhanced forward scattering is noticeable; the haze looks much brighter when the observer looks towards the Sun.

Of course near sunrise and sunset, when the Sun is near the horizon, the beam has to pass through a much greater path length than during the rest of the day. Path length varies as the sine of the solar altitude angle ($\beta$), so if the Sun is directly overhead the optical air mass of the atmosphere equals 1. When the Sun is at $\beta = 30°$ above the horizon the path length is exactly doubled, and at $\beta = 20°, 10°, 5°$ and $2°$ the path is 2.9, 5.8, 11.5 and 28.7 times greater, respectively. With a long path most shorter wavelength light is scattered

away leaving more of the longer wavelengths, rich in oranges and reds to come through. If the air is also polluted only the reds make it through, this is what gives deep red sunrises that end later, and sunsets that start earlier. Indeed, at the height of the Industrial Revolution, in smoky industrial cities, it is reported that the Sun only appeared as a dull red ball, even in the middle of the day, and light levels were poor all day.

Visibility is empirically related to the particulate mass density ($\mu g\ m^{-3}$). An approximate rule-of-thumb is that the visual range (in km) is given by $1000/\rho_{TSP}$, where $\rho_{TSP}$ is the concentration of total suspended particulate matter (TSP). Since typical urban TSP values are about 100 $\mu g\ m^{-3}$ that means visual ranges are only about 10 km (Turco, 2002). In the centre of polluted cities visibility is often between 5 and 10 km, whereas typical rural annual mean visibilities are 35–50 km and in remote natural areas visual range may be > 200 km. Time trends of measured visibility clearly show degradation in urban areas during periods of urban growth and/or increased particulate emission and visibility improvement follows implementation of clean air management. Using the simple rule above suggests that halving the TSP through the implementation of an air management plan can double visibility.

In the 10–15 years after WWII, industrial reconstruction in central Tokyo, Japan, led to an increase in the number of days with poor (< 1 and < 5 km) visibility (Figure 5.22). Following changes to the types of fuel and greater regulatory control there was a marked decrease in days with poor visibility. By the mid-1960s good (> 50 km) visibilities started to increase. Mt. Fuji, about 100 km from central Tokyo, could only be seen on 30 days in 1967, but by 1985 the average number of better visibility days increased to about 80 days a year. More recently, increased photochemical pollution has decreased visibility in Greater Tokyo again, with the magnitude of effects depending on where the monitoring station is located relative to the urban pollution plume.

At night cities 'glow' because of the myriad outdoor artificial sources of light they contain. This fugitive light 'pollution', which is evident to an observer distant from the city, is due to backscatter from particles, haze and clouds over the city. Besides needlessly wasting energy, the backscattered light reduces the visibility of the night sky from the ground. Because of this, astronomical observatories are located well

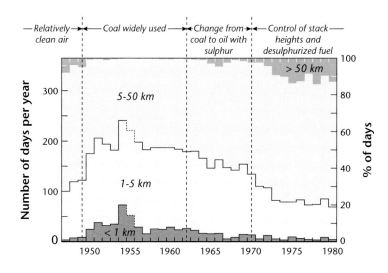

**Figure 5.22** Trends of atmospheric visibility in central Tokyo, Japan, during periods of emission and air management change. Data is missing for 1955 and has been interpolated here (Modified after: Kawamura, 1985).

away from cities. There are also concerns that artificial light at night disturbs circadian rhythms that trigger and regulate biological activities, confuse the navigation of birds and insects and force abnormal plant growth around light sources.

### Spectral Filtering and Directional Effects

The wavelength dependence of scattering and absorption by gaseous molecules and aerosols and their relative abundance in the UBL, combine to impose a spectral filter on the transmission of shortwave radiation. Spectral filtering is not easily characterized for several reasons: the rich mix of aerosols and gases in the UBL each of which absorbs and scatters in different parts of the shortwave spectrum; each city has its own unique mix; and in any one city there are temporal changes in optical characteristics due to atmospheric dispersion conditions and time of day.

Nevertheless some general characteristics can be discerned. In all cities the shortest wavelengths are attenuated most. Therefore at the bottom of the UBL $K_\downarrow$ is most deficient in the ultraviolet (UV, 0.15–0.38 µm), then in the visible (VIS, 0.38–0.7µm) and least in the near infrared (NIR, 0.7–about 3 µm), relative to non-urban air. This filtering effect is greatest at longer path lengths (i.e. larger zenith angles) and least near midday (e.g. Figure 5.23). On the other hand, due to greater scattering, diffuse shortwave irradiance is affected in the reverse sense: enhancement is greatest in the NIR and least in the UV.

The wavelength dependence is particularly important in the UV which can be subdivided mainly on the basis of its biological effects:

- **UV-C** ($< 0.28$ µm) which is extremely hazardous to biological organisms but is completely absorbed by stratospheric ozone ($O_3$) and oxygen ($O_2$) and hence does not reach the ground.
- **UV-B** (0.28–0.32 µm) which is hazardous and contributes to sunburn, skin cancers and damage to materials, crops and marine organisms.
- **UV-A** (0.32–0.38 µm) **exposure** to which is essentially considered 'safe', although it may contribute to immune deficiency.

It follows that the preferential reduction of the shortest wavelengths in the shortwave spectrum is positive in terms of biological harm from radiation, but it arises from the abundance of gases such as $O_3$, $SO_2$ and $NO_2$ that are toxic in other respects. Aerosols also contribute to reduction of UV radiation in polluted atmospheres. The effects are highly variable depending whether the particles mainly absorb (e.g. soot) or scatter (e.g. sulphates). In urban atmospheres enriched with nitrogen dioxide, strong absorption of blue wavelengths light leads to a brownish tint to the sky.

### Urban–Rural Differences in Shortwave Receipt

The gloom of heavily polluted cities has been the stuff of poets and artists for centuries. Conditions were dire

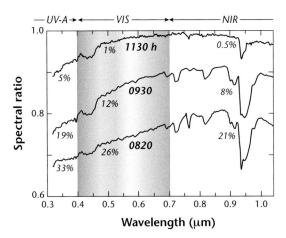

**Figure 5.23** The proportion of shortwave irradiance at a station in Athens, Greece compared to that received outside the city by wavelength (from 0.3 to 1.1 μm) at three times on cloudless days. Notice the reduction is greatest at the longest path length (earliest time) and the shortest wavelengths (Modified after: Jacovides et al., 2000; © American Meteorological Society, used with permission).

in the European Industrial Revolution and this lasted well into the last century. Such conditions still prevail in cities with a high density of heavy industries using old technologies and/or where air management is weak.

Urban-rural differences of shortwave radiation, including the role of spectral filtering are seen in Figure 5.23. It shows $K_{\downarrow U-R}$ measured at two stations, one near the centre of Athens, Greece the other in the less polluted countryside, about 15 km outside the built-up area. In the morning, with a longer path through the polluted UBL, UV wavelengths are reduced at the city site by about 33%, VIS by about 26% and NIR by about 21%. By midday the UV loss is only about 5% and at greater wavelengths depletion is minor ($\leq 1\%$).

Measured urban-rural differences of UV radiation also vary spatially, depending on the location of the station relative to the polluted air mass, which in turn relates to the local ventilation climatology. UV is significantly depleted in the polluted UBL over Los Angeles, United States and the $O_3$ polluted air mass advects inland during the day due to the **sea breeze** circulation (Chapter 12). Peterson et al. (1978) showed the average reduction of UV (0.295–0.385 μm) to be from 11 to 20%, with higher values further inland.

Over the same period the full shortwave spectrum (0.285–2.80 μm) differences were 6 to 8%. Other studies on heavily polluted days showed depletion could be as great as 90%. Stringent controls on urban emissions since the 1970s have greatly improved air quality in Los Angeles.

The range of $\Delta K_{\downarrow U-R}$ between cities is large: from 0 to −33%. Several reasons probably contribute to this variability; some are real UBL differences between city atmospheres whilst others are simply the result of methodological differences and difficulties. The latter include the flaws inherent in using the urban-rural difference approach (Chapter 2): problems using fixed observing sites to assess the properties of polluted air that advects with the wind; separating $\Delta K_{\downarrow U-R}$ due to pollution from those caused by water vapour and cloud; differences in the sample periods used by the studies; and variation of emissions over time due to changes in technology, enforcement practice and a city's economy and population (e.g. Tokyo, Figure 5.22). As a result we do not advise making detailed inter-city comparisons. For example, in virtually all cases the UBL depletes $K_\downarrow$ compared to the non-urban atmosphere. The largest attenuations (say > 10%) are found in cities with significant manufacturing, refining and heavy industrial activities where particulate levels are high. Large attenuations were more commonly observed in studies conducted in the mid-twentieth century when controls on emission sources were weaker or non-existent; there are still cities where this is true. Most modern, less industrialized, cities are characterized by photochemical and fine particulate pollution where reductions of 7% or less are typical. In many cities depletion follows a seasonal variation as do air pollutant concentrations. Those patterns are driven by changes in emissions, especially for space heating, and the local dispersion climatology, which depends on seasonal weather especially **stability** and winds. Over short periods with poor dispersion, depletion of about 30% is still possible in several heavily industrialized cities.

### 5.3.3 Longwave Radiation

Incoming longwave radiation at the surface is often slightly greater in the city. In general, daily $\Delta L_{\downarrow U-R}$ in a large city is typically about 20 W m$^{-2}$, which translates to an increase of 5–10%. The results for Toulouse, France (metropolitan population ~1 M) are a good example of summer values based on

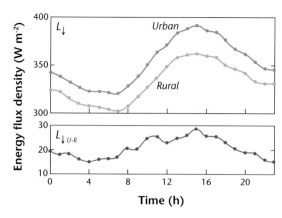

**Figure 5.24** Urban and rural fluxes of $L_\downarrow$ and their difference in Toulouse, France in May-July, 1979 (Source: Estournel et al., 1983; © American Meteorological Society, used with permission).

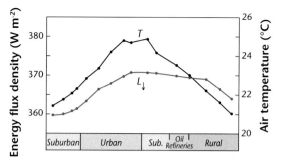

**Figure 5.25** Urban effect on incoming longwave radiation. Transect of $L_\downarrow$ and air temperature across Montreal, Canada, on the nearly cloudless night of June 8, 1970 (Source: Oke and Fuggle, 1972; © D. Reidel Publishing Co., used with permission of Springer).

observations at two fixed stations (Figure 5.24). Values tend to be somewhat greater in the daytime than at night, but with cloudless skies differences can be as large as 40 W m$^{-2}$ at either time of day.

The physical cause of the increase in $L_\downarrow$ is interesting. It was assumed that pollution was chiefly responsible by absorbing much of the outgoing longwave from the surface and re-emitting a significant fraction back (an urban 'greenhouse' effect). However, work in Montreal, Canada, suggests the role of air pollutants is minor compared to that of the heat island effect that results in a warmer UBL that is largely due to convective processes (Chapter 6). So it appears the extra $L_\downarrow$ originates from an atmosphere warmed largely by non-radiative processes. This accords with the observation that transects across a city show the trends of air temperature and $L_\downarrow$ to be similar (Figure 5.25).

On the other hand, results from Hamilton, Canada, suggest an entirely different explanation for the large daytime values of $\Delta L_{\downarrow U-R}$ observed when the city contained large steel mills (now closed). The aerosol laden UBL depleted $K_\downarrow$ and simultaneously increased $L_\downarrow$ (Figure 5.26); when converted into radiative flux densities the shortwave loss is numerically almost equivalent to the longwave gain at the surface. This suggests that absorption of shortwave radiation warms aerosols which emit an equivalent amount of longwave radiation, partly down to the surface and partly to higher layers of the atmosphere. The reason why the effect is so large in Hamilton is thought to be its

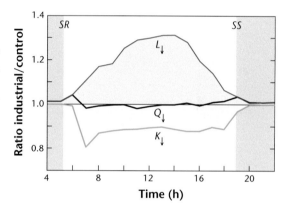

**Figure 5.26** Differences of incoming short- ($K_\downarrow$), long ($L_\downarrow$) and incoming allwave ($Q_\downarrow$) radiation between an industrial site in Hamilton, Canada, and a control (non-urban) site $SR$ – sunrise; $SS$ – sunset (Source: Rouse et al., 1973; © American Meteorological Society, used with permission).

relatively heavy particulate loading and perhaps the heat and cloud plumes from its factories. Hence this may be a relatively anomalous case.

## 5.4 Surface Net Allwave Radiation Budget

### 5.4.1 Urban–Rural Differences of Net Radiation

All terms comprising the surface radiation budget (Equation 5.5) are altered by the surface and

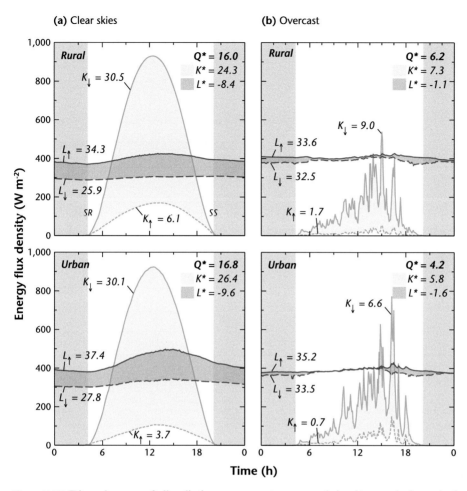

**(a) Clear skies**

**(b) Overcast**

**Figure 5.27** Diurnal course of all radiation components measured simultaneously for a single day above a rural unmanaged grassland surface (top row, LCZ D, 'Va08r' see Table A2.2) and a residential suburb (LCZ 6, 'Va08s', see Table A2.1) of Vancouver, Canada. Left panels **(a)** of each pair on a clear-sky day (June 30, 2009) and the right panels **(b)** on a mostly overcast day with some cloud breaks in the afternoon (June 24, 2009). Numbers are daily totals in MJ m$^{-2}$ d$^{-1}$. The total of the dark grey shaded area is the net shortwave gain $K^*$, the total of the hatched area is the net longwave loss $L^*$.

atmospheric changes wrought by urban development. Figure 5.27 illustrates modification of individual components of the radiation budget for a suburb of Vancouver, Canada, relative to a nearby rural area. In clear conditions (a) $K_\downarrow$ in the urban area is depleted and $L_\downarrow$ is enhanced by the polluted and warmer UBL in comparison to the rural area. The upwelling fluxes $K_\uparrow$ and $L_\uparrow$, which include the effects of the urban surface, are also altered: $K_\uparrow$ is reduced due to the lower urban albedo 0.12 compared to 0.2 in the rural area and $L_\uparrow$ is usually enhanced in comparison with rural values because of the persistent warmth of

the urban surface (Chapter 7). Maximum daytime increase of $L_\uparrow$ is typically ~50 W m$^{-2}$; nocturnal losses are consistently greater by a similar amount (Figure 5.27).

Hence, in both the short- and longwave bands changes to the downwelling flux are compensated in the corresponding upwelling flux. Reduction of the shortwave input due to aerosols, is more than offset by the greater absorptivity of the surface, resulting in a small urban increase in $K^*$ relative to the countryside ($K_U^* = 26.4$, $K_R^* = 24.3$ MJ m$^{-2}$ d$^{-1}$) in Vancouver. Similarly, the increased longwave emission from

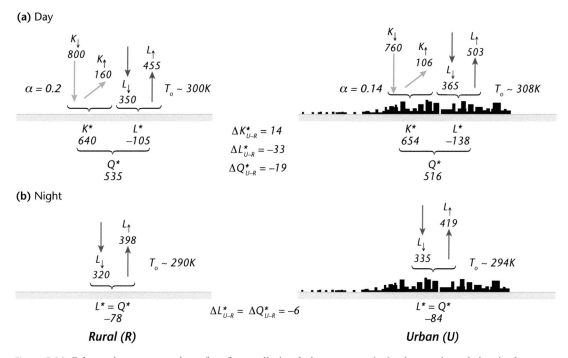

**Figure 5.28** Schematic representation of surface radiation balances at typical urban and rural sites in the mid-latitude at (a) midday and (b) night on a cloudless day in summer. Assumed input conditions: rural $K_\downarrow$ = 800 W m$^{-2}$, urban depletion of shortwave irradiance 5%, constant urban enhancement of $L_\downarrow$ = 15 W m$^{-2}$, UHI$_{Surf}$ = 4 K by day and 8 K by night, albedo and emissivity values: rural 0.2, 0.96, urban 0.14, 0.95, respectively.

the atmosphere is more than outweighed by that from the warmer city, so $L_U^*$ is more negative than $L_R^*$ ($L_U^*$ = −9.6, $L_R^*$ = −8.4 MJ m$^{-2}$ d$^{-1}$). Again, offsetting between $K^*$ (larger in the city) and $L^*$ (more negative in the city) means $Q^*$ differences between the city and the rural area are small (< 1 MJ m$^{-2}$ d$^{-1}$) (Figure 5.27). This is reinforced by the fundamentally conservative nature of $Q^*$ in general. Surfaces with unusually high (low) shortwave absorptivity are hotter (cooler) than average, and consequently $L_\uparrow$ is higher (lower) than average.

Figure 5.28 summarizes the constituent radiative fluxes for a typical urban area in the mid-latitudes on a summer day. It demonstrates that the loss of shortwave input caused by the UBL pollution, can be more than recovered by the city's better absorptivity. Whereas the greater longwave input is more than offset by the extra emission by the UHI$_{Surf}$. Taken together the city absorbs slightly less allwave radiant energy. At night the UHI$_{Surf}$ causes a slightly greater net radiant loss.

Cloud cover exerts strong control on the time series of incoming fluxes and the absolute magnitude of daily totals. In Figure 5.27b urban–rural differences

in cloud cover influence $K_\downarrow$ such that the urban area has a larger relative decrease compared to the rural area than for the clear sky case. Longwave differences are similar to the clear sky case, with greater $L_\downarrow$ and $L_\uparrow$ in the urban area. The net effect is smaller urban $Q^*$, due to differences in cloud cover. When averaged over longer periods, in the absence of systematic urban–rural differences in cloud cover, the differences in urban and rural radiation budgets remains similar to the clear sky case.

That urban–rural differences in $Q^*$ are small has been confirmed in studies across a range of cities. Further, results show that urban–rural differences virtually disappear over the annual period.

Satellite-based studies also confirm the city is hard to distinguish as a distinct $Q^*$ feature compared with its surroundings. Localized spatial differences do occur, for example, between districts dominated by dry surfaces with high albedo (e.g. concrete, bare soil and scrub) and others that are wetter and have a lower albedo (e.g. heavily treed and irrigated areas). However, a survey of $Q^*$ in Vancouver found intraurban spatial differences to be typically < 5%.

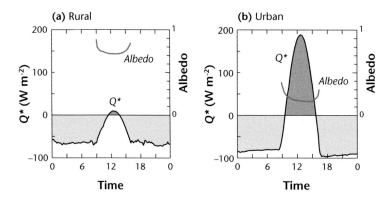

**Figure 5.29** Net allwave radiation ($Q^*$) and surface albedo for a city centre (LCZ 2, average of 'Ba02u1' and 'Ba02u2', see Table A2.1) and a rural site (LCZ D, 'Ba02r3' see Table A2.2) in and near Basel, Switzerland on the clear day of January 2, 2002 with 0.2 m snow cover (Modified after: Christen and Vogt, 2004).

Arnfield (1982) modelled the $Q^*$ of Columbus, United States through the year. He concluded that using combinations of typical urban and rural radiative and geometric properties, together with plausible urban–rural differences in short- and longwave irradiance, can result in *either* an increase or a decrease, in $\Delta Q^*_{U-R}$, and that in a given city the sign of the difference can change with the season.

The greatest urban-rural contrasts arise after snowfall, because of spatial variations in snow accumulation, soiling and removal. The lower urban albedo means $\Delta K^*_{U-R}$ is positive and so is $\Delta Q^*_{U-R}$. In Figure 5.29 fresh snow produced a $\Delta a_{U-R}$ of about −0.6 and the significantly lower urban albedo enabled $Q^*$ gains of ~200 W m$^{-2}$ at midday. At night $\Delta L^*_{U-R}$ is negative, because the **insulation** provided by the continuous rural snow cover reduces surface cooling,

whereas the warmth and patchiness of the snow cover in the city maintains a relatively larger $L^*$.

Desert cities provide an analogous but less strong urban–rural differential. Desert sand, bare soil and scrub have a higher albedo (lower $K^*$ gain) than a city, but because of its lower **thermal admittance** a desert has a higher daytime surface temperature (therefore greater $L^*$ loss) compared to a city with its paved surfaces, buildings and irrigated vegetation. Satellite analysis shows $\Delta Q^*_{U-R}$ as large as 125 W m$^{-2}$ in Dubai, United Arab Emirates (Frey et al., 2007).

As emphasized in Chapter 2, all urban–rural difference measures, are subject to times and places when and where the difference depends more on the properties of the rural 'reference' than those of the city. This is certainly true for $Q^*$, where differences depend on so many surface and atmospheric factors that vary.

## Summary

The receipt of solar (shortwave) radiation at Earth's surface is the primary driver of near surface climates. A city presents a convoluted and multi-faceted surface. This allows radiative exchanges between facets. To unpack the many exchanges it is useful to distinguish between those below roof-level, and those across the top of the urban canopy.

- Much of our understanding of radiation in the **urban canopy layer** (UCL) is based on idealized streets or **canyons** distinguished by their orientation and aspect ratio ($H/W$). Together, these measures regulate access to the solar beam and the extent of multiple reflections among facets. There are large differences of $K^*$ reaching different facets, depending on solar geometry, facet slope and aspect. On average, compared to a flat surface, urban facets absorb less shortwave radiation per unit horizontal area because of the greater surface area created by the canyon configuration.

- The aspect ratio also regulates the extent of horizon screening. This is summarized by the **sky view factor** ($\psi_{i\to sky}$), which decreases as $H/W$ increases; $\psi_{i\to sky}$ partly governs longwave radiation exchanges between a canyon facet and the usually colder sky. Where $\psi_{i\to sky}$ is small, longwave radiation exchanges among the facets dominate, and at night $L^*$ is diminished, which keeps facet temperatures relatively warm.
- At the canyon top, if the contributions of facets are aggregated and weighted according to their area, any distinguishing features of individual facets are eradicated and the pattern of exchanges become similar to those of a flat surface. The effect of **increasing $H/W$ is to lower the albedo** of the canyon and inhibit nocturnal cooling; hence net allwave radiation ($Q^*$) is more positive during daytime and more negative at night.
- At the top of the urban canopy, radiative properties (albedo and emissivity) depend on the relative proportion of roof to plan area. In densely built areas (e.g. LCZ 1) the roof area may comprise the majority of the urban surface and dominate the radiation budget, but in suburban areas (e.g. LCZ 6) the built area fraction may only be small, so the budget of the non-built fraction dominates.
- The urban boundary layer (UBL) is burdened with **aerosols** hence visibility is reduced, **solar receipt at the surface is diminished** and a larger proportion of it is diffuse. Urban–rural differences of $K_\downarrow$ are usually negative, but the values depend on the aerosol content of the UBL. Urban–rural differences in $L_\downarrow$ are positive (typically $< 10\%$), primarily because the UBL is warmer than the rural ABL; the effects of aerosols are secondary. Similarly urban–rural differences in $L_\uparrow$ are also positive, because the urban surface is commonly warmer.
- At the top of the UCL, despite these differences there is usually **little urban–rural difference in $Q^*$**. It is a remarkable finding that, despite all the complex exchanges within the UCL, net allwave radiation at the top of the UCL is little different from that of surrounding regions, at least at the scale of the whole city. Although each term reveals a distinct urban effect, the nature and direction of the changes tend to offset one another. Hence, we must look to the other fluxes of the surface energy balance, such as anthropogenic heat, heat storage change and the turbulent exchanges of heat and water, to explain urban climate effects on heat. They are the focus of Chapter 6.

Knowledge of radiation transfers within the UCL has important applied value to **building energy management**: to assess energy loads on the building envelope; to design passive heating and cooling systems; to ensure access to natural light and; to aid in the optimal placement of solar panels to harvest energy. It is also important in the design of outdoor spaces because issues of **solar access and/or shade** can be critical to ensure whether a place is comfortable for outdoor activities. In addition, the selection of materials with specific radiative properties (such as light- or dark-coloured roofing and road materials) can be effective ways to mitigate thermal concerns at the microscale, and if extended to entire neighbourhoods, may play a significant role in city-scale urban energy and climate management. Some of these ideas are addressed in Chapter 15.

# 6 | Energy Balance

**Figure 6.1** An extreme case of the replacement of natural properties (surface geometry, materials) by built ones (including anthropogenic heat for space cooling) that radically alter the surface water and energy balances. An alley (microscale) in Singapore, which is otherwise renowned for the abundance of vegetation (especially trees), throughout the city (Credit: L. Wee/Getty Images; with permission).

The **surface energy balance** (SEB) is the fundamental starting point if we are to understand and predict surface **microclimates** and climates of the **atmospheric boundary layer** (ABL). It is a statement of the conservation of energy which is applicable to surfaces and volumes at all spatial and temporal scales. It is used here to assess the transfer and storage of energy within an urban system and between that system and the atmosphere. For urban systems, energy balances can be written for individual **facets** (roofs, walls, roads etc.), for urban elements immersed in the urban atmosphere (human body, buildings), for the entire surface-atmosphere interface, or for selected layers of the atmosphere.

Figure 6.1 is an arresting viewpoint from a relatively extreme canyon within an **urban canopy layer** (UCL) in Singapore. The environment is comprised entirely of hard, **impermeable** manufactured materials. The canyon formed by the buildings and road regulates aerodynamic and radiative exchanges with the overlying atmosphere (Chapters 4 and 5). The underlying substrate is sealed so that its water content is not recharged through the surface. Moreover, there is no vegetation so the normal transfer of water from the substrate to the atmosphere via plant roots and stoma cannot occur. These modifications to the **surface cover** are accompanied by the complementary actions of the city's inhabitants. Note the cables and pipes, they are conduits for energy and water supplies to and from the buildings on either side of the street. Whilst there are small openings in the walls that allow ventilation, it is apparent that much of the imported energy is used to create a comfortable indoor climate through air conditioning systems, in what is otherwise a challenging **macroclimate**. However, indoor cooling requires that waste energy is exhausted to the outdoor atmosphere (see the fan-driven exhaust systems on each cooling unit). This image illustrates several of the radical transformations of the SEB that accompanies urban development.

## 6.1 Basics of Energy Transfer and Balance

### 6.1.1 Energy Balance of a Flat Surface

In a nutshell, variations in the climate of a surface and of the ABL are driven by the SEB, which describes the net result of energy exchanges (flux densities in W m$^{-2}$) by **radiation**, **convection** and **conduction** between a facet, an element or a land surface and the atmosphere. Figure 6.2a deals with the simplest case where all heat flux densities

are restricted to the vertical — essentially a one-dimensional view at a horizontally extensive grass-land site. By selecting an extensive site we avoid complications arising from horizontal heat transport from markedly different upwind surfaces.

### The Surface Energy Balance Equation

The relevant flux densities at a non-urban land surface are: the **net allwave radiation** $Q^*$ (see Chapter 5); the ground heat flux density, that transfers **sensible heat** by conduction to the substrate ($Q_G$); and the two turbulent heat flux densities that exchange energy between the surface and atmosphere – the **sensible heat flux density** $Q_H$ and the **latent heat flux density** $Q_E$. Energy conservation means those fluxes must balance at a surface:

$$Q^* = Q_H + Q_E + Q_G \qquad (\text{W m}^{-2}) \qquad \textbf{Equation 6.1}$$

Notice that Equation 5.1 is a genuine statement of *balance* whereas for radiation (Chapter 5) we referred to the disposition at a surface as a *budget*. In a balance the sum of all terms including their sign is zero at all times, whereas in a budget it is more like a bank account, where it is normal to have a surplus or

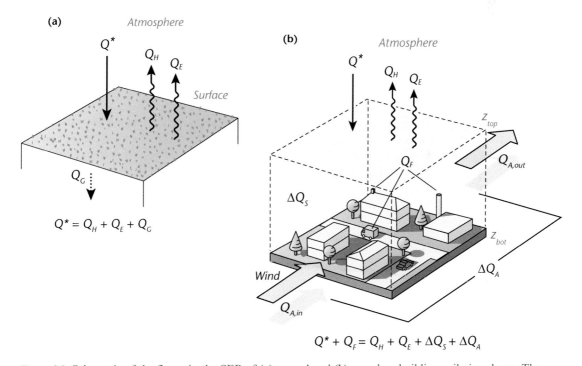

**Figure 6.2** Schematic of the fluxes in the SEB of **(a)** a rural and **(b)** an urban building-soil-air volume. The volume that extends from the top of the RSL ($z_{\text{top}}$) down to a depth where there is no net conduction over the period of interest ($z_{\text{bot}}$). Arrows are drawn in the direction the corresponding flux is considered positive. For $\Delta Q_S$ and $\Delta Q_A$, they are positive if the internal energy of the volume increases. (Modified after: Oke, 1987).

deficit. That is because radiation is only one form of energy, not a complete statement of the total energy of the system.

## Turbulent Heat Flux Densities

$Q_H$ is driven by temperature differences *between* the surface and atmosphere, and minimizes temperature differences *within* the atmosphere by mixing warmer and cooler eddies. $Q_E$ is the consequence of transporting water vapour (a mass **flux density** $E$ of water or **evaporation** in kg m$^{-2}$ s$^{-1}$) and the associated **latent heat** towards or away from the surface. By latent heat we refer to the energy that was used to vapourize the water mass. $Q_E$ and $E$ are linked:

$$Q_E = \mathcal{L}_v E \qquad \left(\text{W m}^{-2}\right) \qquad \text{Equation 6.2}$$

where $\mathcal{L}_v$ is the latent heat of vapourization (2.464 MJ kg$^{-1}$ at 15°C). Once evaporated, the latent heat remains present in water vapour by virtue of it existing in the higher energy state. It is liberated again during **condensation**, if the vapour condenses back into liquid, for example in the process of cloud droplet formation – this could happen far away from the place of evaporation. **Dewfall** is the reverse process to evaporation, it causes negative $Q_E$. In dewfall water vapour is transported to a relatively cold surface where it condenses and releases the same amount of latent heat per unit mass as it locked up in the evaporation process. Smaller, but still significant, amounts of heat accompany the processes of melting and freezing, and the two are summed if **sublimation** occurs (e.g. the liquid state is absent in the conversion). Evaporation removes energy from the local environment, causing the surface and near-surface air to cool; condensation returns it and warms the environment. The changes in air and surface temperature are measurable (i.e. it is sensible), but the temperature of the vapour remains unchanged.

The ratio of the two turbulent heat flux densities $Q_H/Q_E$, known as the **Bowen ratio** ($\beta$), is significant to a surface climate. If $\beta > 1$ it indicates that the surface or system channels more heat into sensible form, which warms the lower atmosphere whereas, if $\beta < 1$ latent heat dominates, which keeps the surface and near-surface air cooler, whilst it adds humidity to the environment.

## Temporal Evolution of the Surface Energy Balance

The pattern of radiation receipts from the Sun sets the fundamental daily and seasonal rhythms of external energy supply (Chapter 5). Equation 6.1 requires that the daytime radiation surplus at the surface ($Q^* > 0$) is conducted in the form of sensible heat into the soil ($Q_G$) or convected by **turbulent** transport into the lower atmosphere ($Q_H$ and $Q_E$). At night all flux densities usually reverse sign. The surface becomes a net emitter of radiation ($Q^* < 0$) and the surface cools, forming a temperature **inversion** in the lowest layer. That heat drain is balanced by the sum of a negative $Q_G$ conducted up from the warmer subsoil, plus negative (and smaller in magnitude) $Q_H$ and $Q_E$ transporting energy from the atmosphere towards the surface. The lowest air layers cool and perhaps dry slightly due to an intermittent flux of vapour onto the colder surface, as dew or frost.

Daytime **turbulent fluxes** and the warming and humidification of the lowest atmosphere, are facilitated by instability that aids mechanical and thermal **turbulence** (Chapter 4), whereas nocturnal stability suppresses convection. There is a phase difference between the daytime course of $Q_G$ compared to those of $Q_H$ and $Q_E$. $Q_G$ typically peaks an hour or two before $Q^*$, whereas the timing of the peak of $Q_H$ and $Q_E$ relative to $Q^*$ and relative to each other, is more complex. The phase of the curves depends on the surface properties (e.g. height of roughness elements, thermal properties) and the state of the ABL (e.g. atmospheric dryness, turbulence) (see Section 6.4.1). Similarly, these controls underlie the asymmetric timing of the resulting daily waves of surface, air and soil temperature.

In summary, the SEB determines the near-surface thermal microclimates of a site, which in turn are controlled by the amalgam of its **forcings** (largely set by the state of the ABL) in combination with the unique mix of radiative, aerodynamic, thermal and moisture properties of constituent surfaces.

## 6.1.2 Energy Balance of Urban Systems and Elements

Urban development of a previously **rural** site leads to significant disturbance of the surface geometry and properties. Natural materials are removed, replaced or modified by a new mix introduced that is often dominated by construction materials. The radiative, aerodynamic, thermal and moisture properties of these materials are radically different to natural ones leading to a greatly modified surface and very different micro- and mesoclimates of the surface and ABL.

This chapter documents these changes in properties, processes and phenomena.

## The Energy Balance of Facets and Facet Combinations

For any urban facet (roof, wall, road, ground) its SEB is similar to the flat surface in Equation 6.1. However, as noted in Chapter 2 when dealing with multiple facets that are linked together to form larger **urban units** (buildings, canyons, neighbourhoods, whole cities) it becomes impossible or inappropriate to deal with individual surfaces; rather we must deal with layers or volumes (Figure 2.16) within which there may be energy sources or sinks. The energy balance of an urban volume is not simply the sum of the energy balances of constituent facets each of which is separately coupled to the atmosphere. The facets are also coupled via radiation, wind and turbulence to each other, which creates feedback and changes their energy balance and there might be internal changes to the energy content of the entire volume, for example, due to warming or cooling of materials, air, etc. This fact is crucial in the development of **numerical models** of the urban SEB. The formulation for such a volume energy balance, that represents the response of the entire **urban ecosystem**, has to provide for three new realities: firstly, new sources and sinks within the volume need to be added; secondly, energy exchange may occur across any or all sides of the volume, not only the top; and thirdly, different facets need to interact. In contrast in Equation 6.1 the surface is considered to be just an interface between two media, it has no mass of its own.

A 3-D treatment greatly complicates the formulation of energy conservation, but it can be made more tractable in practice by defining a conceptual volume, which integrates the entire urban ecosystem. We need to assign bulk values to heat sources within the volume, including their temporal variation if known and give the volume sufficient depth into the ground ($z_{bot}$) that the vertical **heat flux density** by conduction across the bottom is negligible over the period of interest (this will be deeper for longer periods). The balance for such a volume can be written:

$$Q^* + Q_F = Q_H + Q_E + \Delta Q_S + \Delta Q_A \quad \left(\text{W m}^{-2}\right)$$

**Equation 6.3**

where, the new terms are $Q_F$ — heat released inside the volume due to human activities associated with living, work and travel (called **anthropogenic heat flux density**, see Section 6.2), $\Delta Q_S$ – the net **heat storage** change by all the **fabric** of the city including its construction materials, trees, ground and air contained in the volume (see Section 6.3) and $\Delta Q_A$ – the net energy added to, or subtracted from, the volume by wind-borne transport (**advection**, see Section 2.3.2) through any of the volume's sides (simplified to $\Delta Q_A = Q_{in} - Q_{out}$ for any combination of sides and for sensible and latent heat, see Section 6.3.4). In practice field studies and models attempt to avoid $\Delta Q_A$ by selecting or assuming an extensive and relatively homogeneous urban surface, where horizontal differences are negligible.

## Energy Balances of Urban Elements

Analogous energy balances can be written for other elements of urban systems dealt with in this book, such as a building or a human being. The analogous internal heat source for a building is the rate of energy use, including the heat needed to maintain the desired interior climate in the building. In the case of a human being the internal source is the body's **metabolic heat** as it adjusts to maintain a near constant core temperature (Chapter 14). In both cases energy conservation is usually written as the budget of an element, i.e. fluxes (in W) for the entire element, or in the case of buildings the flux per unit volume (W m$^{-3}$). Note, that to convert this to a flux density at the surface of the element, it should be referred to the *total* external area of the building or body, not just the top.

This chapter focuses on non-radiative terms of the SEB. Attention is initially on the anthropogenic heat flux density (Section 6.2), then the transfer of sensible heat to and from storage in urban fabric (Section 6.3) and finally turbulent exchanges of latent and sensible heat between the surface and the atmosphere (Section 6.4). This is followed by examples of urban SEBs at different scales and of urban–rural differences (Section 6.5). The energy balance concept allows us to physically understand, quantify and model thermal, moisture and **precipitation** changes to the urban atmosphere, as discussed in Chapters 7 to 10.

### 6.1.3 Case Studies

Field observations of SEBs gathered in four cities are used here to illustrate the diurnal behaviour of terms (Table 6.1). They are selected because the sites have

**Table 6.1** Surface properties and average energy balance partitioning for summertime, daytime (0900 – 1500h) for sites in four cities (see Table A2.1 for data sources). The residual ($Q_{Res}$) is mainly heat storage change ($\Delta Q_S$) but may also include the anthropogenic heat flux ($Q_F$) and possibly advection ($\Delta Q_A$).

| City<br>Site code [1] | Chicago<br>Ch95 | Tokyo<br>Tk01 | Basel<br>Ba02u1 | Vancouver<br>Va92i |
|---|---|---|---|---|
| **LCZ** | 6 | 3 | 2 | 8 |
| **Land use** | Open low-rise, residential | Compact low-rise, residential | Compact mid-rise, mixed use | Large low-rise, light industrial |
| **Period of observation** | Jun 14-Aug 10, 1995 | Jul 1-31, 2001 | Jun 10-Jul 9, 2002 | Aug 10-25, 1992 |
| **Mean building height** $z_H$ | 5.9 m | 7.3 m | 14.6 m | 5.8 m |
| **Building plan area fraction** $\lambda_b$ | 0.36 | 0.33 | 0.54 | 0.51 |
| **Impervious plan area fraction** $\lambda_i$ | 0.25 | 0.46 | 0.30 | 0.44 |
| **Vegetated plan area fraction** $\lambda_v$ | 0.39 | 0.21 | 0.16 | 0.05 |
| **Complete surface ratio** $\lambda_c$ | 1.74 | | 1.92 | 1.39 |
| **Canyon aspect ratio** $\lambda_S = H/W$ | 1.07 | | 1.30 | 0.57 |
| **Net allwave radiation** $Q^*$ | 545 W m$^{-2}$ | 613 W m$^{-2}$ | 466 W m$^{-2}$ | 451 W m$^{-2}$ |
| **Sensible heat flux density** $Q_H$<br>($Q_H/Q^*$) | 212 W m$^{-2}$<br>(0.39) | 291 W m$^{-2}$<br>(0.47) | 217 W m$^{-2}$<br>(0.47) | 194 W m$^{-2}$<br>(0.43) |
| **Latent heat flux density** $Q_E$<br>($Q_E/Q^*$) | 196 W m$^{-2}$<br>(0.36) | 162 W m$^{-2}$<br>(0.26) | 85 W m$^{-2}$<br>(0.18) | 36 W m$^{-2}$<br>(0.08) |
| **Residual** $Q_{Res}$<br>($Q_{Res}/Q^*$) | 137 W m$^{-2}$<br>(0.25) | 160 W m$^{-2}$<br>(0.26) | 163 W m$^{-2}$<br>(0.35) | 220 W m$^{-2}$<br>(0.49) |
| **Bowen ratio** $\beta = Q_H/Q_E$ | 1.08 | 1.80 | 2.55 | 5.42 |

[1]See Table A2.1 for more details on data sources and measurement site details.

basic commonality: they are all located in temperate climate cities of the Northern Hemisphere; summertime measurements obtained from similar tower-mounted instrument systems above roof-level in the ISL, i.e at the top of the box depicted in Figure 6.2b. They differ mainly by being located in contrasting Local Climate Zone (LCZ) classes, which means their surface properties (cover, structure, fabric, **urban metabolism**) are different (Table 6.1), especially the fraction of vegetation and the compactness of the buildings. The sites include: a well vegetated single-family detached housing area in a suburb of Chicago, United States (Ch95[1], Figure 6.3a); a closely-packed area of residential houses in Tokyo, Japan (Tk01, Figure 6.3b); an area of 3 to 5 storey, attached

apartment buildings that enclose a central area of either courtyards or gardens in Basel, Switzerland (Ba02u1, Figure 6.3c); and a light industrial area with relatively large, detached low buildings and very little vegetation in Vancouver, Canada (Va92i, Figure 6.3d). The general daily course of the SEB terms in Figure 6.3 follow similar patterns to those at non-urban sites, since both are driven by the solar cycle. However, as highlighted here, the magnitude and timing of the flux densities vary significantly between LCZ classes.

## 6.2 Anthropogenic Heat Flux

The anthropogenic heat flux ($Q_F$) is mainly the result of chemical energy or electrical energy that are converted to heat and released to the atmosphere as a result of human activities in a city. This includes

---

[1] Site codes refer to Appendix A2 where details on measurement site characteristics, instrumentation and sources are provided.

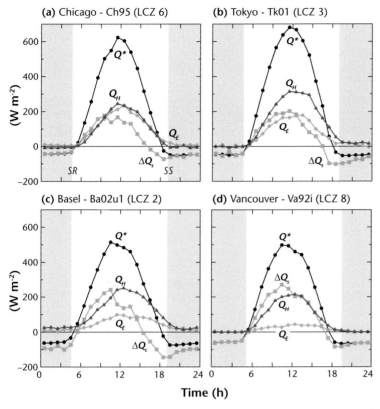

**Figure 6.3** Ensemble diurnal courses of measured net allwave radiation $Q^*$, sensible heat flux $Q_H$, latent heat flux $Q_E$ and heat storage change $\Delta Q_S$ measured in four cities at sites with different land cover and structure. All data are hourly estimates, averaged over several weeks in summertime (Sources: Chicago: Jun 14 to Aug 10 1995, Grimmond and Oke 1999b; Tokyo: 1 to 31 July 2001, Moriwaki and Kanda 2004; © American Meterological Society; used with permission; Basel: 9 to 10 July 2002, Christen and Vogt 2004; © Royal Meteorological Society, used with permission from Wiley; Vancouver: Aug 10 to Aug 25, 1992, Grimmond and Oke 1999b).

energy released during **combustion** of fuels or the energy released by 'consumption' of electricity for human activities such as space heating and cooling of buildings, the energy used in daily living such as lighting, heating water, cooking and the operation of electric appliances and even larger uses such as transport and industrial processing or manufacturing. Energy is also released by the metabolism of the people and animals that live and work in the urban system. These processes essentially convert energy that is usually not accounted for in atmospheric studies, into sensible heat, latent heat or radiation that are then 'injected' into the atmosphere.

The anthropogenic heat flux quantified, on a per-capita basis (in J cap$^{-1}$ y$^{-1}$), allows comparison of the energy efficiency of various cities. Energy use by

individuals varies greatly depending on the climate of the region, the nature of the economy, the modes of transport in use, available energy resources and behavioural/cultural differences. This energy input to the atmosphere is usually expressed as an anthropogenic flux density ($Q_F$, in W m$^{-2}$) at the top of the urban volume (Figure 6.2b) like other SEB terms. To properly model the urban atmosphere it may be necessary to know the variation of $Q_F$ over the course of a day, week or year. These variations are driven by the activity rhythms of people which are only indirectly related to solar cycles.

$Q_F$ is often categorized into three source sectors: buildings, transport and human/animal metabolic output. Energy is transferred from many fuel sources: oil, natural gas, petroleum products, other

**Table 6.2** Energy to heat conversion factors for common fuels and the percentage of energy released as latent heat. Specific energy is energy released per unit mass; energy density is energy released per unit volume. Specific energy and energy density both refer to the higher heating value (HHV), which incorporates both, the energy is released as sensible and latent heat, if all water generated by combustion is injected in vapour state into the atmosphere (Sources: Demirel, 2012 and GREET, The Greenhouse Gases, Regulated Emissions and Energy Use In Transportation Model, GREET 1.8 d.1).

| Fuel | Specific energy (MJ kg$^{-1}$) | Energy density (MJ $\ell^{-1}$) | Energy released as latent heat |
|---|---|---|---|
| **Hydrogen gas** | 143.0 | 0.0108 | 15.5% |
| **Natural gas** | 53.6 | 0.0364 | 9.7% |
| **Liquefied petroleum gas** (LPG) | 46.0 | 26.8 | 8.5% |
| **Diesel** | 45.4 | 38.6 | 6.5% |
| **Conventional gasoline** | 44.4 | 34.8 | 7.7% |
| **Biodiesel** | 42.2 | 33.5 | 7.2% |
| **Fuel oil** (residential heating) | 35.1–36.3 | 38.2–39.5 | 5.7% |
| **Coal** | 14–32 | 20–70 | 4–6% |
| **Wood** (dry) | 18 | | ~5% |

combustibles (coal, wood, peat, dung), and electricity (generated from fuel burning, water-driven turbines, solar, wind or nuclear fission). When a fuel is burnt in air it releases sensible heat, latent heat (water vapour), and other gases. The conversion to sensible and latent heat depends on the type of fuel, the efficiency of the engine and the environmental temperature and humidity. The amount of water vapour can be surprisingly large, for example, burning a litre of gasoline or diesel typically yields a kilogram of water vapour, and the equivalent release for natural gas is even larger. Some idea of the climatic significance of this is gained from the vapour condensation plumes coming from car exhausts and house chimneys on a cold day when the saturation **vapour density** is low (see Figure 8.9). Similarly striking vapour clouds occur above tall buildings due to the exhaust from their heating, ventilating and air conditioning systems or emanating from industrial cooling ponds or cooling towers (see Figure 10.1).

### 6.2.1 Estimating the Anthropogenic Heat Flux

Table 6.3 gives a generalized summary of the magnitude of $Q_F$ at city scale and **local scale**, over annual and shorter periods (for cities in climates with seasons), and according to the density of urban development. Hidden within these aggregate values is considerable temporal variation with time of day and spatially within cities due to land use, or with height. The values in Table 6.3 are not direct measurements, they are *estimates* derived via one of three approaches (Sailor, 2011):

### Top-Down Estimates

This method is essentially a detailed accounting of all known energy use in a region, city or neighbourhood. These data come from reports and databases collected by government agencies, utilities or fuel distributors, which after application of assumed fuel-to-heat conversion factors (Table 6.2) yield a heat equivalent. These are aggregate values, typically only available for longer time periods, and/or for coarser space scales than are required for urban SEB studies. Therefore, **emissions** need to be apportioned in space and time, based on other information: user patterns of activity including commuter and transport traffic by various modes; locations of businesses and manufacturing; work schedules and energy consumption habits of people; the nature of the building stock including its **insulation** efficiency. The process is laborious, and although the total flux can be relatively well quantified for a large geographic area (e.g. a nation) and over

**Table 6.3** Typical values of the anthropogenic heat flux density ($Q_F$) in W m$^{-2}$ organized by scale, season and urban density. Values in brackets refer to the typical ratio of $Q_F$ relative to net allwave radiation ($Q^*$).

**(a)** Cities (Mesoscale)

|  | Annual | | Winter | | Summer | |
|---|---|---|---|---|---|---|
| Large, high density cities | 60–160 W m$^{-2}$ | (> 1) | 100–300 W m$^{-2}$ | (> 10) | > 50 W m$^{-2}$ | (> 0.6) |
| Medium density cities | 20–60 W m$^{-2}$ | (0.4–1) | 50–100 W m$^{-2}$ | (2–10) | 15–50 W m$^{-2}$ | (0.1–0.4) |
| Low density cities | 5–20 W m$^{-2}$ | (0.2–0.3) | 20–50 W m$^{-2}$ | (< 2) | < 15 W m$^{-2}$ | (< 0.1) |

**(b)** Neighbourhoods (Local scale)

|  | Local Climate Zone (LCZ) | Hourly values |
|---|---|---|
| Large dense, city centre | 1, 2 | 100–1600 W m$^{-2}$ |
| Medium dense, city centre | 3 | 30–100 W m$^{-2}$ |
| Low density, open, low-rise | 6 | 5–50 W m$^{-2}$ |
| Heavy industry | 10 | 300–650 W m$^{-2}$ |

long time scales (e.g. a year), **downscaling** to arrive at the spatial and temporal distribution for a city is subject to many uncertainties and assumptions.

### Bottom-Up Estimates

The second approach is to monitor or model typical elements of an urban ecosystem, i.e. buildings, vehicles and people. For example, using consumption meters the energy released by a representative group of building types can be quantified. More typically buildings are modelled using a building energy model (BEM), a tool used in architectural and engineering design. BEMs incorporate both electric and fuel consumption for space heating and cooling, hot water demand, appliances and lighting of a building type, taking into account: its exterior energy balance (including meteorology, solar geometry) internal 3D structure, construction materials and the activity patterns of its occupants. BEMs generate simulations of diurnal, weekly and seasonal energy consumption, which is set by occupant behaviour and meteorological input files, such as hourly weather records assembled for a typical meteorological year (TMY) using data for months that are close to (typical of) the climatological average. The building's energy consumption can be assumed to be released instantaneously to the atmosphere (UCL) or stored temporarily in the building's structure. Running BEMs for a subset of representative building types, together with an inventory of the building stock in the area, can give

spatial and temporal estimates of $Q_F$ for the building sector.

Input from transport can be estimated from information about traffic counts, vehicle types and knowledge of fuel consumption. Estimates need to be complemented by the heat from point sources (e.g. power plants) to scale the values up to total neighbourhood-wide or urban $Q_F$ values.

### Energy Balance Residual Approach

Although $Q_F$ cannot be measured directly, it is known that it is channelled into the other SEB terms, i.e. it augments the radiative ($Q^*$), turbulent ($Q_H$, $Q_E$), advective ($\Delta Q_A$) and/or storage ($\Delta Q_S$) terms therefore, theoretically $Q_F$ can be estimated as a residual. However, this can only be done if there are year-long SEB measurements of $Q^*$, $Q_H$ and $Q_E$ made in the ISL above an **urban canopy** (Section 6.1.2). That is because over an entire year it should be reasonable to assume that storage is close to zero ($\Delta Q_S \approx 0$) and if advection is negligible ($\Delta Q_A \approx 0$) we can solve the SEB as $Q_F = Q_H + Q_E - Q^*$. However, there are several practical impediments to conducting such an analysis. For instance: the cost of undertaking long-term SEB studies is considerable; it is not easy to obtain permission to establish a tower site, or to find a sufficiently extensive site to meet the condition that $\Delta Q_A \approx 0$; and to obtain agreement between the **source areas** of the **radiometers** and the turbulent **sensors** (see Chapter 3). Even if that is done the results apply only to that site and are subject to the several sources of error that normally accumulate in a residual. Therefore, whilst this approach can provide useful comparative data, it is not usually a primary means to obtain spatial estimates of $Q_F$.

In summary, none of the three methods used to obtain the $Q_F$ values in Table 6.3 is simple to implement and all have significant limitations. Nevertheless, the results from high quality studies using these methods converge reasonably.

### 6.2.2 Controls on Anthropogenic Heat Flux

#### Space Heating and Cooling Demand

One of the most basic controls on differences of $Q_F$ is the climatic stress placed on people. That generates the need to achieve thermal regulation, despite the fact that outside it is bitterly cold or stiflingly hot. Clothing and passive design are important but in modern economies it is fuel and electrical energy that powers the active control of buildings, for both space heating and cooling. Heating warms the structure and the internal air, but eventually it drains to the exterior because the heated air seeps through leaks in the building envelope, the windows and doors, out of chimneys, hot water is flushed down the drain, and heat is conducted through walls to the air and the underlying ground. To cool the air in buildings requires energy to power refrigeration units, and to pump warm air through an evaporative cooler or a condenser, with the attendant exhaust of hot and often moist air.

The thermoregulatory effect is demonstrated in Figure 6.4 which shows electrical energy use in Sydney, Australia as a function of ambient mean air temperature. Below a critical temperature (here about 18°C) energy use increases approximately linearly as the temperature drops, because house and office thermostats trigger furnaces or radiators or open the district heating supply valve. People are comfortable and need no active climate control with temperatures between 18 and 24°C, but if temperatures rise beyond that, air conditioning use, and hence energy consumption, increase. Closer inspection will show some of the scatter is systematically due to greater use on workdays and lesser demand on weekends and holidays. The energy consumption variable in Figure 6.4 is not the same as $Q_F$ because it only derives from electrical power use, and does not include the use of other fuels like gas, oil, gasoline, coal and wood. Nevertheless in

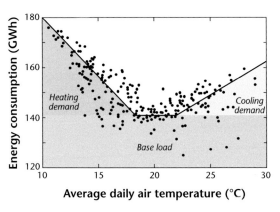

**Figure 6.4** Daily energy demand by Sydney, Australia, as met by electrical power on the grid in 1990–91 (Source: Modified based on data by Pacific Power Inc.).

this city it is likely to be the largest anthropogenic heat source, except for transportation uses.

#### Urban Density and Energy Use Efficiency

Two cities can have very different $Q_F$ values, even though their inhabitants use similar energy per capita. This is so if we compare flux densities, i.e. the energy consumed/released *per unit area*. If the same amount of energy (per building or person) is released in a densely developed area, it generates a larger $Q_F$ than if it is emitted from a dispersed settlement, simply because (with the exception of industrial areas) the density of people is greater in the former. There is a close, approximately linear, relation between urban density (expressed as population density) and $Q_F$ (Figure 6.5).

Since many, but not all, of the most populous cities of the world use most of the global energy resources, it is interesting to know if they are inefficient. In fact, amongst their economic peers and within the same cultural context, their citizens are often amongst the more efficient users of energy; they consume/release relatively less energy on a *per-capita* basis. The primary reasons for this are:

- The greater use of multi-unit apartments in denser cities means they share more walls and roofs, and use less space, which **reduces the surface area to volume ratio** of group housing, which in turn conserves heat.
- The presence of multi-unit buildings in a densely developed city, means a **more efficient land-use**

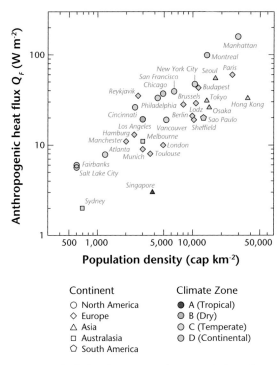

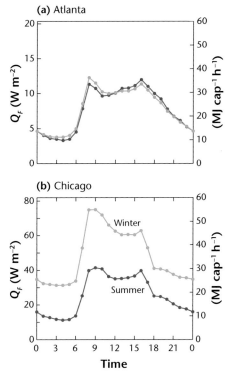

**Figure 6.5** Relation between anthropogenic heat flux density $Q_F$ and population density for cities.

**Figure 6.6** Hourly variation of $Q_F$ in **(a)** Atlanta and **(b)** Chicago, United States in both winter and summer. The right hand axis shows the per-capita equivalent of the energy released (Source: Sailor and Lu, 2004; © Elsevier, used with permission).

**mix**, aggregate travel distance (to work, school, etc.) is shortened, which reduces fuel consumption, especially for transport (see Figure 1.7 and Figure 13.3).
• The high density of commuters in dense cities makes it economic to construct **mass transit systems**, so in addition to walls they also share commuting vehicles.

Dispersed settlements, characterized by sprawl and strict spatial separation between land-uses (residential areas, commercial/industrial parks, shopping-malls etc.), are less energy efficient for the opposite reasons. The North American cities in Figure 6.5 cluster along the top of the relation; this indicates their energy release is anomalously large, for several of the above reasons.

Whilst urban density is undoubtedly the primary reason for the relation in Figure 6.5, macroclimate also plays significant role. The largest energy users at a given density live in cities at high latitude (e.g. Fairbanks, Alaska, United States; Reykjavik, Iceland) and/or in a continental climate (e.g. Montreal, Canada; Chicago and Salt Lake City, United States). The cold winter and/or large summer-winter temperature range in such places imposes greater need for space heating and cooling than in more benign lower latitude or maritime climates (e.g. Hong Kong, China; Singapore; Sao Paulo, Brazil; Sydney, Australia; or Toulouse, France).

### Temporal Variations

There are three main temporal variations of $Q_F$. These are illustrated in Figure 6.6 for the cases of Atlanta, and Chicago, United States. Firstly, in both cities, there is a characteristic double-peaked course in the hourly variation of $Q_F$ through the day. This daily shape contains two elements: the higher values in daytime are due to the cycle of human activities built around the periods for work and sleep, and extra surges of energy due to the morning and evening traffic to and from work. Secondly, what is not shown, there is a difference between energy demand on work-days and on weekends or holidays, which creates a weekly cycle on top of the daily one. Whilst the daily

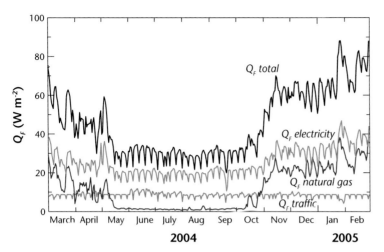

**Figure 6.7** Mean daily variation of different energy sources in Toulouse for a year, as calculated using the inventory approach (Source: Pigeon et al., 2007; © Royal Meteorological Society, used with permission from Wiley).

and weekly variations have approximately the same course in the two cities, the effect of the seasons and their absolute values are very different. There is almost no summer/winter change in Atlanta because the winter climate is sufficiently mild that space heating is hardly needed, whereas Chicago has much colder winters so winter $Q_F$ is approximately double (48 W m$^{-2}$) its summer rate (26 W m$^{-2}$). Both cities have hot summers and use air conditioning. Hence the main reasons for the absolute differences between their annual $Q_F$ Chicago (37 W m$^{-2}$) and Atlanta (8 W m$^{-2}$) are Chicago's winter climate and its much higher population density (Atlanta 1,200; Chicago 4,900 cap km$^{-2}$).

Almost a full set of cycles is illustrated by the mean daily estimates of $Q_F$ in Toulouse, France based on an inventory of electricity, natural gas and vehicle fuel use (Figure 6.7). The annual cycle is very clear as are the weekly cycles shown by the small humps. Notice also that the traffic value varies little through the year whereas the electricity, and especially the natural gas, shows a marked increase in the winter heating season.

Using a database of building energy consumption and a model of occupancy patterns Moriwaki et al. (2008) calculated $Q_F$ in a high-rise district near the centre of Tokyo, Japan. A large fraction of total $Q_F$ is emitted as latent heat not just sensible heat (see also Sailor, 2011). In central Tokyo the seasonal pattern of $Q_F$ peaks in summer, due to the intense use of cooling systems.

### Intra-Urban Variations

Figure 6.8 maps $Q_F$ in the summer and winter for Toulouse, based on a detailed inventory of sources

in space and time. As already seen in Figure 6.7, there is a marked increase in heat output in all districts in the cold season, for space heating. The distribution is closely tied to the location and density of buildings, vegetation and the traffic load on roads. Viewed at the city scale $Q_F$ generally declines along the density gradient of urban development, from centre to periphery. This approximate 'gravity' model form is found in all cities where detailed mapping of $Q_F$ has been conducted.

Less investigated, there is variation of $Q_F$ in the vertical. Tall buildings release $Q_F$ from their sides from ground level to the top. The heating and refrigeration vents and exhaust fans of these towers are located on the roof. Hence large quantities of both sensible and latent heat are injected above the UCL. Emissions from tall chimney stacks and cooling towers are also elevated and localized. This means that to comprehensively understand and model the impact of $Q_F$ requires accounting for the height of release. For example, an aggregated source inventory may indicate large emissions from a district, but if most of the output is located at roof-level or from tall chimneys it may have little impact on the thermal or moisture climate in the UCL.

### 6.2.3 Significance of Anthropogenic Heat Flux

The significance of $Q_F$ in the SEB of a city, and a neighbourhood, can range from dominant to negligible. Clearly in a town located at high latitude where the winter **shortwave irradiance** is very weak or absent

**(a)** Summer

**(b)** Winter

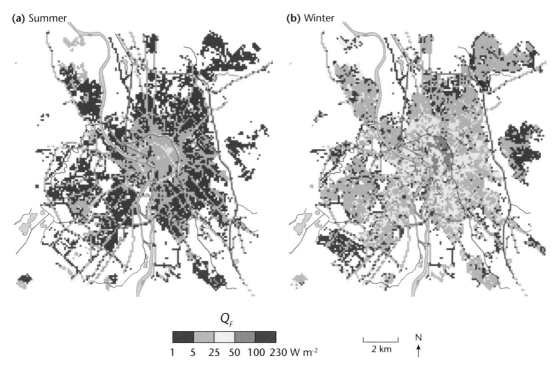

$Q_F$

1  5  25  50  100  230 W m⁻²

2 km    N ↑

**Figure 6.8** Spatial pattern of modeled anthropogenic heat flux density $Q_F$ for Toulouse. Mean $Q_F$ for **(a)** the summer months (June, July and August), and **(b)** for the winter months (December, January and February) of 2004 (Source: Pigeon et al., 2007; © Royal Meteorological Society, used with permission from Wiley).

(Arctic night) $Q_F$ can be the primary input of energy to the balance. It can become very relevant, even dominant in a heavily industrialized zone. Similarly in a city centre, with densely packed high-rise buildings, especially ones with glass facades and perhaps streets heavily used by vehicles, $Q_F$ can be $>$ 100 W m⁻². In winter cities where heating is needed, $K_\downarrow$ is often weak and nights are long, hence the 24-h ratio $Q_F/Q^*$ is greater than unity (Table 6.3). That means anthropogenic heating is larger than radiation in the SEB – a truly human-dominated climate.

Just as external air temperature is a factor in setting the need for space heating/cooling as part of the **urban metabolism**, so the release of energy via $Q_F$ has an impact on the external air temperature. The atmospheric thermal response (increase in air temperature per unit energy flux density, K W⁻¹ m²) depends on the whole SEB, including the ratio of $Q_H$ to $Q_E$ and the mixing efficiency of the air in the RSL. Indications of the direct thermal impact are emerging from numerical simulations that include both the building and RSL energy exchanges, especially if the magnitude of $Q_F$

can be included or removed. In the high-rise core of Tokyo the largest impact of $Q_F$ on external air temperature, is about 2–3 K, at night in the winter. Model results suggest that air temperature is raised by about 0.8 K per 100 W m⁻² of $Q_F$ input.

In summary, the primary underlying factor controlling $Q_F$ is the macroclimate, including the seasonal and daily rhythms it imposes on the need to maintain human comfort (Chapter 14). That basic need usually has to be met through active control of indoor climate. Secondary factors include the economic prosperity, access to resources and cultural-political system in which the city exists. They set the city's **urban form** and infrastructure (transport technologies, housing stock, industry and behaviour). Of particular importance is the value of land and/or development regulations which set the efficiency of the land-use mix, density and transport. Hence in addition to the background climate, geographical, historical and cultural factors substantially shape the energy consumption and anthropogenic heat released by individual cities, all of these contribute to the diversity seen in

Figure 6.5. Energy efficiency becomes even more relevant when dealing with emission of **air pollutants** and **greenhouse gases** (Chapter 11 and 13).

## 6.3 Heat Storage Change

The rate of heat transfer between the exterior and interior of a building and the degree of insulation provided by the building envelope, has been of relevance since the days of cave dwelling and the use of other basic building materials (earth, mud, adobe, stone and rock). The other related property of interest is the retention of heat by the envelope – its heat storage.

Urban development entails the replacement of natural materials (e.g. soils and vegetation) with construction materials (e.g. stone, glass, brick and asphalt). They are selected for several distinct properties (e.g. hardness, lightness, transparency, impermeability). Further, these materials are configured to create buildings (hollow 3D features) that regulate indoor climate by separating it from the outdoor environment. From an SEB perspective the relevant thermal properties of these materials are their ability to transfer and to store heat.

Efficient heat transfer through the building envelope and low heat storage mean a dwelling interior that is strongly coupled its external environment. On the other hand, poor heat transfer and high storage capacity mean weak coupling and a structure that retains heat and gives it up slowly, thereby creating differences between the timing of exterior and interior cycles of daily and longer-term temperature. Further simple controls on the thermal performance of the structure are the thickness of the walls and roof and the degree of insulation provided by the construction technique (e.g. attic space, wall cavities). The corollary is that cities where light structures are favoured are likely to exhibit relatively large diurnal range in surface temperature and weaker heat storage than in cities with more massive structures. The differing thermal performance of these alternatives present a wide range of architectural possibilities, none is best for all situations. Modern construction techniques and new building materials have broken down some of the traditional more simple expectations (Chapter 15).

### 6.3.1 Controls on Heat Storage Change

The earliest work on **urban climates** refers to the expectation that because urban development entails

the wholesale replacement of natural materials by construction materials, the city must be better able to store heat, and that this is a central cause of the urban heat island effect (Chapter 7). Some writers still take it to be intuitively obvious that the thermal properties of urban construction materials are markedly different from those of rural areas, or even that contrasting the properties of concrete with those of soils somehow encapsulates urban–rural heat storage differences. This section confirms that cities have greater ability to uptake and store sensible heat and to delay its release, but it is not sufficient to explain this by simple urban–rural comparison of thermal properties. The full story involves consideration of three sets of intertwined factors: thermal properties, moisture availability and geometric form (structure).

### Thermal Properties

Table 6.4 gives the thermal properties of common materials. Four different properties control distinct heat processes and thermal phenomena:

- **Heat capacity** ($C$, in J m$^{-3}$ K$^{-1}$) is the product of density ($\rho$) and **specific heat** ($c$, in J kg$^{-1}$ K$^{-1}$) of a material (i.e. $C = \rho c$) and expresses its ability to store heat and gives the temperature change to expect due to uptake or release of sensible heat.
- **Thermal conductivity** ($k$, in W m$^{-1}$ K$^{-1}$) is the ability of a substance to conduct a heat flux density (W m$^{-2}$) by passing energy from one molecule to another along a given temperature gradient (K m$^{-1}$). High $k$ simply means that relatively large amounts of heat are transmitted for a given temperature gradient whereas low $k$ implies little heat would travel for the same gradient.
- **Thermal diffusivity** ($\kappa$, in m$^2$ s$^{-1}$) is the ease with which temperature signals are transmitted through the material. In particular $\kappa$ controls the time for temperature changes to travel (i.e. the speed at which temperature waves move) and helps to estimate the depth of the layer involved in thermal changes. Thermal diffusivity is the ratio of the conductivity to the heat capacity (i.e. $\kappa = k/C$). Hence we see the speed of thermal activity is directly related to the ability to conduct, but inversely proportional to the amount of heat needed to warm or cool the intervening substance. Materials with a high $\kappa$ allow rapid penetration of surface temperature changes and a thick layer is involved,

**Table 6.4** Typical thermal properties of rural and built materials (Sources: Oke, 1987; Thornes and Shao, 1991; Crevier and Delage, 2001).

| Material | State | Heat capacity $C$ (MJ m$^{-3}$ K$^{-1}$) | Thermal conductivity $k$ (W m$^{-1}$ K$^{-1}$) | Thermal diffusivity $\kappa$ (m$^2$ s$^{-1}$ × 10$^{-6}$) | Thermal admittance $\mu_s$ (J m$^{-2}$ s$^{-\frac{1}{2}}$ K$^{-1}$) |
|---|---|---|---|---|---|
| **Natural materials** (rural and undeveloped urban sites) | | | | | |
| **Sandy soil** (40% porosity) | Dry | 1.28 | 0.3 | 0.24 | 620 |
| | Saturated | 2.96 | 2.2 | 0.74 | 2,550 |
| **Clay soil** (40% porosity) | Dry | 1.42 | 0.25 | 0.18 | 600 |
| | Saturated | 3.10 | 1.58 | 0.51 | 2,210 |
| **Peat soil** (80% porosity) | Dry | 0.58 | 0.06 | 0.10 | 190 |
| | Saturated | 4.02 | 0.5 | 0.12 | 1,420 |
| **Snow** | Fresh | 0.21 | 0.08 | 0.10 | 130 |
| | Old | 0.84 | 0.42 | 0.40 | 595 |
| **Ice** | 0°C, pure | 1.93 | 2.24 | 1.16 | 2,080 |
| **Water**[1] | 4°C, still | 4.18 | 0.57 | 0.14 | 1,545 |
| **Air**[1] | 10°C, still | 0.0012 | 0.025 | 21.5 | 5 |
| | Turbulent | 0.0012 | ~125 | ~10 × 10$^6$ | 390 |
| **Construction and building materials** in dry state (built sites) | | | | | |
| **Asphalt road** | Range | 1.92–2.10 | 0.74–1.40 | 0.38–1.04 | 1,205–1,960 |
| | Typical | 1.94 | 0.75 | 0.38 | 1,205 |
| **Concrete** | Aerated | 0.28 | 0.08 | 0.29 | 150 |
| | Dense | 2.11 | 1.51 | 0.72 | 1,785 |
| **Stone** | Typical | 2.25 | 2.19 | 0.97 | 2,220 |
| **Brick** | Typical | 1.37 | 0.83 | 0.61 | 1,065 |
| **Adobe** | | 1.50 | 0.57 | 0.38 | 922 |
| **Clay tiles** | | 1.77 | 0.84 | 0.47 | 1,220 |
| **Stone ballast** | 40% void | 1.30 | 0.86 | 0.66 | 1,058 |
| **Wood** | Light | 0.45 | 0.09 | 0.20 | 200 |
| | Dense | 1.52 | 0.19 | 0.13 | 535 |
| **Steel** | | 3.93 | 53.3 | 13.6 | 14,475 |
| **Glass** | | 1.66 | 0.74 | 0.44 | 1,110 |
| **Plaster** | Gypsum | 1.40 | 0.46 | 0.33 | 795 |
| **Gypsum board** | Typical | 1.49 | 0.27 | 0.18 | 635 |
| **Insulation** | Polystyrene | 0.02 | 0.03 | 1.50 | 25 |
| | Cork | 0.29 | 0.05 | 0.17 | 120 |

[1] Properties depend on temperature

whereas low $\kappa$ mean change is restricted to a shallow layer.

- **Thermal admittance** ($\mu$, in J m$^{-2}$ K$^{-1}$ s$^{-1/2}$; also called 'thermal inertia') is a property of a surface (interface), whereas the preceding measures are intrinsic to a volume of the subsurface material itself. In fact $\mu$ has two values one for each substance forming the interface; for the city that usually means the air and construction material or ground. The thermal admittance for air ($\mu_a$) is highly variable as it depends on the thermal diffusivity of the atmosphere and therefore turbulence, but for soil or construction materials the value is more stable, simply involving the other thermal properties, viz: $\mu = \sqrt{kC} = C\sqrt{\kappa}$. The materials are usually just solids, not a fluid.

The units of $\mu$ are difficult to get a feel for intuitively, but it is worth developing an appreciation for this property and the significance of possessing large or small $\mu$. Arguably $\mu$ is the most important thermal property in urban climate. This is because $\mu$ is directly related to both the ability of the system to store heat (e.g. Campbell and Norman, 1998) and the amplitude of the diurnal wave of surface temperature change in response to energy forcing. Further, this property governs heat sharing between the two media. In essence, when $\mu$ of the surface material is large, most of the surface heat input is used to heat the substrate (i.e. it is stored) and relatively little is left to heat the atmosphere. So surfaces with large $\mu$ sequester heat *within* the material and there is relatively small change in the *surface* temperature through the day (i.e. it has a small diurnal range). But surfaces with low $\mu$ store heat less readily; their surface temperature has large amplitude ($\Delta T_0$) and these surfaces shed large amounts of heat to the atmosphere ($Q_H$).

Sensible heat sharing between the substrate and the atmosphere is given by the ratio of the thermal admittances of the two:

$$\frac{Q_H}{Q_G} = \frac{\mu_a}{\mu_s} \qquad \textbf{Equation 6.4}$$

### Urban–Rural Differences in Thermal Admittance

Given its significance in storage and surface temperature we should ask 'do the $\mu$ values in Table 6.4 support the case that thermal properties alone explain the greater heat storage and the heat island that are observed in cities'? Undoubtedly a difference in the $\mu$ of adjacent urban and rural areas could create a distinct urban climate, but assessment is complicated by the large range of $\mu$ values for materials in both areas, and the fact both environments may have canopy layers. The urban canopy has a 'honeycomb' structure, due to the air inside buildings (Chapter 2), both rural and urban areas commonly have plant or tree canopies and can be smothered by snow cover. These canopies contain internal air spaces, which lower the density and heat storage capacity of the system, relative to the properties of the individual solid materials.

Inspection shows the rural $\mu$ range is strongly affected by the state of soil moisture (Table 6.4) or the density of any snow cover. Wet soils have $\mu$ values four to seven times greater than the same soil when dry, and a dense, old snow pack can be more than four times greater than fresh snow. In the city there is an even greater range of values for concrete and to a lesser extent for wood, depending on their density and for asphalt depending on its composition and age. Nevertheless, the temporal variation of $\mu$ is less in urban areas because variations due to soil moisture and snow cover are likely to be less in cities, given the small plan fraction of open areas and building walls remain snow free.

Certainly if we compare expectations when rural soils are dry or there is snow, the city is likely to have a greater $\mu$, but when soils are wet it is quite possible for the difference to be the reverse. In the more typical intermediate case with moist soils, rural $\mu$ are typically about 1,200 to 1,600 J m$^{-2}$ s$^{-\frac{1}{2}}$ K$^{-1}$ and urban values are likely to be similar. By analyzing the amplitude of the diurnal surface temperature wave, measured by thermal **remote sensing** it is possible to map contours of $\mu$ across urban–rural landscapes (e.g. Lewis and Carlson, 1989). Those contours do not show a distinct $\mu$ anomaly that coincides with the location of the urban area. This is suggestive, but not definitive evidence, because remotely measured surface temperatures in cities are biased by under-sampling of vertical surfaces (Chapter 7). However, we can conclude that neither tabulated values nor analysis of remote sensing confirm large urban–rural $\mu$ differences ($\Delta\mu_{U-R}$).

### Moisture Availability

The sensitivity of soil $\mu$ to soil moisture content (Table 6.4) means the variability of rural moisture is capable of creating positive or negative $\Delta\mu_{U-R}$. Further, Chapter 8 stresses the low availability of surface water in urban areas in general and the hydrologic reasons for that (e.g. vertical and sloped building surfaces, efficient water routing by gutters and sewers, 'water-proofing' of paved surfaces, etc.). As a result evaporation is usually less in a city.

### Urban Structure

The characteristically convoluted 3D form of cities, largely created by buildings, increases the **complete aspect ratio** ($\lambda_c$, Section 2.1). This benefits urban heat storage in two ways. Firstly, it increases the absolute **active surface** for energy exchange, compared to typically flatter rural surfaces. In the centre of cities with high-rise buildings $\lambda_c$ can be two or three times larger than a flat plane, the energetic impact of this is augmented by **canyon trapping**, i.e. the geometric effect on shortwave radiation via multiple reflection between canyon facets, and the reduction of **longwave** losses by screening out portions of the cold sky (Chapter 5). These features tend to increase heat absorption compared with a flat plane. Secondly, greater $\lambda_c$ means increased wind shelter (Chapter 4) and decreased sensible heat loss by turbulent transport.

Sugawara et al., (2001) estimated the effective $\mu$ of a district of Sapporo, Japan ($\lambda_c$ = 1 to 3), using remotely-sensed surface temperatures. Signals were corrected for **thermal anisotropy**, but not surface **emissivity**. They found $\mu$ to be 2 to 4 times larger than the plan area weighted mean of the component surface materials.

### Net Effect of Urban–Rural Differences

At the city scale, although urban–rural thermal property differences are typically small, seasonal changes in soil and surface wetness and the insulating properties of canopy layers can alter the sign and size of differences. If the surrounding rural area has dry soils, fresh snow or well-developed vegetation canopies, conditions favour heat storage in the city. On the other hand, if soils are wet, even flooded or the terrain is largely bare rock, differences are likely to be much smaller, or their sign may even be reversed. Similarly, a city situated on the edge of an ocean or large lake has a very large storage body on one, or perhaps several sides (see Chapter 12). Other than these special cases a pervasive feature of cities is their ability to sequester heat efficiently. This is primarily due to the convoluted geometric form of their surface and hence greater active area.

It is not as useful to summarize the situation at the local scale, because the degree of development of individual LCZs vary so greatly; all the way from low density, semi-rural districts to intensely built-up neighbourhoods. In general, the greater the density of development the greater the potential for storage.

### 6.3.2 Estimating and Modelling Urban Heat Storage Change

Similar to $Q_F$ it is not possible to directly measure $\Delta Q_S$ for entire urban canopies. Also similarly there is

no absolute standard against which estimates of $\Delta Q_S$ can be calibrated. At present there are four approaches to its estimation:

- **Energy balance residual approach** – if all other terms in Equation 6.3 have been measured or estimated for a site, $\Delta Q_S$ can be solved as the residual.
- **Thermal mass scheme (TMS)** – uses measurements of temperature change in representative, constituent materials of an urban system, and information about their thermal properties and arrangement, to calculate $\Delta Q_S$.
- **Numerical simulation** – urban canopy and building energy models are able to simulate the heat exchange and subsurface climates of layers and buildings including their heat storage change (see Chapter 3).
- **Parameterization** – description of a system's $\Delta Q_S$ using a model that includes empirical coefficients that need to be determined.

### Energy Balance Residual Approach

Implementation of this approach requires SEB flux densities of $Q^*$, $Q_H$ and $Q_E$ to be measured in the **inertial sublayer** (ISL) above an extensive urban surface with similar properties. This is technically demanding, expensive and site specific. This method is similar to the energy balance residual approach for estimating $Q_F$ (Section 6.1.1), but works on much smaller time scales (hours to days) when $\Delta Q_S$ is substantial in magnitude. The approach is to solve Equation 6.3 for $\Delta Q_S$ (i.e. $\Delta Q_S = Q^* + Q_F - Q_H - Q_E$). To do that $\Delta Q_A$ must be assumed to be small or negligible and $Q_F$ is either neglected, assumed constant or is modelled. As discussed in Section 6.2.1 any residual approach has the weakness that estimates incorporate all the uncertainties and errors inherent in obtaining the other terms.

### Thermal Mass Scheme

From basic concepts of heat conduction and heat storage in a volume $\Delta Q_S$ can be calculated for an urban building-soil-air volume, such as that in Figure 6.2b. Using measurements of temperature change by the constituent materials, and knowledge of their thermal properties and arrangement:

$$\Delta Q_s = \sum_{i=1}^{N} \Delta Q_{s_i} = \sum_{i=1}^{N} \frac{1}{A_i} \int_V C_i \frac{dT}{dt} dV \quad (\text{W m}^{-2})$$

**Equation 6.5**

where the index $i$ identifies $N$ surface types within the urban volume, $A_i$ is the surface area of component $i$ within the system, and the product $C_i(dT/dt)$ is the heat storage change in the urban fabric, which is integrated with respect to the urban volume – $C_i$ is the heat capacity (J m$^{-3}$ K$^{-1}$) of the $i$ materials, $dT/dt$ is the change in temperature over a given time period (K s$^{-1}$) and $dV$ is the volume of the material involved (m$^3$). To be complete, heat summation must include storage change by all components (building, soil, water and air) comprising the UCL. To implement the TMS scheme requires knowledge of the construction of the built volumes, the surface materials and their orientation, mass and abundance within the area of interest, plus a large sample of exterior and interior building temperatures. Usually the mass of air and vegetation is small enough, in comparison with the other components, that they may be neglected, or handled in less detail. The requirements of TMS are relatively onerous, which limits its general use.

## Numerical Simulation

Whilst heat conduction through a single material is relatively simple, that involved in heat loss and gain to and from the interior of heated or cooled buildings, and the ambient environment is relatively complex. **Urban climate models (UCM)** and BEMs are able to numerically solve differential equations for heat conduction in the different facets of an urban canopy and link them to the SEB of the facet and temperature changes.

The heat transfer is often simulated using a **resistance** network approach to handle the layers of different materials and spaces involved in the construction of walls and roofs. The resistance approach considers conduction of heat through materials as analogous with the flow of electrons (current, $I$) in an electrical circuit, as given by Ohm's Law, $I = V/r$ where $V$ is the electrical potential (voltage) and $r$ is the resistance of the wire to the flow. In our case $I$ is the conductive heat flux density through a layer of interest ($Q_G$), $V$ is the driving 'force' produced by the difference of temperatures (e.g. $\Delta T = T_2 - T_1$) across the layer, multiplied by the heat captivity $C$ of the material, and $r$ the resistance:

$$Q_G = C\frac{\Delta T}{r} \quad (\text{W m}^{-2}) \qquad \textbf{Equation 6.6}$$

Resistances have units of s m$^{-1}$. Some workers prefer to use **conductance** ($g$), which is simply the inverse of a

resistance, i.e. $g = 1/r$). Conductance has the more familiar units of velocity, m s$^{-1}$. It is is a measure of the degree to which the layer facilitates transfer of heat ($Q_G$), whereas $r$ indicates its ability to moderate $Q_G$.

For heat transfer through multiple layers in series, resistances are simply additive. For example, to calculate heat conduction through several layers of a wall that uses different construction materials, we write:

$$r_T = r_1 + r_2 + r_3 + \ldots \quad (\text{s m}^{-1}) \qquad \textbf{Equation 6.7}$$

where $r_T$ is the total resistance, and $r_1$, $r_2$, $r_3$ etc. are the resistances of each layer. However, if multiple pathways exist, for example, heat can be conducted into a building through its roof ($r_1$), windows ($r_2$), walls ($r_3$), etc., we must consider parallel routes, so the total resistance of the system is:

$$r_T = \frac{1}{r_1} + \frac{1}{r_2} + \frac{1}{r_3} + \ldots \quad (\text{s m}^{-1}) \qquad \textbf{Equation 6.8}$$

In addition, the resistances need to be weighted by the areal fraction of the different materials. An urban climate model that solves heat conduction using a resistance network approach is the **Town Energy Balance (TEB)** model (Figure 6.9). A facet is modelled as a series of resistances for different layers, whilst heat can move in/out of a building through parallel routes (roof, wall). In TEB, heat capacities, thicknesses and resistances for walls, roof, and road facets must be specified to solve for $\Delta T$. Such models can be forced with observed or simulated meteorological data, and be part of an operational weather forecasting model (see Figure 3.27).

BEMs can simulate heat storage at the building scale, and through aggregation, to larger scales. In addition to anthropogenic heating most BEMs incorporate the radiation loading on the building exterior and the energetic implications of exterior air temperature on the thermal balance of the structure. What are not usually covered are sensible or latent heat exchanges with the environment by turbulent transfer. This is largely because the focus of BEMs is on the internal building climate, and building energy demand.

## Parameterization

SEB observations of surfaces show a strong correlation between $Q^*$ and the sensible heat conducted into the substrate. Provided the material is a solid this holds for both natural and construction materials such as the roof in Figure 6.10. Further, whilst the relation

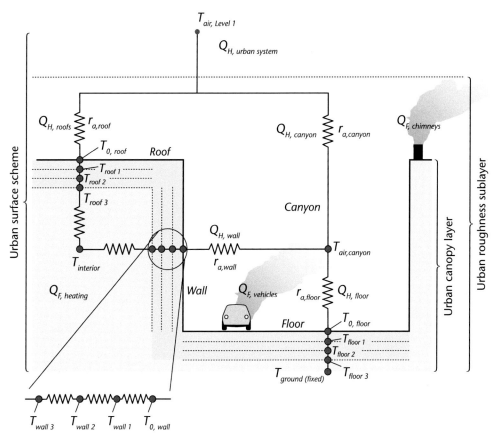

**Figure 6.9** The structure of an urban canopy model, which simulates exchanges at street, wall and roof surfaces representative of parts of a city. The model shown here (TEB) uses electrical analogues that describe the flux of sensible heat in the substrate and air and the associated changes in wall, roof, ground and air temperatures (Modified after: Masson et al., 2002; © American Meteorological Society; used with permission).

is basically linear there is an inertial lag in the conduction, which results in a characteristic diurnal hysteresis loop (e.g. Section 6.1.1 and Figure 6.10b).

To parameterize urban heat storage for individual surfaces, and of most interest here, for a whole building-soil-air volume, the Objective Hysteresis Model (OHM) of Grimmond et al. (1991) uses:

$$\Delta Q_s = \sum_{i=1}^{N} \left( f_i a_{1,i} Q^* + f_i a_{2,i} \frac{\partial Q^*}{\partial t} + f_i a_{3,i} \right) \quad (\text{W m}^{-2})$$

**Equation 6.9**

where, $i$ are different surface types such as roofs, walls, lawns or roads. The $a_1$, $a_2$ and $a_3$ coefficients come from independent empirical studies, such as that in Figure 6.10. They relate $Q_G$ (or $\Delta Q_s$) to $\Delta Q_s$ over one of the $N$ individual surfaces. Equation 6.9 weights

them by $f_i$, which is the surface fraction occupied by each of the $i$ types in the particular area or city under study. To apply the scheme an inventory of the surface cover types and their corresponding $a$ coefficients from the literature are needed (see Grimmond and Oke, 1999b; Meyn and Oke, 2009). OHM has been applied to several cities with modest success and is a module in several empirical urban SEB models.

## Comparison of Methods

All four methods were applied to an area of the old city of Marseille, France (Ma01, see Table A2.1). It is a densely built-up district with narrow streets ($H/W$ = 1.6), buildings with thick limestone walls, sloping tiled roofs and little vegetation ($\lambda_v = 0.14$). The numerical model was TEB (Masson, 2000) and the parameterization scheme was OHM.

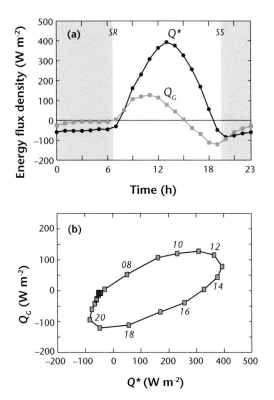

**Figure 6.10 (a)** Ensemble mean (17 summer days ) observed SEB terms for a dry, gravel-covered roof in Vancouver, Canada. **(b)** Scatter plot of same values of $Q^*$ and $Q_G$. Regression analysis of the *hysteresis loop* gives slope ($a_1 = 0.26$), width of *loop* ($a_2 = 0.89$ h) and intercept ($a_3 = -21$ W m$^{-2}$). SR – sunrise, SS – sunset (Modified after: Meyn and Oke, 2009; © Elsevier, used with permission).

The ensemble average results (Figure 6.11b,c) show a similar diurnal course for the different methods. The $\Delta Q_s$ vs $Q^*$ plot for this complex urban district are broadly consistent with results for simpler plane surfaces (e.g. Figure 6.19) including a reversal of sign near sunrise and sunset and a well-defined **hysteresis loop**. The peak is, however, an hour or two later than for flat surfaces which peak before noon. In this chapter we adopt $\Delta Q_{s\,RES}$ to represent storage change.

### 6.3.3 Typical Values of Urban Heat Storage Change

We illustrate characteristics and controls on heat storage using $\Delta Q_{s\,RES}$ results from the four cities noted in Section 6.1.4. After normalizing storage flux densities

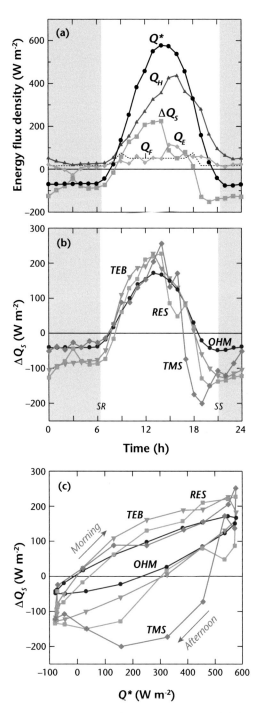

**Figure 6.11 (a)** Ensemble average measured SEB on eight fine summer days in central Marseille (Ma01, see Table A2.1), **(b)** time series of four estimates of heat storage change in the same period, and **(c)** same estimates plotted versus net allwave radiation ($Q^*$), illustrating hysteresis loops (Source: Roberts et al., 2006; © American Meteorological Society, used with permission).

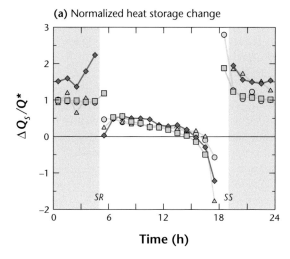

**(a)** Normalized heat storage change

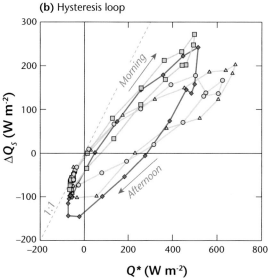

**(b)** Hysteresis loop

◆ Basel Ba02u1 (compact midrise, LCZ 2)
△ Tokyo Tk01 (compact lowrise, LCZ 3)
○ Chicago Ch95 (open lowrise, LCZ 6)
▣ Vancouver Va92l (large lowrise, LCZ 8)

**Figure 6.12** Ensemble diurnal course of **(a)** heat storage change $\Delta Q_S$ normalized by net allwave radiation $Q^*$, and **(b)** the hysteresis loop between $Q^*$ (x-axis) and $\Delta Q_S$ for the four urban sites described in Table 5.1.

thermal instability significantly enhances turbulence, following which $\Delta Q_s/Q^*$ drops in relative importance at all sites, and turns negative (storage release) in the late afternoon, well before sunset. After sunset, $Q^*$ quickly drains energy from the surface and the ratio of the two negative quantities becomes positive. At night the ratio is less well behaved (Figure 6.12a). Some of the scatter is just the instability of ratios between two small quantities, when measurement errors are magnified. In general the nocturnal $\Delta Q_s/Q^*$ is $\geq 1$, i.e. the radiation drain is at least balanced by the release of daytime heat from storage. When $\Delta Q_s/Q^* > 1$ the requirement for balance implies that one or both of the turbulent flux densities is positive (i.e. into the atmosphere) which is not often found at non-urban sites.

The results follow well-defined hysteresis loops (Figure 6.12b) with shapes similar to those of a single roof (Figure 6.10) and the central district of Marseille (Figure 6.11). Based on analysis of similar loops from studies in ten cities (including Ch95 and Va92i), Grimmond and Oke (1999b) conclude there are few obvious relations between the physical attributes of the site and the correlation coefficients ($a_1$, $a_2$ and $a_3$ in OHM, Equation 6.9). But in a general way the slope ($a_1$) appears greatest at the most built-up sites (those with the largest $\lambda_b$ and $\lambda_i$ and least $\lambda_v$) and the phase lag ($a_2$) seems smallest (narrow loop) at dry sites and increases with wetness.

## 6.4 Turbulent Heat Fluxes

Sometimes turbulent heat transfer, or the outcome of its operation, is visible. For example, convectively driven filaments of sensible heat are responsible for the shimmering and mirage-like distortions seen near the surface of very hot roofs, roads and parking lots on sunny days. A similar process underlies the drying of wet surfaces at the end of a rainstorm when they appear to 'steam', or when dewdrops dissipate after sunrise. A visual indication of the vigour of this convection is the rapidity with which a smoke plume expands and becomes diluted with distance after leaving a chimneystack. The plume is advected downwind by the mean flow, but it is turbulence that regulates its **dispersion**. In a turbulent, **unstable** environment the plume widens, its path contorts and it becomes less dense. However, if conditions are **stable** the plume stays in a narrow 'pipe' which meanders gently at the same elevation, but remains dense (see Section

by the corresponding net allwave radiation (i.e. $\Delta Q_s/Q^*$), which aids comparison between different sites, we find the diurnal variation is surprisingly similar (Figure 6.12a).

In the energetically important daytime, storage initially absorbs a large fraction of $Q^*$. By midday

4.1 for discussion of the fundamentals of turbulence and stability).

Turbulent motion is responsible for the transfer of all entities like **momentum**, heat, water vapour, trace gases and air pollutants to and from a surface. Turbulent activity is enhanced over cities because of the mechanical production of turbulence caused by the rough surface (Chapter 4) and the instability arising from the extra warmth of the urban heat island (UHI, see Chapter 7). Here we concentrate on the role of turbulent exchange in the SEB as it facilitates the loss or gain of sensible or latent heat to or from active urban surfaces and areas.

## 6.4.1 Modelling Turbulent Heat Transfer

Turbulent fluxes arise because of the existence of a gradient of a related property in the air (e.g. $\theta_a$, $q_a$ — note **potential temperature** ($\theta$) is used instead of $T$ because height differences may warrant including the effects of pressure change with height) and turbulent motion is set off either by mechanical or thermal production (see Section 4.1.2). The greater the gradient and more efficient the mixing, the stronger is the flux density. Such flux-gradient relations can be expressed in several ways. To simplify here we mostly use simple finite differences (e.g. $\Delta\theta_a/\Delta z$, etc., e.g. see Myrup model, Figure 3.23) or the resistance approach discussed above in the context of heat conduction (Section 6.3.2). For the case of transfer in the atmosphere we can use **aerodynamic resistances**, $r_a$ which depend upon the molecular **diffusivity** of air and the state of turbulence. The resistance approach can be applied to exchanges between a surface and the ambient air at all scales, for example between: a leaf surface and its surrounding air; a wall facet and the adjacent air layer; and whole neighbourhoods and the overlying atmosphere. The full path of a Joule of energy or a molecule of water from its source surface to a level of interest in the atmosphere usually involves several layers, each with a different ability to moderate the rate of transfer.

At the smallest scales mentioned here the resistances relevant for a leaf are those offered by plant stoma to the escape of water and the intake of **carbon dioxide** ($CO_2$), and by the **laminar boundary layer** that adheres to the surface of the leaf. Here our interest begins at the top of that layer where laminar flow transitions into turbulent activity. Extending the electrical analogy, the total path can be analyzed as a

resistance network, as in an electrical circuit diagram. As a flux traverses layers it encounters resistances to its passage. These can be summed in series to obtain the total resistance of the pathway. For example, starting at the top of the laminar layer there are resistances representing the turbulent state of the **roughness sublayer** (RSL), the ISL and the **mixed layer** (ML) within the UBL (see Section 2.2.2). Where there are many such surfaces (trees, rooftops, walls, car parks, etc.) their paths can be summed in parallel, to characterize the flux from an area (e.g. Figure 6.9).

Turbulent fluxes can be presented in both **eddy covariance** and resistance format. For example, the flux density of horizontal momentum ($\tau_0$), given in covariance form in Equation 4.7, can be expressed in resistance form as:

$$\tau_0 = -\overline{\rho w'u'} = \frac{\rho\Delta\bar{u}}{r_{aM}} \qquad (\text{N m}^{-2} = \text{Pa})$$

**Equation 6.10**

Similarly for the turbulent flux densities of sensible heat ($Q_H$)

$$Q_H = C_a\overline{w'\theta_a'} = -C_a\frac{\Delta\bar{\theta_a}}{r_{aH}} \qquad (\text{W m}^{-2})$$

**Equation 6.11**

and latent heat ($Q_E$):

$$Q_E = \mathcal{L}_v\overline{w'\rho_v'} = -\mathcal{L}_v\frac{\Delta\bar{\rho_v}}{r_{aV}} \qquad (\text{W m}^{-2})$$

**Equation 6.12**

where, $r_{aM}, r_{aH}$ and $r_{aV}$ are the **aerodynamic resistances** for momentum, sensible heat and latent heat, respectively. The terms $\Delta\bar{u}, \Delta\bar{\theta_a}$ and $\Delta\bar{\rho_v}$ are the corresponding mean vertical differences of horizontal wind speed, air temperature and atmospheric water vapour density; $C_a$ is the heat capacity of air and $\mathcal{L}_v$ is the latent heat of vapourization. Equation 6.9 can be converted to evaporation, the mass flux of water vapour ($E$) by dividing through by $\mathcal{L}_v$ (see Equation 6.2).

Resistance to the transfer of momentum is about an order of magnitude smaller than for other entities, like heat and water vapour. This is because in addition to **skin drag** (Section 4.2.1) momentum transfer over rough surfaces like urban areas includes **form drag**, due to bluff body effects. Form drag is created by **vortices** and **wakes** which add the transfer of momentum by pressure **perturbations**. There is no parallel process associated with the transfer of heat or water vapour. Hence, the resistances for heat and vapour are greater than for momentum; the difference is termed the **excess resistance** ($r_b$),

$$r_{aV} \approx r_{aH} = r_{aM} + r_b \quad (\text{s m}^{-1}) \qquad \textbf{Equation 6.13}$$

By analogy with the vertical wind **profile**, we conceive of turbulent exchanges occurring at a distance above the solid surface, in other words there are **roughness lengths** for heat ($z_{0H}$) and vapour ($z_{0V}$). The more familiar roughness length for momentum $z_{0M}$ is much larger than the **scalar** lengths. This signifies a shorter distance between the turbulent exchange surface for momentum and the ambient atmosphere, and a smaller resistance. The greater resistance to the transfer of heat and vapour is the equivalent of a larger gap across which the gradients $\Delta\overline{\theta_a}$ and $\Delta\overline{\rho_v}$ occur. At the neighbourhood scale, we extend our model for momentum and place the height of the effective urban surface for $Q_H$ (and $Q_E$) at ($z_d + z_{0H}$ or $z_{0V}$), slightly lower in the urban canopy than ($z_d + z_{0M}$) for $\tau_0$; see Figure 6.13. Kanda et al. (2007) and Kawai et al. (2009) provide relations that can be used to estimate $z_{0H}$ for urban areas at the local scale.

The source of turbulent heat at the active surface is the **available energy**, $A = Q^* + Q_F - \Delta Q_s$; i.e. the other terms in the SEB (Equation 6.3). So the turbulent heat flux densities ($Q_H$ and $Q_E$) are driven by the surplus/deficit in the radiation balance ($Q^*$, Chapter 5), the emissions of anthropogenic heat ($Q_F$, Section 6.2) and the uptake or release of heat by the urban fabric ($\Delta Q_s$) which acts like a rechargeable 'battery' (Section 6.3).

If water is present at a surface, $A$ is preferentially, but not exclusively channelled into evaporation – i.e. it is converted into latent, rather than sensible, heat. Its significant thermal impact, and role in setting $\beta$, means the presence/absence of surface moisture is a powerful tool in the arsenal of urban architects and designers (Chapter 15). Because of these critical aspects of urban energetics and their application to urban design we devote more space to latent than sensible heat transfer. However, that should not be taken to say that $Q_H$ is not of great significance in urban climate, in fact the reduction in $Q_E$ and increase of $Q_H$, relative to the surrounding countryside is almost a distinguishing feature of urban climates, as the following chapter on the UHI shows.

In Equation 6.12 $\Delta\overline{\rho_v}$ is the gradient in atmospheric vapour density, but availability of moisture at a surface is partly regulated by the surface itself. The resistance approach can be expanded to include this physiologic control. The total resistance of the system to the transfer of $Q_E$ then is comprised of both the

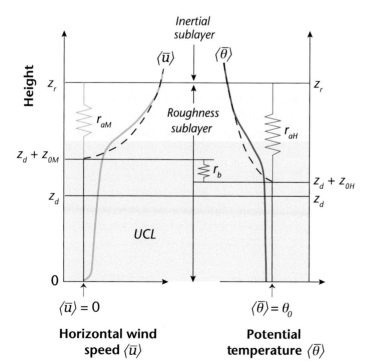

**Figure 6.13** Illustration of the excess resistance concept. The relationship between mean ISL profiles of wind speed ($\overline{u}$) and temperature ($\overline{\theta}$) and the transfer resistances for momentum ($r_{aM}$) and heat ($r_{aH}$). Also shown are the roughness lengths for wind speed ($z_{0M}$) and temperature ($z_{0H}$) and the zero-plane displacement height of a canopy layer ($z_d$). Dashed curves are the hypothetical logarithmic extrapolation of ISL profiles into the UCL, to the mean height of the apparent sinks for momentum and the source of sensible heat. The difference between the two heights illustrates the *excess resistance* ($r_b$) encountered by heat (Modified after: Malhi, 1996).

aerodynamic resistance, $r_{aV}$ which depends on the surface roughness and wind speed, $r_a \propto \bar{u}/z_0$ and the **canopy** or **surface resistance**, $r_{cV}$.

The surface resistance of dry surfaces is approximately infinite (i.e. there is no surface moisture available), whereas at the other end of the spectrum open water and surfaces externally wetted by rain or sprinkling, have values approaching zero resistance (i.e. water is freely available). Moist surfaces like soil and porous construction materials have intermediate values that depend passively on their porosity. The resistance of plants, on the other hand, is actively controlled by their stoma which open and shut in response to light intensity, temperature, leaf water potential, atmospheric moisture and $CO_2$ concentrations. This physiological control affects rates of **transpiration** from vegetation. The total resistance by an urban surface is an amalgam of the resistances of these active and passive elements, weighted by their plan and complete surface fractions.

It follows that to understand the fluxes of water at the local scale in cities requires quantitative knowledge of features like the amount of standing water (lakes, ponds, swamps), completely dry areas (rocks), and both the amount, species and class of plants (i.e. are they are adapted to dry, moist or wet conditions, xeric, mesic or hygric, respectively).

The requirements for evaporation are: a source of water, a supply of energy, a surface-to-air gradient of water vapour and sufficient turbulence to renew the gradient. For application to urban areas assuming $r_a = r_{aH} = r_{aV}$, an appropriate framework for $Q_E$ is the Penman-Monteith equation:

$$Q_E = \frac{[s(Q^* + Q_F - \Delta Q_S) + (C_a\, vdd_a)/r_{aH}]}{\left[s + \gamma\left(1 + \frac{r_{cV}}{r_{aH}}\right)\right]} \quad (\text{W m}^{-2})$$

**Equation 6.14**

where, $s$ is the slope of the relation between saturation vapour density and temperature, $C_a$ is the heat capacity of air, $\gamma$ is the psychrometric constant, and $vdd_a$ is the vapour density deficit of the air. Equation 6.14 encapsulates the main elements driving surface evaporation, viz: the first term in the numerator on the right hand side is the available energy $A$, which is primarily driven by radiation and the second term is the influence of advection (i.e. the delivery and replenishment of drier air to and from the site; its dryness is represented by the deficit, $vdd_a$). Advection is necessary to maintain

a surface-air vapour gradient and to transport vapour between the surface and the ABL. A version which expresses the degree of coupling between the **surface layer** (SL) and the rest of the ABL is:

$$Q_E = \Omega \frac{s}{s+\gamma}(Q^* + Q_F - \Delta Q_S) + (1-\Omega)\frac{(C_a\, vdd_a)}{\gamma\, r_{cV}} \quad (\text{W m}^{-2})$$

**Equation 6.15**

where, $\Omega$ is a decoupling factor defined:

$$\Omega = \left[\left(1 + \frac{\gamma}{s+\gamma}\frac{r_{cV}}{r_{aH}}\right)\right]^{-1} \quad (\text{W m}^{-2})$$

**Equation 6.16**

(McNaughton and Jarvis, 1983). The decoupling factor $\Omega$ modulates the relative significance of surface and atmospheric influences on $Q_E$. It varies with the ratio of the surface resistance $r_{cV}$, to the aerodynamic $r_{aH}$ resistance (see controls, Section 6.3.3). $\Omega$ increases as $r_a$ decreases at rougher sites, where turbulent exchanges are enhanced, and $r_c$ increases at sites where vapour loss is restricted such as bare, sparsely vegetated or otherwise water stressed locations.

In ecosystems with tall elements such as urban areas and forests $\Omega$ is small, typically about 0.01 to 0.2 (e.g. $r_a$ ~10 s m$^{-1}$, $r_c$ ~200 s m$^{-1}$) (Roth and Oke, 1995). This means the surface is well coupled to a deep layer of the ABL so $Q_E$ responds to advection and $vdd_a$. Surfaces like heath land and prairie areas have intermediate $\Omega$ values of about 0.3 to 0.5. On the other hand, ecosystems with short elements, such as low agricultural crops and pasture, have relatively large $\Omega$ of 0.5 to 0.8 (e.g. $r_a$ ~70 s m$^{-1}$, $r_c$ ~50 s m$^{-1}$), meaning the surface is poorly coupled to the atmosphere and $Q_E$ is primarily driven by $A$.

There are significant implications for the diurnal course of SEB terms. Variation of SEB terms at poorly coupled low roughness sites tend to be more in-phase with $Q^*$ than at well coupled locations, such as cities. That being so, it can be anticipated that urban sites will show greater hysteresis between $Q^*$ and the storage and turbulent terms.

## 6.4.2 Measurement of Turbulent Heat Fluxes

Measurement of $Q_H$ and $Q_E$ over extensive, homogeneous rural surfaces requires complex instrumentation and careful field implementation. It is even more challenging over the patchy, inhomogeneous surfaces of cities, where the theoretical requirements of several methods are not met.

### Turbulent Exchange of Individual Facets

Viable measurement solutions to measure $Q_H$ at individual urban facets include: solving for $Q_H$ by residual in the SEB for dry surfaces (roofs, roads, walls) where $Q_E \sim 0$. This strategy requires direct measurement of $Q^*$ (with the radiometer close to the facet so its FOV is just the facet), and of $\Delta Q_s$ (by placing heat flux plates within the fabric). For moist surfaces direct measurement of $Q_E$ (e.g. with a mini-lysimeter) allows the same approach to be used. By monitoring sap flow using heat dissipation methods, it is possible to follow transpiration by individual trees. The method requires careful scaling up from point to whole tree results (Čermák et al., 2004; Smith and Allen, 1998). Estimates for individual canyon facets also require scaling up for each facet and summing to give the SEB of a complete canyon system. Measurement projects of this type are usually undertaken as case studies to give insight into exchanges at the microscale. It is not feasible to apply these approaches at the neighbourhood scale because of the sheer number of microscale facets comprising the system.

### Turbulent Exchange of Urban Systems

The solution at the neighbourhood scale is to take measurements in the ISL over a reasonably homogeneous area where turbulence in the RSL has blended microscale variability (Chapter 3). Measurement of $Q_H$ and $Q_E$ of neighbourhoods encounters several practical and theoretical constraints. Methods such as the aerodynamic, or Bowen ratio-energy balance approaches rely on measurements of vertical gradients of $\bar{\theta}$, $\bar{\rho_v}$ and $\bar{u}$, that incur large errors over urban terrain. This is because to avoid microclimatic effects by roughness elements (buildings and trees) the sensors must be exposed in the ISL (e.g. $> 2z_H$) to avoid microclimatic influences from individual roughness elements (see Chapter 3). At such large heights vertical gradients are very slight (e.g. $\partial \bar{\theta}/\partial z < 0.05$ K m$^{-1}$) and therefore hard to measure consistently and reliably.

Results representative of the local scale can be obtained using the eddy covariance (EC) technique (Figure 6.14). The EC technique is considered to be the standard method for urban SEB flux studies (Feigenwinter et al., 2012). It requires that (a) the instruments are mounted sufficiently high to be in the ISL, and (b) source area analysis of the site shows surface inhomogeneity in the source area of the EC sensors, is not unacceptably large (Figure 3.3). If these

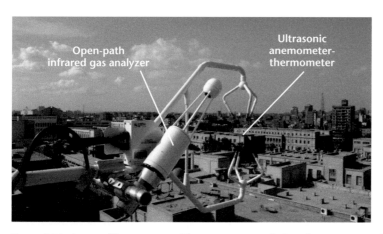

**Figure 6.14** Array of instruments able to measure turbulent fluxes, mounted on a mast over a district of Cairo, Egypt. The ultrasonic anemometer-thermometer is able to obtain fast-response values of three wind components ($u, v, w$), and acoustic temperature ($\theta$). The infrared gas analyzer measures fast-response water vapour density ($\rho_v$) and carbon dioxide density ($\rho_v$) along an approximately 12.5 cm path inside the open frame. The path is tilted to bring it close to the centre of the six anemometer prongs without interfering with airflow. Signals can be analyzed separately or in several combinations to calculate statistics and turbulent fluxes. The wind components can be combined to calculate Reynolds stress using the EC approach. The turbulent fluxes of sensible and latent heat, water vapour and carbon dioxide can be calculated by combining the vertical wind fluctuations with those of the relevant variable (Credit: R. Vogt; with permission).

conditions are met, the results possess spatial validity and the fluxes are essentially independent of wind direction. The EC method employs the covariance relations for the turbulent transport of sensible heat (Equation 6.8 and Equation 6.9). The convention is that fluxes directed away from the surface are positive (opposite to the convention for momentum. Instruments must be able to sense and record the fluctuations of the vertical wind component ($w'$) and of $\theta'$ or $\rho'_v$ with sufficient speed to be considered 'instantaneous' (typically $> 10$ Hz). Similar covariance equations apply to the transport of any scalar (air pollutants, etc.).

### 6.4.3 Controls on Turbulent Heat Fluxes

The magnitude of turbulent fluxes depends on the nature of both the *surface* and the overlying *atmosphere*. Probably the most influential surface property is whether the cover is dominated by vegetation or built materials. Other important surface influences include: surface wetness/dryness, the tightness with which water is 'held' by plants or other surfaces and the spatial arrangement of surfaces. In the atmosphere the main controls are: wind speed and direction, the thermal and humidity structure of the ABL and FA and the likelihood of large-scale advection. Finally, there is the active contribution of humans to the available energy ($A$) via $Q_F$. Each of these controls is discussed below.

### Surface Controls

*Surface water availability*: It is difficult to exaggerate the significance of the availability of surface water (*i.e.* 'wetness'), and the spatial arrangement of the sources of water, to the SEB and the related moisture and thermal climate. In urban areas some water is freely available (e.g. open water of ponds, lakes, rivers, swamps, surfaces wetted by dew, rain, irrigation) whereas some is 'held', more or less effectively, by a surface (e.g. the **stomatal** pores of vegetation or the pores and cracks in construction materials). A first approximation to water availability in cities is the fraction of the surface vegetated ($\lambda_v$). Unless wetted by rain or covered with snow it is usual to assume that buildings and impervious ground surfaces ($\lambda_b$, $\lambda_i$) are dry and therefore are only sources of $Q_H$.

It is also relevant to know the phase of any surface water: i.e. is it liquid, or held in the solid phase as ice or snow? At high latitudes in winter, water can be

present almost everywhere, but since it is in the form of snow or ice cover it is largely unavailable for evaporation and humidification of the air.

*Surface (or canopy) resistance* ($r_c$): This expresses the ability of a system to hold or release entities such as heat, moisture or $CO_2$. Plants play an active role in water exchange via the aperture of the stoma on their epidermis. These pores have guard cells that cause them to open and shut in response to light intensity, leaf temperature, $vdd_a$, $CO_2$ concentration and leaf water content. Thus they act like valves in the process of transpiration. Small holes in concrete, cracks in pavement, soil pores and oily films on open water, passively control the availability of water for evaporation from urban surfaces. $r_c$ values range from very low (zero to ~5 s m$^{-1}$) for open water and rain-wetted surfaces to almost infinitely large for 'dry' built surfaces. For plants under little water stress, $r_c$ values are typically ~50 to 100 s m$^{-1}$. Unless there is dew-wetting of the canopy in the early morning, $r_c$ follows an approximately 'U'-shaped path in daytime. The larger values in the early morning are due to low light intensity, there is a midday minimum (typically $r_c \approx 60$ s m$^{-1}$ for most vegetation including grasslands and forests, but $r_c \approx 30$ s m$^{-1}$ for low crops, Kelliher et al., 1995). $r_c$ values increase again in the evening as stoma begin closing due to lack of light and depletion of soil and leaf moisture.

Given the reduced vegetation cover of cities and the dryness of many urban soils (Chapter 8) it follows that in general urban $r_c$ values are higher than for most rural ecosystems. Values are approximately inversely related to the vegetative cover fraction ($\lambda_v$). The few available $r_c$ values for urban areas show midday values for open low-rise sites vary from $r_c \approx 250$ s m$^{-1}$ in a city with well-watered gardens, to $r_c \approx 1,200$ s m$^{-1}$ in a city with a dry climate and xerophytic landscaping.

*Surface patchiness and advection*: The inhomogeneity of the urban surface creates microscale and local advection. It is particularly evident in cities because, in contrast with most rural areas, the borders between patches of vegetation, building lots and roads are often sharp. At each border there is a transition in surface properties, which in turn creates a gradient in the near-surface microclimate. Depending on the sign and magnitude of the surface contrast, this can assist or hinder a surface flux.

To illustrate, take the case of momentum exchange. If the transition is from a smoother-to-rougher

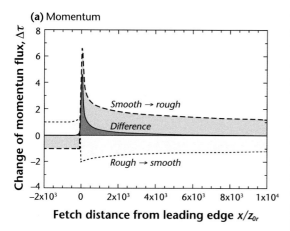

**(a)** Momentum

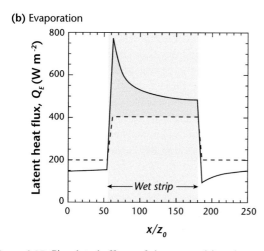

**(b)** Evaporation

**Figure 6.15** Simulated effects of sharp transitions in surface roughness and wetness on surface fluxes as air traverses patches across a landscape. Wind direction left to right in both figures. **(a)** Change in roughness ($z_0$) and Reynods stress ($\tau$), (Source: Schmid and Bünzli, 1995; © Royal Meteorological Society, with permission from Wiley); and **(b)** change of surface wetness and latent heat flux (Modified after: Bünzli and Schmid, 1998; © American Meteorological Society, used with permission). In **(b)** dashed line – for a site where $Q_E$ is fully adjusted to the surface wetness value i.e. there is no advection. Solid line – $r_{c\ dry} = 120$ s m$^{-1}$ for the dry upwind zone and downwind surfaces, but $r_{c\ wet} = 0$ for the wet strip. In both cases $Q^*$ is set at 500 W m$^{-2}$. Darker shading is enhancement and lighter is suppression of $Q_E$ relative to the no-advection case.

surface, the **Reynolds stress** increases sharply to a peak downstream from the **leading edge**, and then relaxes to a new, but greater, equilibrium value (Figure 6.15a) –

this is an example of the **leading edge effect**. The reverse applies in the case of a rougher-to-smoother transition, but notice that the magnitude and rate of adjustment is *different* in the two cases (i.e. smooth → rough is greater than rough → smooth). This is because the adjustment is related to the *absolute* value of the downwind roughness length, which controls the vigour of the turbulence conducting the transition. If there are many such transitions to and forth, differences build up across a patchy cityscape. This differential adjustment causes a net accumulation of the overall flux of momentum at the surface, because adjustment is not complete before the next difference is encountered. This cumulative effect is directly related to the number of borders crossed, i.e. a measure of the 'patchiness' of the area traversed by the airstream.

Urban patches are also likely to possess different moisture availabilities. The patterns of surface cover in cities commonly show completely dry surfaces located immediately next to moist ones (e.g. dry roads, parking lots, buildings bordered by healthy vegetation, moist soils) and even fully wet ones (e.g. irrigated vegetation, rain-wetted surfaces and open water).

The modelling results in Figure 6.15b show the case of moisture leading edge effects. When air traverses from a drier surface (higher $r_c$) to a wetter surface strip ($r_c \approx 0$), evaporation 'spikes' up and then settles back to a new, greater rate over the wet strip (Figure 6.15b). This happens because advection of dry air over a wet strip increases the air-to-surface moisture gradient. This occurs at the windward border of a lawn or park and the size of the effect is related to the strength of the contrast between the wetness of the two surfaces. The opposite happens in a wet-to-drier transition, where advection of moist air over the drier surface reduces the air-to-surface moisture gradient and therefore $Q_E$. But, as with momentum, there is a *difference* between the two cases: the increase in the first type of transition is larger than the decrease in the second type. Therefore, over cities with many patches (multiple transitions), there is a net bias that produces a spatial mean $Q_E$ that can be greater than even that over extensive areas with only wet surfaces.

This largely microscale leading edge effect is augmented by the **oasis effect** at larger scales (i.e. local or mesoscale) that occurs when large moist patches are embedded in much drier, and therefore $Q_H$-rich, surroundings. Being relatively cool the patches induce **subsidence** of drier air from the upper SL or ML down into the SL. There it enhances the air-to-surface water vapour gradient (i.e. $vdd_a$ in Equation 6.14) and drives much larger evaporation than normal. This is made energetically possible by reversal of $Q_H$, i.e. heat is extracted from the warmer air above *towards* the cooler surface below, even in daytime, which augments the local SEB. If despite the greater drying power the wetness of the patches can be sustained, by irrigation using water from artesian, river or reservoir sources, it creates a positive feedback mechanism that can be maintained. The net result is that evaporation rates can substantially exceed the locally available **radiant energy**, (i.e. $Q_E > Q^*$) on both an hourly and daily basis. The dry-moist SEB differences give a hot-cool patchwork of surface temperature ($T_0$), each with sharp edges, that is a hallmark of urban surface thermal imagery (e.g. see Chapter 7 and Figure 7.7).

Both the leading edge and oasis effects are at play in Figure 6.16. It shows evaporation rates both at a tower site in the Carmichael suburb of Sacramento,

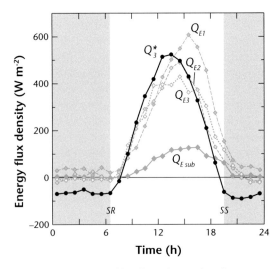

**Figure 6.16** Impact of leading edge and oasis type advection on evaporation rates in Sacramento, United States. $Q_{E\,1-3}$ latent heat flux density measured at lysimeter stations located at, 3.5, 18.5 and 59.3 m downstream from the windward edge of an irrigated park. $Q_{E\,sub}$ are eddy covariance (EC) fluxes measured on a tower site in the same suburb, and $Q^*$ is net radiation at station 3. Data are 3-day averages from the study of Spronken-Smith et al., (2000), where the methods are fully described.

United States (LCZ 6), and from mini-lysimeters installed at three sites in nearby Orville Wright Park, which is irrigated. Two points are clear immediately: first, $Q_E$ at all sites in the park is much greater than in the surrounding suburb and second, rates are greatest near the park border. Sites 1, 2 and 3 are located at **fetch** distances of 3.5, 18.5 and 59.3 m respectively, downstream from the edge of the park. The fraction of radiant heat used in evaporation ($Q_E/Q^*$) at Sites 1, 2 and 3 are 1.69, 1.25 and 1.26 on a daily basis, respectively. In the surrounding suburb the equivalent fraction is 0.41, which is higher than its longer-term average of 0.33, but far below the greatly elevated rates in the park. To give context, park values are about one third greater than even those at a heavily irrigated grass sod farm located outside the city at the same time (not shown). Clearly the park is behaving as an oasis, and the variation of $Q_E$ within about 20 m of the border is typical of a leading edge (compare with Figure 6.15b).

As noted, the available energy, $A$ (usually dominated by $Q^*$) fuels the turbulent fluxes $Q_H$ and $Q_E$. Further, there is a complementary relation between the turbulent terms; dry surfaces favour large $Q_H$ but small $Q_E$; it is *vice versa* for moist surfaces. A reasonable surrogate for that is the fraction of the plan area occupied by vegetation ($\lambda_v$). This is evident in Figure 6.17, which summarizes the relations between the turbulent terms of the SEB across a wide range of cities. It shows the daily total partitioning of the urban energy balance measured at many tower sites around the world. A clear relation arises between the fraction of the available energy partitioned into the latent heat flux (evaporation) ($Q_E/A$) and $\lambda_v$. The corresponding Bowen ratio ($\beta$) is plotted on the right hand scale.

The essence of the relation in Figure 6.17 is simple and direct: as more surface cover is converted from natural (soils and vegetation) to built (construction materials), energy is increasingly partitioned into $Q_H$ at the expense of $Q_E$, which conversely means that $\beta$ increases. The relation cannot be precise because, despite our best endeavours to compare similar sites, results from such a wide range of cities necessarily include large variability in surface and atmospheric controls. Dividing $Q_E$ by $A$ has the useful effect of normalizing the forcing energy, which facilitates comparison of cities from different macroclimates. Table 6.5 puts typical urban $\beta$ values in the context of several other ecosystems. Early musing about

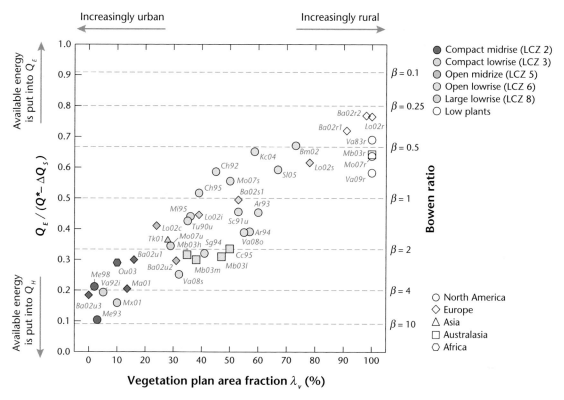

**Figure 6.17** Relation between vegetation plan area fraction and partitioning of daily total turbulent fluxes. The left axis shows partitioning of the daily total available energy ($Q^* - \Delta Q_S$) into sensible and latent heat flux densities ($Q_H$ and $Q_E$ respectively), as a function of the vegetated plan area fraction ($\lambda_v$) of different neighbourhoods. Each point represents a measured multi-week average of the energy balance measured in summer at a mid-latitude site. The right axis shows the corresponding 24-h Bowen ratio calculated based on daily totals of $Q_H$ and $Q_E$. Soil moisture at rural sites ($\lambda_v > 90\%$) is moist and all data are for snow-free conditions. Site codes correspond to those in Appendix A2.

appropriate $\beta$ values to use for urban areas tended towards estimates more typical of desert and semi-arid areas. Clearly Figure 6.17 indicates that evaporation is a significant component in the SEB of most cities.

## Atmospheric Controls

*Water vapour deficit*: ($vdd_a$, defined in Section 6.4.1) is a measure of the amount by which the ambient atmospheric humidity differs from its saturation value, at the same air temperature and this deficit is almost certainly less than its value at the surface ($vdd_0$). In broad terms, the difference between the surface and atmospheric deficits creates the driving gradient for the exchange of water vapour from the surface; hence its significance as the numerator in Equation 6.14.

*Aerodynamic resistance ($r_a$):* the aerodynamic resistances, $r_{aM}$, $r_{aH}$ and $r_{aV}$ are a statement of the degree to which atmospheric turbulence facilitates transport of momentum, heat and vapour to and from a surface. They are related to the wind speed, surface roughness and atmospheric stability. For a given wind speed $r_a$ decreases as roughness and instability increase. In general therefore, cities have relatively small aerodynamic resistance values of $r_{aM}$ ($\approx 25$ s m$^{-1}$ for open lowrise, LCZ 6 and even smaller for rougher sites $\approx 11$ s m$^{-1}$, LCZ 2; Grimmond and Oke, 1999a). For example, $r_{aM}$ values for a mean wind speed of 5 m s$^{-1}$ at 30 m at two sites used in this book are: Vancouver, Canada (Vs09, LCZ 6), $r_{aM} = 17.6$ s m$^{-1}$; and Basel, Switzerland (Ba02u1, LCZ 2) $r_{aM} = 8.1$ s m$^{-1}$. In dry environments increased wind speed encourages turbulent loss ($Q_H$) over storage uptake ($\Delta Q_s$). Therefore in windy, dry

**Table 6.5** Typical values of daytime values of the Bowen ratio $\beta$ for natural and urban systems. Bold values are generic class mean; only for use in descriptive or general studies. Range usually depends on moisture availability. Urban values are for snow-free conditions; adding snow would give a much wider range including larger values (Sources: Bonan, 2002; Eugster et al., 2000; Oke, 1987; Rouse et al., 2008; Valentini et al.,1999; Wilson et al., 2002).

| Surface description | LCZ | Bowen ratio $\beta$ |
|---|---|---|
| Tropical oceans | G | 0.1 |
| Lakes[1] | G | 0.2–1 |
| Wetlands | G | 0.2–0.7 |
| Tropical wet forest | A | 0.1–0.3 |
| Temperate deciduous forest (full leaf) | A | **0.4** 0.2–0.8 |
| Crops | D | **0.3** 0.1 – 1 |
| Temperate grassland | D | **0.4–0.8** 0.5–1.6 |
| Prairie grassland | D | **0.9** 0.3–4 |
| Tundra (inland) | C, D | **1** 0.3–2.5 |
| Temperate coniferous forest | A | **1** 0.8–1.8 |
| Urban 35–75% greenspace | 6, 9 | **1** 0.5–2.5 |
| Urban 25–40% greenspace | 3, 5 | **2** 1.5–3 |
| Urban < 20% greenspace | 2, 3, 8 | **4** 3–8 |
| Semi-arid lands | C, F | **2–6** |
| Sandy desert | F | ~10 |

[1] Large range because: storage can be large (depends on lake size, depth, time of year); wide range of lake temperature; varying flow-through from connected water bodies

cities $Q_H$ is often relatively large and sensitive to change in wind speed. In moist and wet environments sharing is more equal and depends more on surface water availability than wind speed.

Evaporation is not entirely controlled by conditions in the SL and at the surface. Especially in ABLs with well-developed turbulent activity, typical of unstable conditions, the ML and even the free atmosphere can interact with the SL and surface heat fluxes. Thermals originating at the ground can become organized into convection cells that extend over the full depth of the ML (Figure 6.18). Cities are a preferential zone for convective structures, because they promote **convergence** and uplift, due to their greater roughness and thermal effects (Chapter 4). The descending limb of such convection cells often draw down warmer/colder or drier/wetter air from the **entrainment zone** (EZ) or upper ML, and transport it down to the ground. Injections of warmer/drier air increase $vdd_a$ and boost $Q_E$, rather like the oasis effect. Of course if the downdrafts are moister than in the SL they could tend to suppress the surface flux of moisture. In general if the descending air is warmer it suppresses $Q_H$ at the surface, and boosts it if it is cooler.

Full understanding of surface turbulent heat fluxes, especially in an environment like the city, must include attention to total ABL dynamics. This requires knowledge of the vertical profiles of wind, temperature and humidity, an appreciation of the evolving atmospheric circulation and its implications for the march of **synoptic** and mesoscale air masses to and from the site. By the same token numerical models designed to simulate turbulent fluxes in cities, require a scheme that properly couples the SL, ML and FA. If such a model is nested within, or linked to a numerical weather prediction model, it includes synoptic and even global characteristics.

### Anthropogenic Sources of Sensible and Latent Heat

Sensible and latent heat fluxes cover a wide range of values and can be considerably elevated by direct anthropogenic heat releases, which are considered to be part of the available energy ($A$) via $Q_F$ (see Section 6.1, Table 6.3 and A2.8). The additional available energy via $Q_F$ can considerably elevate $Q_H$ (e.g. at major office nodes in city centres, industrial factories, district heating and power plants) and/or $Q_E$ (e.g. fuel combustion, industrial cooling ponds, steel mills). Analogously irrigation can be a significant anthropogenic control of surface water. This aspect is covered in Chapter 8 in relation to its role in evaporation and the **surface water balance** of cities.

### 6.5 Example Energy Balances in Cities

### 6.5.1 Energy Balance of Individual Facets

The SEB of a city is the summation of the SEBs of its many component units (facets, buildings,

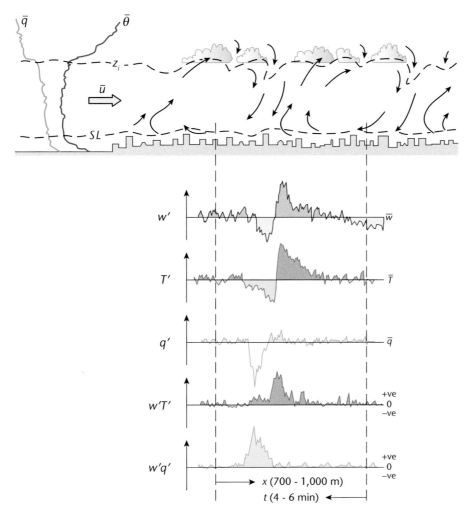

**Figure 6.18** At the top is a conceptual description of enhanced coupling between the SL and the warmer, drier EZ of a UBL during a day with cumulus clouds. Below is a segment of the surface and typical traces found in the SL as a convection cell is advected from left to right, by the wind at about 3 m s$^{-1}$ (Source: Roth and Oke, 1995; © American Meteorological Society, used with permission).

canyons) and their interactions with each other (Chapter 2). The simplest unit is a horizontal, flat, facet with open exposure. Common facets found in urban areas include: flat roofs, exposed sections of road or other paved area, open grassed areas and walls. Here we use these four facets to illustrate the diversity of SEBs in the UCL. Each differs substantially in its geometry, radiative properties (especially **albedo**, $\alpha$), thermal properties (especially thermal admittance $\mu$) and moisture properties (soil moisture content, $S$).

## Roofs

Roofs have excellent exposure to the Sun and sky and are built with surface materials that generally possess low albedo and uniformly high emissivity, except those made of certain metals (Table 5.2). Hence, by day the surface of roofs are strong absorbers of solar radiation producing very high surface temperatures. If there is little cloud, roofs are always free to exchange longwave radiation but at night, with no solar input, and since the sky is almost always much colder than the surface, $L^*$ drives strong cooling of roof surfaces.

In addition to protecting from precipitation and excessive irradiance, roofs are designed to maintain the building's interior climate free from temperature extremes. Materials and construction techniques are selected to ensure the large diurnal range swings of temperature at exterior surfaces, do not migrate indoors. Commonly just beneath the outer tiles, metal sheets or membrane is a layer designed to have low $\mu$, and if there is a space between the exterior roofing and the top floor it contains a lot of still air (e.g. an attic). Hence a roof assembly has poor ability to diffuse or store heat and it is a thermal buffer to prevent unwanted heat gain or loss from the occupied space. Ideally the roof exterior undergoes wide diurnal variation of temperature, whilst the interior range is greatly damped.

Roofs are also fully exposed to precipitation, but are designed to remain dry most of the time. This is possible because they are built of impermeable materials, designed to prevent **infiltration** into the building envelope and to efficiently shed water via gutters and downpipes. Further, their exposure to Sun and wind facilitates evaporation of surface wetness. Hence, everything works to keep roofs dry most of the time.

This characteristic dryness means the diurnal SEB of most roofs does not contain the latent heat flux term (Figure 6.10). By day, hot roofs shed most of their $Q^*$ surplus as $Q_H$. If the urban SL is very unstable this may create micro-**thermals**, heat plumes and a shimmering layer of heat haze above roofs, due to the differing refraction between eddies with different temperature. The discontinuous roof layer located at the top of the UCL is a primary source of sensible heat for the urban SL. It also coincides with the main **shear zone** where the efficiency of turbulent heat transport is greatest (Figure 4.19). Because the roof in Figure 6.10 is covered with a layer of gravel, it has some ability to store heat ($Q_G$). When this heat is released, at about 1500 h, it supports a $Q_H$ that exceeds $Q^*$. At night, once this small heat store is exhausted ($Q_G \approx 0$) the radiative deficit ($Q^*$) is largely matched by $Q_H$ from the slightly warmer air onto increasingly cold roofs.

### Roads

Roads are also dry most of the time, so again only $Q^*$, $Q_H$ and $Q_G$ are involved in the SEB (Figure 6.19a). The major difference between the SEBs of roofs and roads is the much larger $\mu$ of the paving materials (asphalt, concrete, rock cobble stones) (Table 6.4). Even roads made of gravel and compacted soil have elevated thermal properties. As a result more of the

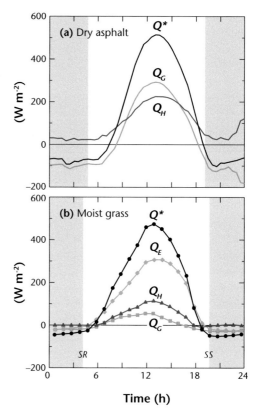

**Figure 6.19** Example SEBs of unobstructed urban facets: **(a)** dry asphalt road near Vienna, Austria on August 11, 1994 (Source: Anandakumar, 1999; © Elsevier, used with permission); **(b)** slightly moist grassed site in an urban park in Vancouver, Canada, for a 4-day period in late July 1992 (Source: Spronken-Smith, 1994; with permission). $Q^*$ and $Q_G$ and $Q_E$ in (b) are measured, but $Q_H$ is solved as the residual of the SEB.

available energy is conducted into the ground below the road, so $Q_G$ is larger by day, and consequently more is released at night (Figure 6.19a) compared to roofs. In daytime $Q_G > Q_H$, but contrary to the roof case, the larger nocturnal $Q_G$ from the substrate is sufficient to usually maintain a positive nocturnal $Q_H$ turbulent output into the lowest air layer over the road at least in the evening and sometimes longer.

### Open Grass

Figure 6.19b shows the SEB of an open stretch of short grass on moist soil in the centre of a large city park. Following the equivalent SEB of largely dry roofs and roads it illustrates the extremely powerful energetic role played by available moisture. $Q_E$ becomes the largest SEB term for this site and the daytime and daily $\beta$ are 0.31 and 0.33, respectively. As with the roof and road SEBs the four flux densities are approximately in-phase, except there is some asymmetry in $Q_E$ in late afternoon (when $Q_E$ becomes approximately equal to $Q^*$, which probably indicates the park is an 'oasis' patch, see Figure 6.16 and discussion in Section 6.4.3). The site is roughly 100 m away from the leading edge yet daytime $Q_E$ and daily total $Q_E$ are 20 to 35% greater than over a comparable extensive rural crop at the same time. Over smaller irrigated patches, such as garden lawns, $Q_E$ is boosted to an even greater extent.

Notice in Figure 6.19 that the diurnal course of the flux densities is approximately in-phase with each other. This is typical of flux partitioning at smooth surfaces, where poor coupling to the turbulent SL results in flux densities that are primarily driven by $Q^*$ (Section 6.4.1).

### Walls

Walls are almost the only vertical urban units. Their SEB depends strongly on orientation, exposure and access to solar irradiance and their **sky view**. The latter is critical, especially if they are embedded in built-up or treed parts of the UCL because then they exchange radiation with other surfaces they can 'see'. This creates complex radiation exchange by reflection of short- or longwave radiation, or emission of longwave radiation (Chapter 5). There is great diversity of wall materials, external cladding (thatch, wood, glass) and paints which create a wide range of albedos (Table 5.1). Walls irradiated when the Sun has low altitude (early morning, evening) can receive remarkably high solar receipt (Section 5.2.2) hence their surface can get very hot for short periods. The sharp difference between the wall surface and air temperatures creates a sheet of rising, hot air next to the wall. Of course at the same time walls at other orientations may be in shade. Heat conduction through walls depends on the materials, wall construction and the number of openings (windows, doors).

### 6.5.2 Energy Balance of Buildings

The SEB of a building depends on its construction (design, materials, insulation), setting (isolated or embedded in an UCL), the weather and climate, season and number of occupants and their uses of the building. An extremely general division of SEB exchanges would include energy gains (space heating/cooling 80%; solar inputs and occupant metabolism 20%) and energy losses (leaks from windows, doors and cracks 60%; conduction through external walls 25%; through roof 10%, down to ground 5%).

The daytime overall $Q^*$ on the exterior surface of an isolated building (Section 5.2.1) is mainly driven by solar irradiance on individual facets and is closely related to the facet albedo. If the building is embedded in an UCL the balance is far more complex, because solar irradiance may be obstructed by other buildings; conversely there is input from **reflectance** from those buildings and the ground. There are analogous exchanges of longwave radiation from the sky and interaction with other buildings. A small amount of solar radiation penetrates window glass.

The main energy input is anthropogenic, imported as electricity, natural gas, coal, wood and other combustible fuel and converted inside to achieve space heating, air conditioning, hot water generation, cooking and the operation of electronic or other equipment. Energy exchanges inside the structure (solar irradiance into windows, heat input to and losses from windows and interior walls, between the walls, ceilings and floors) redistribute energy in a building, and respond to the exterior weather conditions.

The primary heat loss processes are longwave radiation to the sky and other colder surfaces, sensible heat convection from warmed surfaces and chimney gases and a small amount of conduction to the substrate.

### 6.5.3 Energy Balance of Urban Canyons

A dry canyon-plus-roof system is a first order surrogate of the UCL in densely built districts of cities. This geometric form is repeated across urban landscapes, with variations of canyon orientation, aspect ratio, thermal properties and moisture availability. Many numerical urban canopy models use it to calculate energy exchange in a simple 2D canyon-plus-roof system (floor, walls and roof). This representation of **urban structure**, with only three or four facets, can add considerable realism to models, with relatively few additional computing demands (Section 3.3).

Here we demonstrate the nature of the SEB at the canyon top, and by combining it with that of the roof, arrive at an approximation to integrated fluxes from urban districts into the SL. Initially we consider the case of a dry city centre under a cloudless sky. Later we add moisture and vegetation to arrive closer to the more general case of urban neighbourhoods.

### Dry Urban Canyons

By day (Figure 6.20a) solar irradiance largely drives the SEB of the canyon system. Roofs are fully open to radiant receipt but little of the available $Q^*$ is taken into storage ($Q_G$), because of their low $\mu$. Instead the majority of $Q^*$ is convected from the hot roof surface to the cooler air as $Q_H$, and that heat is easily diluted in the atmosphere by turbulence. On the other hand, solar irradiance on surfaces within canyons is more restricted and depends on shadow patterns, which are governed by Sun-Earth geometry, canyon orientation and $H/W$ ratio. Other things being equal, the larger $H/W$, the smaller is average solar access. Within the canyon, multiple reflection between facets increases

overall **absorption** of both short- and longwave radiation (see Section 5.2.2). Because of their relatively large $\mu$ the walls and floor, accept and store heat well during daytime (i.e. large $Q_G$). This means their surface temperatures are much cooler than those on the roof facet. Further, since they are relatively sheltered, $Q_H$ from those facets is also not large. However, even if only for a few hours at a time, a fully irradiated wall may shed enough $Q_H$ to either aid or hinder the cross-canyon vortex (see 6.5.1 and Chapter 4).

In the evening, roofs cool fairly rapidly, expending much of their daytime heat store quickly (Figure 6.20b). Their relative coolness creates a micro-scale inversion at roof-level. In near-calm conditions, at irregular intervals, part of this roof cold air layer may drain, under gravity, into the adjacent streets. This undercuts the warmer canyon air and triggers the latent instability causing it to vent out of the canyon top. On the other hand, within the canyon, whilst $L^*$ is still negative, cooling is weaker than at

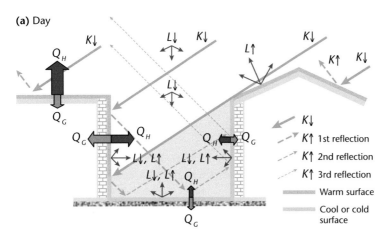

**(a) Day**

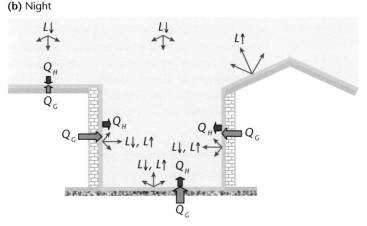

**(b) Night**

Figure 6.20 Schematic depiction of the main energy exchange fluxes comprising the SEB of roof and urban canyon facets **(a)** by day and **(b)** at night.

roof-level, because of the reduced sky view factor and the heat store to be dissipated is larger. The strength of cooling is controlled by the: antecedent temperatures of canyon facets; $H/W$ which sets the sky view and which of the other canyon surfaces can be 'seen'; emissivities of the surface and the effective radiating temperature of the sky (mainly controlled by cloud cover) (Chapter 5).

Figure 6.21a shows modelled energy fluxes for a dry canyon system using the TUF-3D model, as the aggregate fluxes passing through an imaginary plane across the canyon top, i.e.:

$$Q_{\text{top}} = \frac{WQ_{\text{floor}} + H(Q_{\text{wall1}} + Q_{\text{wall2}})}{2H + W} \quad \left(\text{W m}^{-2}\right)$$

**Equation 6.17**

where, $Q$ is any SEB flux. Despite real differences in the timing of radiative forcing on individual canyon facets (noted in Chapter 5), the diurnal course of the combined fluxes at canyon top is fairly smooth, resembling that of more simple horizontal surfaces (e.g. Figure 6.19).

Figure 6.21b adds the calculated roof SEB to that of the canyon top, which effectively approximates the SEB of an array of dry buildings and their streets. The addition of the flat roof does little to change the phase of the curves, but notice that values for the array are generally smaller. This is mainly because $Q^*$ is smaller due to the high daytime roof temperatures which cause large $L^*$ losses and boost $Q_H$ relative to $Q^*$ over the roofs.

Observations in a real-world canyon also show the smoothing effect (Figure 6.22). Close agreement between the calculated and measured values is not to be expected given that the radiative and thermal properties of the materials is not exactly the same, and the real canyon is oriented in only one direction and a

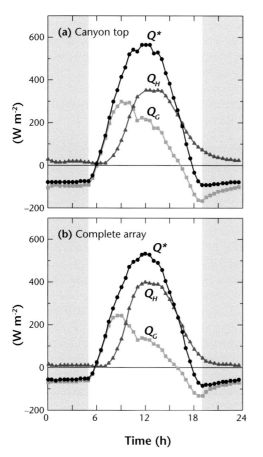

Time (h)

**Figure 6.21** Modelled ensemble average SEB of a grid city of E-W and N-S oriented dry, canyons, $H/W = 1$, $H/L = 0.4$, using the TUF-3D numerical model, with the same input properties used for **Figure 5.8**. **(a)** Modelled SEB at the canyon top level, and **(b)** SEB for the combined canyon top+roof system (i.e. it approximates the SEB at the local (or neighbourhood) scale).

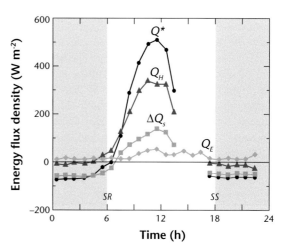

Time (h)

**Figure 6.22** – Diurnal SEB of an urban canyon ($H/W \approx 0.86$) based on the mean fluxes in the period from September 9 to 11, 1973 in Vancouver, Canada. The fluxes on the walls and floor of the canyon are measured, and then combined using the surface area of each as a weighting factor relative to the area of an imaginary horizontal plane at the canyon top (Source: Nunez and Oke, 1977; © American Meteorological Society, used with permission).

small amount of moisture is available at the floor (gravel). Real-world SEBs of urban neighbourhoods (Section 6.5.2) show similar characteristics. These results attest to the compensating nature of radiation processes (reflection, absorption and re-emission) *within* canopies, and the blending power of turbulent mixing *above* them.

### Moist and Vegetated Urban Canyons

The energetically powerful significance of adding patches of water in urban areas, by precipitation wetting or irrigation, or the presence of a low plant cover, and the process of micro-advection have already been mentioned (Section 6.4.3). Here we consider the SEB significance of trees in the UCL. Trees alter the canyon SEB in the following ways: they cast shade on other surfaces; enhance **drag** and hence reduce wind; and add a source of transpiration above ground level. Individual trees are important regulators of street and adjacent building microclimates. In some cities with scattered trees the UCL is rather like a savannah, in more densely treed districts the ecosystem can be officially classed as a forest (Oke, 1989).

The addition of small, widely-spaced trees has relatively little impact on the microclimate of an otherwise mainly built zone such as a street or parking lot. The shade is locally useful in tempering built surface temperatures, but the effect on wind speed is not large and the much-vaunted impact on the cooling of canyon air is often exaggerated. The reasons are several: (a) shading is patchy and limited in area, (b) single trees are easily circumvented by the wind (over, under or around), (c) being elevated the trees are especially well ventilated which is ideal for **evapotranspiration**. The resulting cooling and humidifying effects are quickly diluted into a large volume by mixing and/or replacement of air from above the UCL (Lowry, 1988), and the demand on soil water may be excessive, perhaps leading to the longer-term demise of the trees.

On the other hand, closely-spaced, tall trees lining the centre or sides of streets provide partial or complete shade for the road, pedestrian walk and adjacent buildings, which often is thermally beneficial to pedestrians and the flanking buildings by reducing overall surface temperatures. The air is cooler in such streets due to the combined effects of shade and tree transpiration. In older districts, mature street trees may be taller than the buildings thereby increasing the neighbourhood roughness. Under the canopy, wind speed and exchange between UCL and the layer above the trees are both reduced (see also Section 15.2.6).

Avenues of street trees, or trees in or around parking lots, receive an enhanced sensible heat load due to micro-advection from adjacent dry, paved patches. This effect boosts $Q_E$ (and tree transpiration) such that unless the site is located in the wet tropics or is continuously irrigated, trees need to develop deep tap roots to access sufficient water supplies to prosper.

## 6.5.4 Energy Balance of an Urban Canopy

Figure 6.3 gives the components of the neighbourhood SEB, measured or estimated, in the four cities introduced in Section 6.1.4. The results are long-term average values over several weeks. Due to the averaging, the course of $Q^*$, $Q_H$ and $Q_E$ are relatively smooth. Results for individual days might show more variable traces, e.g. due to changing cloud cover. Comparison between sites is facilitated when the terms are normalized by $Q^*$, the principal forcing term of the SEB (see Figure 6.23a,b).

### Daytime Energy Balance of an Urban Canopy

Figure 6.23b shows that $Q_E$ consumes an almost constant fraction of $Q^*$ at each site in the middle of day, when energy input is largest. The magnitude of the fraction $Q_E/Q^*$ is related directly to the availability of surface moisture at the different sites, which is controlled by the vegetation cover fraction, $\lambda_v$ (see Table 6.1). The lack of change in $Q_E/Q^*$ in daytime is actually a result of competing influences. On the one hand, SL instability and turbulence, and therefore the potential for evaporation, increase through the middle of the day. On the other hand, in the absence of precipitation the act of drying and sometimes partial stomatal closure progressively reduces the availability of surface water to support further evaporation.

The daytime course of $Q_H/Q^*$ follows a different path. The fraction increases towards the afternoon (Figure 6.23a) which approximately mirrors the shape of $\Delta Q_S/Q^*$ at the same time (Figure 6.12a). The course of daytime partitioning *between* $Q_H$ and $\Delta Q_s$ is shaped by the increasing role of convection over that of conduction, as the UBL and ABL grow. But unlike the controls on $Q_E$ there is no intrinsic limitation on the availability of heat at the surface. The daytime variation of $\beta$ in the four cities, and their rank order, are neatly differentiated by $\lambda_v$ (Figure 6.23c; Table 6.1).

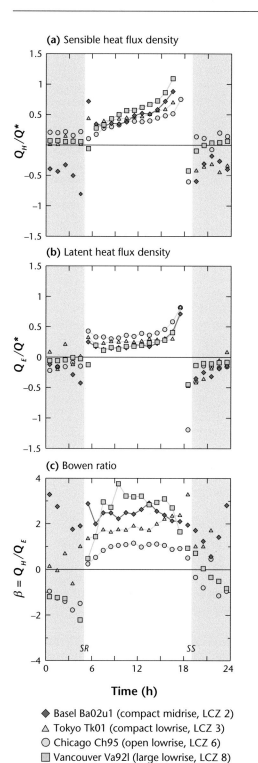

- ◆ Basel Ba02u1 (compact midrise, LCZ 2)
- △ Tokyo Tk01 (compact lowrise, LCZ 3)
- ○ Chicago Ch95 (open lowrise, LCZ 6)
- ▨ Vancouver Va92l (large lowrise, LCZ 8)

**Figure 6.23** Mean diurnal variation of SEB terms for the four case study sites noted in Section 6.1.3. **(a)** The normalized turbulent sensible heat flux ($Q_H/Q^*$), and **(b)** latent heat flux, and **(c)** the Bowen ratio ($\beta = Q_H/Q_E$).

## Nocturnal Energy Balance of an Urban Canopy

Near sunset thermally-driven turbulence drops away and often so does the wind speed and mechanically-driven turbulence. In general, as at most rural sites, a small positive $Q_E$ (evaporation) is maintained, perhaps interspersed with periods of dewfall when latent heat is released and $Q_E$ is negative (see Chapter 8). Dew or frost requires both the surface to cool below the **dewpoint temperature** of the near-surface air, and that there is sufficient turbulence to deliver a supply of water vapour to the surface. Both requirements are more difficult to meet in cities than in rural areas. The main constraint on dewfall is the warmth of built surfaces, with the exception of well-insulated roofs and vegetated surfaces with large sky view (see Section 9.3.1).

Nocturnal $Q_H$ is often distinctly different to the rural case surfaces. Over many urban surfaces (including the four depicted in Figure 6.23) $Q_H$ often does not change sign at night and is positive both day and night, i.e. the transport of sensible heat is into the atmosphere. In Mexico City, at a site in the central city almost devoid of vegetation, daytime $Q_E$ is negligible, about 60% of $Q^*$ is stored in the fabric ($\Delta Q_s$), with the rest convected ($Q_H$) into the SL (Figure 6.24). At night, as the surface starts to cool, the stored heat wells back up, such that the heat release exceeds the net radiative drain and fuels a

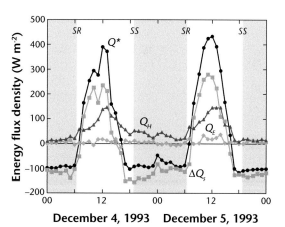

**Figure 6.24** Course of SEB components in central Mexico City, Mexico (Site Me93, see Table A2.1) during a cloudless period in winter 1993. Notice in particular, the large magnitude of both the uptake of storage ($\Delta Q_S$) by day and its release at night. Further, $Q_H$ at night is unusually large and directed upwards (Source: Oke et al., 1999; © Elsevier, used with permission).

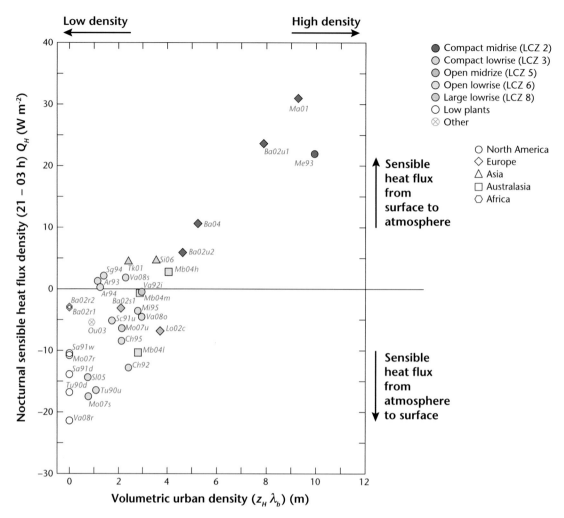

**Figure 6.25** Relation between the average nocturnal $Q_H$ and urban density. For source and site codes see Appendix A2.

substantial $Q_H$ away from the still relatively warm surface.

There is a direct relationship between the magnitude of this positive nocturnal flux and the built density of the city (Figure 6.25). The relationship between nocturnal $Q_H$ and urban density (expressed as the product of building height $z_H$ and **plan area fraction** of buildings $\lambda_b$) is a continuum: from undeveloped rural sites to the centre of large cities. We have noted that in deep dry canyons there can be intermittent venting of heat and a weak convective layer. At more open **suburban** sites, the nocturnal $Q_H$ flux may be positive or negative and the vertical air temperature profile may be adiabatic or inversion (Section 7.4.2). This nocturnal, upward directed $Q_H$ is an almost uniquely urban feature on land. There is a comparable

effect in bodies of warm water which have even higher $\mu$ than cities (Table 6.4). Despite the relatively small magnitude of $Q_H$ compared to daytime values, this upward flux can keep the urban SL **neutral**, even unstable, and well mixed at night, and probably contributes to heat islands of both the surface and UCL (see Sections 7.2 and 7.3).

### Seasonal Effects on Urban Energy Partitioning

The seasonality of the SEB terms is specific to a given location, since it depends on macroclimate and geographic controls. The seasonal phenology of vegetation (leaf on/leaf off) plays an important role, as does the timing and amount of snow cover, and the patterns of city living (e.g. space heating/cooling, garden irrigation).

Snow is unique in that it can suddenly alter the SEB and microclimate of any surface. Fresh snowfall immediately alters the **surface radiation budget** by increasing $\alpha$ and probably $\varepsilon$ (see Section 5.2.4) and modifying surface temperature. Further, a snow layer decreases heat storage (smaller $\mu$), and dampens turbulent transfer (surface temperature is capped at 0°C and frozen surface water is not readily available for evaporation).

Figure 6.26 illustrates the effect of snow on the SEB at an urban site (LCZ 2) in Montreal, Canada, for a period (a) with, and (b) without snow cover. The greater $Q^*$ in the snow-free case is partly due to the later date. Loss of snow cover has two main impacts on the SEB. Firstly, diurnal cycles of $Q_H$ and the residual $Q_{Res}$ ($= \Delta Q_S + Q_F + Q_M$ where $Q_M$ is energy used in snowmelt) changes from being approximately symmetrical and in-phase with $Q^*$ (i.e. like many low roughness surfaces), to asymmetric with the $Q_{Res}$ peak before noon and that of $Q_H$ later in the afternoon (Figure 6.26). The $Q_H$ pattern is typical of most snow-free surfaces discussed before. Secondly, the loss of snow exposes roofs and ground to the air, that boosts the role of turbulent transfer, especially $Q_H$, relative to that of $Q_{Res}$, in the SEB. Hence the loss of the insulating effect of the snow cover is more than matched by greater direct surface-air sensible heat transport. $Q_E$ remains negligible throughout.

### 6.5.5 Urban–Rural Energy Balance Differences

As noted in Chapter 2, the urban–rural difference $(U - R)$ paradigm is often used in **urban climatology** as a surrogate of the impact of urban development. In theory there is no reason to choose a rural zone over an urban one, as the reference. What is critical is that the reference remains a stable and repeatable standard of comparison. Urban–rural differences in SEB terms explain why urban climates and urban climate phenomena, such as the UHI, develop and exist.

#### Seasonal Changes in Urban–Rural Differences

Seasonal differences in the SEB partitioning are shown in Table 6.6 for a suburban-rural site pair in typical mid-latitude climate. $Q^*$ is slightly larger at the rural site, so $\Delta Q^*_{U-R}$ is slightly negative. The main reason is the surface UHI, its warmth underpins a steady net longwave ($L^*$) drain for the city, both day and night and in all seasons. Shortwave differences are small and only operate by day. In all seasons, $\Delta Q_{H\ U-R}$ is positive, but the difference is especially prominent in spring and summer. At that time the drier suburban surface promotes $Q_H$ and generates additional turbulence in the UBL. Conversely $\Delta Q_{E\ U-R}$ is negative in all seasons, especially in spring and summer. These conditions result in mean seasonal $\beta$ between 2 and 3.5 in the city, even though some of the summertime loss of water is compensated by garden irrigation. At the same time, rural foliage is verdant and evaporation keeps $\beta$ between 0.2 and 0.5 from spring to fall. In winter seasonal

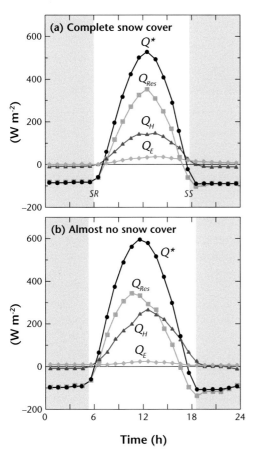

**Figure 6.26** Neighbourhood SEB components at an urban site (LCZ 2) in Montreal, Canada (Site Mo05, see Table A2.1) **(a)** 4-day ensemble average with a complete snow cover on roofs and ground, and **(b)** 3-day average with almost no snow cover. The SEB residual ($Q_{Res}$) includes $Q_F$, $\Delta Q_S$ and the energy involved in snowmelt, $Q_M$ (Source: Lemonsu et al., 2008; © American Meteorological Society, used with permission).

**Table 6.6** Daily totals of the energy balance terms measured over an unmanaged rural grassland site near Vancouver (Va08r, see Table A2.2 for details) and simultaneously at a suburban flux tower within the city (Va08s, Table A2.1) between Oct 2008 and Sep 2009. The suburban $Q_F$ is calculated from building energy models (BEMs), traffic counts and census data. $Q_{Res}$ is the energy balance residual, which is mainly heat storage change ($\Delta Q_S$) but also includes a minor fraction of energy used during photosynthesis and released by respiration and possibly advection ($\Delta Q_A$). $\beta$ is the Bowen ratio (Source: Christen et al., 2013).

| | Net allwave radiation $Q^*$ | | Sensible heat flux $Q_H$ | | Latent heat flux $Q_E$ | | Anthropogenic heat flux $Q_F$ | | Residual $Q_{Res}$ | |
|---|---|---|---|---|---|---|---|---|---|---|
| | Rural | Suburban | Rural | Suburban | Rural | Suburban | Rural | Suburban | Rural | Suburban |
| Spring (MAM) (MJ m$^{-2}$ day$^{-1}$) | 9.10 | 8.94 | 2.64 | 6.07 | 5.16 $\beta = 0.5$ | 3.04 $\beta = 2.0$ | - | 1.09 | 1.30 | 0.92 |
| Summer (JJA) (MJ m$^{-2}$ day$^{-1}$) | 13.44 | 12.53 | 4.40 | 9.56 | 8.39 $\beta = 0.5$ | 3.31 $\beta = 2.9$ | - | 0.74 | 0.65 | 0.40 |
| Fall (SON) (MJ m$^{-2}$ day$^{-1}$) | 3.75 | 3.38 | 0.77 | 2.51 | 3.18 $\beta = 0.2$ | 2.06 $\beta = 2.0$ | - | 1.07 | -0.20 | -0.12 |
| Winter (DJF) (MJ m$^{-2}$ day$^{-1}$) | 0.28 | -0.01 | -0.46 | 1.01 | 1.33 $\beta = -0.3$ | 1.10 $\beta = 0.9$ | - | 1.53 | -0.59 | -0.59 |
| Yearly total (GJ m$^{-2}$ y$^{-1}$) | 2.41 | 2.15 | 0.67 | 1.75 | 1.65 | 0.87 | 0.00 | 0.40 | 0.10 | 0.05 |

mean $Q_H$ and $\beta$ are slightly negative at the rural site, even though evaporation remains a sink, due to rain wetting the surface. $Q_F$ is only calculated at the suburban site because there are no sources in the countryside. Although hourly $Q_F$ is relatively small, its persistence makes it a significant term in the suburban SEB over the year. The residual of the SEB is primarily $\Delta Q_S$; it indicates more is stored at the rural site in spring and summer. This unexpected outcome may simply relate to the fact that any errors in the measured terms accumulate in the residual. The annual mean difference of the residuals, is very small at both sites. This helps give confidence that the SEB is virtually closed and that other terms not accounted for (e.g. advection, chemical storage due to **photosynthesis**) are not large.

### The Role of the Reference Site

The last example of suburban-rural differences in measured energy balance terms comes from Sacramento, United States (Figure 6.27). It illustrates how sensitive differences can be to the choice of the reference site. Synchronous SEB observations were made at a suburban site (LCZ 6) and two rural grassland sites (LCZ D) in the Sacramento region of the Great Central Valley in California: an irrigated sod farm (grows short grass for lawns) and a semi-arid prairie of unmanaged grass ($z \approx 0.20$ m). We use the

suburban site as the reference so as to investigate the influence of soil moisture availability at the rural site.

The differences of $\Delta Q^*_{U-R}$ are relatively small ($< 50$ W m$^{-2}$), mainly negative and follow a similar diurnal course at the two sites, especially at night when differences are only about 10 W m$^{-2}$ (not shown).

Differences of $\Delta Q_{S\ U-R}$ follow similar diurnal courses at both rural sites (Figure 6.27a). Daytime storage at the suburban site is greater than at both rural sites (differences are positive), reaching about 100 W m$^{-2}$ around midday at the wet site. Conversely nocturnal storage release is also greater in the city than at either rural site (differences are negative).

$\Delta Q_{E\ U-R}$ and $\Delta Q_{H\ U-R}$ are large and opposite to each other. At the irrigated (wet) rural site with surface moisture available evaporation is much greater than in the suburban area (differences of $Q_E$ are negative) whereas at the dry site $Q_E$ it is very small over the dried grass prairie compared with that in the city (differences are positive) (Figure 6.27b). $Q_H$ at the (dry) rural site is intense compared with that from the residential zone that includes surfaces like irrigated gardens and parks (differences are negative) (Figure 6.27c).

The main point of this example is that the moisture state of the rural site is sufficient to produce *opposite* differences in turbulent flux density comparisons;

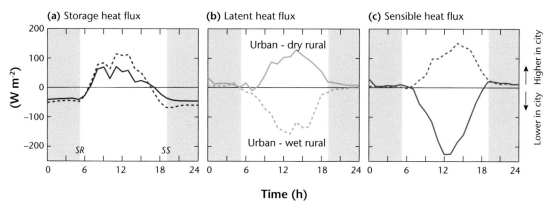

**Figure 6.27** Ensemble mean suburban-rural differences of SEB fluxes observed in Greater Sacramento, 1991: **(a)** Storage, **(b)** latent heat, and **(c)** sensible heat differences between suburban and rural. An open low-rise suburban site (S92u, see Table A2.1) is used for both difference pairs, but there are two rural grassland sites: one 'wet' (S91w see Table A2.2) and one 'dry' (S91d). *Note*: U-R differences are positive when the flux is larger at the urban site (Source: Oke et al., 1998; © American Meteorological Society, used with permission).

indeed it is able to *reverse* the sign of the urban 'effect'. Hence the notion of 'typical' urban impacts depends not just on the LCZ classes of the urban and rural sites, but also on the moisture state of the reference. It follows that the reference site in an urban-rural difference study should be representative of a commonly occurring environment, and its properties must remain reasonably fixed.

## Summary

In this chapter we have applied the principle of energy conservation to representative facets, surfaces and volumes in cities. The urban canopy is complex, owing to its faceted nature. Each facet has different radiative, thermal, moisture and aerodynamic properties that in combination with their geometrical arrangements produce a great variety of energy differences. The following are the main points of significance:

- Facets made of construction materials are generally **dense** and have relatively **high heat capacity** and and a **high conductivity**, closer to those of saturated rather than dry, natural surfaces. As a result, they are comparatively good stores of sensible heat.
- Built facets are usually **dry** and, when wetted, dry quickly. As a result, available energy ($Q^*$) at walls, streets, roofs, etc. is **expended as a sensible heat flux into the atmosphere ($Q_H$) and into the substrate ($Q_G$)**, rather than as a latent heat flux ($Q_E$).
- At the neighbourhood scale the SEB of a representative volume integrates the exchanges of all facets including fluxes due to heat storage change ($\Delta Q_S$) and **anthropogenic heat ($Q_F$)**. $\Delta Q_S$ is heat temporarily stored in building materials and soils that warm up during day/summer and cool during night/winter.
- A feature of the urban SEB is the relationship between $Q^*$ and $\Delta Q_S$; following sunrise, **a large fraction of Q$^*$ is initially transferred into $\Delta Q_S$** in the morning, however after noon, as urban building materials and soils warm up, they reach an equilibrium and more energy is channelled into the two turbulent fluxes $Q_H$ and $Q_E$. At night the turbulent fluxes subside,

the stored heat is released again and in dense urban districts **fuels continued positive $Q_H$ into the atmosphere** during night.

- The proportion of a neighbourhood that is vegetated ($\lambda_v$) regulates the ratio $Q_H/Q_E$, known as the Bowen ratio ($\beta$). Due to an overall reduction in vegetated area ($\lambda_v$) and limited surface moisture, urban surfaces typically channel **more heat into $Q_H$ than $Q_E$** (i.e. $\beta > 1$), but large $\lambda_v$ and irrigation can increase $Q_E$ (decrease $\beta$).
- More generally, the magnitude (and sign) of urban-rural differences in the terms of the SEB critically depends on the nature of the rural surface and its moisture status, because of its greater seasonal variation than in a city.

The distinct **SEB** of cities is **responsible for many observed urban climate effects**, including: enhanced generation of thermal turbulence and initiation of mesoscale wind circulations (Chapter 4), the urban heat island (UHI) effect (Chapter 7), urban-rural moisture differences (Chapter 9), and cloud formation and convective precipitation (Chapter 10). A physical understanding of the SEB at the building-, neighbourhood- and urban scales is germane to intelligent selection and implementation of design strategies to mitigate or enhance urban thermal and moisture climates (Chapter 15).

# 7 | Urban Heat Island

**Figure 7.1** Wide paved surfaces, like this road in an Indian city, that are dark, dry and unshaded can become extremely hot (notice the mirage patches in the foreground). This discourages pedestrian traffic; compare with the canyon in Figure 11.1. Hot urban surfaces contribute to heat islands in surface, air and subsurface temperatures in cities (Credit: Reuters, used with permission).

Decades of research have shown that cities are almost always warmer than their surroundings. This phenomenon, known as the **urban heat island** (UHI), is one of the clearest examples of inadvertent climate modification due to humans. It has many impacts and can be seen for example in the flowering of plants, which occurs earlier in urban areas, in lower space heating costs but higher space cooling requirements in cities, on increased heat stress on human residents in summer and in less dense **fogs** and increased rate of chemical reactions leading to **smog**.

At first glance the UHI appears to be a simple phenomenon with obvious causes many of which were identified nearly two hundred years ago by Luke Howard (1818). The root causes of the UHI were identified in Chapter 6: they are the changes urban development makes to the energy balance of the pre-urban site on which the city is built. Figure 7.1 illustrates one such change – here a paved road from an Indian city provides a dry, dark and unshaded surface that results in an increase in surface temperature relative to pervious natural surfaces. This represents

**Table 7.1** Summary of UHI types, their scales, causative thermal processes, approaches used to model them and direct and remote measurement techniques used to observe them. Also see Section 3.1, Figure 3.9.

| Section | UHI type | Scale | Processes | Models | Direct measurement | Remote sensing |
|---------|----------|-------|-----------|--------|-------------------|----------------|
| 7.2 | **Surface heat island** (UHI$_{Surf}$) | Micro | Surface EB | Surface EB and equilibrium surface temperature | Temperature sensors attached to surface | Satellite/ aircraft sensors |
| 7.3 | **Canopy layer heat island** (UHI$_{UCL}$) | Local | Surface EB and EB of UCL air volume | Canopy and RSL scheme incl. interactions with subsurface and overlying BL | Temperature sensors at fixed points, arrays and mobile in UCL and rural SL | Mini-sodar[1], mini-lidar |
| 7.4 | **Boundary layer heat island** (UHI$_{UBL}$) | Local and meso | EB at top of RSL and BL EB | BL scheme incl. interaction with RSL/surface and free atmosphere | Temperature sensors mounted on aircraft, balloons and tall towers | Sodar, lidar, RASS profiler |
| 7.5 | **Subsurface heat island** (UHI$_{Sub}$) | Local | Subsurface Energy Balance (EB) | Heat (water) diffusion in solid | Temperature sensors within substrate | - |

[1] Sodar does not measure $T$, but can sense temperature structure.

a contribution to a heat island of urban surface temperature; similar warming occurs in the air and subsurface substrate. While the temperatures of an UHI are relatively straightforward to measure, there are several types of UHI each of which is temporally and spatially dynamic which makes it methodologically complex to study. Each responds to a different set of scales, is caused by a different mix of processes, and requires different monitoring schemes to measure it and models to simulate it (Table 7.1). Therefore, one should not refer loosely to *the* UHI as if it is a unitary phenomenon, but rather identify *which* type is being measured, described, interpreted or modelled.

This chapter begins by defining the UHI more precisely and relating these definitions to the temperature changes with time that are linked to the energy balance (EB) processes described in Chapter 6. Then, for each UHI type we consider its spatial and temporal characteristics, how it is caused and how it links to other types of UHIs. To aid understanding, and establish some control, the approach is usually first to show the 'ideal' weather case with clear skies and low winds. Differentiation of thermal microscale and **local-scale** climates is greatest, and therefore most clear, when in absence of clouds, radiative heating or cooling is strongest and when the transport and mixing produced by winds is weakest. Then we add the thermally diminishing effects

of more complex weather conditions and surface controls. The reader will note an uneven treatment of the different UHI types. This reflects the level of study each has received in the literature. We restrict discussion to UHIs in simple physiographic settings, i.e. flat **orography** in the absence of large water bodies and complex land cover (see Chapter 12 for reference to urban heat islands in complex **topography**).

## 7.1 Urban Temperatures and Heat Island Types

Heat islands are a difference in temperature between urban areas and their surroundings. Here we use the following simple classification of UHI types (Oke, 1995):

- **Subsurface urban heat island** (UHI$_{Sub}$) – differences between temperature patterns in the ground under the city, including urban soils and the subterranean built **fabric**, and those in the surrounding **rural** ground.
- **Surface urban heat island** (UHI$_{Surf}$) – temperature differences at the interface of the outdoor atmosphere with the solid materials of the city and equivalent rural air to ground interface. Ideally those interfaces comprise their respective complete surfaces (i.e. $\lambda_c$, see Section 2.1.1).

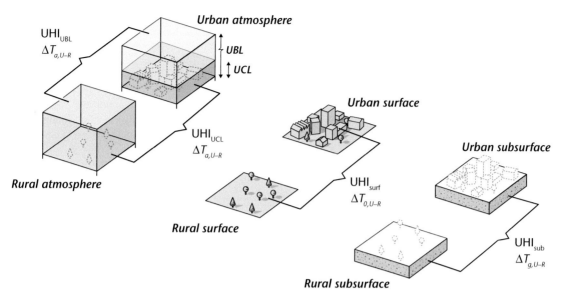

**Figure 7.2** Illustration of the temperature differences forming the four types of UHI: in the UBL (air layer from the ground up to the entrainment zone), the UCL (air layer from ground to about roof level), the surface (the complete surface including ground and all exposed facets of urban elements), and the subsurface (ground surface to depth of active temperature change over period of interest).

- **Canopy layer urban heat island** (UHI$_{UCL}$) – difference between the temperature of the air contained in the urban canopy layer (UCL), the layer between the urban surface and roof level (the exterior UCL), and the corresponding height in the near-surface layer of the countryside.
- **Boundary layer urban heat island** (UHI$_{UBL}$) – the difference between the temperature of the air in the layer between the top of the UCL and the top of the **urban boundary layer** (UBL), and that at similar elevations in the **atmospheric boundary layer** (ABL) of the surrounding rural region.

These four UHI types (illustrated conceptually in Figure 7.2) arise from differences in urban and rural cooling and warming rates at the surface, in the substrate and in the air. Alterations to these rates are caused by changes to the **surface energy balance** (SEB, described in Chapter 6). Here we add the corresponding changes to the EB of UCL and UBL air layers and the underlying substrate (soil).

The term 'heat island' was coined[1] because of the similarity between the spatial pattern of the isotherms

of air temperature in the UCL and height contours of an oceanic island. The analogy is apt for the surface, boundary layer and subsurface heat islands both by day and night, but for the UHI$_{UCL}$ it usually only describes the nocturnal case (e.g. Figure 7.3). The physiographic analogy has been extended to the terminology used for certain UHI features. Air temperatures ($T_a$) at an urban–rural border change abruptly and are called the UHI 'cliff', much of the urban area has relatively shallow gradients which is the UHI 'plateau', and the most densely built-up area (often in the city's commercial core, which may be ex-central to the built-up outline) is the UHI 'peak'. Other areas of particularly dense, perhaps tall, buildings or areas with a substantial **anthropogenic heat flux** ($Q_F$), are often relatively high 'hills' of warmth on the map of $T_a$. On the other hand, districts with relatively little or no urban development including forests, lakes, rivers, golf courses, vegetated parks and other greenspace are 'valleys' or 'pools' of relative coolness within the plateau. Details of the urban thermal landscape at the surface and in the UCL, especially the shape of the UHI outline, are closely tied to the degree of urban construction and human activity. This is particularly true when there is little wind to advect and mix thermal features.

[1] Origin unclear, but used by Balchin and Pye in 1947. Balchin, W.G.V. and N. Pye, 1947: A micro-climatological investigation of Bath and the surrounding district, *Quarterly Journal Royal Meteorological Society*, **73**, 297–323.

**(a)**

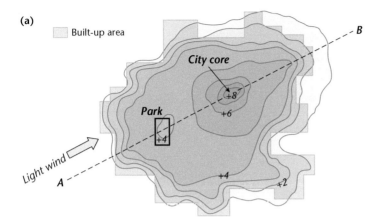

**Figure 7.3** Schematic depiction of a typical $UHI_{UCL}$ at night in calm and clear conditions in a city on relatively level terrain. **(a)** Isotherm map illustrating typical features of the UHI and their correspondence with the degree of urban development. **(b)** 2D cross-section of both surface and screen-level air temperature in a traverse along the line A–B shown in (a).

**(b)**

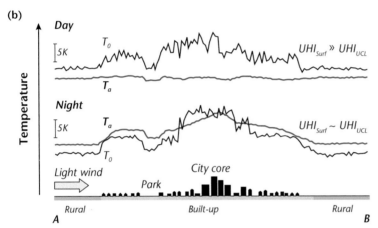

The objective of this section is to make explicit the link between the EB and the temperature changes in rural ($\partial T_R/\partial t$) and urban systems ($\partial T_U/\partial t$) as a basis for understanding how the temperature change affects the UHIs that occur for surface, atmospheric and subsurface temperatures.

### 7.1.1 Surface Temperatures

Every surface possesses a unique SEB for which there is a single temperature at its interface with the air — the surface temperature $T_0$ that satisfies its combination of radiative, conductive and **turbulent fluxes** (see Section 6.1). This temperature is the common boundary in the temperature gradients that generate a **sensible heat flux density** ($Q_H$) upwards into the atmosphere, and similarly conducts a sensible heat flux downward into the substrate ($Q_G$). Equation 7.1 represents the SEB for a surface, as a function of the change of $T_0$ with time for a layer of thickness $z$ with **heat capacity** $C$:

$$C\frac{\partial T_0}{\partial t}\, z = Q^* - Q_H - Q_E - Q_G \quad \text{(W m}^{-2})$$

**Equation 7.1**

It demonstrates that changes to the EB terms or of the thermal properties of the material, will affect $T_0$. Five surface properties exert particularly strong control on Equation 7.1 and therefore on $T_0$: 1. geometric, 2. radiative, 3. thermal, 4. moisture and 5. aerodynamic. Variability in these properties underlies the greater variability of surface temperature compared to air temperature, particularly by day (Figure 7.3).

### Geometric Properties

Figure 7.4 provides the opportunity to examine in more detail surface temperature variability in an urban environment. Geometric properties, including orientation (slope angle and azimuth) and openness to Sun and sky (Section 5.1) provide a strong control on $T_0$. By day the highest $T_0$ occur on **facets** that maximize the local **irradiance** ($K_{\downarrow}$); for example, facets with no

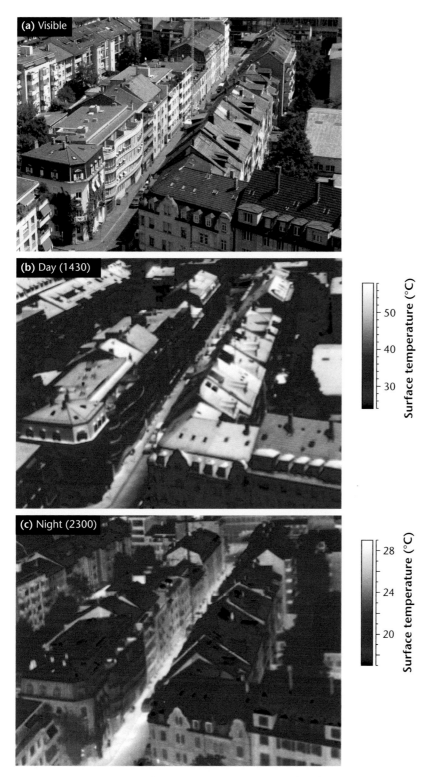

**Figure 7.4** **(a)** Photograph of Sperrstrasse canyon in Basel, Switzerland and adjacent courtyards (Credit: J. Voogt), and **(b)** thermal image of same canyon at 1430 h, and **(c)** at 2300 h on July 12, 2002 (Credit: J. Voogt). Viewing direction is to the NE. No correction for surface emissivity hence metals appear anomalously 'cold' (e.g. tower, skylight flashing, metal roofs).

shade well exposed at small local zenith angles to the solar beam. Note the contrast in $T_0$ for the two sides of the roof gables on the right-hand side of the canyon. Facets with lowest $T_0$ are found in shade (e.g. the southwest-facing walls in the foreground) or on slopes where the local **solar zenith angle** is large.

## Radiative Properties

Radiative properties control the ability to reflect **shortwave** (**albedo**; $\alpha$) and **longwave radiation**, and to emit longwave radiation (**emissivity**; $\varepsilon$ Section 5.1.3). Facets with low $\alpha$ favour shortwave **absorption** and higher $T_0$, whereas those with high $\alpha$ reduce shortwave gain and lead to cooler $T_0$. Two roof facets with nearly the same slope but different colours are visible on the left side of the canyon; that with the lower albedo results in a higher $T_0$. High $\varepsilon$ increases both absorption and **emission** of longwave radiation (Equation 5.2). If a surface is warmer than its surroundings and it has high $\varepsilon$ it will radiate heat more effectively which promotes its cooling. Facets with low $\varepsilon$ retain heat but appear anomalously cold (low **brightness temperature** $T_{0,B}$) when viewed by thermal remote **sensors** (e.g. metal skylight frames in Figure 7.4) because the reflected component of longwave radiation from the colder source is increased but the emitted component of the total outgoing longwave radiation is reduced (Equation 5.12).

## Thermal Properties

Thermal properties of the material, including, **thermal conductivity** ($k$) and heat capacity ($C$) (Section 6.3.1) govern the ability to conduct and diffuse heat into/out of the substrate material. Highest $T_0$ by day occur on facets made of materials with low $k$ and $C$ (i.e. low **thermal admittance** $\mu$), which resist heat transfer into the substrate and instead concentrate it in a thin surface layer that gets very hot (e.g. roofs). At night, roofs become relatively cold as they have a limited store of heat to draw from relative to road surfaces (Figure 7.4c).

## Moisture Properties

The availability of surface and near-surface soil and plant water moisture to evaporate (Section 6.2.1) provides a mechanism for heat loss by **latent heat flux** ($Q_E$) in Equation 7.1. Highest $T_0$ are found generally over dry facets (e.g. concrete, roofs) that do not channel much **available energy** into evaporating water.

Surfaces with access to water (e.g. wet soils, lawns, leaves) experience lower $T_0$ by day due to evaporative cooling and a lower diurnal temperature range.

## Aerodynamic Properties

Finally, aerodynamic properties, especially aerodynamic **roughness length** $z_0$ (Section 4.2.1) and shelter from wind influence $T_0$. Highest $T_0$ occur on facets that are smooth with little **turbulence**, and facets sheltered from the wind, while lower $T_0$ are observed over rough facets well coupled with wind.

There is an almost infinite mix of these properties in an urban area, so spatial patterns of $T_0$ in cities are extremely varied especially by day. The temporal variability of $T_0$ in cities is much greater than that of the air temperature ($T_a$) and that of the subsurface temperature ($T_g$). This is illustrated conceptually in Figure 7.3b and for select measured surfaces in Figure 7.5. During daytime (Figure 7.5), the $T_0$ of all surface facets is above the simultaneously measured canopy layer $T_a$ except transpiring vegetation that remains close to $T_a$. Conversely, at night (Figure 7.5), the $T_0$ of roofs is below $T_a$, walls are substantially warmer than $T_a$, but vegetation remains close to $T_a$ throughout.

Differences between these properties in urban and rural areas thus underlie different urban and rural temperatures as well as the large intra-urban differences in $T_0$ that are observed. The resulting surface temperature difference between urban and rural areas is the surface heat island ($UHI_{Surf}$).

## 7.1.2 Air Temperatures

The warming or cooling of an air layer or volume is explained by the EB of a *layer* or *volume* of air. Energy exchanges in the volume or layer can occur via radiation, **conduction** and **convection**. Molecular conduction in air is negligible compared to the effectiveness of the other terms and can be ignored. Equation 3.4 shows how changes in the internal energy of a parcel of air (temperature changes) can be related to adiabatic expansion/compression and to additions or subtractions of energy from the parcel.

### Air Temperature Change in a Layer over an Ideal Surface

For the simple case of a near-surface air layer over a spatially extensive, flat and relatively homogeneous rural surface where there are no net heat transfers in

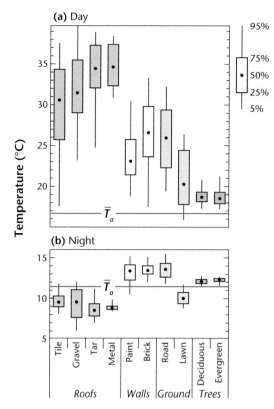

**Figure 7.5** Range and average surface brightness temperatures of various urban facets. Measurements are obtained from an oblique field of view of a thermal camera in Berlin, Germany over 80 minutes during day (around solar noon) and early night (21:00) of April 20, 2009. Values are compared to the simultaneously measured air temperature at screen-level in the UCL ($T_a$) (Source: Christen et al., 2012; © 2011 Springer-Verlag, used with permission from Springer).

the horizontal plane we only need to consider vertical fluxes for different layers (Figure 7.6a). Convection is the main transport process, but unless fog is actively forming (releasing latent heat) only $Q_H$ is relevant for changing $T_a$. It is the change in $Q_H$ with height not its absolute strength that controls temperature changes in an air layer. If inputs of **sensible heat** exceed losses for an air layer, warming takes place (turbulent flux **convergence**). If losses of sensible heat exceed gains then the layer cools (turbulent flux **divergence**).

An atmospheric layer can also both absorb *and* emit radiation; if emission exceeds absorption, the layer cools (radiative divergence). If absorption exceeds emission the layer warms (radiative

convergence). By convention, a convergence is a negative divergence and we hereafter refer only generically to divergence. We rarely consider radiative divergence over thin air layers where the magnitudes of absorption and emission are very small, often too small to be measurable. However, under certain circumstances the divergence of $Q^*$ and $Q_H$ within the layer are capable of modifying $T_a$. An equation that represents the contribution of both terms to the near-surface air temperature change in the layer is:

$$\frac{\partial T_a}{\partial t} = \frac{1}{C_a}\left(\frac{\partial Q^*}{\partial z} + \frac{\partial Q_H}{\partial z}\right)$$

$$= \frac{1}{C_a}\left(\text{div}Q^*_z + \text{div}Q_{H\,z}\right) \qquad \left(\text{K s}^{-1}\right)$$

**Equation 7.2**

where, $\text{div}Q^*_z$ and $\text{div}Q_{Hz}$ denote the change of **net allwave radiation** and sensible heat respectively with height in the layer $z$. Equation 7.2 states that warming/cooling of an air layer $\partial T_a/\partial t$ is directly related to changes in its heat content, which are due to changes of the vertical fluxes of $Q^*$ and/or $Q_H$ with height. Substituting the typical magnitude of $C_a$ (~1,200 J m$^{-3}$ K$^{-1}$) into Equation 7.2, we see that a vertical change of only 0.1 W m$^{-2}$ in radiation and/or sensible heat flux over a distance of 1 m is sufficient to cause warming or cooling of 0.3 K h$^{-1}$.

Figure 7.6a illustrates a nighttime case for a simple rural surface. The blue sphere represents a point a few metres above the surface where, at night under clear skies and very light winds, strong radiative cooling of the layer occurs after sunset due to vertical divergence of $Q^*$. Under these conditions, a surface-based temperature **inversion** develops as the surface cools radiatively to space through the **atmospheric window** (Figure 7.23). The vertical divergence of $Q^*$ is dominated by divergence in the upward longwave radiation (Steeneveld et al., 2010). This effect is maximized a few metres above the surface and decreases above. The vertical radiative divergence is enhanced with larger vertical variations of temperature and humidity. **Turbulent** transfers in this **stable** layer are weak and intermittent but observations suggest an elevated minimum in the sensible heat flux generates a small sensible heat flux convergence in the layer (Garratt and Brost 1981) that accounts for observations of total layer cooling that are less than the cooling rates determined from the radiative divergence.

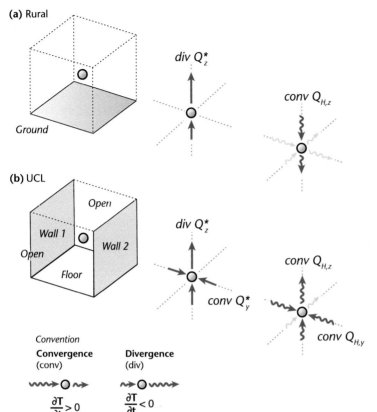

**(a)** Rural

Ground

*div* $Q^*_z$

*conv* $Q_{H,z}$

**(b)** UCL

Open

Wall 1

Open

Wall 2

Floor

*div* $Q^*_z$

*conv* $Q^*_y$

*conv* $Q_{H,z}$

*conv* $Q_{H,y}$

Convention

**Convergence**
(conv)

$$\frac{\partial T}{\partial t} > 0$$

**Divergence**
(div)

$$\frac{\partial T}{\partial t} < 0$$

**Figure 7.6** Heat exchanges that warm or cool a near-surface air volume at: **(a)** an extensive rural site, and **(b)** a street of the urban canopy layer. Conditions are those of an 'ideal' night (cloudless sky, weak wind or calm). The length of arrows is proportional to the relative strength of the heat transport (longwave radiation or turbulent sensible heat). Vector arrows greyed out if difference is minor (convergence or divergence uncertain).

## Air Temperature Change in a Volume

In the **surface layer** an air volume is often horizontally constrained by surrounding surfaces that participate in surface-air exchange. Thus we must consider a *volume* not just a layer. For example, in the UCL, building walls bound the air volume of an **urban canyon** (Figure 7.6b). Expanding Equation 7.2 to represent a volume, retaining the assumption that no phase changes of water are occurring, and adding the effect of **advection** by the mean wind in the x-direction and a term to represent additional heat contributions such as that from **anthropogenic** sources, we can write

$$\frac{\partial T_a}{\partial t} = \frac{1}{C_a}\left(\mathrm{div}Q^*_v + \mathrm{div}Q_{Hv}\right) + \bar{u}\frac{\partial \overline{T}_a}{\partial x} + \frac{\partial S}{c_p \partial t} \quad (\mathrm{K\ s^{-1}})$$

**Equation 7.3**

where the volumetric divergence of net allwave radiation, $\mathrm{div}Q^*_v$ is:

$$\mathrm{div}Q^*_v = \frac{\partial Q^*}{\partial x} + \frac{\partial Q^*}{\partial y} + \frac{\partial Q^*}{\partial z} \quad (\mathrm{W\ m^{-3}})$$

**Equation 7.4**

and similarly the volumetric divergence of turbulent sensible heat $\mathrm{div}Q_{Hv}$ is:

$$\mathrm{div}Q_{Hv} = \frac{\partial Q_H}{\partial x} + \frac{\partial Q_H}{\partial y} + \frac{\partial Q_H}{\partial z} \quad (\mathrm{W\ m^{-3}})$$

**Equation 7.5**

For a simple urban canyon with the canyon axis oriented in the x-direction, the change in the y-direction represents contributions from the canyon walls and the z direction represents contributions from the canyon floor and turbulent mixing from the overlying air layer (that may be influenced by the building roofs). Figure 7.6b shows the nighttime case with clear skies and light winds where convergence of $Q_H$ and $Q^*$ from the canyon walls occur in the y-direction and divergence of both $Q_H$ and $Q^*$ occurs in the vertical direction. The difference between the air temperature in the UCL as governed by Equation 7.3 and that of a layer of corresponding height in the surface layer of the rural surroundings (as represented by Equation 7.2) is the basis for the UHI$_{\mathrm{UCL}}$.

## Air Temperature Change in the UBL

Air temperature changes in the UBL must consider the sensible heat flux contributed from the urban surface below and added via anthropogenic heat flux, and the convective flux of heat via **entrainment** from the **capping inversion** at the top of the UBL (i.e. $\partial Q_H / \partial z$ is important). Advection of heat by the mean wind is also important as UBL warmth is transported downwind. Radiative divergence can occur within the UBL due to **aerosols** and gaseous **air pollutants** that affect both shortwave and longwave radiation.

In the rural boundary layer both the sensible heat flux contributions, especially from below, and radiative properties are different. Thus a temperature difference exists between rural and urban boundary layers that is referred to as the boundary layer heat island ($\text{UHI}_{\text{UBL}}$). Formally we define the $\text{UHI}_{\text{UBL}}$ as the difference between the temperature of the air in the layer between the top of the UCL and the top of the urban-affected ABL – the UBL – and that at equivalent elevations in the ABL of the surrounding rural region.

### 7.1.3. Subsurface Temperatures

The SEB provides a **boundary condition** for the substrate below the surface. In the case of a layer with constant properties below the surface, the right hand side of Equation 7.1 is simplified so that only conductive heat transfers remain ($Q_G$). In a rural area, we can write the one-dimensional heat conduction equation that represents the change of the ground temperature $T_g$ over time. This relates the vertical conductive heat flux in a layer with homogeneous horizontal and vertical properties to the vertical gradient of temperature in the material and the material thermal properties (see Section 6.3).

$$\frac{\partial T_g}{\partial t} = -\frac{1}{C_g}(\text{div}Q_{Gv}) = -\frac{1}{C_g}\frac{\partial Q_G}{\partial z}$$
$$= -\kappa \frac{\partial}{\partial z}\left(\frac{\partial \overline{T}_g}{\partial z}\right) \quad (\text{K s}^{-1}) \qquad \textbf{Equation 7.6}$$

In the substrate below the city surface, we should initially consider the volumetric divergence of the sensible conductive heat flux recognizing that heat may flow into a volume from both the surface and from buried elements of urban infrastructure so that

$$\text{div}Q_{Gv} = \frac{\partial Q_G}{\partial x} + \frac{\partial Q_G}{\partial y} + \frac{\partial Q_G}{\partial z} \quad (\text{W m}^{-3})$$

$$\textbf{Equation 7.7}$$

With increasing depth the primary heat flux is in the vertical direction. At large depths in rural areas, it is expected that the gradient of $T_g$ will match the geothermal heat flux. Not considered here are heat fluxes associated with ground water flows which may be important; addition of an advective term to Equation 7.6 may be used to represent such transfers.

From these equations we see that $T_g$ for both urban and rural substrate materials depend on the material thermal properties of the substrate and by the boundary conditions – specifically the surface conductive heat flux. Differences in thermal properties or in the surface conductive heat flux, which may be enhanced in urban areas due to changes in surface properties and by the loss of heat from the built infrastructure at depth leads to urban–rural differences in $T_g$. The difference in urban and rural temperatures in the ground *under* the city, including urban soils and the subterranean built fabric, and those in the ground of the rural surroundings, defines the subsurface heat island ($\text{UHI}_{\text{Sub}}$).

### 7.1.4 Heat Island Magnitude

Whilst spatial variability exists in both urban and rural temperatures and thus the difference in the two that define a UHI, we commonly seek a single measure for comparison purposes. The **heat island magnitude** for each type of heat island is usually defined as the difference between the maximum urban temperature $T_U$ and a representative temperature of the surrounding rural area $T_R$ over a specified period, i.e.

$$\text{UHI}_{\text{type}} = \Delta T_{U-R} = T_U - T_R \qquad (\text{K})$$

$$\textbf{Equation 7.8}$$

Care is needed in both urban and rural environments to identify representative measurement locations; if spatial variability is large it may be advisable to calculate a spatial mean temperature for each environment. The difference may also be calculated spatially; where a single (possibly mean) rural temperature is used, plotting this difference identifies the spatial structure of the UHI. The **Local Climate Zone** (LCZ) scheme (Section 2.1.4) can be used to assess the thermal difference of urban areas from their surroundings. This scheme accounts for variations in the geometric, radiative, thermal, moisture and aerodynamic surface characteristics that impact both urban and non-urban surfaces and affect their thermal response. The resulting heat island magnitude will depend on the combination of LCZ used to define $T_U$ and $T_R$. Seasonal influences that affect phenology, snow cover and moisture further modulate these differences. This approach can be extended to consider the spatial differences in both urban and rural temperature variability as illustrated in Figure 7.3.

## 7.2 Surface Heat Island

### 7.2.1 Observation

The $UHI_{Surf}$ for large areas of cities is usually estimated from remote observations of $T_0$. The instruments are mounted on an airplane, helicopter or satellite platform. Those observations can provide a spatially continuous image of urban $T_0$ across a city. This helps overcome difficulties associated with using *in situ* sensors to sample adequately the vast range of surface facets comprising an urban system. The observation height and **field of view** (FOV, Section 3.1.1) of the sensor control the resolution of the image. Quantitative assessment of the spatial variation of $T_0$ requires corrections for atmospheric and surface emissivity effects (Section 5.1.3) and ideally for the anisotropy due to the corrugated nature of urban surface structure that creates 3D surface temperature patterns (Chapters 5 and 6). The **thermal anisotropy** of $T_0$ of a 3D surface such as a city, seen from a sensor platform, depends on both the viewing angle *and* the solar geometry. Hence, remotely sensed $T_0$ is a function of both the thermal state of the city, the optical properties of the instrument used to view it and the position of the Sun in the sky (that controls heating rates). Urban thermal anisotropy is greater than for most rural areas; it can exceed 10 K for densely built areas of a city in daytime (Lagouarde et al., 2004, 2010), but is smaller at night or in areas of more open structure. The directional control by the solar beam means that sensor platforms located on opposite sides of the same city, and therefore viewing from different directions, can simultaneously record different magnitudes of $UHI_{Surf}$. Restricting observations to near-**nadir** angles (straight down) gives a more consistent view of an urban area, but it has an inherent bias in favour of $T_0$ of horizontal surfaces. A nadir view omits $T_0$ of surfaces under tree canopies, awnings, etc. and reduces or eliminates the ability to view $T_0$ of vertical surfaces like walls.

Summing and averaging the $T_0$ of the myriad facets at the urban-air interface gives the **complete surface temperature** $(T_{0,C})$ which is the key variable in modelling the SEB, building energy demand and human **thermal comfort** (Voogt and Oke, 1997).

Defining the rural $T_0$ needed for Equation 7.8 can be a challenge because it is quite common to find that the non-urban area shows similar, or greater spatial variability of $T_0$ to that in the urban area. Selection of an appropriate reference $T_0$ should consider surface types, soil moisture, shadows and topographic setting (e.g. elevation, slope, proximity to water bodies) and whether single or multiple temperatures from an image will be used. Imhoff et al. (2010) provide guidance on how to select a reference non-urban $T_0$ from satellite images.

Finally, note that assessment of $UHI_{Surf}$ from orbital (i.e. satellite) platforms depends on the timing of the overpass. This is often at a constant local time of day, which restricts estimates to times of day that may not be ideal, in terms of the daily cycle of heat island development. The magnitude of $UHI_{Surf}$ also depends on whether a single image or a composite over a period of time is used; the latter tends to produce smaller magnitudes.

### 7.2.2 Spatial and Temporal Variability

#### Spatial Morphology under 'Ideal' Weather

Airborne or satellite thermal imagery that includes a complete city region helps visualize a picture that is close to that of the $UHI_{Surf}$ (some surfaces remain hidden). In thermal images recorded from satellite platforms urban areas often stand out from the surrounding landscape, both by day and night. The relative warmth of built surfaces (especially when dry) and their contrast with vegetation and water surfaces, is usually unmistakable.

Take the case of Vancouver, Canada (Figure 7.7). The urban area is bounded naturally: to the north by forested mountain slopes, to the west by ocean and to the south by a bog and farmland. The internal morphology of the built-up area is also evident. Light industrial, warehouses and transportation infrastructure (airports, wide streets) are often relatively hot during daytime. Areas of tightly-packed tall buildings appear slightly cooler because of mutual shade, whereas similarly highly developed, but less closely-spaced lower buildings, appear as warmer nodes. More heavily vegetated areas, especially with tree canopies, have lower daytime $T_0$ (see western sections of the City of Vancouver compared to eastern districts). The coolest

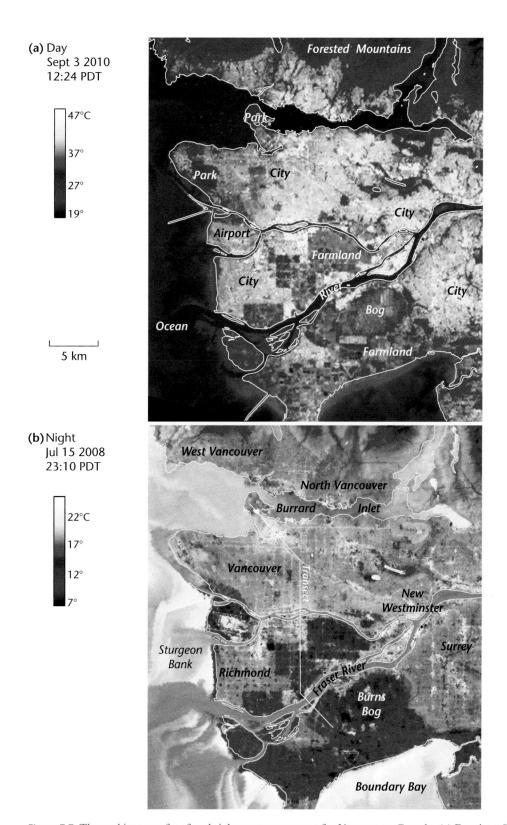

**Figure 7.7** Thermal images of surface brightness temperature for Vancouver, Canada. **(a)** Daytime: September 3, 2010 at 1224 PST. **(b)** Nighttime: July 15, 2008 at 2310 PST. Observations from the ASTER (Advanced Spaceborne Thermal Emission and Reflection Radiometer) instrument on the Terra satellite. Temperatures for Band 13: 10.25-10.95 μm, corrected for atmospheric but not emissivity or surface geometry effects; pixel size 90 m.

surfaces in the daytime are water bodies and areas of well-watered vegetation. To the south in the City of Richmond, strong contrasts in $T_0$ are associated with surfaces of intensive agricultural systems separated from highly developed urban areas by roads which give linear boundaries between the two land uses. These contrasting uses give consistently positive correlations between $T_0$ and the impervious surface fraction ($\lambda_i$), and strongly negative relations between $T_0$ and the vegetated surface fraction ($\lambda_v$). Comparing $T_0$ of urban and rural areas suggests an UHI$_{Surf}$ of about 9–12 K.

The effect of incorporating the **complete urban surface** into estimates of remotely-assessed surface temperature ($T_{0,c}$) by correcting for anisotropy, usually leads to a *decrease* in daytime estimates of urban system $T_0$ by over 3 K (Figure 7.8). This is because the surfaces added are mainly vertical and are typically cooler than the unobstructed horizontal surfaces preferentially seen from above. Such adjustments are rarely undertaken, but the effect can be sufficient to affect interpretation of both the magnitude of UHI$_{Surf}$ (as illustrated in the conceptual Figure 7.9) and the size and sign of surface-air temperature differences.

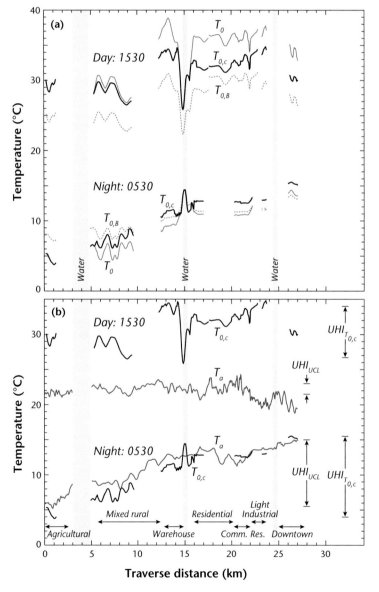

**Figure 7.8** Temperature transects across Vancouver, Canada, of **(a)** surface, and **(b)** air temperature on August, 10/11,1992 at two times of day. Surface temperature measured by an airborne thermal scanner and air temperature measured at 1.5 m by vehicle traverses. Initial brightness temperatures corrected for atmospheric effects and surface emissivity, then corrected for geometry to account for surfaces *not viewed* i.e. those obscured from the scanner FOV (Modified after: Voogt and Oke, 2003).

Further, if $T_0$ is used to drive or evaluate an **urban climate model**, it could mean the observed and simulated surfaces are not adequately matched and therefore nor are their surface temperatures.

The map of a nocturnal $UHI_{Surf}$ (Figure 7.7) is different to that in daytime. The relative warmth of light industrial and warehouse districts often disappears, while commercial and city centre districts emerge as the warmest. The **sky view factor** ($\psi_{sky}$) of the former is large, which assists cooling, the $\psi_{sky}$ of the latter is much smaller. Roads throughout the built-up area remain warm and are more evident than during daytime.

Vegetated but open parks (i.e. those with mostly grass surfaces) which have relatively large $\psi_{sky}$ are relatively cool at night and in marked contrast to surrounding urbanized areas (Figure 7.7). The strong differentiation in temperature associated with tree canopy coverage by day is less evident in the nighttime image. Water bodies and areas of well-watered vegetation become relatively warm compared to other surface types, especially towards the end of the night.

The use of $T_{0,c}$ at night leads to a larger spatial mean temperature compared to that viewed directly from above; this is because roof surfaces are relatively cold while the walls included in the adjusted version remain warm. After adjustment for the complete urban surface area, nocturnal $T_{0,c}$ becomes equal to, or warmer than, air temperature in some of the more densely built parts of the city (Figure 7.8). As a result, $Q_H$ often remains slightly positive at night in densely developed city centres (Figure 6.25).

## Temporal Variation and the Influence of Weather

In cities with distinct seasons the $UHI_{Surf}$ as observed from satellites is largest in the daytime and during summer, but in all seasons the nocturnal UHI is smaller than in daytime and it exhibits less spatial variability (Imhoff et al., 2010, Peng et al., 2011). Seasonal variation of $UHI_{Surf}$ depends on changes in surface properties, especially soil moisture. Wet winters are typical of many mid-latitude cities, this increases soil moisture and creates relatively large soil $\mu$, that in turn reduces nocturnal surface cooling (Chapter 6) resulting in smaller $UHI_{Surf}$.

Nearly all estimates of $UHI_{Surf}$ from satellites or aircraft are made with cloudless skies, for the simple reason that clouds block surface longwave radiation from getting to the sensor. The $UHI_{Surf}$ is reduced in magnitude with high humidity, because it reduces

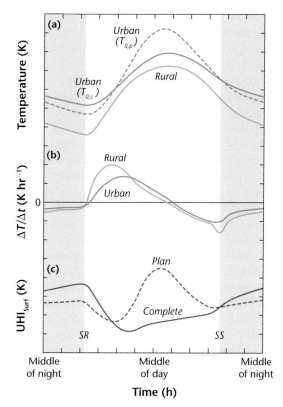

**Figure 7.9** Schematic **(a)** temporal variation of urban complete, urban plan and rural surface temperature on a day with fine weather, **(b)** associated urban complete and rural warming/cooling rates, and **(c)** temporal evolution of the $UHI_{Surf}$ for both the plan and complete surface. Vertical scale units are approximately 5 K, 1 K h$^{-1}$ and 2 K respectively.

differential radiative cooling between urban and rural sites. The effects of wind speed on $UHI_{Surf}$ have not been studied, but higher wind speeds should tend to increase mixing and reduce the $UHI_{Surf}$ both day and night. Antecedent conditions also play a role; cloudy conditions prior to the time of image acquisition can dampen $UHI_{Surf}$, while high soil moisture content, particularly in rural areas, affects $\mu$. This may increase $UHI_{Surf}$ by day, but reduce it at night, because of its impact on the diurnal range of rural $T_0$.

Figure 7.9 shows the expected diurnal evolution of surface temperatures, their heating and cooling rates, and the associated $UHI_{Surf}$ under fine summer weather for a moderately developed urban area (LCZ 2; $\lambda_b = 0.5$; $H/W = 1.25$) with a well-watered rural reference of low plants. These results are based on model simulations from Stewart et al. (2014). The model output has the advantage of directly showing the continuous evolution of $T_{0,c}$ and its heating and cooling rate along with both plan and complete representations of UHI$_{Surf}$. Observations of $T_0$ derived from ground-based upwelling longwave radiation measurements that provide some representation of vertical as well as horizontal surfaces at urban locations show similar features to those described here.

Shortly after sunrise, the unobstructed rural surface begins to warm rapidly as it receives direct solar radiation. The complete urban surface warms more slowly, in part because of its thermal properties, and in part due to shading of some walls and portions of the canyon floor. As a result both the plan and complete UHI$_{Surf}$ decrease to a minimum a few hours after sunrise. The maximum rate of urban warming lags that of the rural surface by 1–2 hours. By late morning, the urban warming rate exceeds that of the rural and UHI$_{Surf}$ increases. Heating rates decrease through the midday period for both urban and rural areas in a similar manner, with the urban rate remaining slightly above that of the rural through this period. Consequently there is a slow increase in the complete UHI-$_{Surf}$. The UHI$_{Surf}$ based on plan surfaces exhibits a maximum at midday due to strong heating of building roofs and of warming at the floors of canyons. Through the mid to late afternoon, a transition to cooling occurs, with the urban surface slightly lagging that of the rural. Cooling rates are greatest just before sunset for the urban area, possibly because of strong heat loss by roofs at this time and shading on parts of the canyon floor and some walls that leads to a shift in earlier peak cooling. In the rural area, the greatest cooling is near coincident with sunset and then declines quickly thereafter. Nocturnal cooling in the urban area is reduced slightly relative to the rural area, this leads to a gradual re-intensification of the UHI$_{Surf}$ for both plan and complete surface representations. Should dewfall or winds occur that differentially disrupt rural cooling, the magnitude of UHI$_{Surf}$ may decrease as rural cooling ceases but urban cooling is sustained by the release of stored heat. In agreement with observed results from Vancouver (Figure 7.8) the UHI$_{Surf}$ using $T_{0,c}$ is warmer than that for plan surface temperature ($T_{0,p}$) at night and cooler by day, with the daytime decrease in $T_{0,c}$ relative to $T_{0,p}$ larger than the night time increase. The midday UHI$_{Surf}$ based on $T_{0,p}$ is greater than the nighttime UHI$_{Surf}$ in general agreement with satellite observations.

The results illustrated in Figure 7.9 are sensitive to rural moisture conditions. For a dry rural surface, the rate of early morning warming is magnified and as a result, a negative daytime UHI$_{Surf}$ or 'cool island' may result. A dry rural surface can also sustain large nocturnal cooling rates that enhance the nighttime UHI$_{Surf}$ for both plan and complete temperatures relative to their daytime values. An extreme case is that of desert environments, for which satellite-scale observations show daytime cool islands in plan $T_0$ and positive night time UHI$_{Surf}$ (Figure 12.9).

### 7.2.3 Genesis of Surface Heat Islands

As noted at the opening of Section 7.1 the spatial pattern and magnitude of $T_0$ is the outcome of the SEB of urban facets, canyons and **neighbourhoods** (Section 6.3), and they are subject to the controls exerted by the surface properties and structure described in Section 7.1.1. Figure 6.20 illustrates the energy exchanges of the main urban facets (roof, canyon walls and floor) in 'ideal' conditions with cloudless skies and weak winds over the course of a day. In this subsection we build on that and show the thermal outcomes of those exchanges. Figure 7.10 illustrates the temporal evolution of surface temperatures for facets in an east-west oriented street canyon along with canyon air and building interior temperatures that complements the spatial variability of temperatures shown in (Figure 7.4, Figure 7.5 and Figure 7.7).

#### Daytime

The combination of low albedo, poor conductivity and dryness makes most roofs very hot on sunny days: roof surface temperatures ($T_{0,roof}$) of 50°C are common, and 60°C is possible, even in temperate cities (Figure 7.4, Figure 7.5, Figure 7.10). Hot roofs are fully exposed to the wind so they are the source of large turbulent sensible heat fluxes ($Q_H$), and thermal plumes and sometimes the extreme thermal gradients immediately above roofs generate a visible shimmering appearance. High $T_{0,roof}$ values contrast sharply with the $T_0$ of most ground surfaces in open, moist rural areas at the same time.

*Within* street canyons the receipt of short- and longwave radiation is disrupted compared with the rural open case. Shortwave radiation input to the walls and floor is subject to shadowing and multiple reflection, whereas longwave exchange comprises absorption

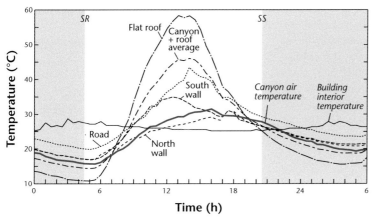

**Figure 7.10** Temperatures observed in an approximately E-W oriented street canyon (see also Figure 7.4) in central Basel, Switzerland on July 8/9, 2002. The day was mostly sunny with some cloud in the afternoon which cleared before sunset. Surface temperatures of the roofs and walls are measured by infrared thermometers. The canyon system average is the temperature derived from a downward-facing longwave radiometer at the centre of the top of the canyon plus the value of the flat gravel roof. Data from site Ba02u1 (Table A2.1) operated during the Basel Urban Boundary Layer Experiment, BUBBLE (Rotach et al., 2005).

and re-emission between the different canyon facets. $Q^*$ absorbed by wall and floor facets is more readily conducted into the buildings and ground than it is into roofs. The large $\mu$ of walls and roads gives them relatively high thermal inertia. Efficient **heat storage** uptake *into* the fabric in the morning means surface warming is delayed in comparison with the country-side. $T_0$ of the canyon floor depends on its thermal properties and permeability and the presence or absence of vegetation. Canyons may include trees and other plants that cast shade and transpire, both of which lead to relative surface coolness within the UCL. So even though individual facets may experience intense heating for short periods and most sur-faces are dry, the average $T_0$ *inside* canyons is usually much cooler than on adjacent roofs (Figure 7.4).

Comparing the daytime $T_0$ of an urban system to a rural environment, the facet surface temperatures are typically ranked as follows: $T_{0,\text{roof}} > T_{0,\text{walls}} > T_{0,\text{floor}} > T_{0,\text{rural}}$. That is just a general ranking, because it also depends on the urban fabric, **urban structure** (e.g. $H/W$), latitude and time of year which control canyon shade patterns, and also the moisture status of the soils and plants at the rural site. If the canyon and roof facets are combined into a single surface temperature for the system, by weighting facets according to their surface area (that is, $T_{0,c}$), the urban system is usually warmer by day than its rural surroundings.

## Night

In the city, roofs have a large sky view (often approach-ing unity) so the net longwave radiation $L^*$ is strongly negative in late afternoon and early evening; this drives strong cooling. Further, the **insulation** typical of roof construction prevents rapid replenishment of heat from the building interior. As a result, $T_{0,\text{roof}}$ becomes the lowest surface temperature in the urban system (Figure 7.4, Figure 7.5, Figure 7.10) so later in the night the magnitude of $L^*_{\text{roof}}$ declines and may become substantially smaller than $L^*_{\text{canyon top}}$.

Inside the canyon, the $T_{0,\text{walls}}$ and $T_{0,\text{floor}}$ are warmer than $T_{0,\text{roof}}$, for several reasons. Canyons are relatively sheltered from wind; $\psi_{\text{sky}}$ is relatively small (because the buildings flanking the canyon obstruct the horizon); and heat storage in the walls and floor retain daytime heat more effectively. Cooling rates near sunset are reduced relative to more open rural sur-roundings (Figure 7.9) and under optimum conditions remain lower than the more open rural area; this leads to a nighttime surface UHI. Under less optimum con-ditions in which rural cooling is reduced, urban areas can maintain significant heat loss due to the accumu-lated heat *from* the fabric; in this case the nocturnal surface UHI may reach a maximum earlier in the evening and thereafter decline.

At night the ranking of facet surface temperatures is typically: $T_{0,\text{roof}} < T_{0,\text{rural}} < T_{0,\text{walls}} < T_{0,\text{floor}}$. This is just the typical ranking, it can be disrupted by the local effects of primary controls like $H/W$ or the state of rural soil moisture. Notice that facets that were hottest by day are usually the coldest at night (Figure 7.4, Figure 7.5, Figure 7.10). This diurnal reversal in the relative temperatures of the facets is a fundamental feature of urban canyons. When these facets are combined into $T_{0,c}$ for the urban system it is always likely to be warmer than $T_0$ of the countryside.

### Controls on Urban–Rural Differences of Surface Temperature

A more quantitative analysis is possible if we reduce the analysis of $UHI_{Surf}$ to those properties most important to its essence. If anthropogenic heat release is relatively small, the controls on the $UHI_{Surf}$, on calm, clear nights, are remarkably few. With no short-wave or cloud impacts on radiation transfer, nor wind speed effects on turbulent mixing and advection, the surface temperature decrease through the night is simply a balance between the strength of the radiation drain ($L^*$) versus the ability of the substrate

heat reservoir to slow the loss down ($\mu$). Rewriting Equation 7.1 we have:

$$C\frac{\partial T_0}{\partial t}z = L^* - Q_G \quad \left(\text{W m}^{-2}\right) \qquad \textbf{Equation 7.9}$$

The course of $T_{0,c}$ with time for this 'ideal' nocturnal case could be solved using a numerical SEB model, which also allows the sensitivity of the $UHI_{Surf}$ to changes in physical properties to be assessed. Figure 7.11 shows the results of a sensitivity analysis and depicts the relative roles of urban–rural property differences in developing a nocturnal $UHI_{Surf}$. The vertical axis is the magnitude of $UHI_{Surf}$ at the end of a 12 h period of cooling (a 'night' starting at sunset), in a temperate climate with the following properties: $L_\downarrow$ = 300 W m$^{-2}$, initial isothermal surface and deep substrate temperature = 17°C, surface emissivity $\varepsilon$ = 0.95 at both the urban and rural sites and rural $\psi_{sky}$ = 1.0, the canyon has a symmetric cross-section and is infinitely long. These results represent a sensitivity test where urban and rural temperatures at sunset are assumed equal, in reality the urban surface is likely to be warmer and therefore $UHI_{Surf}$ is already

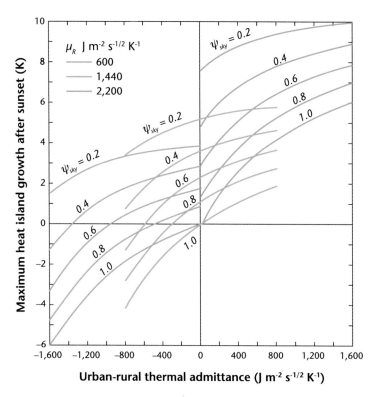

Figure 7.11 Predicted maximum magnitude of nocturnal $UHI_{Surf}$ in 'ideal' weather using a 1-D SEB model as a function of the difference between surface thermal admittance of rural and urban environments ($\Delta\mu_{U-R}$). Variables: sky view factor of canyons in city centre ($\psi_s$), and absolute rural value ($\mu_R$). Urban $T_0$ is represented by $T_{0,floor}$ in this analysis (Source: Oke et al., 1991; © Kluwer Academic Publishers, used with permission from Springer).

positive at that time (Figure 7.9). The three controls that vary are:

- The **sky view factor** ($\psi_{sky}$) for a point on the middle of the floor of the urban canyon (values from 0.2 to 1.0, as noted on each curve). Other things being equal, as the canyon sky view factor $\psi_{sky}$ becomes smaller (i.e. at larger $H/W$ ratios) the UHI$_{Surf}$ always increases.

- The absolute value of the **rural thermal admittance** ($\mu_R$) of the surface in the rural surroundings depends on soil moisture (see Section 6.3.1), as labelled on the three sets of $\mu_R$ curves. The value of $\mu_R$ sets limits on the magnitude of the UHI$_{Surf}$: if $\mu_R$ is large (e.g. waterlogged soils or flooded rice paddy fields; see the left hand set of curves) UHI$_{Surf}$ cannot become large, because rural cooling will be small. Indeed, it is possible that urban thermal admittance is *less* than the rural value so UHI$_{Surf}$ can be negative (a 'cool island'). On the other hand, if $\mu_R$ is small (e.g. dry soils; deserts or snow-covered rural areas; the right hand curves) UHI$_{Surf}$ can be large, because small inertia favours strong rural cooling. Other things being equal, as $\mu_R$ decreases, the UHI$_{Surf}$ always increases.

- The **urban–rural difference in thermal admittance** $\Delta\mu_{U-R}$. To investigate the significance of $\Delta\mu_{U-R}$ look at the range of UHI$_{Surf}$ values covered by any of the $\psi_{sky} = 1.0$ curves (i.e. both environments are flat with no geometry control) – in each of the $\mu_R$ groups the maximum sensitivity is about 6 K. To see the effect of geometry alone, look along the vertical line at zero on the horizontal $\Delta\mu_{U-R}$ scale (i.e. with no inertia effect), and note the UHI$_{Surf}$ at each $\psi_{sky}$. The maximum sensitivity varies from about 4 to 7.5 K in the three $\mu_R$ groups.

Notice that the sensitivity of UHI$_{Surf}$ to changes in $\psi_{sky}$ and to $\Delta\mu_{U-R}$, is of the same order. The most effective combination of the geometry and thermal admittance controls can passively generate an UHI$_{Surf}$ of about 10 K (in the top right corner), even without the influence of anthropogenic heat. If the weather is not 'ideal', so that cloud and wind become significant controls on the UHI$_{Surf}$, they always reduce its magnitude below the values in Figure 7.11. Anthropogenic heat has the potential to override all other controls. More complex urban climate models are able to incorporate these controls, and the role of vegetation, within the canopy.

## 7.3 Canopy Layer Heat Island

The UHI$_{UCL}$ is the most commonly studied heat island. Time (of day and year) and weather (wind, cloud) are strong controls on the UHI$_{UCL}$, the magnitude of which waxes and wanes in response to these modulating influences. Here, we start with the simplest set of controls before introducing others.

For clarity initial discussion focuses on calm, clear nights when radiation cooling is strong and turbulent mixing is weak. This permits sharp temperature gradients to exist, both in the horizontal between different surfaces, and in the vertical above them. UHI$_{UCL}$ development is then subject primarily to local controls and its magnitude is maximized. Given these conditions we look firstly at spatial patterns of temperature (Section 7.3.2, 7.3.3), secondly the cyclical rhythms of time of day and season (temporal dynamics) (Section 7.3.4), thirdly the damping effects of increased wind and cloud (Section 7.3.5), fourthly the impacts of seasonal changes of rural soil moisture (thermal inertia) and finally the seasonal changes of time of sunrise and sunset (Section 7.3.6). Initially it is assumed that $Q_F$ is a relatively insignificant factor.

### 7.3.1 Observation

The canopy layer heat island (UHI$_{UCL}$) is observed using thermometers to measure air temperature ($T_a$) near the ground. Air temperature can be measured in a weather screen or ventilated radiation shield, at one or more sites considered to be representative of urban and rural LCZs (WMO, 2008; Stewart and Oke, 2012). This is called the 'fixed' approach (Section 3.1.3). If the stations have continuous monitoring and recording equipment, temporal variations can be studied. Alternatively, or in addition, a thermometer can be mounted on a vehicle and traversed across a settlement and its surrounding non-urbanized area. This 'traverse' approach (Section 3.1) provides insight into small scale spatial variations of $T_a$. Fixed stations more easily reveal the temporal dynamics, whereas traverses provide insight into the spatial thermal response to variations in **urban form**.

### 7.3.2 Spatial Morphology

The basic structure of the UHI$_{UCL}$ was introduced in Section 7.1. The close relation between urban

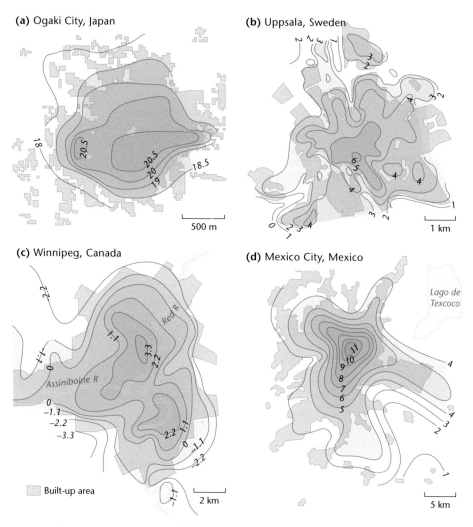

**(a)** Ogaki City, Japan

**(b)** Uppsala, Sweden

**(c)** Winnipeg, Canada

**(d)** Mexico City, Mexico

**Figure 7.12** Isotherm maps of the $UHI_{UCL}$ and their correspondence with the shape and degree of urban development. The size and population of the settlements increases from (a) to (d) – notice the tenfold change in scale (Sources: (a) Takahashi, 1959; (b) Taesler, 1980; (c) Einarsson and Lowe, 1955; (d) Jauregui, 1973).

development and thermal patterns means the $UHI_{UCL}$ of every settlement is unique, modified by the form of its local orography, the distributions of soils, vegetation and water bodies in the rural surrounding plus the patterns and nature of the built forms and functions (Figure 7.12). The unique pattern for every city may explain why so many $UHI_{UCL}$ studies have been conducted. But in truth the isotherm details in and around a settlement are mainly of local value only. The information is useful to establish areas of cool air (e.g. near greenspace, in river valleys, along coastlines) or warm zones (e.g. on slopes, near intense development or sources of anthropogenic heat). Such knowledge may

be useful in applied work on air pollutant **dispersion**, human comfort or energy use. However, rather than dwell on the nearly infinite variety of the $UHI_{UCL}$ in different places, here the focus is on underlying similarities, general characteristics, controls and genesis.

Figure 7.12 gives examples of the horizontal spatial form of nocturnal $T_a$ patterns in four towns and cities from different parts of the world, in nearly ideal weather. The settlements mainly differ in their 'size', measured both by surface area and the number of inhabitants. While the details of the pattern in each settlement vary according to their local of physical and urban geography they all demonstrate cliff, plateau, hill, valley and peak features similar to those in Figure 7.3. That is, the patterns found in smaller towns and cities resemble miniature versions of those of larger cities and megacities. Further, these features exist despite differences of **macroclimate** and culture (including architectural form).

### Intra-Urban Temperature Distributions

Intra-urban open patches and greenspace, especially managed parks, clearly show their presence, often as relatively cool areas at night. The reasons for this depend on the properties of these spaces, including the availability of moisture, the sky view factor and the presence of vegetation and/or water. These properties affect both the daytime heating, nocturnal cooling rates and the *relative* coolness or warmth of the thermal climate relative to that of the surrounding urban area in which it is embedded. The difference, sometimes called the 'park cool island' (PCI) has been studied in the parks of several cities located in different climatic zones. Like the $UHI_{UCL}$, the PCI is usually largest on nights with 'ideal' weather and its influence declines roughly exponentially with distance from the park edge, this limits its effective impact to a distance of about one park width (diameter) into its surrounds. The largest PCI differences reported are about 5 K (Spronken-Smith and Oke 1998, Upmanis et al., 1998). The results from tropical cities, while fewer in number, suggest similar effects to those of mid-latitude parks.

### 7.3.3 Maximum Canopy Layer Heat Island

The maximum UHI for a given city $(\Delta T_{U-R})_{max}$ is most likely to occur on nights with 'ideal' weather when large thermal differences can develop. Whether it is truly close to the largest $UHI_{UCL}$ found in that settlement also depends on the time of year and the soil wetness in the rural area (Figure 7.12). The controls operating to modify air temperature in the UCL are micro- and local scale, including:

- **Street geometry**, because it affects radiation receipt/loss (Chapter 5) and airflow (Chapter 4).
- **Building fabric**, because it affects heat storage/release (Chapter 6) and waterproofing (Chapter 8).
- **Vehicle traffic** and **space heating/cooling**, that release $Q_F$ (Chapter 6).

The power of these controls explains why in 'ideal' weather the location of the isotherms defining the

$UHI_{UCL}$, show such close correspondence with the built-up outline and internal patterns of a city (Figure 7.12).

The strength of the air temperature gradient across the cliff of the $UHI_{UCL}$ is related to the contrast between the rural and urban properties. If the rural-urban transition in thermal admittance and sky view factor is abrupt, so is the gradient in $T_a$. The largest gradients observed are about 5 K km$^{-1}$, which is sufficient to induce thermal breezes across an urban–rural border (Chapter 4). Such contrasts could also create mini-circulations within an urban area (e.g. across the edges of open areas, urban parks, forests, lakes and golf courses).

### Effect of 'City Size'

It is reasonable to expect that the absolute value of the maximum $UHI_{UCL}$ in a city relates to some measure of the 'size' of the city like population, city area or diameter (e.g. Oke, 1973). However, these measures are of less value that than those that describe the physical properties of the **urban canopy**, especially in the urban core. Historically, the most desirable and valuable land in the city was located in the central commercial core, which resulted in greater density of buildings and other infrastructure, taller buildings, more anthropogenic heat, less vegetation and so on. The **canyon aspect ratio** $(H/W)$ and the sky view factor $(\psi_{sky})$ of the street canyons in the city's core are reasonable surrogates of this transformation. Figure 7.13 shows the relation between these two measures and the maximum $UHI_{UCL}$. The results incorporate a wide range of settlements extending from small settlements (villages and towns) to large cities (with several million people). This suggests that $H/W$ and $\psi_{sky}$ are measures of the physical properties of cities that possess a degree of universality for thermal studies.

The magnitude of the largest *observed* $UHI_{UCL}$ that is passively-driven (i.e. not due to larger than normal $Q_F$) is about 12 K. Despite reporting from many $UHI_{UCL}$ studies and the development of megacities this value has not increased over the past four decades. It is possible that this apparent maximum is self-limiting. One potential mechanism for this might be the UHI circulation that is initiated by the large urban–rural differences (Chapter 4). The associated incursion of cool rural air limits the strength of the urban–rural thermal difference. A second possibility is that if the $H/W$ ratio of the urban system exceeds

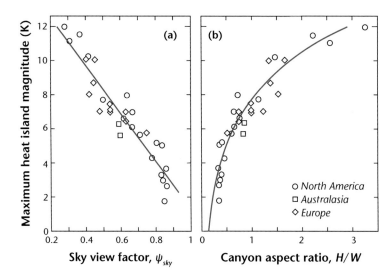

**Figure 7.13** Maximum heat island magnitude in the UCL plotted against measures of canyon geometry in the most built-up part of a city. **(a)** Sky view factor of canyons, $\psi_{sky}$ (Source: Oke, 1981; © Royal Meteorological Society with permission from Wiley), and **(b)** canyon aspect ratio, $H/W$ (Source: Oke, 1987; with permission from Taylor & Francis). The observations come from settlements in temperate regions on a single night with 'ideal' weather. All data gathered by automobile traverse, rural soil moisture relatively dry, no snow cover, no significant space heating or cooling.

some threshold, increased shade within canyons may limit daytime heat storage uptake sufficiently to outweigh the nocturnal thermal benefit of **horizon screening** that reduces longwave radiation heat loss from within the canopy. Since the shading depends on solar angle such competition between the two controls may depend on latitude.

### 7.3.4 Diurnal Variations

#### Warming and Cooling Rates

The largest $UHI_{UCL}$ are likely to occur on calm clear nights following cloudless days with weak winds. This combination produces large amplitude waves of surface and air temperature in both urban and rural areas. By day, solar heating of the surface is strong and in the morning heat sharing favours heat storage over convection; in the afternoon it is the reverse. At night, clear skies enhance the net loss of longwave radiation. That in turn lowers the surface temperature and creates an upward flow of heat from the subsurface heat store. Surface cooling is not effectively opposed by convection from the warmer air above.

In ideal weather, the amplitude of both the urban and rural temperature waves is large but the rural is greater. While daytime air temperature maxima are fairly similar in the two environments the nocturnal minimum in the rural area is much lower (Figure 7.14a). This is the essence of the $UHI_{UCL}$ effect: it is primarily a *nocturnal* phenomenon created

because urban areas fail to cool as rapidly as the rural surroundings in the late afternoon and evening. Hence it is driven by differences in rates of urban warming and cooling (Figure 7.14b) which create the daily variation of the $UHI_{UCL}$ (Figure 7.14c).

A typical daily sequence of **screen-level** air temperatures, warming/cooling rates and the resulting $UHI_{UCL}$ is as follows. Near sunrise rural areas are distinctly cooler, but soon air temperature increases rapidly. This is because rural surfaces are generally open to the Sun and their effects on warming are concentrated in the lowest air layers, where the lingering nocturnal inversion hinders mixing between the ground and the upper ABL until mid-morning. In the same period the warming of the UCL is more sluggish. This is for two reasons: firstly, early in the day canyons in the UCL are largely in shade (Section 5.2.3) and secondly, surface temperatures respond less rapidly due to the high thermal admittance of urban fabric (Section 6.3.1). By midday $T_U$ catches up to $T_R$, the overnight $UHI_{UCL}$ is rapidly eroded, so that by midday it is not uncommon for urban–rural differences to be slightly negative, i.e. $T_R > T_U$; a cool island. After their mid-afternoon maxima both environments cool down. The rural cooling rate is much larger because $\psi_{sky}$ is large and unless rural soils are very wet thermal admittance is relatively small, especially if soils are dry. On the other hand within the UCL it is sheltered, the horizon is restricted, heat is released more slowly from storage and plus there are

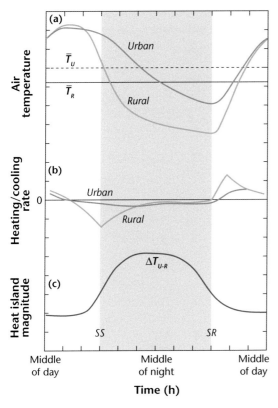

**Figure 7.14** Schematic of **(a)** temporal variation of urban and rural air temperature on days with fine weather, **(b)** the associated warming/cooling rates, and **(c)** the temporal evolution of the $UHI_{UCL}$. Vertical scale units are approximately 2 K for air temperature and heat island magnitude and 2 K h$^{-1}$ for the heating/cooling rates (Modified after: Oke, 1982).

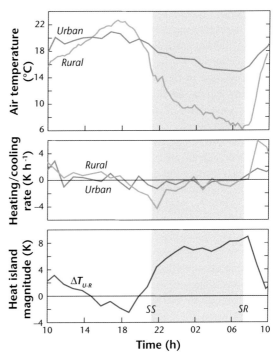

**Figure 7.15** Variation of air temperature at urban (LCZ 1) and rural (LCZ D) sites in Vancouver, Canada on August 10/11, 1992. Note: results are for same day as in Figure 7.8.

anthropogenic heat emissions. If soils are not wet the rural rate peaks near sunset at 3 K h$^{-1}$ or more, while urban rates are typically less than 1 K h$^{-1}$. $UHI_{UCL}$ growth is most rapid in the late afternoon to early evening period. Later in the night the rural rate declines as the daytime heat store is depleted, urban and rural rates become more similar and the $UHI_{UCL}$ usually reaches its peak in the middle of the period of darkness. For the rest of the night rates are similar in the two areas. After dawn the daily sequence repeats, as rapid rural warming quickly erodes the nocturnal $UHI_{UCL}$.

Figure 7.15 is the real world example of this diurnal sequence. The results for Vancouver, Canada, were observed on the same day as the car and helicopter transects of air and surface temperature shown in Figure 7.8. In most respects the thermal dynamics

follow those of the idealized case (Figure 7.14), the main difference being the timing of the UHI peak (just after sunrise in Figure 7.15) which is a local feature. Note that the diurnal course of $T_a$, $\Delta T_a/\Delta t$ and UHI for Vancouver (a temperate mid-latitude city) bears strong similarity with those for Ouagadougou, Burkina Faso in the arid tropics (Figure 9.7). This suggests the ABL physics is similar, it is the boundary conditions themselves (e.g. **surface cover**, ground conditions and weather) that differ.

The temperature transects across Vancouver on the same day show marked differences in the relative thermal responses of urban and rural areas (Figure 7.8). In mid-afternoon $T_a$ across the urban region are relatively similar; there is perhaps a small cool island of ~1 K. Around sunset (not shown), while the air temperature of the city core remained similar to that in mid-afternoon, the rural area cooled rapidly by several degrees and a $UHI_{UCL}$ of about 5 K developed. As night progressed, the city cooled by a few degrees, but rural cooling rates remained much larger, so by the middle of the night there was a $UHI_{UCL}$ of about 9 K.

That remained the case until near sunrise (Figure 7.8). Over the period the central city cooled by about 6 K, whereas the rural area cooled by more than 15 K.

The results in Figure 7.8 also allow us to see how the **lapse rate** in the near-surface layer varies both spatially and temporally. The mid-afternoon transect shows the surface is significantly warmer than the air, in both the city and rural area. Further, $UHI_{Surf}$ is about +8 K but $UHI_{UCL}$ is very small or even negative. Taking surface-air temperature differences $(T_0 - T_a)$ across the region we see large positive (lapse) differences of +11 or +12 K in the city and about +6 K in rural areas. These differences imply instability in the near-surface layer in both environments, and probably proportionally strong upward fluxes of $Q_H$. At night conditions are very different, the air is then warmer than the surface in the rural area, with a fairly consistent inversion of −2 K. In the city there are no strong differences, surface and air temperatures are about the same, but there are patches with weak lapse and even weak inversions. The $UHI_{Surf}$ is about +12 K while the $UHI_{UCL}$ is about +9 or +10 K.

Figure 7.14 makes clear that the switches from day to night and *vice versa* are key events for the $UHI_{UCL}$; sunset is the main period of growth, whereas sunrise signals its imminent destruction. When comparing UHI results from a city in different seasons, or between different cities, it is helpful to normalize the time scales. By doing that these 'triggers' of temporal dynamics act at a similar point in the daily cycle, even though the data are from different seasons or latitudes. In Figure 7.16 we see the $UHI_{UCL}$ results for four cities at different latitudes and of very different size. Initially, the daily courses of their mean raw UHIs appear dissimilar. However, when the amplitude of the UHI peak is set to unity, and time of day is normalized by plotting it as the fraction of the elapsed period of daylight or darkness (darkness extends from 0.5 to 1.5, and daylight from 1.5 to 2 and 0 to 0.5), the daily course of their $UHI_{UCL}$ show a remarkably consistent phase pattern (Figure 7.16).

### 7.3.5 Effects of Weather and Surface State

The controls exerted by weather and surface state, particularly soil moisture but also snow cover on urban–rural differences, are most keenly felt at rural stations. Stations in open country are more sensitive

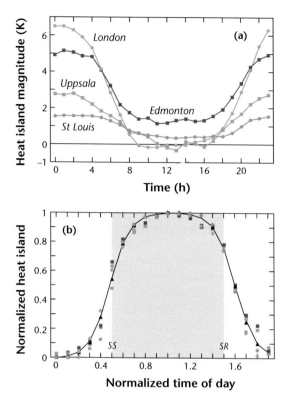

**Figure 7.16 (a)** Diurnal evolution of the $UHI_{UCL}$ in four cities as observed in local time, and **(b)** the same data with UHI magnitude normalized by its amplitude and time by hours of daylight and darkness (Data source: Oke, 1999).

to wind speed because they are in more exposed areas whereas urban ones, especially those in the UCL, are more sheltered. Likewise rural sites can 'see' clouds, or their absence, over most of the sky hemisphere, whereas within the UCL the sky view is considerably reduced. Rural soils are usually pervious and support plants, hence they readily allow both **infiltration** into the substrate and **evaporation** from the soil moisture store. The resulting seasonal variation of soil moisture results in relatively large changes of soil thermal admittance. In the city, where soil and vegetation is sparse, or is sealed over by construction, water is shed rather than stored. Hence the seasonal variation of 'soil' moisture, surface thermal properties and **Bowen ratio** are similarly muted.

### Wind Speed

Weather that is not 'ideal' disrupts the simple daily cycle of the $UHI_{UCL}$ and mutes its magnitude. In his classic pioneering study Sundborg (1950) found the mean daily UHI to be correlated with several weather elements, but the chief ones are wind speed and cloud cover. The physical reason is that wind speed is a surrogate measure of atmospheric transport and mixing; the main drivers of advection and turbulent exchange that limit horizontal and vertical temperature differences. Similarly, cloud cover is a strong control on shortwave and longwave exchange; the main drivers of heating and cooling, respectively. This also explains why **static stability**, which is due to the combined effects of radiation and mixing as measured by the vertical lapse rate of temperature at a rural site, also correlates well with the UHI. Humidity and pressure are less strongly correlated with the $UHI_{UCL}$, than wind and cloud.

To unravel the intertwined effects of wind and cloud we need to establish experimental control. This can be done by restricting analysis of the $UHI_{UCL}$ to the time of the daily maximum (i.e. holding the effect of time of day constant) on occasions when skies are clear (i.e. holding the influence of cloud constant). Now we can see the effect of mean wind speed ($\bar{u}$) without the competing (and confusing) influence of cloud and a fairly simple empirical relation emerges. It applies to both the case of individual nights using automobile traverse observations, and averaged data from fixed stations (Figure 7.17 a and b, respectively). It can be approximated with an equation of the form:

$$\Delta T_{U-R} \propto \bar{u}^{-k} \qquad \text{Equation 7.10}$$

where $k$ is a dimensionless number usually about 0.5 (i.e. approximately an inverse square root function). At the time of the diurnal peak and with no cloud, the curve describes the decrease of $UHI_{UCL}$ from its 'ideal' maximum (e.g. as in Figure 7.13) due to the effects of advection and mixing (i.e. injection of cooler rural air into, or escape of warmer air out of, the city and the lessening of air temperature differences by turbulent mixing). As wind speed increases, at some point the $UHI_{UCL}$ drops below a threshold (say < 1 K), making it unclear if it is more than natural variability. That speed is a matter of interest, but given the asymptotic form of Figure 7.17, is not easily quantified.

The overall shape of the heat island, while strongly affected by a city's internal structure, is also moulded by the wind direction. Figure 7.18 nicely illustrates the effects of both speed and direction on the $UHI_{UCL}$ based on a network of more than 65 stations in London, United Kingdom. Increased wind speed is mostly responsible for the reduction in the UHI magnitude between the two cloudless summer nights; one calm the other with westerly winds at a moderate speed. The overall outline of the $UHI_{UCL}$ on the calm night is almost circular in shape (Figure 7.18) and corresponds roughly to the shape of the built-up area. On the more windy night the outline is 'stretched' eastward, like a plume in the downwind direction (Figure 7.18). In the lee of a city the UHI is advected beyond the built-up limits and onward out into the rural environs. This was not verifiable in London

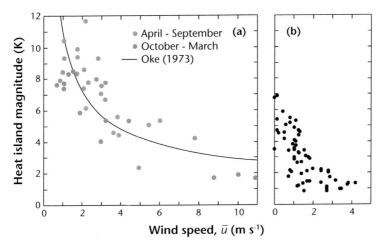

**Figure 7.17** The relation between the regional wind speed and the nocturnal $UHI_{UCL}$. **(a)** In Vancouver, Canada, at the time of maximum heat island magnitude from temperature traverses on cloudless nights (Source: Oke, 1976; with permission). **(b)** In London, United Kingdom the mean for the period 2200–0400 h on all August and September days in 1999, based on the two fixed sites marked in Figure 7.18: U-British Museum, city centre (LCZ 2), and R – rural, park in farmland (LCZ D) (Source: McGregor et al., 2006; with permission).

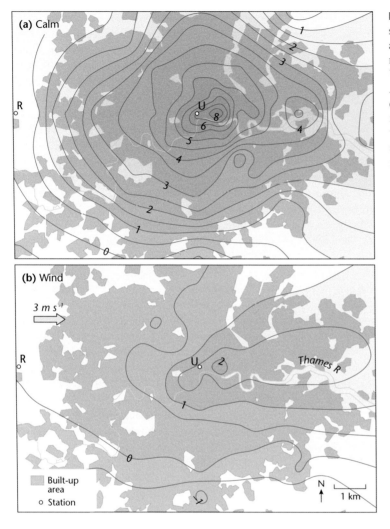

**Figure 7.18** The effect of wind speed in reducing the magnitude and changing the shape of the nocturnal UHI$_{UCL}$ of London, United Kingdom **(a)** On 28 August 1999 at 0100 h in near calm conditions the shape is rounded with its centre just E of the British Museum (located at station U at the centre of the rectangular grid). **(b)** The 'stretched' shape of the UHI$_{UCL}$ on 14 August, 1999 at 2200 h, with a W wind at 3 m s$^{-1}$, preceded by 4 hours with W winds between 3 and 4 m s$^{-1}$, it extends towards the E. Isotherm interval 0.5 K. R is the rural reference station (Source: McGregor et al., 2006; with permission).

## Cloud

Both cloud cover fraction (amount of sky obscured) and cloud type (indirectly related to height) are important modulating controls on the UHI$_{UCL}$. Most stations reporting cloud give the amount (in tenths or octas) but only a few give type (e.g. cirrus, Ci; alto-cumulus, Ac; cumulus, Cu; stratus, St; etc). Hence most studies of the influence of cloud on the UHI, use only cloud cover in their analyses. While amount captures the main cause of the reduction of solar receipt and the trapping of longwave radiation by cloud, it overlooks more subtle effects on the solar streams and certainly misses the significance of cloud type on the nocturnal radiation balance that is central to UHI$_{UCL}$ growth.

The combined effects of cloud type and cover fraction on net longwave radiation cooling at night are given by the Bolz relation:

$$L^* = L^*_{clear}(1 - kn^2) \qquad (W\ m^{-2}) \qquad \textbf{Equation 7.11}$$

where $L^*_{clear}$ is the net longwave radiation with cloudless skies, $k$ is a cloud type factor accounting for the decrease in cloud base temperature with increasing height (e.g. Table A2.3 in Oke, 1987), and $n$ is cloud amount (in tenths). Thus the term in brackets expresses the degree to which the longwave loss is mitigated by the absorption and re-emission of clouds.

Consider the case of complete overcast ($n = 1.0$), if the cloud type is stratus (low cloud base therefore relatively warm) the drain of heat by $L^*$ is reduced by about 90% (average of low clouds – those with bases below approximately 1.5 km) so the UHI potential is severely curtailed. For altus types (middle height) the energy sink is cut by about 75% and with cirrus (high altitude clouds consisting of ice crystals) even with overcast $L^*$ is only reduced by about 25%. So with weak winds the $UHI_{UCL}$ attains large magnitudes, even with complete cirrus cloud cover. This highlights the merit of using *both* type and amount, especially in nocturnal UHI studies; a scheme that neglects cloud type will miss much of the impact of cloud on the $UHI_{UCL}$.

The effects of wind speed and cloud may be combined in an empirical relation to predict the potential magnitude of the canopy layer heat island ($UHI_{UCL,pot}$ — i.e. the probable value if the weather had been 'ideal') from the observed ($UHI_{UCL,obs}$):

$$UHI_{UCL,obs} \propto \frac{UHI_{UCL,pot}}{\bar{u}^{\frac{1}{2}} \left(1 - kn^2\right)} \qquad \textbf{Equation 7.12}$$

### Soil Moisture

As noted previously, the seasonal variation of soil moisture *within* cities is usually not great and therefore is not a major control on the magnitude of the $UHI_{UCL}$. On the other hand, the variation in rural areas has a significant impact on the surface temperature, because of its dominant role in setting the thermal properties of the substrate.

The results in Figure 7.19 show the variation of thermal admittance at a rural station near Vancouver, Canada over a 21 month period, together with the value of the measured nocturnal $UHI_{UCL}$ corrected for the effects of weather using Equation 7.12. The two variables are approximately in anti-phase. This is because in the cold half of the year Vancouver is relatively wet, so soils in rural areas are nearly saturated and possess large thermal admittance ($\mu$). That means rural cooling rates are weak, so the magnitude of the $UHI_{UCL}$ can only be relatively small. In the warmer half of the year, soils dry out, $\mu_R$ drops and the potential for larger magnitude $UHI_{UCL}$ is enhanced. This is another example of establishing control, if the data had not been corrected for the effects of weather the $\mu$ effect would not have been

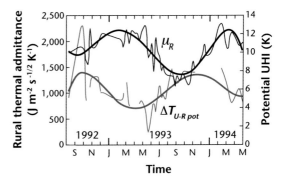

**Figure 7.19** Seasonal variation of rural soil thermal admittance ($\mu_R$) and the nocturnal potential $UHI_{UCL}$ from Equation 7.12 in Vancouver, Canada, for a 21-month period (Source: Runnalls and Oke, 2000; with permission from Taylor & Francis Ltd.).

evident, it would be jumbled in with the effects of all the other controls.

In some cities there is significant anthropogenic control of soil moisture through irrigation of gardens, golf courses, boulevards, etc. This is true of some temperate cities in summer and a few semi-arid or desert cities. The daytime cooling afforded by evaporation, and the reduced nocturnal cooling due to the greater $\mu$, have measurable effects on the surface and canopy UHI. If rural soils remain dry this urban control could increase the nocturnal UHI and create the possibility of a daytime cool island.

### Impact of Weather Effects

Warm and cold air masses, and the **fronts** that mark their edges, regularly move across the landscape. Air masses and fronts are associated with **synoptic** scale systems, like mid-latitude **cyclones** and **anticyclones**, and **mesoscale** systems like **land-** and **sea breezes**, mountain-and-valley winds (see Chapter 12) or cold downdrafts from thunderstorms. When these features cross an urban area they have several possible impacts on an existing heat island. They can completely overwhelm and destroy an UHI and reset the boundary conditions for a new one to develop (e.g. changed cloud cover and winds). They can also create spurious 'heat-' or 'cool islands' when the heat island magnitude is derived from urban-rural pairs that are differentially affected at times when the advective effect is present at one station but not the other. Figure 7.20 shows the impact of a passing thunderstorm on a summer night in Lodz, Poland. The normal development of a nocturnal

UHI from a daytime cool island continues until about 2300 h when there is a rapid drop and then a sudden 'heat' spike – this is the effect of a thunderstorm downdraft of cold air affecting only the rural site, after which the spread of colder air over the wet city keeps the UHI small for the rest of the night. To avoid misreporting urban–rural temperature differences that arise from these advective effects as UHI requires attention to more of the meteorology than just the temperature record. Chapter 12 provides further examples of thermal anomalies that affect the UHI.

### 7.3.6 Seasonal Variations

The synoptic weather at a location delivers a continuously changing mix of the meteorological variables

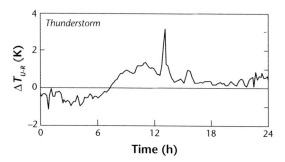

**Figure 7.20** Effect of an evening thunderstorm on urban-rural temperature differences in Lodz on July 23/24, 1998 (Source: Fortuniak et al., 2006; © Springer-Verlag 2005, used with permission from Springer).

that affect urban–rural thermal differences. This modulation of the thermal climate by weather is illustrated by the mean daily $UHI_{UCL}$ in Orlando, United States, as it varies with wind speed and cloud (Figure 7.21). In this study, Yow (2007) used the full Bolz relation described by Equation 7.12. As expected the UHI magnitude is largest when both factors exert little control (calm and clear) and smallest when both are large (windy, cloudy).

In mid-latitude cities the most fundamental actors setting the annual pattern of the $UHI_{UCL}$ are: solar irradiance (including the variation of daylength), **precipitation** and its influence on soil moisture and plant growth, and wind speed and cloud. Only during anticyclonic weather do ideal conditions exist, at other times the weather mix conspires to reduce its magnitude. Hence a city with a population of about 1 M inhabitants, which has the potential for a $UHI_{UCL}$ of say 10 K on an ideal night (Figure 7.13), typically has a mean annual value of only 1 to 2 K.

Figure 7.22 is an isotherm plot of the annual UHI climatology of Basel, Switzerland. The heat island magnitude is muted, because these are averages that include all types of weather conditions in their multi-year records. Nevertheless, they show the daily and seasonal dynamics of the $UHI_{UCL}$ well. The seasonal variation is strongly related to day (night) length. Solar control of warming and cooling clearly sets the timing of the daily UHI. The close spacing of isotherms near sunrise and sunset define the times of most rapid decay and growth of the UHI, respectively. The Basel $UHI_{UCL}$ is small, but does not show

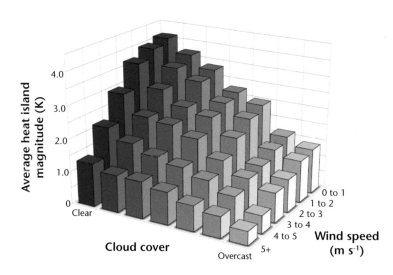

**Figure 7.21** Average nocturnal UHI in Orlando, FL, United States from September 1999 to December 2001. The cloud scale uses the Bolz cloud formula (Equation 7.11) to account for the effects of cloud type on longwave radiation (see Section 7.3.4). Cloud category 1 — clear and 7 — overcast. Wind speed categories are 1 m s$^{-1}$ bins (Source: Yow, 2007; © Blackwell Publishing with permission from Wiley).

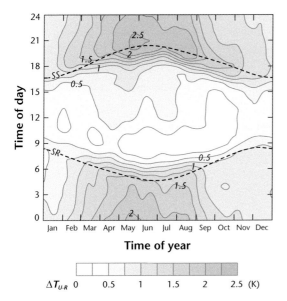

**Figure 7.22** An isothermic plot of the daily and seasonal variation of the UHI$_{UCL}$ (K) for Basel, Switzerland. The UHI uses air temperatures from two fixed stations: Spalenring (urban, 3 m, LCZ 2; 'Ba02u2', see Table A2.1) and Lange Erlen (rural, 2 m, LCZ D; 'Ba02r3', see Table A2.2) averaged for the period 1994–2003. The dashed lines are times of sunrise and sunset (Source: Vogt and Parlow, 2011; with permission).

negative values at any time (Figure 7.22); it is dominantly nocturnal at all times of year and its magnitude is greatest in the middle of the year. This seasonal pattern is probably mainly related to the thermal admittance of the rural soils and the climatology of storms. The summer cooling demand is not great, so $Q_F$ is unlikely to be a major factor.

Urban warmth has several practical implications. For example, in regions with cold winters the UHI lowers the number of **heating degree-days** which often directly translates into fuel savings for homeowners and businesses (e.g. see Figure 7.22). Extra warmth also means the city has a greater number of growing degree-days and frost-free days in a year. This is helpful to plant growth and may result in earlier dates for seed germination, budding, blossoming, leaf-out and fruiting. The city may also expect less fog (Chapter 9) and lower wind chill for those outdoors. Conversely in hot regions the UHI adds to the number of **cooling degree-days** thereby increasing the cost of air conditioning buildings. More importantly

it adds to the burden of hot weather for inhabitants of cities in the tropics, or those subject to seasonal heat waves. For those who have no access to air conditioning the UHI is an added threat to their health. Heat stress is the leading cause of mortality from weather related hazards in the United States greater than more violent phenomena like hurricanes (typhoons), tornadoes and floods. Prolonged exposure to excessive heat is particularly dangerous because the human body requires regular periods of sufficient coolness in which to recover. It is a concern that the nocturnal UHI ramps up in the evening, just when respite is sought after the day's heat (see Chapter 14).

### 7.3.7 Genesis of Canopy Heat Islands

The theory and processes involved in genesis of the UHI$_{Surf}$ are well known (Section 7.2.3). Causes of the UHI$_{UCL}$ are allied closely with those at the surface (Table 7.2) but the big difference is that they deal with a *layer* or *volume* of air, not just the air-surface interface. Naturally the surface temperature ($T_0$) remains critically important to that of the lowest layer of air ($T_a$), however, while the two are closely linked, the relationship is not linear. If $T_a$ is measured at screen-level, then at a city centre station it will be located in a canyon space. To understand air temperature change at any point in the atmosphere we need to analyze the EB of the air volume itself.

Of course it follows that suggestions about the 'cause' of the UHI$_{UCL}$ should also refer to processes in the *air* (Section 7.1.2) not simply those at the *surface*. Hence it is fair to criticize some of the reasoning offered to explain observed UHI in Sections 7.3.4, 7.3.5 and 7.3.6. However, our understanding of UHI$_{UCL}$ causation is not yet complete. Indeed in order to address the question, unlike most other topics in this book, what follows is more speculation than established fact. In rural areas, the daytime difference $T_0 - T_a$ and vertical **profile** of $T_a$ is typical of the **unstable** lower atmosphere in fine weather (Figure 7.23). The corresponding difference and gradient within the UCL is much smaller (Figure 7.23). In the most densely developed parts of a city, *absolute* air temperatures are similar to, or even slightly cooler than, rural air temperatures in the middle of the day. This coolness of daytime $T_a$ in the UCL is thought to be related to surface shading by the convoluted surface form; the greater thermal admittance of the fabric (surface temperature changes are

**Table 7.2** List of potential causes of the heat islands of canopy ($\text{UHI}_{\text{UCL}}$) and urban boundary layer ($\text{UHI}_{\text{UBL}}$). (Sources: Oke, 1982, 1995; Oke et al., 1991; Voogt, 2002).

| Cause | Description of cause |
|---|---|
| **Canopy layer heat island** ($\text{UHI}_{\text{UCL}}$) | |
| **Surface geometry** | (a) Increased surface area ($\lambda_c > 1$)<br>(b) Closely-spaced buildings<br>  – multiple reflection and greater shortwave absorption (lower system albedo);<br>  – small sky view factor ($\psi_{sky} < 1$) reduces net longwave loss, especially at night;<br>  – wind shelter in UCL reduces heat losses by convection and advection. |
| **Thermal properties** | Building materials often have greater capacity to store and later release sensible heat. |
| **Surface state** | (a) Surface moisture-waterproofing by buildings and paving reduces soil moisture and surface wetness.<br>(b) Convection favours sensible ($Q_H$) over latent heat flux density ($Q_E$).<br>(c) *If snow* – lower albedo in city gives relative increase of shortwave absorption compared to rural areas. |
| **Anthropogenic heat** | Anthropogenic heat release due to fuel combustion and electricity use is much greater in city. |
| **Urban 'greenhouse effect'** | Warmer, polluted and often more moist urban atmosphere emits more downward longwave radiation to UCL. |
| **Boundary layer heat island** ($\text{UHI}_{\text{UBL}}$) | |
| **Polluted boundary layer** | Aerosol and gaseous pollutants alter radiation transmission resulting in greater absorption and scattering of shortwave and greater absorption and emission of longwave. |
| **Sensible heat flux** | Greater turbulent sensible heat flux from rougher, warmer city surface. Upward mixing of warmer canopy layer air (i.e. $\text{UHI}_{\text{UCL}}$). |
| **Anthropogenic heat** | Heat injected upward into UBL from chimneys and factory stacks. |
| **Entrainment** | Stronger convection causes greater injection of warmer, drier air from above capping inversion, down into UBL. |

both reduced and delayed), and; mixing by the cross-canyon circulation.

At night in the rural area, on the other hand, where sky view is almost unobstructed, the air-surface interface is freer to cool by loss of longwave radiation. A stable layer forms that resists vertical mixing (Figure 7.23). As a result the lower atmosphere cools in an approximately exponential manner with time as it drains heat from the subsurface store. In contrast, surface cooling in the urban area is relatively weak (Figure 7.9), leading to almost **neutral** stability and gentle mixing. The vertical profile drawn in Figure 7.23 is for a densely built urban district, at less densely developed sites the profile above roof-level is probably closer to neutral, or is even slightly stable. The canyon air volume itself is bounded on three sides by wall and floor surfaces and thus requires the volumetric process introduced in Section 7.1.2 to explain the cooling.

## Heat Flux Divergence/Convergence

Section 7.1.2 introduced the basics of the warming and cooling of air volumes. Equation 7.2 shows that both, $\text{div}Q_H$ and $\text{div}Q^*$ are *potentially* important to air temperature change over a simple rural surface, however, if the air is unstable and well mixed as it is by day, both short- and longwave radiation pass through without sufficient absorption or emission to substantially alter $T_a$ and $\text{div}Q^*$ is weak except in calm conditions when a hot surface can result in long-wave radiative heating of a shallow unstable layer near the ground. Hence, by day when the atmosphere is well mixed, warming is due primarily to $\text{div}Q_{Hz}$. Similarly on most nights, $\text{div}Q_{Hz}$ remains the main determinant of air temperature. But if atmospheric stability is strong, turbulence near the ground is suppressed, strong vertical gradients in the **concentrations** of water vapour, other gases and dust may develop. These are radiatively-active, especially in the thermal

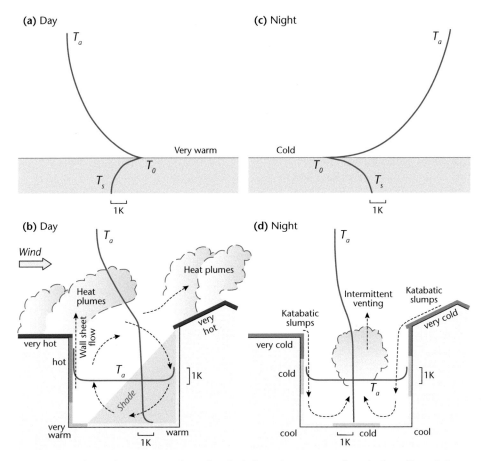

**Figure 7.23** Schematic representation of typical diurnal sequence of vertical profiles of air temperature in the near-surface layer of **(a,c)** an open rural area and **(b,d)** in the UCL and RSL of an urban area in **(a,b)** daytime, and **(c,d)** at night on a day with light winds and little or no cloud.

infrared (Chapter 5). The vertical gradients mean that differential absorption and emission between air layers can occur causing warming or cooling by radiative divergence/convergence. In ideal conditions, strong stability near the ground in rural areas can partially or even completely suppress turbulence and then $\mathrm{div}Q_z^*$ can become a significant, even the main control on the cooling of $T_a$ (Figure 7.6). Under those circumstances, the radiative cooling rate can exceed the actual rate of cooling measured using thermometers $\partial T_a/\partial z_m$ — this condition implies that while $\mathrm{div}Q_z^*$ is cooling the air strongly, $\mathrm{div}Q_{Hz}$ must be simultaneously warming it by convergence (Figure 7.6a). That offset reduces the measured rate of cooling.

Within urban areas, by day it is likely that as in the rural case $\mathrm{div}Q_v^*$ is minimal, therefore $\mathrm{div}Q_{Hv}$ caused by the heat flux convergence from walls and floor (Equation 7.3 and Figure 7.3), and perhaps heat advected from the surrounding UCL (second term of

Equation 7.3) are the drivers of air temperature change. From Equation 7.3 we see that a slower warming rate of the UCL air $(\partial T_a/\partial t)$ during morning, as seen in Figure 7.14 and Figure 7.15 is caused by a reduced convergence of sensible heat in the air layer compared to the rural reference. A reduced convergence of sensible heat can be explained by an overall lower $Q_H$ entering the UCL volume (shading inside the canyon, preferential partitioning of available energy to heat storage in the canyon materials) and/or by an enhanced turbulent mixing of the UCL air with the air above as compared to near-surface layers in the rural terrain (Figure 7.23).

At night, with clear skies and weak winds above roof level, even though the UCL is not as stable as in rural areas the air becomes almost calm due to the shelter provided by the canyon walls. The release of daytime $\Delta Q_s$ from the fabric and the restriction of $L^*$ from the canyon facets (due to their reduced $\psi_{\mathrm{sky}}$), creates much weaker heat loss in the city than in more open rural areas. In ideal conditions experimental observations suggest a rough equivalence between $(\partial T_a/\partial z)_m$ and $\mathrm{div}Q_v^*$ i.e. almost all cooling is radiative, with $\mathrm{div}Q_{Hv}$ being only minor. This may be the energetic basis supporting the largest nocturnal $\mathrm{UHI}_{\mathrm{UCL}}$ (e.g. Figure 7.14). However, corroborating evidence is needed from observation or **numerical modelling** investigations. The situation is further complicated by non-turbulent exchanges such as **katabatic** flows of cold air that are produced on roof tops and which drain into the canyon (Figure 7.23b). These effects are not included in Equation 7.3. In more typical conditions the role of $\mathrm{div}Q_v^*$ will decline and disappear as cloud and/or wind speeds increase leaving $\mathrm{div}Q_{Hv}$ as the only control on canyon air temperature change.

To this point we have considered $\mathrm{UHI}_{\mathrm{UCL}}$ development due to passive cooling differences between urban and rural areas, without major influence by $Q_F$. In Chapter 6 we note that $Q_F$ can add as much as 2 or 3 K to the UHI magnitude of large cities with intensely developed and active central districts. There are also settlements in the subarctic and arctic where $Q_F$ clearly dominates the SEB (see Chapter 12). For example, in mid-winter a town like Inuvik in northern Canada experiences 'arctic night': i.e. it receives no solar input for weeks or months. To heat itself the town depends entirely upon burning imported fuel and a small $\mathrm{UHI}_{\mathrm{UCL}}$ of about 1 K is created. More intense thermal anomalies are likely to exist in the immediate vicinity of heavy industry complexes.

### 7.3.8 Final Remarks

The $\mathrm{UHI}_{\mathrm{UCL}}$ is the most studied heat island reported in the literature. It seems easy and simple to observe. This perception may be responsible for the unfortunate fact that about half of reported studies have significant methodological flaws (Stewart, 2011). Common weaknesses include: poor (unrepresentative) site selection; lack of metadata; too small a sample size; failure to properly control or filter the results for the effects of weather, including frontal activity; or to control for the effects of the seasons on soil moisture and daylength. Attention to the concepts presented here and in Chapter 3 can help to avoid these problems. Probably some of the best conducted $\mathrm{UHI}_{\mathrm{UCL}}$ studies are those in Uppsala by Sundborg, 1951 and the follow-up by Taesler, 1980 both of which include fixed and traverse approaches. Other good examples: fixed – Chow and

Roth 2006; Fortuniak et al., 2006; Holmer et al., 2013; Fenner et al., 2014 and for traverses – Chandler, 1965; Sakakibara and Matsui, 2005.

## 7.4 Boundary Layer Heat Island

Naturally there is continuity between isotherm patterns in the UCL, and those just above, at the base of the UBL (Figure 7.24). Standard climate station data show the location of the $\mathrm{UHI}_{\mathrm{UCL}}$ shifts laterally

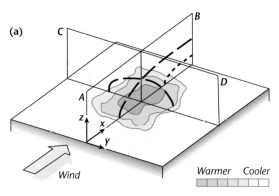

(a)

Wind    Warmer    Cooler

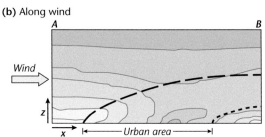

(b) Along wind

Wind    Urban area

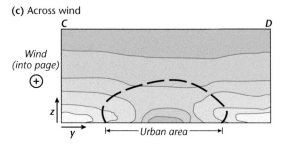

(c) Across wind

Wind (into page)    Urban area

**Figure 7.24** Schematic depiction of a typical $\mathrm{UHI}_{\mathrm{UBL}}$ at night with clear skies and a light wind over a large city in an area without strong topoclimatic control. **(a)** Horizontal distribution of the isotherms of potential temperature ($\theta$) at the top of the UCL, **(b)** vertical section of the $\theta$ distribution in the along-wind direction, and **(c)** across-wind direction (wind direction is into this panel). The dashed line indicates the top of the $\mathrm{UHI}_{\mathrm{UBL}}$ and in **(b)** the top of the rural boundary layer at the lower right.

with the wind (Figure 7.18). In the 1960s it was suggested that at the mesoscale a city can be viewed as a heat source anomaly that creates an **internal boundary layer** (IBL), just like a heated plate in a **wind tunnel**. The simple model of Summers (1964) proposed that the depth of the nocturnal $UHI_{UBL}$ is directly proportional to the strength of the urban heat flux, inversely related to the wind speed, and grows as the square root of the distance of **fetch** from the upwind rural-urban border (e.g. see Figure 7.25). The depth also

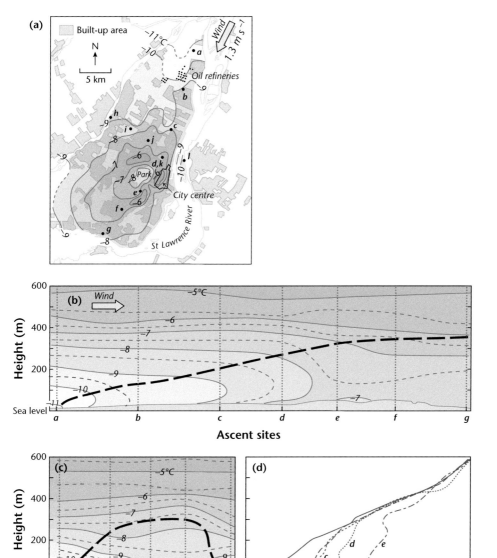

**Figure 7.25** Air temperature distributions in the early morning of March 12, 1968 with wind 1.3 m s$^{-1}$ from the north and north east and 1/10th altocumulus cloud cover. **(a)** The horizontal pattern at screen-level, **(b)** the vertical potential temperature ($\theta$) cross-section almost aligned with the flow, **(c)** the cross-section approximately normal to the flow, and **(d)** vertical profiles of $\theta$ at sites leading from the upwind urban edge to the city centre. Heavy dashed lines are estimates of the top of the thermal UBL. Vertical exaggeration ~12.5 (Source: Oke and East, 1971; © D. Reidel Publishing Co. used with permission from Springer).

depends on the stability of the rural area surrounding the city and the height of any rural inversions. At the downwind edge of the city the warm layer becomes elevated because it is eroded from below by a new IBL forming over downwind rural areas whose cooler air near the ground makes the UBL aloft look like a giant urban 'plume' of warm air spewing from a chimney stack (the city).

This concept was a paradigm shift that subsequently inspired many theoretical developments. UBL growth was verified by several field campaigns but none that show its complete 3D form. The $UHI_{UBL}$ is much less frequently observed because it requires very tall towers, aircraft or balloon measurements (Figure 3.9) and because it is a massive task to fully and quickly sample the volume of air that is thermally modified. Typically that involves a volume of air several tens of km on a side and 1 km deep. Ground-based **remote sensing** using profiling **radiometers** (Table 3.1) can now be used to infer the boundary layer heat island. It is important in $UHI_{UBL}$ measurements to consider the wind direction when identifying relevant rural reference temperatures due to advection of the warmer urban air downwind over rural surfaces.

### 7.4.1 Spatial Structure

Two forms of the UBL are observed:

- The **urban dome** is found with calm, or very weak, airflow. This generates the self-contained UHI circulation discussed in Chapter 4 (see Figure 4.31).
- The **urban plume** form is found in both stable and unstable conditions: The first plume case typically occurs at night, when weak or moderate winds advect stable air across a city. There it is modified from below, by enhanced mixing due to greater surface roughness and the extra heat of the surface heat island (Figure 7.22, Figure 7.26). It could also occur by day if rural areas are cold (frozen or snow covered) so the upwind ABL is stable. Depending on the city's roughness and heat output and the strength of rural stability, this type of UBL is usually shallow (say 50 to 300 m) at its deepest. The second plume case occurs by day when an unstable ABL is advected by moderate airflow across a city with strong sensible heat fluxes. The enhanced **buoyancy** creates vigorous **thermals** that rise easily through the entire **mixed layer** (ML) to the base of

the capping inversion where it spreads out laterally (cf. a forest fire) (Figure 7.27). The extra heat and instability over the city increases the depth of the ML (Figure 7.29 and Figure 7.31).

### Effect of City Size

There is no observational study of the relative magnitude of the $UHI_{UBL}$ in different cities. However, from general reasoning the size of the ML anomaly is probably proportional to that in the UCL unless there are special circumstances, such as a larger than average number of industrial or power production chimney stacks. The horizontal extent of the $UHI_{UBL}$ is related to that of the city, but if it grows into a megacity there seems no reason to anticipate that the magnitude of its heat island will increase greatly. The form of relations relating city size and the $UHI_{UCL}$ all indicate that heat island magnitude does not grow without bounds (Figure 7.13). A megacity's footprint is obviously large, but its vertical extent is usually constrained by a capping inversion and its horizontal persistence as a warm anomaly is relatively rapidly lost downwind due to dilution by turbulent mixing.

### Three-dimensional Structure

A study of the nocturnal $UHI_{UBL}$ of Montreal, Canada, gives the most complete picture of the full 3D temperature field over a large metropolitan area. Even so it is still only skeletal, based on simultaneous observations of the $UHI_{UCL}$ using a car, and the $UHI_{UBL}$ from helicopter soundings. The vertical profiles extend up to about 1 km above ground and include measurements of both temperature and sulfur dioxide ($SO_2$). The 11 sounding sites are arranged to form two cross-sections, approximately at right angles to each other (similar to the Figure 7.25 schematic). This combination of observations gives a 2D horizontal picture of the $UHI_{UCL}$ at screen-level from the car traverses, plus a partial view of the 3D form of the $UHI_{UBL}$, and its relation to the dispersion of air pollutants, using the helicopter profiles and cross-sections.

Results for March 12th, when the weather was favourable for UHI development, show a $UHI_{UCL}$ of about 4 K (Figure 7.25) and a thermal IBL that develops at the upwind rural-urban border, and continues to grow towards the city centre, where the surface is roughest and warmest (Figure 7.25). At the city centre the UBL is about 250–300 m deep

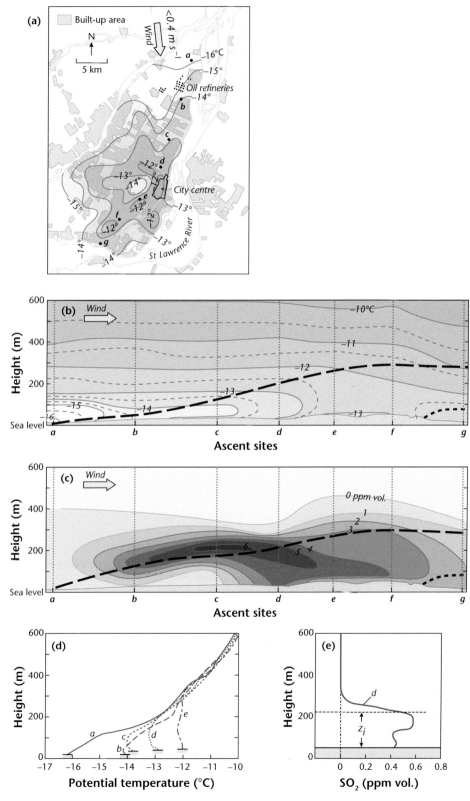

**Figure 7.26** Air temperature distributions at 0700 h on March 7, 1968 with winds of $< 0.4$ m s$^{-1}$ from the north with cloudless sky. (a) The horizontal pattern at screen-level, and (b) the vertical potential temperature ($\theta$)

(the height of the tallest buildings). The vertical cross-section aligned at right angles to the wind direction (Figure 7.25), shows the UBL thins at its edges, towards the river on either side of the Island of Montreal. The heat island is a ridge, elongated along the direction of the wind. The UBL fades out downstream from the centre on this day and there is no urban plume. Warm air advection aloft from the south east makes it difficult to identify the urban effect.

The vertical **potential temperature** ($\theta$) profiles from the upwind edge of the city (profile *a*), to the centre (profile *e*), show how a strong rural inversion is eroded by mixing as it traverses the city. Note that the slope of a $\theta$ profile directly indicates static stability: neutral (vertical), unstable (slope to left), and stable (slope to right). On this day the lowest layers are neutral or slightly unstable, even at the end of the night. Above that the rest of the UBL is slightly stable. This is quite common; the lowest layers of the UBL even support weak convection, generated by a positive $Q_H$ during the night over denser city districts (Chapter 6).

The weather was at least as conducive to UHI development on the morning of March 7th with cloudless skies, very weak airflow near the ground, increasing to 4.5 m s$^{-1}$ at 200–300 m. There was an UHI$_{UCL}$ of about 4.5 K (Figure 7.26) and a rural inversion with similar strength to that on March 12th. The depth of the UBL develops across the city, again extending up to about 300 m near the centre. The layer is unstable in its lowest levels and neutral or slightly stable above (Figure 7.26). A 'rural' boundary layer starts to form near the downwind border of the city (at this time of year the first surface encountered in the lee of the city is the river with ice floes). The shape of the SO$_2$ isopleths are due to the presence of elevated plumes emitted by oil refineries located near the start of the transect (Figure 7.26). After emission the initial **trajectory** of the air pollutants is slightly upward, but when they intersect with the top of the growing thermal UBL (after profile *c*), their trajectory bends downward, causing relatively high concentrations ($\geq 0.4$ ppm vol. SO$_2$) to be mixed down into the

central part of the city (Figure 7.26). The SO$_2$ profile near the city centre (at profile *d*), shows: (i) a narrow layer (between 125–200 m) of high concentration, with a sharp top that coincides with an inversion, and (ii) a layer beneath that coincides with the unstable part of the temperature profile. The layer with high concentrations are the elevated refinery plumes and the unstable layer beneath is a good example of urban **fumigation** (Section 10.2.3), wherein air pollutants are uniformly mixed down to the ground.

Figure 7.27 is a daytime cross-section of the UHI$_{UBL}$ on an afternoon with a UHI$_{UCL}$ of 2 to 2.5 K in St. Louis, United States. Winds averaged through the ABL are 8 m s$^{-1}$, but much less near the surface. Warmer urban air rises easily to the elevated inversion base and cause the ML to be domed, increasing its depth by about 300 m compared with the rural ABL. This is an increase of about 30%; more commonly it increases the UBL depth by 10 to 15%. The warm plume is carried downwind of the city, and remains identifiable for about 35–40 km.

Figure 7.28 shows how the UHI$_{UBL}$ varies with height in three cities, based on measured urban and rural temperature profiles. The mean daytime profile in Christchurch, New Zealand shows a relatively cool island ($\Delta T_{U-R} \leq -0.5$ K) up to about 70 m, above that up to about 300 m, it is warmer in the city ($\Delta T_{U-R} \leq 0.5$ K). At night the heat island extends from the surface up to almost 300 m (Figure 7.28). That compares well with similar profiles in the much larger cities of New York and Montreal. In all three the nocturnal UHI$_{UBL}$ declines approximately linearly with height. The exception is the warm 'bulge' aloft in Montreal, which is due to heat from industrial chimney stacks, i.e. a local effect. In the other two cities at the top of the heat island there is a shallow layer where the city is cooler than at the same height in the countryside. This is a common feature called the 'crossover-effect'. Several explanations for its existence have been suggested but none has been tested.

Figure 7.29 is a summary of the main thermal features of the UBL. It must be stressed that this is

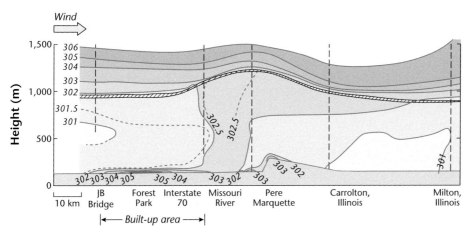

**Figure 7.27** Boundary layer potential temperature ($\theta$) structure in the St. Louis area measured by aircraft profiles and radiosonde flights at 1300 h on August 20, 1974. Hatched line is the mixed layer depth ($z_i$) and shaded plume is 1 K anomaly. Winds from the SE (almost parallel with the cross-section) at 8 m s$^{-1}$. Vertical exaggeration 34 (Source: Shea and Auer, 1978; © American Meteorological Society, used with permission).

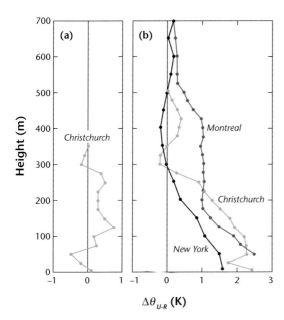

**Figure 7.28** Observations of the vertical $UHI_{UBL}$ based on averaged urban–rural profile *differences*. **(a)** daytime in Christchurch, NZ and **(b)** nocturnal profiles in Christchurch, using an instrumented tethersonde balloon, and New York and Montreal, both from helicopter surveys. Sampling in Christchurch was confined to cloudless weather in both summer and winter, and included 15 nocturnal profile pairs (observed between 2200 and 0100 h), and 10 daytime pairs (between 1100 and 1400 h) (Data sources: Bornstein, 1968; Oke and East, 1970; Tapper, 1990).

the smoothed average view; individual field studies report much more complex conditions than these, including the presence of multiple inversions aloft. This should not be surprising given the many place-specific controls in each city due to their unique orography, distribution of water bodies, urban development and industrial point sources. However, important as these surface controls are, most of the variability in the structure of UBL UHIs is due to weather, especially thermal advection and the changes in temperature structure as air masses and fronts develop, modify and surge across the landscape.

### Dynamic Stability

The most significant implication of the UBL temperature structure is its impact on atmospheric stability, which directly affects the dispersion of air pollutants (e.g. Figure 7.26). While the slope of the vertical profile of $\theta$ indicates static stability, a fuller measure is the **dynamic stability**, which incorporates the influences of both buoyancy and **wind shear** as given by the dimensionless group: $\zeta = (z - z_d)/L$ (Section 4.3.3). Observations of $\zeta$ over urban, **suburban** and rural areas in Basel, Switzerland show significant differences in stability between the three environments (Figure 7.30).

In daytime the atmosphere in the densely built-up inner city is unstable > 95% of the time, whereas in the rural surroundings the equivalent figure is about 80%, but only about 70% in the suburbs. The suburban site is thought to be closer to neutral because the

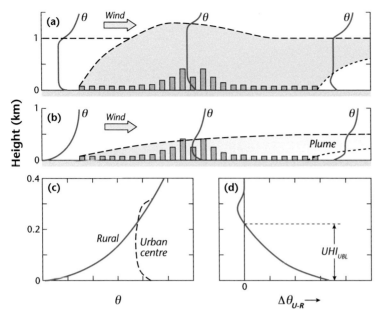

**Figure 7.29** Idealized summary of the thermal structure of the UBL during fine weather, in **(a)** daytime, **(b)** at night. **(c)** Rural and city centre profiles at night define the variation of the magnitude of the UHI with height, and its depth. Temperature profiles use potential temperature ($\theta$) (Source: Oke, 1982).

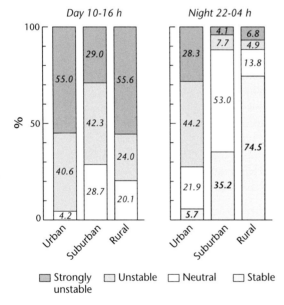

**Figure 7.30** Frequency of dynamic stability classes (stable: $0.1 < \zeta < 10$; neutral: $-0.1 < \zeta < 0.1$; unstable: $-0.5 < \zeta < -0.1$; strongly unstable: $-100 < \zeta < -0.5$) observed in summer in the inertial sublayer above urban, suburban and rural sites in and near Basel, Switzerland (Data source: Christen and Vogt, 2004).

open-set (LCZ types 4–6) roughness of the site encourages mixing more than in the more closely packed city centre (LCZ types 1–3) with its **skimming flow**, or the lower roughness of the countryside.

At night the frequency of stable situations is decreased markedly by the city. It is stable only 5% of the time in the city centre compared with 60% at rural sites (Figure 7.30). Within the city the occurrence of unstable conditions increases with building density. This is consistent with observations of positive $Q_H$ at densely developed urban sites, due to the delayed release of $\Delta Q_S$ and the continued emission of $Q_F$ (Chapter 6).

### 7.4.2 Variability

Diurnal Variation

Typically by day, after sunrise $Q_H$ increases in rural areas and the ML grows rapidly, then it levels off in the afternoon and collapses to zero near sunset, when $Q_H$ changes sign (Figure 7.31). The daily variation of the UBL and the ML follow a similar course, but in the city enhanced production of turbulence (both thermal (surface heating) and mechanical (urban roughness)) can cause the urban ML to grow up to about 20% deeper (Figure 7.31). Areas of enhanced convection are usually patchily distributed across the urban area. Buoyant updrafts bombard the base of the overlying **free atmosphere** (FA) and since the transition layer into the FA is an inversion, the entrainment flux brings heat, and drier, cleaner air back down into the UBL.

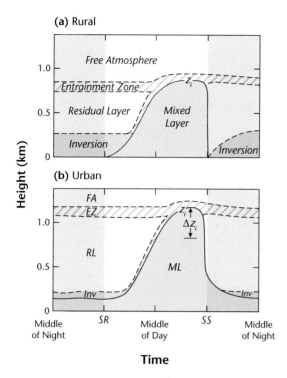

**Figure 7.31** Idealized diurnal development of **(a)** the rural and **(b)** urban boundary layer in fair-weather. Note urban ML is deeper in the afternoon by $\Delta z_i$, and remains as a shallow mixed layer through the night.

Over a rural area at night, the sensible heat flux extracts energy from the lowest layers and a shallow stable **nocturnal boundary layer** (NBL) forms with a ground-based inversion (Figure 7.31). The inversion decouples upper layers from these surface effects leaving a layer of decaying turbulence aloft; this is the **residual layer**. In the evening the most densely built-up districts of cities release $\Delta Q_S$ from the fabric and $Q_F$ from vehicles and buildings, together this is sufficient to support an upward $Q_H$ and a weakly convective lower layer, perhaps about 75 to 150 m deep (Figure 7.31). This weak nocturnal version of the daytime ML sees intermittent 'venting' of heat, from within canyons because they are warmer than the cool layer at roof level (Figure 7.23). The strength and sign of $Q_H$ at night is related to building density (Figure 6.25), so while built-up LCZs have a ML, in less heavily developed zones of suburbs, the ML may be absent and there may be large patches with ground-based inversions. Commonly there are also elevated inversions with several layers in parallel, perhaps

including wave-like structures (see also Section 2.2.2 and Figure 2.14).

### Effect of Weather

To this point we have mostly looked at cases with well-developed heat islands in nearly ideal weather. However, when skies are overcast and/or with strong winds, thermal differentiation of surface local climates is not strong: radiation cooling is weak and enhanced mechanical turbulence thoroughly mixes the ABL and blends urban-rural temperature differences. This effectively obliterates most signs of urban thermal effects. Observations in the ML over St. Louis show, as expected, that as wind speeds in the ABL increase the magnitude of $UHI_{UBL}$ is decreased. The daytime positive anomaly in the upper part of the UBL varies from a typical value of 1 to 1.5 K when speeds are about 2 m s$^{-1}$, but decreases to about 0.5 K at about 8 m s$^{-1}$ (Figure 9.12).

### 7.4.3 Genesis of Boundary Layer Heat Islands

The $UHI_{UBL}$ has connections with the $UHI_{UCL}$, but it is a distinctly boundary layer phenomenon, and is driven by a different, but related, set of processes (Table 7.2). The mix of processes can be divided into 'top-down' and 'bottom-up' groups, to indicate the origin and direction of the **forcing**. The bottom-up set are driven directly by energy fluxes at the top of the RSL (Chapter 6). The top-down set are due to radiation interactions with the polluted ML and convective processes (thermal and mechanical) that originate at the top of the RSL but interact with the **entrainment zone** located at the top of the ML. The two sets of processes are linked and driven by the strength of the surface energy fluxes and pollution in the UBL.

At night, as noted in Section 7.3.2 there are bottom-up processes that originate as terms in the SEB, but they exit the RSL and enter the UBL. Firstly, there is $Q_H$ which continues into the night at densely developed sites. Secondly, there are anthropogenic fluxes of sensible and latent heat from vehicles, domestic chimneys and industrial stacks. Heat from the last two sources is injected directly into the UBL, without passing through the UCL. The latent heat component of these fluxes contributes to atmospheric heating only during **condensation** of the vapour into cloud plumes or other cloud. Thirdly, the greater thermal and mechanical mixing induced by a city redistributes heat that was originally part of the upwind rural inversion. This raises the temperature of the lower, and decreases the temperature of the upper, part of the profile. In summary, the nocturnal UHI$_{UBL}$ is stoked by sensible heat fluxes that converge from above and below: one due to the energy balance of the UCL, the other to advection of the rural inversion across the city.

The daytime UHI$_{UBL}$ is similarly fuelled by bottom-up and top-down processes, except that the strength of the RSL sensible heat flux is much greater and UBL instability assists, rather than limits, the rise of buoyant thermals. The bottom-up heat mainly comes from hot urban surfaces and to a lesser extent from anthropogenic emissions. Daytime thermals have considerable energy that both mix the ABL, and drive entrainment, that eats away at the base of the overlying inversion and generates a downward $Q_H$ (Figure 7.32). The latter heat flux is indirectly linked to the surface one because the greater the strength of thermals created at the surface, the greater is the entrainment flux. A rule-of-thumb is that the entrainment flux is about 20% of that at the surface, when convection is strong. In windier conditions the ratio of the entrainment flux to the surface flux can be even larger. A potential source of additional UBL heating is radiation flux convergence. A polluted UBL contains many particles and gases some of which, such as black carbon, have absorption/backscatter ratios

that favour absorption (Chapter 5), which heat the UBL. This is significant in cities with heavy industries and refineries, or where coal is a main fuel or waste burning is uncontrolled.

## 7.5 Subsurface Heat Island

The surface heat island extends down into the soil, bedrock and **groundwater** beneath the city. Most evidence of urban $T_g$ patterns is obtained from thermometers mounted inside water wells or bore holes that that extend from a few 10s to over 100 m in depth. The placement of sensors is therefore likely to be opportunistic rather than dictated by a scientific plan. Here we define an UHI$_{Sub}$ as the increase in $T_g$ above that found at the same depth in surrounding rural areas, i.e. that due to the background climate plus the geothermal heat flux.

Given the relatively large thermal admittance of soils (Table 6.4) and since heat transfer is essentially limited to conduction, response times are slow. The speed at which a temperature wave travels downward depends on the strength of the vertical gradient of ground temperature ($\partial T_g / \partial z$) and the thermal properties of the substrate. The $T_g$ gradient is sharp near the surface so heat penetrates relatively rapidly, but at greater depths the gradient slackens and the pace of heat transfer slows causing a lag in response to surface forcing ($T_0$). Subsurface spatial and temporal thermal patterns emerge at seasonal, annual and decadal time scales. If temperature measurements have sufficient resolution, waves that originated at the surface can be detected many years later at greater depth (Figure 7.33). That means the history of thermal activity at the surface is preserved, including very long-term events such as changes in regional and even global climate forcing.

A secondary control on $T_g$ is drainage of water from rain, irrigation, rivers and other water bodies. It complicates spatial patterns of $T_g$ because surface infiltration only occurs in pervious areas and it preferentially follows along geologic structures, especially more porous beds. The ground beneath cities is a complex mix of both natural elements (soils, sediments, bedrock) and buried urban infrastructure (below-grade parts of buildings, roads, tunnels, subways, water pipes, sewers, etc.) which may extend tens of metres below the surface and can block or steer water movement.

The temperature of the deep substrate is affected by the natural geothermal heat flux (the weak conduction of heat upwards from much deeper layers in the

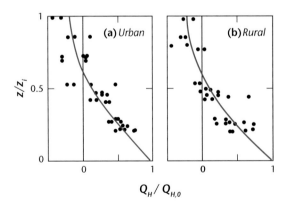

**Figure 7.32** Convergence of measured sensible heat flux $Q_H$ (normalized by its surface value $Q_{H,0}$) in the afternoon urban–rural boundary layers over St. Louis, United States between surface and mixed layer depth $z_i$ Data from several runs with winds between 0.5 and 4.5 m s$^{-1}$ and mixed layer heights between 1.2 and 1.5 km. The line is the average of the data (Source: Hildebrand and Ackerman, 1984; © American Meteorological Society, used with permission).

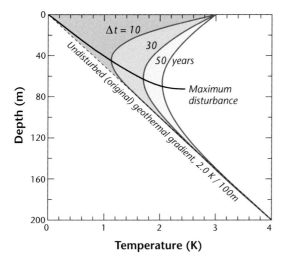

**Figure 7.33** Generalized form of subsurface temperature profiles after 10, 30 and 50 years following a 3 K step increase in surface temperature due to urban warming. Note: depth of maximum thermal disturbance and the original (undisturbed) geothermic profile (slope 20 K km$^{-1}$) (Source: Bodri and Cermak, 2007).

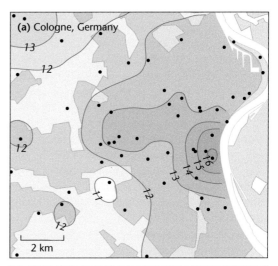

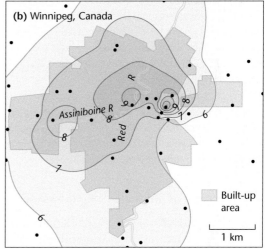

**Figure 7.34** Isotherm patterns of $UHI_{Sub}$: **(a)** at 15 m depth in Cologne, Germany in 2009 (Source: Zhu et al., 2010; © IOP Publishing Ltd. CC BY-NC-SA, used with permission), and **(b)** at 20 m depth in Winnipeg, Canada, in 2007 (Source: Ferguson and Woodbury, 2007; © American Geophysical Union, used with permission from Wiley).

Earth) and the annual average air temperature (i.e. the background climate) because the air and ground systems must come into long-term thermal equilibrium. Close to the surface, smaller scale processes and influences affect $T_g$, but the typical temperature profile of a non-urban environment with uniform surface conditions, shows a linear increase of $T_g$ with depth, due to the geothermal heat flux. Surface warming due to urban disturbance modifies the profile so that near the surface $T_g$ is increased and the sign of the gradient of $T_g$ is reversed (Figure 7.33, Figure 7.34). Measurements of deep soil temperature can be used as a proxy for screen-level air temperature in climate change studies. This is because, as already noted, the two are in long-term equilibrium. If synchronous profiles of soil temperature are available for both urban and non-urban locations, their difference can be used to identify urban-bias (Changnon, 1999).

A characteristic of the $UHI_{Sub}$ is a deeper than normal layer in which $T_g$ decreases, before starting the expected increase with depth, due to the geothermal heat flux. In cities the depth of the affected layer may be several tens, even a hundred metres deep. Several thermal zones can be distinguished (Popiel et al., 2001). Firstly, a shallow near-surface layer (0–1 m) in which the $T_g$ is strongly coupled to surface conditions and is sensitive to short term weather.

Secondly, from about 1–8 m in dry light soils and 1–20 m in moist heavy soils, $T_g$ responds mostly to the seasonal climate and exhibits relatively little variation. Thirdly, below that $T_g$ approximates the average annual value in the air. At yet greater depths $T_g$ becomes almost constant but shows a small, steady increase with depth forced by the geothermal gradient.

There are few observations of $UHI_{Sub}$ compared to those of the surface or the air (Menberg et al., 2013); they consistently occur in large cities, irrespective of their background climate. Magnitudes are typically about 1–5 K, depending on the city and the depth of the measurement (Figure 7.34). The spatial patterns of $T_g$ respond to the surface character; being warmer and more variable beneath more densely built areas and areas with little vegetation, cooler in more vegetated areas (see Figure 7.34). In Cologne, Germany the 15 and 16 K $T_g$ isotherms and in Winnipeg, Canada, the 9 K isotherm west of the river, respectively, are centred on the densest urban LCZ. In Winnipeg the 10 K core of tightly spaced isotherms east of the river, is a local anomaly due to injection of warm industrial water. Because of the time taken for heat to conduct downward, patterns of $T_g$ are a reflection of conditions at the surface several years *prior* to the time when the subsurface observations were made (Ferguson and Woodbury 2007). Extra warmth affects both the solid materials of the substrate and the included water. Since groundwater flows both vertically and laterally it has the ability to deform the shape of the subsurface temperature anomaly into a thermal 'plume' or other shape.

The extra warmth typical of urban soils may increase **respiration** and other processes affecting gas exchanges between the soil and atmosphere. Similarly warmer groundwater alters both the rate of chemical reactions and of microbiological activity.

### 7.5.1 Genesis of Subsurface Heat Islands

Heat transfer processes of the urban SEB (Chapter 6) generate the surface temperature distribution of a city (UHI$_{Surf}$), as outlined in Section 7.1. Surface temperature fluctuations and the difference between them and the $T_g$ of deeper layers, gives a vertical gradient that drives thermal waves and heat fluxes ($Q_G$) into the substrate, by molecular heat conduction.

Cities include two further heat transport processes. First, especially in winter there is leakage of heat from building basements, sewers, and underground pipes (especially if there is a centralized community heating system). Such leakage can be large in older cities where insulation of basements and pipes is poorer. Second, there is downward percolation of relatively warm water from the surface and from buried pipes and sewers, into the soil and groundwater. The temperature of urban surface **runoff** is often elevated, because of the high temperature of roofs, roads and other paved areas. Sewers and water pipes have surprisingly high rates of leakage (Chapter 8). Rapid infiltration of urban runoff can dramatically alter the temperature variation of groundwater seasonally, as can the water and snowmelt from storms. Further, some industrial plants inject warm wastewater into the substrate to use it as a heat sink. This produces anomalous warm spots in the UHI$_{Sub}$ pattern (Figure 7.34b).

The thermal impact of infiltration in natural systems depends on the background climate: if rainfall is concentrated in the warm (cold) season it is likely to warm (cool) the groundwater. However, urban runoff is likely to be warmer compared to that flowing over natural surfaces, in both seasons. **Detention storage** of excess storm water in ponds or basins creates wet spots and the potential for lateral heat transport into adjacent subsurface zones.

### Summary

Urban heat islands (UHI) are the most studied of all urban effects. It is important to be clear about the four different types of UHI, their causation and spatial and temporal behaviour. That knowledge is critical to accurate interpretation of observations, unambiguous communication and comparison of results and provides an intelligent basis to design appropriate policy responses. The four types are:

- The **surface UHI** (UHI$_{Surf}$) has a complex spatial pattern mainly due to the geometry, radiative and thermal properties of surface facets. Airborne or satellite sensors show its magnitude is greatest during clear daytime conditions in parts of a city with little vegetation, or if a large fraction of its plan area consists of roof facets. The magnitude of the UHI$_{Surf}$ is sensitive to the surface cover of the rural surroundings. Use of complete surface temperatures ($T_{0,c}$; i.e. adjusted to account for the full 3D form of the surface) reduces the magnitude of daytime, but increases nocturnal, UHI$_{Surf}$.

- The **canopy layer UHI** ($UHI_{UCL}$) is the difference between the near-surface air temperature below roof level in the city, with that over the non-urban landscape. Its magnitude is **greatest after sunset when air above urban surfaces cools more slowly than air above rural ones**. The sky view factor ($\psi_{sky}$) is inversely related to the rate of nocturnal cooling, so where buildings are tall and streets are narrow (i.e. city centre sites where there is little vegetation), the magnitude of the $UHI_{UCL}$ is strongest. Daytime $UHI_{UCL}$ are usually small or negative.
- The **boundary layer UHI** ($UHI_{UBL}$) is maintained by an enhanced sensible heat flux from the city, which promotes mixing in the lower atmosphere during daytime and sustains it overnight. As a result, the urban boundary layer is warmer than the rural one throughout its depth. Transport by the wind displaces the envelope of urban warmth downwind as a plume well beyond the built-up area.
- The **subsurface UHI** ($UHI_{Sub}$) forms due to transfer of sensible heat from the urban surface and urban infrastructure into the ground. Heat stored in the substrate represents accumulation over a considerable period such that the $UHI_{Sub}$ responds to climatic (and anthropogenic) processes on monthly and decadal and even longer timescales.

The various types of UHI induce other climatic phenomena. The extra warmth shifts the thermal structure of the boundary layer towards neutrality and instability, which encourages vertical mixing. The modified temperature structure also generates country-breezes in an **urban heat island circulation** (UHIC) that draws air towards the city centre at low level where it converges, rises, flows outwards at higher level and sinks away from the city over adjacent rural areas (Chapter 4). Cities are preferred areas for uplift at all times which contributes to urban effects on **cloud formation and precipitation** (Chapter 10). Existence of the $UHI_{UCL}$ affects the heating and cooling needs. This all adds to the thermal stresses on humans during **heatwave** events (Chapter 14) prompting considerable research into ways to mitigate aspects of UHI in cities through urban design measures (Chapter 15). Finally, UHI skew the temperature records of meteorological stations located in urbanizing areas relative to those at undeveloped locations. This is of special concern to those examining the impact of global and regional, rather than local climate change (Chapter 13).

# 8 | Water

**Figure 8.1** The Shibuya River in Tokyo, Japan (Credit: Mhiguera / Wikimedia, CC 2.0).

Access to drinking water is indispensable and a primary determinant of the location of a city. Unfortunately building a city always disrupts the pre-existing drainage patterns and hydrology. Rivers and streams are channelized, diverted or covered over; parts of the pre-urban network are replaced by systems of pipes and **storm sewers** linked to the road and building network. Wetlands are drained or filled-in; much of the low vegetation and tree cover is cleared or replaced by exotic species; natural soils are covered by buildings, concrete or asphalt, which reduces **infiltration** and enhances **runoff**. The great utility of water means that it is added, removed, diverted, constrained, pumped and piped as needed to achieve various practical goals: to prevent flooding, irrigate, supply drinking water, cool buildings, dilute heat pollution, remove waste, generate electricity or operate industrial processes.

Figure 8.1 portrays an intensely urbanized district of a 'compact high-rise' **Local Climate Zone** (LCZ 1) in Tokyo, Japan, where the surface is almost completely sealed. The 'river' channel in the centre of the image occupies a narrow concrete gutter designed to remove storm water from the city as efficiently as possible. Its feeder tributaries are pipes connected to drains that form an artificial drainage network, mostly hidden underground. The walls on either side of the channel offer flood protection and bank control; they permit dense building construction to abut directly up to the river channel. In terms of exchanging water with the urban atmosphere, there are few potential sources of water vapour in this image: the exposed channel water, the scanty vegetation on the banks, the heating systems, and the exhaust vents of air conditioning units. Effectively the transformation of the rural ecosystem to an urban one is almost total. In parts of cities where the building density is lower than in Figure 8.1, greater fractions of the landscape are still able to exchange water vapour from green-space and trees. There are then additional pathways for **evaporation** from both natural (soil moisture, plant leaves) and artificial (spray from sprinklers, spray-wetted surfaces) sources. Indeed some cities are *more* vegetated than their surroundings e.g. in arid areas.

Urban water management is a large field of study that includes water quantity and quality issues (such as the supply of clean drinking water, flood management, waste water treatment). Further, there are economic and political aspects such as water access, governance, social justice and jurisdictional disputes which are beyond the scope of this book. This chapter simply describes the physical effects of cities on surface water. The approach mirrors the discussion of the energy balance in Chapter 6.

## 8.1 Basics of Surface Hydrology and Water Balances

### 8.1.1 Hydrologic Cycle

The global **hydrologic cycle** constantly recycles water mass through components of the Earth-Atmosphere system as ice, liquid water or water vapour. The cycle is powered by the global **surface energy balance** via meteorological processes such as the vapourization of water from moist surfaces (e.g. vegetation, soils, lakes, oceans), vertical mixing, transport by winds, **condensation** as cloud and its eventual return to the surface as **precipitation** in various forms (snow, rain, **dewfall**, fog drip) where it restocks water stores (soil moisture, **groundwater**, snowpacks, glaciers, lakes, oceans). The hydrologic cycle transports water between components of the system: by **convection** and **advection** in the atmosphere, by glaciers, runoff and rivers over the land, and currents in the oceans.

A **surface water balance** (SWB) for a defined portion of the landscape accounts for all water that passes through, or is stored within the volume over a period. An ability to evaluate and predict the terms of the SWB is essential for urban water management. Water in the atmosphere exists as solid (ice, snow, etc.), liquid (e.g. cloud droplets, rain droplets) and gas (water vapour). A full account of the water cycle includes all phases, which depend on temperature

and involve the uptake or release of energy, and hence are linked to the surface energy balance (Chapter 6). For the **atmospheric boundary layer**, the SWB is an essential **boundary condition** of atmospheric humidity (Chapter 9).

The remainder of this section outlines basic concepts and processes related to the surface water balance and the energetic implications. Readers with a good grasp of these concepts may wish to skip to 8.2.

### 8.1.2 The Surface Water Balance

The SWB is easiest to outline for a natural catchment – loosely defined as an area of the landscape that gathers all precipitation falling within its boundary with a depth that extends from the top of any **surface cover** (e.g. vegetation) down to the permanently saturated soil or rock (Figure 8.2a). The topographic irregularities of natural landscapes, in combination with the action of gravity, define catchments just as they define river networks (the areal extent of Figure 8.2a is unable to incorporate sufficient topographic relief to illustrate the effect of elevation differences in the creation of a river catchment).

Most precipitation ($P$) reaching the ground enters the soil and percolates to deeper layers. Some is intercepted by vegetation, where a portion is retained briefly before evaporation returns it to the atmosphere and the remainder reaches the ground. If soil near the surface is unsaturated, water at the surface enters the soil by infiltration the rate of which is governed by two forces: the gravitational potential and the matric potential. The former is due to gravity acting on water in the soil column and is most effective on water held in large pores that are free to drain. The matric potential arises because water is attracted to solid surfaces (e.g. soil particles, stones) where it adheres by surface tension.

Some water moves down into the groundwater zone which consists of an unsaturated (vadose) layer overlying a saturated (phreatic) layer where voids are filled with water. The water table defines the upper level of the phreatic layer. The water table is uneven, generally following the relief of the landscape in muted form. The rate of any horizontal movement of soil water is governed by the slope (gradient) of the water table and the hydraulic conductivity of the substrate. Compact clay has a conductivity of only about 0.0001 m day$^{-1}$, whereas for fine sand it is about 3 m day$^{-1}$ and for coarse sand and gravel it

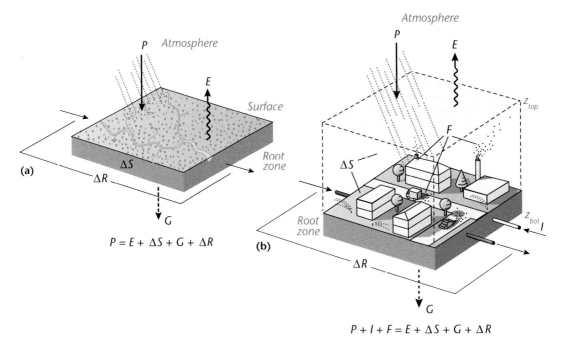

$$P = E + \Delta S + G + \Delta R$$

$$P + I + F = E + \Delta S + G + \Delta R$$

**Figure 8.2 (a)** Surface water balance (SWB) of a natural and **(b)** the equivalent SWB of an urban hydrological unit; e.g. a city and its associated volume of air and moist soil in the root zone. See text for detailed explanation of symbols.

can be as large as 300 m day$^{-1}$. Nevertheless, given a typical water table slope of 0.005, horizontal movement in this zone is very slow (Shanahan, 2009). The substrate can contain many different layers of materials, some of which are nearly **impermeable** and act to confine groundwater thereby forming an aquifer; i.e. a zone of water-bearing permeable rock from which the groundwater can be extracted by drilling a well. The areal extent of the groundwater zone is often larger than that of an individual catchment, so extraction can respond to changes occurring outside the immediate area (Welty, 2009).

Runoff ($R$) is water that moves overland, often exiting the catchment through a stream channel fed by a series of branching tributaries. $R$ comprises water from several sources including: $P$ that falls directly into the channels; overland flow if the rate of $P$ exceeds the infiltration rate; water in the unsaturated layer that migrates downhill into a channel; and groundwater that emerges in a channel where the water table intersects with the terrain. Because water from each source takes a different route (and time) to get to the stream, the flow at any given time represents a combination of that gathered over different time scales. The **base flow** in a stream during dry periods comes from the groundwater store that is recharged

by precipitation events over long periods. The level in the store is established during a dry period when surface runoff is negligible. In urban areas, base flow is usually lower than in **rural** areas because precipitation is diverted to surface runoff; consequently less water infiltrates the soil to recharge groundwater.

$R$ exhibits a distinct temporal response to precipitation depending on the magnitude and intensity of the event, the characteristics of the catchment (the depth of soil, slope, underlying geology, etc.) and any lingering effects due to pre-existing conditions. Following a dry period $P$ may simply replenish depleted stores leaving little available for $R$ but, following a wet period, reservoirs may be full so $R$ responds quickly to $P$. The SWB for a natural catchment is:

$$P = R + E + \Delta S \quad (\text{mm d}^{-1}; \text{ kg m}^{-2}\text{ d}^{-1})$$

**Equation 8.1**

where $\Delta S$ is change in storage including soil water, groundwater and any surface water stores (e.g. lakes, streams, snowpacks, vegetation **interception**). It is positive after a precipitation event (or in the wet season) and negative during a dry period.

Measuring a complete SWB is often a challenge, because of mismatches between the time and space scales at which the different terms of Equation 8.1 operate. For example, $\Delta S$ operates over days or weeks. If the time frame for the analysis is long (many years) it may be possible to capture the full sequence of wetting and drying in a catchment, in that circumstance $\Delta S$ approaches zero. In contrast $P$ occurs in discrete bursts, often intermittently over periods of hours or days, perhaps followed by a month or longer with no input. Evaporation ($E$) is a semi-continuous process, driven by the availability of energy, and the efficiency of convection over periods as short as seconds and plant **transpiration** over hours (Section 6.4). In addition to these mismatches the receipt of rain- and snowfall can be spatially very patchy, even during a single storm.

## 8.2 Water Balance of Urban Hydrologic Units

### 8.2.1 Urban Development of River Catchments

Settlements seek a reliable water supply of sufficient quality to meet their consumption demands; they also seek to remove excess water and wastes to minimize the potential for flooding and to reduce water pollution. Over history the relationship between the growth of a settlement and its water needs follow a common path. Initially, local supplies are used for drinking water; wells are sunk to access groundwater or water is diverted from a lake or river. Groundwater has the great advantage that it often requires little or no treatment before it is drinkable. However, continued abstraction can threaten to deplete the aquifer. Eventually urban growth usually means water has to be imported from distant sources which require construction of an increasingly extensive network of channels and pipes.

As already noted urban development means vegetation clearance, ground compaction and introduction of a more impervious surface cover. These modifications initiate systemic changes to the pre-urban SWB, including reduced rates of infiltration which decrease both soil moisture content and evaporation and a significant increase of the volume and speed of runoff. These changes in $R$ prompt engineering responses such as the construction of large capacity drainage ditches and culverts to avert flooding after major storms. The idea is to convey water quickly to the natural stream network or to **detention storage** in ponds, wetlands or other suitable low-lying ground. Unfortunately stream courses are often modified by

widening or straightening, in attempts to cope with the surge that follows a rainstorm or sudden snowmelt. Further, additional drainage channels including storm sewers are constructed to convey rainwater away even more quickly. Modern cities still need to accomplish these hydrologic imperatives but the best managed water authorities try to preserve or emulate as much of the natural ecology as is using **water sensitive urban design** (Chapter 15).

A modern city has three artificial networks: pipes for drinking water, storm sewers for surface runoff, and sanitary sewers for wastewater. Typically within the city these follow the road network and are buried at different depths – storm sewers are shallowest and sanitary sewers the deepest.

Older cities may have **combined sewers** (a system that has to serve both the sanitary and excess runoff functions) (Figure 8.3). During relatively dry periods combined sewers route all flow (i.e. from both sanitary and any runoff sources) directly to a water treatment plant, if there is one. Such a plant is designed to reduce or ideally remove contaminants. Several levels of treatment could be involved, depending upon the sophistication of the system and the intended end-use of the water (e.g. drinking, industrial processes, irrigation, recreation). However, during large storms when the normal sewer is at capacity, a relief channel comes into play allowing both types of discharge to pass untreated directly to the environment. Clearly this is less than ideal and should only be an emergency measure. Of course it is an unfortunate fact that in many cities of the Less Developed world there is no treatment of water-borne waste and possibly even no sewers.

### 8.2.2 The Water Balance of an Urban Catchment

In addition to the SWB terms of a natural catchment (Equation 8.1), an urbanized one needs to account for the existence of artificial pipe networks and additional sources:

$$P + I + F = R + E + \Delta S \qquad \left(\text{mm d}^{-1}; \text{ kg m}^{-2} \text{ d}^{-1}\right)$$

**Equation 8.2**

where the new input terms are: piped water ($I$) imported from outside the catchment, and water formed chemically when fuels are combusted in air ($F$). Both new terms involve time and space scales that respond to human activities, for example, irrigation or industrial processes. Rather than natural events, these flows depend on decisions exercised by industrial managers

**(a)** Dry weather

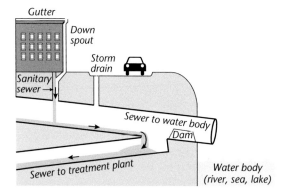

**(b)** Heavy rain

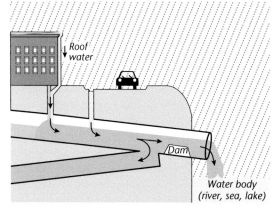

**Figure 8.3** A combined sewer system **(a)** in dry weather, and **(b)** during a heavy rainstorm (Modified after: EPA, 2004).

or householders about whether the water is going to be used to drink, cook, bathe, sprinkle, clean, cool or flush. In Equation 8.2 $R$ includes water carried by sewers as well as by streams and subsurface drainage.

Urban water management is best done, as in rural areas, at the scale of a catchment basin. However, definition of an urban catchment is not straightforward because it often does not have a well-defined perimeter, nor is water channelled to a single basin exit. The catchment's natural drainage is disrupted, supplemented or replaced by artificial pipes, tunnels, gutters and sewers that cross natural divides; therefore it is more realistic to refer to **urban hydrologic units**. Such units have multiple inlets and outlets (some natural, others **anthropogenic**) because the pipe water and sewer networks interweave with each other. Further a unit may include several neighbourhoods constructed in different time periods and serving very different

inhabitants and land uses and can be applied at any scale. A primary aim is to simplify the complexity of an urban system in order to avoid the need to measure, model or predict the water balance of individual urban elements (buildings, trees, lawns, roads), just to describe the integrated response of the unit.

### 8.2.3 Water Balance of an Urban Neighbourhood

If we take a single urban **neighbourhood** as a hydrologic unit then the balance applies to a volume, similar to that for energy (Figure 8.2b):

$$P + I + F = \Delta R + E + \Delta S + G + \Delta A \quad (\text{mm d}^{-1};$$

$$\text{kg m}^{-2} \text{ d}^{-1}) \qquad \textbf{Equation 8.3}$$

The top of the 'box' is ideally in the *inertial sublayer* (ISL) above the tallest elements, where mass fluxes of water vapour can be measured independently of horizontal location. The bottom of the 'box' is the base of the soil layer actively involved in water exchange with the surface over the period of the balance. The layer may not extend all the way to the water table, but should be deep enough to avoid fine-scale near-surface anomalies within the box. If there is significant infiltration from below into the groundwater, that fact must be accounted for in $G$. The balance assumes the 'box' is bounded laterally by similar neighbourhoods, if there are significant discontinuities a term for advection ($\Delta A$) must be added to account for exchange of drier or more moist air with adjacent areas. At this scale, runoff is treated as the *net* contribution of the volume to overland and stream flow, i.e. the difference between runoff that flows in and out ($\Delta R = R_{in} - R_{out}$). Such a volume approach is useful when modelling the urban water balance in a gridded **numerical model** ( Section 3.3).

### 8.2.4 Water Balance of an Entire City

Considering a complete city, including its atmosphere and subsurface, as a single hydrologic unit, Equation 8.3 still applies as long as the volume is considered to extend vertically from the level of permanently saturated ground to the top of the **urban boundary layer** (UBL) (Figure 8.4). In the air there is need to consider advection of water vapour because cities can be drier or wetter than their surroundings; similarly in the subsurface lateral flows of groundwater are possible (e.g. due to a lower

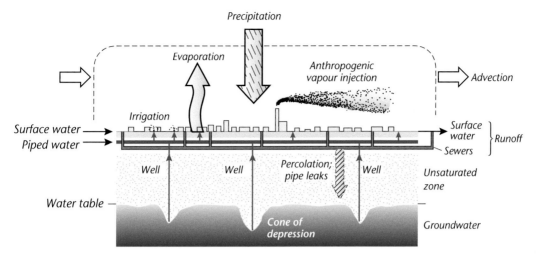

**Figure 8.4** Processes relevant in the water balance of a city.

water table in the city compared to its surroundings). Finally, it may also be necessary to include industrial point sources that can inject large amounts of water vapour directly into the UBL from tall stacks ($F$).

## 8.3 Urban Effects on Water Balance Components

This section outlines the nature of each term in Equation 8.3 and its contribution to the surface and volume balances of water in different urban hydrological units. It starts with the input terms precipitation ($P$, Section 8.3.1), piped water supply ($I$, 8.3.2) and anthropogenic water vapour ($F$, 8.3.3) then the outputs due to urban modified evaporation ($E$, 8.3.4), runoff ($R$, 8.3.5), storage change ($\Delta S$, 8.3.6), groundwater ($G$, 8.3.7) and advection ($\Delta A$, 8.3.8).

### 8.3.1 Precipitation

Evidence suggests that urban areas modify the processes of cloud development, atmospheric electricity and precipitation ($P$ in Equation 8.3 and Figure 8.2b), over and/or downwind of a city (Chapter 10). Here however, $P$ is simply taken to be that set by the regional weather and climate of the location, *including* any urban effects.

#### Distribution of Precipitation within the Canopy Layer

Precipitation at the urban surface is distributed over the **facets** and elements comprising the **urban canopy**

**layer** (UCL). The spatial distribution of deposition is highly variable owing to microscale airflow anomalies that are accentuated in cities (Chapter 4). Technically, we distinguish between $P$ that falls directly on the ground and that intercepted by the roofs and walls of buildings and tall vegetation. In general, initial deposition on the ground is greatest where horizontal transport by the wind is least although local circulations produce complex patterns and wetting under overhangs (roofs, awnings, balconies, trees and bushes) may be close to zero.

Typically, measurement of $P$ in urban environments is mainly by rain and snow gauges at standard meteorological observation stations located near ground level and situated to avoid microscale wind effects. However, the sampling density in standard networks is so low it may be necessary to translate observations from rural climate stations to where they are needed within cities, after due allowance for elevation differences. Measurements made at rooftop sites are of limited value, even misleading, because the catch registered by a gauge is greatly affected by building wind effects. The distribution of snow in an **urban canopy** is more complex because, once deposited it can be drifted by strong winds (such considerations can be incorporated into urban design, see Chapter 15).

#### Interception

The rainfall intercepted by building roofs and walls and tree leaves, branches and stems, is stored temporarily before finding its way to the ground or evaporating.

The fraction of $P$ intercepted is well documented for forest canopies (typically about 10–50% of the total input) but similar field data for urban canopies are almost absent. Nakayoshi et al. (2009) used the COSMO outdoor hardware model (shown in Figure 3.24) of bare concrete blocks in aligned mode with $H/W = 1$ to simulate UCL interception. They found the interception fraction to be just 6%. This comparatively low value is perhaps not surprising given that COSMO has no trees and therefore presents only about 20% of the surface area of a typical forest canopy. The surface area fraction is much greater in real urban environments, especially if they are vegetated or located in densely built areas (e.g. LCZ types 1 to 3) where the roof area may occupy 70% of the plan area.

Generally, a small proportion of intercepted water is retained on the surface before it evaporates. Most roofs remove most liquid water quickly via gutters down to ground level. In some cases this enters the sewer system, in which case roof surfaces behave much as roads and pavements and provide little storage capacity. However, this water can be collected in water butts or cisterns and/or be directed to pervious surfaces such as gardens. Green roofs can be installed to store more water and retard runoff. The amount stored depends on the technology used, which in turn depends in part on the strength of the roof and its slope. These strategies are employed to manage urban runoff generation, diverting it to recharge soil moisture or meet other water uses.

Snow can settle on a surface for just a few minutes or accumulate for extended periods of days and even months before melting. In cities with snowy winters substantial snow layers can build up on roofs and the ground. The amount of water temporarily stored depends on both the depth of accumulation and the snow density. A deep pack may pose hazards. Excessive weight is a structural threat, potentially causing roofs to collapse. Also if a sudden warming releases large amounts of meltwater in a short span, there is danger of flooding which can be heightened if a residual saturated pack hinders surface runoff leading to ponding.

### 8.3.2 Piped Water Supply

Piped water supply ($I$ in Equation 8.3 and Figure 8.2b) supplements water available via precipitation in the SWB. This water usually has to be

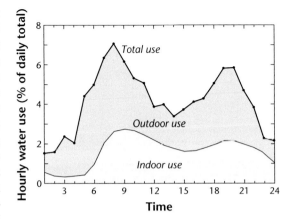

**Figure 8.5** Hourly water use based on average of assessments of 1,188 households in single-family residential neighbourhoods of 12 North American cities. Results gathered during two 2-week periods designed to capture the peak water use in summer and winter (Source: Mayer et al., 1999; © Water Research Foundation, used with permission).

treated to be potable, because it is potentially polluted (e.g. rivers, lakes), contaminated groundwater or rich in salts if the water source is the sea – or a brackish lake. In More Developed economies, this supply follows constrained routes because it is destined for 'consumption' by industry, local government and individual residents. Its path through the urban system is dictated by the network of water mains necessary to deliver it directly to its end-users. Temporal fluctuations in supply are driven by human activity cycles (Figure 8.5).

Per capita water use depends on the availability of water, the nature of the economy, the affluence of the population and how onerous are water charges. New York, United States, an affluent city in a temperate climate, has a drinking water demand of about 500 $\ell$ cap$^{-1}$ d$^{-1}$, which is met mainly by importing water from nineteen reservoirs and three lakes located outside the city. The equivalent demand in Karachi, Pakistan, a less affluent city in an arid climate, is just 165 $\ell$ cap$^{-1}$ d$^{-1}$, much of which is drawn from the Indus River. The level of any water charges also affects the amount used; in the United States it can cost as much to keep the garden looking green in an arid city as it does to heat the house in a cold climate.

The amount of water flow in pipes can be monitored in situ by meters on a routine basis and the data used in support of charges for water supply and/or to assess compliance with regulations governing water restrictions or conservation. To isolate areas of interest as part of a neighbourhood study, gauges can be placed at critical points of the pipe network and in many cities the water use by each house is monitored routinely for billing purposes. The end-use of piped water dictates its pathway in an urban system. The total piped water supply I can be conceptually separated into the following uses:

- Piped water used in **buildings** for residential or commercial purposes, and in most cases later directly channelled into the sewer system ($\Delta R$). A small fraction is evaporated due to cooking, air conditioning, human **respiration** or indoor heating and subsequent venting to outdoors. An assessment of indoor water use in twelve North American cities indicates the following breakdown: toilets (27%), clothes washing (22%), showers (17%), faucets (16%), leaks (14%) and miscellaneous other domestic uses (5%) (Mayer et al., 1999).
- Piped water used for **garden irrigation** substantially affects the urban surface water and energy balances; it has immediate effects on surface and soil water availability ($\Delta S$), and is a critical control on neighbourhood evaporation ($E$) from transpiring plants and evaporation from ground. There is great variation from city to city due to climate and weather patterns. In the North American study of Mayer et al. (1999) the amount of water used for irrigation is positively correlated with the size of a home and households with swimming pools have twice the standard outdoor use. The diurnal pattern of hourly water use by households in North America is remarkably similar (Figure 8.5). Further, the pattern broadly mimics the behaviour of the anthropogenic use of heat (see also Figure 6.6) and the **emission** of vehicular **air pollutants**. Together they illustrate the repetitive daily activity cycle followed by most residents, commuters and workers.
- Piped water used for **industrial processes** is either incorporated into products, or used in processing or cooling. Industrial cooling is often accompanied by release of water vapour ('steam') directly into the urban atmosphere through cooling towers and stacks ($F$).
- Water lost through **leaks** from the pipes during transport increases soil water availability and some may percolate to deeper layers ($\Delta S$).

**Figure 8.6** Automatic sprinkler system for lawn irrigation in the single-family residential district of Oakridge, Vancouver, Canada (see also Figure 8.7) (Credit: A. Christen).

### Irrigation

The piped water used to irrigate is extremely significant in some urban areas. Irrigation is applied in several ways: watering from hand-held cans using water from a tap or stored from household activities, or runoff from roofs, watering from subsurface perforated hoses, drip feed from hoses laid amongst plants, oscillating sprinkler heads or in some area by occasional flooding. Figure 8.6 shows an automatic sprinkler system common in many North American cities. Irrigation practice varies greatly with weather and cultural norms.

Figure 8.7 shows the impact of irrigation on the SWB of a neighbourhood in Vancouver, Canada, which has a temperate climate, with mild wet winters and a short summer drought that gardeners seek to offset by irrigating their gardens (lawns, flowers, produce). In winter (Figure 8.7a) the total piped water demand is small (~1 mm) because the internal budgets of the households (e.g. washing, bathing, toilets) are relatively steady and external use is small, indeed the record indicates almost no discernible response to variations in external weather ($T_a$, $P$).

In contrast, the summer water demand by the neighbourhood fluctuates greatly: an almost eightfold increase over the base flow (Figure 8.7b). Since internal household uses do not vary much seasonally, the likely cause of this variability is external irrigation. Closer inspection shows the water demand is tied directly to the weather; it is positively correlated with increases of $T_a$ (fine and warm weather causes soil

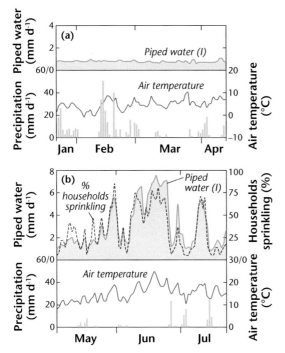

**Figure 8.7** Daily weather and water use by households in mm per day for a single-family residential district of Oakridge, Vancouver, Canada, in **(a)** the colder and **(b)** warmer season (Source: Grimmond and Oke, 1986; © American Geophysical Union, used with permission from Wiley).

drying leading gardeners to increase sprinkling) but retreats close to base flow when there is significant rainfall (sprinkling is curbed when the soil is deemed sufficiently moist for growth). As long as the piped supply remains abundant, outdoor uses flourish, and might even allow the irrigation input ($I$) to exceed $P$ over longer time periods.

Figure 8.8 shows the measured summertime SWB of another **suburban** neighbourhood in Vancouver, Canada. In a normal summer (Figure 8.8a) irrigation water is the largest input to the SWB, indeed larger than $P$. Together with the increased evaporative potential of the fine weather, irrigation supports large $E$, sufficient to almost compensate for the soil drying. During 1992 there was a severe drought that necessitated a regulatory ban on outdoor water use and dramatically changed the SWB (Figure 8.8b). Note, the percentages are of much smaller absolute amounts than usual, hence the apparently *larger* percentage in runoff, was actually smaller than normal in magnitude. The proportion of the SWB given to $E$ dropped compared with a year without a ban. The **Bowen ratio** in 1992, with the ban in effect, was about 3, whereas in a normal year it is usually between 1 and 1.5, this meant that a greater than normal fraction of the **available energy** was channelled into the **sensible heat flux density** ($Q_H$) which directly contributed to already high air temperatures, and synergistically reinforcing the drought.

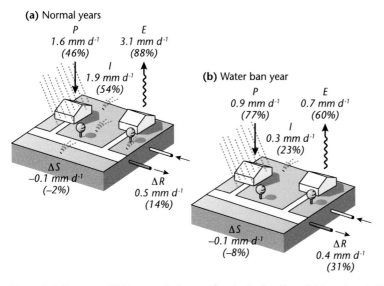

**Figure 8.8** Summer (JJA) water balance of a single-family neighbourhood (Vs, LCZ 6) calculated as **(a)** the average of two years with normal weather and moist soils, and **(b)** a year with severe drought and a ban on outdoor irrigation. Percentages are of total water input ($P + I$).

## Leakage

The water and sewer systems of cities commonly deteriorate and leak, often badly. In addition to simple decay some pipes are especially vulnerable to damage by earth movement: earthquakes, frost heave, subsidence caused by degrading permafrost, collapsing karst, mine operations or groundwater pumping, all of which can rupture pipes and sewers.

Water leaks from pipes can be surprisingly large, partly because a water supply system is under considerable hydraulic pressure (due to gravity and/or pumps), so even small fissures can inject a lot of water into a localized area relatively near the surface. This contributes to soil moisture and groundwater recharge which can be sufficient to *raise* the water table. In cities where the pipes are maintained by a water authority, leakage is typically about 5–15% — individual cities report up to 60% of the total water flow is lost! For example Greater London, United Kingdom, suffers leaks variously stated to be between 28 and 40% of its high quality, potable water. This is partly because about half of the mains water pipes are more than a century old. Losses due to leakage are difficult to assess unless all abstractions are metered. Only that way can water use be properly compared with water supply. Losses are often largest in areas where the building density (and therefore pipe density) is greatest and the pipe system is oldest.

Sewer systems also leak water into the ground but the flow is not usually under pressure, except during storm events. As a result water enters and exits without forcing losses through weak joints. Leakage

of wastewater of course carries the potential to contaminate soils and groundwater. In general, obtaining information about these exchanges is challenging. For example, New York, United States, has a 13,000 km long sewer system 60% of which is a combined sewer system (Figure 8.3). This makes isolating leaks and identifying their source a major headache.

### 8.3.3 Anthropogenic Water Vapour

Water vapour released to the UBL from human activities, including emissions from space heating, air conditioning, industrial processing, cooling ponds, cooling towers, vehicles and human respiration is referred to as **anthropogenic water vapour**, $F$ (Equation 8.3 and Figure 8.2b). Estimates of $F$ rely on inventories of energy consumption and surveys of the number and physical properties of the emitters.

## Combustion

From a balance perspective, it is important to distinguish between water already accounted for as inputs ($P$, $I$) that are vaporized by anthropogenic processes, from those not yet considered. The term $F$ in Equation 8.3 only refers to newly formed water as a result of the **combustion** of wood, natural gas, gasoline and other fuels in vehicles, heating systems and industry. It therefore represents a new source of water in the urban surface balance, as impressively visualized by plumes of condensed vapour formed over chimneys in winter (Figure 8.9). The amount of water vapour formed during combustion is substantial, for example burning a litre of gasoline (about 0.75 kg) yields about

**Figure 8.9** Condensation plumes from houses in a northern city (Rouyn, Canada) where natural gas is burnt in furnaces to provide space warming. This illustrates the anthropogenic water vapour injected into the urban atmosphere (Credit: CMHC, used with permission).

one kilogram of water vapour. Even more striking, 1 kg of natural gas yields more than 2 kg of water vapour.

## Space Cooling

Condensation plumes are evident above tall buildings due to exhaust from their ventilation and air conditioning systems. An air conditioning (AC) system cools outdoor air (often resulting in condensation) before being drawn into buildings and later causing the ejection of warmer interior into the ambient surroundings. Thus, while AC is a source of anthropogenic heat ($Q_F$, Section 6.2), it is not a source of new water; the AC system dehumidifies the building interior while increasing the humidity of the outdoor air. Only when balancing the outdoor air without exchange, do vapour emissions due to space cooling need to be considered as a vapour source. The type of AC system (e.g. vapour compression refrigeration or evaporative cooler) determines the division between emissions of **sensible** and **latent heat**. In a densely built environment (e.g. LCZ 1) in summer when AC use is intense, the accumulated outdoor effects of AC can be significant.

## 8.3.4 Evaporation

The energetics of evaporation[1] ($E$ in Equation 8.3 and Figure 8.2b), including the controls upon it, are also discussed in Section 6.4.3, where it is treated as a turbulent *heat* flux ($Q_E$, in W m$^{-2}$). Here we are interested in the equivalent **turbulent flux** of water *mass* ($E = Q_E/\mathcal{L}_v$, in mm d$^{-1}$). The latent heat of vapourization $\mathcal{L}_v$ is the link between the heat and water balances (Equation 6.2). The equivalence between the two fluxes means evaporation can be measured or calculated using either a heat or mass approach. That duality provides independent means to check the **accuracy** of estimates using the other method.

Direct measurement of $E$ at the city scale is currently only possible from airborne surveys using aircraft fitted with fast-response **sensor** systems flown through the lower ML, or by combining observations from a network of **local-scale** micrometeorological

towers located in representative LCZs. Both are expensive undertakings. Alternatively $E$ can be estimated rather than measured using a combined SEB-SWB model such as SUEWS (Järvi et al., 2011) or a **mesoscale SEB parameterization** scheme (e.g. TEB-ISBA, Lemonsu et al., 2007). Both require input information about urban surface properties and observed weather data, or output from a mesoscale weather forecast model.

Less desirably $E$ can be found as the residual in Equation 8.3; this requires good knowledge of $P$ (precipitation gauges), $I$ (household pipe meters) and $R$ (stream gauge and sewer records) and neglecting the SWB terms $F$, $\Delta A$ and $\Delta S$ (unless there are estimates or field observations). However, this approach is likely to incur cumulative errors of 20–30% in $E$ for catchments (Shuttleworth, 2012). Unfortunately simple equations using standard weather input, like those successful in agricultural applications, do not work in cities because of the complexities outlined here and in Section 6.4. Similarly, measurements using evaporation pans exposed in urban neighbourhoods, where micro-advection is the norm, do not perform acceptably.

## Urban Controls on Evaporation

Observations confirm that by day $E$ is smaller in most cities than in their surrounding countryside (Section 6.5). This is unsurprising given the systematic replacement of evaporation sources, such as vegetation and soils by impermeable paving and building materials. An exception to this generalization is settlement in desert, semi-arid or oasis areas provided they have access to adequate water to support semi-permanent irrigation.

Although the magnitude of $E$ pales in comparison with $P$ and $R$ in wet periods, nevertheless this flux is potentially significant to hydrology between storms. If cumulative $E$ is small because the weather is dull or the inter-storm period is short, soil water remains well stocked, so the next storm has little chance to infiltrate rather it contributes to runoff. However, if cumulative $E$ is large, soils start to dry creating storage capacity so a portion of $P$ can be taken up by the soil column. Vacant storage thereby provides buffering against runoff when the next storm arrives. Enlightened hydrologic managers therefore keep an eye on $E$ and $\Delta S$ on a continuing basis.

In cities where irrigation is significant there can be a close link between piped water supply and

---

[1] Some prefer to retain use of the term *evapotranspiration* which combines evaporation from free water and transpiration from plants. Transpiration is a physiologic process involving the activity of plant stomata mediating the release of plant water as vapour. The end point of transpiration is still evaporation.

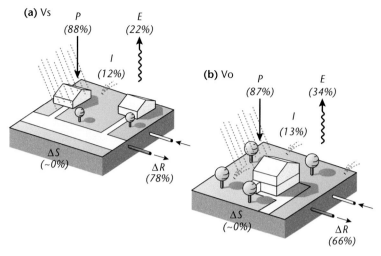

**Figure 8.10** Annual SWB of two residential suburbs in Vancouver, Canada, in 2009 calculated using the SUEWS SEB-SWB model with measured inputs ($P$ and $I$). Percentages are relative to total water input ($P + I$).

evaporation during the growing season (Figure 8.7). Gardeners are particularly sensitive to the state of soil moisture for plant growth; they start or increase watering upon signs of leaf wilt and cease after the arrival of sufficient rain.

Figure 8.10 illustrates SWB differences between two suburban single-family residential districts in Vancouver, Canada (labelled Vs and Vo), in 2009. They both experience approximately the same weather conditions but while both are classed as open low-rise neighbourhoods (LCZ 6), they differ markedly in the density of housing and tree cover. Vs is the same site seen in Figure 8.8 with similar fractions of its plan view occupied by buildings and trees (17% each), whereas Vo has double the tree cover (35%) but about half the buildings (9%). Further, Vs is 87% watered by hand and 12% is unirrigated, whereas 61% of Vo is serviced by automatic sprinkling and only about 1% is left unirrigated. Hence the SWB of the more densely developed Vs site produces about 12% less evaporation from its less vegetated area and its more impermeable surface fraction translates into about 12% more runoff.

### 8.3.5 Runoff

Figure 8.11 illustrates the way four ecosystems are likely to handle the partitioning of $P$. Together they illustrate a typical transition of a landscape from rural to urban uses and the resulting impact on runoff ($R$ in Equation 8.3 and Figure 8.2b). Forest and low plant

environments intercept $P$ fairly effectively and take water into storage. Over the medium term this means water remains available in the soil for $E$. Overall this slows the throughput of water and generates relatively less runoff ($R$). Urban cover types, on the other hand, are much less permeable, hence water is easily repelled, which enhances runoff.

The results in Figure 8.10 follow this pattern, for example, at Vs the greater impermeable surface fraction translates into about 12% more runoff, but 12% less evaporation. As the magnitude of $R$ increases with urban development, so does the ratio $R/E$. The critical control is the total impervious surface cover fraction, which includes both paving and buildings. In Figure 8.10 $R/E$ increases from about 1.9 to 3.5.

In addition to land cover, the connectivity of the river/sewer network and the presence of additional water sources over and above $P$ (e.g. $I$, $F$) exert control on $R$. For example, have rivers been diverted or sewers re-routed to form new network connections or channels?; have sewers been upgraded so leakage and groundwater flow have changed?; is the sewer system combined or separated?

### Discharge Curves in Urban Systems

The main result of urban development on streamflow is threefold (Figure 8.12):

• Following a storm event a much **greater volume** of runoff is generated.

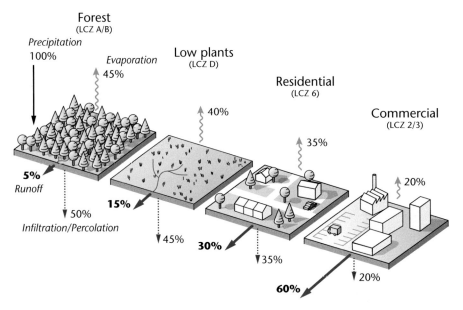

**Figure 8.11** Generalized approximation of the effect of land cover on the partitioning of precipitation (100%) between evaporation, infiltration and percolation and runoff over periods as long as a year, so that storage change in the root zone is negligible.

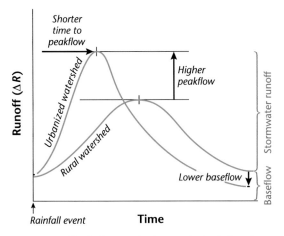

**Figure 8.12** Typical storm hydrograph depicting the relation between stream discharge ($R$, left axis) and time, starting with a major input of rainwater ($P$) due to a storm event. One curve is typical of flow from a rural catchment, the other after urban development. Comparing the two curves, note that after development: the increased value of peak flow; the shortened time before arrival of that peak, and the drop in the baseflow.

- The **time taken** before water input appears in gutters, drains, culverts, streams and rivers is **shortened**.
- The **base flow** is usually **lower** as less water infiltrates into the substrate.

Both the rising and declining limbs of the discharge curve are steeper after urban development than before (Figure 8.12). This is because overland flow is faster over the impermeable and relatively smooth surfaces found in cities (e.g. roofs, roads and paved areas) than those that pass through and across natural soils and vegetation. The runoff, being both greater in volume and faster flowing, is more powerful and increases the chances of flooding. This is especially the case where drainage capacity is exceeded, or channels become blocked by debris rafted along with the surging flow.

### Surface Water Quality

Runoff from urban areas becomes contaminated by flowing over surfaces laden with pollutants such as salt, heavy metals, **particulate matter**, animal faeces, refuse, etc. that are deposited on roads and car parks which are then carried into the drainage system. The contaminants are transported into rivers and aquatic ecosystems downstream. The pollutant load deposited in watercourses is likely to be greatest following dry weather, that permits detritus to accumulate. The resulting 'toxic shock' can cause significant ecosystem damage.

Urban runoff is also relatively warm owing to higher urban surface temperatures (see Chapter 7). For example, a thundershower at the end of a hot day gives a short-term pulse of water that is sometimes several degrees higher than that from rural areas. At the city scale these surges of warm water are added to the continual injection of hot effluent from the outfalls of industrial plants, power stations and sewage. The combination is sufficient to raise significantly the temperature of large rivers as they flow through urban regions. Observations by Clark (1969) show that the pulse from each settlement contributes to an accumulated warming trend in the downstream direction.

After certain household and office uses piped water can be successfully recycled. Except for wastewater generated by toilet flushing, which could contain fecal matter or other pathogens, recyclable water sources include discharge from domestic sources (sinks, baths, washing machines, showers) plus rainwater stored in water butts. Following treatment this 'grey water' can be used to irrigate, recharge soil moisture, wash vehicles and roads. To avoid dispersing pathogens in the air irrigation should be applied beneath the soil surface or via drip feed systems not sprayed into the air. Moreover, greater than 50% of the heat contained in waste hot water can be recovered by a heat exchanger.

### 8.3.6 Change in Storage

Storage change ($\Delta S$ in Equation 8.3 and Figure 8.2b) is the overall change of water mass in the air and soil column per unit area (in kg m$^{-2}$ h$^{-1}$ or mm h$^{-1}$). We might practically divide the change in storage of water within the volume into storage changes of atmospheric humidity (i.e. from the surface to $z_{top}$), the changes in surface water (addition and removal of snow and liquid surface water such as ponds or pools), and changes in soil and substrate moisture from the surface down to $z_{bot}$.

#### Storage Changes in the Air

Within the balancing volume, most water is stored either at the surface or within the subsurface not the air; the typical volumetric water holding typical of air is $10^4$ times smaller than that of soils or substrate. Nevertheless, this mass of water (measured as humidity) fluctuates very quickly due to the action of wind and **turbulence**. For example, $\Delta S$ in an air column within a vegetated UCL can easily equal 10–20% of $E$ over an hour. However, over longer time periods (days, weeks), $\Delta S$ in the air is negligible.

#### Surface and Soil Moisture

Available surface and soil moisture across a city and its change with time is highly fragmented. This is because of the patchiness of urban surface properties, especially the sharp edges of impermeable patches (paving and buildings) and the localized control of irrigation. Another contributor to variability is the construction process, which greatly disturbs urban soils. Digging, laying pipes and the introduction of new materials like gravel and sand destroy the layered structure of a natural soil profile and increase hydraulic conductivity along preferred substrate routes. Finally, horizontal exchange of soil water is impeded by urban infrastructure (basements, tunnels, pipes, utility lines), and complicated by extraneous sources of water from leaking pipes and sewers. This spatial fragmentation is indirectly illustrated by daytime thermal imagery as discussed in Section 7.2.2.

Soil water has relevance in tracking plant-available water and is a major control on $E$. Moreover, the status of soil moisture under permeable surfaces is useful for flood forecasting as it indicates the available storage capacity. Field observation, via ground-level surveys at the city scale, is probably impractical for most applications. Because $\Delta S$ in soils is only a small component of the overall water balance, it is not realistic to solve for it as the residual of the balance, even if all other terms are observed. Crude estimation may be possible using **remote sensing** and algorithms for rural soil types but it is often more promising to use a numerical model that calculates $\Delta S$ as part of a simulated water balance. The highly patchy distribution of soil moisture makes it very difficult to construct a spatial sampling scheme that is representative at larger than the microscale. Given this, spatially coherent estimates are hard to find, and it must be assumed that those available, even at the local scale, possess variability and uncertainty. That understood, the general consensus is that urban soils are drier than their rural counterparts. Irrigated settlements in otherwise dry regions are again an exception.

#### Infiltration

Infiltration is a critical process that controls the movement of water between the atmosphere and the substrate. Soils that are close to saturation or have an impermeable surface cover do not allow significant infiltration so $P$ gathers on the surface or feeds into $R$. As urban development intensifies, the impervious

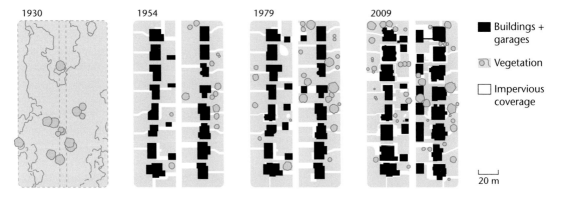

**Figure 8.13** Replacement of vegetation (trees, shrubs) by impervious cover (roads, paths, buildings) due to initial development and subsequent densification in an urban block dominated by single-family residential houses in Vancouver, Canada (based on data from the City of Vancouver).

surface fraction ($\lambda_i$) increases, infiltration to the soil and substrate declines, as does $E$.

Figure 8.13 chronicles changes to the land cover of an **urban block** in a North American suburb over an 80 year period. The area was originally densely forested until the trees were completely cleared in the early decades of the twentieth century. By 1930 it was just covered by underbrush, saplings and an unpaved track. A block of two rows of houses with a road and lanes were built and increasingly developed over the succeeding 80 years. During that period the fraction of pervious cover decreased from 100% at the start, to 83% in 1930, 55% in 1954, 52% in 1979 and 42% in 2009 as houses and garages were added, paved driveways and paths became more extensive.

### 8.3.7 Groundwater

Water that infiltrates but cannot be stored in the near-surface soil layer percolates down towards the water table ($G$ in Equation 8.3 and Figure 8.2b). Many cities have followed a common path to management of water at this depth. Initially, surface paving reduces infiltration and recharge. It follows that water withdrawal by wells depresses the level of the water table (Figure 8.4). As the city matures, investment in public utilities to increase piped water supply slows the extraction of groundwater. Continued urban development reduces infiltration further, so it is reasonable to expect the water table to drop further. However, in many places, leaks from pipes (and excessive irrigation) more than make-up for the loss of natural recharge, and the water table *rises* (Lerner, 2002). Further, some industries

dispose of their 'waste' water, by injecting it directly into the ground, adding to recharge.

This historical path is a good description of the story in London, United Kingdom. The city lies in a basin and until the 1950s it was able to rely on the underlying aquifer to meet most of its water needs. Since then, partly because of legislated regulation, abstraction has nearly ceased. In response, the water table under the city has risen to the point where it threatens underground infrastructure, such as the subway system, which is now below the water table and occasionally experiences flooding.

In other cities groundwater remains a mainstay of the water supply despite unregulated wells causing serious depletion. For example, in New Delhi, India, it is estimated that up to 50% of the water use derives from over 350,000 private wells. In Mexico City, extraction of groundwater over many years has caused the water table to drop so much that parts of the city have sunk more than 8 m since the beginning of the twentieth century, disrupting the foundation of many buildings.

### 8.3.8 Advection

The water balance for an urban neighbourhood, or especially for the UBL, can be subject to advection ($\Delta A$ in Equation 8.3 and Figure 8.2b) The main controls are the physiography and spatial diversity of the urban structure and surface cover. A few well-instrumented studies (e.g. Pigeon et al., 2007) show $\Delta A$ can be significant in cities where there are moisture transition zones (e.g. at a coast, urban edge, or the edge of vegetation/non-vegetated areas).

## Summary

The surface water balance (SWB) is a statement of mass conservation that describes the partitioning of available water. In cities the input terms include precipitation ($P$), dewfall ($E$), piped water ($I$) and water generated by the combustion of fuel ($F$). This water is expended by evaporation ($E$), runoff ($R$), infiltration and soil moisture recharge ($\Delta S$) and possibly advection ($\Delta A$).

- The majority of **piped water** ($I$) is used for drinking water, domestic activities and removing wastes but it contributes relatively little to most urban environments. Nevertheless, the environs of cities have more available water than equivalent rural landscapes due to leakage from pipes that increase $\Delta S$; vegetation irrigation that maintains a greener urban landscape and boosts $E$; and water released by combustion ($F$) which may boost humidity at night, when atmospheric mixing is reduced.
- Increased **impermeability of urban land cover** due to urban development causes $R/P$ to increase and both $E/P$ and $\Delta S/P$ to decrease. Consequently peak flows increase and arrive more quickly in urban than rural catchments.
- Generally, **urban soils possess less available moisture** and the magnitude of $E$ is reduced. In some places, where vegetation growth is supported by garden **irrigation**, the combination of $E$, $R$ and $\Delta S$ can then exceed $P$.

# 9 | Atmospheric Moisture

**Figure 9.1** Visible atmospheric moisture from natural and anthropogenic sources in the early morning in Vancouver, Canada (Credit. A. Clark/Reuters, used with permission).

Figure 9.1 shows early morning **fog** and **condensation** plumes in Vancouver, Canada. In the foreground is advection fog creeping at low level inland from the sea towards the city through an inlet spanned by a suspension bridge. In the middle ground, the fog flows between large and probably warm buildings that both deepen and dilute the fog. Several of the large buildings are emitting plumes of water vapour from roof vents which condense into cloud in the colder ambient air. In the background the advection fog advances across a rural river delta.

As a general rule, densely built urban landscapes are 'arid' by design. Whilst there may be a plentiful water supply, the predominant impervious **surface cover** and the relative lack of vegetation means there is relatively little water available for 'natural' **evaporation**. The direct injection of water into the atmosphere by human activities (air conditioning systems, fuel **combustion**, cooling towers, etc.) partly

compensates, but overall the water content of the urban atmosphere is low especially in daytime, compared with the surrounding environment. At night we'll see the opposite may be true: there can be an **urban moisture excess**. Of course, there are exceptions: some urban landscapes may be more vegetated (or wetter) than the surrounding area because of abundant renewable water supplies (oases, mountain lakes and snowmelt) that are used to support significant irrigation. In practice, the urban effect of humidity (and condensation) depends on a range of factors including background weather and climate, seasonality, surface characteristics (e.g. green fraction) and human factors, specifically the **emission** of heat, moisture and materials. Urban effects on fog are complex due to the competing effects of urban warmth from the heat island (Chapter 7), anthropogenic vapour (Chapter 8 and the present chapter) and extra condensation nuclei from urban air pollution (Chapter 11) all acting within the context of local topographic controls (Chapter 12).

This chapter is concerned with atmospheric moisture in all its states (vapour, liquid droplets and ice crystals). Water can change state with relative ease each involving release or uptake of **latent heat**. Discussion is divided into urban effects on humidity and condensation.

## 9.1 Basics of Atmospheric Moisture

### 9.1.1 Humidity

The moisture of the atmosphere is constantly fuelled by evaporation from the surface. The rate of evaporation from a surface is a balance between competing processes. One is the loss of water molecules from a surface to the air; the other is the capture of water molecules from the air by the surface. Other things being equal, loss depends on a surface-to-air **concentration** gradient of water vapour and the surface temperature. Capture depends on the gradient being in the reverse direction. If loss exceeds capture the process is evaporation, if capture exceeds loss it is condensation (or deposition if it is solid). Condensation also occurs if water vapour attaches to the surface of a particle (**cloud condensation nuclei**) suspended in the air; this forms liquid droplets (e.g. clouds, fog, water vapour plumes from chimneys) or ice crystals on **ice nuclei**. Condensation is favoured on other surfaces if they are cooler than the air (e.g. **dewfall** on leaves, roofs and ground). In moist environments if the concentration in the near-surface air and surface are equal, evaporation ceases and the air is said to be saturated.

Air temperature ($T_a$) is a simple quantity to express and to measure; however, atmospheric water vapour (humidity) is more complicated both to describe and to monitor, because of the compressible nature of the atmosphere (see Wallace and Hobbs, 2006). We can define the following direct measures of atmospheric humidity:

- **Absolute humidity** ($\rho_v$) is the ratio of the mass of water vapour ($m_w$) to the total volume of air ($V$) in g m$^{-3}$ (also called: **vapour density**). The volume of air varies with pressure ($p$). That restricts the use of $\rho_v$ to applications over small height intervals.
- **Specific humidity** ($q$) is the mass ratio (in g kg$^{-1}$) between the mass of water vapour ($m_w$) and the mass of moist air ($m_a$). Specific humidity is a conservative humidity measure. This means for an air

parcel, it does not change even if pressure or temperature changes.
- **Vapour pressure** ($e$) is the partial pressure (in Pa) exerted by water vapour molecules in an air sample.

The maximum water vapour pressure in the air before it causes saturation and condenses on surfaces or condensation nuclei depends upon $T_a$ and $p$. At a given $T_a$ and $p$, the vapour pressure where saturation happens is called the **saturation vapour pressure** ($e^*$). The vapour pressure deficit ($vpd$) is defined as how much more partial pressure can be taken up before saturation occurs, i.e. $vpd = e^* - e$. Figure 9.2 shows $e^*$ as a function of $T_a$ for a pressure of $p = 1,013$ hPa, for example. The graph is drawn with respect to a plane surface of pure water (or ice), whereas in real atmospheres condensation nuclei have curved surfaces and complex shapes and chemistries. In the absence of sufficient condensation nuclei, it is common for air to become weakly **supersaturated** (i.e. $e^* > e$) before condensation takes place (Section 9.1).

Figure 9.2 illustrates definition of **relative humidity** ($RH$) as the ratio (in %) of the actual vapour pressure to the saturation vapour pressure at the given $T_a$ and $p$ (Equation 9.1)

$$RH = 100\frac{e}{e^*} \quad (\%) \qquad \textbf{Equation 9.1}$$

Hence, saturation occurs when $RH$ reaches 100% and $e = e^*$. It is important to note that $RH$ varies with $e$, $T_a$ and $p$. $RH$ is a measure of how close the air is to saturation *at its present temperature,* but it is *not* a direct or absolute measure of moisture content or its vapour deficit. It follows that given its dependence on both $T_a$ and $p$, $RH$ should not be used to measure horizontal or vertical *differences* of moisture content across a landscape, because they involve changes of $T_a$ and/or $p$ (e.g. due to altitude). A map of $RH$ may be able to help gauge how far the air is from becoming saturated, at the temperatures present in the area, but it cannot discriminate whether the air at one site is more or less moist, than at another.

To give an idea of relationships and the magnitudes, at 25°C and a pressure of 1,013 hPa, air with an $RH$ of 50% has the values: $\rho_v = 11.5$ g m$^{-3}$, $q = 9.8$ g kg$^{-1}$, and $e = 15.9$ hPa. For comparison, air at 5°C (with same pressure and $RH$) has much lower values: $\rho_v = 4.0$ g m$^{-3}$, $q = 2.7$ g kg$^{-1}$, and $e = 4.4$ hPa. Despite its dependence on temperature and pressure, $RH$ is a very useful measure to predict the likelihood of condensation (e.g. dew, or fog) and it is also

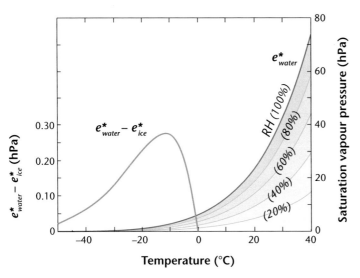

**Figure 9.2** Relations between temperature and both vapour pressure ($e$) (right scale) and relative humidity ($RH$) (noted on curves) over a plane surface of pure water at sea level. At 100% RH, $e$ is equal to the saturation vapour pressure ($e^*$). The single-peak curve at the lower left applies to temperatures below freezing. If the water surface is ice rather than liquid water, the surface saturation value ($e^*_{ice}$) is slightly lower than for liquid water. The curve is the difference between the values for water and ice (on the left axis) (Source: calculated from WMO, 2008, Annex 4B).

significant in relation to human **thermal comfort** and bioclimate (e.g. in sweating, Chapter 14). The last measure used in this book, is the **dewpoint temperature** ($T_d$), which is the temperature (in K or °C) to which a parcel of air must be cooled (at constant pressure) for saturation to occur; only at $RH = 100\%$ is $T_d$ equal to $T_a$.

We should also note that it is considerably more of a challenge to measure the moisture content of air than its temperature. There are about 30 different techniques available to sense humidity (for a review of methods, errors, instrument **exposure** and calibration, see WMO, 2008) and a similar number of instrument types (called **hygrometers**) to measure it each with its own requirements and **accuracy**. This diversity sometimes hampers comparison between different humidity studies.

### 9.1.2 Condensation

Condensation is the opposite of evaporation; instead of the vapourization it leads to deposition of water on surfaces and releases, rather than consumes, latent heat. At the ground's surface (soil, vegetation, roofs) it results in the formation of dew, whereas in the atmosphere water is deposited on nanoscale surfaces such as **aerosols** to form clouds. Here, we mainly consider cloud plumes that form soon after the release of vapour from tailpipes and chimneys into cooler air.

Dew formation occurs on surfaces if their surface temperature ($T_0$) is at or below the dewpoint temperature ($T_d$) of the near-surface air. Dewfall involves the turbulent transfer and **diffusion** of water vapour molecules from the atmosphere onto a surface where they condense as discrete water droplets. At the time of formation there is both a reduction of absolute humidity in the near-surface air (Section 9.2.1) and a release of latent heat of condensation. For dew it is necessary that $T_0 < T_d$, if $T_d$ is below the freezing point the deposit is hoar frost wherein water vapour is deposited directly as ice crystals. Dewfall and hoar are best developed with weak, but not calm, winds because there is need for a flux of moisture onto the surface where condensation occurs (Garratt & Segal, 1988).

Fog is defined as a cloud with its base at ground level that reduces **visibility** to < 1 km. It requires the air to be at or below $T_d$. and include **cloud condensation nuclei** (CCN) onto which water vapour can condense. Visibility is inversely proportional to the extinction coefficient for visible light, which is a function of **liquid water content** and droplet number concentration (Gultepe et al., 2007). Therefore, in urban areas, where the atmosphere has abundant CCN and hence many small droplets, visibility is less at a given liquid water content, than it is in cleaner air.

In the **atmospheric boundary layer** (ABL) the magnitude of the **scalars** $\rho_v$ and $T_a$ are similar, because both are driven by surface **turbulent fluxes** (**evaporation** $E$ for $\rho_v$ and **sensible heat flux density** $Q_H$ for $T_a$). However, whilst all surfaces are sources of heat, not all are necessarily sources of moisture. That fundamental difference plus the spatial variability of soil moisture due to natural differences of soils, vegetative cover and agricultural practices makes urban moisture availability very patchy. This variability is extreme in cities, because unlike most natural ecosystems, dry and wet surfaces can occur immediately next to each other, for example, a lawn next to a road. This fact also contributes to the patchiness of humidity patterns.

In the absence of liquid water, the mass conservation for humidity in the ABL (or for a unit volume in the atmosphere above a surface) can be approximated as:

$$\frac{\Delta \overline{q}}{\Delta t} = \overline{u}\,\frac{\Delta \overline{q}}{\Delta x} + F + \frac{E_0 - E_z}{z} \qquad (\text{g m}^{-3}\,\text{s}^{-1})$$

**Equation 9.2**

which says that changes in mean absolute humidity $\overline{q}$ within the volume over time (g m$^{-3}$ s$^{-1}$) could be caused by moisture **advection** ($\overline{u}$ is the mean horizontal wind speed (in m s$^{-1}$), $x$ the direction along the mean wind), or the addition of water vapour directly, for example, emissions from chimneys and stacks ($F$, in g m$^{-3}$ s$^{-1}$), or by the difference in evaporation between the surface ($E_0$, in g m$^{-2}$ s$^{-1}$) and at the top of the **mixed layer** $E_z$. Equation 9.2 is analogous to Equation 7.6 that describes changes of $T_a$ based on the principle of energy conservation.

## 9.2 Urban Effects on Humidity

In this section, we discuss the moisture characteristics of the urban atmosphere (UBL), how they link to the surface water budget and the **entrainment** processes at the top of the ABL. Discussion is similar to that for temperature (Section 7.3) starting with moisture in the **urban canopy layer** (UCL) (Section 9.2.1), followed by that in the **urban boundary layer** (UBL) (Section 9.2.2), and the genesis of observed moisture effects (Section 9.2.3). However, it is not always possible to standardize and compare urban humidity effects in different cities, as we do for temperature effects (e.g. $\Delta T_{U-R}$). This is mainly because authors use different humidity measures but give insufficient information to convert them to a common measure.

### 9.2.1 Urban Canopy Layer

Treatment of UCL humidity in the literature is less complete than for the **canopy layer urban heat island** (UHI$_{\text{UCL}}$). Indeed, at present, there is insufficient work to establish comprehensive general principles, so here we emphasize recurrent features and draw tentative conclusions.

### Urban Moisture Excess

The similarity of the spatial fields of temperature and humidity in the UCL of Leicester, United Kingdom is evident in Figure 9.3. At the time of these field surveys it was a relatively small city (Population: ~270k) located on fairly level terrain (< 40 m relief). The inner city is densely built-up with narrow streets and little vegetation, with a mixture of industrial and residential uses (LCZ 2 and 3). The suburbs are less dense, largely residential, houses with gardens, parks and open space (LCZ 5 and 6). The night of August 23, 1966, was almost cloudless and nearly calm. A well-developed UHI, with a maximum $\Delta T_{U-R}$ of about 4.4 K, was centred on the inner city (Figure 9.3a). The peripheral UHI 'cliff' generally follows the built-up limit of the city and isotherms inside correspond with breaks of building density and open space.

The corresponding $RH$ field also follows the built-up outline and density patterns inside the city (Figure 9.3b). Notice the strong correspondence between the fields of $T_a$ and $RH$, with a maximum $\Delta RH_{U-R}$ of $-12\%$, again centred on the inner city. Remember that unlike vapour pressure $e$, $RH$ is not a measure of water vapour content, but does show how close the air is to saturation. Whilst the $RH$ of air is high (80%–90% in the countryside), nowhere is it saturated and no condensation (dewfall or fog) was observed that night. The spatial distribution of $e$ in the UCL (Figure 9.3c) follows the density of built structures and land cover (especially at the built-up limit). That results in an urban moisture excess (UME) and maximum $\Delta e_{U-R}$ ($\approx 1.8$ hPa) is found in the city centre.

This nocturnal pattern is fairly typical, especially in larger cities. At first glance this might appear counter-intuitive, since the central areas of cities usually have relatively little vegetation cover. But the explanation is that at night urban air often does have greater moisture content but its vapour is further from saturation owing to higher air temperature (i.e. the UHI).

### Diurnal and Seasonal Patterns in Temperate Cities

From a variety of studies in cities with temperate climates (Köppen types C and D, see Section 12.1), the temporal patterns of urban–rural moisture differences seem to display the following common characteristics:

- They are best displayed in 'ideal' (calm, clear) weather, much like the UHI.

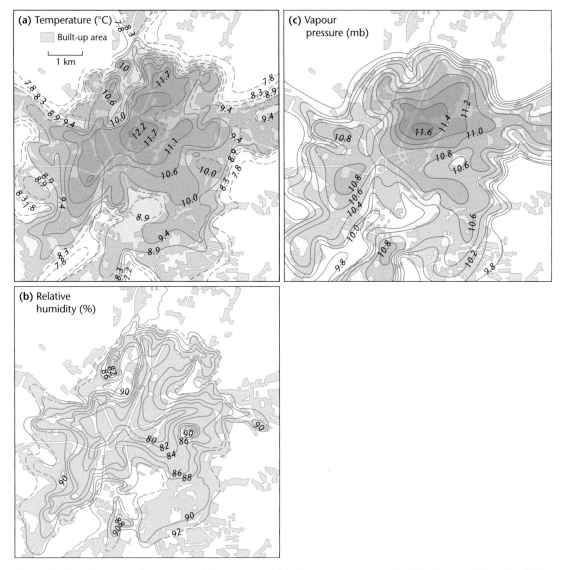

**Figure 9.3** Detailed maps from automobile surveys of **(a)** the air temperature $T_a$ (°C) (Source: Chandler 1976; with permission from WMO) and atmospheric humidity using both **(b)** *RH* (%) and **(c)** vapour pressure, $e_a$ (hPa), in Leicester, United Kingdom on a calm, clear summer night (Source: Chandler, 1967; © American Meteorological Society, used with permission).

- They are the largest and spatially coherent at night but, during daytime they are complex and patchy.
- There is a seasonal shift in the diurnal pattern; in summer, urban air is less moist than in the countryside by day, but more moist at night and in winter a city is commonly more moist at all times.

All of these features can be illustrated using results from Edmonton, Canada (Population: ~480k), which

has a continental location with cold winters (Hage, 1975). Observations show that nocturnal $\Delta\rho_{v\ U-R}$ is positive throughout the year (Figure 9.4a) with largest differences in August and the smallest in the winter and spring. On the other hand, the annual course of daytime $\Delta\rho_{v\ U-R}$ changes from a deficit (city drier) in the warmer part of the year (March to September), to excess in the colder part (Figure 9.4a). In winter there

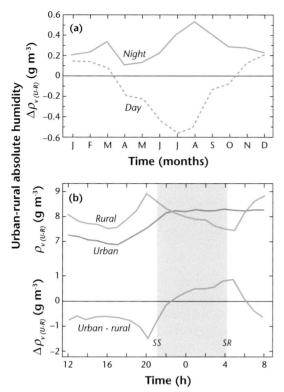

**Figure 9.4** Urban effects on humidity in Edmonton, Canada, based on observations at two airports, one urban the other rural. **(a)** The seasonal variation of the urban–rural difference of absolute humidity ($\Delta\rho_{v\ U-R}$) by day and at night. **(b)** Hourly variation of $\rho_v$ and their difference on cloudless summer nights (Modified after: Hage, 1975; © American Meteorological Society, used with permission).

is little diurnal change in the magnitude of the excess, but by mid-summer there is a strong diurnal shift from a large daytime deficit to an equally large nocturnal excess.

The diurnal pattern of $\rho_v$ at a rural and urban site and the difference ($\Delta\rho_{v\ U-R}$) during cloudless summer nights are shown in Figure 9.4b. The rural site displays a double peak, one a few hours after sunrise and another just before sunset. The early morning peak in moisture near the ground is due to the onset of $E$ into a shallow mixing layer that is capped by the remnant nocturnal **inversion**. As the day progresses the mixing layer deepens, by convective heating, and $\rho_v$ decreases. An hour or two before sunset $Q^*$ turns negative and surface cooling increases **stability**, which stifles **convection** but evaporation continues, so $\rho_v$

suddenly increases. After sundown $E$ slows and is often replaced by intermittent dewfall, which causes partial depletion of $\rho_v$ in the lowest layers.

On the other hand, the daily course of humidity in the city, is a relatively simple wave: after sunrise there is a gradual decline in $\rho_v$ through the day, the trend is similar to the countryside but moisture values are consistently lower until early evening, then urban $\rho_v$ values increase surpassing rural $\rho_v$ by midnight, and remain higher through the rest of the night (Figure 9.4b). Absolute humidity in the city during the daytime is lower than at the rural site, because urban $E$ rates are weaker and convective mixing is greater (Chapter 6). The nocturnal reversal is probably the result of three processes: firstly, weak $E$ usually persists, perhaps intermittently, into the night; secondly, dewfall in the city may be lighter or absent (Section 8.3) and; thirdly, there is often significant input of moisture from combustion sources ($F$). The net effect is that the city is drier by day but moister at night (Figure 9.4b).

Urban–rural differences in dewpoint temperature ($\Delta T_{d\ U-R}$) from Chicago, United States (Figure 9.5), illustrate an annual picture of the temporal patterns that generally agree with the Edmonton results. At night, in all seasons, the city is moister with the maximum UME in late autumn/early winter. The late night maximum is followed by a sharp drop to a moisture deficit or much smaller excess soon after sunrise. There is a deficit on most mornings until one or two hours before noon and in summer, $\Delta T_{d\ U-R} < 0$ through the daytime. Except for those occasions, $\Delta T_{d\ U-R} > 0$, and that is the case for the full 24-hour period in late autumn. Similar patterns have also been found in cities with less extreme winters, like London, United Kingdom (Population: ~8.5M), Gothenburg, Sweden (Population: ~700k), and Christchurch, New Zealand (Population: ~300k).

The recurrent features described here approach a 'standard' pattern for large cities — like the temporal dynamics of the UHI (Figure 7.15). A typical nocturnal UME reaches a maximum magnitude of about 1 hPa (approximately 1 g m$^{-3}$). The highest values in large cities range from 1 to 3 hPa (Holmer and Eliasson, 1999) with extreme values up to 6 hPa. However, there does not appear to be a simple relation between the magnitude of the excess, and the size of the city. This is hardly surprising, given that such differences depend on the humidity of both the city *and* rural environments. Rural moisture properties cover a wide range.

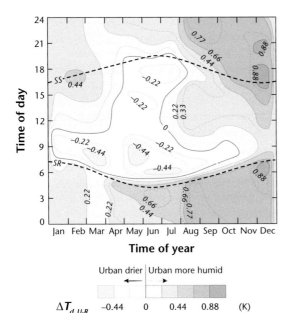

Urban drier | Urban more humid

$\Delta T_{d\ U\text{-}R}$   -0.44   0   0.44   0.88   (K)

**Figure 9.5** Isothermic plot of the daily and seasonal variation of the dewpoint temperature differences ($\Delta T_{d\ U-R}$) between urban and rural stations in Chicago for the period 1963–70. Dashed lines are times of sunrise (SR) and sunset (SS) (Source: Ackerman, 1987; © American Meteorological Society, used with permission).

Whilst most humidity studies show the typical urban effects described above, several find spatial and temporal patterns that deviate from this picture probably for one or more of the following reasons:

- Unusual seasonal patterns of water vapour emissions due to the need for **space heating or cooling** driven by weather conditions.
- Anomalous water vapour emissions near **industries** (e.g. steel mills).
- A particular district is heavily **vegetated**, or conversely, **devoid of vegetation** cover.
- Access to abundant water for garden **irrigation**, whilst the surrounding countryside experiences drought, for example, an oasis settlement, surrounded by desert.
- Being close to a **water body** so there are incursions of air with relatively high humidity.

It is perhaps not surprising to find UCL moisture climates are diverse and variable given this list, and the special properties of humidity outlined at the start of this Section. Lodz, Poland (Warm summer continental

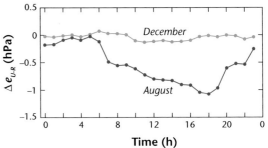

**Figure 9.6** Urban–rural differences of vapour pressure in Krefeld, Germany for a winter and a summer month, based on hourly averages (Source: Kuttler et al., 2007; © Royal Meteorological Society, used with permission from Wiley).

climate; Population: ~0.8M), illustrates the kind of variability that exists. Research on $\Delta e_{U-R}$ found that on some nights, the course of the moisture effect is similar to that described above, but on other seemingly ideal UHI occasions, it evolves very differently. For example the largest positive values $\Delta e_{U-R}$ can occur late in the night followed quickly by the largest negative value (Fortuniak et al., 2006). Why this is so remains unresolved. It is suggested that whilst simple averaging of data from fine days may work well for temperature, it is less appropriate for humidity. Another indication of diversity occurs in towns and small cities: the temporal pattern may or may not follow a course similar to that in larger cities, indeed the UME may be hardly present. In Krefeld, Germany (Maritime temperate climate; Population: ~238k), a city with ~60% open or green space, small UME are found often (~31% of the time) but relatively large (> 0.5 hPa) differences occur on a few (< 5%) occasions. In winter (December) in contrast to the results for large cities, the UME is almost absent throughout the day, and there is virtually no diurnal pattern (Figure 9.6). In summer (August) at night, there is virtually no UME in the mean, however, by day the UME is always negative and by late afternoon urban deficits are large (average about −1 hPa).

These results reiterate that, even for mid-latitude cities where most studies have been conducted, a coherent picture of humidity in the UCL has yet to emerge.

## Cities in Tropical Climates

Humidity effects in tropical cities have been much less studied. Results from cities in tropical wet, and subtropical highland, climates mostly appear to resemble the mid-latitude case (above). Jauregui and Tejeda (1997) studied the spatial and temporal features of humidity in Mexico City (subtropical highland climate; Population: ~21M). In the summer wet season there is a nocturnal UME, but a deficit, by day. In the winter dry season urban–rural differences are smaller, and sometimes insignificant.

Holmer et al. (2013) studied Ouagadougou, Burkina Faso (semi-arid climate; Population: ~1.3M), in the cooler, but still warm period (November–December): days are hot, windy and **unstable**, but nights are cloudless with weak winds and strong stability. Figure 9.7 gives results for very **stable** nights at two sites: one located in the city centre (open-set mid-rise, $\lambda_v \approx 5\%$) it is on a rooftop not in the UCL, the other at a rural station with

shrub-bush vegetation cover, $\lambda_v \approx 44\%$). The diurnal course of specific humidity ($q$), at the urban station, follows a similar path to that found in the mid-latitudes with a minimum in the afternoon, rising rapidly around sunset to values that are sustained until after sunrise. Rural humidity follows a similar daily course but is always greater than in the city, i.e. no UME at any time. Since the rural atmosphere does not become saturated overnight, there is no dewfall to cause $q$ to fall, in contrast to the Edmonton study (Holmer et al., 2013). Humidity at both sites is closely related to changes in stability near sunset and sunrise.

Some arid zone cities can be more moist both day and night (e.g. Tucson, United States, Figure 9.8). The reason for the greater daytime moisture is usually because these cities have sufficient water available to support irrigation. They may be situated on a naturally occurring water source (artesian well, river) or they can import water from mountain or distant reservoirs. Small additional sources include water vapour from air conditioning and from vehicle exhaust. The results for Sacramento, United States in the hot summer climate of the Central Valley of California show the urban–rural difference ($\Delta e_{U-R}$)

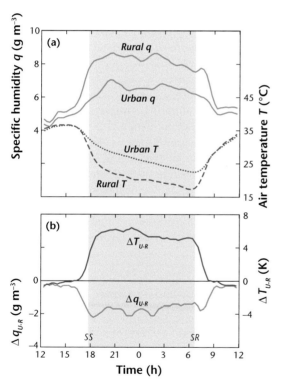

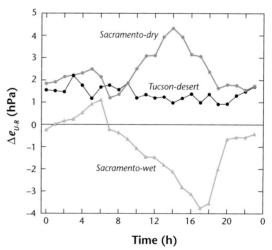

Figure 9.7 (a) Temporal variation of specific humidity and air temperature in Ouagadougou, Burkina Faso, averaged over 5 days: all unstable by day with very stable nights (Source: Holmer et al., 2013; © 2012 Royal Meteorological Society, used with permission from Wiley). (b) Mean variation of urban–rural differences of specific humidity and air temperature (Data source: B. Holmer).

Figure 9.8 Suburban-rural vapour pressure differences from screen-level observations in two cities with hot summers. The Sacramento rural readings are two extensive sites; one is a dry grassland, the other a heavily irrigated farm growing lawn grass. The Tucson rural site is in an extensive desert of creosote bush and cactus (Source: Oke et al., 1998; © American Meteorological Society, used with permission).

depends critically on the choice of the rural station. The urban station is situated in a suburb with patches with irrigated gardens and parks. If the rural station is located over dry grassland (natural to the Central Valley), there is a UME present all day (i.e. like Tucson). But if the rural station is in an extensive area of irrigated short grass the daytime difference is reversed (urban deficit).

Urban effects on the humidity of hot/wet cities are least well documented. As a result there is an insufficient basis to establish general patterns of urban–rural differences.

### Urban Humidity and the Canopy-Level UHI

The dynamics and magnitude of urban–rural humidity differences and the UHI show similarities. The temporal patterns for both are often in approximate phase, their greatest daily range occurs in fine weather and the magnitudes of both are strongest on calm, clear nights in summer. Such conditions in rural areas promote nocturnal cooling, temperature inversions and strong stability. The cooling often results in saturation and perhaps condensation in the air (fog) or on the surface (dewfall). The UHI tends to inhibit saturation by maintaining a weak convective layer so that despite weak evaporation and water injection via combustion, there is little dewfall (Section 9.3). However, any attempts to generalize can only be based on a few studies and there are exceptions to the rule (e.g. Ouagadougou, Figure 9.7).

Similarly, long-term temporal trends in urban humidity may match those of temperature but care must be taken not to draw overly simple analogies. The annual mean *RH* of Tokyo, Japan, fell from 75% (a value close to that of the surrounding rural area) near the start of the twentieth century to 63% by the 1960s, resulting in a rural-urban difference of 12% (Figure 9.9). But, given the definition of *RH*, we cannot be certain to what extent this difference was due to thermal (UHI) or moisture effects. Omoto et al. (1994) devised a means to separate the two factors (Figure 9.9). The visual appearance of the trend at each station suggests *RH* at the urban station decreased from the start, except for the 1940s. The rural location appears fairly stable until the 1970s, after which its *RH* also decreases. The overall impression is that the growth of Tokyo has resulted in increasing 'dryness' (but *RH* only tells us that the city is further from saturation at the ambient temperature). In examining this evidence we should remember

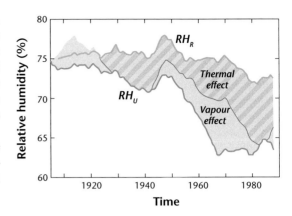

**Figure 9.9** Trends of urban and rural *RH* in the Tokyo region during the twentieth century. Shading identifies the separate contributions to the total difference $\Delta RH_{U-R}$ made by differences of moisture and air temperature, between the urban and rural stations (Modified after: Omoto et al., 1994).

Lowry's framework (Chapter 2); the proper interpretation of climatic trends requires taking into account any trends at other scales that might contribute to the signal. In the present case, since both stations lie in the same region, we may assume them to be approximately equally affected by trends at larger (greater than urban) scales, so trends in their *difference* are likely to be significant. Um et al. (2007), using similar methods to those of Omoto et al. find Seoul, South Korea has experienced similar humidity changes to those of Tokyo in the twentieth century.

Finally, like the UHI, we might anticipate a relationship between the intensity of urban development and the magnitude of any urban effects on UCL humidity but so far evidence is weak.

### 9.2.2 Urban Boundary Layer

The present observational evidence of humidity effects in the ML is based on fewer than ten studies, from about five cities, all in the temperate zone. Observations come from helicopter flights, fixed wing aircraft traverses at a few levels, or profiles from instruments attached to free-flying or tethered balloons. Results from tall towers are indicative but limited to a few sites and unable to characterize the full UBL. As a result available evidence is very limited in its ability to describe the spatial and temporal characteristics of humidity in the UBL and how it differs from the non-urban ABL.

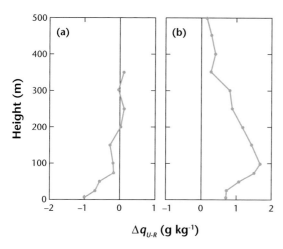

**Figure 9.10** Urban–rural differences in vertical profiles of specific humidity ($q$) in and near Christchurch, New Zealand **(a)** by day, and **(b)** at night (Source: Tapper, 1990; © Elsevier, used with permission).

The study of the vertical variation of specific humidity ($q$) in the UBL of Christchurch, New Zealand, by Tapper (1990) using a mobile tethersonde system gives a rare picture of conditions (details of the field observations are given in Section 7.5.1). Specific humidity is a conserved variable with height, so $\Delta q_{U-R}$ is valid without need for correction. The averaged daytime $\Delta q_{U-R}$ values steadily decrease with height, from about 1 g kg$^{-1}$ at the surface to zero at a height of about 200 m, and remain the same above that (Figure 9.10a), i.e. the UBL is drier than the rural ABL. The corresponding nocturnal profile is positive: the value is about 0.7 g kg$^{-1}$ at the surface, increasing to about 1.7 g kg$^{-1}$ near 100 m, above which it remains positive (a UME) all the way up to about 350 m (Figure 9.10b). This extension to heights above the daytime case might appear unusual. In fact, temperature profiles from the same study show that the nocturnal UHI reaches up to a similar height as the UME, (Figure 7.26a,b). The depth of the nocturnal UME is likely due to mixing over the city and entrainment of moisture from elevated layers down into the UBL.

The typical magnitude and extent of these urban effects on atmospheric moisture in the UBL are illustrated by aircraft surveys over Greater St. Louis, United States (Figure 9.11). Results from the middle of the day (1100–1800 h) in light wind conditions ($< 2$ m s$^{-1}$), with little cloud cover ($< 0.3$) show good

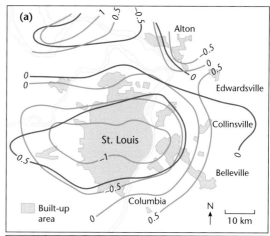

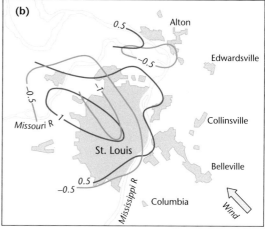

**Figure 9.11** Spatial distribution of anomalies of moisture ($q$, dashed lines) and potential temperature ($\theta$, solid lines) over St. Louis, United States. Composite of aircraft surveys between 150–450 m above ground, each for five days, **(a)** with light and variable winds, and **(b)** with moderate SE winds. Anomalies calculated relative to upwind values at 300–600 m (Source: Dirks, 1974; © American Geophysical Union, used with permission from Wiley).

correspondence between the $T_a$ and moisture distributions in the UBL (Figure 9.11a). This largely non-advective case is a composite created from several days all with light and variable winds, so there is little directional bias. The moisture anomaly shows the UBL to be drier than the upwind rural case, by about 1 g kg$^{-1}$, and warmer by about 0.5 K. On the other hand, the composite with moderate SE winds shows the moisture and heat plumes are elongated and displaced to the NW, by about 10 km (Figure 9.11b).

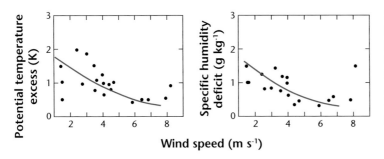

As the UBL wind speed increases, the magnitude of both the moisture and temperature anomalies *decreases* (Figure 9.12). This is characteristic of plume concentrations in general — greater speeds mean forward 'stretching' of the plume, greater mechanical mixing, both of which dilute the plume contents. The height of the urban mixing layer, during these daytime flights, also varies with wind speed. It is about 1.2–1.0 km deep with light winds of 2.5–3 m s$^{-1}$, but it contracts to about 0.5 km with winds of 8 m s$^{-1}$. The dilution of these thermodynamic anomalies, due to the presence of St. Louis, was not detectable beyond ~35 to 40 km downstream.

Almost all results show the humidity climate of the UBL to be an extension upwards of that in the UCL. However, it must be stressed that the sample of cities is very small.

### 9.2.3 Genesis of Effects

Hypothesized causes of urban moisture effects are not fully verified in any city, but piecemeal findings through observation and modelling suggest the following:

#### Urban Canopy Layer

In summer there is an urban daytime moisture deficit and nighttime surplus. The former is mainly a result of lower $E$ in the city due to the greater **impermeable** surface cover (pavement, buildings) and compacted soils. These are compounded by the greater vertical transport and mixing in cities (due to greater roughness and thermal **turbulence**) that enhance exchanges between near-surface air and the drier air above roof level and further dilute the moisture content. At night the UME is probably a result of two factors. Firstly, dewfall may be reduced in the city (Section 9.3) because of the warmth of the UHI$_{Surf}$ and because of emissions of **anthropogenic water vapour** ($F$).

In winter a moisture surplus is found both day and night in cities. It is largely due to anthropogenic emissions from space heating appliances and vehicles.

#### Urban Boundary Layer

There is good agreement between the sign of the moisture anomalies in both the UCL and UBL. This suggests good exchange, linking concentrations between the two layers. Upward transport is usually the more significant by day due to evaporation from the ground and trees, plus $F$ from vehicles, vents and chimneys. Some emissions directly enter the UBL from building vents and chimneys; routes that bypass the UCL. Finally, both by day and at night, there is enhanced entrainment between the turbulent UBL and the overlying **free atmosphere**. The entrainment process is similar to that of **momentum** (Chapter 4) and heat (Chapter 6), it can be a source of extra moisture or dryness for the UBL, depending on the sign of the difference between the two layers.

## 9.3 Urban Effects on Condensation

The processes of fog formation and dewfall are sensitive to microscale variations in surface thermal and moisture properties, both of which can be spatially very variable in cities.

### 9.3.1 Dew

At night canopy-level humidity is often greater in cities (the UME, Section 9.2). Given the relatively small vegetation fraction in cities this might seem counterintuitive. One hypothesis to explain this is that dewfall removes less moisture from the lower urban atmosphere than it does in rural areas. The hypothesized causal processes for less urban dew are:

- **Extra warmth** of the nocturnal **surface urban heat island** inhibits condensation, because reduced cooling of urban surfaces lessens the probability that $T_0$ drops to $T_d$.
- **Greater shelter** within the UCL, compared with rural areas, this lessens the supply of water vapour to surfaces, from above roof level.

Both processes are physically reasonable, but they do not cover all the actors in the cooling/moisture flux picture of cities. Further, not all urban surfaces are warmer. For example insulated roofs are often relatively cold at night (Figure 7.3) and attract copious dewfall, but their areal extent is relatively small. Emission of anthropogenic water vapour ($F$) could partly offset the reduced water vapour supply from above. Richards (2005) concludes the evidence for reduced urban dew is weak.

To put this in perspective, dewfall is a small flux both in mass ($E$) and energetic ($Q_E$) terms. The maximum deposition on grass in rural areas is about 0.5 mm night$^{-1}$ (Garratt, 1992) but more typical rates are 0.05 to 0.2 mm night$^{-1}$. The water is removed from the lowest atmosphere, which causes a drop in **surface layer** (SL) humidity. The maximum rate of dewfall implies a drop in $\rho_v$ of $< 3$ g m$^{-3}$ for a stable layer 100 m thick. However, this loss is partially offset by local advection that can supplement the moisture content. In urban areas, rates of dewfall on some roofs and other urban surfaces with a large **sky view factor**, are similar to those in rural areas, but vertical **facets** receive very little or nothing. Observed urban–rural $E$ differences are typically less than 0.1 mm night$^{-1}$. Over a 10 h night that gives an average flux of about $-7$ W m$^{-2}$. This suggests reduced dewfall is a contributor but is not the sole reason for the observed UME (Section 9.2.3).

The presence or absence of dew in cities is of potential significance in several ways. For example, the extra wetness promotes growth of mildew, moulds and algae on built and plant surfaces. Further, it reduces the efficiency of solar panels, aids deposition of **air pollutants** and the corrosion and formation of rust on outdoor materials. Ecologically dew is a source of drinking water for animals and water for plants.

## 9.3.2 Fog

There are three main paths to fog formation:

- **Cooling the air:** Radiation fog usually occurs at night when radiative cooling of the surface cools

the lowest air layer. Advection fog is created as moist air advects across a cooler surface and is cooled from below. Upslope fog forms when air is forced to rise, over terrain or denser air, causing it to cool adiabatically.
- **Adding moisture:** Frontal fog occurs when warm rain drops evaporate as they fall through a colder layer underneath so that the layer becomes saturated and condensation occurs.
- **Mixing:** If unsaturated air parcels with contrasting temperature and humidity mix, the resulting air mass may become saturated and condensation takes place. Steam fog occurs where cold, dry air flows over a body of warmer water. The near-surface air is both warm and humid, but as it is mixed with the cold air above, the temperature falls quickly bringing the air to saturation.

Ice fog is any fog at temperatures below –30°C. Typically it is found near water vapour sources such as open water (streams, ocean) and combustion exhausts (chimneys, tailpipes, jet engines). The term **smog** was originally introduced to describe a mixture of smoke and fog but now it is extended to include **photochemical smog**, which contains few or no fog droplets (Section 11.3.2). Finally, acid fog forms when marine or rural fog absorbs acidic material from the gas phase in the vicinity of pollution sources, creating a **pH** that is typically about 3 in the liquid phase.

The type of fog is controlled by the **synoptic** weather conditions, which often depend on the season (Tardif and Rasmussen, 2007), and **topography**. In addition, there is close coupling between a fog and the nature of the underlying surface, especially its wetness and ability to cool. As a result, fogs form across a range of geographic scales including hills and valleys, river valleys and basins, lakes and parks.

### Spatial Distribution of Fog in Urban Regions

Fog is sensitive to the special character of **urban climates**, especially effects on airflow, temperature and humidity. For example, UHI warmth can inhibit the formation of nocturnal radiation fog in the centre of cities and erode it in **suburban** and peripheral rural environments. Most significantly, urban fogs depend on the air pollution climate, because condensation requires the presence of CCN around which to coagulate (Chapter 10). Analysis of fog in a given city must account for the composition and transport, transformation, **dispersion** and removal of air pollutants (Section 11.1).

The topographic setting of a city is critical in establishing favourable conditions for fog formation. Many coastal cities are subject to advection fog. This might be due to the clash of warm and cold marine currents, or the existence of upwelling cold water along the coast (e.g. San Francisco, United States and Lima, Peru). The marine fog bank moves on- and offshore in response to the approach of a cold **front** or to **land-** and **sea breezes**. The daytime extent of any incursion into a coastal city is a continual tussle between the forces of motion wafting the fog in-and-out (see Figure 9.1) and the rates of fog production and dissipation. The latter may be enhanced by the warmth of the UHI, solar heating of paved surfaces and mixing over the rougher city.

Similarly, cities located in valleys, basins or near swampy land are prone to radiation fog. The presence of a river, lake or swamp favours humidification of the SL and the **orography** of a valley or basin is naturally suited to cold air drainage and stagnation (Section 12.2). Good examples are Milan and Turin in Italy's Po valley and Chongqing, China, which is situated at the confluence of the Yangtze and Jialing rivers. The Central Valley of California is the site of dense and persistent fogs in winter, especially in **anticyclonic** weather owing to the configuration of the valley and the copious irrigation of the valley floor. Overnight radiation fog thickens and visibilities decline close to zero, but the next day clear patches often form over the largest cities in the valley (Figure 9.13). Other large cities report similar daytime clearing (e.g. Milan and Turin in the Po Valley of northern Italy and Munich in Bavaria; Sachweh and Koepke, 1995).

These clear patches over cities in an otherwise continuous regional fog are probably due to two urban effects. First, the warmth and dryness of the urban atmosphere may cause the fog to thin, which increases solar input at the ground and warming (UHI$_{Surf}$) from below. Second, surface heating generates instability that initiates an urban heat island circulation (UHIC, Chapter 4) that further erodes fog. The edge of a fog bank has its own thermal circulation: a near-surface flow from the colder fog layer towards the warmer ground at its edge. If the fog happens to border a warm city, the subsiding limbs of the UHIC and fog-edge circulations could operate in unison, forming a feedback loop that supports development of urban clear 'islands' (Sachweh and Koepke, 1997).

### Impact of Fog on Cities

Some coastal and high-altitude communities short of potable water exist for long periods above the cloud base. Hence they are regularly exposed to fog-bearing

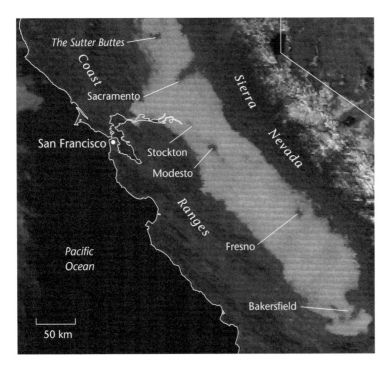

**Figure 9.13** GOES satellite image of the Central Valley of California covered with fog at 1401 PST on 18, December, 1985. The Sutter Buttes in the north rise 600 m above the Valley floor and poke through the fog. The valley is bounded on the west by the Coast Ranges and on the east by the snow-capped Sierra Nevada mountains. Patches clear of fog are located over the major urban centres in the valley (Based on: Lee, 1987).

winds and may be able to gain significant water supplies through fog collection. Fine mesh nets strung up on poles intercept relatively clean water and feed it under gravity to storage tanks. Fog harvesting has proved to be sustainable in several communities along the western coasts of South America and South Africa that otherwise receive scant **precipitation**.

Urban ice fogs are restricted to settlements in areas with very cold winters, such as Siberia, northern Scandinavia and northern North America. Many are based on the extraction and processing of natural resources, especially fossil fuels. They are also transport hubs for shipping and air lines. In winter survival necessitates that buildings are heated. All these activities are the source of air emissions including water vapour. In mid-winter with long nights and air temperature below $-30°C$ air has little capacity to retain water vapour so it condenses, almost immediately, into a cloud composed of droplets or directly into ice crystals. The presence of ice fog at high latitude, when there is little or no natural light anyway, increases the sense of gloom. Further, if toxic exhaust gases or particulates attach to the fog crystals, it makes breathing unpleasant and potentially hazardous.

### Historical Fog Trends in Urban Regions

There are well-defined historical trends in the occurrence and severity of urban fogs. As with temperature trends (Section 13.3.1), the interpretation of long-term fog records requires careful consideration. Fog has several weather controls (e.g. temperature, humidity, wind speed, cloud), each of which follow their own long-term trends in response to urban, synoptic and even global influences. In addition, urban fog depends on air pollutant emissions which depend on a city's transport system; industrial base; types of fuel; and source of electrical power (especially fossil fuels or renewables).

Anecdotal evidence relating the correlation of poor air quality with poor visibility and fog is available for many cities over many centuries. Observations in the late 1800s and early 1900s, mainly in Europe, suggested urban areas experienced increased fog frequency and reduced visibility (Brimblecombe, 2011). This was at the height of the Industrial Revolution when large emissions from intensive coal use produced particulates rich in soot (carbon) and sulphates. Condensation onto a

particle depends primarily on the amount of soluble ions present, but it is initiated on larger particulates at lower values of supersaturation (Section 10.1.1). Hence, in an atmosphere enriched with particles of various sizes, fog formation begins at lower humidities if large particulates are available (Gultepe et al., 2007). Although smaller particles are not sites of condensation (because that requires higher supersaturation), they do contribute to poorer visibility. Once formed, a dense layer of small fog droplets and aerosols reduces receipt of **shortwave irradiance** and inhibits erosion of the fog. This feedback gives a fog that is deeper and with higher liquid water content than would otherwise have developed.

In the late nineteenth century, London was in the early stages of industrial development and particulate pollution dominated because the main fuels were wood and coal. This persisted in only modified form until just after WWII (see sulfur-based fog, Section 11.3). Chandler (1965) differentiated the intensity of the fog in London (as indicated by visibility between urban, suburban and rural locations). The most intensely developed central area showed the highest frequency of light fogs, nearly twice that of surrounding rural areas. However, for denser fogs the frequency in the urban core was decreased, relative to that in the suburbs and countryside. The high frequency of light fogs was attributed mainly to the increased number of condensation nuclei in cities (i.e. greater competition for water vapour), whilst the reduced frequency of denser fogs was considered to be due to the warmth of the UHI and lower urban relative humidities. Another example of spatial differences of annual fog frequency across a city comes from Greater Munich, Germany (Figure 9.14). Over a 30-year period a rural station shows no obvious trend, whereas at the most urban location there is a decline from about 40 to 25 d $y^{-1}$. Interestingly, a suburban station that became increasingly enveloped into the urban area over time, shows a decline in fog from values typical of the rural area at the start, to those closer to the urban area at the end.

Large cities in most of the More Developed economies show a clear downward trend in fog frequency in the last 30 years, based on indicators such as number of fog hours, fog days and visibility. In Western Europe this is correlates with reductions in sulfur dioxide ($SO_2$) emissions, attributable to improved

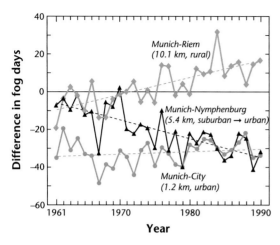

**Figure 9.14** Trend of number of days with fog in the Greater Munich region, 1961–1990. Urban station (Munich-City), is located 1.2 km from city centre; Suburban → urban station (Munich-Nymphenburg) is 5.4 km from centre and is becoming more densely developed; Rural station (Munich-Riem) is 10.1 km from centre (Source: Sachweh & Koepke, 1997; © Springer-Verlag, used with permission).

regulation of sources (Vautard et al., 2009). It is argued that reduced aerosol loads allow increased solar **irradiance** at the surface that contributes to warming. Such changes are most evident in urban areas, and have been ongoing since the 1950s and 1960s. These West European findings agree with results from other advanced economies (e.g. Japan, Figure 5.19).

Shi et al. (2008) summarize the findings of several long-term studies of fog trends in China in the period (1985–2005). They report decreasing fog in many regions, especially in large cities. These findings broadly agree with those from Europe; however, some of the more recently established cities initially showed an increase of fog, up to the 1990s followed by a decline. They postulate this was because UHI effects (which inhibit fog formation) lag behind urban emissions (which enhance fog formation). However, post-2005 conditions suggest the potent link between urban emissions and fog occurrence has resumed. In several large Chinese cities the smog problem has become pervasive and insidious. Pollutants including toxic metals dissolve into the **haze** droplets rendering them harmful to human health. Like the worst of the Industrial Revolution in Europe, human activities in these cities have become a limit to orderly development of the national economy, livability and a healthy population.

## Summary

There has been far less study of the urban effect on atmospheric moisture compared to that on temperature. Atmospheric moisture is a complex variable because its physical state can change relatively readily. Transition between the **vapour, liquid and solid states has energetic implications** due to the attendant uptake or release of latent heat. In its gaseous state, humidity can be represented as an atmospheric property in several different ways, similarly there are many different devices to measure it. As a suspended liquid or solid, it forms a cloud of droplets or ice crystals that can be precipitated or deposited onto surfaces. An unfortunate result of all this variety is that different researchers favour, or only have access to, a particular measure or instrument for their study. A further outcome is that relatively few studies are directly comparable. The following general points can be concluded:

- **Atmospheric humidity** is generally **lower in cities during the daytime** due to lower evaporation ($E$) rates compared to rural areas owing mainly to the smaller fraction of vegetative cover in urban areas.
- Conversely, at **night and in winter generally there is an urban moisture excess** (UME) in mid- and high latitude cities. The UME is probably due to the continuation of weak $E$ into the urban atmosphere that is typically slightly unstable (able to maintain a weak water vapour lapse rate), plus additional water vapour is injected by anthropogenic activities ($F$).

- In some cities, **extensive irrigation** in urban neighbourhoods introduces surface and atmospheric water that raises humidity above values found in the countryside. This effect may occur during a dry spell or where the city is located in an arid climate.
- In the absence of air pollutants (chiefly smoke particles) that can initiate the condensation process at lower relative humidity, the UHI phenomenon **decreases the frequency of fog occurrence** in cities. However, poor air quality can overcome the UHI effect and cause a dramatic increase in fog, particularly where the city is located in geographic circumstances otherwise normally conducive to fog formation.

Urban effects on moisture have obvious relevance to the topics of clouds and precipitation (Chapter 10) but there are also implications for human bioclimates (Chapter 14) and for urban design (Chapter 15).

# 10 | Clouds and Precipitation

**Figure 10.1** Clouds forced by local-scale human activities at the surface. This fumulus cloud, is associated with the emissions of water vapour, heat and aerosols from the Fiddler's Ferry coal-fired electrical generating station near Warrington, Cheshire, United Kingdom (Credit: Getty Images, used with permission).

First principles suggest that cities should be the preferred locations for cloud development, and possibly **precipitation**. The main reasons are:

- Urban atmospheres are rich in **air pollutants**, including **aerosols** (Chapters 5 and 11) which aid cloud formation by providing condensation nuclei.
- The **urban boundary layer** (UBL) is characterized by **convergence** and uplift, due to the effects of roughness (Chapter 4) and the **urban heat island** (UHI) (Chapter 7).

The very visible fumulus clouds developing above refineries, steel and pulp mills, power plants and cooling towers are a concentrated example of **emissions** of heat, moisture and aerosols (Figure 10.1). Moreover, there is ample evidence that emissions from these sources contribute to the downwind development of localized **fog**, persistent local cloud and sometimes even rain and snow. However, if we 'zoom-out' to the city scale, these simple associations are not as clear-cut. In fact, although studies of urban effects on cloud and precipitation have been conducted for well over a century, confirming urban effects on cloud and precipitation has proved to be one of the least well understood, even controversial, topics in **urban climatology**.

Section 10.1 begins with a brief review of the physical basis for droplet/crystal formation and growth in the atmosphere – a reader with a background in meteorology can advance to the following section. Section 10.2 outlines three important methodological challenges that pose important difficulties in the study of urban cloud and precipitation: the basic physical characteristics of clouds and precipitation; inherent limitations to measurement of these phenomena; and the challenge of establishing scientific proof (Section 10.2). We examine the observational evidence of urban-modified cloud and precipitation (Section 10.3) and finally, the various hypotheses being invoked to explain any observed effects (Section 10.4).

## 10.1 Basics of Cloud and Precipitation Formation

Cloud and precipitation formation depends on the presence of aerosols (see Chapter 5) to act as **cloud condensation nuclei** (CCN), they facilitate **condensation** of water vapour and subsequent growth of droplets. However, only a subset of atmospheric aerosols function as CCN and this depends on a number of properties (Wallace and Hobbs 2006) including their size and solubility and if insoluble, whether the particle is wettable (i.e. allows water to spread out in a thin film over its surface). Other aerosols, especially those with a structure similar to that of ice crystals can act as **ice nuclei** (IN); however, these occur much less frequently than CCN. The **concentration** (number per volume), size and composition of aerosols are important determinants of a cloud's properties: such as droplet sizes, droplet concentration, cloud **reflectivity** and the probability of precipitation.

### 10.1.1 Warm Cloud Processes

Rain from warm clouds is most common from clouds at low altitude in low latitude locations or from low-level convective clouds in the midlatitudes in summer. The initial formation of a droplet in a warm atmosphere (i.e. where the temperature is above 0°C) depends on the ambient humidity and size, solubility and **hygroscopic** nature of available condensation

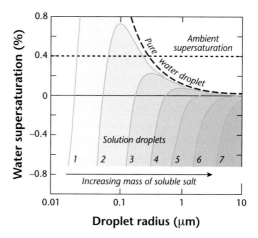

**Figure 10.2** Köhler curves that show the equilibrium supersaturation with respect to a plane surface of water for droplets containing a dissolved salt (solid lines 1–7) and pure water (long-dashed curve) as a function of droplet radius. Curves 1–7 represent increasing concentrations of salt solutions. The short dashed line represents an arbitrary ambient supersaturation of the environment (Modified after: Hobbs, 2000).

nuclei. Once formed, droplet size determines the ability to grow by continued condensation.

### Effect of Humidity on Initial Droplet Growth

Droplets formed around small CCN need an environment with **relative humidity** (*RH*) well above 100% (supersaturation) to grow. On the other hand, droplets that form around larger CCN can grow at smaller supersaturation values, owing to their larger surface area. Figure 10.2 illustrates this so-called curvature effect. The long-dashed line represents the equilibrium supersaturation for pure water droplets. If the intersection of droplet radius with this curve lies below the ambient supersaturation, the droplet might grow.

### Effect of Salt on Initial Droplet Growth

In warm environments, the initial growth of droplets is enhanced by the dissolution of CCNs composed of salt particles. The creation of a weak salt solution lowers the equilibrium **vapour pressure** of the droplet, the so-called 'solute effect'. Hygroscopic aerosols may form small droplets in equilibrium with *RH* well below 100%. The growth of these droplets is described by a Köhler curve (Figure 10.2; curves labelled 1–7), the peak of which marks the critical radius of the droplet. Where the ambient supersaturation (horizontal dashed line in Figure 10.2) is less than the peak supersaturation of the Köhler curve, small droplets on the left hand side of the diagram are limited in size. A small solution droplet represented by Curve #2 will grow until it reaches the intersection of the Köhler curve with the ambient supersaturation. This point marks an equilibrium, above which a droplet would shrink by **evaporation**, and below which a droplet would grow by condensation. These droplets are referred to as **haze** droplets. The equilibrium size of these droplets will shift as the concentration of the droplet solution changes and as the ambient humidity changes. Haze can form at RH well below 100% on hygroscopic aerosols and can impact the receipt of **shortwave irradiance** by the urban surface and alter urban visibility (Chapter 5).

If ambient supersaturation is greater than the peak of the Köhler curve, the droplet can continue to grow, since the equilibrium vapour pressure is now decreasing relative to the environment. Those droplets are said to be activated and will grow to become a fog or cloud droplet. A droplet is more likely to be activated if it is more soluble (e.g. has a higher salt concentration) and/or if the initial aerosol particle size is large. Curves #3–7 of Figure 10.2 represent activated droplets.

### Further Growth by Collision and Coalescence

As a droplet grows larger, the rate of growth by condensation decreases as its radius increases. However, such droplets are still far too small to precipitate, and require further growth mechanisms to get to that point. This occurs in warm clouds by coalescence as droplets collide with each other in a cloud. Ideally two drops merge into a single larger drop after collision, but especially at high speeds they may separate into several smaller drops or disintegrate. The most suitable environment for droplets to grow to raindrop size in warm clouds is a broad distribution of droplet sizes, that includes at least some large droplets, a thick cloud and sustained strong updrafts. Merger of droplets is promoted by favourable electric charges.

### 10.1.2 Cold Cloud Processes

When air temperature drops below freezing within some part of a cloud, both liquid water (supercooled) and ice may be present with the proportion of ice increasing as temperatures decrease. A supercooled

liquid droplet that contains an IN freezes when an ice embryo forming within the droplet reaches a critical size, or a droplet may impact an IN and freeze. Deposition of vapour to ice may also occur directly onto IN. At temperatures below about –35°C, droplets spontaneously freeze, even without the presence of an IN, with the freezing temperature decreasing as droplet size decreases. By –40°C all supercooled water is converted to ice.

### Bergeron-Findeisen Process

In an environment where water exists in all three phases: as vapour, supercooled liquid water and ice, ice particles grow at the expense of liquid cloud droplets because the **saturation vapour pressure** over ice is lower than that over liquid water; the difference is greatest around –12°C (see Figure 10.2). In this environment, liquid droplets evaporate while ice crystals grow as vapour fuses to their surface (Bergeron-Findeisen process). As in warm clouds, growth by vapour deposition is slow for larger particles, so further growth of ice crystals occurs through collision processes that include aggregation (collisions between snowflakes) and riming or accretion (freezing of supercooled droplets onto an ice particle). The growth of liquid water droplets by coalescence is also possible in cold clouds. The resultant precipitation at ground level may be of various types (snow, sleet, rain or freezing rain) depending on the ambient temperature **profile** between cloud base and the ground.

### 10.1.3 Thunderstorms

Thunderstorms are convective storms characterized by strong up- and down drafts, a mix of liquid water and ice, and an electrical charge separation process that causes parts of a cloud (and the ground below) to become electrically charged. Lightning describes the resulting flow of electricity, which occurs along a narrow channel; the rapid and intense heating along this channel and resulting expansion of air produces thunder. Most lightning occurs within or between clouds but the convective system can induce an electrical charge on the ground below the cloud resulting in a cloud-to-ground (CG) electrical discharge. CG lightning is predominantly due to negative polarity (i.e. a negative charge is transferred from the base of the cloud to the ground) but a small fraction of flashes carry a positive charge. In convective clouds that extend well above the freezing level, recirculation of hydrometeors (any product of condensation) up and down within the cloud produces a sequence of freezing and melting that creates hailstones. When the weight of a hailstone is great enough to overcome the updraft strength, it precipitates and a hailstorm is recorded.

## 10.2 The Methodological Challenge

Observing the urban effect on cloud and precipitation is especially difficult for a number of reasons: first, the areas of urban influence are often displaced downwind of the city; second, these areas are dynamic and vary in size, change quickly and are sensitive to the **synoptic** weather controls. To address the challenge we must consider both the nature of the phenomenon and the limitations of our observational systems to detect modifications.

### 10.2.1 The Discrete Nature of Cloud and Precipitation

The range of scales involved in cloud formation and precipitation is large and includes: microscale physical-chemical processes in the atmosphere (e.g. aerosols, phase changes, droplet growth) that result in droplet and ice crystal formation; **local-scale** to regional-scale processes associated with individual clouds and; meso- and large-scale processes that organize clouds within larger synoptic weather patterns.

In fact, for many weather events, the dynamic processes required for sustained uplift occur at scales much larger than that of a city, so that any urban effect may constitute only a small **perturbation**. Moreover, this effect may be confounded by cloud and precipitation patterns created by non-urban controls associated with **topography**, coastal effects, spatial distributions in atmospheric moisture near water bodies and local variations in aerosol loading due to specific industries and practices or seasonal variations due to biomass burning.

In contrast to airflow or temperature, the outcome of these processes is a distribution of cloud and precipitation that is spatially and temporally discrete. This makes it hard to acquire a sample of precipitation or cloud that is adequate for urban studies using individual stations, or even a network of stations. In the case of precipitation there may be long periods with nothing to observe and then a sudden short burst

of activity, similarly it may be dry at one location yet be raining only one kilometre away. Further, there are many different forms of precipitation (e.g. **dewfall**, frost, fog wetting, drizzle, rain, freezing rain, snow, snow pellets, hail) that require multiple and different means of measurement.

### 10.2.2 Observing Cloud and Precipitation

#### Precipitation at Urban Scales

A standard rain gauge receives precipitation via an orifice with an area of 300 $cm^2$. The gauge stores the precipitation 'catch' and the depth accumulated is recorded at regular intervals, usually once every 24 hours (often at 0900 h local time) although some record precipitation at hourly intervals. Modern, automatic gauges employ a tipping bucket mechanism that can capture a time series of receipt over shorter periods but the availability of such time-resolved records is not universal. Hence, most instruments are not designed to capture precipitation events of short duration that are characteristic of convective storms.

While the principle of precipitation measurement appears straight-forward, the catch of the instrument is highly dependent on its exposure to wind, which transports precipitation into the gauge. The presence of nearby obstructions (trees, hedges, buildings) and local topography can modify airflow and affect the deposition of rain and of snow. As a result, seemingly insignificant changes in the microscale positioning of a gauge can have a large effect on its catch. These problems are exacerbated in urban environments where the near-surface airflow is highly disturbed (Oke, 2008).

Rain gauge networks are best used to identify patterns of precipitation receipt that are either associated with a storm that has a spatial extent much larger than the network's station spacing, or those produced by a number of smaller storms over a long period, such as a month or a year. They are of limited value to examine the distribution of precipitation associated with a single event (like a thunderstorm) that has a short duration. Even a dense network may have just one rain gauge per 20 $km^2$, i.e. an areal sampling density of only $1:10^9$. Moreover, rainstorms rarely form where observations are available, usually they are advected in the ambient airflow. That means the network needs to be both dense and extensive, far larger than the extent of the city, if any urban effect is to be detected.

It is clear that much of our historical records of cloud and precipitation are based on observation systems that were never designed to capture the temporal and spatial variations typical of urban scales. **Remote sensing** systems (e.g. weather **radar**, satellite microwave and radar, lightning networks) can help overcome some of these limitations and have recently reinvigorated research into urban effects on precipitation and clouds.

## Cloud at Urban Scales

Information on cloud amount and type is typically recorded by a trained observer who follows a protocol specifying how to scan the atmosphere (to assess cloud cover) and classify cloud types. Observations are made at specific times from a viewing platform that provides an unobstructed view of the sky. Such observations are made at just a few weather stations, often those near airports. This makes it nearly impossible to assess urban influences on cloud. Again, remote sensing systems are useful to overcome these limitations; in particular satellite imagery provides spatially continuous coverage that includes both a city and its surroundings. High resolution satellite-based sensors have a regular daily repeat cycle that permits observations at the same time of day, but limits sampling to a few times per day. Ceilometers and **lidar** systems can help to probe the height of the cloud base from ground.

## The La Porte Anomaly and Scientific Proof

Some of the difficulties in isolating the urban effect alluded to above can be illustrated using the example of the La Porte, United States precipitation 'anomaly', which was extensively studied in the 1970s. Rain gauge records showed the station at La Porte received exceptionally high rainfall relative to surrounding stations between the late 1930s and 1960s. This seemed to correspond with increased storm activity, stream flow and crop losses due to hail damage in the same area. Scientific debate arose concerning both the veracity of the anomaly and, if real, its physical cause. The primary hypothesis forwarded stated that Chicago, United States, located approximately 80 km to the northwest was responsible. But there were concerns about: the quality of the precipitation observations (potential observer error and poor gauge siting); the lack of physical proof of mechanisms to explain any urban influence, and; the absence of corroborative observations of such large positive anomalies from other cities in the region (Changnon, 1980; Lowry, 1998). Alternative hypotheses (e.g. changes in synoptic weather patterns and the influence of nearby Lake Michigan) further confounded picture. During the 1970s the anomaly shifted its location and then disappeared from observations (Changnon, 1980). Unfortunately, the absence of precipitation measurements from over Lake Michigan made it impossible to confirm if the anomaly had simply moved there and outside the observational **domain**. That left the possibility that the anomaly was an urban effect, unresolved.

A modern reassessment of this case study makes progress on this question. Niyogi and Schmid (2014) examined observations from an extended rain gauge network supplemented with precipitation data from radar and simulations using a **numerical model** that incorporates the temporal and spatial distributions of aerosols. The reassessment concludes the anomaly is real and is influenced by both urbanization of the land surface and aerosol emissions. Radar analysis shows an anomaly that is sensitive to wind direction and day-of-week. Storms that travel over a longer urbanized **fetch** show stronger modification, but even storms that do not directly pass over the most highly urbanized region may be still be affected by the aerosol plume from Chicago. The results from the numerical model indicate that growth of the urban area and changes in aerosols can modify the spatial patterns of precipitation both upwind and downwind of Chicago.

The La Porte anomaly demonstrates the difficulty of proving that cities affect precipitation. Given this reality Lowry (1998) advocated four guiding principles when using observations to establish an urban precipitation effect:

1. Design a study that tests explicitly stated hypotheses; this requires removing extraneous influences, such as topography and weather conditions that are likely to confound analysis.
2. Replicate the study in several urban areas to confirm the generality of the results.
3. Given that precipitating systems generate discontinuous patterns in space and time, use observations that can monitor these attributes; conventional observational data that has a poor spatial resolution and only records daily accumulation may be of no value to test some hypotheses.
4. If a study uses standard climatic data, it should be disaggregated into different synoptic weather situations so that the urban effect can be isolated.

## 10.3 Urban Observations

### 10.3.1 Clouds

While it is easy to observe fumulus clouds (Figure 10.1) and associate them with an industrial source, it is more difficult to observe urban scale effects on cloud. One of the most detailed studies of urban cloud and precipitation was **METROMEX** (Metropolitan Meteorological Experiment), a multi-institutional project carried out in and around St. Louis, United States, during the 1970s (Changnon, 1981). It provided detailed observations of clouds and their properties based on measurements made from aircraft and ground-based radars. In the period since this experiment, new remote sensing techniques (e.g. satellite and radar) have added to our knowledge.

#### Cloud Cover

Figure 10.3 shows results from a study using daily afternoon satellite overpasses to observe cloud cover in and around Moscow, Russia, for the spring and summer seasons. Increases in cloud cover are largest over the city centre and extend at least 20 km downwind from it. The average difference for the period March–October was 0.76 tenths and is statistically significant. Similar increases of cloud cover have been noted for other cities, including Tokyo, Japan, and St. Louis, United States, from studies using satellite-based assessments. The temporal development of convective warm season clouds suggests that the urban enhancement is most marked through the middle part of the day.

As clouds develop, ground-based radar can be used to monitor their location and growth through the level of the reflected radar signal. The location of the first identifiable echo return (a radar reflectivity above a defined threshold level) – known as first echoes – gives an indication that precipitation is developing in convective clouds. In METROMEX, the location of first echoes (Figure 10.4) showed enhanced frequency of rain initiation over St. Louis during the summer. Greater densities of first echoes observed over and downwind of St. Louis were attributed mainly to an increased number of clouds over the city (Braham, 1981). In addition, the convective clouds over and downwind of the city had higher concentrations of Aitken condensation nuclei by a factor of up to 3 (Chapter 5; Semonin, 1981a). These concentrations are evidence that the clouds were coupled to the UBL and 'ingesting' urban air that altered the droplet distribution size (Section 10.4).

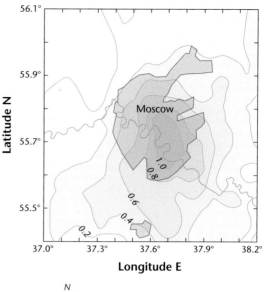

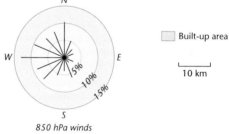

**Figure 10.3** Spatial distribution of the differences in average fractional cloud cover (labelled in tenths) over Moscow, Russia. Differences in fractional cloud cover for March–May compared to the minimum value for the study area. Inset is a wind rose for 850 hPa winds, during the same period derived from NCEP reanalysis data (Source: Romanov, 1999; © Elsevier Science Ltd., used with permission).

When cloud coverage is extensive due to larger scale meteorological **forcing**, the influence of the urban area is much less noticeable. Larger differences are noted under summer conditions, especially those favourable for convective processes that generate cumulus clouds which are more likely to be influenced by surface exchanges. Winter season differences tend to be smaller, but may show the influence of emissions from large industrial sources (that lead to increased cloudiness downwind) or even the UHI (which can reduce the thickness of stratiform type clouds). Clouds in the mid- to upper **troposphere** (altus and cirrus clouds) are mostly forced by larger scale processes and unlikely to

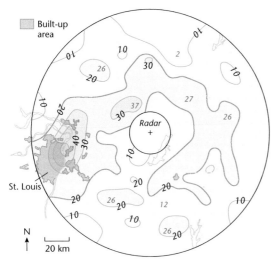

**Figure 10.4** Map of radar first echo densities (in units of numbers of first echoes per 100 square miles (259 km$^2$)) that represent the initiation of convective cloud precipitation from a 44-day data set during the METROMEX project. The radar is located east of St. Louis and is able to observe the atmosphere over the city and that downwind of the city for storms moving in the SW – NW sector (Source: Braham and Dungey, 1978; © American Meteorological Society, used with permission).

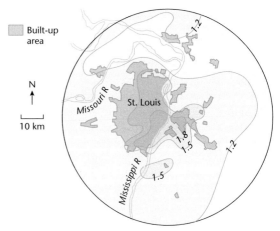

**Figure 10.5** Summer afternoon cloud base heights (km) in the St. Louis area derived from aircraft observations of small cumulus during the METROMEX project (1971–75) (Source: Semonin, 1981a; © American Meteorological Society, used with permission).

be influenced directly by urban areas and their activities.

## Cloud Heights

Measurements of cloud base height during METRO-MEX showed that the average cloud base height for clouds over and slightly downwind of the city was approximately 600–700 m higher than clouds in surrounding **rural** areas (Figure 10.5). The higher cloud base height is attributed to a less humid and deeper UBL. Further, in contrast, cloud bases were lower near major industrial sources of water vapour and CCN.

Radar echoes can also be used to make estimates of relative cloud height. Assessment of the tallest echo from each radar survey represents conditions during the afternoon peak convective rain period. The location of these echoes shows distinct urban enhancement (Table 10.1); under light wind conditions, the tallest clouds are over twice as likely to occur over the urban area. As winds increase, the area of maximum relative frequency shifts downwind of the city and becomes

more muted. The distribution of heights (Figure 10.6) shows a distinctly bimodal distribution for rural clouds. The upper peak represents strong convective clouds that extend through the troposphere whereas the frequency minima between 8–11 km may represent clouds whose heights are restricted by a drier or more **stable** layer in this region (Braham 1981). Urban cloud top heights have a more unimodal distribution with greater frequencies of cloud heights in the range 6–14 km compared to the rural areas. This suggests urban clouds are somehow able to overcome the arresting layer that limits some rural clouds (Braham 1981).

## 10.3.2 Precipitation

Urban effects on precipitation have been studied since the early decades of the twentieth century. Much of the research on European cities is summarized by Kratzer (1956); this includes evidence for the urban effect on both the total amounts over a period of years and the spatial distribution. Moreover, changes to the temporal distribution within a day and during the week, due to urban effect are also hinted at. However, advancement of understanding was limited by observing technologies (reliant initially on simple networks of rain gauges that were often limited to the confines of cities) and lack of numerical modelling tools. More

**Table 10.1** Area-normalized frequencies of the tallest cloud echo of each day measured as part of METROMEX in St. Louis, United States, over the city (a fixed region) and in downwind advection areas that are 1, 2 and 3 hours downwind of the city based on the mean echo movement vector (Source: Braham and Wilson 1978; © American Meteorological Society, used with permission).

| Class | Number of Days | City | Relative Frequency | | | Rural |
|---|---|---|---|---|---|---|
| | | | 1h | 2h | 3h | |
| Echo movement vector $< 3$ m s$^{-1}$ | 22 | 2.55 | | | | 1.00 |
| Echo movement vector $\geq 3$ m s$^{-1}$ | 118 | 1.07 | 1.57 | 1.61 | 0.86 | 1.00 |
| Total sample | 140 | 1.34 | 1.67 | 1.72 | 0.92 | 1.00 |

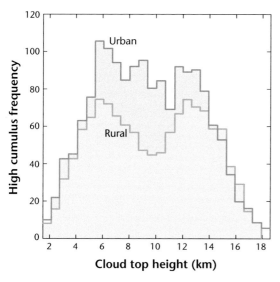

**Figure 10.6** Distribution of the tallest radar echoes of rural and urban convective clouds in the St. Louis, area. Urban clouds are those that occur over the city or in areas downwind of the city subject to advection with less than 3 h travel time and their frequencies are adjusted to account for the ratio of the average size of rural to urban areas (Source: Braham and Wilson, 1978; © American Meteorological Society, used with permission).

recently, new observing technologies, especially those based on remote sensing (Table 3.1) significantly improved capacity to observe the conditions and processes of the atmosphere in and surrounding cities. Numerical models of cloud physics and chemistry plus greatly expanded digital computing resources provide another important advance. This has led to an increase in the number and sophistication of studies (Han et al., 2014; Shepherd 2013).

### Convective Precipitation Downwind of Cities

Precipitation patterns around cities suggest enhancement of precipitation downwind of the urban areas, particularly in the summer season with convective rainfall. Enhancement increases with daily rainfall totals, suggesting that the relative occurrence of heavy precipitation events is magnified in and downwind of cities. This evidence points to urban enhancement of convective systems. Figure 10.7 presents the summer (June, July, August) precipitation totals surrounding St. Louis, United States, measured during METRO-MEX. It illustrates some basic features of this type of urban precipitation pattern. We see higher precipitation totals downwind of the city with the area of enhancement linked to the wind direction near the top of the ABL. Enhancement of rainfall by up to 30–50 cm relative to that over the urban area can be seen which corresponds to an average increase of about 25%. In St. Louis, little effect was noted on winter season precipitation. This distinction highlights the importance of the synoptic and time elements of the Lowry conceptual framework presented in Section 2.6. The complexity of spatial patterns illustrated in Figure 10.7 is also related in part to another aspect of Lowry's framework, namely the influence of topographic/landscape scale effects.

There is some evidence that urban areas can suppress rather than enhance precipitation under certain circumstances. These effects may be influenced by additional aerosols originating from urban areas that act to reduce droplet size in a manner analogous to that of continental vs maritime clouds (Section 10.4.3). Thin, weakly convective clouds that are likely to produce very low precipitation may be more susceptible to suppression as are orographic clouds forming over higher terrain downwind of cities.

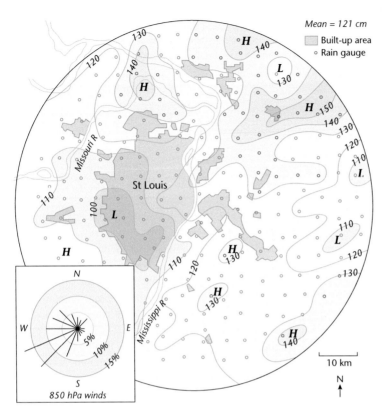

*Mean = 121 cm*

**Figure 10.7** Summertime precipitation totals in the St. Louis, United States, area observed between 1971 and 1975. Inset is a wind rose for 850 hPa winds, during the same period derived from NCEP reanalysis data. These winds are near the top of the BL and commonly represent steering winds for convective storms. The base map reflects the built-up area during the time of the experiment (Source: *Total summer rainfall (cm), 1971–1975,* from *Summary of METROMEX Volume 1: Weather Anomalies and Impacts,* by S.A. Changnon, F.A. Huff, P.T. Schickendanz, and J.L. Vogel, Bulletin 62, Illinois State Water Survey; used with permission).

## Thunderstorms, Hail and Lightning

Observations suggest that the incidence of convective storms that generate lightning, thunder and hail can be increased in the vicinity of cities. In METROMEX, the incidence of thunder was enhanced downwind of St. Louis. These effects were greatest in the late afternoon but increased thunder was also noted at night and early morning in the period 0200–1000 arising from storms associated with synoptic scale events such as cold **fronts** and squall lines. Ashley et al. (2012), hypothesize that under very weak synoptic forcing, thunderstorms are more sensitive to differences in land cover and the presence of cities. Using a 10 year record of radar reflectivities for Atlanta, United States, they conclude that with very weak synoptic forcing the frequency of summer season thunderstorms was higher nearer the city centre. When high and medium radar reflectivities (representing severe and strong thunderstorms, respectively) are considered separately, the high reflectivity category shows a greater increase. Thus, under weak synoptic forcing, cities may be characterized by increases in both thunderstorm frequency and occurrences of intense thunderstorm events (Ashley et al., 2012). The initiation of isolated convective storms, as derived from radar analysis of the Atlanta area under similar synoptic conditions, confirms that more storms form over urban areas, especially in the late afternoon and early evening, with more storm initiation downwind compared to upwind regions (Haberlie et al., 2015). Such storms are of considerable concern because they generate high rates of precipitation that could result in significant **runoff** and flooding, especially in urban areas where there is greatly increased impervious **surface cover** (Section 8.1.9).

Spatial patterns of CG lightning strikes can be assessed using ground-based networks at a spatial resolution of a few km or better; coarser resolution data are available from satellite-based sensors. Counts of lightning strikes can be aggregated over space to create flash densities, or summed to assess event or daily totals. To extract the urban effect, these data must be filtered to reduce the impacts of individual large storms that otherwise tend to overwhelm flash counts. Moreover, CG lightning strikes are sensitive to topographic variations, including elevation, slope and valley position and location relative to water bodies. Westcott (1995) showed that cities bordering the Great Lakes have urban lightning influences reduced or overwhelmed by lake effects. There is some evidence of both storm initiation as determined from radar analysis and lightning patterns being correlated with day-of-the-week (Coquillat et al., 2013, Haberlie et al., 2015), with weekdays having greater lightning activity compared to weekend days. This may be related to weekly variations in aerosol concentrations in and near urban areas.

Ground-based network data show cities increase CG flash densities, especially downwind of the maximum built-up area, which supports the METRO-MEX findings. Urban areas consistently decrease the percentage of positive CG downwind of cities (Stallins and Rose, 2008). This may be related to aerosol size, which has been correlated to CG polarity; larger aerosol sizes lead to a decreasing percentage of positive CG flashes (Stallins et al., 2013). But CG polarity is also influenced by storm type, intensity and stage as well as cloud microphysical conditions that are themselves affected by aerosols. Urban CG enhancement tends to occur on days when thunderstorms are widespread. The increase in flash density is associated with the increased **convection** that created the higher urban clouds shown in Figure 10.6. The enhanced convection should also lead to a higher ice mass in the clouds (Gauthier et al., 2006). Figure 10.8 shows that the spatial area of enhanced flash densities over and downwind of Houston, United States, is also accompanied by a region of higher ice mass as derived from radar observations.

METROMEX also examined hail characteristics in and around St. Louis (Table 10.2). Stations were grouped into regions identified as urban effect and non-urban effect zones. The urban area was found to cause more frequent and more sustained production of hail downwind; this effect corresponds to patterns of crop-hail insurance losses, which has been noted for areas downwind of other cities. The spatial location of hailfall enhancement also coincided with that of thunderstorms, the peak in rainfall and electrical power outages (Braham et al., 1981). Table 10.2 also presents characteristics of hailstreaks created as a column of hail precipitates from a moving storm. In this study, an affected storm is defined as one that passes over or develops over the urban area. The areas and duration

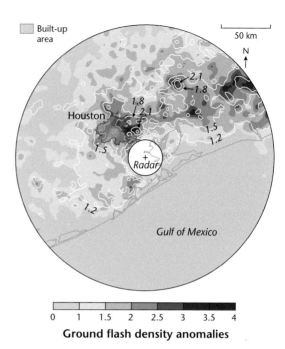

Ground flash density anomalies

**Figure 10.8** Summer season cloud-to-ground (CG) lightning flash density anomalies (shaded contours) for Houston, United States, during the period 1997–2003. Positive anomalies are > 1, negative anomalies are < 1. Anomalies are created by normalizing the flash density for each pixel by the mean for the entire domain. White contours are positive anomalies in ice mass derived from radar analyses at 1.2, 1.5, 1.8, 2.1 and 2.4× the domain mean ice mass (Source: Gauthier et al., 2006; © American Geophysical Union, used with permission from Wiley).

of hailstreaks are only minimally impacted by the urban area, but the duration of hail at individual measurement sites, maximum hailstone diameter and hailstone counts are increased and contribute to the 100% increase in the calculated kinetic energy of hail impact as measured from the size and frequency of impacts on hailpads.

### 10.3.3 Snowfall and Freezing Rain

#### Phase of Precipitation

Snowfall and freezing rain in cities are typically reduced relative to their non-urban surroundings. Table 10.3 gives the percentage reduction of freezing rain and snowfall events for four large cities in the

**Table 10.2** Comparison of hailfall characteristics measured as part of METROMEX in and downwind of St. Louis, United States (1971–1975). Hail from storms that form or pass over the urban area is classified as urban effect; all other storms are classed as no-urban effect (Modified after: Changnon 1978).

| Characteristic | Units | Frequencies | | Increase[1] |
|---|---|---|---|---|
| | | Urban Effect | No-Urban Effect | |
| **Effect Area vs No Effect Area** | | | | |
| Average frequency of hailfall | Counts | 5.5 | 4.3 | +28% |
| Median hailstone diameter | cm | 1.27 | 0.97 | +32% |
| Average hailstones per pad[2] | Count | 372 | 259 | +44% |
| Mean hail impact energy | $J \times 0.1\ m^{-3}$ | 1.45 | 0.74 | +96% |
| **Hailstreaks[3]** | | | | |
| Duration | min | 14.9 | 14.2 | +5% |
| Area | $km^2$ | 37.4 | 36.4 | +3% |
| Average duration at a point | min | 2.9 | 2.1 | +40% |
| Maximum hailstone diameter | cm | 1.41 | 1.15 | +22% |
| Hailstone count per pad | Count | 74 | 54 | +36% |
| Impact energy | $J \times 0.1\ m^{-3}$ | 1.35 | 0.67 | +100% |
| Mean rainfall in hailstreak | cm | 2.05 | 1.69 | +22% |

[1] The percent increase is determined: (Urban effect – no-urban effect) / no-urban effect.
[2] Hailpad measuring area is 929 cm$^2$.
[3] Comparison based on effect versus no effect value for 53 observed hail periods when both effect and no effect hail streaks occurred.

**Table 10.3** Percentage reduction in freezing-rain and snowfall events for four cities in the United States (Source: Changnon 2003; © American Meteorological Society, used with permission).

| City | Reduction in freezing rain days | | Reduction in snowfall events (%) |
|---|---|---|---|
| | Total range (%) | Range with lake/ocean influence removed (%) | |
| New York City | 28–43 | 17–28 | 17 |
| Chicago | 16–42 | 12–31 | 23 |
| Washington, D.C. | 10–27 | 10–27 | 20 |
| St. Louis | 9–30 | N/A | 15–25 |

United States. This climatological study used standard observing stations that met rigorous selection criteria to determine available urban and non-urban control stations. The reductions are attributed in part to the UHI that alters the phase of precipitation in favour of rain over snow or freezing rain. Some tendency for these effects to be most strongly expressed for light snowfalls has been noted (Changnon, 2004). However, given the coastal location of some of the cities, ocean and lake influences which can cause local warmth must also be involved (Table 10.3); this is particularly the case in the early part of the winter season when water-land temperature contrasts are large. The potentially confounding influence of the ocean or lake effect accounts for a third of the decrease in freezing rain events observed in New York City and a quarter of that observed in Chicago (Table 10.3; Changnon, 2003). This underscores the importance of considering the Lowry framework when trying to identify any distinct urban influence (Section 2.4). Additional analysis also showed that the average and maximum annual number of freezing-rain days was reduced in the cities, and that in some the freezing-rain season was shorter relative to that in their surroundings.

## Snowfall Enhancement

The urban effect on winter-time precipitation events over and downwind of cities is not well studied. However, there are observations of snowfall created by major point sources of water vapour and/or pollutant nuclei. Sources have included power generation stations, pulp and paper mills, ore processing and petroleum refineries. Radar and satellite imagery show downwind plumes associated with point sources that result in localized snow accumulation. Figure 10.9 is an example from Dodge City, United States, where a radar-detected plume of snowfall (10.9 a) is linked to emissions of water vapour and IN from a power station and several meat processing plants. The accumulated area of snowfall (10.9b) is wider than the instantaneously observed radar image because it represents the time-integrated snowfall pattern due to small changes in wind direction. Based on several observations these plumes are typically several 10 s of kilometres in length, with precipitation beginning about 5–10 km downwind from the source and

extending 30–40 km downwind. Such enhancements may also occur at the city scale (10.4.1) in cities located within polar regions where the addition of **anthropogenic** moisture, released during the process of burning fuel (Chapters 8 and 11), is large. Radar analysis of snowfall events in the Minneapolis-St. Paul, United States, area mostly show decreases of snowfall downwind of the city centre, but in a few instances enhanced radar reflectivities suggest enhanced precipitation (Perryman and Dixon, 2013).

## 10.4 Hypotheses of Urban Effects

The empirical evidence given above indicates that cities can alter precipitation patterns in specific circumstances, but we have not discussed the mechanisms. Several hypotheses have been proposed which we organize into three categories:

- **Modification of moisture and thermodynamic processes**, such as changes to the **surface energy balance** (SEB) that causes the UHI and initiates a thermally-driven circulation. These can alter atmospheric convection and convergence of winds over and downwind of cities (Section 10.4.1).
- **Modification of dynamical processes** mainly due to the roughness of urban surfaces, which enhances airflow convergence over and downwind of cities

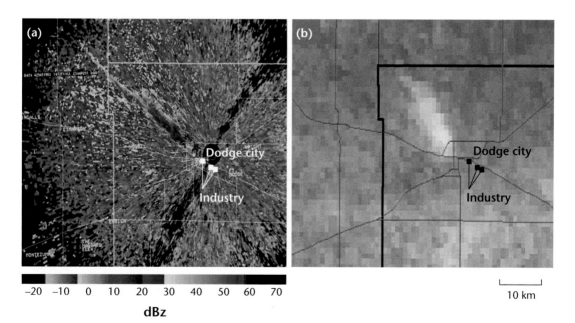

**Figure 10.9** Observed snowfall plume in Dodge City, United States. **(a)** Radar image during the event and **(b)** satellite image showing the resulting area of snowfall (Source: Shepherd and Mote, 2011).

and may act as a diversionary barrier to regional airflow. The effects are to alter precipitation development and its spatial distribution (Section 10.4.2).

- **Modification of microphysical processes** due to ingestion of urban aerosols that modify particle size distributions over cities. This can alter the formation, timing, phase and duration of precipitation (Section 10.4.3).

To test these hypotheses we must consider which cities are most likely to be affected and when effects may be most apparent. Observations (e.g. Section 10.3) show that urban influences are greatest during periods of weather dominated by convection, especially during the warm season when heating of Earth's surface generates uplift. In midlatitudes, frontal precipitation associated with synoptic-scale **cyclones** (with diameters of 1,500–5,000 km) is common, especially in the cold season. The large-scale processes that drive these systems may mask any urban effect and limits the opportunity to test these hypotheses to only a subset of precipitation events. In regions where the climate is influenced by a monsoon with distinct wet and dry seasons, there may also be seasonal limitations on our ability to detect urban influences.

The topographic setting of a city must also be considered as other influences (such as coastal and valley circulations) may also be present, making it difficult to detect the urban influence (see Chapter 12). It is not surprising then that work on the urban precipitation effect has focussed on cities situated in regions of low relief far from major water bodies like Atlanta, Indianapolis, Moscow, Munich, Paris, Oklahoma City and St. Louis.

### 10.4.1 Modification of Moisture and Thermodynamic Processes

#### Summer Season

An **urban heat island circulation** (UHIC) can be initiated when conditions are favourable for heat island formation and regional winds are weak (Chapter 4). The UHIC involves the confluence of winds, low-level convergence and upward motion (Figures 4.30–4.33). We should ask 'are these forcings sufficient to engender urban-induced precipitation?' – both modelling and observational case studies suggest the answer is 'yes'.

The UHIC is sensitive to both heat island intensity and atmospheric **stability**. While strong nocturnal UHI are associated with more stable rural conditions

(Chapter 7), these conditions are not typically favourable for the development of convective storms. Weaker UHI associated with less stable nocturnal cases may have UHIC of similar strength (Vukovich and Dunn, 1978). By day, the warm and less stable UBL also permits development of a strong UHIC that facilitates convection. Numerical models are an important tool for this work, because they allow the conduct of controlled experiments (see Section 3.3) in which the city, topography and rural surface covers can be included or omitted and their effects studied in isolation and atmospheric conditions can be prescribed. This capability is important because the particular geographic setting of each city has an influence on the background climate (Chapter 12); topography in particular plays an important role modifying regional winds and initiating circulations that can affect precipitation formation and distribution.

Models show that the presence of heating from an urban area alone can be sufficient to create an area of upward motion downwind of a city (Han et al., 2014). In Figure 10.10, a city is represented as a circular source of heating with a radius of 10 km in a stably stratified atmosphere with light winds; the simulated landscape is flat with no variations in surface roughness or land cover. The heating source, representing a surface UHI, is shown by a dotted line centred at 50 km in the along wind ($x$) direction. Effects on the atmosphere when it is dry three hours following the start of the simulation are shown in panels (a)–(e).

Urban heating leads to perturbations in the wind field arising from thermally induced gravity waves that in turn give a zone of upward motion, downwind of the city (panels a, b). A plan view of the simulation shows horizontal convergence (panel c) around the zone of upward motion and advected warmth (d) that leads to a broadly v-shaped area of upward motion including a more localized zone directly downwind of the city (e). The strength of the upward motion increases as the heat island intensity (represented as the rate of heating in this model) increases and as the regional wind speed decreases.

When simulations are conducted with the same model but using a moist atmosphere, the results (Figure 10.10 panels f–i) show the UHI can be a trigger for convection leading to both cloud and precipitation downwind of cities. A deep convective cloud appears downwind of the heating centre (f) that leads to localized heavy rainfalls (g). A strong updraft is created, in part by the release of **latent**

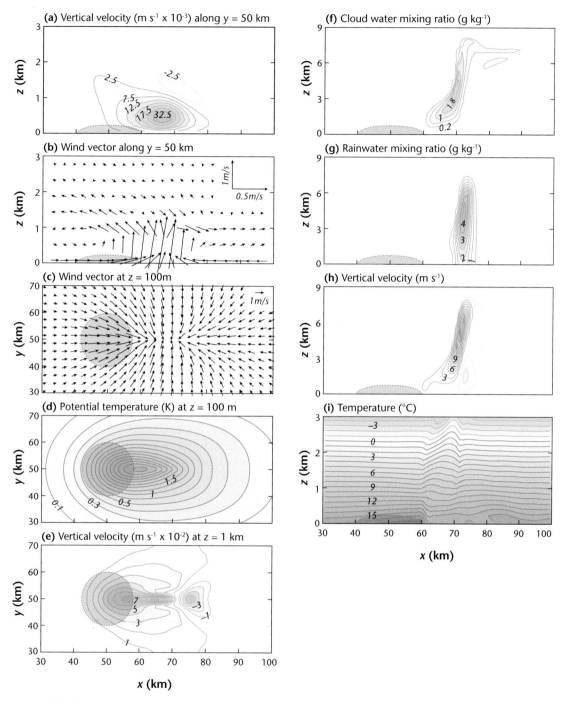

**Figure 10.10** Numerical simulations of the impacts of urban heating (circular dotted region between 40–60 km on the $x$ axis) in a dry (left panels) and moist (right panels) atmosphere. Left panels: **(a)** vertical velocity (m s$^{-1}$), **(b)** perturbation wind vector through and downwind of the urban area along the centreline of the domain, **(c)** plan view of the wind vector perturbation, **(d)** plan view of potential temperature perturbation, **(e)** vertical velocity field at a height of 1 km. Right panels: **(f)** cloud water mixing ratio (g kg$^{-1}$), **(g)** rainwater mixing ratio (g kg$^{-1}$), **(h)** vertical velocity (m s$^{-1}$) and **(i)** temperature fields along the simulation centre line 3 hours into a moist simulation. Mean horizontal wind speed is 4 m s$^{-1}$ and considered to be constant with height for all cases (Source: Han and Baik, 2008; © American Meteorological Society, used with permission).

heat in the moist convection, which is stronger than in the case of dry convection (h). Downwind of the updraft, a downdraft is associated with the precipitation region. The release of latent heat warms the atmosphere, while evaporation of precipitation cools the near-surface layer downwind of the city (i). The effects are sensitive to the moisture, stability, heat island intensity and wind speed. Effects are maximized when heat island intensity is large, stability is weak, moisture is high and wind speeds are low. The patterns of downwind precipitation are also sensitive to the spatial configuration of the heating source (i.e. the UHI).

Another approach is to model specific cases. For example, Rozoff et al. (2003) use a 3-D cloud-resolving model that includes the **Town Energy Balance (TEB) parameterization** (Chapter 3), variations in land cover types and topography. The study examines a single day (June 8, 1999) when the St. Louis region was affected by isolated convective air-mass storms in a moist **unstable** atmosphere with weak synoptic scale forcing. This type of weather is ideal for UHIC development and can initiate low-level convergence. These conditions are similar to those in which METROMEX studies found large increases in urban versus rural differences in air-mass thunderstorms (Changnon, 1981). In the simulation, atmospheric conditions are initialized from the observed weather conditions and the experimental design allowed individual factors related to the UHI convection hypotheses (including urban radiation and energy balance fluxes, urban roughness and the effects of local topography) and their interaction, to be examined. Moreover, although the simulation is restricted to just one day, a pre-screening process was designed to identify ideal weather conditions for evaluation of the UHI effect. As a result the study has wider relevance to other days and other cities that experience similar weather.

The results in Figure 10.11a are from a 'control' simulation set at noon local time on the study day and both the urban area and topography are represented. It shows a daytime UHI with winds from the south-south-east and water vapour mixing ratios are lower over the city. Panels (b) and (c) show the difference in model results when a simulation that does not include the urban SEB is subtracted from one that does. Because the urban SEB is largely responsible for the UHI (Chapter 7), these differences isolate the contributions to precipitation formation from convective

activity that arise solely due to the UHI. In panel (b), urban heat and dry islands can be seen together with evidence of an UHIC (centripetal flows around the border of the urban area). The regional winds (Figure 10.11a) displace the heat island centre downwind (roughly northwards). Convergence in the UHIC is strong over the centre of the heat island (Figure 10.11c) which contributes to rising air downwind of the city, similar to that shown in Figures 4.31 and 4.32a. **Subsidence** on the periphery of the UHIC leads to areas of surface **divergence**. In the region of convergence, convective available potential energy (CAPE) is elevated relative to the simulation that did not include the urban SEB, but it is otherwise reduced primarily because of the drier urban atmosphere but also due to divergence. The net effect of the downwind convergence zone is uplift and condensation that leads to a significant enhancement of precipitation totals (Figure 10.11d) downwind of St. Louis. Separate simulations on the impact of topography show that the downwind enhancement of precipitation was not related to topographic effects on this day.

In the same vein, Dixon and Mote (2003) suggest that at night, convection in a marginally unstable atmosphere with high moisture content could be triggered by the UHI. To test this hypothesis, they examine weather records around Atlanta, United States, from five summers on days when synoptic scale forcing was weak. They find 37 cases that could be attributed to urban-initiated precipitation; most of these events occurred at night and when low-level humidity was high.

To decrease the sensitivity of model results on initial conditions, an ensemble modelling approach can be adopted. This was used by Kusaka et al. (2014) to represent August precipitation in the Tokyo area. The ensemble combines eight simulation years and four sets of **boundary conditions** to generate 32 simulated 'Augusts'. Simulations show that urbanization increases the sensible heat flux, leading to a warmer and deeper UBL that exhibits greater instability relative to the non-urban case. An area of low pressure develops over the urban area producing convergence over the city. Because of the coastal location, this convergence enhances the **sea breeze** and leads to increased horizontal moisture convergence over Tokyo. That increase in moisture exceeds the decrease caused by the drop in evaporation associated with urban development. The net result is a statistically significant increase in August precipitation over the

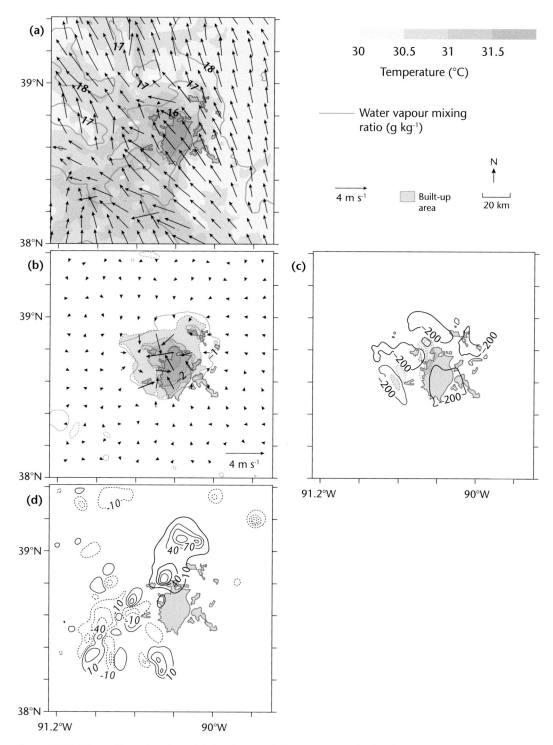

**Figure 10.11** Three-dimensional model simulation of the atmosphere in the Greater St. Louis area for 8 June 1999. The area of urban land cover is shaded grey in panel (d) and has been generalized to match the St. Louis area as shown elsewhere in this chapter. **(a)** The air temperature (shaded), wind vectors and isolines of water vapour mixing ratio at 1800 UTC for the control simulation. **(b)** The difference field at the first model

city and an increase of precipitation intensity, particularly for the most intense events.

### Winter Season

In winter, the additional warmth of the UHI can change the phase of the precipitation type relative to that received in the rural surroundings so that urban locations are more likely to receive rain in place of freezing rain or snow. Moreover, the UHI may also contribute to additional **sublimation** and evaporation of precipitation as it falls through the UBL, thereby reducing the total precipitation received at the ground. Of course, snow cover accumulation is affected by more than urban warmth, soiling of snow cover can also increase melting and sublimation and then there is mechanical removal for snow clearance (Section 5.2.4). Trying to assess urban snowfall by depth measurements across a city after an event can be confounded by these factors (Lindqvist, 1968).

Direct modification of cloud and precipitation formation processes in winter can be seen when cloud develops and/or snow falls in connection with plumes from major point sources of water vapour and/or air pollutants. This occurs in cold and humid conditions often with otherwise clear skies. Koenig (1981) suggests temperatures of –13°C or colder are required for sufficient ice particle concentrations and to provide optimum precipitation growth rates in the plume. Saturation deficits less than 0.5 g m$^{-3}$ provide plume longevity sufficient to allow particles to grow large enough so they precipitate before the plume evaporates. The rate of plume mixing with the ambient air and the humidity characteristics of the ambient air and plume also affect the amount and location of snowfall.

Snowfall enhancement can also occur when pre-existing snow from higher clouds seeds a plume with IN and allows ice crystal growth in the plume by riming, aggregation and the Bergeron-Findeisen process. The combination of potential IN from major emission sources along with additional water vapour

may also provide the conditions for precipitation formation in the plume.

At the city scale, there is little research on urban-induced changes to the formation of winter precipitation. In one winter-time modelling study for Fairbanks, United States, which is located in a high latitude climate zone, Mölders and Olson (2004) tested the effect of increasing the size of the urban area, or increasing the moisture flux, anthropogenic heat and urban aerosols upon a precipitation event. Addition of moisture increased precipitation with enhancements noted closer downwind of the city than is the case with increased aerosols. This finding agrees with results from studies that examine individual point sources of moisture. The modifications led to changes in **buoyancy** that indirectly altered cloud microphysics and precipitation processes and location. It should be noted that in this study even the largest effects are very small (~0.5 mm water equivalent) and would be extremely difficult to discern in an observational study.

### 10.4.2 Modification of Dynamical Processes

The second hypothesized mechanism is the influence of the rough urban surface, which acts to slow the near-surface winds over the city; this leads to convergence upwind and over the city and upward motion but divergence and downward motion further downwind of the city (Chapter 4). Numerical model tests have been used to isolate this influence (e.g. Rozoff et al., 2003). These suggest that roughness effects alone on convergence and divergence are much smaller (an order of magnitude) than those associated with the UHIC. The ability of urban roughness effects alone to generate precipitation therefore seems unlikely (Han et al., 2014) but is constrained by a lack of studies.

### Bifurcation of Thunderstorms over Cities

Some observations suggest that pre-existing thunderstorms approaching a city can be diverted around the

Caption for **Figure 10.11** (*cont.*) level (48 m) due to inclusion of the urban SEB compared to a case with a rural SEB in its place. Air temperature (dashed contour every 0.5°C; dark and light shading when temperatures are below −0.5° and above 0.5°C, respectively), water vapour mixing ratio (solid contour at every 1 g kg$^{-1}$), and wind vectors (m s$^{-1}$ reference vector at bottom of panel). (c) Corresponding difference field showing 48 m divergence [dashed contour every 0.5 ($\times$ 1000 s)$^{-1}$; dark and light shading when divergence is below −0.5 and above 0.5 ($\times$ 1000 s$^{-1}$), respectively] and surface-derived CAPE (solid contour: J kg$^{-1}$). (**d**) Total precipitation as of 0000 UTC 9 June 1999 with contours every 30 mm (Source: Rozoff et al., 2003; © American Meteorological Society, used with permission).

periphery of the city. This is related to the physical **barrier effect** described in Chapter 4. Where the storms are part of a larger synoptic scale feature, for example thunderstorms associated with a cold front, storm cells may bifurcate so that some pass either side of the urban area (Bornstein and Lin 2000). Dou et al. (2015) suggest this process may be related to the anticyclonic turning induced by the rough urban area (Figure 4.27) or from the associated convergence and formation of a high pressure zone on the upwind side of the city in conditions with strong regional winds. Where bifurcation takes place a shift in the precipitation maximum to the lateral and downwind edges of the city and a relative minima over the urban area is anticipated. Lightning flash densities measured near Atlanta from thunderstorms associated with fronts show lower values over the city and larger values towards the city perimeter (Stallins and Bentley 2006).

Figure 10.12 shows time-height profiles of radar echoes from urban and rural clouds during the METROMEX project that did or did not merge. An urban echo is classed as one that at some point in its

life passed over some part of the urban area. The time spent over the urban area is relatively short compared to the total life (on average 35 min for urban merged echoes and 23 min for urban non-merged echoes) so the profiles represent the time that urban echoes are present downwind (and upwind) of the urban area. Considering all echoes, urban echoes tend to grow higher, faster and last longer than rural echoes. When separated into classes of echoes that represent mergers (which is an important mechanism for cloud growth), differences between urban and rural clouds are even more apparent. Urban echoes that merge again are more likely to be higher, grow faster and last longer than rural-merged echoes; this corresponds to the distribution of cloud heights in Figure 10.6. Mergers more frequently occur on the eastern edge of the urban area and downwind of the mean wind direction (Huff, 1978).

Case studies and climatological analyses also suggest that storm cells can be split by the presence of cities. Radar analyses of storm cells in the region of Indianapolis, United States, show statistically significant changes in the distribution of radar reflectivity

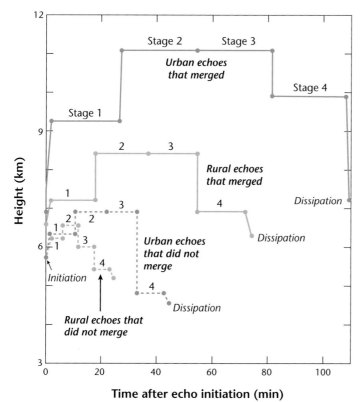

**Figure 10.12** Time-height profiles of radar echoes representing cloud development in and around St. Louis during the METROMEX project (Source: Braham, 1981 and Changnon, 1976; © American Meteorological Society, used with permission).

size and intensity upwind and downwind of the built-up area (Niyogi et al., 2011). Upwind of the city, convective cells were observed to increase in size as they approach the city, with a maximum average size at 40 km upwind of the city centre after which the size decreases towards the city centre (Figure 10.13). Downwind of the city, cell sizes are small but increase at distances greater than 50 km from the urban centre. There were also more cells observed overall in the downwind versus upwind areas. This pattern supports the idea of storm splitting on the upwind side of the city followed by a merger of convective cells and/or reintensification of cells downwind, possibly combined with the initiation of new cells.

The processes associated with storm bifurcation and/or splitting are often difficult to detect amongst other sources of influence on storms, especially topography, land use variation and coastal influences (Shepherd, 2013). Interactions of sea or **lake breeze** fronts with the UHIC (see Chapter 12) can affect the speed of frontal passage. Ganeshan et al. (2013) track propagating storms for coastal cities and suggest the interaction of the UHIC with sea or lake breeze fronts may lead to increased intensity of rainfall over coastal urban areas.

### 10.4.3 Modification of Microphysical Processes

#### Droplet Size Distributions

The droplet characteristics of clouds are greatly influenced by properties of their aerosols. As discussed in Chapter 5, continental atmospheres have much higher aerosol number concentrations than marine atmospheres. As a result, continental clouds have much higher droplet concentrations (Figure 10.14). Therefore for a fixed cloud **liquid water content**, the higher aerosol loading results in a narrower distribution of droplet sizes and a shift in the droplet radius distribution to smaller droplets in continental atmospheres. Similar to this distinction between continental vs maritime aerosol concentrations, urban atmospheres are characterized by much higher concentrations of aerosols and a shift towards smaller sizes when compared to rural atmospheres (Figure 5.19). The impact of urban aerosols on cloud droplet distributions is illustrated in Figure 10.15 from the METROMEX project over St. Louis. At the start of the flight track, distributions have a decidedly rural character, with low concentrations and relatively larger diameters. Further to the northwest along the flight track, distributions become more characteristic of urban air, where concentrations are greater and droplet sizes are smaller. There is also an increase in the

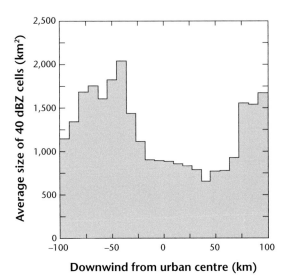

**Figure 10.13** Frequency distribution of the average size of strong radar echo cells with downwind distance from the city centre of Indianapolis, United States (Source: Niyogi et al., 2011; © American Meteorological Society, used with permission).

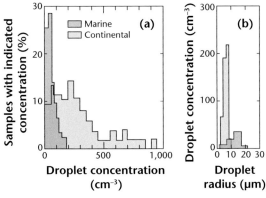

**Figure 10.14** Differences in marine and continental cloud properties. Panel **(a)** shows the percentage of cumulus clouds with a given droplet concentration. Panel **(b)** contrasts the difference in droplet size distributions between marine and continental clouds (Source: Wallace and Hobbs 2006; © Elsevier, used with permission).

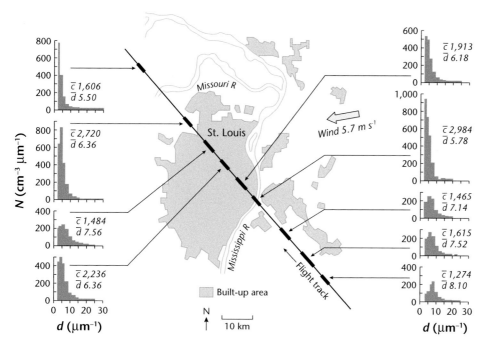

**Figure 10.15** Cloud droplet distributions observed from an aircraft traverse across St. Louis during METROMEX. $\bar{c}$ and $\bar{d}$ represent the mean droplet concentration (cm$^{-3}$) and diameter (μm) respectively. (Source: Braham 1974; © American Meteorological Society, used with permission).

concentration of very large droplets (those with radius $\geq 10$ μm) in the urban-affected distributions.

### Aerosol Invigoration Effect

A conceptual model of the effect of aerosols on cloud and precipitation formation is shown in Figure 10.16. In clean atmospheres (top set) a low concentration of aerosols means there are relatively few cloud droplets and more large drops. The distribution of drop sizes allows for an efficient collision-coalescence process that increases droplet size and rapid precipitation formation. In more polluted atmospheres (lower set), the extra CCN compete for available water content and leads to a greater number of small droplets. In an environment where atmospheric moisture content is limited, such as thin clouds, precipitation suppression has been observed (Rosenfeld et al., 2008). The resulting cloud droplet size spectrum is narrow, which is less favourable for the collision-coalescence process in warm clouds and for riming in mixed-phase clouds, in that case increased aerosols can delay the formation of precipitation sized particles. In cumulus type clouds, this delay results in more liquid water being moved to the top part of clouds where evaporation of

smaller droplets leads to cooling that helps to destabilize the atmosphere thereby promoting the growth of deeper clouds. The occurrence of precipitation is delayed but cloud development is enhanced and leads to a taller cloud with more water, stronger updrafts and a conversion to ice-phase precipitation processes in the upper part of the cloud. Freezing of droplets and the riming process releases latent heat that adds to cloud growth. Stronger downdrafts from the additional evaporative cooling of small droplets also assists with cloud growth by providing uplift to surface level air – a dynamic feedback. This so-called aerosol invigoration effect has the support of both observational and model evidence from the Great Plains region of the United States during the summer, for mixed-phase clouds with low warm bases (Li et al., 2011). Results suggest that aerosols in moist climates may enhance convective clouds and increase summer time precipitation but suppress precipitation from clouds forming in dry environments.

In urban areas, modelling studies provide some evidence for this invigoration effect (e.g. Han et al., 2012), however they only show effects in the presence of an UHI that leads to enhanced convergence

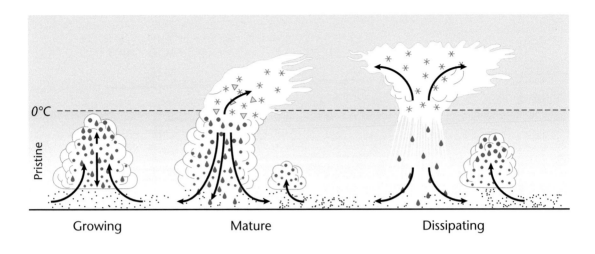

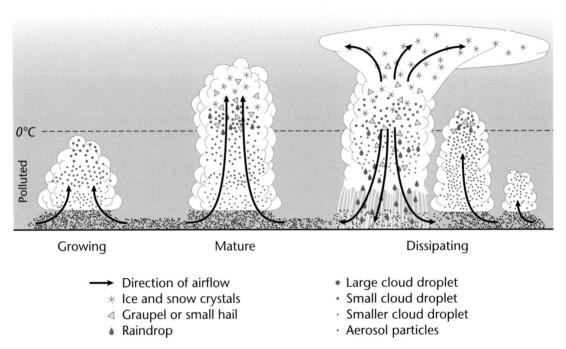

→ Direction of airflow
✳ Ice and snow crystals
◁ Graupel or small hail
💧 Raindrop

• Large cloud droplet
· Small cloud droplet
· Smaller cloud droplet
· Aerosol particles

**Figure 10.16** Conceptual figure of cloud development in clean (top set) and polluted (bottom set) atmospheres showing delayed but invigorated convective clouds in the presence of additional aerosol (Source: Rosenfeld et al., 2008; © American Association for the Advancement of Science, used with permission).

downwind of the city, suggesting that urban land cover, rather than aerosols is the dominant control (van den Heever and Cotton, 2007). In the simulations, cloud development is initially delayed (warm rain suppression) when additional CCN are added to an urban atmosphere, but cloud updrafts eventually become stronger and generate more precipitation

(Figure 10.17). The aerosol loading of the rural atmosphere is also important to the results; where upstream aerosol concentrations are high, model results show low sensitivity to urban additions. Model simulations for Houston, United States (Carrió and Cotton, 2011), show significant intensification of convective clouds downwind of the city as aerosol

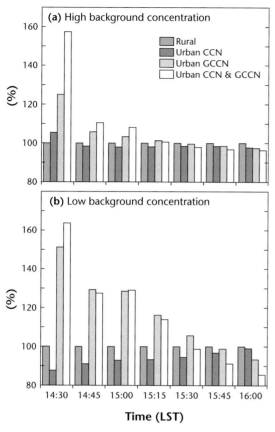

**Figure 10.17** Model simulation of accumulated precipitation downwind of St. Louis, United States, for a case study on 8 June 1999. Results expressed as a percentage of those with only rural concentrations of CCN and GCCN (giant cloud condensation nuclei) for **(a)** high and **(b)** low background rural concentrations (Source: van den Heever and Cotton 2007; © American Meteorological Society, used with permission).

concentrations in the urban atmosphere were increased. However, for higher aerosol concentrations, the efficiency of growth by riming of ice particles was reduced and precipitation efficiency of the clouds decreased. Where convective clouds, that are already raining, approach a city, their interaction with an urban aerosol field may increase the rainfall.

## Thunderstorm Electrification

Increased concentrations of ice and supercooled water droplets enhance thunderstorm electrification. Model results show increased liquid water masses

when aerosols are added to an urban simulation that suggests enhanced lightning in more polluted convective clouds (Carrió et al., 2010). Patterns of CG lightning strikes from ground-based lightning detector networks have shown that individual features such as coal-fired power plants, that are sources of increased aerosol concentrations, also result in an enhanced probability of lightning strikes (Strikas and Elsner, 2013). More supercooled water and stronger cloud updrafts also provide a mechanism for hail stones to grow larger.

## Differences in Urban and Rural Aerosol Emission Characteristics

The sensitivity of precipitation to upwind rural and urban aerosol characteristics may be important to understanding precipitation effects in cities in different locations. Coastal cities or those in regions with relatively clean upwind airsheds may show greater sensitivity to urban precipitation effects (van den Heever and Cotton, 2007). The winter-time model case study of Mölders and Olson (2004) of Fairbanks included tests of enhanced urban aerosols on a precipitation event; they resulted in less precipitation over the city and more further downwind. When combined with a larger urban area, the aerosol influence was reduced and shifted precipitation further downwind.

Urban atmospheres may also receive injections of giant aerosol particles that act as CCN and are important in warm cloud precipitation because they can enhance the collision-coalescence process. In their case study simulations van den Heever and Cotton (2007) found giant nuclei led to strong early enhancement of cloud development and precipitation, but that the associated stronger downdrafts and precipitation also led to an earlier end to the storm.

## Impact of Day-of-Week Changes in Aerosol Emissions

There has long been suspicion that precipitation shows day-of-week influences that originate in the weekly cycles of air pollutant emissions due to the rhythms of urban activities. The underlying hypothesis being that increased emissions of aerosols due to greater urban activities (traffic, industrial activities, etc.) on weekdays relative to those on weekend days can modify the precipitation processes described in Section 10.4.3 sufficiently to lead to measureable differences in precipitation and other related storm characteristics such as cloud top height, hail and lightning frequency. Assessment of such cycles continues, for example using lightning networks. Statistical confirmation of weekly cycles or weekend effects is difficult and requires careful attention to the statistical analysis. It also must recognize the growth of cities through time and the spatial and temporal patterns of contributing aerosol emissions, as these may be possible for the appearance and disappearance over time of weekday–weekend effects over cities and on regional scales (Stallins et al., 2013).

In summary, aerosol effects are very complex and absolute statements regarding their impacts are difficult (Feingold et al., 2009). Aerosols have complex interactions with cloud microphysics, cloud and storm dynamics and radiative heating and cooling. The response of particular storm systems must also recognize the unique storm environment that is due in part to larger scale influences. Urban increases in aerosols appear able to influence cloud dynamics and precipitation formation downwind of cities, but by themselves are not a trigger for convectively-enhanced precipitation.

## Summary

At the outset the urban effect on clouds and precipitation was described as one of the least well understood topics in urban climate. The evidence to date indicates that:

- During the summer season when instability and convective uplift dominate in light regional airflow, **increased cloudiness and enhanced precipitation** is found in and downwind of cities. Cities also **increase the frequency of thunderstorms and hailstorms**. However, when synoptic-scale weather events that produce widespread dynamic uplift occur, urban effects on cloud and precipitation are weaker and more difficult to detect.
- This urban effect on clouds and precipitation is due to the **combined influence of the urban heat island** (UHI)**, surface roughness** and **aerosols**. Together, the UHI and roughness enhance cloud formation and precipitation via low-level convergence and uplift. Under calm or light wind conditions, convergence due to the urban heat island circulation (UHIC, Chapter 4) appears to be most important cause and can lead to precipitation initiation and/or enhancement over and downwind of the city, when atmospheric conditions are suitable for moist convection. With stronger winds, **moving regional storms may bifurcate and/or split** over cities. Under these conditions precipitation may increase on the lateral and downwind side of the city and show a relative minimum over the city centre.
- **Aerosol** number, size, type and chemical characteristics initiate processes that may **enhance, suppress or delay precipitation processes**, especially depending on their size and concentration. It is likely they all contribute to urban effects but that the interactions, leading to enhancement or suppression, within or downwind of cities, are place and climate specific.
- The limited research on cold season urban effects indicates that cities **reduce the frequency of freezing rain and snow**, but additions of water vapour, especially from major point source emissions under very cold conditions, can locally increase clouds and under certain conditions increase precipitation.

Urban influences on cloud and precipitation are most often subtle – the majority of most cloud and precipitation affecting cities is forced by scales much larger than that of the city. Identification of specific urban influences is thus constrained by the synoptic conditions which may or may not permit the urban surface and the UBL to play a role in modifying the formation of precipitation. These conditions vary with season and location. In many cases **urban effects enhance or suppress pre-existing clouds or storms rather than initiate new events**. In all cases, the topographic setting of a city plays an important role in the spatial distribution of precipitation. Finally, the nature of the aerosols in the background air mass, plus those emitted by the urban area, are also important elements of the precipitation formation process. Taken together, we can expect that particular combinations of synoptic conditions, topography and aerosol characteristics lead to many different urban cloud and precipitation outcomes, which makes the identification of urban controls on precipitation a challenging task.

# 11 | Air Pollution

**Figure 11.1** A busy street canyon in Cairo, Egypt (Credit: A. Serrano; with permission).

Ask most people what they regard as the defining characteristic of the urban atmosphere and they refer to its degraded air quality: the distinct odour, reduced visibility or irritating effects in the throat or eyes. Inevitably, the concentration of human activities in cities results in **emissions** that modify the thermal and chemical composition of the urban atmosphere.

Figure 11.1 illustrates the experience of urban living for many; a busy street in Cairo, Egypt, crowds intermingle with cars, buses and motorbikes, each of which injects heat, moisture, noise and a host of **air pollutants** into the **urban canopy layer** (UCL). All of this occurs within a street canyon that confines the air flow and restricts the dilution of air pollutants. Contaminated air drifts into adjacent buildings where it lingers longer and affects the comfort and health of occupants. Although air pollution is not confined to cities,

the density and complex intermingling of emission sources, their proximity to large populations coupled with distinct features of urban meteorology (such as recirculating flows in streets), means it has particular relevance there.

This chapter introduces the topic of air pollution meteorology and climatology as it applies to urban areas. Further, it deals with the resulting effects of urban pollutant emissions on air quality at regional to continental scale. It is a subject with an extensive and lengthy history, much of which overlaps with the broader subject of **urban climatology**. In particular there are special connections to questions of airflow (Chapter 4) and temperature structure (Chapter 7) because urban air pollution depends critically on wind, **turbulence** and **stability** that operate across the spectrum of urban scales to transport, dilute and ultimately remove urban pollutants. We begin as before with a discussion of the basics of air pollution before considering its urban character: readers experienced with the chemical and physical fundamentals of air pollution can skip to Section 11.2.

## 11.1 Basics of Air Pollution

Air pollutants are substances which, when present in the atmosphere in sufficient **concentration**, may harm human, animal, plant or microbial health, or damage infrastructure or ecosystems. The most common air pollutants found in urban air are listed in Table 11.1 along with their potential impacts on human health and ecosystems. Air pollutants occur in all states (gaseous, dissolved in droplets, liquid or solid particulates). In all cases, they are transported by the air flow and mixed by turbulence.

**Air pollution** is the atmospheric condition where air pollutants are present in concentrations that are a concern, or even an immediate danger, for human health, ecosystems or infrastructure. **Exposure** refers to the condition when air pollutants reach individuals (e.g. through breathing), while **dose** is the actual amount taken up by the human body (e.g. mass of a pollutant that reaches the blood vessels of an individual in a given time).

The harm of air pollutants can be immediate and catastrophic. In 1984 a storage tank located on the outskirts of the city of Bhopal, India, ruptured due to an accident when water was entering through leaking valves. The unintentional exothermic reaction caused methyl isocyanate to vapourize and form a highly toxic cloud. Remaining close to the ground, the contaminated plume advected into adjacent shantytowns, poisoning inhabitants. The leak caused the immediate death of ~5,000 people and more than 20,000 in the aftermath. This tragedy was the result of a series of errors. The primary one was the decision to place such a risky industrial facility in a highly populated urban area, where any accident was likely to cause great harm. That was compounded by poor maintenance of the facility and insufficient training of the staff (Varma and Varma, 2005).

Knowledge about the atmospheric **dispersion** of chemicals, and their potential reactions, in urban areas is essential to assess the potential threat posed by such events. For example, **local-scale** dispersion models (Sections 11.2.2 and 11.2.3) can help assess which area is at risk near facilities involving storage,

transport and processing of dangerous goods. Further, their results help to inform the required emergency response actions.

More commonly, the harm by air pollution is due to long-term exposure to elevated concentrations that cause chronic damage. Air pollutants that originate from industrial sources and the many engines and furnaces that burn fuel across a city, contaminate the overlying atmosphere. Under unfavourable meteorological conditions air pollutants may accumulate and progressively degrade air quality in the **urban boundary layer** (UBL). The World Health Organization (WHO) estimates that worldwide, exposure to air pollutants results in about 7 million premature deaths per year (WHO, 2014). About 4.3 million of these are attributed to indoor air pollution in less economically developed countries (LEDC). The remainder are due to outdoor air pollution, which is primarily an environmental health problem in heavily urbanized regions of middle- and high income countries.

Urban air pollution can be managed and reduced by targeting emission processes, for example, by requiring proper technological measures, such as emission control systems, fuel switches, efficient use of resources and enforcing restrictions on the use of vehicles and other sources. Emission control tackles the problem at its origin. Management of air pollution is also achieved by improving the location and height of the point of release and promoting dispersion by proper design considerations (see Section 15.3.3). Weather conditions worsen or improve air quality for a given set of emissions. Grid models that predict airflow and air pollutant dispersion, and that incorporate atmospheric chemistry, can provide forecasts of air quality over cities and inform **mitigation** actions. They can also be used to project potential impacts when siting industries or infrastructure and hence minimize adverse impacts (see Section 11.3.3).

Air pollutants spread far beyond the city limits to affect distant places across a range of space and time scales. In fact, the specific nature of the emissions from a city produce a plume with a unique mixture of chemical constituents. For example, if this includes sulphur dioxide ($SO_2$) and nitrogen dioxide ($NO_2$), given sufficient time they will be transformed into sulphuric and nitric acids, respectively. When they are deposited they damage terrestrial and aquatic ecosystems (acid deposition). Similarly, the emission of **greenhouse gases** (GHGs) from cities change the composition of Earth's atmosphere and alter its radiative

**Table 11.1** Selected major chemicals that are medium to long-lived air pollutants emitted in cities (in alphabetical order), their source, health and environmental effects and scales of impact whether they have are a major contributor (•) to air pollution problems at the given scale or a minor contributor (○).

| Compound | Major emission sources in urban areas | Impact on human health | Impact on environment and infrastructure | Scales of major concern | | | | |
|---|---|---|---|---|---|---|---|---|
| | | | | Indoor | Local | Urban | Regional | Global |
| **Carbon dioxide** ($CO_2$) | The major product emitted during all fuel combustion. Also released by respiration of humans, animals, vegetation and soils. | Even high concentrations not toxic, but high density can accumulate in basements and poorly ventilated rooms and reduce / displace oxygen. | Most important long-lived greenhouse gas – alters global radiative forcing and therefore driver of anthropogenic climate change. | ○ | | | | • |
| **Carbon monoxide** (CO) | A primary pollutant from oxygen-limited (incomplete) combustion in motor vehicles, industrial processes and domestic heating. | Interferes with absorption of oxygen by haemoglobin resulting in oxygen deprivation. Low-to-moderate dosages cause headaches, impair brain functions and reduce manual dexterity. At high concentrations death ensues. Heavier than air. | Affects animals the same way as humans. Contributes to formation of the greenhouse gases $CO_2$ and $O_3$ when oxidized. | • | • | • | | ○ |
| **Halocarbons and halogenated gases** | A class of inert gases including CFC-11, CFC-12 and $SF_6$. Used as insulating gases, fillings or refrigerants in consumer goods, in the electrical, polymer and metal industry. | | Halocarbons and halogenated gases contribute to $O_3$ depletion in the stratosphere. Many are very long-lived (and hence effective) greenhouse gases. | | | | | • |
| **Lead** (Pb) | A toxic metal used as petroleum additive in certain countries and released during combustion. Also industrial sources. | Affects nervous system, kidneys, liver and blood-forming organs. Increases blood pressure and disturbs kidney and reproductive functions. Can cause brain damage, impaired mental development and reduced growth. | Persistent when deposited on terrestrial or aquatic ecosystems. Affects ecosystem functioning, animal reproductivity and accumulates in food-chains. | | • | • | ○ | |
| **Methane** ($CH_4$) | By-product of fuel combustion, and fugitive emissions from leaking of natural gas pipes and pipelines. Anaerobic decomposition (e.g. sewage, landfills) | | Second most important long-lived greenhouse gas. | | | | | • |

| Pollutant | Sources | Human health effects | Environmental effects | | | | |
|---|---|---|---|---|---|---|---|
| **Nitrogen oxides** (NO, NO$_2$) | Primary and secondary pollutant resulting from fuel combustion in motor vehicles, coal-, oil-, and gas-fired power stations, industrial boilers and waste incinerators. Minor emissions occur naturally from soils. | Acute exposure causes respiratory diseases (coughs, sore throats) and at high concentration inflames airways and reduces lung functioning. Can aggravate bronchitis, asthma and emphysema. | Contributes to the eutrophication of aquatic ecosystems. | • | • | • | ○ |
| **Nitrous oxide** (N$_2$O) | Minor combustion by-product of fuel combustion, emitted from soils, primarily managed (fertilized) greenspace. | | Important long-lived greenhouse gas. | | | | • |
| **Non-methane volatile organic compounds** (VOC) | Gasoline vehicle exhausts, leakage at fuelling stations, paint, manufacturing, solvents. Biogenic emissions from trees. | Some compounds are carcinogenic. Selected VOCs cause eye and mucous membrane irritation, others may cause fatigue and difficulty concentrating. | Selected compounds damage vegetation. | • | • | | |
| **Ozone** (O$_3$) | Secondary pollutant formed in urban areas primarily from VOCs and NO$_x$, but also CH$_4$ and CO can play a role in O$_3$ formation at larger scales. | Damages respiratory tract and impairs lung function. Physical activity increases the dosage. Long-term exposure may result in decreased lung capacity and premature mortality. | Higher concentration can damage vegetation leading to reduced plant growth. O$_3$ is a short-lived greenhouse gas changing radiative transfer in the troposphere. | • | • | • | |
| **Particulate matter** (PM10, PM2.5, UFP) | Emitted as primary pollutant during combustion (low-temperature fires, diesel vehicles, waste incinerators, domestic heating and cooking), released during mechanical abrasion (road dust, construction) or formed as secondary pollutant from SO$_x$, NH$_3$ and NO$_x$. | Can reach sensitive parts of the respiratory system. Exposure to fine particulates reduces lung function, increases cardiovascular and respiratory diseases and may cause premature mortality. | Changes radiative transmission in atmosphere (Chapter 5), can impact cloud droplet size distribution (Chapter 10) and alter radiative forcing on global scale. When deposited on snow and ice surfaces, changes their albedo. | • | • | • | • |
| **Polycyclic aromatic hydrocarbons** (PAH) | Benzene (C$_6$H$_6$) and Benzo(a)pyrene (BaP) are the most relevant PAHs and are emitted during incomplete combustion in vehicles, domestic heating, organic material (wood) and oil refining. | PAHs irritate the eyes, nose, throat, and bronchial tubes (BaP). Benzene and BaP are carcinogenic for humans, and harm immune system and central nervous system, blood production. Can cause leukaemia and birth defects. | Many PAHs have acute toxic effects on aquatic life and birds and damage leaves of plants. Selected PAHs bio-accumulate in food-chains. | • | • | ○ | |
| **Sulphur dioxide** (SO$_2$) | Primary pollutant emitted during combustion of sulphur-containing fuels (coal, diesel, fuel oil), and in industrial processing. | Exacerbates asthma causing wheezing, shortness of breath and coughing and inflammation of respiratory tract. Synergistic effects with exposure to O$_3$ and particulate matter. | Damages buildings. Causes acidification of soils and aquatic ecosystems downwind of emissions, can damage forests ecosystems. Contributes to the formation of particulate matter. | • | • | • | |

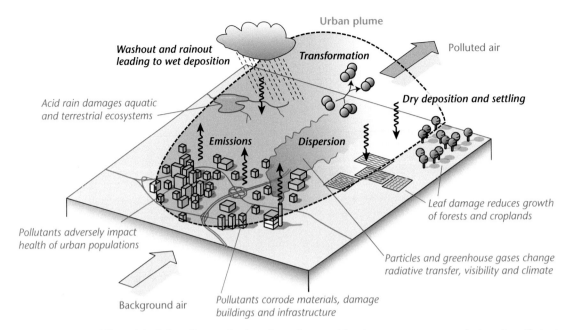

**Figure 11.2** The 'life cycle' of air pollutants in the urban plume, with relevant processes regulating air pollutant concentrations in black and impacts of pollutants in grey italics.

budget, especially that of the **troposphere** (Chapter 13).

Proper understanding and management of air pollution at all scales requires complete understanding of the 'life cycle' of air pollutants at the scale of interest. This includes knowledge of (1) their emission or formation, (2) their dispersion and transport, and (3) their transformation and eventual removal from the atmosphere (Figure 11.2). Sections 11.1.1 to 11.1.3 provide an overview of the relevant processes involved.

## 11.1.1 Emissions

Air pollutants originate from **anthropogenic** and natural sources, although in most cases the anthropogenic sources dominate in an urban environment. **Primary pollutants** are those directly emitted into the atmosphere and retain their chemical character. They arise from chemical processes during fuel **combustion**, or releases such as leakage from pressurized storage or pipes (fugitive emissions), suspension or **evaporation** into the atmosphere. Biological processes also generate emissions that can be relevant in urban air pollution chemistry.

Pollutants formed in the atmosphere from primary pollutants are called **secondary pollutants**. Probably

the best known secondary pollutant is ozone ($O_3$), formed from a suite of reactions of primary pollutants that become chemically transformed in the urban plume. The transformations take time, because they require a specific mix of primary pollutant concentrations and incoming solar radiation (Section 5.3.2). Disentangling the pathway from primary sources to secondary pollutant formation is a challenge that requires **numerical models** that incorporate the appropriate meteorological and chemical processes in the urban atmosphere.

### Combustion

The majority of indoor and outdoor air pollutants are emitted during combustion. Combustion is the sum of chemical reactions between hydrocarbons and atmospheric oxygen ($O_2$) that produce heat and light, and release **carbon dioxide** ($CO_2$) and water vapour into the atmosphere. The dominant hydrocarbons involved are fossil fuels (e.g. natural gas, gasoline, diesel and coal) and biofuels (e.g. wood, bioethanol and biodiesel). If there is an inadequate supply of $O_2$, the combustion process is incomplete and part of the carbon is released as the toxic gas carbon monoxide (CO), instead of $CO_2$. Often furnaces and power plants are equipped with oxygen sensors in the exhaust stack to ensure an adequate supply of $O_2$ to reduce incomplete combustion and CO. If combustion occurs in a very high temperature environment (typical of internal combustion engines), atmospheric nitrogen ($N_2$) reacts with $O_2$ to form nitrogen dioxide ($NO_2$) and nitric oxide (NO). Further, impurities of non-carbon and non-hydrogen elements in the fuel such as nitrogen or sulphur also inject NO, $NO_2$ and $SO_2$ as by-products into the atmosphere. Modern cars are equipped with catalytic converters to partially remove such air pollutants.

Apart from gaseous pollutants, combustion also releases **aerosols** (Chapters 5 and 10). In the context of air pollution, aerosols are usually referred to as **particulate matter** (PM). PM includes both solid and liquid particles in the atmosphere and exists across a wide range of sizes, varying in diameter from $> 100\ \mu m$ to $< 0.1\ \mu m$. Smoke, diesel exhaust, coal fly-ash, mineral dust, paint pigments all contain particulates, mostly made of carbon or silica, but also iron, manganese, chromium, copper and toxic metals such as lead, cadmium, nickel or beryllium. The impact of PM on human health depends partly on its size: PM between $15$–$100\ \mu m$ is trapped in the nose and throat, while that in the $5$–$10\ \mu m$ range reaches into upper parts of the lung, particles smaller than $5\ \mu m$ reach into the fine airways of lungs, and the finest PM ($< 2.5\ \mu m$) can become dissolved in the bloodstream. Air pollution management uses the terms **PM10** for PM $< 10\ \mu m$ (possibly entering the lung), **PM2.5** to quantify PM $< 2.5\ \mu m$ (possibly entering the bloodstream) and ultrafine particulate matter (UFP) is $< 100\ nm$.

### Fugitive and Evaporative Emissions

Fugitive emissions describe the unintended 'escape' of gases held in pressurized containers (e.g. tanks and pipes) that diffuse into the atmosphere. The dense network of natural gas pipes in many urban areas can be a significant source of fugitive hydrocarbon and methane emissions, if it is poorly maintained and leaky. Also, liquefied petroleum gas (LPG) is used as a major energy source for cooking and space heating in cities of LEDC. Leaks of unburned LPG can substantially contribute to air pollutant emissions in cities and play a major role in $O_3$ formation (Blake and Rowland, 1995).

Evaporative emissions occur when liquids, with a low boiling point, evaporate. In cities **volatile organic compounds** (VOC), which have a relatively high **vapour pressure** at ambient temperatures, are a significant concern because they are found in many processes and products. VOCs vapourize from solvents, paints, vehicle fuelling stations, fuel storage facilities, refineries and in various industrial processes. Evaporative emissions increase with the ambient temperature (Rubin et al., 2006).

### Biogenic Emissions

All living organisms, including humans, animals and vegetation, constantly exchange gases as part of their metabolic activity. In the process of **respiration**, humans, plants and microbes release the non-toxic GHG $CO_2$. Some microbes and fungi emit toxins that lead to diminution of indoor air quality; many tree and plant species emit specific VOCs, such as isoprene and more complex terpenes that promote chemical reactions leading to the formation of $O_3$. Emission of terpenes by trees depends on physiological controls, air temperature and light availability and hence exhibits a marked diurnal pattern (Guenther et al., 1993) and it also depends on tree species; substantial contributors are oaks, poplars, eucalypts, pines, sycamore and thuja.

## 11.1.2 Dispersion and Transport

Once released, air pollutants become part of the air. Unlike temperature or pressure, they do not have a direct effect on airflow, therefore they are considered to be passive **scalars** transported by, but not modifying, the airflow. However, it is worth stating that many air pollutants, in particular PM and GHGs, modify radiative transfer processes that impact airflow at longer time scales.

### Concentration and Mixing Ratio

The abundance of air pollutants in the atmosphere is quantified as a concentration $\chi$, i.e. mass per volume of air (e.g. $\mu g\ m^{-3}$). Concentrations are not conservative because they depend on the density of the air, therefore **molar mixing ratios** $r$ (mole of an air pollutant per mole of air, e.g. $\mu mol\ mol^{-1}$) are also used: they are conserved even as an air parcel undergoes thermodynamic changes. For trace gases in the atmosphere, $r$ is typically expressed in parts per million (ppm = $\mu mol\ mol^{-1}$) or parts per billon (ppb = $nmol\ mol^{-1}$). For trace gases, concentration $\chi$ ($\mu g\ m^{-3}$) and molar mixing ratio $r$ (in ppm) can be converted by the ideal gas law:

$$r = \frac{\chi \mathcal{M}_a}{\rho_a \mathcal{M}_p} \qquad (\text{ppm} = \mu mol\ mol^{-1}) \qquad \textbf{Equation 11.1}$$

where, $\mathcal{M}_p$ is the molar mass of the air pollutant (in g mol$^{-1}$), $\mathcal{M}_a$ is the molar mass of dry air (28.96 g mol$^{-1}$), and $\rho_a$ is current (dry) air density (in g m$^{-3}$). Note that Equation 11.1 is only applicable for gaseous pollutants, for example, it does not hold for **PM** or liquid droplets, which can only be expressed as a concentration.

## Dilution

As a scalar, air pollutants are advected by the mean flow and mixed by turbulence, which relocates them and dilutes their concentration. The effect of wind speed on air pollutant concentration is illustrated in Figure 11.3 where a chimney stack emits an air pollutant at the constant rate of one puff every second. If the wind speed ($\bar{u}$) is 2 m s$^{-1}$ there will be 2 m between puffs; but if $\bar{u}$ is 6 m s$^{-1}$ they will be spaced every 6 m. Thus the higher the wind speed, the greater is the volume of air passing the stack per unit time, and the smaller the concentration downwind of the stack exit.

## Turbulent Mixing

Greater speeds also mean greater turbulence that disperses air pollutants. The effect of turbulence depends on the difference between the scale of the pollution (the puff) and the typical **eddy** size (Figure 11.4). Eddies created by mechanical effects have

characteristic sizes that match the dominant roughness elements in a city (buildings, trees). Eddies created by thermal effects have a characteristic size that is set by surface temperature differences. The upper limit for eddies is the **mixed layer** (ML) depth $z_i$.

If the characteristic eddy size is much smaller than the puff, the effect of turbulence causes the puff to grow relatively evenly in the vertical and horizontal directions (Figure 11.4a). For a continuous source, such as the exhaust from a stack, this results in a cone-shaped plume that grows both laterally and vertically downwind of the point of release. This situation is typical of emissions from high stacks over relatively smooth and uniform terrain in the absence of thermal effects. If the typical size of the eddy is close to that of the puff, it becomes stretched and distorted (Figure 11.4b). This is a feature of emissions occurring closer to the urban surface, for example. Finally, if the characteristic eddy sizes are much larger than the puff dimensions, then the entire puff may be simply advected by turbulence without much distortion (Figure 11.4c).

Knowledge of the scale of turbulence, and the overall **turbulent kinetic energy** (TKE) is essential to be able to properly predict air pollutant dispersion in an urban environment. Stability further modifies this situation because the vertical spread may become quite different from the lateral spread. For example, a plume in a **stable** atmosphere is often said to be **fanning**, where lateral dispersion is much larger than in the vertical because vertical movements are suppressed. Conversely, in **unstable** conditions, dispersion happens preferentially in the vertical direction,

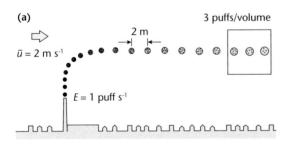

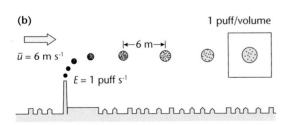

**Figure 11.3** Effects of mean wind on air pollutant concentrations from a source with a constant emission rate (Modified after: Oke 1987).

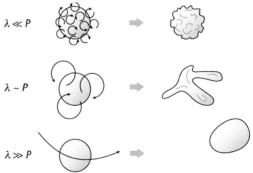

**Figure 11.4** Conceptual effects of turbulence on air pollutant dispersion as a function of the difference between scale of puff $L$ vs characteristic scale of eddies $\lambda$ (Based on a sketch by H.P. Schmid).

causing an effect called **looping**. Urban atmospheres tend towards **neutrality** (see Chapter 7) hence dispersion patterns typical of strong stability (fanning) or strong instability (looping) are relatively rare.

### 11.1.3 Removal and Transformation

Without removal, air pollutants would accumulate in the atmosphere, and concentrations would continuously increase. In fact, this happens to many GHGs that remain for decades to centuries in the atmosphere (Chapter 13). Their accumulation alters radiative transfer and modifies the climate system. Most other air pollutants are removed sooner after emission by four processes: gravitational settling; dry deposition; wet deposition and; chemical reactions and/or decay.

#### Gravitational Settling

Gravitational settling is responsible for the removal of heavier particulates from the atmosphere. The terminal settling velocity $v_s$ (in m s$^{-1}$) of particles determines the rate of removal, as particle mass increases, higher terminal settling velocities will be reached. Most PM $> 10$ μm in diameter settles out relatively rapidly, in closer proximity to emission sources and even strong turbulence is unable to hold these particles in suspension for long. The descent of smaller particles is slowed by turbulence, and those $< 1$ μm in diameter remain aloft for weeks to months (see Figure 5.3). Gaseous pollutants can be adsorbed onto particles that settle gravitationally and be removed with them.

#### Dry Deposition

**Dry deposition** is a transfer process similar to that involved in the vertical transfer of heat, water vapour and **momentum** to and from surfaces. In this process gases and particles are transported by turbulent mixing to surfaces (ground, walls, roofs, vegetation), however to actually reach the surface, they need to cross the thin **laminar boundary layer** (LBL) attached to all surfaces, this is achieved by means of **diffusion**. This can be expressed as a vertical mass **flux density** $F_p$ of an air pollutant (in μg m$^{-2}$ s$^{-1}$) where the underlying surface acts as a pollutant sink. In analogy with the turbulent transfer of heat and water (Section 6.4), the vertical transfer of the air pollutant through the turbulent **boundary layer** can be described by a covariance of vertical wind and concentration:

$$F_p = \overline{w'\chi'} \quad \left(\text{μg m}^{-2}\text{ s}^{-1}\right) \qquad \textbf{Equation 11.2}$$

The vertical flux of an air pollutant $F_p$ in the LBL is due to diffusion:

$$F_p = -k_p \frac{\partial \overline{\chi}}{\partial z} \quad \left(\text{μg m}^{-2}\text{ s}^{-1}\right) \qquad \textbf{Equation 11.3}$$

where the flux is proportional to the molecular **diffusivity** $k_p$ (in m$^2$ s$^{-1}$) and the vertical gradient in the mean pollutant concentration in the LBL. The flux through both the turbulent and LBLs can be expressed as a **deposition velocity** $v_d$:

$$F_P = v_d \Delta \overline{\chi} \quad \left(\text{m s}^{-1}\right) \qquad \textbf{Equation 11.4}$$

The deposition velocity $v_d$ is a **conductance**. Its reciprocal can be written as the sum of various **resistances** (Section 6.4), including the **aerodynamic resistance** $r_a$ for the turbulent layer, the LBL resistance $r_l$, and for gases the **surface resistance** $r_s$ which accounts for the ability of the surface to absorb the gas:

$$v_d = \frac{1}{r_a + r_l + r_s} \quad \left(\text{m s}^{-1}\right) \qquad \textbf{Equation 11.5}$$

For a given atmospheric mean concentration, the rate at which an air pollutant is delivered to the surface is governed by the value of the deposition velocity, which in turn depends upon the state of turbulence and molecular / surface properties. The surface resistance $r_s$ varies with the type of surface. For example, on a leaf surface the stomatal aperture partly controls $r_s$, while over soil it is affected by bacterial activity, and over water by the surface tension. It may also be affected by electrostatic attraction, and chemical reactions between the surface materials and the pollutant.

#### Wet Deposition

Some of the PM, especially if **hygroscopic**, becomes **cloud condensation nuclei** (CCN, see Section 10.1) around which water or ice condenses to form a droplet or ice crystal in clouds. Further, particulates and gases can then collide with existing droplets and ice crystals. This is most efficient for small particles ($< 0.1$ μm) and gas molecules. Molecules and particulates that collide with a droplet are then dissolved up to the limit given by the solubility of the pollutant in water. For example, the solubility of gaseous pollutants such as $SO_2$ and $NO_2$ is higher than it is for $O_3$. This in-cloud pollutant scavenging process is called **rainout** (or snowout).

Eventually, if the droplet becomes large enough, it is precipitated and transports the pollutants with it to the surface. Below cloud level, **precipitation** also cleanses the air by 'sweeping-out' some PM. As drop-lets or ice crystals fall they can collide with particles and carry them to the surface (**washout**). This process is most efficient for larger PM ($> 1$ μm). Altogether, a precipitation event removes pollutants in the form of CCN, dissolved gases and particles involved in colli-sions and entrained during the descent. A rainstorm can transform a murky **haze** into visibly cleaner envir-onment. Note that PM between ~0.1 and 1 μm are not efficiently removed by gravitational settling, rainout or by washout. This explains why particles in this size range remain longer in the atmosphere than larger or smaller ones. Note also that whereas many pollutants are efficiently removed from the atmosphere by wet deposition, they are still deposited on the ground where they can pollute aquatic and terrestrial ecosys-tems ('acid deposition', see Section 11.4.2).

## Chemical Reactions and Decay

The mechanisms discussed so far remove air pollutants by mechanical and microphysical means, but chemical reactions are also crucial in air pollution climatology. While some reactions may aid in the decomposition and decay of air pollutants, others lead to the formation of new secondary pollutants. Some reactions happen dir-ectly in the gas-phase by collision of molecules, while others are more efficient if the air pollutant is dissolved in a liquid droplet or is adsorbed on a particle. Chemical reactions in the atmosphere are generally classified as thermal or photochemical. Thermal reactions involve the collision of two gas molecules. A very common class of thermal reactions transform gases into more oxidized states, such as NO to $NO_2$ and eventually to $HNO_3$ (nitric acid), hydrocarbons can be oxidized to alde-hydes, and $SO_2$ to $H_2SO_4$ (sulphuric acid). Note that in those cases, the higher oxidized products are second-ary pollutants formed by the transformation of primary or already secondary pollutants.

Solar radiation can be absorbed by molecules and they can be split which is called 'photodissociation'. This process is relevant to the destruction of $O_3$, $NO_2$, aldehydes and many other air pollutants including some GHGs. Only **photons** with sufficient energy are able to break molecular bonds. In the atmosphere highly energetic ultraviolet radiation is largely respon-sible for photodissociation. The resulting products are highly reactive radicals, fragments of stable mol-ecules, such as atomic oxygen (O), the hydroxyl radical (OH) or the hydroperoxyl radical ($HO_2$), which immediately react with other molecules.

## 11.1.4 Scales of Air Pollution

Air quality is relevant at many scales of inquiry. Table 11.1 lists common air pollutants, their sources, impacts on human health and the environment gener-ally and identifies the scales at which the problem is greatest. The scale of the pollution is largely dictated by the atmospheric **residence time** (i.e. the average time before removal) and its effects on health and ecosystems.

Figure 11.5 links the residence time of common air pollutants to their characteristic horizontal distance of dispersion. Short-lived radicals such as OH or $HO_2$ have residence times of seconds to minutes, so their dispersion in the atmosphere is very limited (lower left of Figure 11.5). At micro- and local scales, air pollu-tion is characterized by primary pollutants from com-bustion, with considerable small-scale variability due to the specific arrangement of emission sources inter-acting with the complex wind field and/or indoor venting patterns sourrounding buildings, urban blocks and **neighbourhoods**, characteristic for the UCL. Thus air pollutants from one street can be channelled at intersections into adjoining streets, or be trapped in a recirculating eddy at an intersection. Some eventu-ally settle or are captured inside the UCL on building or road surfaces and vegetation and some are swept from the UCL by turbulence into the overlying UBL.

In the UBL, typical air pollutants emitted from individual sources (cars, buildings, industry, etc.) mix and transform over time to form a characteristic urban blend, commonly containing CO, $NO_2$, $SO_2$ and/or $O_3$. Without a regional wind, urban breezes create a polluted **urban dome** (Figure 2.12a). With a regional wind, the contaminated UBL extends down-wind as an **urban plume** (Figure 2.12b).

At the regional to global scale, long-lived species such as GHGs have residence times of years to cen-turies (upper right of Figure 11.5). The individual contributions of cities across the globe become well-mixed in their corresponding hemisphere and add to those from other sources. Since emissions exceed rates of removal, this can steadily increase global back-ground concentrations of certain gases. They are effectively changing **radiative forcing** for the planet; a major cause of anthropogenic climate change (Chapter 13).

The rest of this chapter is organized according to scale. Section 11.2 focuses on pollution at small scales – buildings, single sources and the UCL.

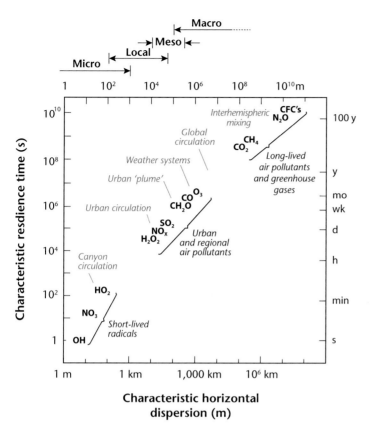

**Figure 11.5** Characteristic atmospheric residence times for selected gaseous air pollutants along with the characteristic horizontal scale of dispersion and the relevant atmospheric phenomena causing dispersion (grey) (Data sources: Seinfeld and Pandis 2006, and Hobbs, 2000).

Section 11.3 deals with the pollution of the UBL and the processes that cause persistent outdoor air quality problems in cities. Section 11.4 discusses regional to continental impacts of urban pollutant emissions.

In this chapter we restrict discussion to simple, generic circumstances, even though most cities are located in more complex terrain (**orography**, shorelines). In those places air pollution is the outcome of interplay between characteristic emissions, local wind systems, **topography** and the background climate. This creates unique challenges to the management of air pollution. These issues are discussed in Chapter 12.

## 11.2 Micro- and Local-Scale Air Pollution in Cities

The urban canopy layer (UCL) is the zone of human occupation and the quality of the air in it is critical in any assessment of the public health consequences. A great many emission sources are located within the UCL where contaminants are injected directly into the indoor or outdoor air, close to breathing level. Here air pollutants encounter a great variety of microscale conditions that inhibit or enhance their dilution. Our goal is to predict their dispersion and venting out of the UCL, where they eventually mix with emissions from sources at or near roof level. At the local scale, the aggregate effect of microscale processes within the **roughness sublayer** (RSL) is to mix all emissions together so they can be considered to arise from a neighbourhood. To this must be added other sources of pollution, such as those from chimney stacks that are emitted directly into the atmosphere above mean roof height. The overall outcome is an urban ISL that is contaminated to a greater or lesser degree with great variability in time and space. From an air quality management perspective, the challenges are to: (i) identify the ambient conditions to which the population is exposed over the course of a day, month, year, or lifetime; (ii) to assess the health consequences and; (iii) to regulate emissions so as to limit exposure. At the outset, it is worth distinguishing between the indoor and outdoor environments, which reflect very different ambient conditions.

## 11.2.1 Indoor Air Pollution

For a great deal of urban living, an indoor environment represents ambient conditions. In fact, indoor air pollution could be of greater concern than that outdoors, simply because most dwellers spend substantially more time inside buildings and vehicles and in many activities (domestic and occupational) the emission of air pollutants first occurs within a confined indoor setting. However, the health consequences of exposure to many of the air pollutants found indoors have not been examined to the same extent as the common outdoor air pollutants.

As a simplified, yet illustrative, model of indoor pollution, the rate of change of pollutant concentration for a given indoor air volume $V$ (in m$^3$) can be described:

$$\frac{\partial \overline{\chi}_i}{\partial t} = \frac{E - D + X(\overline{\chi}_a - \overline{\chi}_i)}{V} \qquad (\mu g\ m^{-3}\ s^{-1})$$

**Equation 11.6**

This is a simple 'box' model as illustrated in Figure 11.6, where $\overline{\chi}_i$ is the time-averaged indoor air pollutant concentration, and $\overline{\chi}_a$ is the air pollutant concentration of the outdoor air (both in $\mu g\ m^{-3}$). $E$ is the sum of all indoor emissions of the air pollutant

of interest (in $\mu g\ s^{-1}$). The major sources of indoor air pollutants are fuel combustion, smoking and cooking, and fugitive and evaporative emissions from furniture, walls and ground in the living space. The venting rate $X$ (in m$^3$ s$^{-1}$) describes the rate of air volume exchanged between indoors and outdoors through windows, doors, chimneys, cracks and by forced ventilation. $X$ depends on the construction details and the outdoor air flow (pressure differences across building). Note that the air volume taken in is about equal to that leaving the building. $D$ is the decay or removal rate (in $\mu g\ s^{-1}$) of the air pollutant of interest in the indoor setting, for example when particulates are removed by gravitational settling, dry deposition or when a compound reacts and /or decays. The simple model presented by Equation 11.6 illustrates that the build-up of indoor air pollutant concentrations is controlled by rates of emission, removal, venting and the volume. Buildings typically enclose a relatively small air volume and its reduced ventilation compared with outdoors means that relatively small emissions can result in the build-up of unhealthy concentrations.

Equation 11.6 also allows us to identify measures to manage and reduce indoor air pollutant concentrations, namely to:

- **Reduce net emissions** $(E - D)$ – for example, through controls on the combustion process, avoidance of certain fuels or construction materials ($E$) and installation of filters and scrubbers ($D$)
- Provide a **larger indoor air volume** ($V^{-1}$) – for example, construct homes with higher ceilings and larger rooms
- **Increase the rate of exhaust** per volume ($X/V$) – for example, by forced ventilation
- **Reduce outdoor air pollutant concentrations**, because this increases the difference $\overline{\chi}_a - \overline{\chi}_i$

Indoor air pollution is of biggest concern in cities of LEDCs, especially where solid fuel combustion (for cooking and heating) occurs within spaces that are crowded and poorly ventilated. However, unhealthy indoor air is not restricted to poor quality housing and low-quality fuel use. It can also be a feature of some modern buildings designed to be energy efficient, these buildings are nearly completely sealed so that indoor-outdoor exchanges are tightly controlled and much of the indoor air is recirculated. In such circumstances, even small pollutant emission rates can result in high concentrations over time.

The list of pollutants found in indoor air is huge – the following sections are an attempt to classify some

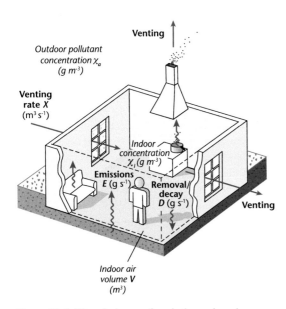

**Figure 11.6** Mass balance of an indoor air volume.

of the most common air pollutants by their emission process – combustion, evaporation and suspension, leakage and biological processes.

## Combustion By-Products

Many indoor air pollutants are, similar to outdoor air pollutants, released during combustion for space heating, hot water generation, cooking and/or smoking. The major air pollutants of concern are the primary pollutants CO, $NO_x$, PM and carcinogenic organic by-products, such as polycyclic aromatic hydrocarbons (PAH, see Table 11.1). To a lesser extent, selected secondary pollutants such as $O_3$ can also be of concern indoors.

Combustion of solid fuels (wood, dung, agricultural residues, coal) often takes place in simple stoves or open fires indoors with an inadequate supply of oxygen, this causes incomplete primary combustion by-products such as CO, PM and VOCs. Exposure to those air pollutants can lead to chronic and acute health risks. Globally, indoor pollution from solid fuel combustion is on the list of the top 10 health risks that cause overall mortality, and it is the single most important environmental health risk (WHO, 2009).

Estimates from dwellings in India suggest indoor exposure to PM is much greater than that experienced outdoors. Table 11.2 shows the ambient levels of PM10 when averaged over 24 hours in 'typical' exposure environments for different income levels in Delhi, India. While all income levels experience approximately similar outdoor air quality, they are distinguished by their different exposure in domestic or occupational settings, and access to affordable fuels. The lowest income class are more likely to cook over open solid fuel fires and occupy small, poorly

ventilated dwellings. Moreover, as cooking is traditionally a female role, the dosage received by females is higher than that of males in the same income group.

For CO, high concentrations are reported inside cars and buses where they are usually 2–5 times those measured in adjacent streets. Inhalation of high concentrations of CO in buildings emitted from gas stoves, vehicles in garages and other combustion engines are the most common means of poisoning – these also occur in cities of more economically developed countries (MEDC). Occasionally deaths from CO poisoning are reported, usually due to improper venting of appliances and because victims are often rendered unconscious in their sleep.

There are well-established dose-response relationship between tobacco smoke and chronic respiratory symptoms — not only for smokers but also for individuals simply exposed to environmental tobacco smoke (i.e. 'passive smoking'). For example, it has been shown that exposure of children to passive smoking in indoor settings increases their risk of developing asthma between 40% and 200% (e.g. Cook et al., 1998). From a public health perspective, passive smoking is the most relevant indoor pollution source in most middle- and high income countries.

## Evaporation and Suspension

In contrast to the chemical process of combustion, a host of other air pollutants are released constantly by the built environment through physical processes, including (i) evaporation of volatile compounds, (ii) suspension of fine particulates, and (iii) diffusion from pressurized natural gas pipes.

Evaporative emissions of VOCs indoors can originate from furniture, consumer products and

**Table 11.2** Exposure to PM10 concentration and time spent in different microenvironment settings as a function of income class in urban India, for the city of Delhi (Modified after: Kandlikar and Ramachandran, 2000).

| Microenvironment | Low income | Medium income | High income |
|---|---|---|---|
| **Cooking** | 1,000–4,000 µg m$^{-3}$<br>1–2 h | 350–600 µg m$^{-3}$<br>2–3 h | 200–300 µg m$^{-3}$<br>1–2 h |
| **Non-cooking** | 650–800 µg m$^{-3}$<br>10–12 h | 200–300 µg m$^{-3}$<br>12–14 h | 200–300 µg m$^{-3}$<br>12–14 h |
| **Occupational** | 650–800 µg m$^{-3}$<br>12–14 h | 200–300 µg m$^{-3}$<br>8 h | 200–300 µg m$^{-3}$<br>8 h |
| **Outdoor** | 200–700 µg m$^{-3}$<br>1–2 h | 200–700 µg m$^{-3}$<br>1–2 h | 200–700 µg m$^{-3}$<br>1–2 h |

construction materials. A prominent example is formaldehyde released from particle board, insulation and furnishings. Many other organic compounds evaporate from adhesives, solvents and paints (e.g. toluene, styrene, alcohols, acetaldehydes and terpenes). In typical indoor air, a few hundred different VOCs are traceable, and their chemical 'mix' is sometimes sensed as the characteristic odour of buildings. The sum of all VOCs is expressed as the concentration of total volatile organic compounds. At the time of initial occupancy, typical office buildings have been shown to contain indoor air with 50 to 100 times higher total VOC concentrations than outdoor air (Bluyssen et al., 1996). Unsurprisingly, it is challenging to isolate health impacts of individual indoor VOCs given the complexity of the total chemical mix – however, effects attributed to selected VOCs in health studies include mucous membrane and eye irritation, fatigue and carcinogenicity.

The most hazardous indoor air pollutant is asbestos, a class of long and flexible or crystalline mineral fibres. Asbestos fibres are released from the wearing down of selected manufactured products and construction materials (pipes, fire retardants, floors, tiles, insulation). Being particulates, asbestos fibres can remain suspended in the indoor and outdoor air for long periods and are transported by airflow through air conditioning systems. As for PM, heavier particles settle more quickly. The major concern is that asbestos fibres are very persistent — they cannot evaporate or dissolve in water and if they reach human lungs, they accumulate and damage the tissue.

## Radon Venting

A unique case of indoor air pollution is caused by radon, an inert radioactive gas. The isotope radon-222 is naturally formed in rocks and soils by radioactive decay of certain minerals containing uranium (e.g. pitchblende). It decays with a half-life of 3.8 days to more stable products that are also radioactive, but not gaseous (lead, bismuth, polonium deposited on particles). Because radon is an inert gas, it can escape easily through cracks and fissures into tunnels, pipes and basements of buildings. Radon-222 accumulates indoors due to low ventilation rates. When inhaled, it deposits its non-gaseous and radioactive daughter products on the lung tissues where they remain, irritate cells and can cause cancer. After tobacco smoke, radon inhalation in indoor settings is the second most important cause of lung cancer worldwide, and is responsible for 3 to 14% of all mortalities (WHO, 2009). Radon isotopes also account for roughly half of the general population's exposure to ionizing radiation. There are large regional differences, due to varying near-surface geology and differences in building construction practice.

The micro- and urban meteorological aspect of radon pollution concerns its transport from the soil to the indoor space and eventually to the outdoor atmosphere (Nazaroff, 1992). Typical concentrations of radon-222 in soil pores result in radioactivity of several 10,000 Bq m$^{-3}$ in the ground. The escape of radon to the overlying atmosphere depends on soil porosity and the pressure gradients in the soil. Although the airspace in soil pores is in constant exchange with the outdoor air, in most cases the soil airspace is renewed very slowly because of the less permeable organic top-soil and the presence of sealed surfaces (asphalt, concrete) in urban areas that form a 'lid' over the bedrock. If radon reaches the outdoor atmosphere, it is quickly dispersed and diluted. The radioactivity in the outdoor space near the ground is small compared with that of soil (2–10 Bq m$^{-3}$). Nevertheless, radon-222 in buildings is a problem; and on average it reaches radioactive levels between 20–400 Bq m$^{-3}$, but in unfavourable cases can exceed 10,000 Bq m$^{-3}$, which is highly unhealthy. Buildings with poorly sealed basements are a particular concern. Firstly and obviously, the small volume and restricted venting of buildings prevents effective dispersion compared to the outdoor atmosphere. Secondly, in buildings with basements, the top-soil has been removed hence the less permeable lid is missing, in particular if the basement has a natural floor. Also a generally lower soil moisture underneath buildings means higher permeability for gases in the soil. Thirdly, the airflow around the building envelope causes pressure differences (see Section 4.2.1) that affect the pressure in the soil and building interior causing preferred venting of radon from underneath buildings. The negative pressure in buildings caused by active ventilation and heating can further 'suck' radon-222 from the subsurface through cracks in the bottom slab or walls into indoor spaces. Radioactivity in homes caused by radon has hence been shown to depend on outdoor wind speed and heating demand (Miles, 2001).

## Biological Air Pollution

Biological contaminants, namely bioaerosols including fungi, mites, amoebae, bacteria and viruses are present and transported in indoor air. Molds and microorganisms cause infections and allergic, and hypersensitivity reactions. Mycotoxins and certain VOCs released by these organisms are suggested to be compounds capable of causing such symptoms. Humans and animals are the source of bacteria, and viruses that spread through the ventilation systems of larger buildings.

$CO_2$, released by both, human respiration and during fuel combustion, is routinely used as a reliable indicator of biological indoor air pollution, especially for emissions originating from building occupants. Although not a health concern, concentrations of $CO_2$ more than ~1000 ppm in buildings are considered an indication of poor ventilation and indoor air quality (Zhang and Smith, 2003) and is usually accompanied by feelings of stuffiness, tiredness, headache and loss of concentration and productivity. Relatively simple sensors that monitor $CO_2$ can be used in combination with active control systems to increase/decrease the inflow of outside fresh air into larger buildings.

Fortunately, all indoor air pollution problems (combustion, evaporation, radon and biological contaminants) can be mitigated by adequate technical, hygienic and design measures. In this respect, overcoming educational and economic barriers to implementation is key.

## 11.2.2 Outdoor Air Pollution in the Urban Canopy Layer

Compared to indoors, air pollution climatology in outdoor urban spaces between buildings is more complex for several reasons. Firstly, emission sources in an urban system are in multiple locations. For example, emissions could originate from buildings, mobile vehicles near ground level and even from sources above roof level (e.g. chimneys) that might be drawn down into the UCL, in addition there could be biological contributions by plants and trees. Secondly, airflow itself follows complex patterns, so dispersion in the UCL must be treated as fully 3-D (Section 4.2). Thirdly, the movement of people (i.e. the 'receptors') through a street system exposes them to an extraordinary variety of transitory ambient conditions.

Figure 11.7 illustrates the exposure of a pedestrian to ultrafine particulates over a period of 30 min when walking through the UCL in central London, United Kingdom. Exceptionally high exposure occurs at intersections, traffic crosswalks and when passing people smoking. On their daily commute to and from work people tend to follow preferred routes of this kind. Knowledge of repeated exposure to air pollutants along such routes can be of value in evaluating exposure, threats to health and how to manage air quality at this scale.

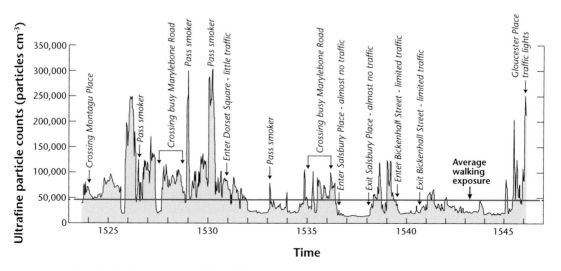

**Figure 11.7** A pedestrian's exposure while walking through a complex urban concentration field of ultrafine particulate matter in London, United Kingdom (Data source: A. Robins).

In many cities, air pollutant concentrations in the UCL are dominated by emissions from traffic. The primary pollutants created during the combustion of gasoline, diesel and biofuels (CO, $NO_x$, PM, PAHs, VOCs, and in some cases lead from leaded gasoline) arise near street level but their fate very much depends on airflow patterns in the UCL. Buildings and trees produce considerable mechanical turbulence that enhances mixing and dilutes concentrations. Canyon airflow includes channelling and recirculation vortices including possible thermal effects due to patterns of sunlight and shadow created in the UCL. Given these circumstances, measuring and modelling air quality to which the urban population is exposed in the UCL is a difficult task, however, without such assessments it is not possible to manage emissions to ensure a public health. Observations of air quality must be of sufficient density to ensure the spatial and temporal variability of air pollutant concentrations in the UCL is resolved. Hence, observations must consider problematic hot-spots and episodes (that may be localized or short-lived).

## Monitoring Outdoor Air Quality

Air pollution monitoring is often carried out using an observation network consisting of fixed stations deployed in the UCL to capture the average and the extreme ambient concentrations. These stations can be of two types:

- **Urban background stations** that represent 'typical' concentrations most people experience. The stations are usually placed in courtyards, parks, or other places away from significant pollution sources (such as major roads). They capture a spatially averaged signal representative of the lower end of concentrations typifying the local scale.
- **Roadside stations** are often used to measure pollution 'hot-spots'; those places where concentrations are expected to be highest and to which significant numbers of people are regularly exposed (e.g. when driving or walking along streets or living in buildings flanking the street). They are typically located on a pedestrian walkway along selected busy streets, close to the traffic lanes and with the **sensor** inlets at 1.5 to 3 m above ground. Similar stations may be placed in the vicinity of large industrial plants.

Such networks of fixed stations are often supplemented by mobile or short-period fixed observations, to either ensure the network observations are

sufficient or to monitor the effects of changes in the urban environment (e.g. new traffic regulations or large-scale construction projects).

Despite these efforts, observations alone are not sufficient to assess air quality across a city. Numerical modelling at the microscale is necessary to complement an observation network and assess the representativeness of its measurement sites. Microscale models can also guide air quality management strategies at the scale of streets and urban blocks (e.g. traffic routing and building density).

## Air Pollution in Urban Canyons

Air quality in streets has received particular attention because of the ubiquity of these exposure environments, their relatively confined setting and the intensity of activities near ground level that generate air pollutants. The effects of traffic emissions in long and relatively narrow (small aspect ratio) streets have been well studied because the limited mixing can result in poor air quality (see Section 4.2.3 and Figure 11.1).

Air pollution in a street canyon is controlled by the emission strength, the exchange of air within the UCL and its interactions with the air above roof level. Air pollutants may be transported along a canyon or laterally into intersecting streets, courtyards and adjacent buildings or mixed with the air above roof level. Where street width is narrow and the buildings on either side form an almost unbroken wall, lateral dispersion within the UCL is restricted. Such street canyons experience **skimming flow** (see Figure 4.9c) meaning that above-roof winds skip along without penetrating into the UCL. Instead, UCL air and its contaminants are recirculated internally by a helical **vortex. Canyon aspect ratio** $H/W$ and roof shape greatly affect the vertical ventilation of such canyons (see Section 4.2.3).

Figure 11.8 shows the spatial variability of the traffic-related pollutant benzene, a carcinogenic PAH emitted during combustion of gasoline (see Table 11.1). Measurements were taken at various locations in the busy 4-lane canyon 'Rue de Rennes' in Paris, France and reflect the average molar mixing ratios over five consecutive weekdays. The canyon has an aspect ratio of about 1 and is bounded on either side by rows of uniform 6–7 storey buildings. Two additional measurements were taken at a distance of about 300 m on either side of the canyon at urban background locations (greenspaces). The observed benzene mixing ratios are substantially higher in the

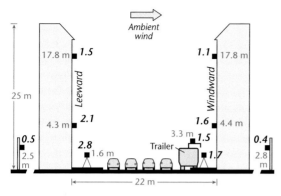

**Figure 11.8** Average weekday mixing ratios of Benzene in ppb (bold numbers) measured at different locations and heights in a regular street canyon (Rue de Rennes, Paris, France) and adjacent urban background sites within 300 m on either side (gap in surface). Data from July 19–23, 1999 (Modified after: Vardoulakis et al., 2002).

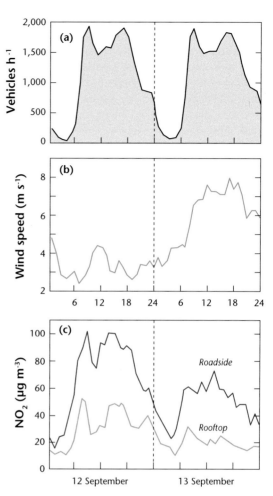

**Figure 11.9** Relations between **(a)** traffic intensity, **(b)** wind speed and **(c)** $NO_2$ concentrations on two consecutive weekdays in 1997, measured in a busy city street of Copenhagen, Denmark. The concentration measurements were made simultaneously at the roadside (solid curve) and on a nearby rooftop (Source: Fenger, 1999; © Elsevier Science Ltd., used with permission).

canyon than at the urban background sites. Also there is clear cross-canyon asymmetry of pollutant mixing ratios: when the dominant above-roof winds are nearly perpendicular to the canyon axis, values on the leeward wall (left) of the canyon are higher than at the same height on the windward wall (right). At the leeward wall the helical flow pattern in the canyon causes updrafts of polluted air, whereas on the windward wall cleaner air from above roof level is entrained (see also Figure 4.11a and b). Measured mixing ratios decrease with height in the canyon; higher values are measured deeper in the canyon (closer to the source of emissions) where the legal standard is exceeded. In fact, highest mixing ratios are reported on the pedestrian walkway on the leeward side, which is immediately downwind of the road lane. This location may also experience corner vortices that further trap and recirculate air pollutants.

Temporal variations of emissions and of the flow field combine to produce characteristic pollution concentration cycles inside street canyons. Figure 11.9 shows the relation between traffic, winds and $NO_2$ mixing ratios over two days in a street canyon in Copenhagen, Denmark. Emissions into the street are governed by the traffic density, which shows a diurnal 'wave' pattern associated with a daytime peak that is shifted towards evening. Superimposed on the wave crest are two peaks associated with commuter traffic.

The $NO_2$ concentration in the street reflects the emissions pattern but is moderated by wind. The stronger winds on the second day have the effect of ventilating much of the contaminated air from the street, reducing mixing ratios. Note that in both cases there is a vertical gradient in $NO_2$ with maximum values near the source, at street level.

The pattern of pollution dispersion within such ideal street canyons is well-established and it has allowed the development of simple microscale air pollution models to predict concentrations inside canyons in response to street dimensions, above-roof wind direction and the rate of emissions. For example, early street canyon pollution models such as that of Johnson et al. (1973) use simple box approaches and empirical relations based on measurements to predict concentrations for receptors on the leeward and windward sides of a street canyon. Concentrations are directly proportional to the number of vehicles, and inversely proportional to the ambient wind speed and are scaled by the dimensions of the canyon. Their model adjusts wind speed for traffic effects and accounts for traffic-produced turbulence (see Chapter 4).

More sophisticated models combine box-models with plume dispersion models. As an early example, Yamartino and Wiegand (1986) developed a plume dispersion model that decomposes the above-canyon flow into two segments: a cross-canyon component (which drives a vortex) and an along-canyon component. Each flow component regulates distinct dispersion patterns of vehicle emissions that results in canyon zones where advective or turbulent processes dominate. Along the windward wall a narrow channel of fresh air from above the canyon is drawn into the street. As it descends it mixes with a proportion of recirculating canyon air, in the lower half of the street, the mixed flow crosses the vehicle-generated pollutant plume to the lee side of the street and then upwards. At the canyon top a portion of the contaminated air is advected into the above-canyon flow, while another portion diffuses and recirculates.

Today, a large number of operational street canyon air pollution models are available for use in regulatory applications (Vardoulakis et al., 2003). They can all produce time series of air pollutant concentrations within canyons based on a limited number of simple inputs (traffic volume, fleet composition, canyon geometry, meteorological data).

## Air Pollution and Dispersion in Complex Geometries

In many other cases, the outdoor airspace in the UCL cannot be adequately described using simple 2-D canyon dispersion models. In such cases, more sophisticated simulations are possible using **computational fluid dynamics** (CFD) simulations of the flow field inside canyons (Li et al., 2006) or neighbourhoods

with real buildings resolved (Tseng et al., 2006; Giometto et al., 2016). Those approaches are flexible, and solve Reynolds Averaged Navier-Stokes Equations (Chapter 3) or use **Large Eddy Simulations** (LES) to incorporate the effect of the **urban structure** on the flow. CFD models can track air pollutant concentrations at each grid node as passive scalars or coupled with chemical reactions. CFD approaches have high computational needs and their use to date is mostly restricted to research, and to selected subsets of cities, rather than for general regulatory purposes in air pollution management.

The population density of cities comes with increased **vulnerability** to accidental and intentional airborne release of hazardous gases and materials such as those in industrial and transportation accidents or by acts of warfare or terrorism. In such cases, knowledge about the dispersion, potential reactions and transformation of chemicals, and settling or deposition is required to identify areas of greatest impact, for example, to prioritize areas for evacuation or other emergency measures.

Over open and flat terrain, knowledge of wind direction and speed is usually sufficient to make a rough estimate of zones of danger after an accidental release. Observations on site, with emergency weather stations or using operational weather data, might be appropriate to estimate probable spread. Not so in urban areas, where complex flow in the UCL renders point measurements of wind nearly useless and weather forecast models are too coarse to resolve local flow features. Figure 11.10 depicts the above-ground plume released during a tragic subway fire in the city of Daegu, South Korea. Although the plume is easily identifiable and coherent, it experiences complex behaviour, changing its direction as it is first channelled along one street canyon and then helically turns by 270° as it rises above the intersection and mixes with the ambient flow. The case also illustrates the interconnectivity between different air spaces in cities, including underground space (tunnels), indoor space and the outdoor air: all are affected but to different degrees.

Figure 11.11 illustrates modelled zones of danger from a hypothetical release at ground level in Portland, United States (a) with and (b) without the effect of buildings near the release point considered. The regular arrangement of streets and buildings in (a) causes preferential channelling, which redirects the

**Figure 11.10** Plumes of smoke emerge from a subway fire in Daegu, South Korea, and illustrate the complex behaviour of air pollutant plumes in the urban canopy layer (Credit: © Press Association; used with permission).

plume centreline considerably. The presence of tall buildings greatly enhances both the lateral and vertical mixing of the hazardous material. The plume shape close to the source is completely different in the two cases, as are the predicted zones of danger. Even far downwind, on the East side of the Willamette River, the plume is broadened and shifted northwards due to diversion along the street grid.

The two examples in Figure 11.10 and Figure 11.11 illustrate that considering complexities of airflow in the UCL are crucial for successful emergency response modelling, if both the source and receptors are in the canopy. However, this requirement comes with substantial costs. Resolving buildings in modern CFD-based models is (i) computationally expensive, (ii) requires knowledge of the actual urban morphometry to be incorporated into the model and (iii) requires

specialized infrastructure and personnel to run such complex models. Emergencies, on the other hand require fast action, and can happen at unpredictable places where geographical data concerning building details and topography and the necessary computational personnel might not be available.

For emergency response, predictions of the flow field and the plume spread in a building-resolving mode are needed very quickly. With current computing power this is unfortunately not yet feasible. And for most cases, the uncertainties about location, quantity, timing and the material released into an urban environment might be much larger than any fluid dynamical errors in models. Nevertheless, for selected subsets of cities and industrial infrastructure, simplified models are run operationally to assist emergency response actions. Such models may use operational

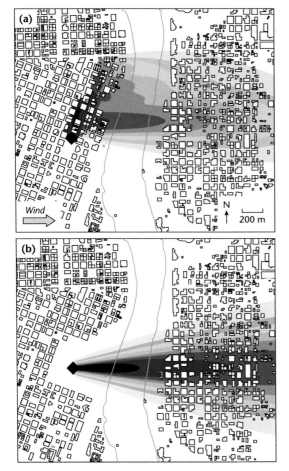

**Figure 11.11** Simulated plume dispersion from a hypothetical ground-level source located in central Portland, United States, **(a)** with and **(b)** without the effects of buildings near the release incorporated (Modified after Brown, 2004).

wind measurements from weather stations to determine the **boundary conditions** or they are informed by coarser-scale weather forecast models. That way they can predict the flow field a few hours into the future, then in case of an emergency the user can insert a suitable source and quickly determine its spread in the pre-calculated wind field. To overcome the computational load (and time required) to model the details of the flow around buildings, some schemes use coarse resolution models or run a suite of simulations under different meteorological conditions in advance and later use databases to interpolate from the pre-simulated cases (Brown, 2004).

Another application of urban-scale dispersion models is to allow authorities to be better prepared in the event of an accidental or intentional release of harmful substances. For this purpose, urban dispersion models can be run in a 'backward mode' to determine areas where a potential release of chemical, biological and radioactive material might affect people, ecosystems or critical infrastructure. In doing this the plume spread is inverted in time (but airflow is not) and so a model can predict risk zones, i.e. areas where a potential release would effectively impact these sensitive targets in a matter of minutes. Policing and surveillance can consequently focus and block-off the main risk areas during large sporting events, other large gatherings of people and protect sensitive infrastructure in cities.

### 11.2.3 Air Pollution from Elevated Point Sources

Above the UCL, most of the multitude of emission sources from the urban landscape can be treated as if they arise from lines (e.g. top of canyons or highways) or areas (e.g. dwellings and vegetation). However, there are also large point sources that can inject air pollutants directly into the **surface layer** above the UCL. These are mainly tall chimneys associated with industrial operations or power plants. The height of the stack is designed to ensure that contaminants are diluted by mixing in the UBL and extreme concentrations in the UCL are prevented.

To model elevated point sources well above mean building height ($z_H$), the specific arrangement of buildings and roughness elements at the urban surface can be neglected and the **urban canopy** can be treated simply as an extended rough surface and parameterized with an integral aerodynamic **roughness length** $z_0$ and **zero-plane displacement** $z_d$ (Section 4.3.2). It is useful to discuss this type of pollution here first, using a simple Gaussian plume model, because it illustrates how atmospheric processes govern the dispersion and dilution of contaminants. This is followed by discussion of urban-specific effects on plume dispersal.

#### Gaussian Plume Models

These are deterministic models that are applied to 'ideal' situations where there are continuous emissions from a point source over extensive, uniform roughness where the plume contents are usually inert. Although initially developed for open flat terrain, they can be modified for urban situations and even incorporate some UCL effects. Gaussian plume models predict the steady-state time-averaged concentration $\overline{\chi}$ at a location (given by the co-ordinates $x, y, z$) downwind from a point source. While they can assess the average concentration over a long time period, they cannot predict peak concentrations experienced at shorter, turbulent time scales. Under favourable conditions, they apply to distances in the range from a few hundred metres to a few kilometres downwind from an emission source.

Figure 11.12 illustrates the Gaussian plume model concept. A plume emerges from an elevated point source (a stack) that is continuously emitting air pollutants at a constant rate $E$ ($\mu g\ s^{-1}$). As the pollutants enter the urban atmosphere, they form a plume that is advected by the mean wind and mixed by turbulence, which regulates its spread. If one were to take a cross-section of the plume at different distances from the source, it would show that the mean pollutant concentration, averaged over a considerable time (to eliminate the random nature of turbulence), is greatest in the centre of the plume and diminishes towards its edges. This 'bell-shaped' distribution is well described by the Normal or Gaussian probability distribution in both the horizontal and vertical, although at different rates

in the two directions. The shape emerges as the core of the plume is gradually mixed by turbulence, causing the peak concentration to fall as the cross-sectional area of the plume increases. The mathematical description of such functions allows us to model the dispersion and calculate the mean concentration outdoors $\overline{\chi}$ (in $\mu g\ m^{-3}$) of an air pollutant at any point $(x, y, z)$ downwind of a stack with height $H$ (in m):

$$\overline{\chi}(x, y, z) = \frac{E}{2\pi\hat{\sigma}_y\hat{\sigma}_z\overline{u}} \exp\left(-\frac{y^2}{2\hat{\sigma}_y^2}\right)\left[\exp\left(-\frac{(z - H)^2}{2\hat{\sigma}_z^2}\right)\right.$$

$$\left. + \exp\left(-\frac{(z + H)^2}{2\hat{\sigma}_z^2}\right)\right] \quad \left(\mu g\ m^{-3}\right)$$

**Equation 11.7**

where $\overline{u}$ is the mean wind speed (m s$^{-1}$), here assumed constant across the depth of the plume. The spread of the plume is regulated by $\hat{\sigma}_y$ and $\hat{\sigma}_z$ which are the lateral and vertical standard deviations of the concentration field, respectively, which increase with distance from the source. They can be related to the lateral ($\sigma_v$) and vertical ($\sigma_w$) standard deviations of turbulent fluctuations (in m s$^{-1}$). For example near the source (< approximately 1 km), the plume spreads linearly with distance $x$:

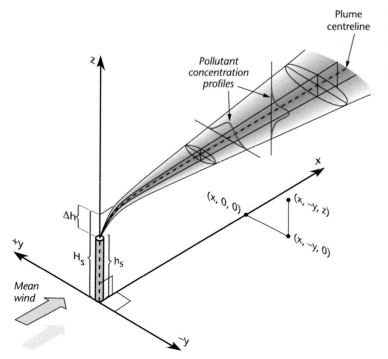

Plume centreline

Pollutant concentration profiles

**Figure 11.12** Air pollutant dispersion from an elevated point source, modelled using a Gaussian dispersion model (Modified after: Turner, 1970).

$$\hat{\sigma}_y \approx \frac{x\,\sigma_v}{\overline{u}} \text{ and } \hat{\sigma}_z \approx \frac{x\,\sigma_w}{\overline{u}} \qquad \textbf{Equation 11.8}$$

Far from the source (> approximately 1 km), the plume spread changes to square-root with distance. The last term on the right-hand side of Equation 11.7 is included to account for the increased concentration at positions downwind of the point at which the plume has already 'touched' the ground. In this formulation it is assumed that all of the pollutant is folded back up into the atmosphere and none is deposited, an assumption called 'eddy reflection'.

In most cases, we are interested in concentrations at ground level, where people live. For this, Equation 11.7 can be simplified to predict concentrations at ground level (i.e. $z = 0$).

$$\overline{\chi}(x, y, 0) = \frac{E}{\pi \hat{\sigma}_y \hat{\sigma}_z \overline{u}} \exp\left(-\left(\frac{y^2}{2\hat{\sigma}_y^2} + \frac{H^2}{2\hat{\sigma}_z^2}\right)\right) \quad (\mu g\ m^{-3})$$

$$\textbf{Equation 11.9}$$

It is worth pointing out some of the controls on the predicted pollution concentration at any location in these formulations. First, $\overline{\chi}$ at any point is directly proportional to the source strength (i.e. $\overline{\chi} \propto E$) and inversely related to the mean wind speed (i.e. $\overline{\chi} \propto \overline{u}^{-1}$, see Figure 11.3). Second, $\overline{\chi}$ is inversely related to the TKE (i.e. $\overline{\chi} \propto (\hat{\sigma}_y \hat{\sigma}_z)^{-1}$, which is controlled by stability). Instability encourages vertical spread, whereas stable conditions have a restraining influence. Third, $\overline{\chi}$ at ground level at a given distance downwind is decreased by raising the effective stack height $H$. Other things being equal, higher point sources cause lower concentrations at the ground downwind, because turbulent mixing will have had longer to dilute the plume contents. Tall stacks are therefore usually of help in combating poor air quality near the point of release, although the total mass emitted does not change. The effective height of release can also be increased if the effluent emerges at a high velocity, and at a temperature well above that of the environmental air temperature so that the plume possesses **buoyancy** and rises higher. The plume then ascends well above the stack exit before bending-over and proceeding downwind.

Generally, Gaussian dispersion models try to account for urban effects by adjusting $\hat{\sigma}_y$ and $\hat{\sigma}_z$. Compared to typical **rural** areas, $\hat{\sigma}_y$ is about two times larger, and $\hat{\sigma}_z$ is 2 to 5 times enhanced over a city. Many Gaussian dispersion models include specific urban parameters and may have modifications to represent urban effects on flow and dispersion. Processes considered are lateral shifting of the plume centreline due to plume channelling along the dominant canyon axes in relation to mean wind speed (Theurer et al., 1996), simple building and **downwash** effects, and traffic-produced turbulence (Section 4.2.3). Some models allow users to input building dimensions to calculate the trapping of air pollutants in building **wakes** and urban canyons.

## Plume Dispersal Characteristics in the Urban Boundary Layer

At the scale of plumes from elevated sources, cities are rarely homogeneous and changes in roughness and thermal surface characteristics are the norm. Figure 11.13 illustrates the behaviour of various plumes that are within, or cross transitions between, rural and UBLs. At the left a rural stack plume is emitted into a stable nocturnal **inversion** layer, with very little dispersive capability in the vertical. It can stay as a narrow concentrated 'pipe-like' plume with little dilution in any direction or it can oscillate lazily sideways forming a 'fan-shaped' thin sheet (plume form: fanning). Upon entering the UBL the same plume encounters greater turbulent activity, due to the UHI and increased roughness. As a result it is rapidly mixed, especially down towards the RSL and surface, it fumigates the UBL (plume form: fumigation). An elevated plume emitted fully within the UBL usually enters an approximately neutral, well-mixed layer and disperses roughly equally well vertically and horizontally (plume form: coning). At the right, downwind of the city, an elevated source is

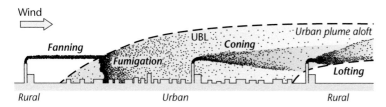

**Figure 11.13** Plume behaviour after emission from elevated point sources (stacks) in the vicinity of and within cities on nights with strong rural stability (little cloud and weak airflow) (Modified after: Oke 1987).

able to disperse easily in the lower part of the UBL but is largely prevented from mixing down due to the newly developing rural stable layer (plume form: lofting). For surface-based receptors the fanning case is temporarily good, but in the morning the surface inversion is eroded and fumigation happens. For as long as it lasts, which often is not very long, the lofting plume form is best for surface-based receptors such as people.

## 11.3 Urban-Scale Air Pollution

At the scale of a city with its many emission sources individual smaller plumes lose their identity and contribute to a more general contamination of the whole UBL. The efficiency of turbulence at this scale produces a relatively homogeneous mélange of contaminants that fill the depth of the ML.

Table 11.3 lists typical long-term (annual) average concentrations of major air pollutants in remote, rural, and urban atmospheres together with current annual and short-term guideline values for each air pollutant as set by the WHO. WHO guidelines offer advice to authorities in charge of air quality to protect human health, they are appropriate for all global regions. Individual countries can set their own standard values for management and enforcement. The concentration of air pollutants in the UBL of cities is substantially greater than in the ABL of remote areas. What these figures do not capture is the temporal variation that is largely a result of changes in the state of the atmosphere and in rates of emission (winter vs summer, weekends vs weekdays, and diurnal **profiles** of traffic and human activity). The following section

first examines controls on urban-scale air pollution before considering the relevant chemical processes that lead to secondary pollutants. Finally, the role of urban-scale air pollution modelling as an operational forecasting and management tool is considered.

### 11.3.1 Meteorological Controls on Air Quality in the Urban Boundary Layer

For illustrative purposes, we formulate a simple model to predict the average concentration $\bar{\chi}$ (in μg $m^{-3}$) of an air pollutant in the UBL (Figure 11.14). Here the UBL is divided into boxes arranged sequentially along the path of the ambient wind. The top of the boxes corresponds to the height of the ML ($z_i$) (Section 2.2.1). The urban surface emits air pollutants from below into the boxes at an average flux density $F_p$ at the surface (μg $m^{-2}$ $s^{-1}$). Pollutant removal from the volume is possible via settling and deposition at the surface (Section 11.1.3), through chemical reactions, or through **advection** by the mean wind (m $s^{-1}$) averaged over the depth $z_i$. For simplicity, removal and chemical reactions are neglected so the conceptual model applies only to relatively unreactive air pollutants that do not settle or are deposited in the urban area.

Further — to simplify discussion — air entering the **domain** from the left is assumed to be clean, the pollutant is thoroughly mixed over the depth $z_i$. Lateral mixing, perpendicular to the mean flow direction, is considered to be instantaneous. In this situation we can write that any distance $\Delta x$ (m) from the upwind boundary of the city the concentration of an air pollutant $\bar{\chi}_{(x)}$ is given from:

**Table 11.3** Typical long-term (annual) average concentrations and guidelines for annual concentrations and short-term maxima of major urban air pollutants (Sources: WHO, 2006, GAW – Global Atmosphere Watch Program Database).

| Pollutant | Remote background ABL | Typical rural ABL | Typical UBL (μg $m^{-3}$) | WHO annual guideline | WHO short-term guideline |
|---|---|---|---|---|---|
| **CO** | 60–150 μg $m^{-3}$ | 150–500 μg $m^{-3}$ | 200–1,500 μg $m^{-3}$ | - | - |
| **NO$_x$** | ~ 3 μg $m^{-3}$ | 3–40 μg $m^{-3}$ | 20–200 μg $m^{-3}$ | 40 μg $m^{-3}$ | 200 μg $m^{-3}$ (1-h mean) |
| **O$_3$** | 40–70 μg $m^{-3}$ | 30–100 μg $m^{-3}$ | 20–90 μg $m^{-3}$ | - | 100 μg $m^{-3}$ (8-h mean) |
| **PM2.5** | | | 5–30 μg $m^{-3}$ | 10 μg $m^{-3}$ | 25 μg $m^{-3}$ (24-h mean) |
| **PM10** | | | 10–50 μg $m^{-3}$ | 20 μg $m^{-3}$ | 50 μg $m^{-3}$ (24-h mean) |
| **SO$_2$** | 0–0.5 ng $m^{-3}$ | 0–20 μg $m^{-3}$ | 0–50 μg $m^{-3}$ | - | 20 μg $m^{-3}$ (24-h mean) |

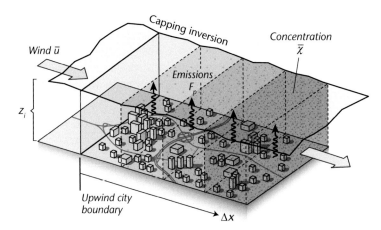

Concentration
$\overline{\chi}$

**Figure 11.14** A simple model of air pollutant concentration in the UBL as a function of distance downwind from the city boundary $\Delta x$, emission strength $E$, mean wind speed and height of the UBL $z_i$ (Modified after: Oke 1987).

$$\overline{\chi}_{(x)} = \frac{F_p \Delta x}{\overline{u} z_i} \qquad \left(\mu g\ m^{-3}\right) \qquad \textbf{Equation 11.10}$$

This relationship, although very simplified, reveals a number of fundamental dependencies and suggests potential air quality management options:

- The stronger the flux of air pollutants at the surface ($F_p$, i.e. rate of emissions per unit area), the greater the resulting $\overline{\chi}$ in the UBL. Although obvious, this reminds us that emission reduction is the single most effective management action to improve urban air quality, and it represents a direct approach to addressing the problem at the source.
- The concentration is inversely related to the depth of the UBL ($z_i$). When $z_i$ is small, it defines a shallow UBL so $F_p$ is mixed into a smaller volume and $\overline{\chi}$ increases. This depth cannot be regulated, but $z_i$ is generally larger during daytime, and lower at night. Further, $z_i$ is larger over cities than rural areas owing to increased roughness (Chapter 4) and the UHI (Chapter 7). This should benefit urban air quality.
- $\overline{\chi}$ increases linearly with distance of travel ($\Delta x$) across the urban area as the wind advects air pollutants from upwind sources resulting in higher concentrations downwind. This implies that to reduce population exposure to degraded air quality in the city, additional polluting industries and infrastructure should, in most scenarios, be sited on the dominant downwind side of the urban area. However, if the less frequent flow from the opposite direction typically coincides with strong stability and low $z_i$, this can cause industrial emissions to flow back into

the city and cause unfavourable concentrations (see also Section 15.4.3).
- $\overline{\chi}$ is inversely related to mean wind speed ($\overline{u}$). Higher $\overline{u}$ means greater flushing and lower concentrations. This is analogous to the case of a single plume, as illustrated in Figure 11.3. Although over the depth of the UBL $\overline{\chi}$ cannot be controlled, there are two management implications: (i) $\overline{\chi}$ will be worse when $\overline{u}$ is small, and (ii) cities situated in wind-sheltered topographies (valleys, mountain basins) are generally more prone to periods of high $\overline{\chi}$ compared to those exposed to strong winds. There are also design considerations at the local- to mesoscale to maintain higher winds and disperse air pollutants.

The **ventilation factor** ($V_f = \overline{u} z_i$) combines the two atmospheric controls in Equation 11.10 and provides a simple means to assess the capacity of the box to disperse and dilute emissions. Generally, periods of poor air quality occur when $V_f$ is smallest, and emissions $E$ are largest. $V_f$ tends to exhibit strong diurnal and seasonal variations related to UBL development and **synoptic** meteorological controls. Other things being equal, $\overline{u}$ is characteristically stronger in daytime and so is $z_i$, due to convective growth of the ML (Figure 2.12). So, the components of $V_f$ are correlated such that the best environment for dispersion usually occurs by day and the poorest by night.

### 11.3.2 Smog

**Smog** refers generally to the phenomena of degraded air quality on urban to regional scales. The air is visually changed appearing variously dirty, hazy, discoloured and may have a noticeable odour. Periods when smog persists and air quality in the UBL progressively deteriorates, to the point where it becomes a public health issue, are termed 'air pollution episodes'. There are two classes of smog.

## Sulphur-Based Smog

The term 'smog' was originally coined to describe a mixture of smoke and fog that characterizes the UBL in cities with high PM and $SO_2$ emissions derived from the burning of coal, solid fuels and sulphurized petroleum. This sulphur-based smog is characteristic of urban systems with a high density of emissions from open fires, furnaces and uncontrolled emissions from industry and power generation that use these fuels. In a very humid atmosphere particulates grow in size by absorption of vapour and some form CCN on which droplets form. The primary pollutant $SO_2$ is absorbed into droplets, where it oxidizes to sulphur trioxide ($SO_3$), which reacts with water vapour in the presence of catalysts to form sulphuric acid ($H_2SO_4$). Since acids have an affinity for water they allow droplets to grow further hygroscopically and cause haze. Conditions ideal for the formation of sulphur-based smog occur in mid- to high latitudes in winter when emissions for space heating are substantial.

The London, United Kingdom smog episode in December 1952 is a case in point. Under cold and stable weather, fuel consumed for space heating resulted in high emissions of $SO_2$ and PM. Chimney smoke from myriad domestic fireplaces, mainly burning coal, fumigated the UCL across the city. Such was the density of the smog that sunshine could not penetrate to warm the underlying surface, inhibiting mixing and growth of the ML by thermal turbulence – which along with weak winds caused a very low $V_f$. The air pollutant concentrations exceeded modern standards by an order of magnitude for several days and resulted in 12,000 'excess' deaths (Bell and Davies, 2001) (i.e. when compared with the expected number of deaths in this period).

Such dramatic air pollution episodes have played an important role in the history of air pollution control. They provided the rationale for intensive investigation of air pollution meteorology and chemistry, which had hitherto received little attention. Also, they are often the inspiration for the legislation that underpins air quality management. Emission control and switching fuel types are the most successful

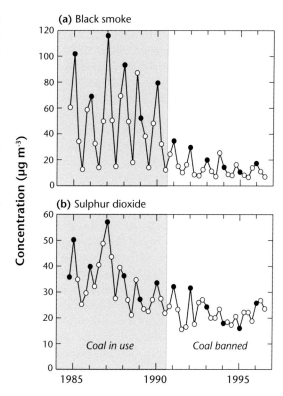

**Figure 11.15** Seasonal mean black smoke concentrations ($\mu$g m$^{-3}$) from September 1984 to 1996 in Dublin, Ireland. The shaded area indicates times when the marketing and sale of poor quality coal was still permitted, while the white area is after its ban. Black circles represent winter data (Source: Clancy et al., 2002; Reprinted from *The Lancet*, with permission from Elsevier).

management strategies for sulphur-based smog. Figure 11.15 shows seasonal average concentrations of $SO_2$ and PM measured in Dublin, Ireland, over a twelve-year period. In the first half of this series, the use of poor quality bituminous coal to heat homes that was still widely common in the high winter season caused frequent sulphur-based smog episodes. In 1990, the city banned the marketing, sale and distribution of bituminous coal, which saw a dramatic improvement in air quality as it did in London. An epidemiological analysis of death rates indicates about 243 fewer cardiovascular and 116 fewer respiratory deaths per year after the ban on coal sales (Clancy et al., 2002). Even today, sulphur-based smogs remain common in many cities where coal is a primary fuel.

## Photochemical Smog

Serious air quality episodes have plagued the metropolitan area of Los Angeles, United States, since the 1940s. At first, industrial sources were blamed, and the air pollution was attributed to sulphur-compounds, but in the early 1950s, chemist Arie Haagen-Smit showed photochemical processes are relevant when emissions from gasoline combustion are converted into the secondary pollutant ozone ($O_3$). This **photochemical smog** needs strong sunlight for certain chemical reactions to occur. Photochemical smogs occur in many urban areas, but because of the requirement of strong **irradiance** their occurrence outside the tropics tends to be restricted to the summer season. Under typical anticyclonic conditions with clear skies that characterize photochemical smogs, air quality is further degraded because low wind speeds and **subsidence** inversions cause a low $V_f$. The net result is a hazy atmosphere with a brownish colour, greatly reduced visibility and distinctive odour. Photochemical smog can irritate the eyes and the respiratory tract (e.g. reduce lung capacity, aggravate asthma) and cause plant damage (e.g. injure tissue and reduce yield).

The principal emissions responsible for the formation of $O_3$ in photochemical smog are $NO_x$ and VOCs. Internal combustion engines using gasoline are the primary source of these emissions. VOCs are also emitted through fugitive and evaporative emissions from solvent use and from the biosphere (Section 11.1.1).

The formation of ozone from $NO_x$ in the UBL is governed by the cycle in Figure 11.16 in the shaded box of the top part. The energy supplied by photons of UV radiation can photodissociate $NO_2$ into nitric oxide (NO) and an atomic oxygen radical (O):

$$NO_2 + \hbar v \xrightarrow{j_1} NO + O. \qquad (R1)$$

Here, $\hbar v$ is the energy of a photon, where $\hbar$ is the Planck constant and $v$ is its wave frequency (in $s^{-1}$). The photolytic rate $j_1$ of reaction R1 (in $s^{-1}$) is directly proportional to the intensity of UV radiation in the wavelength range 0.37–0.42 μm. R1 is hence limited to daytime situations, and $j_1$ is usually strongest at noon. The oxygen radical resulting from R1 is highly reactive and combines immediately with the ambient molecular oxygen ($O_2$) to form ozone ($O_3$):

$$O + O_2 + M \xrightarrow{k_2} O_3 + M \qquad (R2)$$

where $M$ is another molecule involved in absorbing some of the surplus energy (not in Figure 11.16); in most cases this is simply the abundant $O_2$ or $N_2$. The $O_3$ formed then reacts with NO to yield again $NO_2$ and $O_2$:

$$O_3 + NO \xrightarrow{k_3} NO_2 + O_2 \qquad (R3)$$

$k_2$ and $k_3$ are the kinetic rates for $O_3$ formation and $O_3$ destruction, respectively (in $s^{-1}$). The cycle described by reactions R1 to R3 leads to a theoretical equilibrium between $NO_2$, NO and $O_3$, which is a photolytic steady-state called the Leighton relationship:

$$[O_3] = \frac{j_1[NO_2]}{k_3[NO]} \qquad (\mu mol\ mol^{-1}) \qquad \textbf{Equation 11.11}$$

The Leighton relationship results from the assumption that $O_3$ and NO are formed and destroyed continuously with no *net* production over time. The ratio of $[NO_2]$ to $[NO]$[1] sets a limit on $[O_3]$ and is therefore a key parameter to assess ozone formation potential.

Predicted urban $[O_3]$ based on the Leighton relationship would yield only ~0.01 μg m$^{-3}$ under typical urban situations, but observations show values can reach 100 μg m$^{-3}$ (see Figure 11.18). Such high concentrations cannot be explained by the Leighton relationship alone, they are due to a complementary set of reactions that convert NO to $NO_2$ without depleting $O_3$. This is enabled by the simultaneous presence of VOCs in urban air. Select VOCs can oxidize NO to $NO_2$ hence leaving the $O_3$ previously formed by R1 and R2 unaffected and reduce the probability of R3. This is illustrated by the set of reactions in the lower part of Figure 11.16 (in red) which can run several times (hence they are a catalytic cycle) and replenish the $NO_2$ reservoir without destroying an equivalent amount of $O_3$.

For example, in reaction R4, a methyl radical ($CH_3$) is formed from the simplest hydrocarbon, methane ($CH_4$), reacting with the hydroxyl radical (OH). The methyl radical then reacts with oxygen in R5 and the resulting methyldioxy radical can oxidize NO to $NO_2$ and form a methoxide radical (R6). That radical then reacts with $O_2$ (R7) to form two hydroperoxyl radicals ($HO_2$) and formaldehyde (HCHO).

---

[1] Square brackets are a short-hand notation for the concentration of the pollutant.

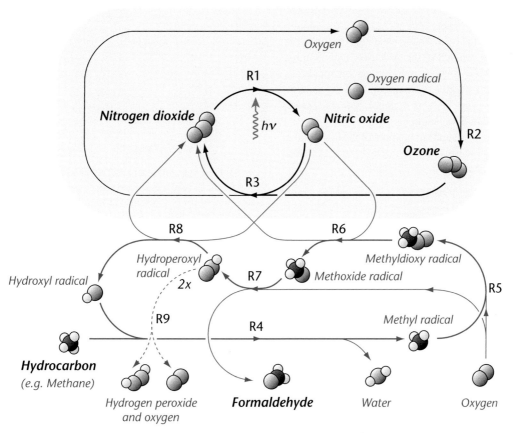

**Figure 11.16** Selected chemical reactions (R1 to R7) relevant in photochemical smog formation that lead to high $O_3$ concentrations. Catalytic cycles converting hydrocarbons (bottom, in grey) can interrupt the steady-state Leighton relationship between $NO_2$, NO and $O_3$ (in the light blue box). Reactions R6 and R8 lead to destruction of NO without involvement of $O_3$ and consequently cause a build-up of $O_3$. Although the cycle is illustrated using methane ($CH_4$), it can be substituted by other hydrocarbons.

At that point the reactions can follow one of two paths. If R8 occurs, $HO_2$ will oxidize another NO molecule and produce a hydroxyl radical (OH):

$$HO_2 + NO \rightarrow OH + NO_2 \qquad (R8)$$

that in turn reacts with another hydrocarbon (through R4) thus closing the catalytic cycle and allowing R4 to R8 to repeat. The net effect of this catalytic cycle is to destroy hydrocarbons while creating formaldehyde that increases its concentration in photochemical smog and oxidizing NO:

$$CH_4 + 2O_2 + 2NO \rightarrow H_2O + 2NO_2 + HCHO$$
$$(R4\text{-}R8)$$

Increasing concentrations of formaldehyde in urban air, as a result of this cycle, are partly responsible for the distinctive odour of photochemical smog. Note that R4 to R8 do not consume OH, rather OH is 'recycled' – it is the catalyst, and none of these reactions require UV radiation, so they can occur day and night.

Alternatively, the two $HO_2$ molecules, formed in R7, can combine via R9 to create hydrogen peroxide ($H_2O_2$) and oxygen:

$$HO_2 + HO_2 \rightarrow H_2O_2 + O_2. \qquad (R9)$$

This breaks the catalytic cycle and terminates the role of the hydrocarbon molecule in ozone formation.

Figure 11.16 shows only one possible pathway, based on the methyl radical ($CH_3$). Similar reaction cycles are possible with longer and more complex hydrocarbon radicals. A relevant case is the acetyl radical ($CH_3CO$), which can fuel the cycle with additional $CH_3$ and form the peroxyacetyl radical ($CH_3COO_2$) that reacts with NO to form the vapour of peroxyacetyl nitrate (PAN), which also builds up in urban air as a product of the cycle over time. Similar to formaldehyde, PAN is an important compound in photochemical smog accompanying high $O_3$ concentrations. Like $O_3$, PAN is a major eye and throat irritant and damages plants.

In summary, to build-up high concentrations of the secondary pollutants $O_3$, formaldehyde and PAN and in the UBL we need:

- **High [$NO_x$],** derived primarily from vehicle emissions in urban air.
- **Reactive VOCs** – also emitted from vehicles plus evaporative emissions that are temperature dependent from other anthropogenic and biogenic sources.
- Sufficient **high-energy UV radiation** to enable a substantial $j$ in R1, this is typically found in **anticyclones** with ample solar irradiance.

Only if all three prerequisites are fulfilled, will the secondary pollutants $O_3$, formaldehyde and PAN be formed. Photochemical smog is hence characteristic for urban areas with $NO_x$ (primarily vehicles) *and* VOC emissions (vehicles, fugitive and biogenic emissions) that build-up under anticyclonic conditions with ample irradiance under weak winds.

## Temporal Dynamics of Photochemical Smog

The daily change in both emissions due to human activity (e.g. traffic patterns) and of solar UV radiation combine to create characteristic traces of both the primary and secondary pollutants involved in photochemical smog. The typical course of measured air pollutant concentrations over a summer day is illustrated in Figure 11.17 in the city of Essen, Germany. The early morning rush-hour adds [$NO_x$], in particular [NO] which results in a molar [$NO_2$] / [NO] ratio of ~1. The abundant [NO] means the net formation of [$O_3$] remains low during rush hours. With increasing **shortwave** irradiance, [$O_3$] forms at higher rates and when emissions cease in the later morning, the reservoir of [NO] is used up quickly. By about 1100 h, the build-up accelerates and high [$O_3$] is present in the urban atmosphere during the afternoon, while primary pollutants are lower. The afternoon is also the time of highest surface and air temperature, so anthropogenic and biogenic fugitive emissions of VOCs, that are temperature dependent, peak. A second burst of $NO_x$ emissions is caused by

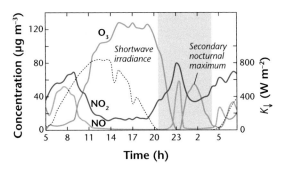

**Figure 11.17** Traces of shortwave irradiance, NO, $NO_2$ and $O_3$ concentrations measured at an urban background station in Essen, Germany on June 26 and 27, 1995 (Source: Kuttler and Strassburger, 1999; © Elsevier Science Ltd., used with permission).

the evening rush-hour, but this is not clearly visible in Figure 11.17 because of the strong mixing of the UBL at that time. The onset of darkness inhibits further formation so [$O_3$] decays in the evening and is completely destroyed by 2300 h. Interestingly, there is a secondary nocturnal maximum of [$O_3$] at 0200 h, associated with an increase in wind and mixing. Such nocturnal secondary maxima are a common feature and cannot be explained by chemical formation, because there is no shortwave irradiance to drive R1. Instead, such secondary maxima are due to air masses still rich in [$O_3$] that are transported back into the city and down to the surface. These nighttime increases are generally attributed either to mixing of [$O_3$] rich air from the **residual layer** downwards, when the nocturnal inversion temporarily breaks down, or by advection in from rural areas outside the city, by the **urban heat island circulation** (UHIC, Section 4.4). In both cases, the rural boundary layer and the residual layer generally have a higher [$NO_2$] to [NO] ratio, which causes [$O_3$] to be less quickly depleted overnight.

## Intra-Urban Patterns of Photochemical Smog

Figure 11.18 shows midday [NO], [$NO_2$] and [$O_3$] measured by a mobile laboratory that traversed through different neighbourhoods of Essen. The values are averaged over many summer days. Data has been grouped into that measured along motorways, main and secondary roads, and in residential and green areas. Along motorways abundant primary pollutants inhibit $O_3$ accumulation. There, [$NO_x$] is

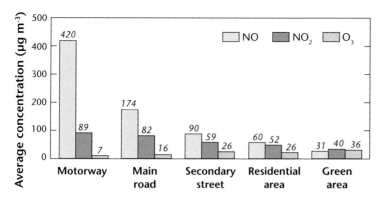

**Figure 11.18** Average concentrations of NO, NO$_2$ and O$_3$ measured with a mobile laboratory in various parts of the city of Essen, Germany. The values are averages over six different traverses (Source: Kuttler and Strassburger, 1999; © Elsevier Science Ltd., used with permission).

high and the molar [NO$_2$] / [NO] ratio is low (~0.1), which means that [O$_3$] once formed is immediately depleted by [NO] (Equation 11.11). In places where primary pollutant emissions are lower, for example along less busy secondary roads, the higher molar [NO$_2$] / [NO] ratio (~0.4) and plentiful [VOC] causes [O$_3$] to climb, yet highest [O$_3$] is found in residential and green areas of the city where molar [NO$_2$] / [NO] is ~0.7. On regional scales, photochemical smog often causes [O$_3$] to peak in rural areas several 100s of kilometres downwind of cities. These regional patterns in the urban plume are discussed in more detail in Section 11.4.1.

## Implications for Air Quality Management

The reaction pathway taken after R7 (that is, whether R8 or R9 dominates), is an important 'switch' in the cycle, and depends on the concentration of NO. If there is a plentiful supply of [NO] but little [VOC], R8 is more likely to happen, the cycle continues and more ozone is formed. In the reverse situation where there is little [NO] but plentiful [VOC] there are more HO$_2$ radicals in the atmosphere, R9 dominates and the cycle terminates. Other reactions, such as between HO$_2$ and the methyldioxy radical, also end the reaction sequence. These 'side tracks' have implications for regulating O$_3$ concentration, which is a function of both VOC and NO$_x$ concentrations (Figure 11.19). Generally, with higher VOC and NO$_x$ concentrations, concentrations of O$_3$ increase. The maximum [O$_3$] concentration that can be reached, however, depends on the ratio of [VOC] to [NO$_x$]. At intermediate ratios (4:1 to 10:1) conditions for O$_3$ formation are most favourable. A closer look at Figure 11.19 shows two particular régimes, depending on the ratio [VOC] to

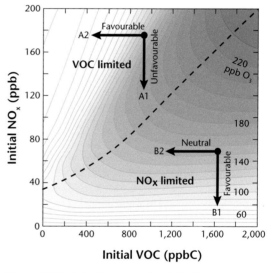

**Figure 11.19** A typical ozone isopleth diagram showing the maximum urban ozone mixing ratio (isolines in ppb) as a function of initial VOC mixing ratios (x-axis in ppb carbon) and initial total NO$_x$ (NO + NO$_2$ in ppb). The dashed line separates the NO$_x$ limited régime (lower right) from the VOC limited régime (upper left). For trajectories A1, A2, B1 and B2 see text (Modified after: Seinfeld and Pandis, 2006 and Sportisse, 2009).

[NO$_x$], that are relevant to the management of O$_3$ in urban environments:

• **VOC limited régime** (also called: high [NO$_x$] régime): The catalytic cycle R4 to R8 flourishes, consuming some of the abundant NO and hence inhibiting partially the destruction of O$_3$. The high availability of [NO$_x$] means R9 is unlikely to happen. [$O_3$] is most sensitive to the availability of VOCs. Higher VOC emissions would mean more $O_3$, while increasing $NO_x$ emissions alone would not change much.

- **$NO_x$ limited régime** (also called: low [$NO_x$] régime): The lack of NO means reaction R9 is more likely to happen, and the catalytic cycle of the VOCs is less effective. At the same time the limited [$NO_x$] means little $O_3$ destruction and production by reactions R1 to R3. [$O_3$] is most sensitive to the availability of $NO_x$. Increasing $NO_x$ emissions would therefore cause more $O_3$, while higher VOC emissions alone would not yield much change to $O_3$ concentrations.

The UBL above city centres tends to have a VOC limited régime because there are high [$NO_x$] emissions from traffic. Here a reduction of $NO_x$ alone may actually worsen $O_3$ pollution (path A1, in Figure 11.19), and a reduction of VOC, or both $NO_x$ *and* VOC in the city is the proper action to reduce [$O_3$] (path A2). But reducing VOC emissions is difficult to implement because there are multiple small-scale sources (e.g. solvents, paints, trees) that are not as easily regulated as vehicles, which are the main source of $NO_x$.

Often the highest [$O_3$] are found in rural areas downwind of cities (Figure 11.22), which are $NO_x$ limited régimes, with ample VOCs (partially derived from biogenic emissions). Here, reducing emissions of $NO_x$ upwind will often lower [$O_3$] (path B1, in Figure 11.19). Reducing VOC emissions alone will have little effect (path B2).

This situation shows the counter-intuitive effect of certain actions: reducing $NO_x$ emission (e.g. by traffic restrictions) benefits communities downwind, but may worsen conditions in the city centre if not accompanied by a simultaneous decrease in VOC emissions. The specific situation depends on the regional emission composition, climate and wind field – so management actions to reduce $O_3$ do not necessarily affect all parts of an urban region the same way and need to be planned carefully. Hence, management of air quality involving photochemical smog is usually guided by numerical model simulations.

### 11.3.3 Modelling Urban Air Pollution

Numerical air quality models at urban- to regional scales are necessary tools to forecast air quality and to guide emission management strategies. Generally such models combine a representation of atmospheric dynamics, atmospheric chemistry and urban emission inventories in a physically-based form to predict air pollutant concentrations in space and time.

### Modelling Atmospheric Dynamics

Gaussian plume models (Section 11.2.3) do not take into account the complex flow field and spectrum of time scales involved in urban-scale air pollution. **Eulerian** and/or **Lagrangian** approaches (Section 3.3.1) are needed to model the dispersion and trans-formation of emissions at urban scales. An early attempt to run time-dependent 3-D air quality fore-casts for a city was the pioneering 'Urban Airshed Model' developed for the Los Angeles Metropolitan Area in the early 1970s (Scheffe and Morris, 1993). The model solved atmospheric dispersion in a simple Eulerian grid, using conservation equations for several air pollutants and selected key chemical transformations. Today, a large number of such **photochemical grid models** (PGM) are available, and at much finer resolution, and with improved sophistication. PGMs can model airflow and pollution dispersal not only for a domain of a single city, but for entire urbanized regions and continents. PGMs also account for radiation exchanges using the specific wavelength bands needed for photochemical reactions, and predict temperatures that control meteorological processes and the rate of chemical reactions. PGMs are used in prognostic modes to forecast pollutant concentrations, to inform regulatory measures (i.e. identify effective emission reduction scenarios) and for research purposes.

### Modelling Atmospheric Chemistry

For computational reasons, it is not possible to incorporate the thousands of reactions of each possible air pollutant. In particular the treatment of the large variety of VOC species is a challenge in PGMs. Therefore, VOCs are often grouped into simple classes of similar reactivity, which are usually olefins, paraffins, aldehydes and aromatics. These classes still require several hundred chemical equations to be solved at each grid node and time step of the model. The corresponding kinetic and photolytic rates for individual reactions are based on many laboratory chamber measurements and are functions of temperature and radiation that are taken from the dynamical part of the model.

## Emission Inventories

The description of emissions are the single most sensitive input to modern PGMs. Emission inventories are geographical databases that describe the location, timing, quantity and chemical composition of primary pollutants emitted in a city or region. The inventories are coupled to the model as one of the lower boundary conditions. Emission inventories are needed to model primary pollutant concentrations and therefore are also critical to the accurate prediction of secondary pollutants.

For convenience, we should first categorize emissions into three major physical processes: (i) combustion and industrial; (ii) evaporative and fugitive emissions, and; (iii) biogenic emissions (Section 11.1.1). Combustion sources are further separated into: traffic (road, rail, ship, air); domestic and commercial buildings due to space heating and cooling, hot water demand and appliances, and industry and power generation. The emissions from each of these sources follow distinctive diurnal, weekly and seasonal patterns associated with the rhythms of the **urban metabolism** (e.g. commuting to and from work, heating and cooling at work and home and economic production cycles). Sources of evaporative and fugitive emissions (as well as biological emissions) are more diffuse in nature and are controlled in some cases by temperature and other environmental or technological controls. In particular, evaporative and biogenic VOC emissions are difficult to estimate.

Urban-scale emission inventories are constructed based on top-down or bottom-up approaches. These are directly analogous to the means by which the **anthropogenic heat flux** ($Q_F$) is estimated (Section 6.2). The top-down approach uses fuel consumption statistics available at coarse (e.g. national or regional) scales and downscales them to a specific urban area. Because the exact location of fuel combustion is not known or is not accessible, these data are distributed spatially and temporally using proxy data, such as population density. Similarly, yearly data may be decomposed to a finer temporal scale using traffic counts, or seasonally using **heating degree-days** (HDD). The strength of the top-down modelling approach is that total emissions over longer time scales are quite robust. However, there is uncertainty about downscaling, i.e. when and where emissions are released in a region or within an urban area.

The bottom-up approach derives emissions from knowledge about a small number of typical urban elements such as buildings, vehicles or trees. Emissions from those generic 'elements' of the **urban metabolism** can be modelled using sub-models such as building energy models, traffic flow models or plant physiological models. Total emissions from an area are aggregated to the scale of interest (grid cell, block, neighbourhood, city) using a known spatial distribution of those elements. Many cities have access to detailed databases or geographic information systems, which allow detailed identification of buildings, roads or green space, thereby enabling and assisting the modelling of emissions. A bottom-up approach has the advantage of functionally and/or physically modelling the urban metabolism, associated fuel combustion and air pollutant emissions. This may allow the construction of inventories for scenarios that reflect plans for changes in **urban form**, the mix of land uses and technology. However bottom-up inventories are sensitive to the modelling assumptions, and small errors in the sub-models propagate into larger uncertainties at the urban scale.

Emission inventories should also separate between emissions that happen within the domain or grid cell of interest (local emissions) and emissions caused by the domain or grid cell but released outside (external emissions). The concept of external emissions is illustrated by the consumption of electricity for heating and lighting. Electricity may be generated outside the grid cell or area of study and be delivered to the city via a grid system. Finally, there are many urban-located functions, such as air and sea transport that generate emissions but are difficult to attribute to any one place.

In practice, scaling bottom-up inventories to match fuel consumption data from top-down approaches at a coarser scale can be a successful strategy to benefit from the advantages of the two emission inventory approaches, i.e. realistic total regional emissions as well as a functional understanding of the system in terms of its detailed spatio-temporal distribution.

Figure 11.20 shows a detailed emission inventory for $NO_x$, $SO_2$, VOC and PM10 derived by a combination of top-down and bottom-up methods for the greater London, United Kingdom area. The maps illustrate the characteristic geographic distributions resulting from the relative importance of emission sectors. The inventory map for $NO_x$ (Figure 11.20a)

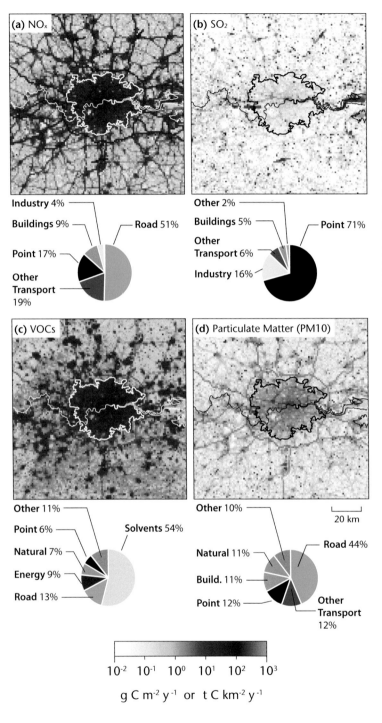

**Figure 11.20** Maps of modelled primary pollutant emissions for the Greater London Area, UK at 1 km$^2$ resolution for the year 2009. The maps show: **(a)** nitrogen oxides, NO$_x$ expressed as NO$_2$, **(b)** sulphur dioxide, SO$_2$, **(c)** volatile organic compounds, VOCs (excluding methane) and **(d)** particulate matter < 10 μm in diameter. The charts underneath of each map describe the contribution of various sectors to the total emissions in the map domain (Source: Maps drawn based on data provided in UK's National Atmospheric Emissions Inventory, NAEI, Department for Environment, Food and Rural Affairs, UK Government).

reflects the road network with motorways and arterial roads being the 'hot-spots'. In contrast the emission map of SO$_2$ (Figure 11.20b) only weakly identifies with the urban area. Emissions are tied to a few strong point sources for fuel processing, ship tracks, rail and air transportation, such as four major airports. Emissions of VOCs are even more diffuse in space (Figure 11.20c) and their variation is related to various land cover – primarily settlements (solvent use). Particulates (Figure 11.20d) again reflect a complex mixture, where the road network is a primary location for emissions.

The link between fuel/energy use and emissions is provided by **emission factors** that relate the mass of air pollutant released to the mass of fuel burned (or to distance driven). These are necessary for both top-down and bottom-up inventories yet are often associated with high uncertainties. They depend on the fuel type and its composition, on the design of the internal combustion engine, and the conditions during the combustion process (temperature etc.). Depending on the combination, a range of values is possible. There are reports and online databases which summarize emission factors for specific sources, practices and economic systems.

## 11.4 Regional and Global Effects of Urban Air Pollution

Full treatment of urban air pollution chemistry and meteorology requires that we cannot simply look at a city in isolation. Many air pollutants emitted in cities are transported far beyond the city's limits where they are still able to affect human health and ecosystems, or even cause large-scale and long-term disturbance to the global climate system. Managing medium to long-lived air pollutants involves regional (transboundary) and global (international) efforts and solutions.

### 11.4.1 Urban Plumes

At the **meso-** to **macroscale**, cities appear as large 'point' sources with their plume extending many hundreds to thousands of kilometres downwind. The persistence of such plumes depends on both the residence time of its constituents (see Figure 11.5) and the degree of mixing that a plume experiences. If conditions are suitable, an identifiable mass of 'urban' air can be transported over great distances. Stohl et al. (2003) employed a numerical model to trace the precise origin of air pollutants reaching Europe that were emitted over North America. The study identified contributions from a number of sources, for example the urban plume produced by the metropolitan area of New York was clearly distinguishable.

### The Extent of Individual Urban Plumes

Recent developments in satellite **remote sensing** make it possible to map air pollutant concentrations in urban plumes. Figure 11.21a shows a composite of column total $NO_2$ recorded by means of passive satellite spectroscopy over the Middle East (Beirle et al., 2011). The column total is expressed as the number of molecules from the ground to the top of the troposphere per column ground area ($m^{-2}$), but note that most $NO_2$ is found in the **atmospheric boundary layer**.

The Middle East, due to the low frequency of cloud cover, is a preferred region for satellite-based remote sensing of air pollutants. Due to its relatively short atmospheric residence time, the mapped $NO_2$ is found in proximity to heavily urban and industrial centres. The case of the Riyadh, Saudi Arabia — an isolated city, dominated by emissions from oil-related industries — is enlarged in Figure 11.21b. Average satellite-measured column totals of $NO_2$ are shown for a 375 by 375 km region centred on Riyadh and conditionally sorted by wind direction. The maps illustrate the extent and persistence of Riyadh's urban plume affecting areas downwind changing with wind direction. The $NO_2$ plumes in all cases have a limited geographical extent of about 200 km, beyond which they approach background concentrations. Generally, the extent of the plumes is determined by source strength, wind speed, mixing and the rates of deposition and chemical transformations for an air pollutant.

Air pollutants in urban plumes undergo transformation as the polluted air travels downwind from a city. Tulet et al. (1999) studied the photochemistry of the plume of Paris, France, during a four-day air pollution episode in July 1996. Hourly $O_3$ concentrations in the core area of Paris and upstream from Paris were considerable, but not extreme ($< 100\ \mu g\ m^{-3}$), however, rural stations located 25 to 110 km downwind of the city experienced seriously high $O_3$ concentrations ($> 200\ \mu g\ m^{-3}$, see Table 11.3). The study linked the high rural $O_3$ concentrations to emissions in Paris, by numerical modelling of the airflow and chemistry with a PGM at the regional scale (Figure 11.22). $O_3$ precursors, such as $NO_x$ experience highest concentrations over the city and! about 50 km downwind in the plume (Figure 11.22a). During transport by the plume,

**(a)**

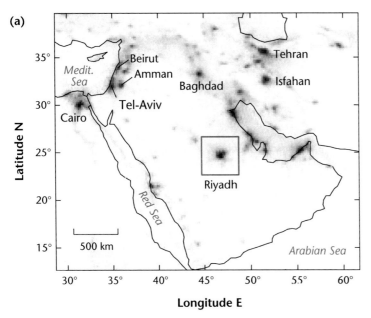

**Figure 11.21** Composite of average total column NO₂ measured by satellite (Ozone Monitoring Instrument on the 'Aura' spacecraft) at a spatial resolution of ~15 × 25 km, **(a)** the Middle East under low wind conditions. **(b)** Enlargement of the city of Riyadh with NO₂ plumes for different wind directions (Source: Beirle et al., 2011; © American Association for the Advancement of Science, used with permission).

**(b)**

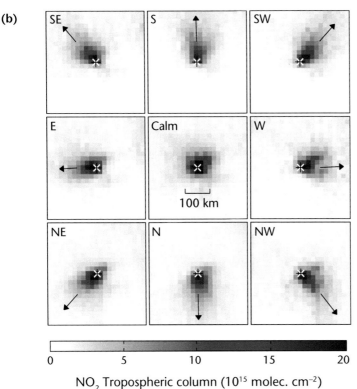

NO$_2$ Tropospheric column ($10^{15}$ molec. cm$^{-2}$)

precursors are chemically transformed to O$_3$, which peaks about 100 km downwind of the city centre (Figure 11.22b). The modelled plume processes are supported by measurements — there was a time-lag in observed O$_3$ maxima between measurement sites closer to Paris and those further away. The lag was between the peaks of the diurnal trace of anthropogenic emissions due to the timing of the rush hours

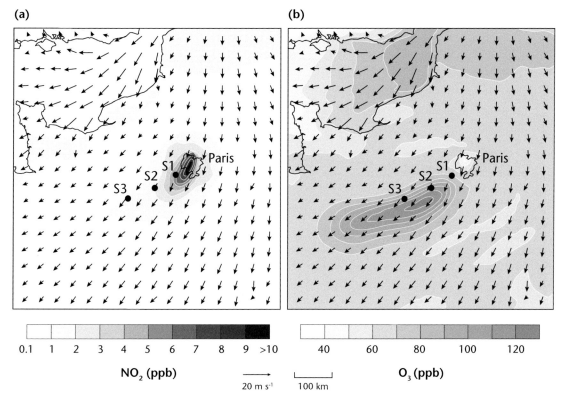

**Figure 11.22** Modelled urban plume downwind of Paris, France, in the evening during an episode in July, 1996. **(a)** Nitrogen dioxide ($NO_2$) and **(b)** ozone ($O_3$) (Source: Tulet et al., 1999; © Elsevier Science Ltd., used with permission).

and the time taken for the wind to advect the plume. Sites further downwind from Paris experienced $O_3$ maxima later in the day.

## Urban–Rural Differences in Air Pollution

Although plumes of individual, isolated cities, can be identified, quantified and modelled, in most cases air pollution at the regional to continental scale is the consequence of many emission sources, where air pollutants from local sources are superimposed on those from distant urban and non-urban sources. Figure 11.23 summarizes air pollutant concentrations measured at thousands of air quality stations across Europe. The stations are classified into 'roadside' (adjacent to a major road or highway, Section 11.2.2), 'urban' (located away from traffic but in urban and **suburban** locations) and 'rural' (away from major population centres and emission sources). The curves in Figure 11.23 show the frequency distribution (histograms) of the annual average concentrations for

each class. Most stations are located in central Europe, so there is a spatial bias to results.

For the three primary pollutants shown, $NO_x$ (Figure 11.23a), CO (Figure 11.23b) and PM10 (Figure 11.23c) roadside station concentrations are generally highest, followed by urban background and lowest at rural stations. Compared to the hemispheric background, all three primary pollutants are elevated, even in remote rural areas – in part because rural areas are not emission free, and in part because they may be located downwind of urban centres and are affected by their plumes. Note that histograms for roadside, urban and rural sites separate best for $NO_x$, followed by CO, while for PM10 the differences are relatively small and the histograms mostly overlap. This is explained by the different atmospheric residence times of the species (Figure 11.5). $NO_x$ has a short residence time of only ~4 hours in an urban plume and hence urban–rural differences are large. CO has an average residence time of weeks to a month

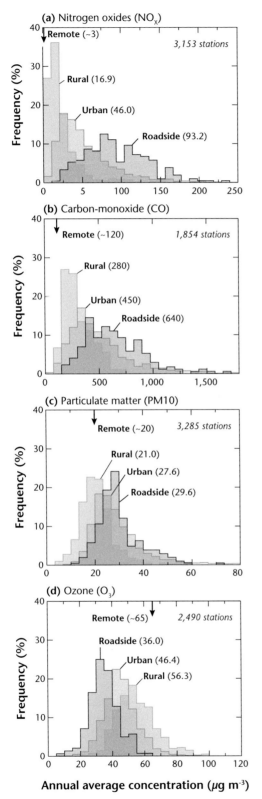

**Figure 11.23** Histograms of average annual concentration for **(a)** NO$_x$, **(b)** CO, **(c)** PM10 and

and downwind areas are more affected by urban emissions. In the case of fine particulates, their long atmospheric residence time (see Figure 5.3), and also the fact that rural areas generate additional emissions (agriculture, dust) explains the tightest overlap between the three histograms.

The case of tropospheric O$_3$ (Figure 11.23d) is the reverse of the others – the ensemble of roadside stations experiences *lowest* concentrations. While O$_3$ is also formed from the primary pollutant NO$_2$, high [NO] immediately reduces O$_3$ back to O$_2$ and NO$_2$. Consequently, [O$_3$] is lower at urban sites where primary pollutants from vehicle emissions are present, while highest values are measured at rural and remote sites. Rural sites do not emit substantial amounts of NO$_x$, but that advected from urban plumes to those more remote areas, along with considerable VOC concentrations, causes the build-up of high O$_3$ concentrations without being depleted by NO. Background O$_3$ concentrations in the troposphere away from urban centres have actually increased by a factor of 3–4 over the last century, due to the increase in emissions of the main precursor pollutants NO$_x$ and VOCs.

### 11.4.2 Effects of Urban Pollutants on Ecosystems Downwind

Air pollutants in urban plumes not only contaminate the atmosphere, but also adversely affect downwind aquatic and terrestrial ecosystems, and structures when they are deposited. Common impacts on downwind ecosystems include water and soil acidification, lake eutrophication and both fertilization and damage to agro- and forest ecosystems.

### Acidification

Absorption of CO$_2$ into cloud droplets causes rain to be acidic — even in the most remote regions. In a pristine

Caption for **Figure 11.23** *(cont.)* **(d)** O$_3$ measured at roadside, urban background and rural sites in the European Union for the period between 2002 and 2012. Values in brackets show median values for all stations in a given category. The remote arrow denotes Northern Hemispheric background concentrations in the PBL measured at a remote site (Mace Head, Ireland). Only years with more than 75% of data coverage at each station are included (Data sources: European Environment Agency, AirBase v. 7.0 and from the Global Atmosphere Watch Program).

atmosphere, rain droplets have a **pH** of about 5.6. How-ever, in an urban plume, in addition to $CO_2$, $SO_x$ and $NO_x$ are also absorbed into droplets and form sulphuric and nitric acid. This further lowers the typical droplet's pH, which can reach values as low as 2, and quite often ranges between 3 and 4 in urban plumes. If such highly acidic droplets (pH < 5.6) are deposited on the ground, built structures or vegetation, they cause damage. This is the problem of acid deposition, which is loosely referred to as 'acid rain'. In addition to deposition by rain, acid deposition includes droplets reaching the surface as acid fog or snow or gravitational settling or dry deposition of acidic particles. Acidification has been observed in many parts of the world, especially downwind from

major urban-industrial regions in Eastern North Amer-ica, Europe and Asia. The acidification of soils and aquatic ecosystems can mean profound changes. Acids on surfaces and soils can corrode materials and dissolve certain rocks (e.g. limestone), further it retards decom-position of organic material in soils and affects plant health.

Figure 11.24, shows the average contribution of major urban centres in Asia to sulphur deposition on terrestrial and aquatic ecosystems. Although the urban imprint is most dominant in close proximity to urban regions, the contribution of cities is detect-able at the continental scale and affects aquatic and terrestrial ecosystems thousands of kilometres

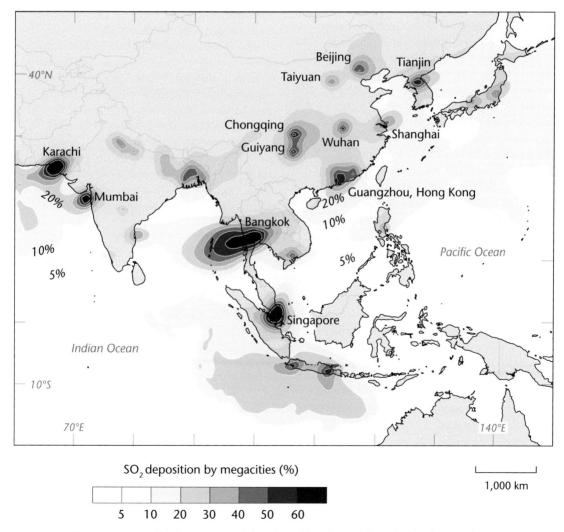

SO$_2$ deposition by megacities (%)

5   10   20   30   40   50   60

1,000 km

**Figure 11.24** Percentage contribution to total sulphur deposition due to $SO_2$ emissions from major urban centres in Asia averaged over the period 1975 to 2000 (Modified after: Guttikunda et al., 2003; © Elsevier Science Ltd., used with permission).

downwind, including the Pacific Ocean. $SO_2$ emissions from fuel combustion have been successfully reduced in many parts of Europe and North America, by abandoning sulphur-bearing fuels (Figure 11.15), so $NO_x$ and ammonia ($NH_3$) are now the principal acidifying components in urban plumes.

### Effects on Vegetation – Fertilization, Eutrophication and Damage

Nitrogen is the limiting plant nutrient in mid-latitude soils, that restricts the growth potential of many terrestrial ecosystems. Inadvertent fertilization by deposition of elevated levels of $NO_x$ and $NH_3$ (or nitrogen in the form of particles) can promote vegetation growth in cities and downwind from them. The typical deposition of N downwind of cities ranges between 10 and 50 kg N $ha^{-1}$ $y^{-1}$. This can cause changes in vegetation composition, or for aquatic ecosystems can cause eutrophication (excess nutrient loading).

Trusilova and Churkina (2008) modelled the resulting changes to plant growth in ecosystems downwind of urban areas in Europe using a biogeochemical land-surface scheme. They found that both, the elevated $CO_2$ in urban plumes and the inadvertent fertilization by enhanced nitrogen deposition increased growth, and consequently the $CO_2$ uptake by vegetation. On the other hand, certain air pollutants such as $O_3$ or PAN; (see Section 11.3.2) are phytotoxics, that cause physiological stress, damage to leaves and reduce growth.

### Physical Changes in Weather and Climate

Of all air pollutants, PM has the strongest impact on physical processes in the atmosphere downwind of cities where they produce a multitude of modifications. Firstly, PM changes the radiative transfer of the atmosphere by reducing shortwave and enhancing longwave radiation fluxes (Section 5.3). These effects are transferred downwind by the regional-scale plume with their magnitude depending on concentration, which in turn, depends on the residence times controlled by the size distribution of the emitted PM, the ambient wind speed and wet deposition. Secondly, PM affects droplet formation and hence cloud cover, cloud lifetime and precipitation. The role of PM as CCN, the effects of number densities and composition on cloud-life-times, and precipitation downwind of cities, are discussed in Chapter 10.

Some urban air pollutants have exceptionally long residence times in the atmosphere, so they become relevant at the global scale. These include GHGs, which accumulate in the troposphere, alter the radiation balance and contribute to global-scale climate changes. These include increases in tropospheric temperatures, changes in precipitation patterns, reduction in sea ice extent and the rise of global sea level (IPCC, 2013). The contribution of cities to global GHG emissions and the consequences of global climate changes on cities are discussed in Chapter 13.

### Summary

Good air quality is a very valuable asset, but maintaining it is a challenge for cities where people are proximate to multiple sources of air pollutant emissions, often in confined spaces like buildings or street canyons. In general terms, **air quality is controlled** by the **strength and density of emission sources**, the **ability of the atmosphere to mix** and advect air pollutants and the **chemical and physical transformation** and **removal of air pollutants**. These controls are specific to individual cities and the climates they occupy.

- The intensity of emissions and the suite of air pollutants in the urban atmosphere depend mainly on the nature of the **local economy**, the **energy base** that underpins it, and the **technology** used. The first determines the relative contributions and intensity of emissions from residential, commercial, industrial and transportation sectors, while the type of fuel (coal, oil, gas, etc.) controls the character of the emissions. Emission control technology, if present, can help prevent or remove unwanted combustion products.
- The form and design of the city in conjunction with its economic functions determines the **geography of emissions**. Air pollutants emitted within buildings, or within the urban canopy layer (UCL), can become highly concentrated if ventilation is weak. Exchanges across the top of the UCL, in addition to emissions above this level, contribute to air quality in the urban

boundary layer (UBL) where they can cause chemical reactions that create **secondary pollutants**, sometimes at considerable distances downwind from major urban emission areas.

• **Wind, turbulence and stability** play a critical role in determining the **mixing and dispersion of air pollutants** at all scales in the city. At the microscale the pattern of wind within the UCL is managed by urban form that helps circulate and channel air pollutants. Within the UBL, wind and the **mixed layer height** regulate the volume of atmosphere into which air pollutants are mixed. Aspects of the urban climate such as the urban heat island can modify these climate controls.

The focus of air quality management is mainly to **regulate emissions**, although **ensuring their mixing and dilution through design**, is an ancillary option. Regulation can be achieved indirectly through urban planning and design (see Chapter 15) and directly though 'command-and-control' legislation. Decisions taken to improve air quality must account for the full life cycle of air pollutants in a broader geographic setting, including areas downwind of the city (see Chapter 12).

# 12 | Geographical Controls

**Figure 12.1** An aerial view of suburban Phoenix, United States, and an approaching haboob generated by downdrafts from nearby convective storms on August 11, 2015 (Credit: J. Ferguson/Chopperguy; with permission).

To this point in the book we have focussed on identifying, isolating and quantifying the magnitude and spatial extent of the distinct urban effect on climate by removing extraneous influences such as topographic variations and weather events. However, most cities are found in complex landscapes and experience changeable weather and climate that can modulate or even overwhelm the magnitude and timing of the urban contribution. For example, in Chapter 11, we concentrated on air pollution with a focus on the **emission** of **air pollutants** from urban sources (vehicles, industry, etc.) but there are natural causes of poor air quality in many cities. Figure 12.1 shows a dust storm generated by the downdrafts from a convective system (a haboob) as it approaches a **suburban** neighbourhood (LCZ 6) in Phoenix, United States, which is located in a hot desert environment. The strong **turbulence** created by downdrafts has lifted dry soil to hundreds of meters in the atmosphere and the billowing mass of suspended particulates moves downwind. The arrival of such storms will have a dramatic (if short-lived) impact on aspects of air quality; an exceptional haboob event in July 2011 generated peak **PM10** and **PM2.5** hourly values of 1974 and 907 μg m$^{-3}$, respectively, which are far in excess of levels associated with typical urban sources of pollution (Raman et al., 2014).

The structure of this Chapter is based on the modified analytic framework of Lowry's presented in Section 2.3.1. This states that the measured value of a weather variable $V_M$ (such as air temperature, humidity, wind speed, etc.) at a station is assumed to consist of three contributions: the background 'flat-plane' value of the variable due to the **macroclimate** of the region ($V_B$); the departure from $V_B$ due to landscape or local climate effects such as relief or water bodies in the vicinity ($V_L$) and the departure from $V_B$ due to the effects of human activities, including urban effects ($V_H$). While most of this book is concerned with quantifying $V_H$, this Chapter deals with the geographical controls that govern the interplay between $V_B$, $V_L$ and $V_H$. These controls include latitude, the distribution of land and water, the general circulations of the ocean and the atmosphere and **orography**. Here we study **urban climate** effects across latitudes, in maritime and continental settings, and introduce how $V_H$ superimposes with $V_L$ in coastal and mountain cities. We also introduce some aspects of extreme weather and how these singular events affect (and are effected by) urban areas; some of the concepts introduced here, such as return periods, are relevant when considering the impacts of global climate change on cities (Chapter 13).

## 12.1 The Macroclimatic Context of Cities

Climate is the typical sequence of weather events found at a place, often represented in terms of statistics of climatic elements (irradiance, wind, temperature, precipitation, etc.). At a global scale the temporal and spatial patterns of these elements are largely controlled by latitude, the relative influence of land vs ocean, and orography. The statistics of these elements (means, variation, extremes) are used to classify a place into macroclimate types.

### 12.1.1 Urban Population and Macroclimates

Approximately one-half of the world population live in urbanized areas at an average population density greater than 260 Inh. km$^{-2}$; in fact, 95% of the world population reside on just 10% of the global land area. Although historic, economic and cultural factors play a role in the establishment, growth and decline of cities, clear patterns between macroclimates and populated areas exist. Despite modern technologies and a global economy, macroclimates still pose limits to human settlements.

### Global Distribution of Irradiance and Population

The main driver for macroclimates is the globally uneven receipt of solar **irradiance** (Table 12.1). Land areas in the tropical zone ($\phi < 23.5°$), which are the home of 35% of the world population, receive the most **radiation**. Radiation receipt in the tropical zone is further differentiated — the land areas in the equatorial zone ($\phi < 10°$), despite the most favourable solar geometry, receive less $K_\downarrow$ (6.7 GJ m$^{-2}$ yr$^{-1}$) than areas further south or north. Here, the intertropical **convergence** zone causes deep **convection** and convective clouds reflect and absorb part of the $K_\downarrow$. Surface irradiance is greatest in the zone $15° < \phi < 20°$ with a zonal average of 7.6 GJ m$^{-2}$ yr$^{-1}$. The subtropical zones ($23.5° < \phi < 38°$), where 42% of the world population live in both hemispheres, receive on average the same $K_\downarrow$ as the equatorial land surfaces, namely 6.7 GJ m$^{-2}$ yr$^{-1}$. As we move to higher latitudes, annual $K_\downarrow$ decreases rapidly. The midlatitude zone ($38° < \phi < 66.5°$) receives only half of the $K_\downarrow$ available in the tropical zone (Table 12.1),

**Table 12.1** Global distribution of population and irradiance by geographical zones and hemisphere for ice-free land surfaces (Data sources: Population and land area-ISLSCP II Global Population of the World Dataset at 0.5° resolution; List of 100 most populated cities from Demographia's online World Urban Atlas, 2015; Irradiance-NASA Surface meteorology and Solar Energy (SSE) Release 6.0 July 1983–June 2005).

| Geographical zone | Fraction of global land area[1] | Fraction of global population | Number of cities within the top 100 most populated cities | Average annual irradiance (GJ m$^{-2}$ yr$^{-1}$) | Variability of monthly irradiance (% of annual total)[2] |
|---|---|---|---|---|---|
| **Tropical zone** ($\phi < 23.5°$) | 37.8% | 35.1% | 35 | 7.15 | 12% |
| **Subtropical zone** ($23.5° < \phi < 38°$) | 24.7% | 42.4% | 43 | 6.71 | 28% |
| **Midlatitude zone** ($38° < \phi < 66.5°$) | 32.9% | 22.5% | 22 | 4.34 | 58% |
| **Polar** ($\phi > 66.5°$) | 4.5% | 0.04% | 0 | 2.93 | 100% |
| **Northern hemisphere** ($0° < \phi < 90°$N) | 73.8% | 88.5% | 86 | 5.10 | 42% |
| **Southern hemisphere** ($0° < \phi < 90°$S) | 26.2% | 11.6% | 14 | 6.87 | 18% |

[1] Land area excludes areas with permanent ice.
[2] Ratio of standard deviation of monthly average irradiance to annual average irradiance.

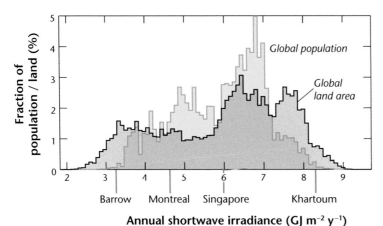

**Figure 12.2** Relative distribution of ice-free land area and global population in comparison to annual average shortwave irradiance $K\!\downarrow$ (Data sources: Population and land area-ISLSCP II Global Population of the World Dataset at 0.5° resolution, Irradiance-NASA Surface meteorology and Solar Energy (SSE) Release 6.0 July 1983–June 2005).

nevertheless, many regions sustain large populations up to $\phi \approx 55°$; overall 22% of the world population lives in this zone. Here, part of the energy to maintain warmer air temperatures is transported from low to high latitudes by ocean currents and wind systems.

Figure 12.2 shows that for selected regions, the distribution of $K\!\downarrow$ imposes a constraint to human settlement: about 10% of the global ice-free land area receives less than 3.5 GJ m$^{-2}$ year$^{-1}$, and only 1% of the world population live in those areas. The resulting cold temperatures and the extreme seasonality in those subpolar and polar climates create climate challenges for sustaining cities and settlements (see Chapter 15). At the other extreme, 6% of the global land area receive more than 8 GJ m$^{-2}$ year$^{-1}$ yet is home to less than 1% of the world population. Those areas of highest annual $K\!\downarrow$ are located in deserts where the lack of available water generally poses a limit on large-scale urban development.

Both annual total $K\!\downarrow$ and its seasonal and diurnal distribution change greatly with latitude. The last column in Table 12.1 illustrates the seasonality of the **shortwave** receipt in each geographical zone, expressed as variability (standard deviation) of the monthly $K\!\downarrow$ relative to the annual average. The seasonality increases from only 12% in the tropics to 100% at the poles.

Figure 12.3 shows Sun path diagrams for four selected cities / settlements to illustrate the effect of latitude on solar geometry. These diagrams show the relative position of the Sun with respect to **solar zenith angle** (distance from origin) and **solar azimuth** (direction relative to origin). The diagram for Singapore ($\phi = 1°$N, Figure 12.3a) illustrates a typical equatorial

case, where days are of equal length across the year, and the Sun passes close to the **zenith** from east to west. Moving northwards, the asymmetry becomes evident for Khartoum, Sudan ($\phi = 16°$N) still in the tropical zone with the Sun in the zenith twice in the year (Figure 12.3b) and midlatitude Montreal, Canada ($\phi = 45°$N, Figure 12.3c). Both, the **solar altitude** and the length of daylight start to vary with season as one moves away from the Equator. Differences are minimal in low latitudes (between 11 and 13 h in Khartoum), but become substantial in midlatitudes (between 9 and 15 hours in Montreal). In the case of Barrow, United States ($\phi = 71°$N) the Sun never sets from May 16 to Jul 27 (polar day) and there is polar night from Nov 18 to 25 Jan. The solar altitude during the year cannot exceed $(90°-\phi) + 23.5°$ so in Montreal the solar altitude reaches a maximum of 68.5° in summer, and 42.5° in Barrow (in the summer solstice). Note also the azimuthal changes of the relative solar position with latitude: whereas in Singapore, the Sun rises and sets within a narrow sector of ±23.5° from E and W, this sector widens to ±24.5° in Khartoum and ±34° in Montreal. In Barrow, above the polar circle, the Sun can rise anywhere from 0° to 180° and set anywhere from 180° to 360°. These variations in the Sun's path have a profound impact on solar access in urban areas (see Section 5.2) and must be accounted for in **climate sensitive urban design** and planning (Chapter 15).

## Global Distribution of Temperature, Precipitation and Population

Figure 12.4 shows the global distribution of ice-free land area as a function of the two major climatic

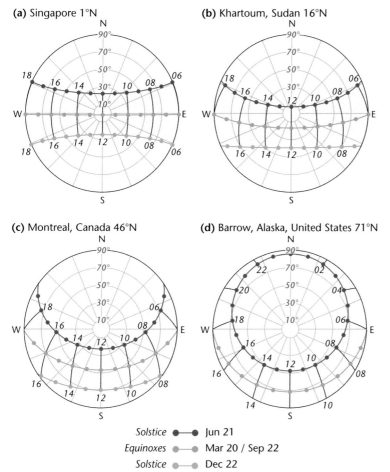

(a) Singapore 1°N

(b) Khartoum, Sudan 16°N

(c) Montreal, Canada 46°N

(d) Barrow, Alaska, United States 71°N

| Solstice | ●━━● | Jun 21 |
| Equinoxes | ●━━● | Mar 20 / Sep 22 |
| Solstice | ●━━● | Dec 22 |

**Figure 12.3** Sun path diagrams for cities in different latitudes in the Northern hemisphere. The diagrams show the relative position of the Sun with respect to solar zenith angle (distance from origin) and solar azimuth (direction relative to origin). Numbers refer to times of day in local apparent time. The shaded area is the range of possible solar positions at the given latitude.

elements, annual mean **precipitation** $P$ and annual mean air temperature $T_a$. The contours show the abundance of land area for a particular combination of $P$ and $T_a$. Superimposed on this parameter space are the population distribution and the locations of major urban areas. Note much of the global ice-free land mass is sparsely populated: less than 1% live in areas where $T_a$ is below 0°C (comprising one-fifth of land area) and; 7% live in areas where $P$ is less than 300 mm (comprising nearly one-third of land area). Cairo, Egypt, which is sustained by the fresh water of the transcontinental river Nile is a notable exception. On the other hand, there are no equivalent occupation limits at the highest annual temperatures. Also excess precipitation does not seem to limit settlements anywhere globally, except in cold continental climates ($T_a$ < 8°C). Here two factors play a role. First, an excess precipitation can cause a longer snow cover in winter.

Along with lack of sunlight (due to cloudiness) this might further impose limits on irradiance that affect local agriculture, human comfort and put high demands on infrastructure. Second, many of the regions with cold temperatures and high precipitation are associated with complex terrain (Andes, Western North America, Himalaya) where precipitation is enhanced by orographic uplift and settlements are limited by the orography.

### World Population in Different Macroclimatic Zones

The two distinct clusters of dense occupation and urbanization evident in Figure 12.4 correspond to temperate midlatitude and tropical climates. Nearly 40% of the world population lives in temperate macroclimates with $T_a$ between 5° and 18°C and annual precipitation from 300 to 1,500 mm (16% of land area). The tropical macroclimate sustains 34% of

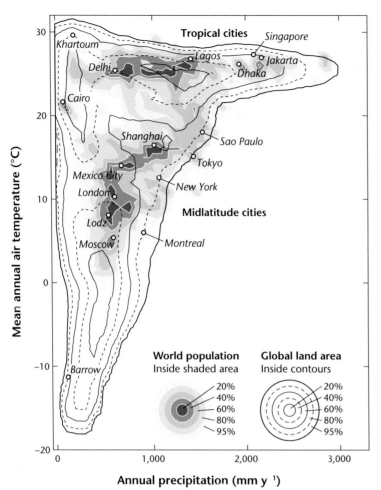

**Figure 12.4** Map of global population distribution (shaded) and distribution of global land area (contours) in the parameter space of annual precipitation in mm yr$^{-1}$ and annual mean air temperature in °C. Data sources: Population and land area-ISLSCP II Global Population of the World Dataset at 0.5° resolution, Precipitation: Calculated from GPCC Full Data Reanalysis Version 6 at 1° resolution, Temperature: Calculated from NOAA/NCEP GHCN CAMS monthly temperature fields at 0.5° resolution.

the global population and has annual mean temperatures from 22° to 28°C and precipitation > 500 mm (23% of land area). It is this second cluster where cities are expected to grow most rapidly in the coming decades (see also Figure 1.3). A more detailed examination of the macroclimate context of a city should account for the seasonal covariation of climate elements. Köppen's climate classification system categorizes climates into types based on statistical relationships between annual and seasonal air temperature and precipitation. These relationships were derived to match the global distribution of vegetation and despite its limitations, it will provide a useful framework for examining how macroclimatic processes influence the magnitude and character of the urban effect over the course of a year.

The Köppen scheme (Kottek et al., 2006) identifies five macroclimatic types: tropical (A); arid (B); mild

temperate (C); continental temperate (D) and; polar (E). Figure 12.5 shows the global distribution of population colour-coded according to their Köppen type. The map shows that the largest cities are concentrated in A, C and D climates, where available water and temperatures suitable for growing crops are relatively abundant. For example, the mild temperate (C) climates, which cover only 17% of the Earth's land area, sustain 45% of the global population. Historically, the C climates have proven to be suited to intensive agricultural production during summertime that can support large populations. By comparison, 29% of the land area in Köppen's classification is attributed to arid (B) climates and is occupied by only 15% of the world population. Similarly, there are even fewer people (and few cities) that are located in continental temperate (D) and polar climates (E) that make up 30% of the world's land area. While Moscow, Russia, and Montreal,

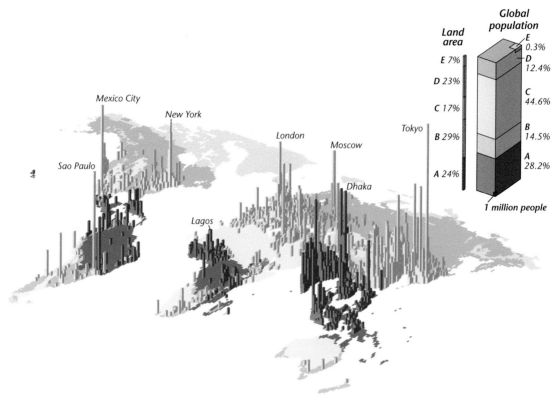

**Figure 12.5** Map of global population distribution coloured by Köppen climates at $1 \times 1°$ resolution. Population is only shown as a bar on the map in cells where population exceeds 500,000. The key indicates the percent of land area and the percent of global population in each macroclimate. The volumetric key for global population is to scale with the bars on the map (Data sources: Gridded Köppen climates based on Kottek et al., 2006, Gridded Population from based on A. L. Brenkert, Oak Ridge National Laboratory).

Canada, are examples of large cities situated in D climates, there are no cities of this size in E climates.

## Seasonality of Temperature and Precipitation

These basic climate types can be subdivided to account for seasonality in precipitation firstly, and temperature secondly. The average monthly temperature and precipitation for selected cities are presented in Figure 12.6 in the form of simplified Walter-Lieth climate diagrams (Walter and Lieth, 1967). This form of climograph used primarily in biogeography gives an indication of the growth conditions and the **surface water balance** at a place based on the close relationship between air temperature and **evaporation**. The temperature and precipitation axes are aligned at 0°C and 0 mm, respectively, as evaporation at subfreezing temperatures is minimal. The relative scaling of the axes is based on a crude relationship that a change in air temperature of 1°C increases evaporation by ~2 mm. If the precipitation curve exceeds the temperature curve a moisture surplus is indicated, the reverse implies a moisture deficit (in other words, more evaporation could occur if there was sufficient precipitation).

The climate of Singapore (1°N, 108°E, Figure 12.6a, Sun path diagram in Figure 12.3a) is an example of a fully humid tropical climate (Af, also 'rainforest climate'); it has a small annual temperature range (2 K) and receives nearly 2,100 mm of precipitation annually with no dry period. By comparison, Brasilia, Brazil (16°S, 47°W, Figure 12.6b), is further from the Equator and has a typical seasonal tropical wet and dry climate (Aw, also 'savannah climate'). Khartoum, Sudan (16°N, 33°E, Figure 12.6c, Sun

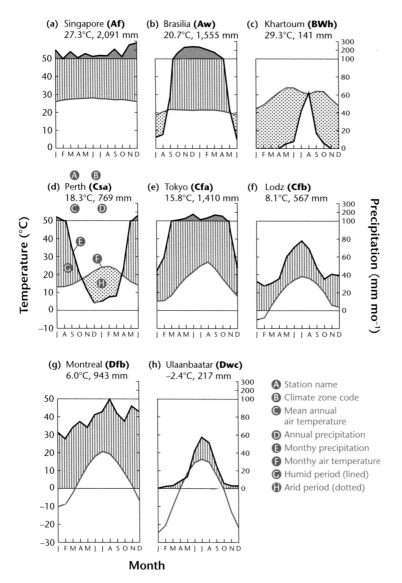

**Figure 12.6** Walter-Lieth ecological climate diagrams for the cities of **(a)** Singapore (1°N), **(b)** Brasilia, Brazil (16°S), **(c)** Khartoum, Sudan (16°N), **(d)** Perth, Australia (32°S), **(e)** Tokyo, Japan (36°N), **(f)** Lodz, Poland (52°N), **(g)** Montreal, Canada (46°S), **(h)** Ulaanbaatar, Mongolia (48°N). For a detailed discussion see text.

path diagram in Figure 12.3c), has a hot arid climate (BW); this climate is characterized by intense **direct-beam irradiance**, a large diurnal temperature regime and a small precipitation receipt over the year that limits vegetative **surface cover**. Note that despite their similar latitudes, Brasilia and Khartoum have markedly different climate regimes.

Midlatitude climates (types C and D) are generally found at higher latitudes. Perth, Australia (33°S,

116°E, Figure 12.6d), has a Mediterranean climate (Csa) with summers that are warm and dry and winters that are cool and wet. Tokyo, Japan (36°S, 140°E, Figure 12.6e), has mild climate with no dry season (Cfa). Lodz, Poland (52°S, 19°E, Figure 12.6f) also has a C climate but given its higher latitude has a lower annual average temperature and sub-freezing monthly average temperatures in January and February. Finally, Montreal, Canada (46°N, 74°W,

Figure 12.6g), and Ulaanbaatar, Mongolia (48°N, 107°E, Figure 12.6h) both have D climates with several months where the average monthly temperature is below freezing. Ulaanbaatar's climate is both cooler and drier owing to its position, far inland and at elevation (1338 m); air temperature is above freezing for just six months of the year.

### 12.1.2 Urban Effects in Different Macroclimates

Although the processes governing the urban climate effect are universal, its magnitude, timing and relevance must be evaluated against the macroclimate context. For example, if a city were to enhance annual precipitation by 50 mm per year, the result may have limited impact in a wet-tropical climate with 3,000 mm per year, but could be a significant climate modifier in a semi-arid macroclimate. Similarly, raising the surface temperature in a very cold environment subject to permafrost by a small amount could have profound implications on near-surface soil stability while the same increase in a warm climate may not have dramatic consequences.

The macroclimate context is even relevant for assessing the urban contribution to air quality. For example, Beijing, China (40°N, 116°E) lies on the Great North China Plain and the Yanshan Mountain lies to the west and north. It has a cold temperate climate (Dwa) with a dry winter (caused by the Siberian **Anticyclone**) and a wet and warm summer (Asian monsoon). Beijing experiences extraordinarily poor air quality that is largely the result of intensive emissions sourced within the urban area. However, during spring it can also experience poor air quality that is natural in origin; on occasion a strong cold

front embedded in the prevailing northwest airflow excavates **particulate matter** (PM) from the Gobi desert and transports it southward, into the city. The distinctive origins of the PM is illustrated in Figure 12.7, which shows two graphs illustrating PM and sulfur dioxide ($SO_2$) during winter and spring 2002. While PM can originate from both **anthropogenic** fuel emissions and natural processes, $SO_2$ is solely an effect of anthropogenic coal emissions. During the winter period the values for both air pollutants are positively correlated suggesting the same source, that is, coal burning within Beijing itself. However, during the spring period the arrival of the dust storm is accompanied by strong winds that ventilate the urban area, diluting the urban-sourced air pollutants (falling $SO_2$) but increasing the natural contribution to PM.

### 12.1.3 The Urban Surface Energy Balance in Different Macroclimates

The **surface energy balance** (SEB) provides the most comprehensive evaluation of energy exchanges at a surface and is ideally suited to examining the physical basis the urban climate in different macroclimates. Figure 12.8 shows SEB terms measured or approximated over residential areas of four cities situated in different macroclimates. Shortwave irradiance, the primary driver of net allwave radiation ($Q^*$), is chiefly a function of both day-length and solar zenith angle which are both controlled by latitude and moderated by cloud cover. Together with the availability of water, $Q^*$ controls the **available energy** for **sensible** ($Q_H$) and **latent heat flux densities** ($Q_E$). Note that heat storage changes in the urban **fabric** and ground ($\Delta Q_S$) over the course of a year should be

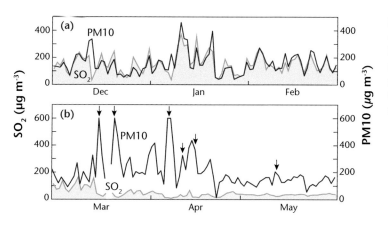

Figure 12.7 $SO_2$ and PM10 concentrations in Beijing, China, in (a) and (b) winter and (c) and (d) spring, circles indicate reported dust storms (Source: Guo et al., 2004; © 2003 Elsevier, used with permission).

approximately zero (that is, positive and negative values cancel each other), while the **anthropogenic heat flux** ($Q_F$) follows mainly the pattern of space cooling and heating (Section 6.2).

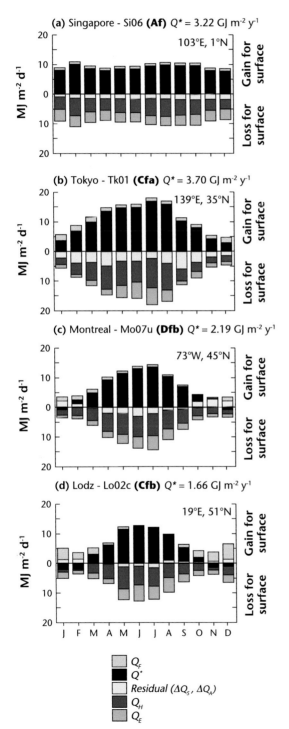

**(a)** Singapore - Si06 **(Af)** Q* = 3.22 GJ m⁻² y⁻¹

**(b)** Tokyo - Tk01 **(Cfa)** Q* = 3.70 GJ m⁻² y⁻¹

**(c)** Montreal - Mo07u **(Dfb)** Q* = 2.19 GJ m⁻² y⁻¹

**(d)** Lodz - Lo02c **(Cfb)** Q* = 1.66 GJ m⁻² y⁻¹

A typical residential area (LCZ 3) in Singapore (Af, Figure 12.8a) experiences very little seasonal variation in all terms of the SEB; the majority of $Q^*$ (3.22 GJ m$^{-2}$ y$^{-1}$) is partitioned into $Q_H$ (the **Bowen ratio** is $\beta = 1.35$) and both $Q_F$ and $\Delta Q_S$ are relatively small. This is because Singapore's tropical climate experiences little change in solar irradiance through the year (see also Figure 12.3a). The fact that $Q_H > Q_E$ suggests that the urban effect has been to limit the availability of water for **evapotranspiration** ($Q_E$) by replacing vegetated with impervious land-cover.

At higher latitudes, the seasonal variation in solar input is revealed by the monthly bars (indicating the magnitude of fluxes), which are shortest in the winter months and longest in the summer months. For a comparable residential area (LCZ 3) in Tokyo, Japan (Cfa, Figure 12.8b), $Q^*$ over the course of the year is actually higher than that at the Singapore site. During the summer months, the available energy is 1.5 times that at the Singapore site but it is half of that available in winter. $\beta$ exceeds 1 during the year and $Q_F$ is relatively small but is larger in the colder winter months, because energy release for home heating in winter is greater than that used in summer for cooling.

These same patterns are accentuated in a residential area (LCZ 2) in Montreal, Canada (Dfb, Figure 12.8c), where $Q^*$ over the urban area becomes negative in winter and heat is withdrawn from storage to compensate; in other words, the urban substrate is cooling in winter. Finally, at a typical city centre location (LCZ 2) in Lodz, Poland (Cfb, Figure 12.8d), the annual $Q^*$ is only 1.66 GJ m$^{-2}$ y$^{-1}$ however this varies from −2.09 (January) to 12.73 GJ m$^{-2}$ day$^{-1}$ (June). $Q^*$ is negative from November until March and there is strong input of energy by anthropogenic space heating during this time.

**Figure 12.8** Monthly surface energy balances for urban residential areas in different macroclimates. The bars for each month are divided by the flux density and are organized according to whether the flux density is a gain (above) or a loss to the surface (below zero line). These data come from sensors positioned on towers above the city that measure radiative and turbulent terms directly ($Q^*$, $Q_H$ and $Q_E$). The studies used different approaches to calculate the anthropogenic heat flux ($Q_F$) (Data sources: (a) Roth et al., 2017; (b) Moriwaki and Kanda, 2004; (c) O. Bergeron / I. Strachan, pers. comm.; (d) Offerle et al., 2005b).

Figure 12.8a–d presents a consistent picture of how the urban energy balance responds to the seasonality of mainly $Q^*$ over the course of the year. In the case of Singapore where the energy inputs are consistent, $Q_F$ is relatively small. At higher latitudes, where the winter is more severe, the seasonal pattern in $Q^*$ is compensated by energy added for heating. In other places (or other parts of the city) we might expect to see different patterns. For example, in densely built commercial areas that rely on mechanical air conditioning, $Q_F$ may be larger during the warmer months.

### 12.1.4 The Urban Heat Island in Different Macroclimates

In the absence of details on the energy exchanges in different climates the widespread study of the urban heat island (UHI) can provide some insights into how the urban effect can be modulated in different macroclimates. Fundamentally, the UHI is evaluated as a temperature difference between the city and the surrounding non-urbanized landscape; the various types and the processes responsible have been discussed in great detail in Chapter 7. Here we will remark on the role of the macroclimate in regulating the magnitude and direction of these differences.

#### Surface Urban Heat Islands

The differences in surface temperature between a city and the surrounding area are controlled to a considerable degree by the characteristic thermal properties of the natural landscape and their change with seasons (see Section 7.2.2). Imhoff et al. (2010) used satellite-derived land surface temperatures to analyse the diurnal and seasonal variations in the **surface urban heat island** (UHI$_{Surf}$) for thirty-eight of the largest cities in the US. Each city area was assigned to a naturally occurring habitat or biome, which can be linked closely to climate type.

Figure 12.9 shows that largest UHI$_{Surf}$ values were observed in the summer daytime for cities located in the temperate forest biome. By comparison, the winter UHI$_{Surf}$ was greatest in the daytime for the tropical and subtropical biome. Interestingly, desert cities showed a negative daytime UHI$_{Surf}$ in summer which may be due to irrigation and the cultivation of ornamental and shade vegetation in urban areas. However, at night, the UHI$_{Surf}$ for desert cities reverses sign and becomes positive.

#### Canopy Layer Urban Heat Islands

The **canopy layer urban heat island** (UHI$_{UCL}$) is measured as a difference in near-surface air temperatures ($\Delta T_{U-R}$); in the mid-latitudes it is typically strongest at night in the densest built of cities where the rate of surface cooling after sunset is slowest. However, in Barrow, United States, which has an extremely cold (ET) climate, the observed UHI$_{UCL}$ is strongest during cold and calm winter days, when there is no solar forcing but the heating demand is greatest. Here, it is the loss of heat from buildings (that is, $Q_F$) that controls the magnitude of UHI$_{UCL}$ (Figure 12.10). Moreover, the heat loss to the near-surface atmosphere in Barrow is enhanced by its layout which consists of dispersed buildings that are elevated above the ground so as to limit heat loss to the soil and protect the stability of the underlying permafrost layer.

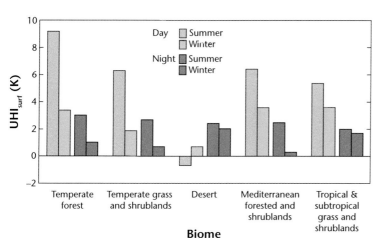

**Figure 12.9** Composite UHI$_{Surf}$ from satellite (MODIS) data for US cities in different biomes. UHI$_{Surf}$ is difference between average temperature of the urban core and surrounding rural vegetated surfaces. Daytime and nighttime results are for 1330 LST and 0130 LST respectively (Source: Imhoff et al., 2010; © Elsevier, used with permission).

Figure 12.11 shows the magnitude of the monthly UHI$_{UCL}$ in selected tropical and subtropical cities. At each place there is a strong negative relationship with precipitation, which is due to a number of factors, most especially the impact that rainfall has on the thermal properties of **rural** soil and vegetation, increasing the **thermal admittance** of the rural surface. Consequently, $\Delta T_{U\text{-}R}$ is greatest when the landscape

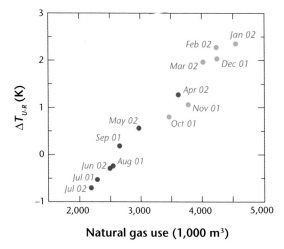

**Figure 12.10** Monthly urban heat island ($\Delta T_{U\text{-}R}$) related to natural gas use in Barrow, Alaska (Source: Hinkel et al., 2003; © Royal Meteorology Society, used with permission from Wiley).

around the city is driest and cools more quickly. Even in Gaborone, Botswana, the driest city, the meagre rainfall received during winter reduces the canopy layer UHI magnitude.

## 12.2 Topography

**Topography** is a description of Earth's features at a place; these features include orography (e.g. mountains, plains and valleys) and variations in the type and state of surface cover (e.g. forest in leaf, cropped grass, etc.). Topography influences climate at all scales; the impact of very large-scale features like the Himalayas or the Indian Ocean are already embedded in the background macroclimates that affect cities, as discussed above. In this chapter, our focus is on smaller-scale topographic features that are larger than cities and modify aspects of this macroclimate. At a smaller-scale still, the influences of topographic features within cities (e.g. hills and green parks) are considered to be part of the urban climate and are discussed elsewhere.

In general, because of common historic origins, cities occupy similar topographic configurations, such as along coasts and/or in river valleys or basins and/or next to mountains (Small, 2004). Table 12.2 provides a simple categorization of these topographies and summarizes their influence on airflow, air temperature and potential air quality. Here, we dis-

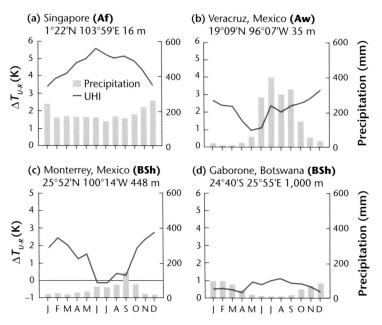

**Figure 12.11** Seasonal variation of mean monthly nocturnal canopy-level heat island intensity ($\Delta T_{U\text{-}R}$) and climatological mean monthly rainfall for various cities in Tropical and Subtropical climates. Values are measured at the time of daily minimum temperature with the exception of Singapore where value is at time of maximum heat island intensity (Source: Roth, 2007; © Royal Meteorology Society, used with permission from Wiley).

**Table 12.2** Common topographic settings of cities and associated meso- and microclimatic effects (Modified after: Wanner and Filliger, 1989).

| Orographic setting | Mechanical | Thermal | Air quality |
|---|---|---|---|
| **Mountain top or ridge** | Wind speed increased due to elevation; cloud and precipitation enhanced due to uplift. | Under clear conditions higher solar irradiance and longwave radiation loss. Higher diurnal and seasonal temperature range. | Good air quality when windy but potential for photochemical smog on clear and calm days. |
| **Slope or slope terrace** | On lee (windward) side shelter (exposure) from (to) ambient winds. Affected by strong downslope winds in certain circumstances. | If the aspect is toward (away) from the Sun, the surface and adjacent air is warmed (cooled) causing upslope (downslope) winds. At night, higher slopes cool fastest and downslope winds result. | Generally good air quality unless located in recirculation zone or close to the inversion level. |
| **Basin or valley** | Effect depends on geometry of valley in relation to ambient wind direction. If perpendicular (aligned) to the valley axis, the valley may be sheltered (exposed). | Cross-valley winds develop during daytime (nighttime) owing to anabatic (katabatic) flows on valley sides. Katabatic system is stronger in shallow nighttime surface layer. | Potential for very poor air quality due to pooling of cool air on valley floor at night. |
| **Base of slope** | On the leeward side foehn winds may occur depending on local topography. On windward side, flow may be blocked against barrier. | Daytime (nighttime) anabatic and katabatic flows. | On the windward side cool air become blocked; on leeward side some conditions can create a recirculation eddy and poor air quality. |
| **Coastal** | On-shore winds generated by storms are much stronger than those off land for the same conditions. | Under calm and clear regional conditions, a coastal breeze circulation is established that brings air on-shore (off-shore) during daytime (nighttime). | Daytime air pollution possible when sea breeze conditions dominate and cause a strong, near-surface inversion. |

cuss the influences of topography on airflow by: obstructing and redirecting the ambient airflow (mechanical effects) and; modifying the diurnal SEB and generate thermal circulations (thermal effects). The latter is of particular interest when studying the urban climate effect as the circumstances that accentuate the topographic influence (i.e. is, calm and clear weather) also magnify the intensity of the **urban heat island circulation** (UHIC, see Section 4.2). However, while the thermo-topographic circulations reverse direction from day to night, the UHIC always acts in the same direction, towards the city centre. As a consequence, these circulations can either oppose or re-inforce each other, which modulates other urban effects on air temperature and the dispersal of air pollutants. Naturally, examining the effects of each of these circulations in isolation is difficult and requires clever observational design and/or **numerical modelling**.

In the following, we deal firstly with orographic influences (Sections 12.2.1 and 12.2.2) associated with hills/mountains and valleys/basins and then with coastal influences (Section 12.2.3). Finally, we consider the interplay of topographic influences that may be difficult to disentangle but are the norm for many cities (Section 12.2.4).

## 12.2.1 Cities Affected by the Mechanical Influences of Orography on Airflow

The influence of orography on the ambient airflow experienced by a city depends both on the physical dimension and geometry of the relief and the ambient weather conditions. Here we divide airflow effects into flow over and around isolated hills, over ridges and along valleys.

### Urban Climates Affected by Flow over Hills

The effect of mountains and ridges on the flow depends on the character of the approaching air mass (its velocity and atmospheric **stability**) and the dimensions of the obstacle presented to the airflow, primarily its height ($H$, in m). To flow over a hill, part of the kinetic energy of horizontal motion must be transformed into uplift (geopotential energy); if the atmosphere is **unstable**, vertical displacement occurs easily but, if **stable**, vertical motion is inhibited.

The effect on flow within a stable atmosphere depends on the relative magnitudes of the upward **forcing** and atmospheric resistance in relation to the height of the hill. These variables can be combined into a single dimensionless value,

$$Fr = \frac{u_o}{N_{BV}H} \qquad \textbf{Equation 12.1}$$

The Froude number ($Fr$) is the ratio of upstream wind speed ($u_o$ in m s$^{-1}$) to the Brunt-Väisälä frequency ($N_{BV}$) and obstacle height ($H$); $N_{BV}$ is a measure of stability in an unsaturated atmosphere,

$$N_{BV} = \left(\frac{g}{\theta_0}\frac{\Delta\theta}{\Delta z}\right)^{0.5} \quad (\text{rad s}^{-1}) \qquad \textbf{Equation 12.2}$$

where $g$ is the gravitational acceleration (9.8 m s$^{-2}$), $\Delta\theta/\Delta z$ is the increase in **potential temperature** with height (in K m$^{-1}$) and $\theta_0$ is the potential temperature at the surface (in K). $N_{BV}$ is equal to infinity in a **neutral** atmosphere and is not applicable to an unstable atmosphere. Most work on airflow over complex orography considers the implications of flow in stable atmosphere because at scales of mountains, the **troposphere** is usually slightly stably stratified.

Further, under stable situations air quality is often of greatest concern (Figure 12.12):

- When **$Fr$ is close to zero** (Figure 12.12a), the atmosphere is very stable and airflow is weak relative to the height of the obstacle. As a result, the approaching air mass does not have sufficient energy to flow over the hill and flows around, transporting some air pollutants to the lee side; if a city is located upwind of the mountain, the blocking will hinder the **advection** of air pollutants and worsen air quality here.

- At higher $Fr$ values (**0 > $Fr$ > 1**, Figure 12.12b) a larger fraction of the approaching air mass is displaced over the hill and standing waves develop downwind of the hill crest as the displaced air oscillates around its upstream level. Increasing $Fr$ means weaker upwind blocking and improved ventilation of air pollutants from upwind cities. Downwind cities do not experience unusual poor air quality.

- When **$Fr$ approaches 1** (Figure 12.12c), an substantial part of the atmosphere passes over the mountain, and the wavelength of the standing wave grows until it is roughly equal to the dimension of the obstacle. If this happens, then recirculating, standing vortices (rotors) can develop downwind of the obstacle that can trap air pollutants and cause serious **air pollution** episodes. Note that the potential for poor air quality has shifted from upwind ($Fr$ close to zero) to downwind ($Fr = 1$).

- When **$Fr$ exceeds a critical value of 1**, **flow separation** occurs near the crest and a recirculation **vortex** develops on the lee side adjacent to the slope (Figure 12.12d); air pollutants released by a city in the lee can be transported against the regional flow near surface and recirculate back into the urban area. However, further downwind, outside the recirculation zone, the near-surface flow is **turbulent** and well mixed.

### Urban Climates Affected by Flow over Ridges

If an orographic barrier is elongated perpendicular to the oncoming airflow, then the exchanges between the windward and lee sides are restricted to flow over the ridgeline. In these circumstances the differences between the weather experienced on either side can be dramatic. Figure 12.13 shows three examples that show the interaction between flow over a barrier juxtaposed with vertical temperature **profiles**.

Figure 12.13a shows the creation of a hydraulic jump as the air that is displaced by the ridge forms a deep wave that 'breaks' on the lee side. The

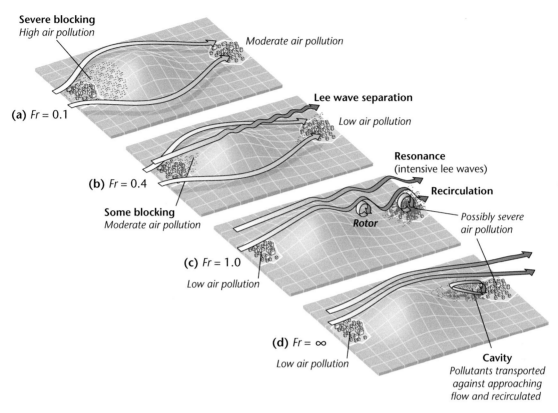

**Figure 12.12** Schematic effect of the Froude number (*Fr*) on the flow around and over a hill. Impact of the atmosphere on air pollution in cities upwind and downwind of the hill, assuming emissions are the same (see text for details).

circumstances for its formation require a large **pressure gradient force** across the ridgeline and a **capping inversion** that is less than 2*H*. As a consequence, the vertical displacement of air is constrained as it flows over the mountain and it must accelerate. This acceleration may be sufficient to disrupt the mountain wave and transform it into hydraulic flow: air flows rapidly downhill along the lee slope in a thin layer; on reaching the base of the slope, it slows abruptly causing convergence and rapid vertical expansion, known as a hydraulic jump (Figure 12.13a).

The urban wind experience depends upon its location *viz-a-viz* the hydraulic flow. Boulder, United States, for example, is located on the eastern slopes of the Rocky Mountains and experiences hurricane force winds and considerable damage if atmospheric conditions are right. Similarly, the bora describes another very strong and cold wind along the Adriatic coastline; here the source is an extensive pool of cold air that has formed in an inland elevated basin. Storms periodically disturb this pool, forcing air through gaps in the Dinaric Alps downhill to the coast. In populated places along the affected coastline (such as Trieste, Italy) this wind causes damage to buildings and creates considerable discomfort; the response can be seen along paths where railings are provided to aid pedestrians.

Figure 12.13b illustrates the situation where a deep pool of cold air is lodged on the lee side of the barrier and is capped by an **inversion** layer. This may be the result of cool air 'pooling' in a wide valley as the surface cools overnight or as a result of advection of cold air on regional scales. The windward air is displaced over the barrier but despite cooling with height remains warmer than the surface layer on the lee side. As a result the cold pool remains substantially undisturbed; any emissions into this layer do not get diluted and air quality is degraded. If the cold pool has formed as a result of nocturnal cooling, then the conditions are unlikely to persist; surface warming during the daytime initiates convection and the cold pool is eroded from below. However, a mixture of processes that include **synoptic**-scale **subsidence** and advection in complex terrain can result in 'persistent'

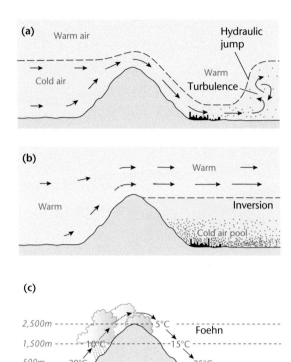

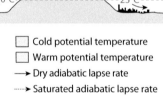

☐ Cold potential temperature
☐ Warm potential temperature
→ Dry adiabatic lapse rate
┈┈➤ Saturated adiabatic lapse rate

**Figure 12.13** Thermal effects affecting the flow over a mountain ridge with a city in the lee. In all cases there is a synoptic (background) wind from left to right. (a) with an upwind cold air pool, (b) with a downwind air pool and (c) with the added effect of latent heat, causing foehn.

pooling. This is the case in Figure 12.14, which shows a part of Santiago, Chile. The city is located on the floor of an elevated basin. During the summer, the **mixed layer** height over the heavily urbanized basin is about 800 m thick but in winter it shrinks to just 200 m (Munoz and Undurraga, 2010). Under clear skies, cooled air from the surrounding slopes flows downhill and 'pools' over the city, creating a near-surface inversion (Romero et al., 1999). The result is an atmosphere that is confined horizontally (by the dimensions of the basin) and vertically by a capping inversion. Emissions of air pollutants from the underlying urban surface accumulate in the basin and, although the intensity of emission is comparable to other cities of

similar size, Santiago experiences much poorer air quality, in particular in the winter months when the stagnant atmosphere limits **dispersion**. A similar situation is found in the basin of Salt Lake City, United States, which experiences poor air quality and **fogs** during wintertime pooling events (Lareau et al., 2013).

Figure 12.13c shows the development of a foehn wind as a deep stable layer approaches the mountain barrier. The oncoming air is forced to rise, cool, condense and precipitate on the windward side; while moving upslope, it cools at the dry $(0.01 \text{ K m}^{-1})$ and then saturated $(< 0.01 \text{ K m}^{-1})$ adiabatic **lapse rates**. On the lee side the air descends and, after evaporating any remaining cloud droplets, warms at the dry adiabatic lapse rate. In other words, for the same elevation change, the descending air (depleted of some of its moisture content) will be warmer than the ascending air. As a result, the air that arrives at the base of the lee-side slope is both warmer and drier than that which ascended the mountain. There are many winds of this type affecting cities and populated areas; in North America the chinook forms when westerly airflow brings moist air from the Pacific over across the Rocky Mountains. These winds occur more frequently in the winter months where their arrival causes a rapid increase in temperature and widespread snowmelt in colder climates.

Of course, each orographic configuration is unique and the interaction with climate/weather may produce distinct regional-local outcomes that may combine elements of the simple situations presented above. For example, Flamant et al. (2006) report on situations in Alpine valleys where both cold air pooling and foehn winds occur (Figure 12.13b and c); however, the foehn is too weak to either erode (through mixing) or dislodge the former. As a result, the foehn flows over the cold pool, producing a shallow inversion that may trap pollutant emissions close to the ground.

## Urban Climates Affected by Channelling

The mechanical effects of orography include redirecting airflow through gaps in mountains and ridges and along valleys. As a consequence, near-surface flow in a valley may bear little relationship to the direction of airflow aloft, which is governed by the interaction of the pressure gradient force, **Coriolis force** and friction force (see Equation 3.2). Within the valley, flow is directed from high to low pressure along the valley axis and its velocity will vary with the cross-sectional area, moving faster where the valley narrows and *vice versa*.

**Figure 12.14** Air pollution in Santiago, Chile. The city occupies an area over 500 km$^2$ on the floor of an elevated basin (500 m a.s.l) bounded on the east by the Andes Mountains (over 5,000 m) and on the western side by the Coastal Range (up to 2,000 m) (Credit: Christian S.; with permission).

In the coastal urban basins of southern California (e.g. Los Angeles, United States), a strong, warm and very dry wind (known as the Santa Ana) causes wind damage, elevated dust levels and considerable physical discomfort for humans. Moreover, the desiccating effect of the dry air affects water use in cities, and increases the risk of fires that are spread by the strong winds. About 20 Santa Ana events occur each year as a result of synoptic-scale circulation systems that generate off-shore winds that disrupt a pool of cool air that has formed over an elevated inland desert (Hughes and Hall, 2010). As the cold air sweeps towards the coast, it has to pass through narrow gaps in the terrain causing the air to accelerate as it descends and warms.

Channelled flow can transport urban pollutants over great distances and if the conditions inhibit vertical mixing, **urban plumes** experience little dilution. For example, the Mistral is an off-shore wind that occurs during winter when an anticyclone over northern Europe causes a stable air mass to flow from the North across Central Europe. As the flow encounters Alps, some flow is diverted to the west and is channelled to the Mediterranean coast along the Rhone river valley. In these circumstances **secondary pollutants** such as ozone ($O_3$) can be funnelled to the coast along valleys (see Section 11.3 on $O_3$ formation). Corsmeier et al. (2005) examined an anomalous $O_3$ peak in the coastal atmospheric boundary layer associated with a Mistral event. The analysis showed that the $O_3$ peak was caused as air exited the Rhone river valley and moved rapidly downhill to the coastal plain. The quick descent resulted in hydraulic flow that thinned the surface layer and elevated $O_3$ **concentration**; as the air moved off-shore, the hydraulic jump that followed quickly mixed $O_3$ into a deeper layer.

## 12.2.2 Cities Affected by the Thermal Influences of Orography on Airflow

Thermal effects on airflow are characterized by a reversal in wind direction over the course of a day and are driven by surface heating and cooling in response to the path of the Sun and terrain geometry. These thermo-topographic circulations can be classified into four types (slope, cross-valley, along-valley and mountain-plain), which are interconnected. Thermal effects are strongest when there is a small regional pressure gradient and skies are mostly clear. During the daytime, differential surface warming results in warming of the near-surface air and instability:

- Winds that move warm air up along slopes during day are called **anabatic winds**.
- On a larger scale, the air in the valley is replaced by air that moves up large valleys from plains to mountains. Those are called 'valley winds'.

At night the process is reversed: the surface cools, air becomes negatively buoyant and slides down the orography.

- Winds that move cold air down a slope during night are called **katabatic winds**.

- On a larger scale, cold air pooled in a valley floor by many katabatic winds can flow down the valleys into plains during night. Those winds are called 'mountain winds'.

The nature of airflow during the daytime depends on solar exposure/shading, which dictates where surface heating and cooling occurs. For example, in an east-west oriented valley outside of the tropics, where one side will be in shade throughout the day, anabatic winds will form on the sunlit slope and draw air from the valley floor. However, at night, katabatic winds will form on both slopes. As a model for discussion, we consider a symmetric valley where the sloped surfaces have identical energy balances.

### Interplay of Urban Effects and Slope Winds

Many cities (and parts of cities) are situated in valleys and experience thermal circulations. The nighttime winds (katabatic winds, mountain winds) have received particular attention as they are associated with inversions that can result in poor air quality. Figure 12.15 is a conceptual diagram that shows the interaction between city and slope wind circulations that form under ideal weather conditions. The UHIC

causes air to converge on the city centre, rise and spread outwards aloft eventually sinking outside the city. In this context however, the UHIC is superimposed by slope winds, which complicate matters. During the daytime, the anabatic winds on both sides will draw air from the valley floor, where it is replaced by descending air in the centre. When fully developed, the circulation will consist of two cells that occupy each half of the valley. The anabatic winds draw air out of the city and counteract the UHIC. In Figure 12.15a the valley-scale circulation dominates, drawing air out of the city near the surface and mixing it through the depth of the valley.

By contrast, the nocturnal UHIC and katabatic wind systems reinforce each other (Figure 12.15b). The katabatic winds from both slopes converge at the mid-point of the valley, where the air lifts, spreads out at height and completes the circulation. In the absence of a city, the nighttime circulation produces an inversion layer above the floor of the valley that deepens as the cold air from the valley sides accumulates. However, the development of an inversion layer in the valley is impeded as the urban surface is comparatively warm and maintains a positive lapse rate near the surface; as a result, the inversion that does

**(a)** Daytime

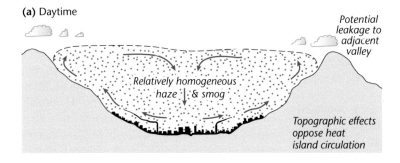

*Relatively homogeneous haze & smog*

*Potential leakage to adjacent valley*

*Topographic effects oppose heat island circulation*

**(b)** Nightime

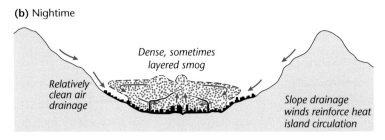

*Dense, sometimes layered smog*

*Relatively clean air drainage*

*Slope drainage winds reinforce heat island circulation*

**Figure 12.15** A cross-section of thermal circulations in a valley and interactions with an urban heat island circulation (UHIC). During daytime **(a)** the UHIC is counteracted by the anabatic flows so that urban air is drawn out of the city at the sides and then circulated into the valley atmosphere. At nighttime **(b)**, the katabatic flow is in the same direction as the UHIC but brings cool air into the city at its sides. The vertical development of the circulation is restricted by the near-surface inversions so that the urban effect is constrained.

form is weaker and higher when a city is present in the valley. In this depiction, the cross-valley and UHI circulations are finely balanced; a strong UHIC could draw in cold air far into the city that would cool the urban surface and extinguish the UHIC itself. Similarly a large UHIC within a relatively small valley could overwhelm the katabatic flows by thoroughly mixing warmed urban air throughout the valley.

The interplay of surface energy exchanges and atmospheric circulation are usually subtle, making their effect difficult to observe and better suited to numerical modelling. Figure 12.16 shows the results of an experiment using a two-dimensional **mesoscale** model to examine the development of circulation systems in a symmetric valley 300 m deep, 10 km wide and oriented east-west. This configuration is based on the situation of a large industrial city (Lanzhou, China), which occupies the valley floor. A series of simulations were performed for calm and clear synoptic conditions during December when air quality is usually poorest. Figure 12.16a shows the valley circulation in the late afternoon when the upslope winds are strongest: the anabatic winds have values of 0.25 m s$^{-1}$ and the return flows aloft have their core at about 1 km above the valley floor. The effect of the city is to weaken the valley circulation by inhibiting the development of the upslope winds, which are 0.2–0.4 m s$^{-1}$ lower when the city is present (Figure 12.16b). As a result, the ventilation of the city is weakened and contributes to poor air quality. On the other hand, the nocturnal simulation found that the katabatic winds were sufficient to 'suffocate' the development of the UHIC and trap air pollutants closer to the surface, thereby worsening air quality.

## Interplay of Urban Effects with Mountain and Valley Winds

Where the orography in a valley itself has a significant slope, the slope winds become part of a larger network of connected flows forming valley winds (day) and mountain winds (night). Finn et al. (2008) examined the dispersion of plumes at night using **tracer** gases under ideal conditions in Salt Lake City, United States, which sits at the base of the Wasatch Mountains on a gentle sloping plain. Tracers were released at ground level under conditions when there was weak synoptic forcing. The plume drifted slowly downhill in a very thin layer that was highly stratified and decoupled from the overlying flow; in fact, conventional airflow measurements (most made at heights from 7–23 m in the city) provided little information

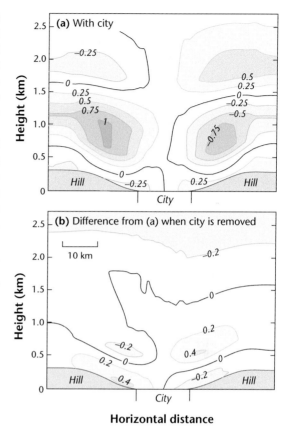

**Figure 12.16** (a) A cross-section of the modelled circulation system at 1700 h in an east-west oriented valley circulation system with a city on the valley floor. Isotachs indicate wind velocity (m s$^{-1}$) and the sign show direction: positive values (green) represent flow from the right and negative values (blue) flow from the left. (b) The difference in horizontal wind speed and direction when the city is removed (Source: Savijärvi and Liya, 2001; © Kluwer Academic Publishers, used with permission from Springer).

on the movement of the plume, which was guided along minor topographic indentations towards the lowest point.

The interplay between valley and mountain winds with urban circulations has not received much attention; following from Figure 12.16a we might expect that a UHIC impedes the development of valley and mountain winds in the vicinity of the city. Piringer and Baumann (1999) observed the development of valley and mountain winds around Graz, Austria, which sits in the Mur river valley. Instruments located on tethered balloons (see Figure 3.12b) were used to

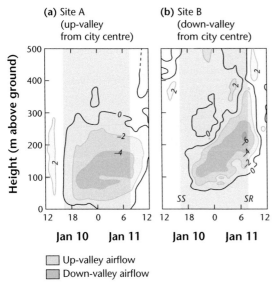

**Figure 12.17** Time height cross-sections of airflow at two stations in a valley based on tethersonde ascents over a 24 hour period during winter characterized by light regional winds. Isotachs indicate wind velocity (m s$^{-1}$) and the sign shows direction: positive values (green) represent flow from the right (up valley) and negative values (blue) from the left (down valley). The city is located between sites A and B (Source: Piringer and Baumann, 1999; © Springer-Verlag, used with permission).

measure the vertical profile of wind speed at site A (about 4 km up valley of the city centre and outside the urban area) and at site B (2 km down valley of the city centre but within the urban area) in calm conditions over a 3-day period.

The time-height cross-sections of valley flow at these two locations are shown in Figure 12.17. At site A, the mountain wind starts at about 1800 h; the mountain wind begins just above the ground and extends upward to over 250 m; it strengthens during the night with a maximum at about 150 m and reaches the ground after midnight and is replaced by valley winds at about 0900 h the next morning. At site B the nighttime mountain wind starts at 2000 h and is located well above the ground whereas the near-surface wind is up valley by day and night. The difference between the two sites is attributed to a UHIC that is centred on Graz, which draws air into the city centre from both sides. At site B, the valley wind is enhanced

by the UHIC during day but at night, the mountain wind is opposed, hence the near-surface flow is upslope by day and night.

Although the effects of orography have been categorized into mechanical and thermal effects, in reality they can interact. Li et al. (2015) presents the meteorological conditions that cause heavily polluted days (HPD) in Urumqi, China, during winter, which is located at the northern side of a mountain gap at 800 m asl; the gap links two basins located to the north and south/south-east of the city. During HPD, the city can experience both mountain/valley breezes and foehn airflow that combine to limit vertical mixing. During the daytime, surface heating drives a valley wind that draws air upslope from a cold air pool that forms in the northern basin. When a foehn occurs, warmer air descends downslope from the south-east and meets the colder valley breeze at the surface, forming along a narrow transition zone (a mini-front). This position of this front moves up- and downslope depending on the strength of either wind, and sometimes divides the city into a warmer and colder side (albeit the air temperatures on either side are well below freezing). Above the surface, the foehn flows over the cold air at the ground creating an intense near-surface inversion that traps air pollutants surface emissions and results in high concentrations of air pollutants.

### 12.2.3 Cities Affected by Coastal Wind Systems

The contrasting properties of water and land surfaces at the coastal interface have a significant impact on the climates of coastal cities, which usually experience higher average winds and a smaller annual temperature range than cities further inland. The former is a result of the lower surface roughness of water so that the on-shore winds will usually be much stronger than off-shore winds for the same pressure gradient. The latter is a result of differences in the radiative, thermal and convective properties of water and land that result in profound differences in the respective SEBs, which initiates thermal circulations known as land- and sea breezes.

### Cities Affected by Land and Sea Breeze Circulations

Other things being equal, when regional airflow is weak and skies are clear, strong daytime irradiance causes the land surface to warm more quickly than the adjacent water. The differential heating of the overlying air mass results in: a pressure gradient aloft from land to sea and an off-shore wind at elevation. The shifting air mass above creates a pressure gradient from sea to land at the surface that generates a near-surface on-shore flow. Conservation of mass means that vertical flows upward (downward) over land (water) compensate to form a closed circulation. The on-shore flow of cool maritime air is termed a **sea breeze**; its advance is marked by a shallow **sea breeze front**, which progresses inland over the course of the day. Ahead of the front, similar to larger scale cold fronts, warmer air is displaced vertically by a few hundred meters; behind the front a near-surface inversion is created as cool air underlies the warmer displaced air. At night, differential surface cooling produces a pressure gradient in the opposite direction and an off-shore breeze develops (**land breeze**); convergence and uplift occurs off-shore and; a return flow occurs aloft.

The daytime on-shore flow has received considerable attention in **urban climatology** as it has significant implications for air quality of coastal cities owing to its association with a near-surface inversion. Yoshikado (1990) observed the development of a sea breeze circulation across Tokyo, Japan, by taking vertical profiles at varying distances inland (up to 80 km) from the coast (Figure 12.18). At 0930 h the incursion of the sea breeze is evident as a shallow layer close to the shore; the nighttime, off-shore breeze can be seen in a layer about 600 m deep further inland (Figure 12.18a). By noon, the sea breeze front has passed the centre of Tokyo and its speed of movement has increased; the remnants of the land breeze now occupies a weak recirculating zone just ahead of the front (Figure 12.18b). At 1300 h near-surface wind speed has increased further and the convergence zone is readily identified more than 40 km inland from the shore (Figure 12.18c) and its progress continues throughout the afternoon reaching near 80 km inland by 1500 h (Figure 12.18d).

A coastal city can affect the development and progress of the sea breeze circulation by virtue of its relative warmth (which generates the UHIC) and roughness. During the daytime, the advance of the sea breeze front inland is aided by the UHIC, which draws air towards the city centre. The movement of the sea air near the surface is slowed when it encounters the rough urban surface and the front steepens as a consequence. Once the sea breeze reaches the city

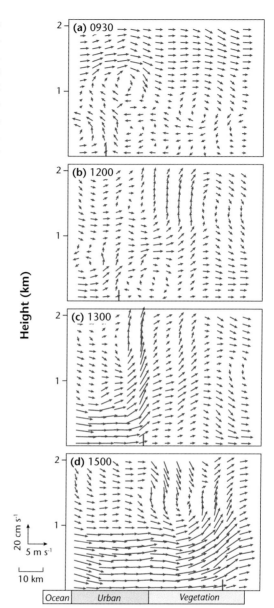

**Figure 12.18** A vertical cross-section of winds across the Kanto Plain on 31st July 1986 at **(a)** 0930, **(b)** 1200, **(c)** 1300 and **(d)** 1500 h. The red vertical line identifies the position of the sea breeze front at the surface. (Source: Yoshikado, 1990; © American Meteorological Society, used with permission).

centre, it is now opposed by the UHIC which further slows its further progress inland.

The roughness effect has received some attention because the sea breeze can act as a natural 'air-

conditioner' for cities during hot weather, when
access to the cooler sea air can offer respite. Where
buildings are small and streets are wide and oriented
perpendicular to the coast, the urban landscape
offers little impediment to the movement of the sea
breeze front passage but tall, closely-spaced buildings and narrow streets can block or greatly lessen
its advance. These effects can be seen in
Figure 12.19a, which shows a sea breeze front
moving inland in the Greater New York, United
States area. The detailed analysis was possible
because of a dense network of sites that measured
the near-surface and boundary layer winds and temperature. The effect of the city is shown by the
distortion of the isochrones that depict the time of
passage. The front crosses approximately normal to
the local coast but because of the shape of the
coastline it moves both westward and northward.
The front starts inland between 1000 and 1100 h
but as it encounters higher urban density, it slows
down (the isochrones are closer). At 1600 h the front
arrives at the southern tip of Manhattan, the most
densely built-up and roughest area and becomes
distorted as it moves inland at different rates. By
1900 h the front breaks through the city. The extra
**drag** of the city causes a steepening of the sea breeze
front; city versus no-city simulations of sea breezes
in New York City show the head of the inflow
bulges upward as it enters the densest urban area
so it is 2 to 3 times deeper than if the city is absent
(Thompson et al., 2007).

Examining the urban impact on development of the
sea breeze is ideally suited to numerical modelling
where experiments with and without the presence of
the city can be conducted. In one experiment, Yoshikado (1992) found the presence of Tokyo enhanced
development of the sea breeze circulation by enhancing the thermal difference between the land surface
(UHI) and the adjacent water body. As a result the
modelled updraft in the convergence zone reached
300 m higher in the experiment when the city was
included. Freitas et al. (2007) also explored the relationship between the sea breeze and a UHIC based on
the situation of Sao Paulo, Brazil, which is located in a
hilly environment more than 50 km from the Atlantic
coast. However, in order to isolate the interactions
between these thermally-driven circulations, simulations were performed with no orography present.
Figure 12.20 shows the flow patterns at 1700 h along
a vertical cross-section extending from the coast

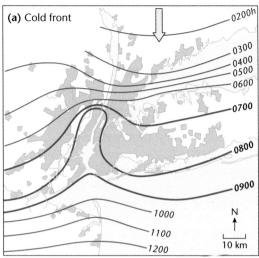

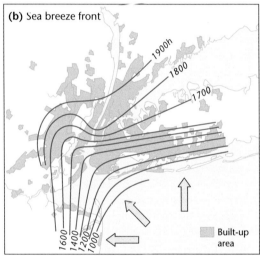

**Figure 12.19** Effect of New York City on **(a)** inland
penetration of the sea breeze in the afternoon of
9 March, 1966 (Source: Bornstein and Thompson,
1981; © American Meteorological Society, used with
permission), and **(b)** passage of a cold front across the
city in the morning of March 11, 1966. (Source: Loose
and Bornstein, 1977; © American Meteorological
Society, used with permission).

inland, through and beyond the city when the sea
breeze circulation is well formed. When no city is
present, the circulation extends horizontally 30 km
inland and vertically to nearly 2,000 m. When Sao
Paulo is included, the model simulates a daytime
UHIC that intensifies the sea breeze circulation and
draws it further inland more quickly. In fact, comparing the city to the no-city simulation, the sea breeze

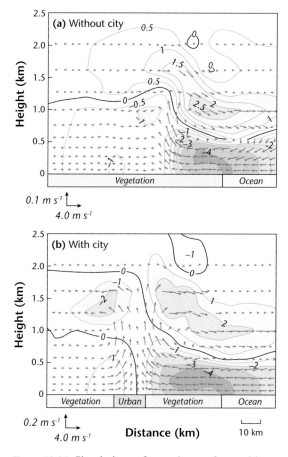

**Figure 12.20** Simulations of a sea breeze front with and without a city (Sao Paulo). Vectors correspond to the vertical and sea breeze wind components. The intensity of these two components is presented in the bottom left corner. Contours represent the magnitude of the sea breeze (in m s$^{-1}$) (Source: Freitas et al., 2007; © Springer Science +Business Media B.V., used with permission).

reaches the location of the city centre two hours earlier in the city case. However, upon reaching the centre of the UHIC, further progression inland is stalled and the combined circulations transfer moisture and heat vertically. After 2100 h the UHIC has weakened and the sea breeze continues inland but at a slower rate than in the no-city case.

### 12.2.4  City Airflow in Diverse Topography

Many cities are subject to several interacting topographic influences that operate at different spatial and temporal scales depending on the background

climate and weather. Figure 12.21 is loosely based on the physical geography of Los Angeles, United States, which is located on a coastal plain adjacent to a significant orographic barrier. Under calm and clear background climate circumstances the thermal circulations associated with coastal, orographic and urban influences combine to create a complex air quality problem. In the morning (Figure 12.21a), a shallow sea breeze front advances from the coast into the urban area. By this time, morning traffic has emitted **primary pollutants** into the overlying atmosphere. As the sea breeze front moves inland, it pushes polluted air near the surface ahead of it and siphons a plume of urban air back towards the ocean at height. At the surface, air quality improves dramatically as the front passes but there is a corresponding deterioration in the zone ahead as emissions from traffic in the coastal urban areas are added to local sources.

By early afternoon (Figure 12.21b) the sea breeze front has arrived at the base of the mountain barrier where air pollutants are drawn upslope by anabatic winds. The primary pollutants emitted by vehicles and industry have now begun to change into secondary pollutants (e.g. ozone and **PAN** in **photochemical smog**, see Section 10.3.2). Note that peak concentrations of secondary pollutants are consequently found both in the trapped layer above the advancing sea air, and on the slope downwind of the city. The former has become isolated from fresh emissions at the urban surface that would accelerate the destruction of $O_3$. At night (Figure 12.21c), a weak land breeze pushes the polluted air back out to sea, so the following day the aged $O_3$, which has been formed during the preceding daytime, returns onto land.

The background climate is itself subject to change that may enhance or diminish these topographic influences. For example, Lebassi et al. (2009) analyzed temperature data for stations in southern California for the summer period 1970–2005 and found warming (cooling) trends at high-elevation inland (low elevation coastal) stations, especially in the daily maximum values. The apparent contradiction is attributed partly to inland warming that has strengthened the land-sea thermal gradient and increased the strength of sea breezes, causing cooling in the near coast zone.

### 12.3  Synoptic Controls

Weather is distinguished from climate largely on the time scale of analysis. Whereas climate is the

**(a)** Morning

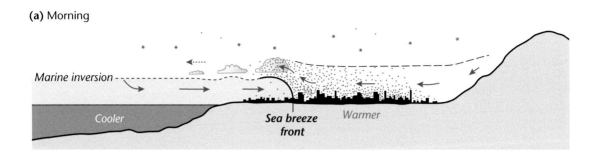

**(b)** Afternoon

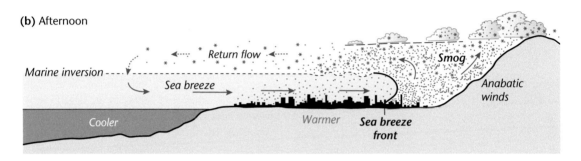

**(c)** Night

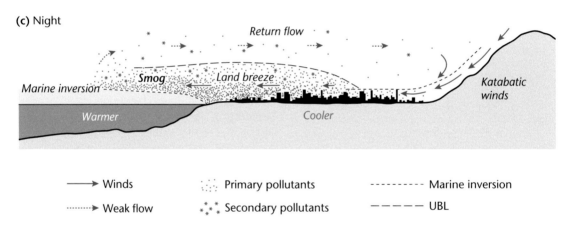

| → Winds | ⋮ Primary pollutants | - - - - - - - Marine inversion |
|---|---|---|
| ·····► Weak flow | ∗∗∗ Secondary pollutants | — — — — UBL |

**Figure 12.21** The development of a sea / land breeze system for an urbanized coastal plain adjacent to a mountain range. The diagrams show a time series of winds and air pollutants over the course of a warm summer's day.

composite of weather on scales from seasons to decades, weather refers to the short-term (hours to days and weeks) variations in the state of the atmosphere. It is these variations, which are usually the outcome of large-scale atmospheric motions acting in concert with topography that regulate the magnitude of urban atmospheric effects. Thus, for example, the calm clear conditions that allow strong UHI formation are associated with anticyclones; it follows then that the UHI climatology for a given city

depends partly on the frequency of anticyclonic conditions. In fact, many distinct urban effects (including poor air quality) are clearest during types of weather when **macro-** and mesoscale influences are weak and local surface-air exchanges dominate the response of weather near the ground. When regional winds are strong and skies are cloudy, the urban effect on many variables (apart from wind, naturally) is muted and non-urban weather controls are dominant. Of particular concern is where weather is extreme (i.e.

damaging to property or injurious to health) and the city can act to either ameliorate or enhance conditions.

## Return Periods of Events

Periods of unusually high/low values of any atmospheric property recorded over a period of time can be described as extreme (in a statistical sense); by definition then, what is unusual depends on what is normal for that climate. **Return periods** describe the average period of time that elapses between events (droughts, rainstorms, floods, wind speed, etc.) of a given magnitude. These are calculated from available historical weather records that are taken to represent climate at a place. This information can be used as the basis to examine natural hazards and assess the city's **exposure** to risk; as the historical record evolves (or climate changes), return periods may change. With few exceptions however, the weather events that generate hazards are much larger than cities and urban planning/design is needed to reduce the associated risks.

## 12.3.1 Storm Systems

A storm may be defined as an individual atmospheric system with associated winds, precipitation, cloud etc. For our purposes here we categorize storms into three types associated with decreasing intensity and increased size:

- **Tornadoes**: rotating columns of air connected to cumuliform clouds with diameters of ~1 km, lifetimes of $< 1$ h and winds exceeding 29 m s$^{-1}$.
- **Tropical cyclones**: typhoons and hurricanes with diameters of ~500 km, lifetimes of 7–10 days and maximum winds over 33 m s$^{-1}$.
- **Midlatitude cyclones**: associated with air masses and frontal systems; diameters of up to 5,000 km, lifetimes of 5–7 days and typical maximum wind speeds between 17 and 33 m s$^{-1}$.

Given the scale of, and energy contained in, larger atmospheric phenomena such as tropical cyclones, the urban influence may be very small outside of the city. However, within the settlement the urban landscape can modulate aspects of the timing, duration and magnitude of storm-related hazards such as wind damage. Moreover, as storm systems have distinct geographical distributions, some cities are more exposed to the associated hazards than others.

## Tornadoes

Tornadoes can form both as isolated storms and as part of larger mesoscale convective complexes; the tornadic system itself may be embedded within a larger scale organized circulation system, such as a midlatitude storm. There are few records of tornadoes that have crossed urban areas, but this may simply reflect the small size of urban areas when compared to the scale of these synoptic-scale systems and the nature of these storms which are small and short-lived. Nevertheless, cities and their residents can be vulnerable to the extreme winds owing to the dense concentration of people and of potential debris; so even if the hazard is a rare occurrence the risk may be great. During a tornado outbreak in May 1999, one storm that passed through the suburban area of Oklahoma City, United States had a maximum wind speed value of 135 m s$^{-1}$ at 32 m above the ground; the vortex itself was 350 m wide and caused catastrophic damage and loss of life (Wurman et al., 2007). While a rougher urban surface slows the ambient airflow, it also makes it more turbulent and potentially more damaging to structures. Building codes that specify construction materials and design standards are a common planning response.

Some research has indicated that the occurrence, duration and track of tornadoes can be affected by the presence of a city however this may be limited to weaker events. Fujita (1973) studied statistics for Chicago, United States, and Tokyo, Japan, and concluded that as cities grow the frequency of tornadoes in the centre decreases whereas the frequency in areas on their periphery increases. Indeed he speculated that this 'peripheral tornado belt' was related to the UHI with a mesoscale thermal front over the inner suburbs that could create significant **vorticity**. In a laboratory experiment he found the extra friction and warmth of the centre reduced tornado strength. However, there is insufficient observational data to test these ideas.

## Tropical Cyclones

Tropical cyclones form and move over tropical oceans, move within easterly winds and at maturity generate hurricane force winds ($> 33$ m s$^{-1}$). Although mature tropical storms occur relatively infrequently and are relatively small in diameter, many major coastal cities (such as Taipei, Taiwan and Hong Kong, China) are exposed to the associated hazards which

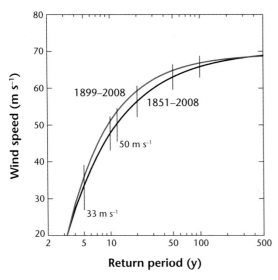

**Figure 12.22** Estimated windspeed (m s$^{-1}$) as a function of return periods for hurricanes passing within 100 km of the centre of Miami, United States. The solid line shows the return-level estimates for the period 1851–2008. The 90% confidence intervals are shown as vertical lines. The dotted lines show the estimates based on the shorter period 1899–2008 (Source: Malmstadt et al., 2010; © American Meteorological Society, used with permission).

include strong winds, prodigious precipitation and storm surge. Even cities located at higher latitudes (such as New York) can experience the effects of the hurricanes that recurve, moving from the Tropics into the mid-latitudes. There is no evidence that cities have an impact on tropical cyclones however, decisions on the location and type of urban development can dramatically affect exposure to the hazard and the associated risks.

Evaluating the nature of risk is a key to its management; which needs an estimate of the statistical character of the hazard. Figure 12.22 shows the return periods for wind speeds associated with hurricanes passing near Miami, United States, based on a model reanalysis of hurricane paths; note that, given the unpredictable movement of these storms, a 100 km 'buffer' was taken as a measure of the hazard. The analysis shows that winds of 33 m s$^{-1}$ have a return period of just 5 years and even exceptional winds (~50 m s$^{-1}$) have relatively short return

periods when compared to building lifetimes (< 20 years). These winds represent maximum values over water and will diminish inland due to surface roughness but buildings near the coastline will experience the brunt of the storm. In addition to strong winds, coastal areas may also experience wind-driven flooding as the storm pushes sea water inland (known as a 'storm surge'). This figure also shows that return period estimates are based upon the available record, which may be limited and give a false sense of security.

## Midlatitude Cyclones and Fronts

Much of the day-to-day weather of the mid-latitudes is regulated by the passage of cyclones that draw polar and tropical air masses into a low pressure centre. These air masses converge along distinct pathways that are separated by fronts which are transition zones between air masses; the passage of a front is often accompanied by a rapid change in weather conditions that affects wind, temperature, precipitation, etc. Again, the main influence of cities is largely contained to the near-surface layer, especially within the **urban canopy layer** (UCL).

When fronts move across an urban area, they can completely overwhelm and destroy urban effects. For example, Figure 12.23 shows the effects of synoptic fronts passing over Wroclaw, Poland. The effect on the UHI$_{UCL}$ is assessed using three stations: rural R1 12 km west of the centre, U in the city centre, and rural R2 18 km east of the centre. In the first panel a cold front arrived from the west at about 1100 h when there was no significant UHI$_{UCL}$ present. It introduced an increase in wind speed from 2 to 6 m s$^{-1}$, a change in wind direction, intense showers and a 7 to 8 K drop in air temperature in only 1 to 1.5 h which was felt at the three stations in succession. Using R1 as a rural reference the front produced a difference U–R1 with a 5 K *warm* anomaly peaking at 1220 h, but if R2 is the reference U-R2 has a 4 K *cool* anomaly peaking at 1250 h. Considering the temporal record of these differences without knowing a front had passed, it appears like a short-lived heat island, or cool island. In fact no urban effect was involved. On the other hand, the second panel shows how arrival of a nocturnal warm front over the same city caused the *destruction* of an existing UHI$_{UCL}$. Warm fronts are not as abrupt as cold ones, but in this case at

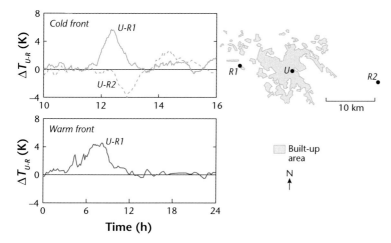

**Figure 12.23** Effects of a daytime cold front on 16 July 2001, and of a nighttime warm front on 11/12 April, 2001, passing the city of Wroclaw, Poland on air temperature differences. Stations: R1, R2 – rural (agricultural, LCZ D), U – urban (LCZ 2) (Source: Szymanowski, 2005; © Springer-Verlag, used with permission).

about 2030 h the 4 K $UHI_{UCL}$, that had started to develop as usual, was eliminated by widespread air mass replacement, for the rest of the night.

Large metropolitan areas can retard the movement of fronts and may trap a layer of air near the ground, which can be over-ridden by the air mass behind the front. Figure 12.19b shows the passage of a weak cold front moving southward over New York, United States, as detected in surface observations: note the close packing of isochrones on the upwind side of Manhattan (where the building density is greatest) as the frontal movement slowed to 1 m s$^{-1}$, just 25% of the upstream value. The rapid southward movement on the east side of the city is attributed to the presence of the UHI, which accelerated the movement of the front to 6 m s$^{-1}$ in this area. However, subsequent analyses of the surface and boundary layer observations suggested that the surface front never passed through the UCL in Manhattan. Instead, the upper part of the front proceeded above the city leaving a stagnant 'bubble' of air below (shown as the loop formed between the 0800 and 0900 isochrones in Figure 12.19b). Once across the city, the upper level front rejoined the segments of the surface front that flowed around Manhattan. The remnant bubble of air in Manhattan eventually dissipated (Gaffen and Bornstein, 1988).

### 12.3.2 Floods

Flooding in cities can occur as a result of intense and/ or long-lasting rainfall events (pluvial), river discharge exceeding channel capacity (fluvial) and sea-level rise (coastal). London, United Kingdom, represents an excellent case study as it typifies the situation of many urban areas. It developed on the floodplain of the Thames River, which experienced flooding as a result of both fluvial events and tidal surges. A series of flood defences accrued over time in response to individual events. A flood management system was put in place after an exceptional tidal surge event in 1953 caused by a large storm in the North Sea. The centrepiece of this system is the Thames Barrier, a series of gates that can be opened or closed to manage water levels. In concert with other barriers and 337 km of walls and embankments these defences protect a floodplain in which 1.25 million people live and work. The design standard allowed for a 0.1% chance of flooding in any year while accounting for a sea-level rise of 8 mm yr$^{-1}$; its estimated lifespan is 50 years and it has permitted urban development on land that would flood naturally on a regular basis (Lavery and Donovan, 2005).

The design capacity of urban storm-water systems must cope with events that might reasonably be expected, such as a hundred year rainstorm.

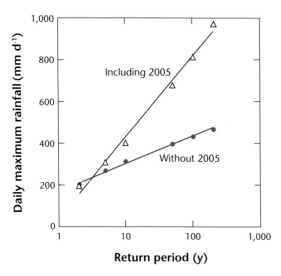

**Figure 12.24** Return periods for rainfall events at Santa Cruz Airport, Mumbai, India, which shows the impact of including the extreme 2400h precipitation event of July 2005 on return period estimates (Data source: Hallegatte et al., 2010).

Figure 12.24 shows a graph of return periods for rainfall events of varying intensities in Mumbai, India, a city of over 18 million. Its macroclimate is monsoonal and its annual precipitation of over 2,000 mm is concentrated in the summer months: an average of over 680 mm is received in July alone, often in the form of intense rainstorms. Yet, a single event can modify the statistics upon which these design curves are based. Over a 24-hour period on 26–27 July 2005 an intense thunderstorm moved over the city producing a prodigious downpour. At the nearby airport, the official meteorological station recorded 944 mm in 24 hours! The result was widespread flooding that left over 400 dead, damaged 175,000 houses and 40,000 commercial premises and disrupted the functioning of the city for weeks afterwards. The inclusion of this event in the historical record produces a new design curve for maximum 24-hour rainfall receipt (Hallegatte et al., 2010).

## 12.3.3 Heatwaves

There is no absolute definition of a **heatwave**. It refers to a sustained period of unusually high temperatures and therefore depends on the macroclimatic

background: a heatwave in Moscow, Russia, is different from one in Brasilia, Brazil, for example. Most often, it is used to refer to instances where relatively high temperatures (often accompanied by high **relative humidity**) cause **thermal stress** and discomfort in the population leading in some cases to hyperthermia and death. The atmospheric conditions that are commonly associated with heatwaves are caused by anticyclones that suppress uplift (clear skies) and generate little wind. Coincidentally, these are the same circumstances that generate strong surface and canopy-level UHIs (Chapter 7). In fact, there is strong evidence that the magnitude of the UHI is enhanced during heatwave events, which increases heat-related impacts in cities (Li and Bou-Zeid, 2013).

The nocturnal $UHI_{UCL}$ is of particular concern during heatwaves as it may hinder sleep and the body's recovery process. During August 2003, Western Europe experienced an exceptional period of dry, warm weather associated with a persistent anticyclone. The strong solar radiation and absence of rain resulted in very high surface and air temperatures that were well above the climate norm. France, in particular, experienced a severe heatwave from 4th to 18th August in which average daily temperatures exceeded 35°C and nocturnal minimum temperatures stayed > 20°C. During the same period the number of excess deaths[1] reached nearly 15,000, which were attributed to the hot weather. The impact was greatest in cities, where the population is concentrated but identifying the contribution of the urban climate effect is not easy owing to the number of confounding variables (e.g. age, poverty, and housing quality). The catastrophic outcome of this event was partly attributed to lack of preparation, which was explained in terms of the rarity of the event. However, it is only recently that the statistics of heatwave events have been examined with a view to determining their return periods. A recent study suggests that the meteorological conditions that underpinned this event have a return period of about 50 years (Charpentier 2011); large-scale climate change (Chapter 13) may act to shorten this period in the future and is a concern for urban planning, design and operations (Chapter 15).

---

[1] Excess deaths are calculated as the number of deaths in a period compared to the number of expected deaths, based on normal conditions.

**Summary**

The magnitude and relevance of any urban climate effect has to be put into the context of the background climate of a city, which is a product of geographical controls that act at a hierarchy of spatial and temporal scales and include the **macroclimate**, the **topography** (land-sea distribution, orography), and at finer scales, **synoptic** phenomena (day-to-day weather patterns).

- The **latitude** of a city primarily affects diurnal and seasonal patterns of daylight and the intensity of solar radiation receipt at the surface. Latitude controls the extent and timing of shadow patterns and is a critical variable in explaining seasonality, the rates and range of surface heating and cooling, and translates to building energy demands.
- Additionally proximity to oceans, the global atmospheric circulation patterns and resulting precipitation distributions have created favourable **macroclimates** for settlements. Climatically, there are two clusters where global population is geographically concentrated: **mild midlatitude seasonal climates** (about 40% of world population), and **tropical climates** (about 35% of world population).
- The **topographic setting** of a city can exert a strong influence on regional weather systems, providing shelter/exposure from/to strong winds, greater/less precipitation receipt, well/ poorly ventilated air basins, and **diurnal wind systems that recirculate urban air pollutants**. Site selection is a fundamental decision for buildings, neighbourhoods and cities that greatly affects the urban climate; wise (or fortunate) site selection can ameliorate many of the undesirable aspects of urban climate and protect from extreme events.
- **Synoptic** phenomena, such as midlatitude frontal systems or hurricanes are so powerful that urban-scale modifications cannot compete with their energy, yet their **passage through large urbanized areas can be retarded or slowed**. Extreme weather events are, by definition, unusual for a given place however their effect may overwhelm urban infrastructure, disrupt city functions and cause loss of life. Even though a hazard may be rare, exposure to the event and population vulnerability result in a higher significant risk in urbanized areas, that must be managed.

# 13 | Cities and Global Climate Change

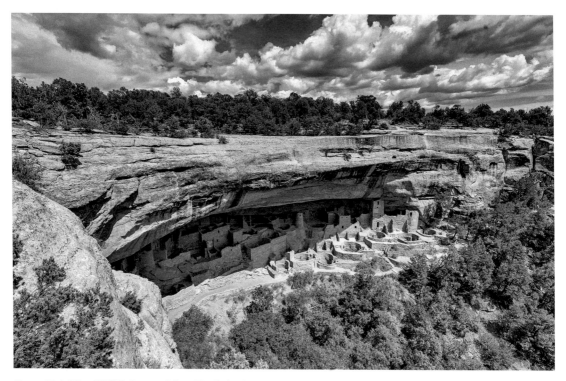

**Figure 13.1** The Cliff Palace at Mesa Verde in the Four Corners region of the Southwest United States. This is one example of many substantial settlements in the region that were abandoned in the thirteenth century (Credit: US National Park Service).

Urban areas are set within existing **macro**- and **mesos-cale** climates and many have developed systems for coping with seasonal variations in weather extremes, based on past experiences of events. However, while cities occupy fixed positions in space the climate system can undergo shifts over time that changes the statistics of climate. Historically, there is evidence that climate change has had significant consequences for settlements, leading in some cases to abandonment. Figure 13.1 shows the remains of a large settlement located in the Four Corners area of North America that archaeological evidence indicates was abandoned by the end of the thirteenth century. Its builders and occupants were known as the Anasazi, whose diet relied on maize, a crop that requires precipitation in the winter and summer. Tree ring evidence from the

region, which can be used as a proxy for precipitation, indicates that the abandonment occurred during a multi-decadal period of drought. It is hypothesized that the sustained reduction in precipitation reduced food production and placed great stress on the population. Although socio-demographic events may have been the proximate cause of migration from the area, the deteriorating environmental conditions played an important role (Benson et al., 2007).

Modern cities are less vulnerable to the effects of regional scale climate changes as their **urban metabolism** does not rely solely on the productivity of their regional hinterland. Rather, they are sustained by exchanges of fuel, food and materials over great distances. As the urban population (and per-capita consumption of natural resources) has increased, cities have progressively overcome regional impediments and their global impact has grown. Global urban coverage remains very small, between 2–3% of the Earth's landmass according to some estimates and less than 1% according to more restrictive definitions of what constitutes 'urban' (Schneider et al., 2009). Yet these places are associated with intense resource use and generate directly more than 50% of the **emissions** contributing to **anthropogenic** global climate change (GCC, see Figure 1.6). Paradoxically, the preferred locations of cities (mostly at low elevation, near a coast and in a river basin), increases their **exposure** to some projected changes such as sea-level rise and more intense precipitation events. In addition, existing hazards, such as **heatwave** events are projected to occur more frequently, which will add to existing warming that occurs in urban areas.

Not surprisingly then the role of cities has become a central part of the debate on managing GCC and their policy responses to the projected changes are categorized into two types. **Mitigation** actions seek to reduce the net emission of the **air pollutants** causing GCC through making cities more energy efficient, for example. **Adaptation** measures are focussed on ensuring that cities can cope with GCC. Many of these responses correspond with climate-based initiatives that focus on climate management at urban scales that are discussed in Chapter 15. This chapter provides the scientific basics to understand, measure and quantify the impact of cities on GCC and to determine and project the impact of GCC on cities.

## 13.1 Urban Impacts on the Global Climate System

Global urbanization in tandem with population growth, land-demand, energy consumption, industrialization and technology development are the main drivers of GCC. Increasing atmospheric **concentrations** of air pollutants – **greenhouse gases** (GHG) and **aerosols** – and large-scale changes in land use and land cover have perturbed the energy balance of the entire **troposphere**. The net effect is that we observe trends towards higher air temperatures over land areas, warmer ocean temperatures, sea-level rise and reductions in annual ice and snow-coverage; it is projected that these trends will continue and accelerate in the coming decades and that contrasts in precipitation between wet and dry regions and between wet and dry seasons will increase (IPCC, 2013).

The changes in the tropospheric energy balance can be quantified as **radiative forcing** (RF) in W m$^{-2}$. The concept of RF is used by climatologists to compare various anthropogenic and natural drivers of GCC and how they affect the troposphere. RF is defined as the change in **net allwave radiation** ($Q^*$) at the top of the troposphere (tropopause) after a **perturbation** relative to the undisturbed value before. It is estimated that the anthropogenically forced change to the global RF is between +1.1 and +3.3 W m$^{-2}$ relative to that in 1750, i.e. before the Industrial Revolution (IPCC, 2013). This RF is comprised of contributions from a number of sources, some positive (warming) and some negative (cooling). The strongest positive forcing is due to GHG (mostly $CO_2$) while aerosols, which in addition to their impact on **radiation** also affect clouds, have had a net cooling effect. **Albedo** change due to land use change, largely the conversion of forests to crop and grazing lands, has also caused a small negative RF. By comparison, natural variations in RF since 1750 due to changes in solar **irradiance** and volcanic eruptions that inject aerosols are estimated to have had small warming and cooling effects, respectively. The RF estimated for solar irradiance since 1750 is just +0.05 W m$^{-2}$, with a **precision** of ±0.05 W m$^{-2}$. The impact of volcanoes is linked strongly to the nature and timing of the eruption that can affect the composition of the stratosphere; however, the long-term RF is equally small.

Estimating the contribution of cities to the global RF values is not an easy task (it depends on how one assigns 'responsibility') but as concentrated areas of energy use they are foci for emissions of GHG and of aerosols. Folberth et al. (2012) provided a preliminary evaluation of the global RF for 34 megacities using a climate model. The model was run using a gridded global inventory of GHG and other climate-active air pollutants (such as aerosols) compiled for 2005. Subsequently it was run after the contributions of these megacities, estimated at between 3–12% of the total emissions, have been removed. The results suggest that the global RF that can be attributed to direct emissions from these cities is +0.15 W m$^{-2}$ and that this is comprised of a warming effect (+0.15 W m$^{-2}$) due mainly to $CO_2$ (+0.12 W m$^{-2}$) and a small cooling effect due to aerosols.

### 13.1.1 Land-Cover Change

The direct effect of urbanization on global land-cover change can be evaluated by using the **surface energy** balance (SEB, see Section 6.1); at a planetary scale, including oceans, the yearly and spatially averaged SEB is (Kiehl and Trenberth, 1997),

$$Q^* - (Q_E + Q_H) = 102 - (78 + 24) \quad (\text{W m}^{-2})$$

**Equation 13.1**

Note that, over a year or multiple years, the storage heat flux ($\Delta Q_S$) is negligible. This yields a planetary **Bowen ratio** of ~0.31, which is dominated by the ocean surface that covers 71% of the planet. The impact of cities can be estimated by evaluating their impact on the SEB and considering their relative area of the planet.

There is little difference between $Q^*$ over urban and **rural** areas over monthly or annual time scales (see Section 5.4.1), hence we may surmise that the direct RF associated with the change from natural to urban land cover is small. Cities are distinguished by the relative magnitude of the **sensible** ($Q_H$) and **latent heat flux densities** ($Q_E$). The $\beta$ value for cities is generally greater than 1 (see Section 6.4). However, as the area of the planet that is urbanized is very small ($< 1\%$), the direct effect of the urbanized landscape on the global SEB is negligible (Lamptey, 2010). The injection of anthropogenic heat by the urban metabolism, although its effects can be dramatic on the urban scale (Section 6.1), is also of minor global importance (Crutzen, 2004).

Of course, the global effects of cities are not confined to the urbanized areas alone (see also Section 1.2 and Figure 1.2). Satisfying the metabolic needs of cities has caused vast land areas to be converted from native vegetation cover to agriculture, mining, plantations, rangeland, and reservoirs; in addition assimilating urban wastes creates direct land changes (landfills), changes to aquatic systems (changes of sediment load and nutrients of coastal waters, algae blooms) and contamination to snow and ice surfaces (e.g. black carbon on snow). Moreover, air pollutants from cities that are advected over the surrounding landscape will affect their productivity (see Section 11.4).

### 13.1.2 Greenhouse Gases

GHGs absorb **longwave** radiation and contribute to the energy balance of the troposphere by re-emitting part of the absorbed radiation back to the surface and converting another part to **sensible heat** in the atmosphere (the 'greenhouse effect'). Each GHG absorbs longwave radiation in distinct parts of the electromagnetic spectrum (Figure 5.2).

The most important natural GHG by far is atmospheric water vapour, which is largely responsible for the planet's favourable climate that otherwise would be about 18 K cooler. Smaller changes in water vapour concentrations, such as those caused by urbanization and associated anthropogenic water injections (Section 8.1), will not change the RF of the global troposphere significantly. This is because the capacity of water vapour to absorb more longwave radiation is very limited given its current concentration. However, in spectral bands where water vapour is not effective in absorbing radiation, other GHGs (even those present in only trace amounts) can be very efficient in absorbing longwave radiation. If their **mixing ratios** increase, they can further narrow previously transparent bands of the longwave spectrum and change the RF of the entire troposphere, especially in the **atmospheric window** region (see Figure 5.2). The most relevant GHGs, in terms of their potential to change the RF by the troposphere and their emission rates due to human activities, are listed in Table 13.1. Also listed are their pre-industrial and current molar mixing ratios, typical **urban boundary layer** (UBL) mixing ratios, and **residence times**. Their dominant emission sources and processes in urban environments are given in Table 11.1.

### Carbon Dioxide

Carbon dioxide ($CO_2$) is the single most important anthropogenically-enhanced GHG with a direct anthropogenic RF of 1.94 W m$^{-2}$ (Table 13.1). If current rates of increase persist it will change the RF of the troposphere by up to 5 W m$^{-2}$ by the end of the twenty-first century. Average molar tropospheric mixing ratios of $CO_2$ have risen from 280 ppm in pre-industrial times to ~400 ppm at the time of writing this book. The major source of $CO_2$ is the **combustion** of fuels that contain fossil carbon (C), which has been sequestered by vegetation over millions of years. In the process of fuel combustion, that carbon is being released to the atmosphere as $CO_2$ over a very short period. Between 2002 and 2011, emissions due to fossil fuels was about 8.3 petagrams (Pg) C y$^{-1}$; a petagram equals $10^{15}$ g. Between 30–40% of these fossil fuel-related global $CO_2$ emissions originate directly from within urban areas, and over 70% are caused by the energetic needs of the urban population (Satterthwaite, 2008). Another 1.0 Pg C y$^{-1}$ are

**Table 13.1** Selected mixing ratios and properties of long-lived greenhouse gases (GHGs) that are enhanced by anthropogenic activities or are entirely synthetic. For emission sources and processes see Table 11.1 (Sources: NOAA ESRL, Global Monitoring Division, ORNL CDIAC 10.3334/CDIAC/atg.032 April 2016, and IPCC, 2013).

|  | Carbon dioxide ($CO_2$) | Methane ($CH_4$) | Nitrous oxide ($N_2O$) | Halogenated gases |
|---|---|---|---|---|
| Pre-industrial tropospheric background concentrations | ~280 ppm | ~700 ppb | ~270 ppb | 0 ppt (nonexistent) |
| Tropospheric background concentrations in 2016 | 400 ppm | 1,834 ppb | 328 ppb | 232 ppt (CFC-11) 516 ppt (CFC-12) 8.6 ppt ($SF_6$) |
| Typical observed mixing ratios in the urban boundary layer | ~400–450 ppm | ~1,950–2,100 ppb | ~320–400 ppb | - |
| Atmospheric residence time | 100–300 y | 12 y | 121 y | 45 y (CFC-11) 100 y (CFC-12) 3,200 y ($SF_6$) |
| 100-year Global warming potential per mass relative to mass of $CO_2$ | 1 | 28 | 265 | 4,660 (CFC-11) 10,900 (CFC-12) 23,500 ($SF_6$) |
| Anthropogenic radiative forcing (RF) on global troposphere in 2016 | 1.94 W m$^{-2}$ | 0.50 W m$^{-2}$ | 0.20 W m$^{-2}$ | 0.34 W m$^{-2}$ |

emitted due to land-cover changes such as biomass burning and loss due to deforestation that are also partly linked to the process of urbanization. Of this enhanced anthropogenic $CO_2$ flux of 9.3 Pg C y$^{-1}$, about 2.5 Pg C y$^{-1}$ is taken up by the world's oceans and by enhanced growth of the land vegetation (2.6 Pg C y$^{-1}$). The remaining 4.3 Pg C year$^{-1}$ is currently accumulating in the atmosphere causing increases in total tropospheric $CO_2$ mixing ratios of ~2 ppm y$^{-1}$ (Le Quéré et al., 2013).

To manage and reduce emissions from cities we must know how much $CO_2$ is released during combustion of various fuels. This can be estimated using **emission factors** for $CO_2$, which describe the mass of $CO_2$ emitted per energy released or per mass of fuel (Table 13.2). Not all fuels have the same energy efficiency; coals, followed by gasoline, yield the highest $CO_2$ emissions, while less $CO_2$ is emitted when burning natural gas to supply the same energy. Emission factors for a given fuel can also differ slightly depending on its geographic origin and how it is refined and burned. In comparison, combustion of hydrogen gas does not result in any $CO_2$ (and only creates water vapour) and, from a direct emission perspective, is the only entirely clean fuel in Table 13.2. It is further notable that wood and other

biofuels do release $CO_2$, but as their carbon content has been sequestered in recent times and will be replenished by new growth they do not release new $CO_2$ to the climate system; in other words, biofuels just accelerate the carbon cycle. Nevertheless the processing, transport and distribution of biofuels also requires energy usually provided in part by fossil fuels, and their growth causes land-cover changes. The emission factors in Table 13.2 account only for direct emissions of $CO_2$ and do not include other GHGs emitted as by-products and $CO_2$ emissions that were generated during production, processing and transport of the fuels.

$CO_2$ is further exchanged between urban and other land surfaces and the atmosphere by **respiration** and **photosynthesis**. These two biological processes are relevant when discussing the potential for direct carbon sequestration of urban vegetation (see Section 13.2.3), or when assessing the change in productivity of ecosystems downwind of cities affected by air pollutants in the **urban plume** (see Section 10.4).

### Methane

Methane ($CH_4$) is the second most relevant GHG in the atmosphere with a direct anthropogenic RF of

**Table 13.2** Typical $CO_2$ emission factors for common fuels sorted by increasing $CO_2$ emission per energy released in the combustion process. Specific carbon or $CO_2$ emission is the emission per kg fuel burned (Sources: Demirel, 2012, NAEI, 2012).

| Fuel | $CO_2$ emission per energy g $CO_2$ MJ$^{-1}$ | Specific $CO_2$ emission kg $CO_2$ (kg fuel)$^{-1}$ | Specific carbon emission kg C (kg fuel)$^{-1}$ |
|---|---|---|---|
| Hydrogen gas | 0 | 0 | 0 |
| Natural gas | 51.3 | 2.75 | 0.75 |
| Liquefied petroleum gas | 65.4 | 3.01 | 0.82 |
| Diesel | 69.5 | 3.15 | 0.86 |
| Conventional gasoline | 74.3 | 3.30 | 0.90 |
| Biodiesel | ~75[1] | 3.17[1] | 0.87[1] |
| Fuel oil (residential heating) | 90.4 | 3.23 | 0.88 |
| Coal | 85–200 | 2.75 | 0.75 |
| Wood | 90–110[1] | 1.7–2.0[1] | 0.45–0.55[1] |

[1] No fossil fuel-release C that has been assimilated by plants in modern times.

currently $+0.49$ W m$^{-2}$ (Table 13.1). Methane is emitted as a result of biogenic and non-biogenic processes and once released, most is gradually transformed into $CO_2$ through oxidation; although it has a relatively short lifespan, its spectral properties make it a more powerful GHG than $CO_2$; 70% of all global anthropogenic $CH_4$ emissions are caused by biogenic (microbial) processes that take place in rice paddies, ruminants (i.e. livestock), landfills and sewage treatment plants. The remaining 30% arises from non-biogenic sources such as fossil fuel mining and burning, biomass burning and from leaks in natural gas pipelines. Globally, mixing ratios of $CH_4$ have more than doubled from pre-industrial times mainly as a result of changing farming practices and food production but a proportion is directly attributable to processes in cities due to fossil fuel distribution, combustion, landfills and wastewater treatment. Wunch et al. (2009) used the correlation between $CO_2$ and $CH_4$ emissions observed in Los Angeles, United States, to suggest that the urban contribution to global anthropogenic methane emissions may be 21–34%.

Direct emissions of methane in urban areas can be attributed to three sources. First, leaks in urban distribution networks of natural gas, which contains $> 80\%$ methane, can be significant. Second, burning gasoline,

natural gas, oil and coal is not completely efficient and produces traces of $CH_4$. Third, anaerobic microbial respiration (in particular a problem in landfills and sewage plants) causes substantial emissions of $CH_4$ in urban areas. Globally, the $CH_4$ emissions from landfills and wastewater account currently for about 90% of waste sector emissions, or about 18% of global anthropogenic $CH_4$ emissions (Bogner et al., 2008). However, emissions from landfills vary considerably based partly on their organic content and the management. Many cities put infrastructure in place to harvest $CH_4$ emissions from landfills for energy production.

### Other Greenhouse Gases

Other GHGs responsible for GCC include nitrous oxide ($N_2O$), ozone ($O_3$) and halogenated compounds (CFC, HFCs). Nitrous oxide, with a direct RF of $+0.20$ W m$^{-2}$, is emitted in cities as a by-product of fuel combustion, primarily from vehicles with catalytic convertors. Additional emissions come from heavily managed urban greenspace, which use fertilizers. Globally, molar mixing ratios of $N_2O$ in the troposphere have risen from 270 ppb to $> 320$ ppb (Table 13.1). Ozone ($O_3$) near the ground is a short-lived **secondary pollutant** with a RF of $+0.40$ W m$^{-2}$

and is closely associated with vehicular emissions in cities (see Section 11.3.2). Also a number of synthetic halogenated compounds (e.g. CFC-11, CFC-12 and $SF_6$) originate from industrial and consumer sources. They are essentially inert, hence have very long atmospheric residence times and are highly effective GHGs (Table 13.1). Although emissions of many of these gases are diminishing over time, measurements downwind of New York, United States, continue to find elevated values in air masses that can be sourced to distant urban areas (Santella et al., 2012).

### Equivalent Carbon Dioxide Emissions

To compare the impact of different GHGs on the global climate system, the metric of **global warming potential** (GWP) is useful. GWP is a factor that defines the RF per mass of a GHG relative to that of $CO_2$ for a given time horizon, e.g. over a 100 year time frame (Table 13.1). For example the 100-year GWP of $CH_4$ is 28, which means over 100 years 1 kg $CH_4$ is as powerful in changing the tropospheric RF as 28 kg of $CO_2$. For comparison metrics, and in emission inventories, GHG emissions of a mixture of $CO_2$, $CH_4$, $N_2O$ and other GHGs can be expressed in **equivalent carbon dioxide emissions** ($CO_2e$). $CO_2e$ is obtained by multiplying the emitted mass of all GHGs by their corresponding GWP for a selected time horizon. For example, if a city emits 3,000 kt $CO_2$ $y^{-1}$, 6 kt $CH_4$ $y^{-1}$ and 1 kt $N_2O$ $y^{-1}$ then this results in 1 × 3,000 + 28 × 6 + 265 × 1 = 3,433 kt $CO_2e$ $y^{-1}$, where 28 and 265 are the GWPs for $CH_4$ and $N_2O$ respectively (Table 13.1).

## 13.2 Greenhouse Gas Emissions from Cities

### 13.2.1 Tracking Carbon in Cities

By far the most important GHG emitted from cities is $CO_2$. Hence a more detailed study of the origin and fate of $CO_2$ in cities helps us to understand and mitigate emissions. For any given system, carbon, as any other biogeochemical relevant element, is conserved. Hence it can be tracked in an urban system by quantifying inputs and outputs and storage changes within the system to quantify and track $CO_2$ emissions (Figure 13.2). Although total carbon is conserved the chemical forms of carbon can change in chemical reactions. It is useful to separate carbon inputs and outputs of an urban system into lateral fluxes and vertical fluxes (Churkina, 2008).

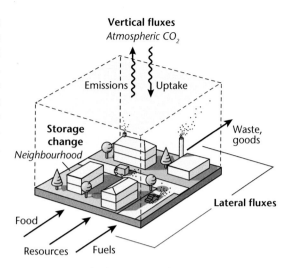

**Figure 13.2** Vertical and lateral fluxes of carbon in and out of an urban system (Modified after: Kellett et al., 2013; Crown Copyright © 2012 Published by Elsevier B.V., used with permission).

Lateral fluxes (Figure 13.2) are entirely anthropogenic processes and are essential for the functioning of a modern city. As part of the urban metabolism carbon is imported in a variety of forms. These include fossil fuel (gasoline, diesel, natural gas), food, and construction material or goods (timber, plastics). There are also exports of carbon in the form of goods, and solid or liquid waste products that remove carbon from an urban system. Lateral fluxes typically transport carbon as solid or liquid organic carbon not gaseous $CO_2$ – an exception is carbon contained in natural gas entering the city through pipelines.

Vertical fluxes (Figure 13.2) refer to the exchange of carbon between the urban surface and the atmosphere and result from three specific processes: combustion (C), respiration (R), and photosynthesis (P). Combustion and respiration inject carbon in the form of $CO_2$ into the atmosphere, while photosynthesis captures atmospheric $CO_2$ and transfers it back to carbohydrates and other organic carbon in biomass of urban vegetation and soils. All three processes can be quantified as mass flux densities (kg C $m^{-2}$ $y^{-1}$).

### Combustion

In urban areas, a large variety of fuels are used, including fossil fuels (coal, oil, diesel, gasoline and natural gas) and biofuels (wood, bioethanol) that, when combusted, release $CO_2$ (Table 13.2). Most fuel

**Table 13.3** Total estimated greenhouse gas emissions for cities and metropolitan regions (Data source: Kennedy et al., 2010).

| City | Administrative unit | Population in millions | Total GHG emissions (Mt $CO_2e$) | Per-capita GHG emissions (t $CO_2e$ cap$^{-1}$) |
|---|---|---|---|---|
| Athens, Greece | Metropolitan Region | 3.99 | 41.5 | 10.4 |
| Barcelona, Spain | City | 1.61 | 6.8 | 4.2 |
| London, UK | Greater London | 7.36 | 70.7 | 9.6 |
| Madrid, Spain | Comunidad de Madrid | 5.96 | 41.1 | 6.9 |
| Oslo, Norway | Metropolitan Region | 1.04 | 4.0 | 3.5 |
| Paris, France | City | 2.13 | 11.1 | 5.2 |
| Rotterdam, Netherlands | City | 0.59 | 17.6 | 29.8 |
| Los Angeles, United States | County | 9.52 | 124.0 | 13.0 |
| New York City, United States | City | 8.17 | 85.8 | 10.5 |
| Toronto, Canada | Greater Toronto Area | 5.56 | 64.5 | 11.6 |
| Sao Paulo, Brazil | City | 10.43 | 14.6 | 1.4 |
| Bangkok, Thailand | City | 5.66 | 60.6 | 10.7 |
| Beijing, China | Beijing Province | 15.81 | 159.7 | 10.1 |
| Tokyo, Japan | Metropolitan Government | 12.68 | 62.1 | 4.9 |

consumption data are available at coarse (national, regional) scales at aggregate time scales (annual at the most) so that estimating $C$ at detailed spatial and temporal scales in cities is not simple. The methods to downscale fuel combustion are similar to those employed to estimate the **anthropogenic heat flux** (see Section 6.2) or emission inventories for air pollutants (see Section 11.3.3) and may be based on bottom-up or top-down procedures. Unfortunately, comparing combustion-derived GHG emissions of different cities (and establishing the per-capita flux) is more difficult still because information is usually collated for administrative units that rarely correspond to the built-up area. Moreover, there are methodological issues such as, how to account for a remote power plant that consumes fossil fuels and causes $CO_2$ emissions yet delivers the energy to a city via an electricity network.

These complications are made clear in Table 13.3 where total and per-capita emissions of GHGs for 14 cities are listed, based on fuel consumption data. The modal range lies between 10 and 20 t $CO_2e$ per capita but there are places with far higher and lower emissions that reflect the activities in the city and the source of energy. For example, Oslo, Norway has low per-capita emissions (3.1 t $CO_2e$ cap$^{-1}$ y$^{-1}$) that reflect its source of energy (hydropower and nuclear) and low level of industrialization whereas Rotterdam, Netherlands has a large oil refining industry explains its high value (30 t $CO_2e$ cap$^{-1}$ y$^{-1}$). Few cities have comparable data so useful relationships are difficult to establish with certainty. In Figure 13.3, the per-capita $CO_2e$ emissions for an unrepresentative mix of ten cities are based on end-use activities (that is, the values represent demand within the city even though some emissions may arise outside its administrative

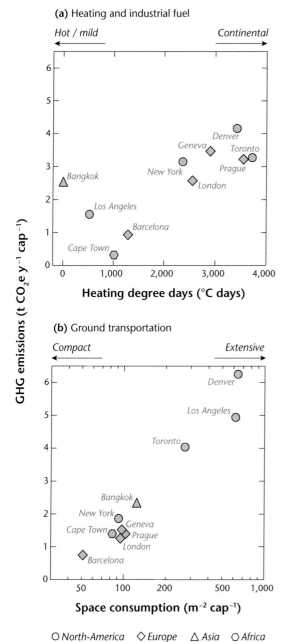

**Figure 13.3** Per-capita GHG emissions from various world cities expressed as equivalent carbon dioxide emissions ($CO_2$e). **(a)** GHG emissions from heating and industrial fuel against heating degree days (using a base temperature of $18°C$) and **(b)** GHG emissions due to transport as a function of space consumption (the inverse of population density) (Data source: Kennedy et al., 2009).

boundaries). Figure 13.3a plots the emissions resulting from heating and industrial fuel use against **heating degree days**. For most of these cities, energy demand is driven by the need for heating in domestic and commercial buildings and hence by the macroclimatic setting. Bangkok, Thailand being tropical, has no heating degree days rather its energy needs are related to industrial demand. The relationship between energy demand for ground transport and the population density (expressed in terms of area per person) is clearer (Figure 13.3b), suggesting that the urban density and metabolism are closely related and that denser cities have less per-capita emissions for transport.

## Photosynthesis

Photosynthesis is the biological process of carbohydrate formation through assimilation of atmospheric $CO_2$ in the chlorophyll-containing tissues of plants and requires daylight, i.e. **photons** in the range of **photosynthetically active radiation** (PAR, $\sim$400 – 700 nm). The rate of photosynthesis ($P$) for a given plant is controlled by the quantity and distribution of PAR, but also canopy-layer air temperatures, vapour density deficit of the air, water stress of the plants, and $CO_2$ mixing ratios of the ambient air. Vegetation (trees, lawns, gardens etc.) in cities is expected to show generally higher $P$ per leaf area compared to unmanaged native species, or crops in rural areas. The urban environment provides more water (from irrigation) and warmer ambient air temperatures (the urban heat island) that advantages plant growth and extends growing periods. Nevertheless the differences greatly depend on the land-cover change, **macro-** and **microclimates**, and also the availability of radiation in the **urban canopy layer** (UCL).

## Respiration

Respiration is the reverse biological process to photosynthesis and describes the oxidation of carbohydrates and other organic carbon by living organisms that releases $CO_2$ from plants, animals, humans and waste into the atmosphere. It unlocks metabolic energy to maintain living functions. Since respired $CO_2$ was captured by plants during photosynthesis the $CO_2$ released by respiration has previously been in the atmosphere. Biologically, we distinguish between autotrophic respiration (vegetation) and heterotrophic respiration of living-forms that consume biomass to live (microbes, humans, animals). The rate of respiration ($R$) in an urban area can be alternatively separated into soil respiration ($R_s$), above-ground plant respiration (leaves, stems, autotrophic, $R_a$), and human and animal respiration respiration ($R_m$)

Major controls on the rate of respiration in urban soils ($R_s$) include soil temperatures, soil water content and soil organic carbon and the supply of organic matter through litter and clippings. $R_s$ generally increases with increasing soil temperature. There is also an optimal range of soil water content, below which water is limiting and above which oxygen limits microbial activity (anaerobic conditions) and hence $R_s$. Irrigation, fertilization and elevated subsurface temperatures due to the urban heat island generally cause an increase in $R_s$ in urban areas. Typically, urban soils respire up to $R_s = 10$ µmol m$^{-2}$ s$^{-1}$ (about 0.4 kg m$^{-2}$ y$^{-1}$) where they are not paved or built-upon. Koerner and Klopatek (2002) compared $R_s$ in rural and quasi-urban lots (golf courses) in the desert climate of Phoenix, United States, and found that urban soils with high irrigation and intensive management showed up to ten times higher $R_s$ compared to the native soils in the city's dry hinterland.

Human respiration ($R_a$) is regulated by human activity cycles; typical human respiration rates reported for adults range between 70 and 100 kg C y$^{-1}$ cap$^{-1}$. Human respiration can be a significant component of the total emissions of an urban **neighbourhood**. For example it makes up 6% of daytime emissions of $CO_2$ in central Mexico City (Velasco et al., 2009). On the whole, neither soil or human respiration add new carbon to the global climate system, rather they cycle what has been assimilated by plant photosynthesis.

### An Urban Carbon Cycle?

Figure 13.4 is an attempt to quantify annual lateral and vertical fluxes, along with typical pools (stores) of carbon, for an urban residential neighbourhood in Vancouver, Canada. Arrow width denotes the magnitude of the carbon transported per time and area in kg C m$^{-2}$ y$^{-1}$. The area of the pools depicted is proportional to the amount of carbon located in the urban system. In this example, $+6.7$ kg C m$^{-2}$ y$^{-1}$ is imported into the system by lateral fluxes and about 90% (5.9 kg C m$^{-2}$ y$^{-1}$) of this carbon leaves the system by way of $CO_2$ to the atmosphere. Only a small part exits the system as a lateral output in the form of waste ($-0.8$ kg C m$^{-2}$ y$^{-1}$). The main message of this diagram is that pathways of carbon in the urban metabolism are more linear than cyclic. More than 95% of the

carbon in this urban system has been imported laterally from outside the system and a large fraction is fossil carbon. This is a fundamental contrast to the carbon cycle found in most other terrestrial ecosystems in which nearly all stored carbon is taken up by plants through $P$ from the atmosphere and is returned to the atmosphere by $R$ within the same system. In most terrestrial systems not dominated by human activity, lateral fluxes are minor. In contrast, carbon in an **urban ecosystem** at the scale of cities, neighbourhoods and buildings must be approached as an open system with throughput rather than cycling.

### Distinguishing among Sources of Atmospheric Carbon

Identifying biogenic and anthropogenic sources of carbon in urban ecosystems can make use of isotopic composition of the gas under study. For example, several isotopologues of $CO_2$ and $CH_4$ exist in the atmosphere, which are variants where for example the dominant carbon isotope $^{12}$C is replaced by the stable $^{13}$C or the radioactive $^{14}$C (radiocarbon). The radiocarbon $^{14}$C isotope is a powerful indicator of the origin of the $CO_2$ or $CH_4$ in the urban atmosphere. The radioactive $^{14}$C is constantly formed by cosmic rays in the higher atmosphere and is then taken up by plants and cycled through the biosphere (plants, soils, animals, humans) and decays with a half-life of 5,700 years. So the amount of $^{14}$C left in a C-containing compound is proportional to the time since the carbon was assimilated by vegetation. The typical turnover rate in the biosphere is short compared to its half-life, so for example $CO_2$ originating as respiration will contain a fraction of the original atmospheric $^{14}$C. However, $CO_2$ originating from combustion of fossil fuels contains virtually no $^{14}$C, because the carbon was taken up by plants millions of years ago and all $^{14}$C has long since decayed. The isotope method allows us to separate between $CO_2$ that is formed by combustion of fossil fuels (no $^{14}$C) vs that originating from biological respiration (contains some $^{14}$C). This fraction of $^{14}CO_2$ relative to the dominant $^{12}CO_2$ is termed $\Delta^{14}$C. In the urban atmosphere $\Delta^{14}$C is lower than in rural or pristine atmospheres. For Krakow, Poland, Kuc et al. (2003) reported a distinct seasonality in $\Delta^{14}$C with a local minima during winter months when the combustion of fossil fuels due to space heating is increased. Higher fractions of $^{14}CO_2$ in the UBL are observed during summer months when respiration dominates.

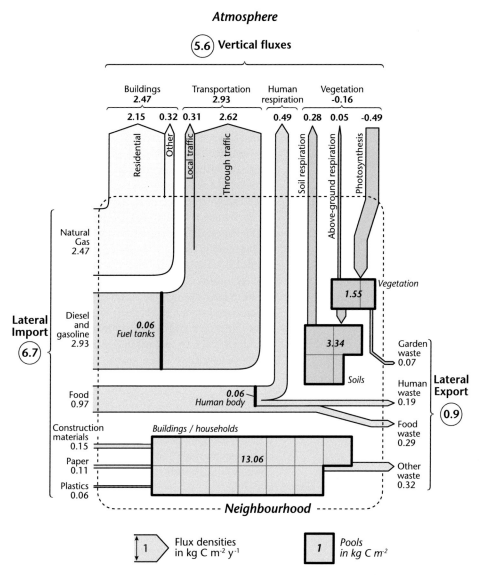

**Figure 13.4** Fluxes and pools of carbon cycled through a typical urban ecosystem representing a residential neighbourhood in Vancouver, Canada. The width of arrows corresponds to the magnitude of fluxes, the size of pools is proportional to the carbon stored (Source: Kellett et al., 2013, Crown Copyright © 2012 Published by Elsevier B.V., used with permission).

Isotopes can be used to study carbon cycling in urban ecosystems as they may act as a **tracer** that follows the path of carbon via different fluxes and pools. As an example, Figure 13.5 shows $^{14}C$ in plant material of urban grasses adjacent to a major highway in Paris, France. The graph shows that plants near the highway have less $^{14}C$ in their plant material compared to plants further away. This is because the grass

assimilates more fossil fuel $CO_2$ near the highway (up to ~12% at the edge), where proportionally more $CO_2$ comes from fuel burning (which is $^{14}C$ free). The contribution of the fossil fuel source declines with distance from the highway as proportionally more background $CO_2$ is assimilated.

Similar considerations can be applied to other isotopes in GHGs such as the ratio between $^{12}C$ and

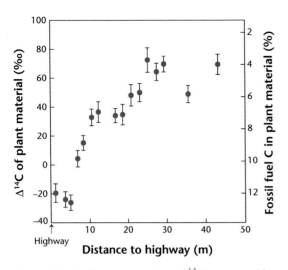

**Figure 13.5** Radiocarbon content ($\Delta^{14}C$) measured in grass biomass as a function of distance to an urban highway in Paris, France. The right axis shows the estimated fraction of the carbon in plant biomass that is originating from fossil fuel emissions at the highway (Data source: Lichtfouse et al., 2005).

$^{13}C$ which allows separation of $CO_2$ from gasoline and natural gas combustion. Studying multiple isotopes at the same time helps us to constrain uncertainties about the sources of GHGs in urban environments (e.g. Djuricin et al., 2010).

### 13.2.2 Directly Measuring Greenhouse Gas Emissions from Cities

There are several approaches available for mapping emission and uptake in urban systems, which can be sorted by scale Christen (2014):

- At the **microscale** we can determine emissions from individual urban elements or track plumes of emitters in the UCL.
- At the **local scale** using flux measurements in the **inertial sublayer** (ISL) on towers, we can determine the GHG emission profiles of urban neighbourhoods.
- At the **city or regional scale,** using airborne platforms located in the **mixed layer** (ML) of the UBL or satellite-based systems.

#### Measuring Emissions from Urban Elements and in the UCL

Bottom-up emission inventories at urban scales rely on emission factors that express the amount of GHG

emitted per mass of fuel burned or energy converted (see Section 11.3.3). If consumption data are available for the contributing sources (vehicle type, heating units, etc.), then fuel use can be converted into the equivalent $CO_2e$ emissions (Table 13.2); it is important to state however, that these factors often refer to measurements made under controlled, laboratory conditions and that the actual emissions may be higher under normal operation. For biological fluxes of $P$ and $R$, ecophysiological or allometric measurements of urban vegetation may help to identify the emission and sequestration rates, which is relevant to assess the potential for GHG emission to be offset by uptake by urban vegetation (Weissert et al., 2014).

The intense GHG emissions in cities result in higher mixing ratios in the UCL compared to remote locations (Table 13.1). Figure 13.6 shows $CO_2$ mixing ratios measured on board a mobile system traversing Phoenix, United States, through the centre of the Metropolitan region extending into rural (desert) background on both sides. There is a clear relation between the measured mixing ratio of $CO_2$ and urban density, similar in form to the **canopy layer urban heat island** ($UHI_{UCL}$, e.g. Figure 7.3); note however that enhanced urban $CO_2$ is not a significant contributor to the $UHI_{UCL}$ (Balling et al., 2001). Some have mooted that cities might be considered 'laboratories' in which to study future global climate scenarios owing to the combination of elevated $CO_2$ concentrations and higher temperatures due to the $UHI_{UCL}$ (Section 7.3), but their role as analogues is very limited since other atmospheric constituents and their dynamics are unlikely to mimic future urban or rural atmospheres.

Mobile traverse systems have been used to map individual plumes or identify areas of high emissions of GHGs released primarily by traffic in the UCL (Lee et al., 2017). A similar approach has been used to identify and map leaking natural gas pipe infrastructure in cities by Phillips et al. (2013). They used a mobile gas analyzer to examine leaks in natural gas pipelines through excessively high concentrations of $CH_4$ measured in the UCL on a vehicle platform as they drove across Boston, United States. They identified and mapped more than 3,000 leaks in the system, most of which were associated with older cast iron pipes that had deteriorated over time.

#### Measurements of GHG Fluxes above the Roughness Sublayer

Much of our direct knowledge of the dynamics of urban GHG fluxes comes from instruments located

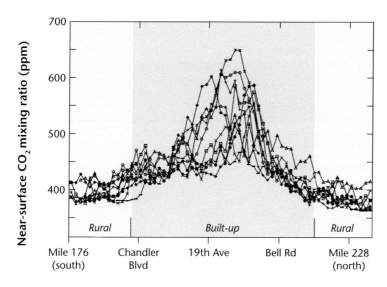

**Figure 13.6** Early-morning mixing ratios of the greenhouse gas $CO_2$ measured on ten weekday car transects across the Phoenix Metropolitan Area, United States (Source: Idso et al., 2001, © Elsevier Science, used with permission).

on towers within cities that employ the **eddy covariance** approach (EC), which has been described previously (see Section 6.5 for determination of $Q_H$ and $Q_E$). These data can be used to validate emissions based on inventories. For example, an EC system (Figure 6.14) to measure the vertical flux of $CO_2$ ($F_{CO_2}$) is composed of a fast **anemometer**, which provides the fluctuations of the vertical wind ($w'$) and a co-located gas analyzer that provides molar concentration fluctuations of $CO_2$ or any other long-lived greenhouse gas ($\chi'$). The flux is obtained from the covariance of $w$ and $\chi$:

$$F_{CO_2} = \overline{w'\chi'} \qquad (\mu mol\ m^2 s^{-1}) \qquad \textbf{Equation 13.2}$$

EC systems must be operated in the inertial sublayer (ISL) above the surface of interest and the flux observations respond to emissions from a turbulent **source area** upwind (see Section 3.1.1 and Figure 3.3). The resulting molar **flux density** $F_{CO_2}$ represents the net effect of combustion, respiration and photosynthesis, for a portion of the urban landscape. Guidelines for performing EC measurements of GHGs in cities are discussed in Feigenwinter et al. (2012).

Table 13.4 lists measured annual totals of $F_{CO_2}$ for ten selected long-term EC towers in urban areas, where at least one full year of measurements is available. Those measurements should not be seen as representative of an entire city; some of these EC towers have busy arterial roads within their source area, while others are located in remote residential areas

with little or no through traffic. Nevertheless, all sites indicate that urban surfaces are – unsurprisingly – emission sources of $CO_2$. The measurements from all mid-latitude sites indicate significantly higher $F_{CO_2}$ in winter and lower $F_{CO_2}$ during summer, which is a combination of two controls: the increased heating demand in winter (combustion) and increased photosynthesis by urban vegetation in summer (sequestration). However, the former dominates because $CO_2$ uptake by urban vegetation is in nearly all cases not strong enough to offset anthropogenic $CO_2$ emissions even in summer.

The diurnal and seasonal dynamics of measured $F_{CO_2}$ are illustrated for two contrasting EC tower sites in Figure 13.7. Figure 13.7a shows flux measurements over a highly vegetated residential **suburban** neighbourhood in Baltimore, United States, while Figure 13.7b is over a densely developed inner-city neighbourhood in Basel, Switzerland. The suburban Baltimore site shows a consistent carbon sink during summer, with a net $CO_2$ uptake by urban vegetation during daytime. This site is characterized by 54% tree coverage, and the absence of any significant road traffic in the source area makes uptake by $P$ the dominant driver during summer days. Nevertheless, compared to a typical daily net-uptake by grassland and forest cover in the region, uptake is small, and suggests residual emissions from traffic-related fuel combustion even in summer (Crawford et al., 2011). The Basel site is dominated by traffic, which peaks in

**Table 13.4** Measured seasonal and annual daily average $F_{CO_2}$ at selected urban eddy covariance (EC) towers. Codes refer to Table A3.1 where additional information on sites, surface cover and bibliographic sources are listed (Source: Christen, 2014).

| Site | Code / Source[1] | LCZ | $\lambda_b$ | Daily $F_{CO_2}$ (g C m$^{-2}$ day$^{-1}$) | Daily $F_{CO_2}$ (g C m$^{-2}$ day$^{-1}$) | Annual $F_{CO_2}$ (kg C m$^{-2}$ year$^{-1}$) |
|---|---|---|---|---|---|---|
| **Mid-latitude cities** | | | | Summer | Winter | Year |
| Baltimore – Cub Hill, United States | Bm02 | 6 | 14% | −0.9 | +3.8 | **+0.35** |
| Montreal – Roxboro, Canada | Mo08s | 6 | 12% | +0.5 | +8.2 | **+1.48** |
| Swindon, UK | Sw11 | 6 | 44% | +0.6 | +9.5 | **+1.84** |
| Melbourne – Preston, Australia | Mb03m | 6 | 46% | +5.9 | +7.6 | **+2.31** |
| Lodz – Lipowa, Poland | Lo06 | 2 | 30% | +6.0 | +9.9 | **+2.94** |
| Tokyo – Kugahara, Japan | Tk01 | 3 | 33% | +6.1 | +13.6 | **+3.35** |
| Basel – Klingelbergstr, Switzerland | Ba04 | 2 | 38% | +6.8 | +10.6 | **+3.63** |
| Essen – Grugapark, Germany | Es07[1] | 3 | 59% | +5.9 | +19.5 | **+3.94** |
| Montreal – Rue des Écores, Canada | Mo08u | 3 | 27% | +9.8 | +23.4 | **+5.64** |
| Vancouver – Sunset, Canada | Va09s | 6 | 29% | +16.4 | +21.4 | **+6.71** |
| **Subtropical and tropical cities** | | | | Dry season | Wet season | Year |
| Singapore – Telok Kurau, Singapore | Si06 | 3 | 39% | +5.10 | +4.31 | **+1.79** |

[1] See Appendix A2 for details on sites and data sources.
[2] Urbanized wind sector only.

the morning and evening. It has even higher emissions in winter, due to space heating. Only in summer at mid-day, is uptake by urban vegetation just able to offset fuel emissions, causing $F_{CO_2}$ to be near zero.

Establishing general relationships between $F_{CO_2}$ and aspects of **urban structure** and metabolism is hampered by the paucity of measurements at this time but some relations can be abstracted from existing data. Figure 13.8 shows summertime $F_{CO_2}$ data for 20 urban EC towers plotted against building density ($\lambda_b$) which reveals a strong positive correlation. Only the EC towers located in the highly vegetated, low-density suburban areas in Montreal, Canada, (Mo08s), Baltimore, United States (Bm02), and a

park in Essen, Germany (Es07[p]), show weak negative values where $P$ offsets local anthropogenic emissions. Note this relationship does not mean that per-capita emissions also increase with increasing $\lambda_b$, in fact the evidence for population density (Figure 13.3b) suggests the opposite.

Control of emissions by the urban metabolism is evident in Figure 13.9, which shows monthly measured $F_{CO_2}$ from a residential site in Vancouver, Canada, against heating degree days. There is a clear positive, linear relationship between heating demand and measured $F_{CO_2}$. This is because the majority of homes combust natural gas to meet their heating needs; the resulting emissions are picked-up by the EC system. If local emission sources in the tower's

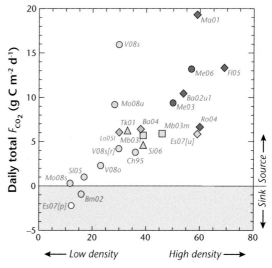

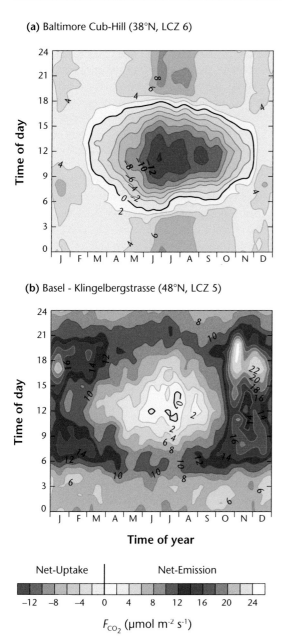

(a) Baltimore Cub-Hill (38°N, LCZ 6)

(b) Basel - Klingelbergstrasse (48°N, LCZ 5)

**Figure 13.8** Measured averaged summertime $F_{CO_2}$ from 20 flux towers in urban environments vs the plan area fraction of buildings. Site codes refer to the list of experimental sites compiled in Table A2.1, where additional information and bibliographic sources are given. Sites Si06, Me03 and Me06 show values from the dry season instead of summer (tropical and subtropical cities) (Source: Christen. 2014; © Elsevier, used with permission).

**Figure 13.7** Isopleths of measured $F_{CO_2}$ at two urban flux towers as a function of time of year ($x$-axis) and time of day ($y$-axis). Positive values indicate net-emissions of $CO_2$, negative values are a net-uptake of $CO_2$. The averages are based on continuous long-term measurements over 5 years each (Data sources: Crawford et al., 2011, Lietzke et al., 2015).

source area are known, $F_{CO_2}$ measured on flux towers can be a powerful tool to validate fine-scale emission models and infer emission factors (Velasco et al., 2009; Christen et al., 2011).

EC methods can be also applied to other GHGs. Gioli et al. (2012) examined both $F_{CO_2}$ and the flux of methane ($F_{CH_4}$) in Florence, Italy over a period extending from late winter to late spring, corresponding to the end of the heating season. Urban emissions of $CH_4$ are strongly associated with the type of fuel (i.e. natural gas) used to meet energy needs and the type of land use (e.g. landfills). The emission sources for both $F_{CO_2}$ and $F_{CH_4}$ included roads and domestic heating and, in the case of the methane flux, leakage from the natural gas network that provides the fuel for heating. $F_{CO_2}$ averaged 68 g $CO_2$ m$^{-2}$ d$^{-1}$ over the period with a distinct descending trend as the weather

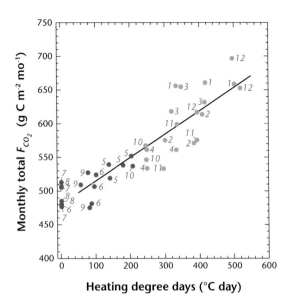

**Figure 13.9** Measured monthly averages of $F_{CO_2}$ in a residential area of Vancouver, Canada, vs heating demand expressed as heating degree days (HDD). The small numbers refer to the month of data (Source: Christen et al., 2011; © Elsevier, used with permission).

warms. By comparison, $F_{CH_4}$ averaged 189 mg $CH_4$ $m^{-2}$ $d^{-1}$ but showed no temporal pattern as it was estimated that over 80% of the observed $F_{CH_4}$ came from leaks in the network where the natural gas was under pressure and continued to be a source outside the heating season.

### GHG Measurements in the Urban Boundary Layer

To estimate GHG emissions of an entire city means integrating the sources and sinks across the urban landscape. This could be done if the contributions of the constituent neighbourhoods were measured using a network of EC towers that captures the spatial and temporal character of emissions. However, it could also be evaluated by measuring GHG mixing ratios within the ML. Acquiring information in this layer is not routine; the total column concentration of GHGs can be estimated using spectrometers but elucidating the vertical details requires *in situ* measurements made on balloons or airborne platforms. Typically, the results of observational field campaigns are used in combination with flux tower measurements and atmospheric models (e.g. box models such as those used for air quality studies, see Section 10.3) to get an insight into the net sources and dilution of urban scale

emissions. As an example, Font et al. (2015) explored $CO_2$ concentration in the UBL created by the Greater London, United Kingdom urban area and attempted to attribute contributions from the surface flux ($F_{CO_2}$), **advection** and **entrainment** using surface and airborne observations and a boundary-layer model. The study illustrated the difficulty of capturing emissions that vary over space and time across a city and are modulated by the UBL which varies in its depth and direction with weather conditions.

On the other hand, the contents of the urban plume can be sampled downwind of a city and the urban contribution assessed against the background concentration if that is known. Figure 13.10 shows aircraft measurements of $CO_2$ and of $CH_4$ made downwind of Indianapolis, United States, within its urban plume (see Figure 2.11); flight paths were oriented perpendicular to the plume, which was sampled at different heights and the data interpolated to provide cross-section maps. Measurements were made during conditions likely to reveal a strong urban signal hence flights were made near the middle of the day during the non-growing season and under light wind conditions. On 21 April 2008 the presence of the plume is readily detected against background concentrations of 399.0 and 1.91 ppm for $CO_2$ and $CH_4$, respectively. However, while the map of $CO_2$ concentrations shows a relatively simple pattern with a single core area of high values, the pattern for $CH_4$ has multiple peaks reflecting its more complex spatial emissions. The researchers conclude that the methane emissions originate from non combustion sources such as leakage from natural gas networks, landfills and wastewater treatment plants. Generally, identifying sources of $CH_4$ can prove a major challenge; in the Los Angeles basin, for example emissions from local petroleum production are intermixed with those linked to imported fossil fuels (Wennberg et al., 2012).

### 13.2.3 Urban Land Cover and Greenhouse Gas Exchange

The term 'carbon pool' refers to locally stored deposits of carbon in an urban ecosystem (shown as boxes in Figure 13.4). Interestingly, while vegetation and soils are important, the largest pools are the buildings and infrastructure that include carbon pools associated with wood in buildings and furniture (Churkina et al., 2010). Trees in cities can also play an important role (Weissert et al., 2014). For example,

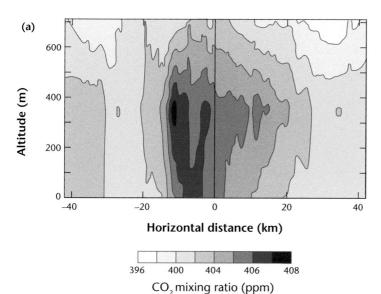

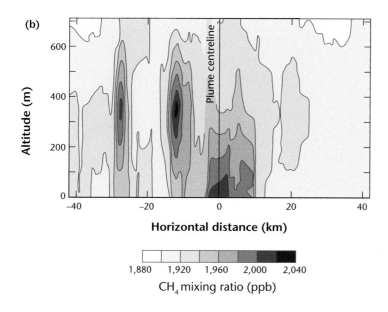

**Figure 13.10** Interpolated **(a)** $CO_2$ and **(b)** $CH_4$ based on aircraft measurements made downwind of Indianapolis, United States (Source: Mays et al., 2009; © American Chemical Society, adapted with permission).

Nowak and Crane (2002) estimated that urban trees in the US store about 700 million t C and sequester 22.8 million t C $y^{-1}$, which equates to about 5.5 months and 5 days of national emissions, respectively.

Although the rates of photosynthesis ($P$) and respiration ($R$) in a vegetated urban system can be considerable (see Section 13.2.2), the annual **net eco-system productivity** (NEP = $P - R$, in g C $m^{-2}$ $y^{-1}$) of plants and soil is usually small. Peters and McFadden (2012) measured NEP on a per-area basis in Minneapolis–Saint Paul, United States, and show that irrigated turfgrass has a nearly twofold higher NEP of 211 g C $m^{-2}$ over the growing season (Apr–Nov) than nonirrigated turfgrass with 115 g C $m^{-2}$. Also ever-green coniferous urban trees showed higher NEP (603 g C $m^{-2}$) than deciduous vegetation (216 g C $m^{-2}$). Generally, NEP in an urban ecosystem is less than that of managed agricultural, forest or natural ecosystems. Although urban vegetation is more efficient in both uptake and release of $CO_2$, the

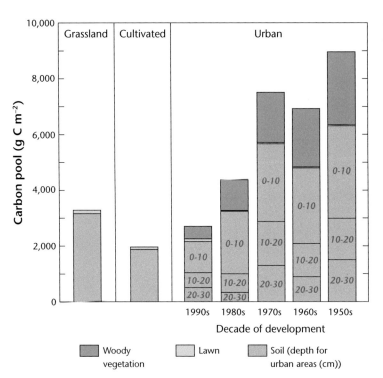

**Figure 13.11** Carbon pools in soils and vegetation in the Denver-Boulder metropolitan area, United States (Source: Golubiewski, 2006; © Ecological Society of American, used with permission from Wiley).

fraction of vegetation coverage expressed as $\lambda_v$ is substantially less than 1 (Figure 2.5).

In the process of land-use conversion, carbon pools can be depleted or enriched. For example, when an urban area is newly developed, soil (along with its organic carbon) and vegetation are disturbed and some may be removed or replaced, releasing stored carbon to the atmosphere. Golubiewski (2006) studied biogenic carbon pools in the Denver–Boulder metropolitan area (United States) which is situated in a semi-arid continental climate. Carbon stored in soils, lawns and woody vegetation was quantified on rural and residential lots of various ages. With increasing age of a suburban development, from the 1990s back to the 1950s, urban trees formed growing C pools (Figure 13.11). The growth in biomass is assisted by intensive irrigation in this climate. Although carbon pools experience an initial decrease following development, due to top-soil removal and vegetation changes, on average after 20 years, organic carbon in soils recover to the initial values of native grassland despite the reduced soil cover. After 50 years, the sampled urban soils contained about twice the organic carbon of native grasslands. Compared to fossil fuel

emissions the size of carbon pools in urban biomass and soils is small, and cannot offset fossil fuel emissions; in this study the carbon sequestered over ~50 years is the equivalent of half the annual fossil fuel emissions from the region. Nevertheless, the assessment of the size of carbon pools is important to quantify the potential for (and limits to) carbon sequestration in urban ecosystems. It further allows estimation of emissions related to changes of urban land cover and form (i.e. development, redevelopment, removal of soil, increase in vegetation, etc.).

## 13.3 Global Climate Change in Urban Environments

The increased concentration of GHGs in the atmosphere has caused a myriad of changes to the behaviour of the climate system at global and regional scales, the best known of which is a warming trend. The evidence for GCC is summarized at regular intervals by the Intergovernmental Panel on Climate Change (IPCC). Based on a review of scientific studies and simulations, the IPCC concluded that warming of the system is unequivocal and that human activity is extremely likely to be the dominant cause since the middle of the twentieth century. Moreover, even if GHG concentrations in the atmosphere were to remain at current levels, the climate system would continue to change as it adjusts to the additional RF of the gases and irreversible land-cover change, hence both adaptation and mitigation are needed to address the projected climate changes (IPCC, 2013). These projections and their implications have a particular relevance to cities for two main reasons: first, being densely occupied places with considerable cultural and capital investments that are worth protecting (adapting) and second, they are the foci of the majority of greenhouse gas emissions (mitigation). In addition, cities usually have strong management systems that can devise and implement policies more effectively than those at other scales.

Here we examine the likely consequences of GCC for cities and consider the responses in Chapter 15. However, we initially discuss how best to disentangle the effects of global and urban climate change processes in the air temperature record, which may confound our analysis of the problem and lead to inappropriate policy responses.

### 13.3.1 Monitoring Climate Change in Urban Environments

The climate changes recorded in (or near) cities contain signals associated with global, regional and urban influences that may vary differently over time (see Lowry's framework, Chapter 2). Of all measured variables, the urban influence on air temperature is most easily identified, yet identification of the urban contribution to the measured signal (and removing it if required) is a major challenge.

Some idea of the problem posed is presented in Fujibe (2011), who examined linear trends in the hourly temperature records for 546 stations in Japan for the decadal period March 1979 to February 2008. The population density in the year 2000 in the immediate area around each station was used as a surrogate for urbanization. The background climate trend was estimated using stations based in areas with low density ($< 100$ cap. $km^{-2}$) at 0.3 K per decade on the assumption that these stations were located in non urban areas and captured the global signal. The decadal rate was generally greater at stations with higher densities but to assess the degree of urban influence, the difference between the hourly temperature trends recorded at

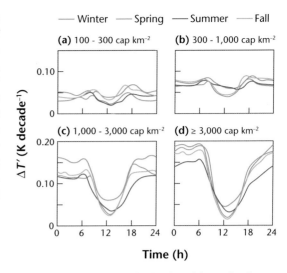

**Figure 13.12** Differences in the decadal trends of hourly temperatures observed at urban stations when compared with non-urban stations (that is, those in areas with population densities $< 100$ cap $km^{-2}$). The urban stations are categorized according to four categories of population density (Source: Fujibe, 2011; © Royal Meteorological Society, used with permission from Wiley).

'urban' ($\geq 100$ cap $km^{-2}$) and background ($< 100$ cap $km^{-2}$) stations ($\Delta T'$) was calculated. In Figure 13.12 values of $\Delta T'$ associated with categories of urbanization (based on population density) are averaged and plotted by hour and season. There are small seasonal differences but the overall pattern is clear: values of $\Delta T'$ are greatest for nighttime hours and for higher densities (that is, more intense urbanization). These distinct diurnal pattern follow the typical canopy-level UHI phenomenon described in Chapter 7.

If **metadata** (see Chapter 3) exists that can describe the landscape around the station and how it has changed over time, it may be possible to distinguish the sources of climate change and adjust the record to remove the urban effect. Figure 13.13a shows air temperature for central Tokyo in the twentieth century; the upward trend is striking but until the effects of regional and global processes are accounted for, we are unable to confidently attribute the warming to urban effects alone. The urban effect of Tokyo is estimated by subtracting the observational record obtained at Mito, a much smaller town within the region (about 100 km to the north-east), from the Tokyo record (Figure 13.13b). Since both records are likely to be

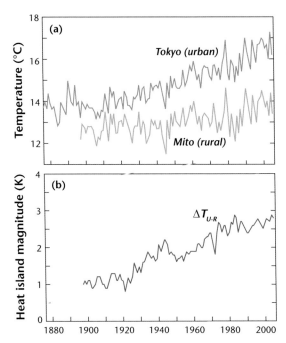

**Figure 13.13** Long-term variation of **(a)** annual mean air temperature for central Tokyo and Mito Observatory, Japan, and **(b)** their difference. The central Tokyo site has moved three times during the period of the record but Mito has never been moved (Source: Kataoka et al., 2009; © 2008 Elsevier, used with permission).

similarly affected by larger-scale **forcing** their difference can be expected to isolate more local scale controls on climate. However, the difference doesn't strictly isolate the urban temperature effect because other local effects are present. For example, the 'rural' nature of Mito degraded over time — at the start of the record Mito was open farmland, but by the 1980s the station was encroached upon by the expanding town. The site still remains fairly open so the difference signal is likely to be dominated by the urban effect of Tokyo.

Ideally the global air temperature record should represent the full diversity of Earth's landscapes. This could include urbanized areas, but no greater than their areal representation on the surface over time. Further, the record should extend over a long time period. Understandably, but unfortunately, apart from a few rural sites most of the long instrumental air temperature records come from places that are now cities (e.g. Berlin, Germany: started 1701; St. Petersburg, Russia: 1743; Boston, United States:

1743; Stockholm, Sweden: 1756; Toronto, Canada: 1770). As a result, there have been many attempts to first, assess the degree of urban 'contamination' of station records and second, to 'correct' the station records so that the urban influence is removed but this is not a simple process (see introduction to Chapter 12). Notwithstanding the need to adjust urban temperature records to assess GCC at a place, urban 'bias' is not responsible for the global warming signal.

### 13.3.2 Projecting Future Climates in Cities

Our knowledge of the future consequences of GCC is based on projections made with **general circulation models** (GCMs) combined with future fuel emission scenarios. Most projections are based on coupled atmosphere-ocean GCMs that simulate the fluxes of energy, mass and **momentum** across the ocean-atmosphere and land-atmosphere interface and resolve the dynamics of the atmosphere, oceans and ice-sheets in a three-dimensional manner. Each successive generation of GCMs has improved in tandem with our understanding of the Earth-Atmosphere system and growth of computer power. The most developed of these models include descriptions of the global biosphere, snow and ice-feedbacks, volcanoes and atmospheric chemistry. Some models (Earth System models) are now able to incorporate biogeochemical cycles (such as the carbon cycle) into their simulations, which allows for more realistic examination of feedbacks with the biosphere.

Simulations by these models compare favourably with the historical record of global air temperatures and help separation of anthropogenic from natural climate drivers. However, all models must strike a balance between the number of processes included and the time and space resolution used in simulations (see Chapter 3). For example, in the Hadley Centre Global Environment Model (HadGEM2) a grid cell at the Equator covers an area (208 × 139 km) much larger than a city; hence, they cannot make climate projections at city scales, nor resolve the localized urban effects discussed in this book (radiation, energy, water and **air pollution** dynamics).

#### Future Emissions

Future changes in global climate due to human activities depend on the evolution of GHG emissions, which are unknown and depend on both economic

**Table 13.5** Future climate scenarios based on representative concentration pathways (RCP). The RCP label indicates the approximate RF in 2100, compared to that in 1750; thus RCP4.5 causes a 4.5 W m$^{-2}$ forcing. The accumulated carbon emissions from 2012–2100 under each RCP and the estimated $CO_2$e in 2100 is shown in the adjacent columns, respectively. Finally, the likely range of increased surface temperature (K) and sea level (m) is shown for two periods in the future (Source: IPCC, 2013).

| Scenario | Carbon Emissions (Gt) 2012–2100 | CO$_2$e 2100 (ppm) | 2046–2065 | | 2081–2100 | |
|---|---|---|---|---|---|---|
| | | | (K) | (m) | (K) | (m) |
| **RCP2.6** | 270 (140–410) | 475 ppm | 0.4–1.6 | 0.17–0.32 | 0.3–1.7 | 0.26–0.55 |
| **RCP4.5** | 780 (595–1,005) | 630 ppm | 0.9–2.0 | 0.19–0.33 | 1.1–2.6 | 0.32–0.63 |
| **RCP6.0** | 1,060 (840–1,250) | 800 ppm | 0.8–1.8 | 0.18–0.32 | 1.4–3.1 | 0.33–0.63 |
| **RCP8.5** | 1,685 (1,415–1,910) | 1,313 ppm | 1.4–2.6 | 0.22–0.38 | 2.6–4.8 | 0.45–0.82 |

growth (which is currently based primarily on fossil fuels) and the success of mitigation efforts. To cope with this uncertainty, models employ a range of curves that reflect possible paths of GHG emissions, known as **representative concentration pathways** (RCPs) into the future (Table 13.5). These are distinguished by their RF of all contributing perturbations in 2100 compared to 1750. In the 5th Assessment report of the IPCC (2013) four RCPs were used to represent potential emission scenarios from now until 2100, these included (Van Vuuren et al., 2011): one mitigation scenario that results in a low RF (2.6 W m$^{-2}$); two medium stabilization scenarios (RF = 4.5 and RF = 6.0 W m$^{-2}$) and; one high emission scenario that produced the highest RF (8.5 W m$^{-2}$). As a result, the simulated climates are referred to as projections (rather than predictions) and are derived by running several different GCMs using the same RCP. The likely range of outcomes include global surface temperature increases of between 0.3 to 4.8 K and sea-level rises of between 0.26 to 0.82 m by 2100, depending on the RCP (Table 13.5).

### Downscaling Projections for Cities

Since GCMs developed to assess GCC cannot simulate processes or outcomes at the scale of cities the global scale results must be downscaled. Dynamical downscaling uses a **numerical model** designed for smaller scales and nests it within the larger-scale model, which provides the **boundary conditions**. Statistical downscaling establishes a relationship between variables at these different scales using historic data, which is then used to estimate local-scale values based

on simulations at the global scale. Neither method is ideal: the dynamical approaches accumulate errors because models, each with their own errors, are run in series, and the statistical approaches are forced to assume that current statistical relationships will hold well into the future.

Downscaling can yield insights into likely changes that can assist decision-making at the urban scale but, of the two approaches, nested models provide a much richer understanding of the nature of climate change at city scale. For example, Oleson et al. (2011) used an **urban canyon** representation to describe city landscapes and capture processes responsible for **urban climate** effects. The surface cover of the Earth within the larger grid of a climate model was divided into subgrids that described natural surface cover and cities; an urban subgrid accounted for roofs, walls (sunlit and shaded) and pervious and impervious ground. Similarly, Masson et al. (2014) employed downscaling and at the same time considered how cities could evolve (representing different urban emission scenarios) over the timeframe of climate changes. This type of work is just beginning and is likely to form the basis of urban scale climate projections in the future. Here, we describe the likely impacts of GCC on cities in general terms.

### 13.3.3 Impacts of Global Climate Change on Cities

Table 13.6 lists recent trends in, and projections for, extreme weather and climate events. These phenomena include warmer air temperatures by day and night

**Table 13.6** Examples of possible impacts of GCC due to changes in extreme weather and climate events, based on projections to the mid- to late twenty-first century. These do not take into account changes or developments in adaptive capacity. The likelihood estimates in column two relate to the phenomena listed in the first column (Data sources: IPCC, 2007 & 2013).

| Phenomenon and direction of trend | Evidence of change | Evidence of human contribution | Likelihood of further change by late twenty-first century | Impact on cities (industry, settlement and society) |
|---|---|---|---|---|
| Warmer and/or fewer cold days and nights over most land areas. Warmer and/or more frequent hot days and nights over most land areas. | Very likely | Very likely | Virtually certain | Reduced energy demand for heating; increased demand for cooling; declining air quality; reduced disruption to transport due to snow, ice; effects on winter tourism. |
| Warm spells/heat waves. Frequency and/or duration increases over most land areas. | Medium confidence | Likely | Very likely | Reduction in quality of life for people in warm areas without appropriate housing; impacts on the elderly, very young and poor. |
| Heavy precipitation events. Increase in the frequency, intensity and/or amount of heavy precipitation. | Likely | Medium confidence | Very likely over most of the mid-latitude land masses and over wet tropical regions | Disruption of settlements, commerce, transport and societies due to flooding: pressures on urban infrastructure; loss of property. |
| Increases in intensity and/or duration of drought. | Low confidence | Low confidence | Likely on a regional to global scale | Water shortage for settlements, industry and societies; reduced hydropower generation potentials; potential for population migration. |
| Intense tropical cyclone activity increases. | Low confidence | Low confidence | More likely than not in the western North Pacific and North Atlantic | Disruption of urban infrastructure by flood and high winds; withdrawal of risk coverage in vulnerable areas by private insurers; potential for population migrations; loss of property. |
| Increased incidence and/or magnitude of extreme high sea level. | Likely | Likely | Very likely | Costs of coastal protection versus costs of land-use relocation; potential for movement of settlements and infrastructure; also see tropical cyclones above. |

over most land areas; more frequent and/or longer heat wave events; changes in precipitation statistics and in the intensity and/or duration of droughts and high sea-level events. Each statement is expressed in terms of likelihood, which expresses the confidence in the statement based on the empirical and model evidence. So for example, there is confidence that a warming of global near-surface temperatures since 1950 has taken place (very likely) and that humans have contributed (very likely); moreover, there is high confidence (virtually certain) that this trend will

continue into the late twenty-first century. By comparison, regarding tropical storm intensity, there is low confidence in identifying trends, attributing human responsibility or projecting changes. Overall, projections show that the statistics (means and variability) of the current climate, including the **return period** of extreme events, will change.

## Hazard, Exposure, Vulnerability and Risk

It is worthwhile distinguishing among various terms that are used to assess the impact of GCC on cities

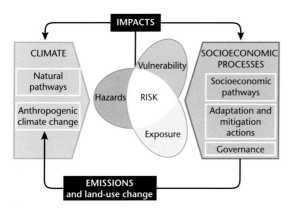

**Figure 13.14** Illustration of the core concepts of the WGII AR5. Risk of climate-related impacts results from the interaction of climate-related hazards (including hazardous events and trends) with the vulnerability and exposure of human and natural systems. Changes in both the climate system (left) and socioeconomic processes including adaptation and mitigation (right) are drivers of hazards, exposure, and vulnerability (Source: IPCC, 2014).

and their response. Natural hazards refer to phenomena that can damage infrastructure or cause loss of life, different hazards have an impact based on their intensity, scale and duration. **Exposure** describes the extent to which a population or a place is in harm's way, while **vulnerability** describes the degree of preparedness to that hazard. **Risk** lies at the intersection of these properties (Figure 13.14). Although two cities may occupy low-lying coastlines and be exposed to a threat of high sea-levels, they may be distinguished by their vulnerability. While one may be devastated by such an event, the other, which has prepared for such an eventuality (with evacuation procedures or protective barriers, for example), recovers quickly. In this case, the latter city has a better **resilience**. Figure 13.14 illustrates that the risks associated with climate-related impacts are an outcome of both climate variability and socioeconomic processes, which regulates the anthropogenic causes of climate change, namely GHG emissions and land-use changes.

The key risks identified by the IPCC (2014) as a consequence of projected GCC include:

1. Risk of death, injury, ill-health or disrupted livelihoods in low-lying coastal zones with large urban populations due to storm surges, coastal flooding and sea-level rise.

2. Systemic risks due to extreme weather events leading to breakdown of infrastructure networks and critical services such as electricity, water supply, health and emergency services.

3. Risk of mortality and morbidity during periods of extreme heat, particularly for vulnerable urban populations and those working outdoors.

4. Risk of food insecurity and the breakdown of food systems linked to warming, drought, flooding and precipitation variability and extremes, particularly for economically and socially disadvantaged populations.

5. Risk of loss of rural livelihoods and income due to insufficient access to drinking water and reduced agricultural productivity.

6. Risk of loss of marine/terrestrial and coastal/inland water ecosystems, biodiversity and the ecosystem goods, functions and services they provide for livelihoods.

Not surprisingly, many of these individual risks overlap spatially in urban areas where heat stress, extreme precipitation, inland and coastal flooding, air pollution, drought and water scarcity have an impact upon people, assets, economies and ecosystems.

It follows of course that risk will be greatest where there is a geographical correspondence between projected hazards, the exposure of population (and infrastructure) and vulnerability. Not surprisingly then, these risks vary between and within cities based on the geographical context (Chapter 12), urban layout and socioeconomic differences (Figure 13.14). Thus, while warmer temperatures may reduce the demand for heating in cities that experience harsh winters (and reduce cold-related deaths) it will surely increase the demand for cooling in cities with warm summers (and increase heat-related deaths). Moreover, variations across neighbourhoods associated with **orography**, the nature of urban development (e.g. land use, land cover, urban structure), socio-demographic profile, etc. create spatial patterns of risk within cities.

Tentative assessments at these scales are only just beginning to appear, owing to the absence of comparable data on cities and the imprecise nature of projections. For example, Table 13.7 estimates the exposure of large port cities (in terms of population and assets) to a 1 in 100 year surge induced flood event for the current climate and for the climate projected for the period 2060–2070. The projection accounts for population growth, sea-level rise, land subsidence and

**Table 13.7** The top 20 world port cities ranked by population exposure under the current climate and a future climate scenario; [D] indicates the city is situated on a river delta. Population in millions and assets in billions of US dollars (Source: Nicholls et al., 2008).

| City | Population in 2005 | Current climate | | Future scenario | |
|---|---|---|---|---|---|
| | | Exposed population | Exposed assets | Exposed population | Exposed assets |
| Mumbai, India | 18.2 | 2.8 | 46 | 11.4 | 1,598 |
| Guangzhou, China [D] | 8.4 | 2.7 | 84 | 10.3 | 3,358 |
| Shanghai, China [D] | 14.5 | 2.4 | 73 | 5.5 | 1,771 |
| Miami, United States | 5.4 | 2.0 | 416 | 4.8 | 3,513 |
| Ho Chi Minh City, Vietnam [D] | 5.1 | 1.9 | 27 | 9.2 | 653 |
| Kolkata, India [D] | 14.3 | 1.9 | 32 | 14.0 | 1,961 |
| New York-Newark, United States | 18.7 | 1.5 | 320 | 2.9 | 2,147 |
| Osaka-Kobe, Japan [D] | 11.3 | 1.4 | 216 | 2.0 | 969 |
| Alexandria, Egypt [D] | 3.8 | 1.3 | 28 | 4.4 | 563 |
| New Orleans, United States [D] | 1.0 | 1.1 | 234 | 1.4 | 1,013 |
| Tokyo, Japan [D] | 35.2 | 1.1 | 174 | 2.5 | 1,207 |
| Tianjin, China [D] | 7.0 | 1.0 | 30 | 3.8 | 1,231 |
| Bangkok, Thailand [D] | 6.6 | 0.9 | 39 | 5.1 | 1,118 |
| Dhaka, Bangladesh [D] | 12.4 | 0.8 | 8 | 11.1 | 544 |
| Amsterdam, Netherlands [D] | 1.2 | 0.8 | 128 | 1.4 | 844 |
| Hai Phong, Vietnam [D] | 1.9 | 0.8 | 11 | 4.7 | 334 |
| Rotterdam, Netherlands [D] | 1.1 | 0.8 | 115 | 1.4 | 826 |
| Shenzen, China | 7.2 | 0.7 | 22 | 0.7 | 243 |
| Nagoya, Japan [D] | 3.2 | 0.7 | 109 | 1.3 | 623 |
| Abidjan, Cote d'Ivoire | 3.6 | 0.5 | 4 | 3.1 | 142 |

increased storminess but does not address adaptation measures to reduce vulnerability. For the current climate Mumbai, India, and Miami, United States, have the greatest population and asset exposures, respectively; note the distinction between cities in less and more economically developed countries on these measures. Under projected future climates, the population and asset exposure to hazards increase substantially for a number of reasons but the main driver is socioeconomic. Other drivers, such as GCC and land subsidence due to **groundwater** mining exacerbate the

socioeconomic changes but play a smaller role in increasing exposure.

## Amplification or Reduction of the Urban Heat Island?

The projected warming of near-surface air temperatures due to GCC could have implications on the magnitude of the UHI effect and the frequency of heatwaves in cities. However, we cannot assume that GCC simply raises the background temperature and that the UHI effect is additive. This is because the definition of the UHI is based on the *difference* in temperatures between urban and adjacent non urban environment and its magnitude is modulated by conditions in both environments and by the background climate, all of which may change in the future.

McCarthy et al. (2010) employed a GCM that had the capacity to include fractions of different land covers; an urban land-surface scheme was included to account for the physical presence of cities and their anticipated anthropogenic heat flux $Q_F$. A series of experiments were conducted using different atmospheric mixing ratios of $CO_2$ and the canopy-level UHI was estimated as the difference in the near-surface temperature over the urban area, when compared to that over the dominant non-urban cover within each grid cell. The results showed that increasing the global $CO_2$ concentration from 323 to 645 ppm enhanced this UHI by less than 0.5 K, substantially less than the warming associated with

the global driver (3 K). However, results also indicated a significant increase in very warm nocturnal temperatures in urban areas, which could cause heat stress and have serious public health consequences.

There have been studies about expected GCC impacts on individual cities. For example, Lemonsu et al. (2013) downscaled GCM projections to the Paris region using the **Town Energy Balance** (TEB) model to represent the urban area. The study considered the change between the current climate and that for the end of the twenty-first century using high emission scenarios. The results show a significant temperature increase ($> 2$ K in both minimum and maximum daily temperatures) but increases were larger in rural areas, so the magnitude of the $UHI_{UCL}$ actually decreased. The cause of this was the projected reduction in precipitation in the region, which dried soils, reduced their **thermal admittance** and made the rural area more thermally responsive (Chapter 7).

## Summary

Anthropogenic global climate change (GCC) is caused primarily by two drivers: **global land-cover change** and the emission of **greenhouse gases** (GHG, especially carbon dioxide ($CO_2$), methane ($CH_4$) and nitrous oxide ($N_2O$). Cities play an important role primarily as concentrated areas of fossil fuel based energy use that are responsible for a large fraction of anthropogenic $CO_2$ emissions as a waste by-product. Cities are also sources of short-lived air pollutants and aerosols that change radiation transfer, clouds, precipitation, properties of snow etc. downwind of cities. The expected outcomes of near-future (50–100 years) GCC include higher air temperatures over land areas, warmer ocean temperatures, reductions in annual ice and snow-coverage, consequent sea-level rise and increased contrasts in precipitation patterns. As a consequence, aspects of the background climate, within which cities are embedded, will change. The extent and magnitude of these changes is projected to vary geographically and depend upon actions to limit GHG emissions. In summary:

- Cities contribution to GCC is **mainly a result of GHG emissions from energy consumption**, rather than direct land-cover change within urban areas. Nevertheless, the indirect land-cover change outside cities — to feed, supply power, supply resources and for transport of urban populations is substantial. Therefore evaluating the contribution of cities has proved problematic owing to the difficulty in attributing GHG emissions based on place of production or place of consumption. About 30–40% of all fossil fuel related global $CO_2$ emissions originate directly within urban areas, while over 70% are caused by the energy needs of urban populations.
- Emissions of GHGs within a city **elevates concentrations of GHGs in the UBL and in urban plumes**. Many GHGs have long atmospheric residence times so that the contents of these plumes mix into the global atmosphere and increase global GHG concentrations. On an

urban scale, the elevated $CO_2$ in the urban atmosphere does not contribute significantly to the boundary layer urban heat island.

- More **compact and densely settled cities generally generate less GHG per capita**. Consequently, global mitigation policies to reduce emissions in cities should not only focus on technology and fuel-switches, but also pay attention to the potential for moderating the city contribution through a more efficient urban form, transport and land-use mix.
- Cities are preferentially located at lower elevations close to coasts and alongside rivers, which **places them at particular risk from some of the projected climate changes**, such as sea-level rise. To reduce vulnerability cities will need to adapt, by modifying urban form and functions.
- The projected increases in global air temperatures will **enhance the effects of the urban heat island** (and increase the risk of heat wave events) although it may not affect its magnitude substantially. Local mitigation of the UHI will be needed to offset the impact of GCC.

# 14 | Climates of Humans

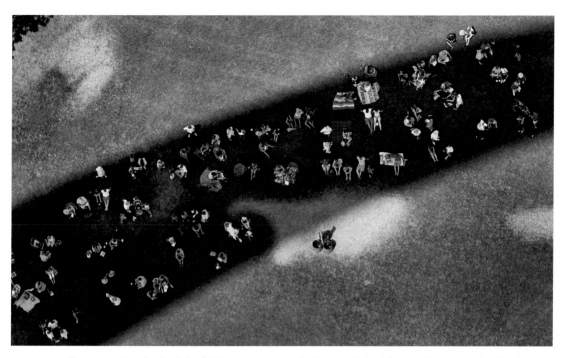

**Figure 14.1** On a very hot day in July, 2012, park users in St. Louis, United States, chose to sit in the shadow of the Gateway Arch, a slender sculpture that generates a narrow band of shade, rather than the adjacent sunlit grass (Credit: St. Louis Dispatch).

A commuter who walks to work leaves the comfort of home and embarks on a journey that may take her through areas that are: shaded or sunlit; next to walls that are warm; across grass that is cool; under trees that offer shelter from the rain; around a building corner where the wind suddenly accelerates; along roads where polluted air is inhaled and so on. As the individual encounters these different **microclimates**, the body responds by becoming warmer/cooler, shivering or sweating and reacts by changing pace, adjusting clothes, leaning into the wind or avoiding roads that are busy and polluted. Where individuals are stationary, they will tend to locate themselves in more favourable microclimates, such as the shelter of a building during windy weather. Figure 14.1 is a remarkable illustration of this behaviour in hot weather. Individuals have decided to avoid sunshine and gather in a narrow band of shade, exposed to the breeze off a nearby river; on the date of this image (July 4th, 2012) the city was in the middle of a **heatwave** and for the previous six days the maximum air temperature was $\geq 38°C$. Persistent high temperatures can place great stress on the ability of the body to regulate its internal temperature, which is necessary for survival. The response of those outdoors here is to cluster within the shade, avoiding the nearby sunlit grass.

This chapter focuses on the bioclimates of humans with particular regard to the climates experienced in cities and on the outdoor thermal and wind environments. Air quality has been discussed in Chapter 11. Initially, we present the basic biophysical systems of humans and how they respond to variations in the ambient environment before discussing bioclimate and its parameters in more detail.

## 14.1 Basics

Humans make deliberate and imperceptible adjustments to an environment, which is in constantly in flux. One of the objectives of building and urban design is to manage this environment to minimize stress.

### 14.1.1 Managing Heat

Humans have a near constant deep body temperature of 37°C that is maintained by the body's thermoregulatory system, which ensures **homeostasis** under normal conditions. Thermoregulation incorporates a set of physiological and behavioural processes that manage heat exchange between the body and the ambient environment such that the net gain or loss of energy is close to zero. This is a complex feat that regulates the disposal of heat generated by internal metabolic processes in response to external environmental conditions including **radiation, sensible** and **latent heat** exchanges.

For an average adult at rest the heart beats 72 times a minute, causing 5,600 cm$^3$ of blood to flow through the body's circulatory system. This system is comprised of the pulmonary and systemic circuits; the former brings blood to the lungs for oxygenation and the latter transfers oxygenated blood to the body's tissues. This blood is carried via arteries that branch successively into smaller blood vessels until finally oxygen transfer occurs across the semi-permeable boundaries of capillaries into the adjacent tissue. While the blood leaves the heart through the aorta with radii of 10 mm at a rate of 0.3 m s$^{-1}$, it moves progressively slower as the blood vessels divide (becoming smaller) and the friction imposed by the vessel walls increases. The de-oxygenated blood is carried back to the heart in the veins.

The circulatory system also transfers heat from the body core to skin surface where heat exchange takes place either directly with the ambient atmosphere or indirectly via a layer of clothing. Thus, the energy balance of the body must account for heat exchanges internally (within the body) and externally (with the ambient environment).

The body can exert a considerable degree of control over the types and magnitudes of external exchanges through both automatic physiologic responses and voluntary actions. The former includes regulating blood flow, rates of sweating and metabolic activity.

These can be complemented through a range of actions that require decisions, such as moving to a more comfortable place or altering clothing. Of course, spaces can also be designed to manage the ambient conditions (see Chapter 15); the most elaborate of these is the construction of buildings for shelter and comfort.

### 14.1.2 Maintaining Postural Balance

Humans control their postural balance while standing, walking or running by integrating the sensory, central nervous and muscular-skeletal systems. This process is complicated by the relatively small base (two feet) and location of the centre of body mass at approximately two-thirds of body height. The effect of the atmosphere on balance is related to wind, both its magnitude and variability.

The force ($F_B$) exerted by the wind on a person is related to the square of the mean wind velocity ($\bar{u}$),

$$F_B = \frac{\rho_a \, \bar{u}^2 \, C_D A_f}{2} \quad (\text{kg m s}^{-2} \text{ or N}) \qquad \textbf{Equation 14.1}$$

The drag coefficient ($C_D$) is a function of body shape and for the upright person is about 1.15 and 1.0 for front and side winds, respectively. However, $C_D$ varies according to the nature of clothing and whether it 'hugs' the body or flaps about it. The area of the average body, projected in the direction of the wind ($A_f$) has a value of 2.4 m$^2$ and 1.6 m$^2$ for front and side winds, respectively.

An average adult pedestrian will begin to feel the force of a breeze at 6 m s$^{-1}$, which for a wind at the back corresponds to 62°N. Note that the force increases with the square of velocity; at 12 m s$^{-1}$ walking becomes difficult and at 18 m s$^{-1}$ an adult can be blown over. To adjust to this force, pedestrians lean into the breeze (see also Figure 4.1). In constant airflow, the angle ($\theta$) required can be estimated from

$$\theta = \arctan\left(\frac{F_B}{m\,g}\right) \qquad (\text{degrees}) \qquad \textbf{Equation 14.2}$$

The upright stability of the body is due to its mass ($m$) times the acceleration due to gravity ($g$) but when the mean wind $\bar{u}$ exceeds 15 m s$^{-1}$ an adult pedestrian must lean 8° into a headwind, which is potentially unstable. However, in **turbulent** airflow where speed and direction change rapidly, maintaining stability at a much lower mean wind speed is more difficult (Bottema, 1993).

## 14.2 The Human Energy Balance

The energy balance of a human must account for: the transfer of heat from the body core to the skin surface and; between the skin surface and the ambient environment. The latter may be modulated by clothing.

### 14.2.1 Internal Energy Exchanges

The heat generated when metabolized food is chemically converted to fuel both internal (e.g. breathing) and external (e.g. running) physical activity is termed **metabolic heat**. While some of this energy is expended as mechanical energy (such as walking uphill), the great majority ($> 95\%$) is converted to heat that must be dissipated to the ambient environment. The magnitude of the metabolic rate depends strongly on the level of exertion (Table 14.1). The lowest value occurs when an adult is sleeping (70 W) and is about 100 W when awake and comfortably at rest; however, it can be as high as 800 W during intense physical activity. Shivering describes involuntary muscular activity

**Table 14.1** The metabolic rate ($Q_M$ in Equation 14.5) associated with different levels of physical activity (Source: ASHRAE, 2009).

| Activity | Metabolic rate | |
|---|---|---|
| | **(W)** | **(W m$^{-2}$)** |
| **Resting** | | |
| Sleeping | 70 | 40 |
| Seated, quiet | 110 | 60 |
| Standing relaxed | 130 | 70 |
| **Walking on a level surface** | | |
| Pace 0.9 m s$^{-1}$ | 210 | 115 |
| Pace 1.2 m s$^{-1}$ | 270 | 150 |
| Pace 1.8 m s$^{-1}$ | 400 | 220 |
| **Office** | | |
| Writing | 110 | 60 |
| Walking about | 180 | 100 |
| Lifting/packing | 220 | 120 |
| **Occupational** | | |
| Cooking | 170–210 | 95–115 |
| Housecleaning | 210–360 | 115–200 |
| Handling 50 kg bags | 420 | 235 |
| Pick and shovel work | 420–500 | 235–280 |
| **Leisure** | | |
| Dancing | 250–460 | 140–255 |
| Tennis | 380–490 | 210–270 |
| Basketball | 520–880 | 290–440 |

initiated when the body loses heat too quickly and increases the metabolic rate.

Metabolic heat is transferred to the body's outer surface through the respiratory and circulatory systems. Rates of breathing vary with levels of exertion and oxygen demand (Table 14.2); an adult male at rest exchanges about 0.7 m$^3$ h$^{-1}$, so if the air temperature were 20°C, the sensible heat exchange would average just 4 W. Exhaled breath is also close to saturation at 37°C (about 44 g m$^{-3}$) but the heat transfer would depend greatly on the **relative humidity** of the ambient atmosphere; if the relative humidity were just 40% at 20°C, about 20 W would be transferred by breathing at rest. Naturally, these exchanges would increase in the same ambient conditions if the individual is engaged in heavy work and breathing more often (~3–4 m$^3$ h$^{-1}$). However, typically the dominant means of exchange internally is by blood flow,

$$Q_{\text{core}\rightarrow\text{skin}} = \frac{v_b C_b \left( T_{\text{core}} - \overline{T}_{\text{skin}} \right)}{A_{\text{body}}} \quad (\text{W m}^{-2})$$

**Equation 14.3**

where $v_b$ and $C_b$ are the flow rate (m$^3$ s$^{-1}$) and **heat capacity** of blood (3617 J m$^{-3}$ K$^{-1}$), $T_{\text{core}}$ is the temperature of the body core and $\overline{T}_{\text{skin}}$ is the average skin temperature. When the body is at rest and comfortable, $\overline{T}_{\text{skin}}$ is 34°C (about 3 K lower than $T_{\text{core}}$) and blood flow to the skin is between 3300 to 8300 mm$^{-3}$ s$^{-1}$ hence, between 36 and 90 W is exchanged.

One of the first responses of the body to excessive heat loss/gain is through **vasomotor control**, which regulates blood flow to the skin organ and thereby manages $\overline{T}_{\text{skin}}$. **Vasodilation** causes the blood vessels

**Table 14.2** Human inhalation rates by activity level (Modified after: Moya et al., 2011).

| Level of exertion | Resting (m$^3$ h$^{-1}$) | Light (m$^3$ h$^{-1}$) | Moderate (m$^3$ h$^{-1}$) | Heavy (m$^3$ h$^{-1}$) |
|---|---|---|---|---|
| Adult female | 0.3 | 0.5 | 1.6 | 2.9 |
| Adult male | 0.7 | 0.8 | 2.5 | 4.8 |
| Average adult | 0.5 | 0.6 | 2.1 | 3.9 |
| Child 6 years | 0.4 | 0.8 | 2.0 | 2.4 |
| Child 10 years | 0.4 | 1.0 | 3.2 | 4.2 |

**(a)** Cool conditions   **(b)** Warm conditions

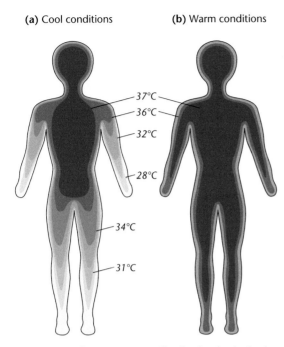

37°C
36°C
32°C
28°C
34°C
31°C

**Figure 14.2** The temperature distribution in the body during cool and warm conditions. In cool conditions, the warmest temperature (37°C) is confined to the head and trunk only. The temperature of the subcutaneous tissue for the hands and feet is < 28°C. In a warm environment the warm core temperature is found over much of the body (Modified after: Mount, 1979).

to expand, increasing the rate of blood flow to the skin, while **vasoconstriction** does the opposite. In this way the body can manage heat loss by altering the **insulation** provided by the body tissue and regulating $(T_{core} - \bar{T}_{skin})$. The effects of vasomotor control can be seen in Figure 14.2 which shows the temperature distribution in the body under cold and warm conditions. Note how much cooler the limbs become when the environment is cool and the body retains heat in its core; $\bar{T}_{skin}$ varies considerably, it is lowest where the potential heat loss to the ambient environment is greatest and the distance from the core is furthest.

### Body Area

The intensity of exchanges at the body's outer surface (that is the energy **flux density**) is expressed in relation to the surface area of the body, which can be estimated using the DuBois relationship,

$$A_{body} = 0.007184 m^{0.425} H^{0.725} \quad (m^2)$$

**Equation 14.4**

where $m$ is mass (kg) and $H$ is height (m) of the body. The surface area of an upright average adult male is about 1.8 m$^2$ based on a weight of 70 kg and a height of 1.65 m. Adult males are typically both taller and heavier (by up to 10%) than female; equivalent surface areas for a typical female and 10-year old child are 1.6 m$^2$ and 1.1 m$^2$, respectively. However, there is considerable variation across ages and cultures. In many studies the human form is represented by a cylinder with a height of 1.65 m and a radius of 0.12 m; this simplifies calculations and allows research on heat transfer from simple geometric shapes to be transferred to the study of human-environment exchanges.

The shape of the body and its parts (segments) plays an important role in managing heat exchange by controlling area of the outer surface exposed to the environment. For example drawing in the limbs and curling up converts the body shape from cylinder (area of about 1.8 m$^2$ for an adult male) into a more sphere-like shape (area of about 0.9 m$^2$); as the volume of the body remains the same, the effect of decreasing the outer surface is to minimize heat loss. Similarly, extending the limbs maximizes surface area and heat loss. The appendages (arms, legs, fingers and toes) are ideally suited to heat loss as they have a high area to volume ratio. As a result surface heat exchange is large but their capacity to store energy in the enclosed mass is small. Consider the implication for **heat storage**: the area to volume ratio for the torso is 17 but is 135 for the hands so for the same rate of heat loss the hands will cool eight times more quickly. Not surprisingly, it is the appendages (especially the fingers and toes) that experience the greatest temperature fluctuations and in extreme cold climates are the first to experience frostbite as the body withdraws blood flow in an effort to maintain core temperature.

### 14.2.2 External Energy Exchanges

The energy exchanges with the ambient environment occur across the outer surface of the body. These can be expressed in the form of an energy balance similar to those presented elsewhere in this book,

$$Q^* + Q_M = Q_H + Q_E + Q_G + \Delta Q_S \quad (W\,m^{-2})$$

**Equation 14.5**

where each term is expressed as an average over the body's outer envelope, which may be the skin or the clothing surface. This energy balance is distinguished by internal energy supplied by the body's metabolism ($Q_M$), which varies according to the level of physical activity (Table 14.1). The healthy body regulates these fluxes so that heat storage is minimal ($\Delta Q_S \approx 0$). This control is necessary to ensure a near constant core temperature of approximately 37°C; the maximum allowable deviation of $T_{core}$ is approximately ±3 K, with a greater tolerance on the cold side.

The sensible heat exchange via **conduction** ($Q_G$) is normally a relatively small term as typically just a small proportion of the body's surface area is in contact with a solid surface. However, there may be situations where the body is prone or immersed in water and this term is significant and must be included. For our purposes here it may be ignored.

### Insulation/Resistance and Heat Transfer Coefficients

In biometeorological studies, energy fluxes employ insulation/**resistance** terms and transfer coefficients; the former are used to account for clothing which has a limited depth and the latter for exchanges with the ambient environment.

The effect of clothing is to trap a layer of still air against the skin surface, which impedes the transfer of sensible and latent heat (i.e. it offers resistance). A typical clothing ensemble (trousers, shirt and sweater) provides a thermal insulation ($I_{cl}$) of about 0.155 K m$^2$ W$^{-1}$. The equivalent thermal resistance can be assessed in relation to the properties of still air; at 20°C and 100 kPa, air has a heat capacity ($C_a$) of 1220 J K$^{-1}$ m$^{-3}$ and a thermal resistance ($r_a$) of 470 s m$^{-1}$ (Monteith and Unsworth, 2008), so a 1 mm layer has an insulation value of 0.0385 K m$^2$ W$^{-1}$ and the typical clothing ensemble is the equivalent of a cushion of 4 mm of still air. The resistance to **evaporation** provided by clothing (kPa m$^2$ W$^{-1}$) varies between 0.15 and 0.30 for common ensembles (Parsons, 2003).

At the outer surface of the body, energy exchanges with the ambient environment may be expressed using transfer coefficients (W m$^{-2}$ K$^{-1}$) and an appropriate gradient, for example the **sensible heat flux density** ($Q_H$) can be expressed as:

$$Q_H = h_c \Delta T \quad (\text{W m}^{-2}) \qquad \textbf{Equation 14.6}$$

where $h_c$ is the convective heat transfer coefficient and $\Delta T$ is the difference in temperature between the **active surface** (here skin or cloth) and the ambient air. Values for these coefficients are established using

thermal manikins to represent the human body (and its segments) in various postures and clothing ensembles. These manikins can be exposed to ambient conditions in a controlled laboratory environment where radiation, air temperature, humidity and airflow can be regulated and heat loss measured.

### 14.2.3 The Radiation Budget

The radiation budget of an individual is the same as that for the **surface radiation budget** of any natural surface, but its application to humans is made more complicated by body shape,

$$Q^* = (S + D)(1 - \alpha) + L_\downarrow - L_\uparrow \quad (\text{W m}^{-2})$$
$$\textbf{Equation 14.7}$$

The direct **shortwave** radiation ($S$) received by an individual will depend on the area of the body as viewed from vantage point of the Sun. The intercepted solar radiation on an upright individual can be estimated by calculating the shadow area ($A_s$) generated by a cylinder that has the approximate dimensions of the human body,

$$A_s = (2rH)\cot\beta + \pi r^2 \quad (\text{m}^2) \qquad \textbf{Equation 14.8}$$

The controlling variable here is the **solar altitude** ($\beta$), which varies with latitude, time of year and time of day. If the Sun were directly overhead ($\beta = 90°$) then the only area in shade is directly underneath the body ($A_s = \pi r^2$) and all the available solar radiation would fall on the crown of the head and the shoulders. By comparison, when solar altitude decreases, the shadow area increases and $S$ is spread over the illuminated limbs and torso. Moving from sunlight into shadow and *vice versa* will change the body's energy receipt significantly.

**Diffuse irradiance** ($D$) can be treated as though it originated from a notional 'hemisphere' that surrounds the body (Figure 14.3) where $D$ is a function of the strength of the radiation source and the proportion of the flux leaving that source incident on the body surface (that is the **view factor**, $\psi$). If one were to treat these radiation sources as simply the sky and the surrounding surface environment, then

$$D = D_{sky} + D_{env} = D_0\,\psi_{sky} + K_\uparrow\psi_{env} \quad (\text{W m}^{-2})$$
$$\textbf{Equation 14.9}$$

where $D_0$ represents diffuse radiation received on flat surface with $\psi_{sky} = 1$ and $K_\uparrow$ represents the

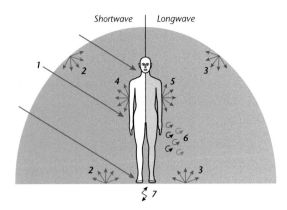

Shortwave | Longwave

**Figure 14.3** Energy exchanges at the surface of the human body. These include: (1) direct shortwave radiation that impinges on the sunlit part of the body; (2) diffuse shortwave radiation that originates from the sky as a result of scattering and from the ground as a result of reflection; (3) diffuse longwave radiation that is emitted from the sky vault and from the ground; (4) reflected shortwave radiation that is controlled by the albedo of the clothed body; (5) emitted longwave radiation which is a function of surface temperature; (6) convective heat loss by sensible and latent heat exchange with the ambient air that is partly a function of wind speed and; (7) conductive heat exchange with the ground through physical contact.

reflected radiation from the environment that impinges on the body. For an individual standing on an extensive horizontal surface $\psi_{sky}$ and $\psi_{env}$ will be approximately equal but this will not be the case in densely built urban environments. An individual walking through a city street intercepts $K_\uparrow$ from a variety of **facets** of a wide range of **fabrics** (glass, asphalt, concrete, grass, etc.) each with a unique solar radiation receipt and **albedo** ($\alpha$). Integrating the diffuse radiation from all of these sources is a complex task but tools are available (e.g. Matzarakis et al., 2010).

The albedo ($\alpha$) of skin regulates the **absorption** of solar radiation. $\alpha$ depends on pigmentation and skin blood flow. The spectral reflectivity of skin at visible wavelengths (0.4–0.7 µm) varies from 0.15 to 0.35 as pigmentation changes from dark to light and; between 0.8 and 1.2 µm, but increases sharply to 0.6 between 0.8 and 1.2 µm before falling at longer wavelengths. However, $\alpha$ can be readily changed through choice of clothing.

## Longwave Radiation

Longwave radiation can be treated in the same way as diffuse shortwave radiation,

$$L_\downarrow = L_{\downarrow sky} + L_{\downarrow env} = L_{\downarrow 0}\,\psi_{sky} + L_\uparrow \psi_{env} \quad (\mathrm{W\,m^{-2}})$$

**Equation 14.10**

where $L_{\downarrow 0}$ represents **longwave** radiation receipt at a surface with $\psi_{sky} = 1$ and $L_\uparrow$ is that emitted by surrounding facets. $L_{\downarrow env}$ is a function of facet **emissivity** and surface temperature; at night, the contribution of $L_{\downarrow env}$ in cities is especially noticeable owing to the **surface urban heat island**. The human body reflects a very small proportion of $L_\downarrow$ (less than 3%) and consequently is a near perfect emitter at these same wavelengths.

## Net Radiation

The diversity of the radiation environment is apparent in Figure 14.4, which shows two urban landscapes that are distinguished by the vegetative **surface cover**, the construction materials and built geometry. The patterns of sunlight and shade and the differing **surface energy balances** result in a great diversity of microclimates that are evident in the surface temperatures. Figure 14.4a and b shows a plaza where there is little shade and the surrounding surfaces have a nearly uniform temperature which is close to that of the people. Figure 14.4c and d shows a more complex urban setting comprised of warm (road and sunlit facades) and cool (grass and shaded pavements) surface that create a heterogeneous environment; the radiation experience of an individual will vary greatly depending on their position in this landscape.

For many practical purposes it can be useful to summarize the external radiative environment using a **mean radiant temperature** ($T_{\mathrm{MRT}}$), the surface temperature of a perfect emitter (**blackbody**) that generates the same radiation as that absorbed by the body,

$$T_{\mathrm{MRT}} = \sqrt[4]{\left[(K_\downarrow \alpha) + (L_\downarrow \varepsilon)\right]/\sigma} \quad (\mathrm{K})$$

**Equation 14.11**

Here, the albedo ($\alpha$) and emissivity ($\varepsilon$) parameters are those for outer (clothed) surface of the body and $\sigma$ is the Stefan-Boltzmann constant. The measurement of $T_{\mathrm{MRT}}$ is done using a globe thermometer that consists of a conventional thermometer housed in a spherical casing that has specific radiative properties. The resulting temperature is converted to $T_{\mathrm{MRT}}$ by solving the surface energy balance of the spherical enclosure.

**(a)**

**(b)**

**Figure 14.4** Visible and infrared (thermal) images for two urban landscapes in Berlin, Germany illustrating the radiation environment (Credit: F. Meier; with permission).

In indoor situations where there is no shortwave radiation component, and little variation in surface temperatures, whole body **net allwave radiation** ($Q^*$) can be obtained from

$$Q^* = f_{cl}\, h_r (T_{0,cl} - T_{\mathrm{MRT}}) \qquad (\mathrm{W\ m}^{-2})$$

**Equation 14.12**

where $T_{0,cl}$ is the surface temperature of the clothing and $T_{\mathrm{MRT}}$ is the mean radiant temperature. The coefficient of radiative heat transfer ($h_r$) is taken as 4.7 W m$^{-2}$ K$^{-1}$ for the whole body and $f_{cl}$ is a clothing area

factor. The latter is a simple ratio of the clothed surface area to that of the nude body ($A_{\mathrm{body}}$); for standard work clothing $f_{cl}$ equals 1.31. If ($T_{cl} - T_{\mathrm{MRT}}$) were 10 K an individual in a typical work setting will lose 61 W m$^{-2}$.

### 14.2.4 Sensible Heat Flux

The sensible heat flux density ($Q_H$) occurs by breathing and by convective exchange at the skin surface, although the latter is modulated by clothing.

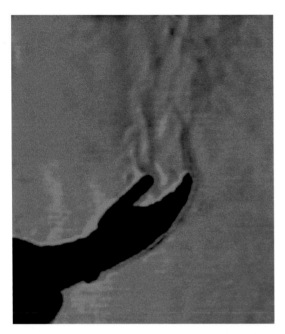

**Figure 14.5** A Schlieren image, which shows distortions in optical depth produced by air currents reveal the nature of thermal turbulence generated by the extended hand in a still atmosphere (Credit: G. Settles; CC3.0).

The breathing rate is a function of activity levels (Table 14.2) but even an individual engaged in moderate activity (exchanging 2.1 $m^3$ $h^{-1}$) in an environment at 20°C will lose about 6 W $m^{-2}$ or < 5% of the metabolic energy generated. So, the majority of sensible heat transfer occurs via the outer surface,

$$Q_H = f_{cl} h_c (T_{cl} - T_a) \quad (\text{W m}^{-2}) \quad \textbf{Equation 14.13}$$

where $h_c$ is the coefficient of heat transfer; its value varies according to: the magnitude of the temperature difference, which regulates thermal turbulence production (see Section 4.1.2) and; ambient airflow ($\bar{u}$), which causes mechanical turbulence production.

Thermal turbulence is generated as the human body warms adjacent air which becomes buoyant and rises forming a vertical stream of heated air (Figure 14.5). If there is little air motion thermal turbulence dominates but is not an effective means of heat loss. Increasing airflow disturbs the air around the body and generates mixing. Mechanical turbulence dominates once the ambient airflow exceeds > 0.2 m $s^{-1}$. Work on heat exchanges using a thermal manikin suggests that $h_c = 10.3\bar{u}^{0.6}$ that is suited for airflow $0.2 < \bar{u} < 0.8$ m $s^{-1}$ in typical indoor situations (de Dear et al., 1997). Note that the

value of $h_c$ is not linear, at 0.2 m $s^{-1}$ it equals 3.92 but at 0.8 m $s^{-1}$ it equals 9.01 W $m^{-2}$ $K^{-1}$; in other words, even low airflow dramatically improves heat loss. Consequently, simple appliances like electric fans can make individuals feel more comfortable in hot weather.

Net allwave radiation and sensible heat fluxes at the exterior of clothing can be combined (Parsons, 2003) by summing the heat coefficients for each and creating the operative temperature ($T_o$),

$$Q^* + Q_H = f_{cl} h (T_{cl} - T_o) \quad (\text{W m}^{-2})$$

**Equation 14.14**

where,

$$T_0 = \frac{h_r T_{\text{MRT}} + h_c T_a}{h_r + h_c} \quad (\text{K}) \quad \textbf{Equation 14.15}$$

and $h = h_r + h_c$. Similarly, the 'dry' heat exchange between the skin and $T_o$

$$Q^* + Q_H = \frac{(T_{skin} - T_o)}{I_{cl} + \left(\frac{1}{f_{cl}h}\right)} \quad (\text{W m}^{-2})$$

**Equation 14.16**

where the effects of clothing are captured by a single insulation value ($I_{cl}$).

### 14.2.5 Latent Heat Flux

Latent heat exchange ($Q_E$) occurs via **respiration** and via the skin. Exhaled air is close to saturation at core temperature (with a **vapour pressure** (e) of 91 kPa). If the ambient air temperature is 20°C and a relative humidity of 50%, then $e_a = 16$ kPa; a breathing rate of 2.1 $m^3$ $h^{-1}$ would exchange approximately 75 g of water vapour, which would represent just 30 W $m^{-2}$.

At the skin surface, evaporation occurs by **diffusion** through the skin membrane and directly as water is secreted by the sweat glands. The former account for a small proportion of heat loss by evaporation, estimated at 6%. Regulatory sweating is the main avenue of $Q_E$ but estimating its magnitude is complicated as the availability of water at the skin surface is governed by involuntary action,

$$Q_E = \frac{w(e_{skin} - e_a)}{I_{(v)cl} + \left(\frac{1}{f_{cl}h_v}\right)} \quad (\text{W m}^{-2}) \quad \textbf{Equation 14.17}$$

where, $e_{skin}$ and $e_a$ are the vapour pressure at the skin and the ambient air, respectively; the former is treated as the saturation value of $e$ at skin temperature. The insulation value of clothing with regard to vapour ($I_{(v)cl}$) is typically 0.015 m$^2$ kPa W$^{-1}$ for work clothing. The latent heat transfer coefficient ($h_v$) is linked with that for sensible heat ($h_c$), via the Lewis Relation with a value of about 16.5 K kPa$^{-1}$ for typical indoor conditions (Parsons, 2003). Hence, if airflow ($\bar{u}$) equals 0.5 m s$^{-1}$, $h_c$ is approximately 6.80 W m$^{-2}$ K$^{-1}$ and $h_v$ equals 112.13 W m$^{-2}$ kPa$^{-1}$. The availability of water is represented by skin wettedness ($w$), which is calculated from the residual in the energy balance equation (Equation 14.5) when $\Delta Q_S \approx 0$. Essentially, $w$ is the ratio of the latent heat required to balance the equation compared to the maximum allowable latent heat exchange in the ambient conditions (Parsons, 2003).

Regulatory sweating is the main means by which the body manages its heat exchanges at high ambient air temperatures. Approximately 85% of human skin can be made 'wet' through the excretion of water through sweat glands onto the skin surface. Over the trunk and limbs there are 1–2 glands per mm$^2$, for the palms of the hand and sole of the feet there are 20 per mm$^2$. However, the maximum sustainable heat loss through $Q_E$ is about 380 W m$^{-2}$ (equivalent to about 1 kg h$^{-1}$ or 2.5% of the body's water content) if it is supplemented by water intake. A loss of 10.5 kg of water for a 70 kg man will cause circulatory failure and death; even half this loss results in headaches and loss of concentration.

## 14.2.6 Clothing

Exchanges between the skin surface and the atmosphere are modulated by clothing, which acts as a new 'surface' that intervenes between the skin and the ambient environment. Clothing modifies the shape of the body and acts as a barrier to solar radiation receipt by reflecting, absorbing and transmitting proportions of the intercepted flux. The magnitude of each depends on the incident flux and the colour and density of cloth. Clothing also exchanges longwave radiation with both the environment and the body it encloses. It also impedes the **turbulent fluxes** by introducing an additional resistance to transfer. Most clothing is designed to protect the body from excessive sensible heat loss and this is managed through the choice of clothing (the insulation values of various garments are listed in Table 14.3). Note that the values change with both the nature of the material

**Table 14.3** The dry thermal insulation values ($I_{cl}$) for individual clothing garments and for selected work clothing ensembles (Source: Parsons, 2003). These data are also shown in the form of the equivalent depth of still air, 1 mm has a value of approximately $3.85 \times 10^{-2}$ K m$^2$ W$^{-1}$.

| Garment | Insulation $I_{cl}$ (K m$^2$ W$^{-1}$) | Depth of still air (mm) |
|---|---|---|
| **Individual clothing layers** | | |
| Underwear (e.g. underpants, T-shirt, slip) | 0.03–0.10 | 0.15–0.52 |
| Footwear (e.g. socks, slippers, boots) | 0.02–0.10 | 0.10–0.52 |
| Shirts/Blouses (e.g. short and sleeve shirt, sweatshirt) | 0.15–0.30 | 0.78–1.55 |
| Trousers (e.g. shorts, trousers, overalls) | 0.06–0.28 | 0.31–1.45 |
| Sweaters/Jackets | 0.20–0.35 | 1.03–1.81 |
| Dresses/skirts | 0.15–0.40 | 0.78–2.07 |
| Outdoor clothing (coat, parka) | 0.55–0.70 | 2.84–3.62 |
| **Clothing ensemble** | | |
| Underwear with long sleeves and legs, shirt, trousers, jacket, socks, shoes | 0.155 | 5.17 |
| Underwear with short sleeves and legs, shirt, trousers, jacket, thermojacket and trousers, socks, shoes | 0.225 | 7.50 |
| Underwear with long sleeves and legs, thermojacket and trousers, parka with heavy quilting, socks, shoes, cap, gloves | 0.395 | 13.17 |

and the extent of the body that it covers. The overall effect of clothing on the human energy balance is evaluated as a product of the ensemble, that is the shoes, socks, undergarments, trousers, shirt, etc. that fully clothe the person.

The properties of traditional clothing usually reflect the climatic environment of the wearer. In hot climate where overheating is an issue, the objective is to limit heat gain and maximize heat loss. Where it is hot and humid, clothing that offers the least resistance to the transfer of heat and moisture is desirable; ideally, the fabric has a wide weave and is loose around the body to allow for circulation. In hot and dry climates, the primary purpose of clothing is to protect the body from solar radiation, which is best achieved through materials with a high **reflectivity**. Although it may appear counterintuitive, darker clothing can achieve the same effect if there is an additional layer beneath the outer layer, so that the heated fabric is not in contact with the skin surface. In these climates it is the head and shoulders that receive the brunt of solar radiation during the hottest period of the day so head cover is needed.

In colder climates, the emphasis shifts towards limiting heat loss so clothing ensembles that provide higher resistance values are desirable. In climates that are both cold and windy, heat loss is accentuated as the laminar layer adjacent to the skin surface is shallow and offers limited protection. Layers of clothing act to increase the depth of this layer by trapping air close to the skin. The coats of animals that have thick furs are also employed for the same purpose.

## 14.3 Thermal Stress and Body Strain

The terms in the energy balance equation (Equation 14.5) adjust continuously as the environment to which the body is exposed changes and the body responds. Altogether, there are six variables that regulate these exchanges: radiation, air temperature, humidity and wind speed, which represent the exposure environment and; metabolism and clothing, which represent the response of the individual. The combination of the environmental variables imposes a **thermal stress** to which the body responds, resulting in **thermal strain** as it seeks to ensure that $\Delta Q_S$ is close to zero. For example, in warm weather, low wind speed and a high relative humidity causes considerable stress to which the body responds by sweating. The degree of strain may be evident by the gleam of sweat that has been excreted onto the skin surface but has not evaporated resulting in a feeling of discomfort.

To understand the relationship between the stresses and strains, it is helpful to consider the case of a person placed in a controlled indoor setting (a climate chamber). In this setting, the ambient air temperature, humidity and velocity are fixed, the surrounding wall surfaces have a uniform temperature equal to the air temperature (that is, $T_{\mathrm{MRT}} = T_a$) and the activity levels and clothing of the occupant are prescribed. In these conditions, the physiologic response of the

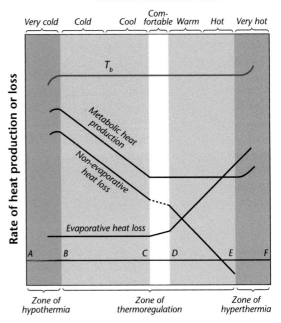

**Thermal sensation scale**

Figure 14.6 Simplified relationships between energy exchanges and a changing environment as represented by an environmental temperature, which represents the ambient conditions. To maintain a constant deep body temperature of 37°C the body must manage its exchanges. When the body is comfortable (C-D), the body has little difficulty in disposing of the metabolic heat ($Q_M$) it generates. When the ambient conditions cool (C to B) the body generates energy ($Q_M$) to compensate for radiative and sensible heat loss ($Q^* + Q_H$). When the ambient conditions warm (D to E), heat loss from the body via ($Q^* + Q_H$) diminishes and the body responds by increasing heat loss through regulatory sweating ($Q_E$). The thermal sensations associated with these changes are described in terms of varying degrees of cold or warm. At the extremes (B to A and E to F), the body cannot maintain a constant deep body temperature and experiences hypothermia or hyperthermia (Modified after: Mount, 1978).

occupant can be evaluated by modifying the temperature of the chamber only (Figure 14.6). A lightly clothed and sedentary individual when exposed to an environment with a wind speed of 0.1 m s$^{-1}$, a relative humidity of 50%, and $T_a$ of 20°C experiences little strain. In these circumstances, the body does not sweat, $Q_E$ is minimal and sensible heat loss ($Q_H$) dominates. Skin temperature is about 33°C. These conditions may be described as comfortable as most will express satisfaction with the environmental conditions.

## Cold

Reducing $T_a$ induces the thermal sensation of cooling and the body responds by managing heat loss while increasing heat generation. The first response will be to employ vasoconstriction to reduce skin temperature, which depresses $Q_H$ and $L_\uparrow$, and the environment feels cool. If $T_a$ continues to fall, it begins to feel cold ('goosebumps' appear on skin) and the body will take actions to increase $Q_M$ such as shivering to offset heat losses. The temperature of the skin and subcutaneous tissue cools unevenly in response to heat loss (Figure 14.2): the lowest temperatures are found distant from the core, especially in extremities such as toes and fingers. A further temperature drop will be sensed as very cold: eventually the body will not be able to sustain heat loss ($\Delta Q_S < 0$) and $T_{core}$ starts to fall leading to hypothermia.

## Warm

On the warm side of comfort, the body must dispose of heat so internal heat generation ($Q_M$) is maintained at a base level. Vasodilation can enhance $Q_H$ and $L_\uparrow$ by raising $T_{skin}$ but this has a maximum value of 36°C, which limits the effectiveness of this response to conditions that feel slightly warm. The most effective mechanism for coping with warm and hot conditions is regulatory sweating and evaporation ($Q_E$). Eventually, increases in $T_a$ will result in a net heat gain ($\Delta Q_S > 0$) as the body's ability to dispose of heat has been overwhelmed. These very hot conditions cause $T_{core}$ to rise, which increases the metabolic rate and internal heat generated; if this continues the thermoregulation system fails and hyperthermia ensues.

The circumstances that govern the onset of hyper- and hypothermia will vary from individual to individual, depending primarily on their age and health. However, from a public health perspective, there is a clear relationship between mortality and air temperature. Figure 14.7 shows a U-shaped curve that captures the statistical relationship between daily mortality and daily maximum air temperature for Manhattan, United States, based on historical records. Notice that when temperature lies within the

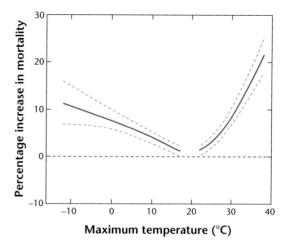

**Figure 14.7** Exposure–response curve for temperature-related mortality in Manhattan, New York, United States. The solid line shows the central estimates. The dashed line shows the 95% confidence intervals (Source: Li et al., 2013; Reprinted by permission from Macmillan Publishers Ltd: Nature Climate Change, © 2013).

zone from 17.2 to 21.7°C, there is no discernible effect but outside these bounds, mortality is correlated with the magnitude of the deviation. This pattern is repeated in other places but the zone of no impact shifts higher (lower) in warmer (cooler) climate, which is taken to reflect a degree of adaption by the population to the climate at a place (Keatinge et al., 2000).

## 14.4 Thermal Comfort and Its Assessment

**Thermal comfort** is defined as *the condition of mind that expresses satisfaction with the thermal environment and is assessed by subjective evaluation*[1]. Thus a comfortable individual *feels* neither too warm nor too cold and has no desire to either alter their clothing/ activity levels or to modify the environment to which they are exposed. Crucially, this definition highlights the role of psychology in assessing one's thermal state. It follows then that although thermal equilibrium may be a necessary condition for comfort, it is not a sufficient one. Moreover, deviations from comfort engender thermal sensations of 'warming' or of 'cooling' that is both a function of biophysical parameters (such as sweat rate or skin temperature) and individual

[1] ANSI/ASHRAE Standard 55-2010, Thermal Environmental Conditions for Human Occupancy

preferences, which may be associated with cultural contexts, thermal expectations and acclimatization. Much biometeorological research has focused on linking these expressed thermal sensations to measures of strains (body responses) and of stresses (environmental conditions).

### 14.4.1 Indoors

There has been exhaustive research carried out on comfort for the purpose of finding the ideal indoor climate that is suited to a particular occupation (such as office work) and the energy needed to create and sustain these conditions. This research is based on two distinct approaches that record the thermal sensations of individuals exposed to different climate conditions.

### Climate Chamber Research

A climate chamber is an enclosed space within which all aspects of the ambient climate (surface and air temperature, relative humidity and airflow) can be managed precisely (Figure 14.6). Subjects are exposed to regulated climate settings wearing specified clothing ensembles and engage in activity levels that exemplify a given work setting. The physical responses and thermal sensations of the subjects are recorded in response to environmental settings. In these controlled settings, the measured physical responses (e.g. skin temperature) and expressed thermal sensations (e.g. warm or cold) are remarkably robust; they have provided a biophysical basis for delimiting conditions that are considered to be universally comfortable.

The widely employed **Predicted Mean Vote** (PMV) model of Fanger is based on this research. PMV is based on a derivation of the energy balance which quantifies the deviation of the body's net energy flux in the exposure environment from that if the body experienced minimal strain (i.e. comfort) doing the same activity and wearing the same clothes (van Hoof, 2008). This deviation is correlated with the thermal sensation expressed as a 'vote' by subjects to create a linear scale. PMV values range from very cold (−3) to very hot (+3); these scores identify the thermal sensation expressed by the majority of those exposed to these conditions (Table 14.4). For example, PMV values of ±0.5 are comfortable, as less than 10% of the population should perceive those conditions as either too warm or too cool. Figure 14.8a shows this zone plotted on a psychrometric chart, with air temperature

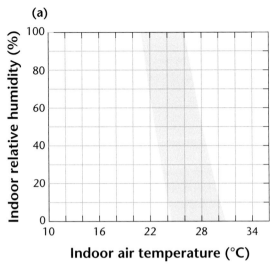

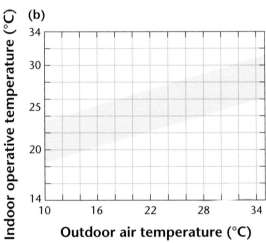

**Figure 14.8** Comfort zone in blue, representing 90% acceptability, plotted according to ASHRAE standards: **(a)** based on Predicted Mean Vote (PMV) values of ±0.5 and plotted according to air temperature and humidity (i.e. a psychrometric chart) and **(b)** based on the adaptive method and plotted according to indoor operative temperature and prevailing mean outdoor temperature. The targets shows the same set of circumstances on each (Source: CBE Thermal Comfort Tool for ASHRAE 55 http://comfort.cbe.berkeley.edu/; CC3.0)

and humidity. The value of this work is that it outlines universal comfort zones suited to particular work practices. For example, a design climate suitable for an office is one in which wind velocity is low ($< 0.2$ m s$^{-1}$), air temperature is between 20°C and 24°C and relative humidity is 40–60%.

**Table 14.4** Ranges of the thermal indexes Predicted Mean Vote (PMV) and physiological equivalent temperature (PET) for different grades of thermal perception by human beings and physiological stress on human beings (Modified after: Matzarakis et al., 1999).

| PMV | PET | Thermal perception | Grade of physiological stress | Universal Thermal Climate Index (UTCI) (°C) range | Physiological responses |
|---|---|---|---|---|---|
| | | Very cold | Extreme cold stress | < 40 | Decrease in core temperature |
| −3.5 | 4 | | | −27 to −40 | Shivering, average skin temperature will fall below 0°C if exposure is sustained |
| −2.5 | 8 | Cold | Strong cold stress | −13 to −27 | Face temperature < 7°C (numbness), core to skin temperature gradient increases |
| | | Cool | Moderate cold stress | 0 to −13 | Vasoconstriction, exposed skin temperature < 15°C |
| −1.5 | 13 | Slightly cool | Slight cold stress | +9 to 0 | Localized cooling, need for gloves |
| −0.5 | 18 | | | | |
| | | Comfortable | No thermal stress | +9 to +26 | Comfortable, sweat rate < 100 g h$^{-1}$ |
| 0.5 | 23 | | | | |
| 1.5 | 29 | Slightly warm | Slight heat stress | +26 to +32 | Slight heat stress |
| | | Warm | Moderate heat stress | +32 to +38 | Positive change in rate of sweating, and of skin temperature |
| 2.5 | 35 | Hot | Strong heat stress | +32 to +38 | Sweat rate > 200 g h$^{-1}$ |
| 3.5 | 41 | | | +38 to +46 | Small core to skin temperature gradient (< 1 K). Sweat rate increase (> 650 g h$^{-1}$ at limit) |
| | | Very hot | Extreme heat stress | > 46 | Increase in core temperature |

### Naturally Ventilated Buildings

A complementary approach measures the environmental conditions and the thermal sensations of subjects in actual work settings selected to represent different climate settings, cultural contexts and building types. This work has shown a significant difference between the responses of individuals in office buildings that have heating and cooling systems from those that are naturally ventilated (de Dear and Brager, 2002). While the responses of the former group conform closely to the results of climate chamber work, those of the latter are closely linked to outdoor air temperature. These differences are explained by a process of cultural and psychological acclimatization that adjusts the human 'thermostat' so that individuals are prepared to accept lower/warmer temperatures in winter/summer and adjust accordingly. In strongly managed climates (represented by the climate chamber approach), these adjustments do not occur; thus effect of managing climates to narrow parameter

values is to heighten expectations. The key finding from this work has been establishing a link between the outdoor temperature, the indoor temperature and expressed satisfaction (Figure 14.8b). This approach has informed the development of models of adaptive comfort, which allow for greater thermal tolerance by building occupants; this work has significant implications for assessments of building energy demand (Halawaa and van Hoof, 2012).

### 14.4.2 Outdoors

Although there have been many attempts to transfer the knowledge and techniques of indoor comfort to the outdoor environment, the process is fraught with difficulty for a number of reasons.

Firstly, the outdoor environment is a great deal more variable over space and time so that conditions are rarely static. As a consequence, the body's biophysical systems do not come into equilibrium with the environment and are constantly adjusting. This process is readily observed in Figure 14.9, which shows the ambient environment of an individual on a path (a **Lagrangian** perspective) through the urban landscape on a warm summer day (Nakayoshi et al., 2015). Note the great variation in air temperature experienced by the pedestrian outdoors when compared to that recorded at a nearby meteorological station and the impact of moving indoors; $T_a$ (and $T_{MRT}$) rises and falls by up to 12°C (20°C) instantaneously on leaving/entering buildings. The changes in wind, humidity and $T_{MRT}$ are equally dramatic; the reader should compare this thermal experience with the outdoor air quality experience depicted in Figure 11.7. Where the indoor space is air-conditioned, the environment is static and there is little temporal variation in variables but when outside, microscale spatial and temporal variation in all of the weather elements is the norm.

Secondly, the activities of those in the outdoors (walking, sitting, running, etc.), their demographic make-up and their clothing decisions are far more diverse (Figure 14.10). This means that it is not generally possible to create outdoor climates that meet narrow objectives. On the other hand, the climate expectations of individuals (that is the psychological component of comfort) are far less stringent; a draught may be unacceptable inside a building but a moderate wind may be perfectly acceptable and even desirable when outdoors. A design goal for outdoor urban spaces is to moderate, to varying degrees, its undesirable properties. So, in a cold and windy environment, one would seek to provide shelter from the wind. Although this is unlikely to make the space comfortable, it will make it less stressful and extend the period of usage.

### Measurement

Acquiring information on the human climate outdoors is extremely difficult. Ideally one needs to gather information on radiation, wind, temperature and humidity in the exposure environment but this is complicated by the great number of microclimates in cities. In addition, it is necessary to link these to indicators of the biophysical and psychological response of the person, which depends on activity levels, clothing, age, health and so on.

Figures 3.10a, 3.11b and 3.11c show instrument systems that are designed to make observations at pedestrian height; note that radiation instruments are arranged to provide information on the exchanges at the sides of an imaginary box. Each instrument combination is oriented perpendicular to the others and records short- and longwave radiation. Typically, these systems are located in an environment that is likely to see public use and a sample of people are interviewed to provide data on clothing and activity levels and their personal assessment of the thermal environment. This information may be supplemented by observations of the public use of the space. Subsequently, this response information is correlated with the environmental data to determine temporal and spatial patterns of use. Figure 14.11 shows a different approach, which is made possible by technological advances that allow instruments to be miniaturized and placed directly onto the human body. Rather than relocating the instruments to the places where outdoor users are located, the user becomes the platform for measuring the ambient environment. Even still, it is necessary to record the thermal sensation of the wearer to assess comfort and stress (Nakayoshi et al., 2015).

### Modelling

The PMV model predicts thermal sensation based on the deviation between the energy exchange that is required to solve the energy balance in given conditions against that required during comfort. It was designed for indoor spaces and its relations are based on empirical studies carried out with subjects in a climate chamber. More general **numerical models** of the human energy balance can simulate the thermophysical responses of the human (e.g. skin temperature and sweat rate) to environmental stimuli that change rapidly; these transient conditions require non-steady state responses. Typically, these models segment the body into parts that are represented by simpler shapes (e.g. a sphere to represent the head and cylinders of different dimensions to represent the torso, arms, legs, fingers and toes) for which view factors and exchange coefficients are known. Each segment is connected via a modelled circulatory system and the energy exchange from the core to the outer skin surface is simulated at a series of nodes that represent layers of tissue. Clothing is included as a new surface layer and the energy exchange between the node at the skin surface and that at the cloth surface is simulated. These models can be used to create *rational* indices of environmental stress by comparing ambient conditions against a standard (see next section).

Human energy balance models are distinguished by their sophistication in representing aspects of the

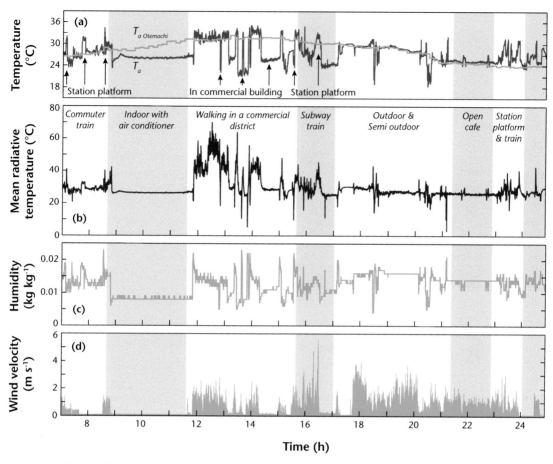

**Figure 14.9** The ambient environment as recorded by an instrumented pedestrian (Figure 14.11) when walking through an urban environment. The variables recorded are: **(a)** air temperature as recorded at a nearby weather station ($T_{a\text{ Otemachi}}$) and by the individual ($T_a$); **(b)** mean radiant temperature ($T_{MRT}$); **(c)** humidity (as mixing ratio) and; **(d)** wind speed (m s$^{-1}$). The shading indicates the indoor, outdoor and semi-outdoor environments experienced during the walk (Data source: M. Nakayoshi; with permission).

energy balance. The Munich Energy-Balance Model for Individuals (MEMI) solves three sets of equations: the energy balance of the whole body; the heat flux from the core to the skin surface and; the heat flux through the layers of clothing. More sophisticated models have realistic descriptions of the body form and of internal transfers. As an example, the UCTI-Fiala model treats the body as composed of 12 compartments: head, face, neck, shoulders, thorax, abdomen, upper and lower arms, hands, upper and lower legs and feet. The transfer of heat within and between body compartments is calculated at 187 nodes that allow for detailed simulation of the variations in skin temperature associated with asymmetric heat gains and losses at the skin surface (Fiala et al., 2012).

One of the obstacles to using such models is the availability of data on the ambient environment. Observations from conventional meteorological stations will often have little in common with microclimate circumstances experienced in cities among buildings. Computer models are ideally suited to dealing with these situations by allowing routine calculations of solar access, and **sky view factor** in urban areas using building morphology data (e.g. Lindberg, 2007). The RayMan model simulates the short- and longwave radiation absorbed by an individual located in complex urban setting. It does this by dividing the urban landscape into surface elements, computing the radiation flux originating for each and calculating the view factor between the recipient and the source of radiation. The results of the

**Figure 14.10** A typical outdoor urban environment (Curitiba, Brazil) which illustrates the varied use of outdoor spaces and micro-climatic experiences. While some are walking or sitting in sunlight, others are situated in shade, some next to fountains (Credit: G. Mills).

computation are converted into a mean radiant temperature ($T_{MRT}$) suitable for biometeorological purposes (Matzarakis et al., 2010).

### 14.4.3 Thermal Indices

A great number of indices have developed that link thermal responses to measures of ambient stresses and body strains (Epstein and Moran, 2006). Some are based on readily available meteorological data so that they have the advantage of ease of calculation; others are based on measures of thermal strain such as skin temperature or sweat rate. The most comprehensive are based on the human energy balance. For practical purposes, many of these indices are calibrated against a suite of ambient climatic, activity and clothing conditions that represent an imaginary setting in which only air temperature is allowed to vary. The 'equivalent' air temperature is calculated that would exert the same stress (or cause the same strain) in these circumstances as the conditions to which the body is currently exposed; this provides a single measure of the thermal environment.

#### Direct Indices

These are based on routinely available meteorological data and are of value in assessing extreme temperatures that may have public health consequences. However, these indices do not provide any detailed examination of the factors contributing to thermal stress and are consequently of limited diagnostic value. As examples the Humidex and Wind Chill Index

(WCI) for warm and cold conditions, respectively, are used by the Canadian meteorological service,

$$\text{Humidex} = T_a + 0.5555\,(e - 10.0) \quad (^{\circ}\text{C})$$

**Equation 14.18**

$$\begin{aligned}\text{WCI} = {} & 13.12 + 0.6215\,T_a - 11.37\,\bar{u}^{0.16} \\ & + 0.3965\,T_a\,\bar{u}^{0.16} \quad (^{\circ}\text{C})\end{aligned}$$

**Equation 14.19**

where $e$ is vapour pressure (hPa) and $u$ is wind speed (in km hr$^{-1}$) measured at 10 m above the ground (that is, a conventional weather station). Both indices produce values that can be linked to a temperature scale and include just two of the environmental variables governing either warm stress (temperature and humidity) or cold stress (temperature and wind). On a warm (30°C) and very humid (30 hPa) day, the Humidex would be close to 40, indicating that the effect on a person is nearly the same as if the air were dry and $T_a$ almost 40°C; the associated advice is 'great discomfort, avoid exertion'. By contrast, WCI is applicable when $T_a \leq 10°\text{C}$ and $V \geq 5$ km hr$^{-1}$ and again indicates the equivalent temperature in calm conditions that would produce the same effect.

#### Rational Indices

These are based on the energy balance (Equation 14.5) which is a complete description of the biophysical processes that underpin thermal state of the body. It can be applied generally once the meteorological inputs (wind velocity, air temperature and humidity

**Figure 14.11** An outdoor Lagrangian measurement system suitable for evaluating the biometeorological environment. The antennae attached to the hat measure radiation, air temperature and humidity and wind speed; the instruments attached to the skin surfaces record skin temperature and heart rate and; the data-loggers recording the ambient environment and thermophysical responses arc attached to the body (Credit: M. Nakayoshi; with permission).

and radiation terms) are available. This model is best applied to indoor situations where environmental and behavioural circumstances can be treated as constant and as a consequence, the body can be assumed to be in thermal equilibrium. It is less suited for outdoor conditions that are transient in nature, the radiation environment is far more complex and the response of individuals is variable.

The physiological equivalent temperature (PET) describes the exposure environment in terms of the air temperature that would be required in reference conditions (using the MEMI model) to produce the same thermal response (Höppe, 1999). These reference conditions correspond to a person located indoors, wearing office attire and engaged in light work. In this setting, $T_a = T_{MRT}$, $V$ is 0.1 m s$^{-1}$ and $e$ equals 12 hPa (relative humidity of 50% at 20°C). In essence, the model is solved for the actual exposure conditions, the calculated mean core and skin temperature are re-entered as fixed values for the reference indoor conditions and a new $T_a$ is obtained that ensures a heat balance; this is the PET. Similarly the **Universal Thermal Climate Index (UTCI)** is based on the UTCI-Faisla model (Krzysztof et al., 2013) which in addition to its more complex treatment of the human energy balance selects an outdoor reference where the individual is walking, $T_a = T_{MRT}$, $V$ is 0.5 m s$^{-1}$ and relative humidity is 50% ($e$ is capped at 20 hPa).

## 14.5 Wind and Comfort

Apart from its role in thermal comfort, the mechanical effect of wind itself can generate discomfort, depending on the outdoor activity (ASCE, 2004). Typically, the attributes of wind are expressed in terms of the mean wind speed ($\bar{u}$) and its fluctuation ($u'$), where the latter are fluctuations of between 1 and 5 seconds,

$$u = \bar{u} + u' \quad (\text{m s}^{-1}) \qquad \textbf{Equation 14.20}$$

Typical gust conditions ($\hat{u}$) over a period of time are evaluated using the statistics of wind at a place,

$$\hat{u} = \bar{u} + f_p\,\sigma_u \quad (\text{m s}^{-1}) \qquad \textbf{Equation 14.21}$$

where $\sigma_u$ is the standard deviation of $\bar{u}$ and $f_p$ is a peak factor; if $f_p = 0$ then gusts are ignored, at $f_p = 1.5$ we can expect the gust speed to be exceeded 10% of the time and, at $f_p = 3.5$, just 0.01% of the time. Discomfort can occur as a result of either strong winds or gustiness so it can be useful to derive a gust equivalent speed ($u_{GEM}$) that can be compared directly with $\bar{u}$,

$$u_{GEM} = \hat{u}/f_{p,\text{fixed}} \quad (\text{m s}^{-1}) \qquad \textbf{Equation 14.22}$$

where $\hat{u}$ is based on a gust factor of 3.5 (i.e. a 3-second long gust exceeded about once every 5–10 minutes) and $f_{p,\text{fixed}}$ is a representative fixed peak factor (1.85). If either $u_{GEM}$ or $\bar{u}$ exceed some established criterion, then the overall assessment is uncomfortable. Table 14.5 provides wind ranges for different activities that would be acceptable if met more than 80% of the time; note that the acceptable ranges increase with activity level, that is, the conditions comfortable for walking may be uncomfortable for sitting.

## 14.6 The Urban Effect on Human Climates

From the perspective of human health and well-being, it is conditions within the **urban canopy layer** (UCL) that are most relevant. Here the ambient environment is dynamic and highly variable over very short distances (Figure 14.4 and Figure 14.9). The following text focuses on the outdoor environment and thermal and airflow effects of cities and is arranged according to the scale of the urban effect in this layer.

### 14.6.1 Microclimates

Given the variety of urban microclimates, most systematic work has been done on common urban configurations such as streets, parks and plazas. This research relies on the methods described above; here

**Table 14.5** Wind criteria (applicable to both $\bar{u}$ and $u_{GEM}$) for different levels of activity based on a 20% probability of exceedance (Source: ASCE, 2004; © ASCE, with permission).

| Activity | Comfort ranges (m s$^{-1}$) |
|---|---|
| Sitting | 0–2.5 |
| Standing | 0–3.9 |
| Walking | 0–5.0 |
| Uncomfortable | > 5.4 |

we present some examples of this work and leave general discussion of design to the following chapter.

### Thermal Environment

The role of street geometry on the climate experienced by pedestrians can be examined using Figure 14.12 which shows the cross-section of canyon that is partly in shade. While $T_a$ is nearly constant owing to mixing (see Figure 7.23), the surface temperatures of the facets vary considerably. The surfaces in shadow are relatively cool, whereas those in direct sunlight are warm. On the shaded side of the street, a pedestrian receives no direct shortwave radiation and the bulk of the intercepted longwave radiation will be sourced from the shaded side of the street, which occupies the larger view factor. However, all these exchanges change as the individual crosses the street to the sunlit side, receives **direct-beam irradiance** and intercepts more of the radiation emitted by the warmer surfaces; the $T_{MRT}$ captures the radical change in the radiation environment when crossing from the shaded to sunlit part of the street. The PET index follows the path of $T_{MRT}$, illustrating the importance of the radiation environment to human (dis)comfort in the outdoors.

Johansson (2006) measured surface and air temperature, wind and humidity in typical streets that characterize Fez, Morocco (33°N), a city that experiences hot summers and cold winters. The old city has a compact design where narrow streets (high $H/W$) separate buildings (similar to that of Marrakech, see Figure 5.1) but the newer part has a more dispersed urban design and is characterized by lower $H/W$ ratios. Figure 14.13 shows PET values calculated for

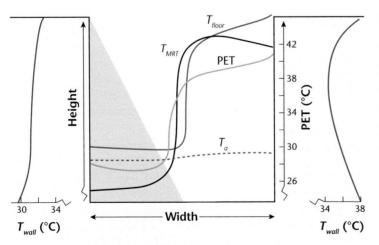

**Figure 14.12** A cross-section of an urban canyon showing simulated variations in wall ($T_{wall}$) and floor ($T_{floor}$) temperatures; air temperature ($T_a$) and mean radiant temperature ($T_{MRT}$). Note the impact of sunshine and shade (in grey) on the surface values (Source: G. Jenritzky; with permission).

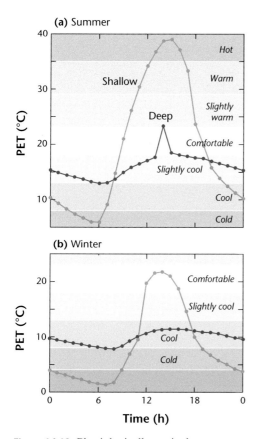

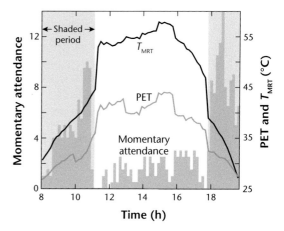

**Figure 14.14** Assessment of the influence of daily shadings pattern on human thermal comfort and attendance in Rome during summer period (Source: Martinelli et al., 2015; © Elsevier, used with permission).

**Figure 14.13** Physiologically equivalent temperature (PET) at the mid-point of a deep and shallow street canyon for summer and winter periods in Fez, Morocco (Source: Johansson, 2006; © 2005 Elsevier, used with permission).

two streets (one deep and one shallow) that represent these neighbourhoods, respectively. During the summer period the deep canyon remains comfortable throughout the day. Here the surface facets remain in shade and surface temperatures are relatively low, hence the low PET. By comparison, the shallow canyon experiences very high PET values and offers little shade from the Sun. At night, the pattern is reversed with higher PET values in the deep canyon. During winter, the same patterns occur but, as the winter climate is cool, access to sunshine is desirable and the shallow canyon is more comfortable during the daytime. On the other hand, the deep canyon retains heat at night and remains relatively warm.

Parks and plazas are both places set aside for outdoor public use and have received some attention. Figure 14.14 shows the results of a study on the public

use of a square in central Rome, Italy (Martinelli et al., 2015). The researchers divided the square into subareas based on the pattern of shade as it moved over the course of a sunny and warm August day. Those in place (not walking through) for 10 minutes or more (momentary attendance) were counted according to their position in the different subareas. The results show a clear correspondence between the choices of those in attendance, the shaded period and recorded $T_{MRT}$ and PET. Evidently, most preferred shade whenever possible and relocated when it was no longer available at that location.

## Wind Environment

The overall effect of cities is to reduce the mean wind speed while simultaneously increasing gustiness at pedestrian level but the wind environment in any given location is controlled by the interaction between the ambient airflow and the immediate geometry of the built environment. Tall buildings in particular present problems as they displace faster moving air from above to ground level and can cause problems for vehicles and pedestrians in their vicinity (Section 4.2.4). Figure 14.15 shows the effects of **turbulence** on the gait of a pedestrian near a tall building. As the wind shifts the body must redistribute its weight by altering footstep patterns to maintain balance.

Although each building has a unique impact on airflow, Figure 14.16 shows a range of building types

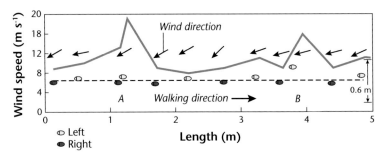

**Figure 14.15** Observed footsteps and wind speed at 1 m height near a high rise building. The distance between the steps reflects the difficulty that the pedestrian experiences while walking in a turbulent urban environment (Source: Murakami, 1982).

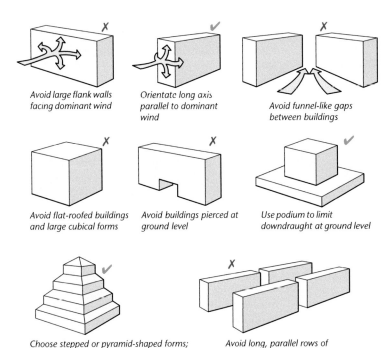

**Figure 14.16** Guidelines for reducing the wind sensitivity of buildings (Source: Littlefair et al., 2000; © IHS, reproduced with permission from BRE BR 380).

and arrangements and their wind effect. In general, tall buildings that present a wide face to the oncoming flow should be avoided; similarly, long smooth-faced buildings aligned along the path of airflow offer no shelter. A common solution to such problems is to raise the tall tower on a larger base (podium) that protects the wind environment at street level. The separation distance between buildings may also be relevant as airflow will accelerate as it squeezes through the gap; large slab-like faces with narrow separation gaps that channel airflow are problematic. For these types of situations, wind shelter at the ground in the form of windbreaks is often needed to make the outdoor space useable. On the other hand, tall buildings that taper with height or present a narrow face to the oncoming flow offer less resistance to above roof winds and are less likely to cause unfavourable winds at the ground.

## 14.6.2 Neighbourhood

Given the diversity of urban microscale environments, generalizations at the scale of a neighbourhood are not simple. A neighbourhood may be described as an urban landscape with a degree of homogeneity in terms of building dimensions, separation distances, **impermeable** and vegetative cover, etc. (see Chapter 2).

**Table 14.6** Frequency of occurrence of climatic event days in the LCZs (see Figure 2.9) in Oberhausen, Germany during the measurement period 1 August 2010–31 July 2011 (Modified after: Müller et al., 2014).

| Local Climate Zone (LCZ) | $T_{a\,(\max)} \geq 30°C$ | $T_{a\,(\max)} \geq 25°C$ | $T_{a\,(\min)} \geq 20°C$ | $T_{a\,(\max)} \leq 0°C$ | $T_{a\,(\min)} \leq 0°C$ |
|---|---|---|---|---|---|
| **LCZ 2**: Compact mid-rise | 3 | 35 | 2 | 12 | 51 |
| **LCZ 3**: Compact low-rise | 3 | 24 | 2 | 12 | 55 |
| **LCZ 5**: Open mid-rise | 3 | 26 | 1 | 9 | 60 |
| **LCZ 6**: Open low-rise | 3 | 29 | 0 | 8 | 69 |
| **LCZ 8**: Large low-rise | 3 | 25 | 2 | 12 | 59 |
| **LCZ 9**: Sparsely built | 3 | 19 | 0 | 6 | 61 |
| **LCZ A**: Dense trees | 2 | 16 | 0 | 5 | 64 |
| **LCZ D**: Low plants | 3 | 23 | 0 | 5 | 72 |

Different neighbourhood types have a 'typical' mixture of microclimates (e.g. streets, plazas, parks, gardens, etc.) that distinguish it from other neighbourhoods. Moreover, as these types are often associated with particular functions such as commercial or residential activities, they will have distinct patterns of indoor and outdoor use.

### Thermal Environment

The design of a neighbourhood affects every meteorological variable that influences human biometeorology; the magnitude of the urban effect on each is closely related to characteristics of the urbanized landscape and the 'natural' landscape that it replaced. However, it is important to note that the body responds thermally to the integrated urban effect rather than any one variable in isolation. Thus, for example, raising $T_a$ and lowering $\bar{u}$ together reduces heat loss, which may make the neighbourhood more comfortable (if cold) or less comfortable (if already warm); similarly, raising $T_a$ while lowering $T_{MRT}$ (through shading) may have little net impact.

Table 14.6 shows the thermal climate of Oberhausen, Germany, over a year (1 August 2010 to 31 July 2011) based on observations made at eight stations representing different **Local Climate Zones** (see Figure 2.9). Note that, in general the urban areas experience warmer air temperatures by day and by night in the summer and fewer cold days during the winter. Figure 14.17 shows the diurnal pattern of PET curves for July 11, which was especially warm. Note the mid-rise compact LCZs, which represent the higher building density with the least vegetation exert

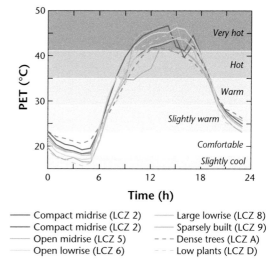

Compact midrise (LCZ 2)  Large lowrise (LCZ 8)
Compact midrise (LCZ 2)  Sparsely built (LCZ 9)
Open midrise (LCZ 5)  – – – Dense trees (LCZ A)
Open lowrise (LCZ 6)  – – – Low plants (LCZ D)

**Figure 14.17** Diurnal variation of physiologically equivalent temperature (PET) values at the climate stations of the monitoring network in Oberhausen, Germany on 10 July 2010, based on measurements of the station monitoring network. (Source: Müller et al., 2014; © The Authors 2013, open access from Springer).

the greatest thermal stress during the day. Although vegetation (to provide shade and evaporation) were important regulators of PET, the greatest influence was near-surface wind speed.

### Wind Environment

More so than any other atmospheric property the statistics of air velocity are extremely variable in time and space within the UCL. In general, the degree of shelter at ground level is closely related to the packing density of buildings (see Figure 4.24). Where buildings are closely spaced, compactly organized and of uniform height, the space below roof level may be aerodynamically sheltered from the ambient airflow above roof level. However, where buildings are widely spaced, aligned along flow paths or of variable heights faster moving eddies from above roof level can be drawn into the UCL where they may pick up loose debris, and cause discomfort for citizens.

### 14.6.3 Urban Scale

At this scale, the average **urban climate** effect is to raise air and surface temperature (the urban heat island), lower relative humidity (largely as a result of air temperature changes) and retard airflow. With the exception of air quality, the urban effect can be beneficial depending on the background climate; when the climate is cold and windy, cold stress and building heating needs can be lessened but, where the climate is already warm and humid, heat stress (and building cooling demand) will increase. However, it is concerns for urban warming that has dominated research. This can be explained by: the increased energy demand for building cooling systems, even in cool climates; the emergence of very large cities in tropical climates where the background climate is already warm and; projections of global warming and increased frequency of heatwave events (see Section 12.3.3 and Table 13.6).

Managing the urban heat island generally is discussed in the following chapter, but it is worth reiterating the integrated nature of the environmental variables that create outdoor heat stress. As an example, the development of Hong Kong as a very densely built and occupied city in a warm and humid tropical setting has added to thermal stresses that residents would experience naturally. The pattern of building has worsened the climate by creating a line of closely spaced tall building that are parallel to the coastline and obstructs the movement of **sea breezes** into the city. These thermal circulations are strongest during the same calm and clear weather associated with the UHI and could provide respite from stressful conditions by bringing cooler and cleaner air into the city. To strike a balance between urban development and environmental concerns, the city has a development strategy that aims to create ventilation corridors between buildings that offer passage for coastal air (Ng, 2009).

## 14.7 Indoor Climates

Although the focus of this book has been on the climate outdoors, there is an intimate link with the indoor climate of buildings, which are an outcome of internal and external energy gains and losses (Section 6.5.2). The internal gains are associated with the metabolic heat released by occupants and that added by appliances (e.g. computers), lighting and heating systems. The external gains are those that enter the building via the envelope and include: radiation transmitted through windows and sensible and latent heat transfer across the wall and roof facets via conduction and through openings via convection. Maintaining a desirable indoor climate involves managing these exchanges each of which has distinct diurnal and seasonal patterns associated with occupation patterns. However, the same outcome could be achieved by different means (envelope design, material choice, energy demand and so on), each of which will have a different impact on the outdoor climate. Figure 14.18 shows a large low-rise building that has a retail function and is located in a hot and arid environment. Part of its energy management strategy is to reduce solar gain via the large roof surface, by increasing its albedo (a **cool roof**) while at the same time converting some of the solar energy to electricity. The combined strategy reduces the cooling energy demand within the building and provides an alternative and renewable source of energy. However, this example also shows the limits of such actions at a building scale, a theme that is picked up in the following Chapter 15.

**Figure 14.18** A large retail shop in Arizona, United States, that has taken steps to reduce its energy demand by using a white-coated roof to reflect solar radiation and installing solar panels to generate a portion of the building's energy needs (Source: Walmart; CC2.0).

## Summary

Concern for the climates of humans is at the heart of much of urban climate research, much of which is concerned with the application of scientific knowledge to the design of cities. The basic energy balance that underpins our understanding of human climates is discussed extensively in this book: it includes the **radiative** and **convective exchanges at the skin surface** and accounts for **heat transfer within the body**. However, the involuntary responses of the human system to these exchanges and the role of psychology in regulating thermal sensitivity distinguish the human climate from those of inanimate objects.

There are **six variables that** govern **the energy balance terms**. The environmental variables include the wind velocity, air temperature, relative humidity and radiation. The latter is encapsulated by the **mean radiant temperature** ($T_{MRT}$), which expresses the short- and long-wave radiation absorbed at the outer surface of the body. The terms that are intrinsic to humans are levels of activity and of clothing. Overall, the body must achieve balance so that the deep body temperature is constant.

- The combination of environmental variables results in **thermal stress**, which causes the body to respond by experiencing strain. The zone of minimum stress (and strain) is referred to as comfortable. Cold thermal stress causes the body to generate internal heat while limiting heat losses. Warm thermal stress causes the body to maximize heat loss.
- There are a great **number of thermal indicators** which are based on simple measures of the environment and of thermal stresses but the most useful indicators are based on the energy balance that link exchanges to physical (dis)comfort.
- Cities create myriad microclimates within the UCL. Indoor microclimates tend to be less variable and exposure can be treated as a steady state problem for occupants. **Outdoor microclimates are extraordinarily diverse** in space and time and **exposure is transient in nature**. A variety of tools have developed to understand the complex relationship between humans, climate and the use of outdoor spaces but much is still unknown.
- The creation of **comfortable indoor spaces** is a primary concern for building engineers and decisions on how to best achieve this **has ramifications for urban energy use** (Chapter 13) and the **anthropogenic heat flux** (Chapter 6). Sustainable architecture attempts to achieve this goal through good design that makes best use of the background climate resources. Even still, there is an ongoing debate on the importance of psychological controls in determining the best indoor climate for a given place and purpose.

All of the environmental variables relevant to human climates are modified in the city. While some of these changes may be beneficial, others are deleterious (e.g. dangerous winds, poor air quality). The thermal effect of the city has received most attention: the urban heat island raises surface and air temperature and in hot climates (or weather) will add to thermal stresses but, in cold climates (or weather) may be beneficial. Managing the urban climate through design solutions for the benefit of humans is the focus of Chapter 15.

# 15 | Climate-Sensitive Design

**Figure 15.1** An aerial view of a small park (St. Stephen's Green) in Dublin, Ireland (Credit: Irish Air Corps).

A primary reason to study **urban climate** effects is to apply the knowledge acquired either to plan new settlements that are climatically-sensitive or to address climatic problems in existing settlements. For the most part, urban climate effects are an unintended (even if predictable) outcome of countless decisions that have shaped particular **urban form** (e.g. street layout, building density and character, **surface cover** mix) and **urban function** (e.g. transport modes and networks, dominant fuel and energy systems and land use mix). However, as we have seen, the climate in the **urban canopy layer** (UCL) consists of a great variety of **microclimates** that can be managed through purposeful design.

Figure 15.1 shows a small (9 ha) green park located in the city centre of Dublin, Ireland. Originally used as pasture for nascent settlement, it became a public park in the nineteenth century and

was designed to provide a 'natural' oasis within the city centre. It contains a number of features that allow it to create a microclimate that is distinct from those created by the surrounding **neighbourhood**. These features include: open spaces that are exposed to the sky vault; a grass surface that is permeable and can store water; tree cover that can provide shade and shelter; vegetation and water features that evaporate water and experience moderate temperature changes. Note also that the park boundary is vegetated and separates the enclosed space from much of the noise, heat and pollution generated by the traffic that surrounds it.

Parks in cities represent conscious attempts to create pleasant outdoor spaces by using design tools (such as trees and water). **Climate-sensitive urban design** (CSUD) extends this perspective to the whole city. In this chapter, we use the evidence presented in previous chapters to extract general principles and provide guidance on urban planning and design from a climatic perspective. We might begin by asking: what makes a well-planned and designed city from a climatic perspective?

- The city is **efficient in its use of resources** (land, energy materials, water, etc.) so as to minimize its global and regional impact (e.g. **emission** of **air pollutants** and **greenhouse gases**, water degradation, waste generation).
- City neighbourhoods are **designed to improve the microclimates** surrounding buildings and their environs (or at least not worsen them).
- People and infrastructure are **protected from extreme weather events** by considering current and future climate variability and extremes.

We return to these points at the end of this chapter once we have outlined some basic principles of CSUD and discussed their application at different urban scales. The hierarchy of urban scales emphasized earlier in this book (see Chapter 2) is central to understanding the character and impact of climate effects and the potential for purposeful urban planning and design.

## 15.1 Basics of Climate-Sensitive Planning and Design

Here, we use the term 'planning' for decisions about the spatial organization of urban functions (e.g. residential and commercial) and the term 'design' for decisions about aspects of urban form at one scale that create the physical character of neighbourhoods appropriate to their function (e.g. single family dwellings or high-rise buildings for residential areas). The urban design tools discussed here refer to decisions on the creation or redesign of aspects of urban form (the cover, **fabric** and structure) and function, rather than the legislative tools that can support these decisions.

### 15.1.1 Political Context and Policy Mechanisms

Planning and design practices are embedded within overarching systems of governance that can be categorized by scale into global, region-national, city, neighbourhood and plot/building practices. The actors in these systems include governments, private and public land developers, public representatives, professional planners and designers and citizens (Figure 15.2). As examples: at global scale the climate

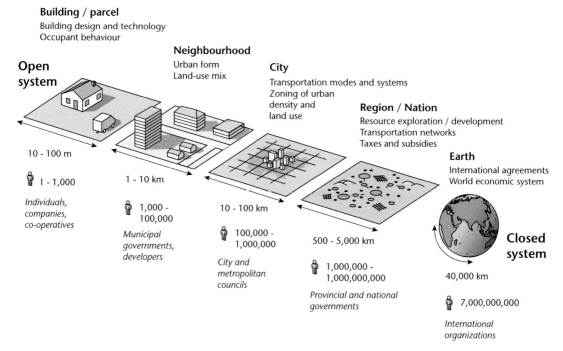

**Figure 15.2** The hierarchies of scales at which actors (whether individuals or organizations) make decisions affecting changes in urbanized landscapes.

context is established by various international agreements, such as the United Nations Framework Convention on Climate Change (UNFCCC) which requires ratifying parties to limit emissions of greenhouse gases (GHG); at regional and national scales there are equivalent policies that stipulate legally binding air quality standards; at city-scale policies to manage extreme events like heat waves and floods are enacted; at neighbourhood scale, decisions on the amount and character of green space and its distribution that affect local climate are made and finally; at the scale of the plot/building owners and occupiers make choices on landscaping, building **insulation** and so on to manage microscale climate and building energy use. Of course, there is great variation in planning systems internationally. An additional complication is that administrative boundaries rarely coincide with the contiguous urban landscape and an individual city may be 'fractured' into several political territories which may have different planning powers and pursue different agendas. These circumstances make the pursuit of common climate policies across an urban landscape a challenge.

Policies usually employ either top-down or bottom-up mechanisms. Top-down mechanisms include laws (ordinances at a city scale), legal commitments and guidelines used to enact decisions made by political institutions. Examples of ordinances include zoning, where land is set aside for certain functions such as the establishment of warehouse districts close to transport lines or the separation of residential areas from noxious or potentially dangerous facilities, like waste landfills or a railway transporting explosive or toxic products. **Zoning** includes specifications on the types of buildings (heights, dimensions, construction codes), road widths, etc. that result in certain urban parameters which affect climate. Bottom-up mechanisms are those associated with individuals (developers, homeowners and so on) who act within the given legal/political context. Their actions may be incentivized by institutional policies such as inducements to improve building insulation through an energy rating scheme, planting trees and maintaining greenspace, etc. The respective roles of institutions versus individuals varies considerably with the political context. A large proportion of modern urbanization in less wealthy regions however occurs as informal settlement, outside a legal context, where land ownership and/or urban services are not guaranteed or well regulated, as a result urban patterns emerge organically.

### 15.1.2 Climate Assessments

Given the range of climatic issues that could be considered in any specific urban design project, it might be helpful to restrict ourselves to one initially, that of the outdoor thermal climate experienced by humans. In Figure 15.3 a notional air temperature ($T_a$) that represents the integrated effect of **radiation**, wind, air temperature and humidity on an individual (similar to that shown for indoor conditions in Figure 14.6) is plotted for a time period. Four scenarios are depicted:

- $T_a$ falls within an acceptable range over the entire time period;
- $T_a$ is too high over the entire time period;
- $T_a$ is too low over the entire time period;
- $T_a$ is too high for a while and too low at others.

Obviously the first instance requires the least intervention and the primary purpose of design should be to ensure the natural environment is not made less desirable by urban development. The second and third situations also offer a distinct design objective, that of warming or cooling throughout the period. The final situation requires cooling at some times and warming at others.

Although this simple example focusses on just one variable and its variation with time, it illustrates the dynamic nature of climate stresses and the need for prior climate assessment to guide design interventions.

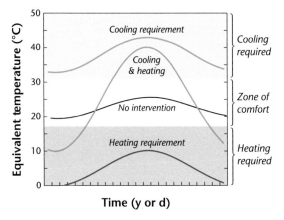

**Figure 15.3** A simplified cross-section of temperature against time. Equivalent temperature represents the aggregate effect of the environmental factors (air temperature, humidity, wind etc.) that influence the thermal environment of humans. The time axis represents hour of day or month of year. The white zone in the centre represents an area of minimal stress.

For other stressors, like air quality or precipitation, the patterns will be different but the principle remains the same; only by identifying the climate issues can informed decisions be made.

## Data Sources

Lowry (1988) describes readily available meteorological data to planners and designers as 'answers to questions they hope someone will ask' and emphasizes the importance of obtaining information that is relevant to the design project at hand. Ideally, weather and climate information should:

- be available for the site under evaluation or a comparable site nearby;
- include the relevant climatic elements (wind, temperature, precipitation, radiation etc.);
- be of sufficient duration that appropriate climatic averages and extremes can be determined; and
- be of sufficient temporal resolution that diurnal/seasonal stresses can be evaluated.

This means acquiring information on the appropriate climatic elements measured at the desired temporal resolution, for a sufficient period of time at a representative site.

For many design projects, such as evaluating periods of heat stress, it is the combined effect of several atmospheric properties (**mean radiant temperature**, air temperature, humidity and airflow, see Chapter 14) that is relevant, rather than just one. While some climatic elements may be readily available others may have to be estimated. Even when just one element is needed, the temporal resolution is important; for example, while the average monthly wind speed may be useful to select potential sites for wind turbines, data on gusts of short duration are needed to assess risks to pedestrian safety. Finally, the microscale context, where observations are made, is critical to assessing their usefulness at another location (see Section 3.1.2). Despite their limitations, meteorological observations made in controlled circumstances are invaluable and may provide input to a simulation model that could extend their usefulness.

## Compiling Climate Information

Given the great range of potential climate data the challenge of assembling and analyzing meteorological data can seem insurmountable. The essence of climate analysis is to distil the data available to a useable form using statistics (e.g. means and variances) and diagrams (e.g. wind roses). For urban planning purposes this information is often best placed in map form so that climate data appears alongside other environmental data and landuse/cover information (Bitan, 1988).

This approach is the basis of an urban climate programme that is exemplified in the governance of planning and design in Stuttgart, Germany, which since 1938 has formally incorporated climate information into its urban planning process (Hebbert and Webb, 2011). Historically, Stuttgart experienced very poor air quality due to its industrial economy, topographic setting and background climate that combine to create frequent near-surface **inversions** and poor air quality. However, the city also experiences nocturnal **katabatic** flows that represent a potential climate resource to ventilate the urban area. Identifying and protecting the sources of these winds and their flow routes provided the rationale for managing urban land use and cover using climate information, which is used to delineate areas with distinct climatic outcomes or **climatopes** (Section 2.1.4). The general methodology has proved an effective way of structuring climate information in a way that planners can incorporate into practice (Ng and Ren, 2015). However, identifying climatopes and assessing their climatic value requires considerable investment.

## 15.1.3 Guiding Principles

Before discussing climate information and its value in planning, it is worth emphasizing some general principles to place urban climate knowledge in context.

### The Background Climate Matters

Each city, by virtue of its geographic location and topographic situation, inherits and modifies a distinct climate (see Section 12.1). The most obvious way to minimize urban effects is to reduce the intensity of urban development by modifying aspects of a city's form and functions. The magnitude and timing of urban effects on climate depends on both the background climate (i.e. the climate that would exist in the absence of the city) and the nature of development at that site. This means that to a considerable degree, general principles of good design practice must be modulated to fit each city. Figure 15.4 illustrates this idea using mean monthly air temperature and humidity for standard meteorological stations (i.e. not accounting for the urban effect) in different cities

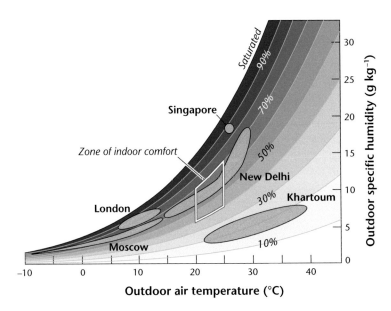

**Figure 15.4** The climates of selected cities expressed in terms of the monthly air temperature and humidity. The area in the centre of the diagram represents a zone of indoor comfort that might be a desirable objective.

and a standard zone of indoor comfort. For example: Singapore is wet and humid throughout the year; London, United Kingdom and Moscow, Russia, are cool; Khartoum, Sudan, is warm and dry and New Delhi, India, is warm at one time of the year and cool at another. Clearly, distinct climate types give rise to different heating and/or cooling and/or (de)humidification needs if we adopt this standard.

For some projects, general information on the background climate and the local **topography** is sufficient to make judgements on urban layout. Figure 15.5 shows two designs for a new low-rise housing development planned in Tahoua, Republic of Niger, in the Sahel region, south of the Sahara desert in north Africa. Here, annual precipitation is between 250 and 750 mm, of which 60–80% arrives in a four month period between July and August, when rainfall can be intense and cause flooding. During the dry season the Harmattan (a hot wind) blows from the Sahara bringing sand and dust and raising air temperatures to ~40°C. Urban design should provide shade and protection from the wind and the dust it carries. Moreover, urban development should avoid steep slopes (> 8%) so as to limit erosion during rainfall events.

Figure 15.5a shows the official plan, which imposes a grid layout that facilitates traffic movement but ignores both topography and climate. Note the logic of the grid means that the edge of the settlement is treated the same as its interior (so little communal shelter is provided). The roads are long and wide (between 10 and 50 m depending on purpose) and many are aligned along the path of the Harmattan. There is limited use of green space and of trees to provide comfortable microclimates. The overall design is poor from a climatic perspective as it provides little protection from the wind, dust and heat and creates a potential flood hazard. By comparison, the alternative (Figure 15.5b) accounts for **orography** by avoiding steeper slopes and designing roads to follow contours. The result is a more compact urban form with narrower streets (to provide shade) and reduced channelling of strong winds. Narrow gaps in the built fabric at the edge of the settlement maximize shelter. In addition, the road design impedes the flow of air through the settlement and tree planting is used as a design tool to provide both shelter and shade, and opportunities for **detention storage** of water, in specific areas.

### Climate Issues Are Best Considered at the Earliest Opportunity

Once a landscape has been developed for urban purposes (roads have been laid and buildings constructed), much of the urban climate effect is in place and the opportunities for change are greatly limited. Buildings have a life-span of decades whereas building plots and road networks can persist for hundreds of years. As a result, decisions on urban layout have long-term implications for servicing the metabolic

**(a)** Official plan

**(b)** Climate sensitive alternative

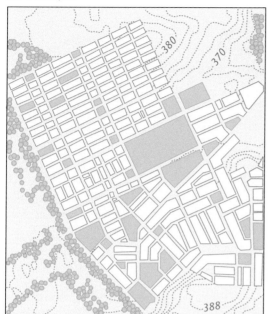
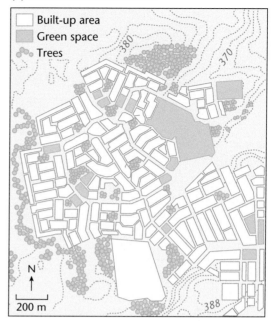

Built-up area
Green space
Trees

N

200 m

**Figure 15.5** Alternative layout plans for the Koufan extension zone in Tahoua, Republic of Niger. The official plan **(a)** is based on a generic grid design **(b)** an alternative plan based on local climate and topography (Modified after: Herz, 1988).

requirements of neighbourhoods and for waste generation (see Section 1.1.2). In fact, once *in situ* it may be easier to change the associated functions than alter form. Where the city is already in place, identifying and protecting natural resources (e.g. river floodplains, ventilation pathways, etc.) through appropriate zoning and landuse management should be a priority.

The process by which climate information can be brought into the design process can be illustrated using the case of a proposed new settlement to be constructed near the Arctic Circle, where protection from the elements is paramount (Figure 15.6). Low winter temperatures coupled with strong winds can make outdoor exposure hazardous so shelter is essential and time spent outdoors should be minimized – hence buildings should be close to one another. At the same time, given the relatively short periods of daylight around in winter, it is desirable that individual dwellings have solar access for as long as possible. Finally, the proposed site is situated near a very large water body so it receives most of its

precipitation as snowfall, which can accumulate if the urban layout hinders drift by wind. The design solutions that maximize solar access (Figure 15.6a), provide wind shelter (Figure 15.6b), and control snow accumulation (Figure 15.6c) are distinctly different. The solar solution recommends building plots on streets oriented east-west and sufficiently wide to guarantee periods of direct sunlight. The exposure solution generates a design that includes a long building (for communal activities) that acts as a windbreak to the north of the settlement. Finally, the stricture on snow accumulation produces a street pattern aligned along the path of airflow. A composite design tries to combine these individual solutions.

While the discussion above is theoretical, many of the design principles are evident in Fermont, Canada (52.5°N), a small (~3,000 residents) mining town constructed in the early 1970s. The design emphasizes wind shelter and access to sunshine as a response to its severe climate, which includes a long winter of 8 months when the average temperature is below

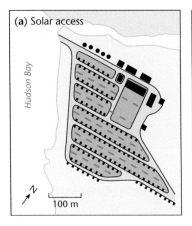 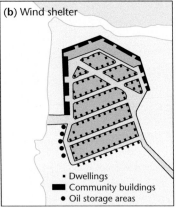 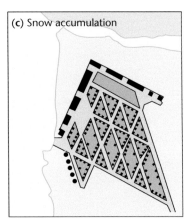

**Figure 15.6** Three town plans for a high-latitude settlements developed to meet different climate objectives with regard to **(a)** solar access, **(b)** wind shelter and **(c)** snow accumulation. A final plan should be a synthesis of the three alternatives through negotiation and resolution of conflicts (Modified after: Zrudlo, 1988).

**Figure 15.7** Fermont, Canada (52.5°N), photographed in 1970s, just after it was founded (Credit: Quebec Cartier Mining Company; with permission).

freezing and it receives snow 80 days a year. Note the placement of a tall (50 m high) communal building (1.3 km long) that protects the lower buildings in its lee from cold northerly winds (Figure 15.7). The retention of plantations of coniferous trees within the settlement also slow the near-surface airflow and snow drifting.

Of course, this example illustrates a process that is difficult to repeat; the settlement is both relatively new and located in an extreme climate where climate concerns dominate. In most cities, much of the urban development has already occurred and the challenge may be to decide how best to permit future building while preserving the climate resources that remain.

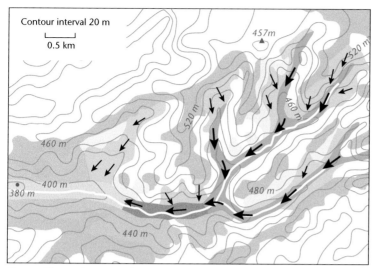

**Figure 15.8** The zones of restricted and banned construction in an alpine valley which is a source of katabatic winds for the city of Graz, Austria (Modified after: Lazar and Podessor, 1999).

Zones of banned construction

◤ Areas of cold air production with down-valley winds
◤ Areas of cold air production with down-slope winds

Zones of restricted construction

◤ Generally sections of local climate with moving cold air
☐ Zone above but with lower wind, already existing and dense development
☐ High contribution to the production of cold air
▨ Already existing and dense development on the slope
▨ Relatively advantageous conditions of temperature and ventilation
☐ Forest

Figure 15.8 shows a landuse management map for an alpine valley in the vicinity of Graz, Austria which is based on an examination of where cold air forms on the ridges and the route that katabatic flow takes as it flows downhill into the city. The map is used to restrict urban development primarily so that the associated clean (and cool) air has an unimpeded passage into the city where it can ventilate the settlement and improve its air quality.

### There Is No Single Best Design that Meets All Climate Objectives

Urban design must attempt to accommodate many, often conflicting demands, such as the desire for shade to manage building energy demands and for sunshine in adjacent outdoor spaces. As a consequence, either some compromise is reached or the impact of one parameter is considered to dominate and over-ride other concerns. For example, there is a considerable body of information on street climate that can be placed in the context of urban climate parameters, such as street orientation and the **canyon aspect ratio** $(H/W)$.

For a given orientation, the value of $H/W$ regulates access to sunshine (daytime warming), radiative heat loss at night (nocturnal cooling), shelter from the ambient wind and street ventilation (air quality). While a small $H/W$ value allows access to sunshine and ventilation, it also promotes night-time cooling and provides little shelter. On the other hand, a high $H/W$ value provides shade, limits night-time heat loss, provides shelter and weakens ventilation. On the basis of the available information (see Figure 4.24, Figure 5.12 and Figure 7.13), Oke (1988b) found that a range of $H/W$ values between 0.4 and 0.6 could satisfy the need to provide wind shelter and retain some of the heat island effect while simultaneously maintaining solar access and allowing the **dispersion** of air pollutants at street level. Similarly Fathy (1986), writing on managing building climate in Cairo, Egypt (30°N), identifies an east-west street orientation as best for solar access but that a northeast-southwest orientation is better for using wind to ventilate the building. However, he decides that solar considerations should dominate building design as indoor ventilation could be achieved via 'wind-catchers' on the roof that would be oriented in the direction of the wind.

## The Same Objective May Be Achievable by Different Strategies

It may be possible to create a desirable climate outcome in a number of different ways. For example, cooling the near-surface atmosphere in a hot and dry climate, where water is scarce can be achieved by reducing the adjacent surface temperature using a combination of interventions such as shading, **albedo** control and enhanced **evaporation**. However, not all interventions are equally effective.

As an example, Shashua-Bar et al. (2009) evaluated the efficiency of shade and evaporative cooling (via vegetation) to create a more pleasant microclimate by conducting an experiment in two semi-enclosed courtyards over a 45-day period during summer, when daytime temperatures exceeded 30°C. Over this period, different strategies were tested using combinations of trees, grass and shading mesh to test their impact on near-surface air temperature. As the experiment was conducted over 45 days, the temperature data were normalized by comparison with a nearby meteorological station. Figure 15.9 shows the hourly air temperature differences for each strategy when compared to the air temperature measured over bare soil. The most effective scheme (up to 2 K cooler) uses trees that provide both shade and evaporative cooling. Using grass alone was largely ineffective and the use of mesh shading produced a counterintuitive warming effect. Overall, trees represent the most efficient use of scarce water resources. While the mesh provides shade it does not cool the atmosphere, and although the grass provides evaporative cooling in the absence of shade its impact is diminished.

## Urban Design Options May Be Restricted by Prior Decisions

The existing urban fabric reflects past decisions on the road network, land parcel sizes, buildings, green

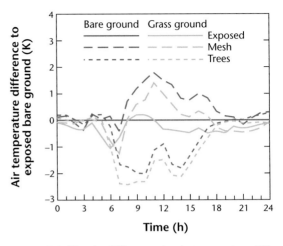

**Figure 15.9** Hourly differences in air temperature (K) between courtyards treated with mesh and grass **(a)** or trees and grass **(b)**, relative to the base case ('exposed bare') (Source: Shashua-Bar et al., 2009; © 2009 Elsevier B.V).

spaces, etc. that can restrict design opportunities by limiting access to natural resources, such as airflow. Ordinances can protect 'rights' to some climate/ weather resources such as daylight and sunshine. Knowles (2003) proposed a simple methodology to derive a solar 'envelope' for urban land parcels (which have irregular sizes and shapes) based on shared access to sunshine for specified periods of the day/ year. The envelope describes a 3-D volume shaped by the Sun's path such that a building constructed within this volume will not cast a shadow over a neighbouring parcel during the specified period. Similar rules can protect the availability of sunlight in outdoor spaces (Bosselman, 1995) and indeed, an inverse strategy could be used to shade outdoor spaces in hot climates (Emmanuel, 2007). Managing other outdoor weather elements, such as wind, can prove more difficult. Downtown San Francisco, United States, which is characterized by high-rise buildings, has a wind ordinance that specifies criteria for outdoor comfort in seating and pedestrian areas and for safety in all areas. Each criterion is expressed in terms of an equivalent windspeed limit (see Section 14.5), and a maximum amount of time per year that the limit may be exceeded. Developers must demonstrate through modelling and measurement that any proposed new building in the downtown area will meet these criteria (Arens et al., 1989).

## Planning Criteria Can Generate Different Urban Forms

Planning ordinances can stipulate neighbourhood and building criteria that must be met while allowing considerable freedom to design different built forms. For example, Figure 15.10 shows three design solutions that meet the criterion of 75 dwellings of the same size per hectare: high-rise (Figure 15.10a) occupies a small plot by arranging the dwellings vertically; low-rise (Figure 15.10b) fills the space with terraced houses and; mid-rise (Figure 15.10c) creates a single **urban block** with an internal courtyard. Each design solution results in different urban parameters (e.g. **plan area fractions** of buildings ($\lambda_b$), impervious ($\lambda_i$) and vegetative cover ($\lambda_v$) and mean building height) that generate distinct microscale climate impacts. Figure 15.10a will cast a long and narrow shadow at higher latitudes and generate strong surface winds that may be problematic. At the same time it makes efficient use of urban land and encloses the same volume within a smaller envelope. Option (b) partitions the entire area into small dwelling units with accompanying garden plots. It affords a great deal of freedom to the

occupants and the terracing limits heat loss. On the other hand, much of the area is occupied by road surfaces that are needed to access the entry point of each building. Finally, the design of option (c) lies between these two in terms of built cover and **impermeable** cover fraction but less area is available for private gardens.

Note that eight towers of the dimensions shown in Figure 15.10a would dramatically increase the floor area ratio ($\lambda_{floor}$) and create 600 dwellings per hectare. This change would modify some parameters, such as **sky view factor** ($\psi_{sky}$, Section 2.1.3) and have a dramatic impact on climate near the ground. However, it would also produce the same fractional coverages as Figure 15.10b, in other words, it could have the same effect on the ratio of **runoff** to precipitation ($R/P$) while increasing the intensity of landuse. The wider climatic impacts of different urban forms depend on their areal extent and what is needed to meet their metabolic needs (energy, water and so on). These ideas are explored later in the chapter when we consider the evidence that supports policies for creating more compact cities with higher built densities.

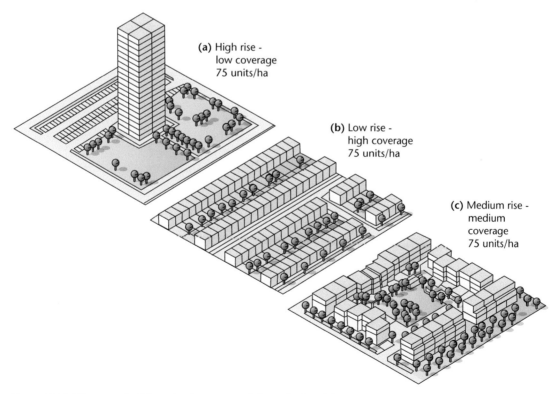

**(a)** High rise - low coverage 75 units/ha

**(b)** Low rise - high coverage 75 units/ha

**(c)** Medium rise - medium coverage 75 units/ha

**Figure 15.10** The same building density can result in radically different building and neighbourhood outcomes (Modified after: The Urban Task Force, 1999).

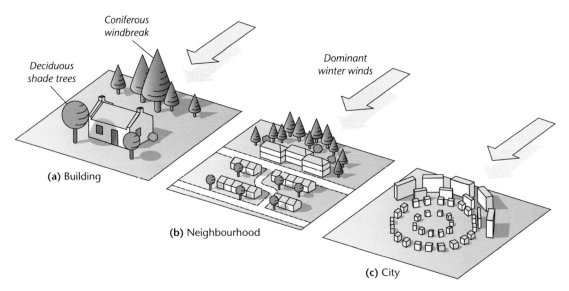

**Figure 15.11** At higher latitudes, a design strategy to ensure access to the Sun and for shelter from cold winds are often compatible at different scales: **(a)** building (Modified after: Akbari et al., 1992); **(b)** neighbourhood (Modified after: Littlefair, 2011) and **(c)** city scale (Modified after: Givoni, 1998).

### There Is Often a Best Scale for Implementing a Tool

Although many design tools can be applied across several scales they may be most effective at a given scale. For example, urban air quality management requires city-scale actions to regulate emissions. Interventions at a street or block scale designed to improve air quality at the microscale by enhancing ventilation, are of limited value by comparison.

Some climate objectives are achievable across the full hierarchy of urban scales. A case in point at high latitudes is design to ensure protection from cold winds yet preserve solar access (Figure 15.11). In this situation, the directional attributes of solar access (heat gain) and cold winds (heat loss) are usually exclusive. In the Northern Hemisphere the strategy is to provide shelter from cold northerly winds and allow access to airflow and sunshine. At the building scale, rows of evergreen trees with canopies that extend to the ground are located to the north whereas obstructions to the south are discouraged (Figure 15.11a). Deciduous trees with elevated canopies can provide shade in summer and solar access in winter and do not hinder near-surface airflow. At the neighbourhood scale, taller buildings to the north of the development can provide wind shelter for smaller buildings to the south (Figure 15.11b). The taller buildings have access to sunshine during winter

months when the **solar altitude** ($\beta$) is low, but do not cast shadows over other buildings. The lower buildings that lie to the south are positioned to maximize solar access and minimize heat loss and east-west oriented streets with terraced dwellings are ideally suited to achieve this. This general approach could theoretically be applied at a city scale (Figure 15.11c) but the neighbourhood scale is probably most efficient because it makes best use of the mutual benefits that can be provided by the building layout, if variations in dimensions and placement are strategically employed.

## 15.2 Design Interventions at Different Scales

The magnitude of climate impacts are related to the scale and type of surface modification. In this book we have discussed the urban surface across a hierarchy of urban scales — **facets**, urban elements (buildings, trees), streets, urban blocks, neighbourhoods and city — that affect the urban atmosphere at the microscale, **local scale** and regional scale (Figure 2.3). Facets are arranged to create buildings, streets, parks and gardens, each of which creates distinctive adjacent microclimates. The typical assemblage of elements forms urban blocks that then aggregate into neighbourhoods and their microclimatic diversity. The impact of neighbourhoods on the atmosphere above roof level is to create a distinct **internal** boundary layer that mixes with other neighbourhood boundary layers. The totality of neighbourhoods comprises the city-scale, where the pattern and size of neighbourhood types contributes to the formation of an **urban boundary layer** (UBL) and the character of the downwind **urban plume** (Figure 2.12).

Table 15.1 places many of the planning/design decisions and their climate effect within this scale hierarchy. It is important to recognize the mutual dependence of outcomes at one scale on another; it is quite possible that well-intended policies designed to achieve a climate outcome at one scale are negated or even counteracted at another scale unless this dependence is taken into account (Mills et al., 2010). As an example, Figure 14.18 shows a large low-rise building that has a retail function and is located in a hot and arid environment and uses a **cool roof** to limit indoor heat gain. This roof will not produce hot air plumes and will contribute to urban heat island (UHI) **mitigation** but the limits of building-scale actions are also clear. The attached asphalt car park counteracts the cool roof effect by raising daytime surface and air temperatures and enhances an already thermally-stressful outdoor climate. Moreover, the facility itself is mainly accessible by private vehicles, which are a source of **air pollution**. Although the climatic efficiency of the individual building is apparent, it is part of a neighbourhood with significant deleterious climate impacts. Hence, to create the well-planned and designed city described in the bullet points opening to this chapter requires a multi-scale perspective.

Here, we assume that there is a system of planning that operates at city and neighbourhood scales. Land-use/cover management at the city scale controls the overall extent of the urban landscape and the location, type and areal size of neighbourhoods and their juxtaposition. In a planned settlement, zoning is used to partition the landscape into areas designated for particular functions, such as residential neighbourhoods; the associated regulations specify the types of buildings (dimensions, materials, etc.), built density, road width, proportion of greenspace and so on. Urban design takes place at the neighbourhood scale in this logic, where regulations give rise to distinct urban forms. It is at this scale that decisions on layout, properties of individual buildings, street landscaping and so on are made that create microclimates.

In the following, we start with a top-down perspective, moving from the city to the neighbourhood scale where the context for CSUD at smaller scales is

established. We then shift to a bottom-up perspective and 'build' the city starting with facets and creating elements, streets and urban blocks. This approach allows us to juxtapose city-scale objectives for resource efficiency and safety with smaller scale objectives for more comfortable and healthy microclimates. At the outset, it is worth stating that many of the policies for CSUD are compatible with **water-sensitive urban design** (WSUD), which has the primary goal of offsetting the deleterious impact of urban development on the quantity and quality of runoff (Section 8.3.5). However, because the surface water and energy balance equations share a common term (evaporation $E$ and **latent heat flux** $Q_E$, respectively), any change to atmosphere-surface exchanges of water affects both. A key measure of the performance of WSUD is the proportion of precipitation ($P$) that is converted to runoff ($R$) following a rainfall event (Figure 8.12). We will make these links explicit in Section 15.3.

## 15.2.1 Cities

At the city scale it is the aggregate impact of the entire urban landscape (rather than its parts) that is regulated. Ideally, as much of the natural hydrologic, atmospheric and coastal resources is retained where possible during urban development. For example, river banks and coastal areas at **risk** of flooding should be set aside for robust uses such as parks and playing fields and where there are well defined airflow routes into the urban area they should be maintained as ventilation pathways. Recreating or substituting these resources at a later stage may be prohibitively expensive.

Where urban development does occur, the available evidence indicates that, all other factors being equal (population, economy and climate) cities that are densely occupied and built (i.e. compact cities) are more efficient (see Section 6.2.2). Evidence is especially strong for transport costs (Figure 1.7 and Figure 13.3b) as a less extensive urban area shortens distances and the higher population density facilitates the operation of efficient mass transit systems (Kenworthy, 2006). Consequently, on a per capita basis, residents consume less area, use private vehicles less frequently, consume less energy and have lower GHG emissions. Of course, increased density may also reduce access to solar energy at building facades and streets, limit ventilation (increase shelter) in the UCL, reduce the green area fraction and enhance the UHI.

Table 15.1 A list of design tools that can be used to improve some aspects of the urban climate. These tools are organized according to scale (left column) and can be used within the context of land-use zoning and ordinances that specify standards for buildings, roads, neighbourhood functions, etc.

| Urban climate scale / Decision-maker | Restrictions | Design tools | | | |
|---|---|---|---|---|---|
| | | Energy use | Water | Wind | Air quality |
| **Urban Facet** *Individual owners, property developers* | • Building codes<br>• Building layout and lot size<br>• Building function<br>• Time of occupation | • Orientation, transparency, reflectivity and porosity<br>• Cool roofs<br>• Cool paving<br>• Green roofs and walls | • Green roofs<br>• Rainfall harvesting | • Porous walls<br>• Openable windows<br>• Roof shape | • Green walls<br>• Reactive materials, adhesive cover |
| **Building / lot** *Individual owners, property developers* | | • Passive design<br>• Landscaping | • Green fraction<br>• Water retention<br>• Irrigation control | • Aerodynamic building form<br>• Sheltering using trees<br>• Sheltering using building elements (e.g. podiums) | • Energy use and fuel source<br>• Emission location (e.g. stacks) and controls, location of air intake<br>• Parking restrictions<br>• Building form and porosity |
| **Streets and blocks** *Municipal government, urban developers, public bodies* | • Controls on building height and plot sizes<br>• Rules on road dimensions and street function(s) | • Shading (street dimensions and orientation) | • Permeable paving<br>• Vegetated medians and paths<br>• Rain gardens<br>• Water features | • Street dimensions, curvature and orientation<br>• Street layout (e.g. Covered footpaths, overhangs)<br>• Trees | • Alternative modes of transport and restrictions on private vehicles<br>• Vegetation for screening pedestrians<br>• Street dimensions and street orientation to enhance dispersion |
| **Neighbourhood** *Municipal government, urban developers, public bodies* | • Zoning regulations on neighbourhood functions<br>• Topographic context | • Building density<br>• Population density<br>• Street length and connectivity<br>• District heating<br>• Renewable energy generation | • Limiting permeable surface cover<br>• Linear parks along river courses<br>• Use of retention ponds<br>• Irrigation management | • Neighbourhood layout (street grid, height variation, density, staggering) | • Mixed land-use developments that shorten commute distances<br>• Availability and accessibility of public or efficient transport |

Table 15.1 (*cont.*)

| Urban climate scale *Decision-maker* | Restrictions | Design tools | | | |
|---|---|---|---|---|---|
| | | Energy use | Water | Wind | Air quality |
| **City** *City and metropolitan councils, regional, national and supra-national governments* | • National and international agreements and commitments<br>• Climate and topographic context<br>• Existing built infrastructure and land-cover | • Compact city: building and population density<br>• Use of renewable energy | • Flood protection by hard defences or land-use management<br>• Efficient use of water resources<br>• Groundwater management | • Ventilation pathways (e.g. greenways) | • Placing pollution sources strategically with regard to population exposure<br>• Compact city policies<br>• Emission reduction by emission controls on vehicles<br>• Provision of mass transit<br>• Use of renewable energy<br>• Ventilation pathways |

**Figure 15.12** An oblique aerial view of two contrasting residential landscapes.
**(a)** L'Exiample, a neighbourhood of Barcelona, Spain, a compact mid-rise neighbourhood (LCZ 2) with a density of 35,000 persons $km^{-2}$ (Credit: Wikipedia CC).
**(b)** suburb of Miami, United States, representing an open low-rise neighbourhood (LCZ 6) with a density of 1,500 persons $km^{-2}$ (Credit: Getty images).

All of these may be undesirable outcomes. The 'best' design will meet the need for efficiency at the urban scale while addressing the microscale climate impacts that more compact cities generate.

To illustrate, Figure 15.12 shows two contrasting residential neighbourhoods: Figure 15.12a is a compact mid-rise part of Barcelona, Spain with a population density of 35,000 persons $km^{-2}$ and Figure 15.12b is an open low-rise suburb of Miami, United States, with a population density of 1,500 persons $km^{-2}$. Clearly, the microscale urban climate effect (on solar access, wind, air temperature, etc.) will be far more intense in the Barcelona but for the same population this design occupies a much smaller area with other efficiencies, such as shared heating between buildings and lower public transport costs. Miami on the other hand has a low urban effect but requires an extensive supporting infrastructure in the form of roads, water and sewer pipes, electric and communication cables and so on.

Martilli (2014) explored the relationship between urban compactness, energy use, air quality and outdoor heat stress using a **mesoscale** model that included both an urban canopy model and a building energy model. Numerical simulations were performed for 22 idealized cities using an array of identical buildings (20 m on a side) that housed 10 million people. The settlement was made more compact by simply changing the population density (to reflect realistic values) so that higher densities resulted in taller buildings more closely spaced to one another. The model was run for summer and winter in a mid-latitude climate and in a hot dry climate. The results suggest that the most compact cities perform best for building energy consumption for space heating/cooling but that the intermediate densities perform best for outdoor comfort and health. At higher densities, air quality is made worse because although vehicle emissions are reduced, the effects are offset by lower ventilation rates from the canopy. In addition, outdoor heat stress was enhanced as higher density enhanced the UHI. The results also show that including vegetation is positive in most situations and never negative. Although this work is abstract, it raises important issues about the interaction between scales and achieving the best overall city plan.

## 15.2.2 Neighbourhoods

Here, we define neighbourhoods in terms of **Local Climate Zones** (LCZ) that categorize landscapes based on a typical range of values for surface cover, **urban structure**, fabrics (radiative and thermal properties of construction materials) and for **anthropogenic heat flux** (Figure 2.9 and Table 2.2). These values provide useful guidance on climate impact of different types of neighbourhoods and the potential for design interventions.

The availability of water at the surface is linked to plan area fraction of vegetation ($\lambda_v$), low values mean that a greater proportion of precipitation either runs off quickly or lodges on the surface (Figure 8.12). Similarly, the relative dryness of most urban neighbourhoods means that much of the available natural energy is transferred to the atmosphere as a **sensible heat flux** ($Q_H$) rather than evaporation ($Q_E$). This relationship is approximately linear (Figure 6.17) so that when $\lambda_v$ exceeds 0.5 (and is not under stress), $Q_E$ is dominant, otherwise $Q_H$ dominates. So, compact LCZs with little vegetation (low $\lambda_v$) and intensive energy use preferentially heat the overlying atmosphere and contribute to the boundary layer UHI. Similarly, the built fraction ($\lambda_b$) and frontal area fraction ($\lambda_f$) exert a strong control on urban surface roughness ($z_0$), which regulates exchanges between the street level air and that above roof level. For simple building layouts (Figure 4.9 and Figure 4.15): $\lambda_b < 0.5$ either

has no effect or enhances exchange and; $\lambda_b > 0.5$ results in lower exchange and sheltering, which can worsen air quality. Where there are high values of $\lambda_b$ and also a need for canopy ventilation, CSUD should take account of wind direction and $\lambda_f$ and vary building heights to promote mixing.

While each neighbourhood has a distinct local impact its contribution to the overall impact of the city depends on both the intensity of urbanization (Figure 15.12) and its areal extent. Each city is comprised of a variety of neighbourhood types that reflect its historical development, advances in transportation systems and its topographic setting. To illustrate, Figure 15.13 shows the distribution of LCZ types in a typical European city (Dublin, Ireland), which is situated on a low-lying plain with few physical impediments to horizontal expansion. In the city centre, which is mixed use, the LCZ type is compact mid- and low-rise, but in much of the surrounding residential area the landscape is open low-rise. Where there are major transportation facilities (airports, major roads), large low-rise associated with light manufacturing and warehouses are located. The pattern of LCZ types reflects its historical development which began as a settlement on a river about 900 AD. Until the mid-eighteenth century it occupied a small area ($< 10$ km$^2$) around the original settlement. Population growth in the nineteenth century especially and improved transportation infrastructure, such as railroad and tram, allowed the city to expand outwards along transport corridors and link some smaller towns outside the original city.

Continued population increase and the demand for adequate housing during the early twentieth century saw the city expand as low-rise compact housing close to the city centre. The availability of private vehicles and of other forms of road-based public transport allowed the development of open low-rise neighbourhoods and the expansion of the city in all directions. Currently, the residential population density in the city centre area (LCZ 2 and 3) is 7,000 to 10,000 persons km$^{-2}$, which is twice that of the mostly low-rise **suburban** area. However, bear in mind that much of the population is mobile on a daily basis. In this city, the centre remains the focus of work and commerce so commuter traffic flows become progressively more intense towards the centre. While most people living in the city centre walk or bike to work, the majority of those outside use private cars or public transit. The magnitude of the urban effects at the

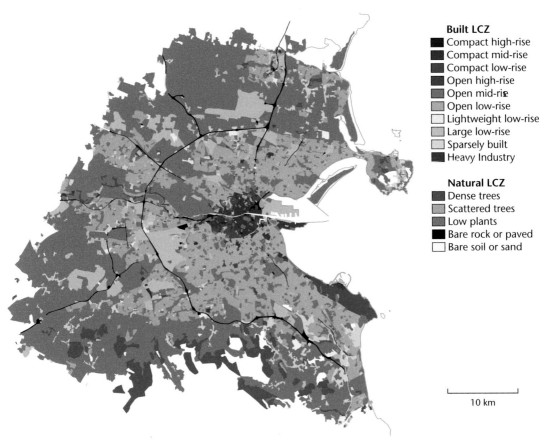

**Figure 15.13** Local climate zone map of Dublin, Ireland (Credit: P. Alexander; with permission).

microscale are greatest in the city centre where $\lambda_b$, $\lambda_i$ and $Q_F$ are highest and $\lambda_v$ is lowest; this is where air quality is poorest and UHI is largest (Alexander and Mills, 2014). Conversely, the magnitude of the aggregate contribution of the open low-rise neighbourhoods to global climate change is largest, mainly owing to GHG emissions from transportation sources.

At this point in the discussion we switch to the individual elements that make-up the neighbourhood.

### 15.2.3 Facets

The fundamental unit of an urban landscape is the facet, which are homogenous, planar surfaces described by their geometry (aspect and orientation) and fabrics (material composition and surface properties); see Table 5.1, Table 5.2 and Table 6.4. In the present section we focus on facets made of fabrics

used for pavement and buildings, which are selected for their strength and durability among other attributes. Here, we have divided the manufactured facets into those for outdoor paving and for buildings, which serve different purposes. What is common is the use of albedo, vegetation and/or permeability as a means of managing the climatic impacts of selected facets. The wider scale impact of each of these interventions depends on their areal extent. As most observational work has been confined to small plots, estimating the potential neighbourhood (or city) benefits are based either on extrapolation to the available facet area or physically-based **numerical modelling**.

### Outdoor Paving

Most construction materials are designed to be impermeable so when wetted, water either lodges on the surface and is evaporated ($E$), or is drained away by runoff ($R$). The underlying soil is partially 'sealed' over so its moisture content is neither recharged (by percolation through the surface) nor withdrawn (by vegetation), hence there is minimal storage change ($\Delta S$). Once dry, these facets are sources of almost zero evaporation and surface temperature ($T_0$) is largely related to albedo ($\alpha$).

Cool pavements are designed to lower $T_0$ using materials and coatings with relatively high $\alpha$ values. Using such materials means more solar radiation is reflected and less is absorbed thereby lowering the daytime $T_0$ and the heat transferred both to the overlying air and stored in the substrate. A freshly laid asphalt road may have $\alpha$ values as low as 0.05 that after use will increase to 0.18 (see Table 15.1), consequently $T_0$ can exceed 60°C on a hot day. Covering the asphalt with a white gravel topping can raise its $\alpha$ to 0.3–0.45 while replacing it with white concrete can raise $\alpha$ to 0.6–0.7 (Santamouris et al., 2011). There are drawbacks however: **reflectivity** tends to reduce over time as deposits dull the surface and; highly reflective surfaces create glare that can create a visual nuisance. Further, be aware that the reflected energy may fall on another surface and contribute to its energy gain.

Permeable paving describes materials that are either porous (i.e, there are gaps in the aggregate material to allow water percolation) or blocks with interstitial spaces filled with porous materials that can support grass (Figure 15.14a and b). While they can provide atmospheric benefits, their primary value is as components of WSUD as they reduce the magnitude and timing of runoff following a rain event and can improve water quality (Scholz and Grabowiecki, 2007). Because they can retain and lose water by evaporation, these surfaces also can moderate surface and air temperature, although some of the benefits are due to albedo changes. The selection of an appropriate paving system depends on a number of issues including the purpose of the paving (pedestrian, parking, traffic, etc.), the climate and the soil.

**Figure 15.14** Facet interventions to manage water and energy exchanges. **(a)** Permeable paving using modular blocks (Credit: B. Wenk). **(b)** Permeable car park with a grass cover (Credit: A. Christen). **(c)** White painted rooftops or cool roofs New York (United States). (Credit: White Roofs Project). **(d)** An extensive green roof on Chicago City Hall, United States, as part of a cool roof project (Credit: Tony the Tiger, July 2008).

Generally block arrangements are best for light use activities rather than major roads that require considerable strength to sustain traffic. Moreover, porous materials can lose their capacity to uptake water over time as debris fills the gaps. Takebabayasi and Moriyama (2009) examined the thermal performance of 36 plots representing different car park treatments in Kobe, Japan, over a two month summer period. They found that the permeable paving was cooler than traditional asphalt cover and that the level of cooling (up to 20 K difference at mid-day) was related to the green area fraction. However, they estimated that extending this cover to an urban neighbourhood with 6% car park cover would have a marginal impact on the near-surface UHI.

### Building Facets

Building facets are arranged to create the envelope that encloses an indoor space and the fabric can include natural (e.g. mud, wood and stone) and manufactured (glass, brick, steel, etc.) materials. These are selected in combination to provide supporting strength, imperviousness, regulate heat exchange and admit light to the interior. The climatic effects of built structure are discussed in the next section, whereas here the focus is on the properties of the facets which are selected primarily to control the indoor climate. Of the building facets, the roof has received the most attention from the perspective of climatologists for a couple of reasons. First, the roof is usually the lightest and least insulated facet and is a major source of unwanted building heat gain and loss (see Section 6.5.1). Second, where building density is high the aggregate roof area is large and this facet offers the potential for managing heat exchanges and rainfall runoff. However, one should consider the height of the building relative to the roof area if the purpose is to manage the indoor climate — the evidence suggests that the building effect of roof intervention is limited to the top floors.

Cool roofs use reflective coatings to reduce roof surface temperatures and moderate heat transfer into the building. They are also used as means of mitigating the UHI directly by reducing the surface-air heat exchange and, indirectly by reducing cooling energy demand. Typical building wall materials have albedoes between 0.2 and 0.6 but the roof surface particularly often has much lower $\alpha$ (Table 5.1). Figure 15.14c shows rooftops in New York, United States, which are typically black but have been painted white to deal with high summertime temperatures.

There are a variety of surface coatings that can create cool roofs, some of which are designed to have varying **emissivity** and reflective properties at different wavelengths. Synnefa et al. (2006) compared the thermal performance of tiles coated with different paints during summer in a Mediterranean climate. The coolest tile had a white acrylic elastomeric coating which was manufactured to be highly reflective (0.8) at visible wavelengths and an efficient emitter (0.9) of **long-wave** radiation; it was up to 20 K cooler in daytime when compared to a dark tile. Other coatings were designed to have high reflectivity in the near-infrared, outside the visible range, which limits glare and allows non-white cool roofs. However, the benefits of coatings tend to diminish over time as air pollutants in the urban environment 'dulls' the white surfaces. Of course in some cities darker building surfaces may be preferable, especially during the winter when energy gain is welcome. In fact, an ideal building envelope would change its reflectivity with the seasons, to shed or absorb the Sun's energy as needed (Landsberg, 1973). Thermochromic coatings that can alter their reflectivity as a function of temperature are a potential solution.

Green roofs are used both as cool roofs and to attenuate building runoff but their performance depends both on the nature of the vegetative cover and the climate to which it is exposed. Extensive green roofs consist of a lightweight structure with a thin soil layer that supports short flowering plants (e.g. sedums) that are adapted to harsh environments and drought. These systems can be applied to roofs that are sloped (up to ~45°) and do not impose a significant additional weight on the building. By comparison, intensive green roofs allow for deep rooted vegetation (including trees) and need a far more substantial growing substrate (> 150 mm) and are confined to strong roofs with a slope of less than 10°. Figure 15.14d shows an extensive green roof on a tall building (66 m) in Chicago, United States, which has a policy to support city greening as part of both CSUD and WSUD.

The cooling function of green roofs for buildings is due to the shade and/or insulation they provide for underlying building surface. In hot, dry climates, where irrigation is not provided, a green roof may provide a marginal cooling benefit compared with a cool roof. Coutts et al. (2013) examined an extensive green roof in Melbourne, Australia and found little evidence for evaporative cooling except during irrigation periods. In wetter climates the cooling potential can be larger but the roof will also impede night-time heat loss, which may be desirable so the net benefit for indoor climates can be limited. In summary then the impact of the green roof on the thermal performance of buildings is largest where the roof is poorly insulated, which tends to be older buildings in modern cities.

The ability of green roofs to store water and attenuate runoff following a precipitation event is measured depends on the intensity of the rain, the storage capacity of the roof and its prior water state. In moderate rainstorms of short duration the green layer may retain all of the water, which is then released to the atmosphere via evaporation. However, in heavier rainstorms or those of longer duration, the system can become saturated so there must be a way to remove excess water. An examination of published research on the observed runoff to precipitation ratio ($R/P$) for green roofs in mid-latitude European climates found that performance varied from 91% to 15% for traditional roofs and intensive green roofs, respectively (Mertens et al., 2006). Again, the potential benefits depend on the extent and nature of the intervention and the climate context; Mertens et al., (2006) estimated that applying extensive green roof systems to 10% of roofs in Brussels, Belgium, could reduce runoff in the city centre following a rain event by just 3.5%.

Cool walls and green walls have much the same effect for the building climate as roof treatments but their outdoor impact is different, as they affect the street climate directly. Cool walls will reflect solar radiation but this energy will fall on other surfaces and can increase the heat load of those outdoor. Green walls have no equivalent soil substrate to which the plants are anchored. For the building, these will moderate the underlying surface temperature. For the outdoors, the green cover will lower the surface temperature to which pedestrians are exposed and may help improve air quality by trapping air pollutants in their vicinity (Pugh et al., 2012). They will also intercept precipitation but their potential for storage is very limited.

## 15.2.4 Buildings

A primary role of buildings is to ensure an indoor climate suited to the needs of the occupants by managing energy gains and losses across the envelope.

Ideally, a building regulates these exchanges using sustainable design techniques to maximize natural energy gains and losses and so minimize the internal energy inputs required; these techniques include site selection, envelope design (its size, shape and orientation), selection of construction materials (fabrics) and landscaping. Where passive energy gains and losses are not sufficient the building has a heating and/or a cooling demand (see also Figure 6.4). Satisfying this demand by conventional fossil fuel energy systems (combusted in the building or via electricity) is a significant contribution to the **anthropogenic** fluxes of heat, moisture, air pollutants and GHGs in cities. Heating, ventilation and air conditioning systems account for about half of building energy use, and building energy use in many modern cities accounts for 40% of total energy use (Pérez-Lombard et al., 2008). Naturally, moderating building energy demand is a key part of efforts to mitigate global climate change.

### Climate Controls on Individual Buildings

The most significant external heat gain for a building is usually solar **irradiance**, which is distributed unevenly over the surface during the day and annually (Figure 5.8). However, the total gain can be managed through the design of the building envelope. In hot climates at tropical latitudes, a tall rectangular building oriented east-west intercepts less solar radiation than a low building encompassing the same volume, because it minimizes the area of the envelope most exposed to direct solar radiation (i.e. the east and west walls and roof). Similarly, in high-latitude cooler climates, increasing the area of the Equator-facing wall maximizes gain.

The climatic controls on building heat losses are ambient temperature and wind speed. The latter causes dynamic pressure differences across the envelope of the building, which in turn cause exchange of air between indoor and outdoor through ill-fitting windows and doors. While air temperature cannot be managed at this scale, wind can be slowed and/or re-directed. In cold and windy climates, buildings would be sited in sheltered areas or protected by a line of trees and hedges (a windbreak) to calm airflow (Figure 15.11). In hot climates, however the passage of air through the building is often desirable hence walls have multiple openings and few internal obstructions to maximize cross-ventilation. In hot and arid climates, where the outdoor air may be both hot and dust laden, wind devices on the roof of buildings allow cleaner air to be channelled into the building.

Figure 15.15 shows a variety of building designs with features explicitly linked to their climate context. Figure 15.15a and b show traditional architectural responses to cool and warm climates. Figure 15.15a shows a cottage that is situated in a relatively cool, wet and windy climate and is designed to limit heat loss: the roof is thatched and the stone walls are thick to insulate the interior and; the openings (doors and windows) are small to limit ventilation and the loss of heat generate indoors and the chimney is needed to expel smoke from a fireplace. Figure 15.15b uses a similar approach to do the opposite in a hot and arid climate: the facets of these buildings are comprised of dried earth (adobe) and have few openings to limit heat gain to the interior and the large feature on the roof is a wind-catcher designed to admit clean air. Ideally, this air is cooled by evaporation before it enters the living space. Modern materials can permit different designs to achieve the same effects. Figure 15.15c is located in a hot and humid climate; like Figure 15.15b it seeks to minimize heat gain but in contrast it also maximizes heat loss. The form of the building is tall and rectangular and is oriented to limit solar gain and maximize wind exposure. The wide wall facet is porous and allows airflow to pass through the narrow section of the building. Figure 15.15d is also located in a cool climate similar to Figure 15.15a but modern materials allow extensive use of glazing while limiting **conduction**. It has been designed in layers with a glazed outer envelope that allows plentiful light into a vegetated interior space that has enclosed office spaces.

## Buildings in Cities

In cities, building design needs to account for the urban climate effect (i.e. the altered air and surface temperatures, wind, air quality, etc.) and the effects on

**Figure 15.15** Buildings and climate: **(a)** A traditional thatched cottage in a cool maritime climate (Ireland) (Credit: G. Mills). **(b)** Traditional adobe buildings in a hot and arid climate (Iran) (Credit: R. Hassanzadeh). **(c)** A modern tall building suited for a hot and humid climate (Singapore) (Credit: Katmorro). **(d)** A modern glass framed building in a cool maritime climate (the Netherlands) (Credit: G. Mills).

direct-beam irradiance of neighbouring buildings. The latter includes shadowing and **horizon screening** (see Section 5.2.3), which affects access to daylight and sunshine. It is a simple matter of geometry (Figure 15.16) to ensure equitable access to solar energy while at the same time increasing building density. Knowles (2003) has shown that solar

Conventional layout                  Modified layout

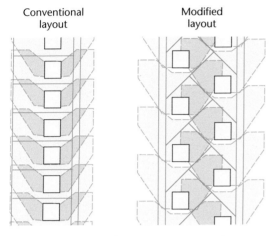

**Figure 15.16** The shading patterns generated throughout the year between 0900 and 1500 h by **(a)** aligned and **(b)** staggered building cube layouts oriented north to south at 50°N. The staggered array increases built density and ensures a more equitable access to sunlight (Modified after: V. Matus, 1988)

geometry can be used in conjunction with variable plot sizes and shapes to ensure the same outcome. Of course, the same technique can also be used to provide shade within the **urban canopy**. At low latitudes this requires highly compact designs to compensate for the high solar altitude (see for example Figure 5.1).

Ensuring access to airflow for building ventilation is more difficult to resolve. Increased building density generally slows the near-surface airflow but the effect depends on the layout of the building array projected in the direction of the wind (see Sections 4.2.2 to 4.2.4). In other words, for the same building density ($\lambda_b$), the frontal area fraction ($\lambda_f$) regulates the degree of shelter within the neighbourhood. Buildings that are aligned along the path of the wind minimize $\lambda_f$ and allow air to be channelled deep into a neighbourhood (Figure 15.17). However, these are very general guidelines only, wind at street level in neighbourhoods varies with layout, the spacing between buildings, small variation in building heights and the presence of trees and vehicles. Also note that poor air quality at street level may mean that ventilation into the building from the street side may not be desirable (e.g. in the case of Figure 11.1). A vegetative screen close to where air is drawn into the building can lower air temperature and remove some of the larger airborne **particulate matter** (PM).

Figure 15.18 shows a number of building designs that illustrate design principles of buildings in an urban context. Figure 15.18a is an aerial image of

**(a) Aligned**                  **(b) Staggered**

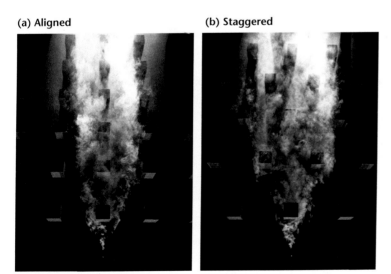

**Figure 15.17** The impact of an **(a)** aligned and **(b)** staggered grid of cubes on airflow near the ground. While the staggered grid obstructs flow through the array and spreads the plume horizontally, the smoke in the aligned grid is more confined (compare with Figure 4.15) (Source: Hall, et al., 1997; with permission).

**Figure 15.18** Examples of climate-sensitive buildings in cities: **(a)** a solar neighbourhood in Freiburg, Germany (Credit: A. Glaser; CC3.0); **(b)** a sustainable neighbourhood (BedZED) in London, England (Credit: T. Chance; CC); **(c)** a building in Fukuoka, Japan shaped by solar geometry (Credit: K. Mabuchi) and **(d)** A street arcade in Barcelona, Spain (Credit: G. Mills).

a settlement where each rooftop has a photovoltaic array tilted in the direction of the Sun at noon. Critically the heights of building are regulated so that each array is not shaded by other buildings. Figure 15.18b shows a set of terraced residences that are designed to make passive use of natural resources. Buildings are aligned east-west and the terraces are sufficiently separated to ensure solar access to each dwelling. The south-facing façade of each building is glazed and the distinctive feature on the roofs are wind cowls that orient themselves in the direction of the wind to ventilate the indoor air. Figure 15.18c shows an entire building designed to maximize solar gain on its sloped south-facing façade. Figure 15.18d illustrates how buildings can also be used as shading devices. In this case, a partially covered walkway (arcade) in a Mediterranean climate provides protection from the Sun and allows ventilation in summer yet shields pedestrians from rain in winter.

## Tall Buildings

These buildings deserve special mention because they can have significant impacts at ground level, especially in relation to solar access and wind. The shadowing effects of a tall building depend on latitude and season. When $\beta$ is less than $45°$, a building casts a shadow longer than its height but the wider overshadowing effect depends on the time of day and the shape of the building as 'seen' by the Sun. The impact of overshadowing on building energy demand depends partly on the purpose of the building and the timing of heating and cooling energy needs (Section 6.2.2). Energy use in commercial buildings is highest during daytime, while that for residences is largest during the morning and evening hours. For office buildings where cooling demand is high, the mutual shade provided by buildings will reduce energy use whereas the loss of sunshine for residences will increase energy use for heating. Residential low-rise and mid-rise compact neighbourhoods that are adjacent to, and poleward of, high-rise neighbourhoods will experience the greatest impact. Conversely, at low latitudes where heat gain is an issue, tall buildings may be well suited to residential use as the bulk of solar radiation is intercepted by the east and west walls and roof; an east-west oriented rectangular-shaped building will minimize solar gain. In addition the shadowing effect, while limited in area will be beneficial for affected buildings and outdoor spaces.

Tall buildings in isolation and in certain configurations can have dramatic impacts on near-surface wind in their vicinity by deflecting and/or channelling airflow (Section 4.2.4). As a guideline, Bottema (1999) suggested from a wind perspective, that, for a given building floor space, the best wind climate is usually the design with the lowest building height and that *ad hoc* interventions at street level, like windbreaks, are less effective than altering building size and/or shape (Figure 14.16) Where tall buildings are clustered together, the net effect can be to substantially reduce near-surface airflow and limit the ventilation of the UCL. This can be beneficial where shelter is desirable but may also worsen air quality and impede the natural ventilation of buildings (Figure 12.20). Conversely, isolated tall buildings within a low-rise urban landscape can be used to improve ventilation at street level, promote vertical mixing and air pollutant dispersal.

### 15.2.5 Streets and Urban Blocks

Streets and urban blocks provide the framework for the layout of buildings in a neighbourhood. In the simplest road system, streets are arranged as a regular grid forming uniform blocks, which are partitioned into individual land parcels. More complex road systems form irregular block patterns of varying sizes. The form and function of streets and intersections can exert considerable influence on the formation of microclimates, especially where neighbourhoods are compact (LCZ 1–3) and the road and adjacent buildings form **urban canyons**. Whereas the primary objective of buildings is the creation of a suitable indoor climate, streets fulfil many functions, the most obvious of which is transportation by various modes, but also serve as usually the public outdoor spaces of a city.

Figure 15.19 illustrates four types of street form that typify different neighbourhood types. Figure 15.19a shows a road in an open low-rise residential neighbourhood (LCZ 6) common to the suburbs of cities in North America, Europe and Australia. Here the built density is low, individual buildings occupy large plots and are separated from the road by driveways and pavements. The remaining figures show neighbourhoods where buildings are sufficiently close to create urban canyons and there is little vegetation present. Figure 15.19b shows a terraced street oriented north to south in a mid-rise open neighbourhood (LCZ 5) that was designed originally for residential use. Despite the wide street, at this latitude (53°N) shadow lengths are long for most of the day and the buildings are often in shade. To compensate for the loss of light, windows are larger on the lower floors. Figure 15.19c and d are both streets in compact high-rise neighbourhoods (LCZ 1) but they are distinguished by the road layout and degree of compaction. The commercial buildings in Figure 15.19c are laid out on a grid road system with large footprints and wide separating boulevards to accommodate traffic. On the other hand, buildings in Figure 15.19d are mixed use (residential and commercial), have small footprints and are very closely spaced. The road itself is a narrow, sinuous channel cut though the built fabric with a small exposed sky vault overhead.

### Radiation Exchanges

The height to width ratio ($H/W$) and orientation are critical street parameters that govern radiation exchanges. In grid road systems, roads are at right angles to each other and the exposure of the block is similar to that of the cube-shaped building (see Section 5.2.2). Where the grid is oriented in the cardinal directions, streets are oriented either north-south or east-west. While an east-west street orientation may be best for building energy gain, it also means that one side of the street is in shade throughout much of the day. Outside of the Tropics this side of the street will be in shade year round. By contrast, the shade switches from one side to the other in a north-south oriented street. The extent of shade depends on the dimensions of the canyon. For example, an east-west street with a $H/W$ of 1, will be in shadow whenever $\beta \leq 45°$; thus, for latitudes greater than 45°, and east-west street facet will be in shade for 6 months of the year. Simply re-orienting the grid system off a north-south axis produces a more equitable distribution of sunlight at street level.

There are a number of ways to design streets to provide shade, such as: increasing $H/W$ by increasing the height of buildings or narrowing the street width;

**Figure 15.19** Street examples that exemplify neighbourhood types: **(a)** typical suburban street (LCZ 6) in Putnam, Australia (Credit: J.W.C. Adam; CC3.0); **(b)** a mixed use residential/commercial street (LCZ 3) of terraced buildings in Dublin, Ireland (Credit: A. Christen); **(c)** a compact high-rise (LCZ 1) commercial street in downtown Los Angeles (Credit: G. Mills) and **(d)** a narrow and sinuous mixed-use street in Hong Kong (Credit: G. Mills).

using arcades (Figure 15.18d); planting trees and; using canopies and awnings when required. Of course, designing the street itself as a shading device requires considerable forethought. Figure 15.20 shows a cross-section of a traditional street in Jaisalmer, India (26.9°N), which is located in a hot and dry desert. It has a high solar altitude in summer daytime and dust storms. In response, all major streets are oriented east-west (at right angles to the direction of dust storms). Whether deliberate or not, the street form that has emerged is narrower near roof level and provides shelter and shade at ground level.

Street geometry also modifies the receipt of **diffuse irradiance** by limiting access to the sky vault, which can be measured with the sky view factor ($\psi_{sky}$). As $H/W$ increases, $\psi_{sky}$ decreases so solar receipt is partly made up of radiation reflected from other facets in the street (Figure 5.9 and Figure 5.12). However, this source does not compensate for the loss of light from part of the sky. More so than sunlight, access to daylight is often enshrined in law, because of concerns about the internal lighting of buildings. In New York City the 1916 Zoning Resolution ensured access to daylight at street level while allowing tall buildings.

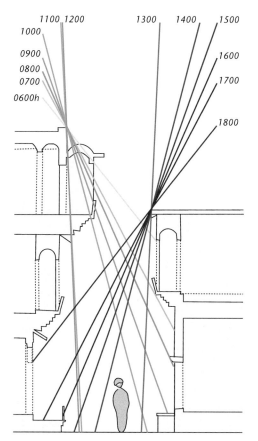

**Figure 15.20** A typical street section in Jaisalmer, India (27°N), showing shadow projections at different times of the day at the summer solstice (Source: Krishan, 1996; © Elsevier Science, used with permission).

It did not limit building height but did preserve a portion of the sky vault defined by the angle subtended at the mid-point of the street. In response, the characteristic skyscraper form that resulted was setback from the street, had a wide base and decreasing floor area with height (Kwartler and Masters, 1984).

Street design affects both heat gain and loss at street and wall facets and is a major factor in the formation of the surface and canopy-level UHIs. During conditions conducive to UHI formation, $\psi_{sky}$ regulates heat loss to the overlying sky and heat exchange between canyon facets (Figure 7.13). Streets with high $H/W$, typical of centre city urban development usually exhibit the strongest UHI effect. Typically these neighbourhoods also have the lowest plan area fraction of vegetation ($\lambda_v$). A higher $\lambda_v$ generally lowers

temperatures in the UCL such that daytime heating is less, and the nocturnal UHI is reduced. Increasing $H/W$ also reduces solar gain at facets within canyons, so that surface and air temperature are cooler as consequence before the nighttime cooling process begins (see Figure 14.13).

### Airflow

Street geometry also regulates airflow near street level by managing interaction with the above roof (ambient) airflow (see Section 4.2.3, Figure 4.11). Limiting this exchange reduces mean windspeed but may also dampen the vertical exchange of air pollutants emitted by traffic. For simple long and symmetric street canyons, the key parameters are $H/W$ and the orientation of the street in relation to ambient airflow. When ambient flow is perpendicular to the street axis, a value of $H/W \leq 1$ for symmetric streets offers shelter at ground level but this may be undesirable if there are sources of street-level pollution. On the other hand, if ambient flow is parallel to the street axis, air pollutants are moved downwind, but there is little shelter.

Again, where the road system is a grid the airflow through the streets can be imputed from the work on building cubes (e.g. Figure 15.17). To ensure that airflow penetrates deeply into the neighbourhood the grid should be oriented parallel to the desired wind direction and streets on this axis should be wider (in other words minimizing $\lambda_f$). Naturally, orienting the grid to maximize $\lambda_f$ will obstruct flow. More generally the road system forms a network and air pollutants may be channelled through it following the mean above-street airflow; at intersections where air pollutants from traffic along the merging streets converge, and vehicles idle, air quality may be poor (see Figure 4.14).

### 15.2.6 Trees

Of all the urban elements, vegetation is the most versatile in the service of climate modification. It may be used to control radiation exchange, airflow, ventilation of air pollutants, evaporation, temperature, erosion, runoff and noise levels. Trees may emit specific air pollutants, but they also promote deposition and adsorption of air pollutants. The climate impact varies with plant architecture (canopy form, foliage density, branch and root systems) and physiology, which depend on plant species, age and health.

Hedges and shrubs have a limited vertical extent and the canopy encloses the entire woody skeletal structure; they are often used to provide wind shelter near the ground and to act as noise barriers. Trees are taller and have extensive canopy crowns that are often displaced from the surface, exposing the trunk.

A major climatic distinction between species is whether they are deciduous (seasonal) or evergreen. In urban environments, which are harsh, particular species are selected for their tolerance to drought, poor air quality or diminished light (Brown and Gillespie, 1995). Figure 15.21 shows examples of streets in compact mid-rise neighbourhoods with climates that are substantially modified by landscaping:

Figure 15.21a and b shows street surfaces with significant vegetative cover that separates the moving traffic from the pathways and car parks on either side of the street; Figure 15.21c places the trees along the centreline of the street with traffic on either side and; Figure 15.21d uses the tree-lined and grassy median of a road way for a light-rail transport system.

### Radiation

Trees cast shadow patterns that are speckled with light that has filtered through the leafy canopy. Deciduous species transmit 10–30% (50–80%) of solar radiation in summer (winter) while coniferous species transmit 10–30% year round. Tree species and planting can be

**Figure 15.21** Examples of vegetation in streets situated in compact mid-rise (LCZ 2) neighbourhoods: **(a)** Lyon, France (Credit: G. Mills); **(b)** Singapore (Credit: M. Roth; with permission). **(c)** Gothenburg, Sweden (Credit: A. Christen) and **(d)** Milan, Italy (Credit: G. Mills).

**Figure 15.22** The seasonal character of deciduous trees can transform streets and their climates. In a mature suburban development in Vancouver, Canada, the contrast between **(a)** winter (February) and **(b)** summer (August) at the same time of day (0900 h) is stark (Credit: A. Christen).

used to regulate solar access for a selected surface at specific times of the day and year. For example, an elevated canopy can allow solar access when the Sun is low in the sky (during morning hours) while providing shade during the hottest part of the day. This may be an ideal strategy to shade buildings (or even air conditioning systems). Typical heights of urban trees ($< 25$ m) generally limit their shade potential to low-rise structures (or the lower floors of high-rise buildings) and the ground (e.g. Figure 15.19c).

In high latitudes, where it is desirable to have shade during summer months but solar access during the winter, deciduous trees are ideal. Figure 15.22 shows the dramatic impact of canopy cover along a street in a low-rise open neighbourhood (LCZ 6). When the trees are in leaf, the canopy encloses the road providing plentiful shade but also limiting ventilation to the air above the UCL. Without leaves, the woody skeleton still provides some limited shade but the sky view from street level is much greater. Figure 15.23 shows the impact of leaf canopy on solar access.

### Temperature and Humidity

Trees modify air temperature and humidity both directly as air passes through the canopy and indirectly as a result of shading the underlying surface. The direct effect is linked to the process of **photosynthesis**, which exposes the moist interior of the leaf and permits evaporation. The rate of evaporative cooling is related to the moisture deficit of the adjacent air, relatively dry air passing through a canopy is both cooled and humidified but the magnitude of these changes depends on both moisture deficit and the rate

of airflow (Section 6.4). At low wind speed the effect is greatest but the volume influenced is small; increasing wind speed replenishes the air moving through the canopy and increases the volume affected but reduces the impact. Lowry (1988) estimated that a stand of trees in a city street could yield a potential cooling of 30 K h$^{-1}$ assuming an evaporation rate of 70 W m$^{-2}$ but only if the air in the street remains in place. If a moderate rate of ventilation (**advection**) is applied, this value reduces to just 0.3 to 0.6 K h$^{-1}$.

The cooling effects of trees in hot climates can improve outdoor **thermal comfort** significantly. Figure 15.24 shows the results of an experiment that observed $T_a$ and $RH$ of the near-surface air along a tree-lined avenue under near calm conditions in a hot and dry climate. The cooling impact of the trees on $T_a$ (and $RH$ as a consequence) is closely related to the area of shade. The weak advection confines the primary cooling effect of the trees to the immediate surroundings. The physiological impact on pedestrians is greater still because trees reduce the radiation load on the body directly through shade and by lowering the mean radiant temperature ($T_{MRT}$) of the surroundings. In another climate (or during different weather), the thermal impact of the same tree-lined avenue may be reduced, for example in cloudy conditions the impacts of shade and solar interception by the canopy are less or in humid conditions evaporation is suppressed and in windy conditions effects are diluted.

### Airflow

As roughness elements trees impose a **drag** on airflow but unlike most buildings their porous nature, slows,

**(a)** Leaves-on                              **(b)** Leaves-off

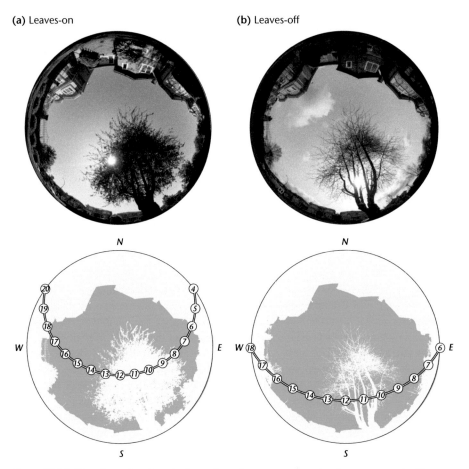

**Figure 15.23** Sky views showing the interplay of tree canopy and solar geometry in Vancouver, Canada (49°N). During summer **(a)**, much of the solar beam from after 0800 to before 1300 h is intercepted by the leafy canopy but during winter **(b)** most of the solar beam reaches the underlying surface (Credit: A. Christen).

rather than blocks airflow. Although it may appear counterintuitive, this property of porosity makes trees (and hedges) more effective as wind shelters. The wind, when it encounters an isolated solid object, such as a wall or a building, causes high pressure differences between windward and leeward, consequently, wind is intensely accelerated around edges and roofs and quickly recovers downwind. On the other hand, the porous nature of leafy canopies creates smaller pressure differences so the recovery rate is slower and the sheltering benefits extend further downwind. Flow through foliage also shifts kinetic energy from larger to smaller eddies, which reduces large gusts and is good for shelter but not for transporting undesirable properties such as heat or air pollutants away (Krayenhoff et al., 2015).

Landscape design for shelter must consider both the type of vegetation (canopy architecture and its seasonal response) and its placement. A canopy that extends to the ground is best where shelter is needed near the ground surface, but an elevated canopy can permit welcome breezes at this level. The aerodynamic properties of an evergreen canopy remain similar through the year whereas a deciduous canopy provides significant slowing and shelter only in summer (Figure 15.22). In low-rise neighbourhoods where trees can be taller than buildings, the overall aerodynamic roughness changes significantly with season, and so does the vertical wind **profile**. If planted in sufficient density (and in leaf) the canopies may merge so that airflow near the ground is almost separated from the overlying air. In mid-rise neighbourhoods

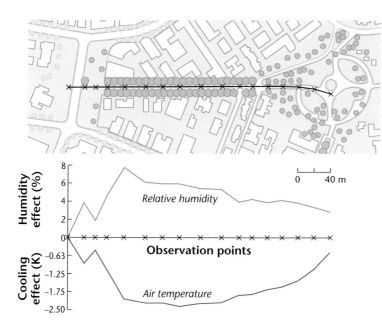

**Figure 15.24** The humidity and cooling effects along Hayeled Avenue in Tel Aviv, Israel. The lower graph shows deviations from a reference point and the symbols correspond to measurements made at different points along the avenue. The values represent observations at 1500 h made on five occasions (July 2, 5, 10, 12 and August 18). All observations were made during calm conditions (Source: Shashua-Bar and Hoffman, 2000; © Elsevier Science, used with permission).

(LCZ 2 and 5), mature trees are often about the same height as buildings and 'cushion' their effects on wind. In streets a tree canopy disturbs circulations that might otherwise form (Section 4.3.2). If the canopy coverage fills the space between buildings ventilation of street air is greatly restricted (Gromke and Ruck, 2008). The result may be to degrade air quality at street level because ventilation of traffic emissions is disrupted. In high-rise neighbourhoods (e.g. Figure 15.19c) trees are relatively short elements with little overall influence on the wind, but they can provide a buffer against strong winds and gusts at ground level caused by diversion of winds from aloft.

### Air Quality

Leaves actively exchange gases and materials with the ambient atmosphere. As a result they can improve air quality by removing some air pollutants, but also their own biogenic emissions can engage in reactions leading to the formation of **secondary pollutants** and **photochemical smog**.

Air pollutants are removed by **dry deposition** and **wet deposition** (see Section 11.1.3). Dry deposition of airborne materials onto a leaf surface is a function of the **deposition velocity** $v_d$, the value of which depends on atmospheric conditions, and the character of the pollutant and the character of the leaf surface (i.e. described by the **surface resistance** in Equation 11.5). While some of the deposited pollutants adhere on the leaves, a portion can be re-suspended or washed off during rainfall. Exchange of gases through open **stomata** further allow the transfer of common air pollutants to the plant interior (e.g. ozone, nitrogen oxides and sulfur dioxide). However, in this process the health of plants can be damaged. For example, excessive sedimentation of **PM** on leaves may prevent proper functioning of stomata and limit their ability to use solar energy. Urban trees, especially those planted on busy streets, must be selected to tolerate poor air quality and drought-like conditions. Species selection should also consider that tree and plant species emit trace gases (i.e. VOCs) that, together with air pollutants from fuel **combustion** react to form secondary pollutants such as $O_3$ or PAN that then cause photochemical smog (see Section 11.3.2).

The benefits of urban trees to overall air quality can be estimated from tree canopy cover, which is a measure of the total leaf area exposed to the atmosphere. Nowak (2006) reports on the results of a model that estimates air pollutant removal based on canopy cover, local climate and ambient air quality. The average pollution removal was calculated at $10 \text{ g m}^{-2}$ for selected cities, but the highest estimate was for Beijing, China ($27.5 \text{ g m}^{-2}$), owing to its poor air quality. Of 13 US cities examined with canopy cover of between 11 and 36%, the average improvement provided by tree canopies during daytime was estimated as: 0.64% (**PM10**), 0.62% ($O_3$), 0.61% ($SO_2$), 0.40% ($NO_2$), and 0.002% (CO).

At the global scale the GHG **carbon dioxide** ($CO_2$) is considered a pollutant even though it does not have direct health consequences for humans (Table 11.1). Cities are significant direct sources of $CO_2$ and trees can play direct and indirect roles in managing urban emissions (Section 13.2.2). However, this ability depends on size of canopy: a large, healthy tree that may have matured over several decades can sequester over $90 \text{ kg C yr}^{-1}$ while a small tree with a smaller canopy may sequester just less than 2% of this value. The potential to offset urban $CO_2$ emissions is shown in Figure 13.7 where measurements in a low density suburb with mature trees is a net carbon sink in summer months when the trees are actively growing. However, at an urban scale, this direct role is very limited. Nowak (2006) calculated that the annual sequestration by Chicago's urban forest, which consists of over 50 million trees, at over 140,000 t C. Although this is an impressive number, it is the equivalent of just one week's emission by transportation in the city! The indirect role of trees in this role may be much larger by reducing building cooling (shading) and heating (shelter) demand. Moreover, by improving outdoor spaces to extend their use the burden on energy systems to maintain indoor climates may be lessened.

### 15.2.7 Gardens and Parks

A considerable portion of many cities consist of managed natural surfaces such as lawns, gardens, parks, urban forests and green recreational facilities. Like other vegetation the climatic response of these surfaces depends on the type of vegetation and the character of the underlying soil, including its moisture content. They create a microclimate in much the same way as **rural** surfaces of the same type but they are distinguished by their areal extent and shape and especially the level of management (e.g. grass cutting, pruning and irrigation). In urban landscapes characterized by low vegetation cover,

parks are often considered a climatic 'oasis', offering respite from the dominantly impervious, surrounding urban environment (e.g. Figure 15.1). During daytime, a large, well designed park provides access to sunshine and airflow and, if landscaped consists of a variety of microclimates suited to different activities. In vegetated parks where water is available the surface and air temperature is lower than a similarly exposed paved facet. At night the effect is revealed by near-surface air measurements wherein parks appear as cool and humid islands within the canopy-level UHI (Figure 7.3).

Parks and gardens affect (and are affected by) the surrounding urban area through advection. In fact, even when there is little regional air movement, small differences in the nocturnal air temperatures can cause cooler air, formed in a park to penetrate into the surrounding neighbourhood. Figure 15.25 shows the distribution of air temperature and **relative humidity** in the vicinity of a large ($6.86 \text{ km}^2$) green park in Mexico City on a clear evening several hours after sunset. Notice that the coolest nocturnal air temperature (and highest relative humidity) is located in the centre of the park, distant from the edges. However, the park effect is seen to extend beyond the boundaries into the surrounding area to a distance approximately equal to the width of the park.

Although there is significant literature on the urban climate effects of parks, the diversity of park types (e.g. grass vs forested or dry vs irrigated) results in a range of microclimates that defies simple generalization (Bowler et al., 2010). In the mid-latitudes, green parks are clearly identified as cooler areas within the urban fabric, however the extent to which they interact with the surrounding urban environment depends on the form of the park (layout, vegetation, and management), its urban context and orographic setting and the ambient airflow. Overall, the literature suggests that the impact of green parks on surrounding areas is limited in extent and depends on the length of the perimeter. As a guide, if parks are used to moderate urban climate effects beyond the park boundaries, several smaller distributed parks are preferable to a few large isolated parks.

Vegetated surfaces can provide other benefits and in dry climates can protect the soil from erosion and the subsequent deposition of dust. Jáuregui (1990/91) describes the impact of draining a large lake ($85 \text{ km}^2$)

**(a)** Temperatures (°C)          **(b)** Relative humidity (%)

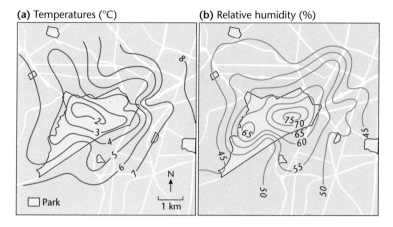

☐ Park

N
↑

1 km

**Figure 15.25** Night-time **(a)** air temperature and **(b)** relative humidity in Chapultepec Park in Mexico City on December 3, 1970. Measurements were made over the period 0530 to 0700 h under calm conditions with clear skies (Source: Jáuregui, 1990/91; © 1991 Elsevier B.V., used with permission).

at the edge of Mexico City allowing airflow to transport dust from the exposed lake bed across the city. In the 1970s a re-vegetation project began that planted native grasses and created a large artificial lake, covering more than 80% of the former lakebed. The result has been a dramatic drop in the frequency of dust storms.

### 15.2.8 Water as a Design Feature

Like vegetation, the climatic effects of water can be incorporated into urban designs at all scales. For neighbourhoods located along a coast, advection from the water body generally brings faster, cleaner and more humid air inland. In winter (summer) the onshore winds are usually warmer (cooler) that over the land. Moreover, when the regional weather is fairly calm and clear, the differential response of land and water to the diurnal cycle of heating and cooling drives a land-sea (or lake) thermal circulation. At the surface, the **sea breeze** brings cooler air off the water during daytime and the flow reverses at night (**land breeze**, see Section 12.2.3).

From the perspective of thermal comfort and clean air the design of neighbourhoods in coastal zones must be aware of these breezes, their climatic benefits and challenges. In order to allow thermal effects to extend into the city, streets should be wide and oriented perpendicular to the coast, and built infrastructure should not completely obstruct onshore flow through the canopy. Of course the reverse applies if shelter is desirable. Smaller lakes

and ponds within a neighbourhood have similar effects but the wider climatic impact on a neighbourhood is limited. Keep in mind that land-sea thermal circulations are closed circulation systems in which air pollutants (including secondary ones) are recycled rather than vented away from the city (see Figure 12.22)

Rivers can also provide climate services if their benefits are integrated into the design of a city and its neighbourhoods. Besides the obvious function of removing water, if the floodplain is substantially retained, the river system can provide many ecosystem services in addition to managing water quantity and quality. Wider channels can provide a framework for linear parks that cut through dense urbanized landscapes and provide a link to the natural landscape outside the city. The impact of these linear parks depends on how adjacent neighbourhoods are designed to allow advection. Figure 15.26 shows a linear park in Singapore designed around the path of the Kallang River, which divides high and mid-rise neighbourhoods. The park was redesigned to allow the river, which was previously enclosed in a narrow concrete channel, to meander and form a channel that could cope with higher runoff levels. The park also provides a route for air to flow into the city.

At a microscale, water is used as a design tool mainly to cool surfaces and the adjacent air. Figure 15.27 illustrates the use of water at neighbourhood and street scales. Figure 15.27a shows a linear park created around the Cheonggyecheon River

**Figure 15.26** Aerial view of Bishan Park, Singapore, which was redesigned to increase the capacity of the Kallang River to cope with runoff from the urbanized landscape (Credit: Atelierdreiseitl; CC3.0).

**Figure 15.27** Water as an urban design feature: **(a)** The Cheonggyecheon River in Seoul, South Korea (Credit: Stari4ek, CC3.0); **(b)** a manufactured lake within the Barbican development in London, United Kingdom (Credit: G. Mills); **(c)** a fountain embedded in the pavement of Nimes, France (Credit: G. Mills) and; **(d)** a water spray is used in conjunction with an awning to provide a cool airspace immediately outside a shop in Florence, Italy (Credit: J. Voogt).

(Seoul, South Korea), which was until recently entombed beneath a major highway, that has since been removed, shifting traffic elsewhere. Figure 15.27b shows a mixed-use (residential and cultural use) urban environment that is intensively developed but incorporates a pond as an outdoor design feature. Figure 15.27c shows a water feature that is a component of a tree-lined avenue used as a cooling area on hot days. Finally, Figure 15.27d shows a very small scale cooling system attached to an awning that delivers a very fine water mist to cool the air in the immediate vicinity of a shop. The system generates a spray at regular intervals because the effect is short-term.

Small ponds and fountains are most often used in parks and courtyards where people are expected to sit, walk and relax. The impact of a water feature depends on the meteorological conditions, especially wind speed and humidity. The evaporation rate is greatest if there is plentiful solar radiation, a large surface-air humidity gradient and a **turbulent** atmosphere. A cooling strategy that relies on water features should also consider how best to manage the circulation of air to maximize its potential for moderating the microclimate. In an urban setting, courtyards and small plazas can provide shelter to limit the volume of air affected so that the impact of the feature is maximized. In these circumstances, fountains that produce a spray of water are more effective means of cooling as the water droplets evaporate more quickly than a stagnant pool. In outdoor parks, without shelter, the wider spatial impact of small water features is very limited (Figure 15.28).

One of the advantages of using water for microclimate control is that it can be applied at a particular time and place as needed (e.g. Figure 15.27c and d). Where water is available it can be spread on street facets to cool the ground and the overlying air. In an experiment conducted on a Paris, France, street during over several hot days in July, about 1 mm was spread over the asphalt surface at frequent intervals. The pavement watering scheme was found to reduce surface temperatures by 2–4 K in the morning and 6–13 K in the afternoon (Hendel et al., 2015). An advantage of this intervention over 'cool' pavements is that a wetted surface does not increase reflection but rather diverts available energy into evaporation. This system requires considerable manual effort and a supply of water, but can be an effective way to reduce the surface and air

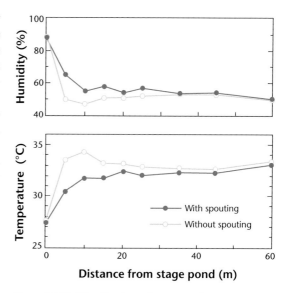

**Figure 15.28** Distribution of average air temperature and humidity on the leeward side of a water feature (between 1400 and 1500 h) located in a park in Osaka City, Japan. The feature consists of stone pillars (2 m tall, 0.5 m wide) that mimic a waterfall and is accompanied by a spray. Measurements were made on July 30–31 1992 during average summer conditions-clear days with a breeze at 2.5 m s$^{-1}$ west north-westerly sea breeze. The cooling effect spreads to a distance of 35 m (Source: Nishimura et al., 1998; © Elsevier Science, used with permission).

temperature in streets. In hot climates where water is a scarce resource, the challenge is to achieve the maximum effect for the least amount of water; this usually means using evaporation cooling in confined, shaded environments such as the courtyards of buildings.

## 15.3 The Well-Planned and Designed City

Our stated goals for a well-planned and designed city from a climatic perspective at the start of the chapter included: efficient use of resources; improvement, or at least not degrading the local climate compared to pre-urban development and; protection of people and infrastructure from extreme weather events. Although the focus here is on CSUD, many of the tools discussed in Section 15.2 have wider environmental benefits for water management especially and can be linked with WSUD (Coutts et al., 2012).

## CSUD and WSUD

WSUD incorporates a raft of strategies that are designed to manage water quantity and quality (Table 15.2). Some WSUD features are designed to capture and store water for future use in the house or to be applied to green landscapes (e.g. rainwater harvesting) and others for reducing the magnitude and improving the quality of urban runoff following a rain event (e.g. swales and biodetention systems). Many WSUD features have already been introduced as CSUD tools including: green roofs (Figure 15.14d); filter strips (Figure 15.21d); trees (Figure 15.21a) and; permeable paving (Figure 15.14a). The effect of WSUD tools on climate depends on how that water

**Table 15.2** A list of WSUD design tools and their description. The collection mechanism is classified into point (P), lateral (L) and surface (S) and the dot symbols indicate its likely valuable contribution to delivery of design criterion (Source: Table 7.1 in Woods-Ballard et al., 2015).

| Tools | Description | Collection mechanism | Peak runoff rates | Small events | Large events | Water quality |
|---|---|---|---|---|---|---|
| Stormwater harvesting systems | Systems that collect runoff from the roof of a building or other paved surface for use | P | | • | • | |
| Green roofs | Planted soil layers on the roof of buildings that slow and store runoff | S | | • | | • |
| Infiltration systems | Systems that collect and store runoff, allowing it to infiltrate into the ground | P | • | • | • | • |
| Proprietary treatment systems | Subsurface structures designed to provide treatment of runoff | P | | | | • |
| Filter strips | Grass strips that promote sedimentation and filtration as runoff is conveyed over the surface | L | | • | | • |
| Filter drains | Shallow stone-filled trenches that provide attenuation, conveyance and treatment of runoff | L | • | • | • | • |
| Swales | Vegetated channels (sometimes planted) used to convey and treat runoff | L | • | • | • | • |
| Biodetention systems | Shallow landscaped depressions that allow runoff to pond temporarily on the surface, before filtering through vegetation and underlying soils | P | • | • | • | • |
| Trees | Trees within soil-filled tree pits, tree planters or structural soils used to collect, store and treat runoff | P | • | • | | • |
| Permeable pavement | Structural paving through which runoff can soak and subsequently be stored in the sub-base beneath, and/or allowed to infiltrate into the ground below | S | • | • | • | • |
| Attenuation storage tanks | Large, below-ground voided spaces used to temporarily store runoff before infiltration, controlled release or use | P | • | | | |
| Detention basins | Vegetated depressions that store and treat runoff | P | • | • | | |
| Ponds and wetlands | Permanent pools of water used to facilitate treatment of runoff – runoff can also be stored in an attenuation zone above the pool | P | • | | | • |

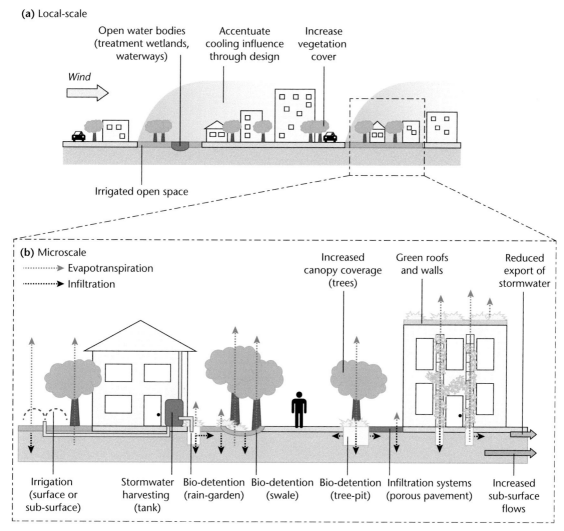

**Figure 15.29** Schematic representation of widespread implementation of storm water harvesting and water-sensitive urban design (WSUD) at the microscale in the restoration of a more natural water balance, along with increased vegetation cover. This enhances urban evapotranspiration and shading resulting in local-scale cooling effects that can improve human thermal comfort (Modified after: Coutts et al., 2012; © The Authors, used with permission).

is employed, that is whether it is stored and sealed, or applied to the surface or added to the soil, where it can be taken up by vegetation. The role of vegetation in both WSUD and CSUD is an important link and taken as a whole is often considered as 'green infrastructure', which places it alongside other types of infrastructure that support the functioning of the urban system. Ideally, the application of tools to achieve the CSUD goals is complemented by WSUD (Figure 15.29).

Figure 15.30 places the desired urban outcomes within a schematic that shows the overlapping nature of climate changes at the global/regional scale and urban scales. The main driver of urban climate effects are decisions to alter the urban footprint, built layout and infrastructure, population densities, economic structure and so on. This diagram emphasizes the systemic connections between components of the urban system and climate and regional/global climate change. Climate-based policy to address impacts and

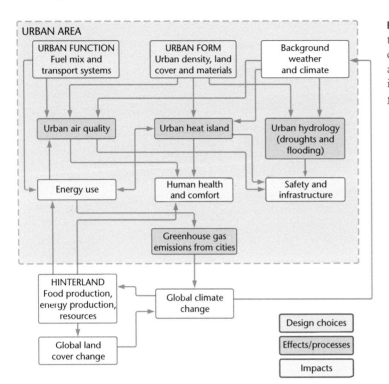

**Figure 15.30** A schematic diagram that links urban planning and design decisions on urban form and functions to their climatic impacts at urban, regional and global scales.

their drivers need to consider the global and urban climate system simultaneously.

### 15.3.1 Resource Use

A core measure of urban sustainability is the efficiency of the resource use required to build its form and maintain its functions through **urban metabolism** (Section 1.4). Indicators of efficiency include per capita energy, water, food and land consumption and waste generation (including air pollutants and GHGs). There is strong evidence that cities which are more compact, that is more densely built and occupied, are intrinsically more efficient (OECD, 2012). The evidence for this is especially strong for energy use and pollution associated with transportation (Kenworthy, 2006). As public transit systems become economically feasible, private vehicle use becomes prohibitive and the average commuting distance decreases. Ideally, a city is designed at the outset to be compact because rebuilding a city, once a low-density layout is established, is very difficult to achieve.

Table 15.3 lists five major policies to make cities more compact. At the outset, note that the policies are metropolitan-wide and apply to the contiguous urban area, rather than just one part. The policies emphasize delineating the urban–rural edge and reusing land and retrofitting buildings within the existing area when possible. Intensification of the existing urban form to increase population density is encouraged through the creation of compact mixed-use neighbourhoods at the nodes of mass transit systems. Critically, good design is required so that the negative impacts of increased compactness are minimized. The latter include increasing vegetative cover and creating high-quality public spaces to encourage outdoor use.

### 15.3.2 Comfort

Comfort describes environmental conditions where an individual *feels* no need to adjust their environment. In the strictest sense, the concept of comfort has a limited value to outdoor conditions where there is limited control on the meteorological stresses. Given the range of weather combinations, it is not possible for a single outdoor space to provide a universally acceptable microclimate. However, good design should enable extension of the period of outdoor use through:

**Table 15.3** Recommendations for compact city policies (Source: OECD, 2012).

| Compact City Policies | |
|---|---|
| **Policy strategies** | **Policy sub-strategies** |
| Set explicit compact city goals | • Establish a national urban policy framework<br>• Encourage metropolitan-wide strategic planning |
| Encourage dense and contiguous development at urban fringes | • Increase effectiveness of regulatory tools<br>• Target compact urban development in green-field areas<br>• Set minimum density requirements for new development<br>• Strengthen urban–rural linkage |
| Retrofit existing built-up areas | • Promote brown-field development<br>• Harmonize industrial policies with compact city policies<br>• Regenerate existing residential areas<br>• Promote transit-oriented development in built-up areas<br>• Encourage intensification of existing urban assets |
| Enhance diversity and quality of life in urban centres | • Promote mixed land use<br>• Attract residents and local services to local centres<br>• Promote focused investment in public space and foster a sense of place<br>• Promote a walking and cycling environment |
| Minimize adverse negative effects | • Counteract traffic congestion<br>• Encourage the provision of affordable housing<br>• Promote high-quality urban design<br>• Encourage greening of built-up areas |

- **Adaptive and responsive urban form and fabrics** – create urban spaces that respond to expected changes in weather and seasons through the use of deciduous trees, intelligent materials, awnings, umbrellas, water spraying, and consideration of solar position in the design process.
- **Diversity in urban form** – design that creates a variety of microclimates within neighbourhoods to offer choices to people under different weather and seasons. For example simultaneously create microclimates that provide shelter (or ventilation), sunshine (or shade), protection from precipitation, etc. based on the expected use of spaces and weather variability.

There are several CSUD books (e.g. Emmanuel (2005), Erell et al. (2012), Lenzholzer (2015)) that discuss how best to design urban spaces to create more desirable climates. Examples of interventions using mechanical and natural tools to manage climate in outdoor urban spaces are shown in Figure 15.31. Each intervention reflects the specific environment and the intended use. Figure 15.31a shows a small square that is used for open air markets and public events in a cool, wet and windy climate. While the small space provides shelter from

the wind, the key to extending its use is protection from rain, which is provided here by large umbrellas that are inverted when opened so as to funnel water to the substrate. Sometimes interventions are needed simply to counteract other urban effects. Figure 15.31b shows a street in a mid-latitude climate with cold winters (note the enclosed pedestrian bridge connecting two buildings in the background). The metal sculptures that line these streets are designed to provide wind protection for pedestrians in a high-rise environment, by reducing gusts generated by the tall buildings. For outdoor spaces that have seasonal use, naturally responsive tools are ideal. Figure 15.31c shows an outdoor restaurant during summer where an expansive leafy canopy provided by deciduous trees creates a pleasant shaded area suited to sitting. The street shown in Figure 15.31d employs a different strategy in a Mediterranean climate characterized by high daytime temperatures during the daytime. Outdoor **thermal stress** can be reduced by providing shade and removing obstacles to airflow. The wide unobstructed pedestrian street shown here allows breezes and the sheets of fabric suspended at roof level provide shade that can be removed when no longer needed.

**Figure 15.31** Managing outdoor climate using adaptable urban forms: **(a)** Outdoor umbrellas to provide shelter from rainfall in Dublin, Ireland (Credit: G. Mills); **(b)** Metal 'trees' located on a pedestrian street in Calgary, Canada, to reduce wind gusts created by tall buildings. (Credit: G. Mills); **(c)** Trees provide an extensive canopy for shade in an outdoor eating area in Munich, Germany (Credit: A. Christen); **(d)** A shading canopy formed by fabric suspended above a pedestrian street in Malaga, Spain (Credit: M. Chernov; CC3.0).

Inevitably, design also depends on the primary function of a street (e.g. residential, commercial-office and commercial-shopping) and the best design for one is not necessarily the best for another. In Table 15.4 we show how different design tools can be used to address a number of climate issues at a street scale. While some tools (e.g. runoff and temperature control), are complementary others (e.g. better ventilation to improve air quality and increasing built density to reduce energy use) may be in opposition to each other. Moreover, the appropriate design must be fitted to the climate context; while solar access is desirable in cool climates, shade is wanted in hot, arid settings.

### 15.3.3 Air Quality

Poor air quality is an almost ubiquitous feature of city climates globally. More so than other urban issues, there is a long history of urban air quality management using a wide variety of planning tools. The total public risk associated with a particular air pollutant depends on several factors (Joumard et al., 1996), including the strength of emissions, air pollutant dilution, physical and/or chemical transformation, toxicity and the population exposed. Although chemistry and toxicity cannot be managed, we can manage for emissions, dilution and the location of population relative to emissions.

### Emission Management

The most direct way of addressing air pollution is to reduce the magnitude of emissions, i.e. their intensity $(m^{-2})$ and rate $(s^{-1})$. Available tools include: technological controls on the efficiency of emission drivers (e.g. vehicle engines, building insulation and manufacturing processes); regulations on the use of fuel type and quality, of materials, of chemicals, etc.; controls on behaviour (e.g. use of open fires and of private vehicles). If this is not possible, the location and

**Table 15.4** Examples of poor and good practices in street design. Note that the 'best' design may vary from street to street depending on the desired objective.

| | | |
|---|---|---|
| **Air Quality control**<br>The simplest means of improving air quality is to reduce emissions. Poorer street level air quality occurs in poorly ventilated streets with heavy traffic. Wider streets and well positioned trees can improve ventilation and screen pedestrians and buildings from emissions. | ✗<br>Chimney  Air conditioner<br>Traffic | ✓<br>Tree  Window |
| **Solar control**<br>Street geometry regulates access to Sun. Trees and arcades can provide additional shade outdoors. Solar gain to buildings can be managed with trees and awnings. Where solar gain is desirable, overshadowing should be minimized. Rooftops provide ideal location for solar gain if not overshadowed. | ✗ | ✓<br>Solar panel<br>Arcade  Awning |
| **Runoff control**<br>Runoff can be managed by increasing the capacity to detain rainwater, adding vegetation and enhancing the permeability of street facets. Green roofs can provide insulation for buildings and some evaporative control but may require that roofs are strengthened to cope with additional weight. | ✗<br>Rapid runoff from impervious facets<br>Storm sewer | ✓<br>Green roof<br>Water storage<br>Permeable paving |
| **Temperature control**<br>Daytime surface temperature can be managed through shading of facets and controlling reflectivity. Fountains can be employed when needed to provide evaporative cooling. Green facets will reduce surface temperature and trees can provide both shade and | ✗<br>Dark roof poor insulation | ✓<br>Reflective roof  Fountain  Green wall |
| **Wind control**<br>Wind shelter within the urban canopy is controlled by ambient wind velocity and street design. When airflow is perpendicular to the street axis, tall buildings will draw faster moving air to ground level while closely spaced buildings of even height provide shelter. | ✗<br>Asymetrical street geometry. Windward building tall. No within street obstructions | ✓<br>Buildings the same height, H/W ≤ 1<br>Trees retard airflow |

timing of emissions provide further control (e.g. regulations on stack heights and operating times).

## Managing Population Exposure

Although there is no ability to control population **exposure** directly (apart from extreme events where an 'evacuation order' is put in place), considering population distribution relative to significant point sources can be an effective means to improve air quality. It is common practice to situate large polluters away from populated places so as to limit exposure but it is important to take meteorological conditions into account when making such decisions. For example, although the prevailing wind for most mid-latitude cities is westerly (suggesting that pollution sources should be positioned on the east side of the city), the weakest winds and most stable weather are often associated with easterly winds. In other words, placing polluting facilities on the eastern side of settlements may actually increase the risk to the population because the extreme events (the highest **concentration** of air pollutants) often correspond to infrequent easterly winds.

### Managing Dilution

The two most effective means of managing risk are to reduce emissions and/or limit population exposure by separating movements of people and traffic. Nevertheless, dilution management provides a third option to augment efforts of the other two. Urban form affects the venting, and hence dilution and transformation of air pollutants at different scales. At a regional scale, where there is the right mix of **primary pollutants**, air temperature and ample **irradiance**, secondary pollutants (e.g. $O_3$ and PAN) can create photochemical smog (Section 11.3.2); in these circumstances the warming effect of cities enhances the production of secondary pollutants. However, the UHI also promotes **convection** and consequently vertical mixing, deepens the depth of the UBL and dilutes the concentration of primary and secondary pollutants in the overlying air. In calm conditions, the UHI may be sufficiently strong to generate a thermal circulation that causes the advection of cleaner air from rural areas into the city.

The impact of urban planning and design is greatest on the UCL, below roof level, where pollution sources (e.g. vehicle exhausts and domestic fires) and human receptors are co-located. At this scale the outdoor exposure to air pollutants depends on the microscale location of emission sources and the most used pedestrian routes (Figure 11.7). Enhancing ventilation at street level can be achieved by modifying the built geometry by:

- Increasing the **spacing between buildings**;
- **Decreasing the height** and increasing the height variability of buildings; and
- Aligning the **street in the direction of the prevailing airflow**.

However, such changes to the physical geometry of streets and their layout is usually not an option, but limiting traffic flows and widening pathways to limit exposure may be more practical. Similarly removing obstacles that impede airflow can assist mixing, for this reason in Hong Kong, signage that protrudes from building facades is regulated. Trees can also obstruct flow but may be effective in scavenging selected air pollutants from the atmosphere. Thus, trees with extensive canopies that cover a street may limit ventilation and increase near-surface concentrations, but a line of trees or hedges that divide pedestrians from traffic could act as a protective screen.

### 15.3.4 Weather Extremes

Figure 13.14 provides a useful framework to discuss urban climate risks. It depicts risk as the intersection of three overlapping issues: hazard, **exposure** and **vulnerability**. While hazard represents the objective statistics governing the reoccurrence of climate events that could result in risks to an urban population and infrastructure, its magnitude is regulated by exposure to the hazard and the ability to respond (vulnerability). Applying these ideas to urban climates means that the following workflow should be applied:

1. Identify potential weather **hazards;**
2. Identify population **exposure** to the hazards, i.e. neighbourhoods and outdoor spaces where risk of the identified hazards is greatest; and
3. Select proper responses that reduce **vulnerability** to the hazards appropriate to the scale.

Here we concern ourselves with two hazards (flooding and heat stress) where the urban effect generally accentuates the risk by exposing vulnerable populations and enhancing the hazard itself.

### Flooding

There are three types of flooding that affect cities: coastal (due to sea-level rise, or storm surge); fluvial (river) and pluvial (precipitation). The first can only be addressed directly using barriers or simply not building on (or withdrawing from) coastal areas likely to be flooded. Within cities, these areas should be parklands set aside to provide a buffer zone between the settlement and the coast. Where building has occurred, the architectural response is to raise buildings so that the ground floor is above the likely flood water level and/or convert the ground floor of buildings to robust uses, such as carparks.

**Figure 15.32** An aerial view of Dallas, United States, and the Trinity River parkway. The parkway has levees on either side to protect the city from flooding but the park has been designed as a series of wetlands to accommodate water during heavy rainstorms when the drainage capacity of the river channel is exceeded (Credit: G. Paccaloni; with permission).

The urban effects on pluvial and fluvial flooding are correlated because both are enhanced by impervious surface cover that either causes water to pool on the surface or deliver it into nearby streams too quickly, overwhelming the drainage capacity of the channel. One way to decrease the vulnerability is to enhance the capacity of the drainage infrastructure to cope with intense precipitation events by increasing the size of pipes and channels. Another approach is to avoid building in areas likely to flood by ensuring suitable buffer zones around river channels. Naturally, this is easiest to do at the outset, because restoring river systems in built-up areas requires very significant intervention. Figure 15.32 shows an aerial image of downtown Dallas, United States, and the Trinity River which is managed within a levee system that encloses the river channel and a set of wetlands (WSUD) that can manage excess water. The linear park shown in Figure 15.26 is an example of WSUD at a neighbourhood scale.

Figure 15.33 shows a variety of techniques that are used to manage water in urban environments. Figure 15.33a and b shows hard engineering approaches that may be needed to cope with excess runoff in built landscapes. The other two figures show examples of WSUD management approaches by creating more natural channels including detention storage (Figure 15.33c) or by replacing paved surfaces

with permeable facets (so-called 'rain gardens', Figure 15.33d). The latter represents a microscale intervention that could solve a specific problem but would only have a significant impact if replicated on a wide basis.

A major urban control on urban flooding is the degree to which permeability of the pre-urban landscape is compromised. In a densely built urban area (e.g. LCZ 1–3), the total impervious cover can approach 100%, so almost all precipitation is converted to runoff. WSUD techniques improve the hydrological response of these neighbourhoods through widespread microscale interventions to increase permeability (Figure 15.29). The design challenge is to transform existing built landscapes where space is often at a premium. If the roof area forms a significant proportion of the surface cover it provides some capacity to store water for later evaporation. However, storing substantial amounts of water on roofs is usually not feasible owing to its considerable weight. This is especially true for large, low-rise neighbourhoods (e.g. warehouses). In such places, increasing storage capacity and permeability at ground level are the only ways of reducing $R/P$. However, minimizing the total impervious fraction ($\lambda_b + \lambda_i$) of a city also means reducing the area occupied by extensive low-density neighbourhoods (e.g. Figure 15.12b and Figure 15.19a). While these neighbourhoods have a

**Figure 15.33** Examples of urban runoff management: **(a)** Conventional flood protection along a river estuary subject to fluvial and coastal flooding in Dublin, Ireland (Credit: G. Mills); **(b)** a concrete channel used to facilitate the runoff of excess water in Nimes, France (Credit: G. Mills); **(c)** a managed urban river in Vancouver, Canada, that provides a riparian bank environment and green embankments (Credit: A. Christen) and; **(d)** a small rain 'garden' along a busy street in Washington DC, United States, that allows water to permeate the surface cover (Credit: G. Mills).

high vegetative fraction, the total impervious area is much larger for the population that they accommodate. Stone (2004) examined suburban developments of different built densities in the United States and found that houses on larger plots were placed furthest from the road network and required lengthy driveways that increased the total impervious cover. Simply limiting the size of building plots was suggested as an effective means of controlling impervious cover.

## Heat Stress

Heat stress (see Section 14.3) in cities is a particular concern because urban effects commonly raise the temperature and reduce wind speed. The UHI is in part caused by increased impermeable cover, the properties of construction materials, the three-dimensional built geometry and the direct addition of heat by humans. These changes create four distinct types of UHI (at the surface, in the UCL, in the UBL and in the subsurface), each of which is linked to its own set of primary drivers which result in distinct temporal patterns (Chapter 7).

At the outset, it is important to remember that the magnitude of a UHI is assessed as a *difference* when compared with the equivalent temperature representing the same environment in the absence of urban development. Hence, the significance of the UHI should be judged against this reference. For cities in cool climates (and cold weather) the UHI reduces the heating load (i.e. a benefit) but in warm climates (and hot weather) it increases the cooling load (i.e. a cost). For places that experience a seasonal climate, the relative benefits and costs vary through the year. During heatwaves, the urban effect increases the thermal stress on the population, results in higher energy use in buildings and poorer air quality (Section 12.3.3). For many cities, climate change will mean more frequent heat wave events of greater intensity (Section 13.3.3, Table 13.6). Counteracting the UHI must consider: the type of UHI and its drivers; the temporal and spatial pattern of the UHI and; the risk it poses.

Strategies to reduce the magnitude of the UHI focus mainly on raising the albedo of urban facets and/or increasing the vegetative fraction (EPA, 2009). Cool roofs are a relatively simple and pragmatic way to lower roof temperature, limiting building heat gain, reduce the thermal stress indoors and reduce additional heat emissions outdoors due to space cooling systems in the building. The roof surface also presents a cooler surface for the above roof air and contributes to reducing the boundary layer heat island. Cool pavements could have a similar outcome within the urban canopy by lowering street temperatures during daytime, however, the reflected radiation will impinge on buildings and pedestrians

and increase their heat gain. Modelling experiments to calculate the UHI in Sacramento, United States, by Taha (2008) indicates that increasing albedo does have a significant impact on the daytime surface (and canopy level) UHI but little influence on the nocturnal canopy level UHI.

To address the canopy-level UHI the focus has to shift to microscale decisions about buildings and adjacent outdoor spaces. An obvious but expensive solution to reduce thermal stress is to simply cool indoor spaces in the same way one heats indoor spaces in cold climates. Of course, the waste heat generated by air conditioning systems contributes to the outdoor heat conditions to which the building's climate control system is responding (see Section 6.2.2). However, in response to a public health crisis arising from a heatwave event, establishing cool indoor spaces for public access is a viable short-term solution. Modifying the outdoor environment is more difficult for obvious reasons. Immediate outdoor responses include providing shade, and using pavement watering and water sprays in accessible places. For especially vulnerable populations that cannot afford indoor cooling systems, access to cooled public buildings, energy credits and/or the provision of air conditioners may be needed. More substantive changes to the urban fabric (e.g. permeable paving and reflective coatings), to surface cover (increased vegetation cover) and to the geometry of the urban landscape are needed to reduce the magnitude of a canopy-level UHI permanently. Of these options, changing the layout of buildings and their dimensions is the most dramatic, but it could boost the near-surface speed, improve air quality, increase natural ventilation in buildings and reduce heat stress.

## Summary

The climate effects of cities are a result of **urban form** (extent, materials, building dimensions and density, etc.) of the landscape and **urban function** (energy, water and material use through the urban metabolism). At its best, **climate-sensitive urban design** (CSUD) can **moderate climate risks, limit resource use and emissions and extend periods of outdoor comfort**. However, with the possible exception of air quality, there are no universal rules for CSUD. What is required is the judicious application of tools based on an understanding of the drivers of urban climate effects; these drivers are linked to location (latitude and topographic setting), regional weather and climate and the prior urban development. To summarize:

- The background climate, hydrological and topographic setting of a city provide the context for CSUD. **Mapping** the temporal and spatial properties of **climatic elements** is an important step in identifying potential **resources and hazards**. This requires meteorological observations that are relevant to the task at hand and modelling tools to generate useful information to guide design and planning.
- Scale is critical to the assessment of climate impacts and appropriate intervention options. To be effective intervention policies must be coherent across these scales. At the **city scale**, the overall areal extent of the city, the proportion occupied by each neighbourhood type and population density influences urban metabolism and the magnitude of urban climate effects. At a **neighbourhood scale**, the distribution (and proportion) of surface cover that is built-upon, impermeable and vegetated are key indicators of the impact on local climate and hydrology and of the potential for intervention. Within the **UCL**, it is decisions on individual buildings (and facets) and intervening spaces (streets) that matter most for human comfort.
- Modifying urban functions to regulate urban metabolism is the most effective way of **managing energy use and waste emissions**. Decisions on urban form can complement these actions by promoting more compact urban layouts that are designed to provide comfortable and healthy microclimates;
- Creating a comfortable outdoor microclimate for all users in all climates is neither feasible nor desirable. Rather, the goal is to design urban spaces that are **responsive** to different climate drivers so that periods of discomfort are reduced. This need is greatest in densely populated, compact and **multi-functional urban settings** where people can choose preferred outdoor spaces rather than retreat indoors to avoid urban climate impacts.
- For existing urban form, **increasing vegetative cover** and incorporating natural landscape features into urban design are the best means of managing urban climate effects at all scales ('green design'). Modifying the **properties of facets**, such as increasing albedo and increasing permeability, can address selected issues but must be extended across the urban landscape to have wide effect.

# Epilogue

*Urban Climates* appears at an important stage in the growth of an **urban climate** science; it has overcome the somewhat fitful initial progress that characterized its early history and now is maturing rapidly as interest in the subject matter grows across various disciplines. The need for such a science is undeniable, given the imperative to manage climate change and improve the lives of many urban inhabitants. There remain, however, knowledge gaps and conceptual barriers; a non-exhaustive list includes:

- A lack of observations in upper parts of the **urban boundary layer**, which links the city-scale with regional and global climates.
- Integration of urban processes and climates with the facts of ongoing urban development within regional and **global climate models**.
- A paucity of information on the **climate of settlements outside regions of the developed world**, especially those located in **tropical climates**.
- A lack of case-study examples that demonstrate how **urban planning and design interventions** impact urban climate effects.

To date, perhaps the most difficult hurdle to overcome is the integration of this knowledge into **climate-sensitive urban design** and planning practice. As the urban climate field has developed, the nature of evidence has become more sophisticated and increasingly inaccessible to a wider audience. A major challenge for **urban climatology** is how best to translate faithfully scientific evidence into tools that are readily usable by designers and urban and regional planners. A broader goal is to integrate climate knowledge into a holistic model of the urban system that incorporates its natural and human components.

Our understanding of urban climate effects has come about as a result of a combination of observations and theoretical developments, both underpinned by technological advances. This pattern is likely to continue in the future as instrumentation and computational power becomes more accessible, less expensive and more versatile allowing investigation and monitoring of urban climates that cannot be explored using conventional methods. Consider, for a moment, the potential of wearable devices to gather personal climate **exposure** using **Lagrangian** methods and crowd-sourcing; a fleet of mobile sensors on cars, bikes, drones and even birds that relay information about the urban atmosphere in real-time; smart indoor and outdoor **sensor** networks to provide real-time data about climate, energy and human behaviour; **remote sensing** devices with great spatial and spectral resolution that can be mounted on various airborne platforms to 'see' parts of the urban landscape that were previously inaccessible. In other words, we will soon have a capacity to gather large datasets at all scales relevant to urban climate. Nevertheless, these atmospheric data will only lead to new or enhanced knowledge if their deployment is well planned and usefully integrated into existing frameworks and they mesh seamlessly with the capacity to assimilate it into predictive models.

# A1 | History of Urban Climatology

This book has compiled and structured a wealth of **urban climate** research generated over many decades. Much of this work incorporated and built upon earlier work that is rarely recognized or acknowledged. In this brief appendix we provide a 'pen portrait' of the development of **urban climatology** by identifying key periods and people.

The scientific study of urban climatology dates from the early nineteenth century. Its development can be conveniently divided into four periods:

- prior to 1930 the field was characterized by **pioneering climatographies** of selected cities and weather elements;
- from 1930 until about 1965 urban research was greatly influenced by the growth of **micro- and local climatology** which provided greater insight into climatic differentiation and new field techniques;
- from 1965 until about 2000 it experienced an explosive increase in research interest including closer **links with meteorology**, and the **emergence of physically-based models** of the urban atmosphere;
- at the start of the twenty-first century urban climatology and meteorology matured into a **predictive science**.

Recently, as in many of the natural sciences, the literature in urban climate has expanded almost exponentially. Before 1930 there were typically less than ten new publications per year, in 1965 there were still less than 30 per year but at the end of the twentieth century it grew to about 140 per year. In addition, fields like urban **air pollution**, urban wind engineering, urban hydrology and building climatology had their own extensive literatures. A fairly complete bibliography of urban climatology can be created by combining: Brooks (1952), Kratzer (1956) (Figure A1.2), Chandler (1970) (Figure A1.2), Oke (1974, 1979, 1983 and 1990), Jáuregui (1993, 1996) (Figure A1.3) and most recently the World Meteorological Organization (WMO)/International Association for Urban Climate (IAUC) Bibliography of Urban Climate (www.urban-climate.org).

The state of the field and its research has been reviewed on several occasions, including by: Kratzer (1937, 1956), Landsberg (1956, 1962, 1970, 1981) (Figure A1.2), Munn (1966), Oke (1974, 1979, 1980), Terjung (1974), Yoshino (1975), Chandler (1976) (Figure A1.2), Lowry (1977) (Figure A1.3), Oke, (1987, 1988), Chow and Chao (1985), Fezer (1995), Yoshino and Yamashita (1998), Helbig et al., (1999), Lowry & Lowry (2001), Arnfield (2003), Erell et al., (2011) and Cleugh and Grimmond (2012).

## A1.1 Pioneer Urban Climatographies (Prior to 1930)

Early urban dwellers were aware the atmosphere was affected by the presence of the city, especially visible phenomena such as **fog** and smoke. By the seventeenth century the largest cities of Europe such as London and Paris were sufficiently polluted to warrant regulation of smoke-producing factories, but the growth of industry during the Industrial Revolution combined with domestic wood and coal burning led to the murky, depressing state of urban air vividly described in Britain by the novels of Dickens and the paintings of Turner. In heavily industrialized areas the Sun appeared as a dull red ball at times when the countryside upwind was sunny and bright.

It was not until the development of meteorological instruments such as the thermometer, barometer and **anemometer** that more subtle aspects of urban climate could be investigated. The first to do this scientifically was Luke Howard (Figure A1.1). He was a chemist and gentleman scientist with a keen interest in meteorology and a perceptive observer of phenomena in the field – he devised the classification of cloud types used today. He qualifies as the Father of Urban Climatology for his observation, interpretation and reporting of air temperature and other meteorological variables in London (at the Royal Society), and at a number of sites, in what was then countryside to the northeast. He reported his findings in two early volumes entitled *The Climate of London* (Howard, 1818, 1820) and later in a much expanded second edition of three volumes (Howard, 1833). Despite what we would now consider to be unrepresentative **exposure** of thermometers he

Luke Howard

Émilien Renou

Viktor Kremser

Julius von Hann

**Figure A1.1** Portraits of some of the more influential scientists in the pioneering phase of climatographies (Credit: Library of the Central Institution for Meteorology and Geodynamics, Vienna).

observed, and correctly recognized, the anomalous warmth of the city (both diurnally and seasonally) that we now call the **urban heat island**. Further, he correctly hypothesized almost all of the causes now considered responsible (Chapter 7). It was a most auspicious start to the scientific study of the urban atmosphere (Mills, 2008). Noah Webster (1799), the lexicographer, was one of the first to write about the relative warmth of cities. He reported ice formation at rural sites when air temperature thermometer readings in New York City were about 8°F (~4.4°C) above freezing, but Howard was the first to provide scientific documentation.

The latter half of the nineteenth century saw the rapid expansion of meteorological observation networks around the world, especially in Europe. This provided the essential data for climatological analysis and city climatographies. The thermal climate of several of the great cities of Europe were studied in this period including Paris by Renou (1855, 1868;

Figure A1.1), Munich by Wittwer (1860), Berlin by Kremser (1886; Figure A1.1) and a comparison of several European cities by von Hann (1885; Figure A1.1). Similar work was begun in North America and Japan in the 1900–1920 period. By the last decade of the nineteenth century air pollution and its effects on urban fogs and the reduction of sunshine became a focus, especially in England. A trend of increasing fog frequency was established and it was recognized that winter fogs and air pollution were sufficient to cut urban sunshine duration by one half. Studies of urban effects on other climatological elements followed; including the decrease in air pressure by von Hann (1895; Figure A1.1), changes in **precipitation** by Hellmann (1892), and the reduction of both humidity and wind speed both first noted by Kremser (1908 and 1909, respectively; Figure A1.1).

## A1.2 Advances in Micro- and Local Climatology (1930 to 1965)

Interest in climates existing *within* the **synoptic** station network were sown by pioneers like Homén of Finland and Kraus in Bavaria at the beginning of the twentieth century followed by Schmauss and Geiger (Figure A1.2) in Germany and Schmidt (Figure A1.2), Lauscher (Figure A1.2) and Steinhauser in Austria. Work in urban areas was greatly aided by the invention of the mobile station by Schmidt (1927). He revealed small variations of climate by attaching instruments to a vehicle and recording the response while traversing a route to obtain cross-section **profiles**, or the spatial distribution of one or more weather elements. This simple technique provided considerable insight into details of urban climates and created a large number of case studies in different cities. German-Austrian workers of this period also began studies of basic urban climate processes. For example, Lauscher (1934; Figure A1.2) laid the foundation to understand the effect of building/street geometry on the exchange of **longwave** radiation and Albrecht (1933) did the same for the circulation of winds.

These developments in the 1930s led to similar work in North America and Japan, but World War II caused a temporary decline in research. The pace accelerated after the war in response to military interest in research into **dispersion** of gases in urban areas. Climate information was also incorporated in planning cities during the reconstruction period, which

Wilhelm Schmidt    Albert Kratzer    Åke Sundborg    Tadeshi Kawamura

Frederick Lauscher    Rudolf Geiger    Tony Chandler    Helmut Landsberg

**Figure A1.2** Noted climatologists in the development of micro-, local- and urban climatology (Credit: Photos supplied by each individual).

was accompanied by rapid housing development, industrialization and environmental deterioration (e.g. in Germany, Japan, Russia and England). This period produced some of the most detailed urban climatographies of specific cities including thermal studies in Bath, United Kingdom (Balchin and Pye, 1947), Uppsala, Sweden (Sundborg, 1951; Figure A1.2), San Francisco, United States (Duckworth and Sandberg, 1954), Ogaki City, Japan (Takahashi, 1959), Linz, Austria (Lauscher et al., 1959), Vienna, Austria (Steinhauser et al., 1959), Kumagaya City, Japan (Kawamura, 1964; Figure A1.2), and London, United Kingdom (Chandler, 1965; Figure A1.2).

Sundborg's (Figure A1.2) work was particularly outstanding: he devised a rigorous statistical method to classify the representativeness of stations along a traverse route; he was the first to predict **heat island magnitude** in terms of synoptic weather conditions; and most significantly he was the first to couch causation of the heat island within a **surface energy balance** framework. Duckworth and Sandberg were the first to fully study the vertical structure of the heat island, and the studies of Takahashi and Kawamura (Figure A1.2) heralded attempts to relate physical features like urban **fabric** and structure to heat island morphology.

Parallel work on urban effects on humidity, wind, fog, precipitation and solar radiation were pursued, but the heat island remained the main attraction. Work was characterized by increased description of urban climates in the **urban canopy layer**, attempts to statistically relate distributions to features of the urban landscape and weather controls and initial attempts to probe above roof level. At the end of this period there was a growing literature on types of urban effects in different places but no physically based theory to link climate outcomes to landscape characteristics. Tropical cities, however, remained little studied until Jáuregui's pioneering work (Jáuregui, 1973; Roth, 2007)

Peter Summers        Alan Davenport

William Lowry        Stan Changnon Jr

Zhou Shuzhen        Ernesto Jáuregui

**Figure A1.3** Urban climatologists in the post 1965 period (Credit: Photos supplied by each individual).

## A1.3 Towards a Physical Climatology of Cities (1965 to 2000)

The choice of 1965 as a break-point in development is arbitrary, but the research at the end of the 1960s was markedly different from that at the beginning. This transformation is attributable to several factors including increased involvement by meteorologists (who applied their training in atmospheric physics)

and the emergence of a generation of physical climatologists. The latter were interested in going beyond description and statistics to a climatology rooted in general principles based on the **surface energy balance** and **surface water balance** framework – this is *physical climatology*. It was applied to agricultural, forest and marine environments as well as cities (e.g. Munn, 1966). The approach required new instrumentation capable of estimating the fluxes of energy, mass and **momentum** and relied on developments in electronic data capture and storage. Moreover, computer technology became more widely available and it was possible to check theories through **numerical modelling**.

This period was also marked by growing concern over the state of the environment. The plight of cities was foremost because air quality became unacceptably poor. These problems, created by poorly or unregulated industrial and domestic **emissions**, became common knowledge because of the shocking death toll of earlier air pollution disasters in London, United Kingdom (1952), and Donora, United States (1948). In many large cities like New York and Los Angeles, the poor air quality was persistent and palpable. This helped to generate interest in urban meteorology, especially the nature of urban winds, **turbulence**, the urban heat island and **mixed layer** depth, because of their significance in the transport and dispersion of **air pollutants**. Concerns covered the spectrum from impacts on pedestrians in vehicle-clogged streets up to the consequences of long-range transport downwind including deposition and impacts of acid rain. Increased research funding led to an outburst of publications at all scales. They included emphasis on basic physical processes in the urban atmosphere such as **radiation**, turbulence, **evaporation**, cloud microphysics and atmospheric chemistry.

A seminal contribution to the conduct of urban climatology was made by Lowry (1977; Figure A1.3) who forwarded a methodological prescription concerning the assessment of urban effects on climate and cautioned about the way errors can be incurred in urban climate analysis (Chapter 2). The notion of the urban atmosphere as a developing **boundary layer** was introduced by Summers in 1964 (Figure A1.3). It was given observational support by field projects that probed the vertical structure of temperature and air pollutants using airborne surveys and balloons. Large-scale observational campaigns were initiated including the Urban Air Pollution Dynamics Program in New York (Davidson, 1967), **METROMEX** in

St. Louis (Changnon et al., 1981; Figure A1.3) and the Complete Atmospheric Energetics Experiment (CAE-NEX) in Zaporozhye, USSR (Kondrat'yev, 1973). METROMEX deserves special mention because of the scale of the project and the number and range of experiments, many of which have never been repeated. It also brought new meteorological expertise to bear on urban climate questions including cloud physicists, statisticians, radar meteorologists and numerical modellers (Changnon et al., 1977 and Ackerman et al., 1978).

First steps towards modelling the processes responsible for urban climate effects include: the numerical heat island models of Summers (1964; Figure A1.3) and Myrup (1969); the scaled hardware model of Davis and Pearson (1970); the **wind tunnel** work on turbulence around buildings and the urban wind profile by Davenport (Figure A1.3) and; numerical models of radiation transfer in the polluted **urban boundary layer** (Atwater, 1971). These early modelling endeavours also deepened understanding of the heat island and its genesis (Arnfield, 2003). On the other hand, attempts to measure the budgets of energy, mass and momentum remained weak because of lack of appreciation about the role of scale and because instruments were often deployed with poor exposure and used inappropriate methods. In fact, it was not until the 1980s that reliable surface energy balance measurements were obtained.

Although considerable progress was being made in understanding the urban climate effect, much of this work was confined to universities and institutions in wealthy economies. Elsewhere, individuals ploughed a lonely furrow by conducting fundamental observational work in often difficult circumstances. Two particular examples come to mind because of their impact on the field. Chow Shuzhen (Figure A1.3) studied the urban climate of Shanghai and co-authored the first book on urban climate in China. Ernesto Jáuregui (Figure A1.3) did ground-breaking work on Mexico City and highlighted for many the profound environmental changes taking place there. Even in the wealthy economies, there were very few centres of urban climate research and progress was inextricably linked to the foresight of a few individuals.

A major impediment to the study of urban climate effects was how to cope with both the heterogeneity of the urban landscape and varying scales of urban effects. The absence of a framework meant that observational data and numerical simulations were often incompatible and the fundamental controls that regulate urban climate effects could not be isolated. Oke (1976) was the first to make explicit the distinction between the **canopy layer urban heat island** and **boundary layer urban heat island** and identify the urban **roughness sublayer**, which separated the scales of the governing processes. The use of the street canyon had already been used in air quality studies (e.g. Dabberdt et al., 1973) but it quickly became a focus of study in energy balance measurement (Nunez and Oke, 1977) and modelling (Arnfield, 1982). Moreover, a simple descriptor of street geometry (the ratio of building height to street width, $H/W$) was shown capable of summarising decades of canopy-level UHI research (Oke, 1981). This ratio captured the combination of thermal and radiative controls that were largely responsible for the magnitude of the UHI and flow regimes in canyons.

Identifying an equivalent model to represent types of urban landscapes that would allow representative observations above the urban canopy layer proved more difficult to resolve. While boundary-layer theory had been developed for extensive homogeneous surfaces, it was not clear whether these concepts could be applied to urban landscapes characterised by great heterogeneity. Turbulence and **turbulent flux** studies using towers with **eddy covariance** instruments mounted well above the roughness sublayer showed that spatial estimates of fluxes can be obtained reliably. Work by Roth and Oke (1993a and 1993b) and Schmid (1994) provided essential steps in answering this key question and showed that urban turbulence is very similar to that over more homogeneous extensive terrain. That work paved the way to recommend guidelines for the exposure of instrument systems in urban settings (Oke, 2004).

The work that emerged in this period established a solid foundation for the measurement and modelling of the key **turbulent** exchanges of momentum, heat, water vapour and air pollutants between the urban surface and the urban boundary layer. By the end of the period, observation and numerical models of urban climates were developing apace. Observational projects allowed coherent information on urban fluxes to be obtained so they could be linked to physical measures of the underlying surface and create robust **parameterizations** (Grimmond & Oke, 1995, 1999). Further, computer advances permitted development of more sophisticated models including incorporation of canopy-level processes in predictive boundary layer models that are forced by the output of weather forecast models (e.g. Masson, 2000). Overall there was an increase in work on meteorological processes, expansion of vision to include the entire boundary layer, and attempts to produce process-response models that link cause and effect.

## A1.4 Consolidation and Prediction (Since 2000)

At the end of the twentieth century the fields of urban climatology and meteorology blossomed into the form illustrated in this text. The community of scientists sensed this and the International Association for Urban Climate (IAUC) was formed. It is the first organization dedicated solely to furthering work in the field of urban climate and fostering cooperation between all interested scientists and practitioners; it now has over 1,000 members.

The descriptive phase of the subject is largely complete, except that modern measurement and logging systems continually increase the ability to probe more fully and rapidly and the international expansion of the field is providing insight into previously uncharted urban environments. There is consolidation of understanding phenomena and processes and the development of powerful models to simulate the workings and climates of urban atmospheres. The more cohesive community is now able and keen to pool resources. Even in one city, the urban climate spans several space and time scales which make it impossible for one research group to adequately sample comprehensively. The realization of this spawned several coordinated international field observation and modelling projects such as ESCOMPTE in Marseille, France (Mestayer et al., 2005), BUBBLE in Basel, Switzerland (Rotach et al., 2005), CAPITOUL in Toulouse, France (Masson et al., 2008), Joint URBAN 2003 in Oklahoma City, United States (Allwine et al., 2004), and EPiCC in Montreal and Vancouver, Canada. The valuable data generated by these field campaigns helped to develop and validate realistic urban systems within weather and climate models. That made it possible to compare the relative performance of different models against high-quality field observations (e.g. the International Urban Surface Energy Balance Model Comparison project, Grimmond et al., 2010).

The application of urban climate expertise is still young and under-developed relative to the basic science, but the basis of a predictive science is now largely in place. This has been aided by remarkable advances in technology that includes: the advent of computers able to deal with huge amounts of data, including distributed **sensor** systems and crowdsourcing of data, and can process these data at great speed and store it efficiently; the development of simulation models able to resolve the detailed effects of buildings on flow and radiation; the availability of geoinformation regarding many surface properties of whole cities and **remote sensing** and satellite systems that can sense the surface from above and the urban atmosphere from above or below. New forecast models enable national meteorological services to issue urban weather forecasts at less-than-whole city scales. Further, emergency measures organizations can obtain intra-urban predictions of pollutant **concentrations** due to release of toxic materials by industrial or transport accidents or terrorist activity. Urban hydrology, wind engineering and architecture similarly are able to access more short-term and small-scale weather information or use more sophisticated prediction models in support of better water management and planning. Strategies for **climate sensitive urban design** can now be supported with quantitative assessments and projections of likely outcomes and impacts.

# A2 | Site Codes and Data Sources

**Table A2.1** List of urban micrometeorological measurement sites/programmes referred from within tables and figures in this book. The dots in the highlighted columns indicate whether the site has published results on wind profiles and atmospheric turbulence (T), terms of the urban energy balance (E), greenhouse-gas fluxes (G), or aerosol and air pollutant fluxes (A) (Source: Modified and expanded based on the IAUC Urban Flux Network and Grimmond and Christen, 2012).

| Code | City, country site-name | Lon./lat. | Köppen climate zone | LCZ[a] | $\lambda_b$ (%) | $\lambda_v$ (%) | $\lambda_i$ (%) | $\lambda_{oth.}$ (%) | Land-use[b] | $z_H$ (m) | T | E | G | A | Measurement period | Selected references |
|---|---|---|---|---|---|---|---|---|---|---|---|---|---|---|---|---|
| Ar93, Ar94 | **Los Angeles**, United States ('Arcadia') | 118.050°W 34.133°N | Csb | 6 | 25 | 53 | 18 | 2 | R | 5.2 (±0.2) |  | • | • |  | 07/1993–08/ 1993, 07/1994 | Grimmond et al. (1996), Grimmond and Oke (2002) |
| Ba95 | **Basel**, Switzerland ('Basta') | 7.600°E 47.565°N | Cfb | 6 | - | - | - | - | R C I | ~24 (±6) | • |  |  |  | 07/1995–02/ 1996 | Feigenwinter et al. (1999) |
| Ba02sl | **Basel**, Switzerland ('Allschwil') | 7.562°E 47.556°N | Cfb | 6 | 28 | 53 | 19 | 0 | R | 7.5 | • | • |  |  | 06/2002–07/ 2002 | Christen and Vogt (2004) |
| Ba02ul | **Basel**, Switzerland ('Sperrstrasse') | 7.597°E 47.566°N | Cfb | 2 | 54 | 16 | 30 | 0 | R C | 14.6 (±6.9) | • | • | • |  | 06/2002–07/ 2002 | Christen and Vogt (2004), Vogt et al. (2006) |
| Ba02u2 | **Basel**, Switzerland ('Spalenring') | 7.576°E 47.555°N | Cfb | 2 | 37 | 31 | 32 | 0 | R C | 12.5 (±5.4) | • | • |  |  | 08/1994–09/ 2002 | Christen and Vogt (2004) |
| Ba02u3 | **Basel**, Switzerland ('Messe') | 7.601°E 47.563°N | Cfb | - | 100 | 0 | 0 | 0 | C O | 18.8 (±6.3) | • | • |  |  | 06/2002–07/ 2002 |  |
| Ba04 | **Basel**, Switzerland ('Klingelbergstrasse') | 7.580°E 47.562°N | Cfb | 2 | 38 | - | - | - | R S O | 13.8 | • | • | • | • | 01/2004–* | Lietzke and Vogt (2013) |
| Bj05 | **Beijing**, China ('IAP 325m tower') | 116.371°E 39.974°N | Dwa | 1 | - | - | - | - | R C S O | ~30 | • | • | • | • | 01/1991–* | Song and Wang (2012) |
| Bm02 | **Baltimore**, United States ('Cub Hill') | 76.521°W 39.413°N | Cfb | 6 | 16 | 68 | 15 | < 1 | R | 5.6 |  | • | • | • | 01/2002–* | Crawford et al. (2011) |
| Cc95 | **Christchurch**, New Zealand ('Beckenham') | 172.640°E 43.564°S | Cfb | 6 | - | - | - | - | R | N/A |  | • |  |  | 08/1995 | Spronken-Smith (2002) |
| Cc96 | **Christchurch**, New Zealand ('St Albans') | 172.647°E 43.521°S | Cfb | 6 | 22 | 56 | 22 | 0 | R C | 6.1 |  | • |  |  | 01/1996–02/ 1996, 07/ 1997–08/1997 |  |

Table A2.1 (cont.)

| Code | City, country site-name | Lon./lat. | Köppen climate zone | LCZ[a] | λ_b (%) | λ_v (%) | λ_i (%) | λ_oth. (%) | Land-use[b] | z_H (m) | T | E | G | A | Measurement period | Selected references |
|---|---|---|---|---|---|---|---|---|---|---|---|---|---|---|---|---|
| Ch92 | **Chicago,** United States ('Suburban') | 87.800°W 41.950°N | Dfa | 6 | 33 | 44 | 22 | 0 | R | 6.7 (±0.5) |  | • |  | • | 07/1992 | Grimmond and Oke (1995) |
| Ch95 | **Chicago,** United States ('Dunning') | 87.795°W 41.949°N | Dfa | 6 | 36 | 39 | 25 | 0 | R | 5.9 (±1.3) |  | • | • | • | 06/1995–08/1995 | Grimmond et al. (2002), Grimmond and Oke (2002) |
| Dn01 | **Denver,** United States ('South Denver') | 105.013°W 39.659°N | Dfb | 8 | 30 | 36 | 30 | 4 | R C I S O | 7 |  | • | • | • | 05/2001–09/2007, 01/2011–* |  |
| Ed00 | **Edinburgh,** United Kingdom ('Nelson Monument') | 3.183°W 55.954°N | Cfb | - | - | - | - | - | R C I | 10.8 |  | • | • |  | 10/2000–11/2000 | Nemitz et al. (2002) |
| Es07 [p/u] | **Essen,** Germany ('Grugapark') [c] | 6.993°E 51.431°N | Cfb | 5/B[c] | 59 / 12 | 22 / 52 | 19 / 29 | 0 | R C | 15 | • |  | • | • | 09/2006–11/2007 | Kordowski and Kuttler (2010), Weber and Kordowski (2010) |
| Fl05 | **Florence,** Italy ('Ximeniano') | 11.256°E 43.774°N | Cfb | 2 | 69 | 2 | 29 | 0 | R C | ~25 |  | • | • | • | 09/2005–12/2005 | Matese et al. (2009) |
| He05 | **Helsinki,** Finland ('Kumpula') | 24.961°E 60.203°N | Dfb | 2/B | 14 | 46 | 40 | 0 | R I | 20 |  | • | • | • | 12/2005–* | Vesala et al. (2008) |
| He10f | **Helsinki,** Finland ('Fire Station') | 24.945°E 60.165°N | Dfb | 2̃ | 57 | 7 | 35 | < 1 | C | 22 | • | • | • | • | 06/2010–01/2011 | Nordbo et al. (2013) |
| He10h | **Helsinki,** Finland ('Hotel Torni') | 24.939°E 60.168°N | Dfb | 2̃ | 55 | 3 | 42 | 0 | C | 24 | • | • | • | • | 09/2010–* |  |
| Kc04 | **Kansas City,** United States ('Greenwood') | 94.346°W 38.851°N | Dfa | 6̃ | 30 | 58 | 12 | 0 | R | 6.9 |  | • |  | • | 08/2004 | Balogun et al. (2009) |
| Ld09 | **London,** United Kingdom ('KSS') | 0.116°W 51.512°N | Cfb | 2 | 36 | 7 | 36 | 21 | S O | 22.8 |  | • |  | • | 10/2009–* |  |

| Code | Location | Lat/Long | Köppen | | | | | | | $z_H$ | | | | Period | Reference |
|---|---|---|---|---|---|---|---|---|---|---|---|---|---|---|---|
| Lo02c, Lo05l | **Lodz**, Poland ('CBD', 'Lipowa') | 19.447°E 51.762°N | Cfb | 2 | 31 | 30 | 39 | 0 | R S O | 10.6 | • | • | • | 11/2000–12/2002, 06/2002, 2006–09/2010 | Offerle et al. (2005b), Offerle et al. (2006), Pawlak et al. (2011), Fortuniak and Pawlak (2015) |
| Lo02i | **Lodz**, Poland ('Industrial') | 19.411°E 51.801°N | Cfb | 8 | 17 | 39 | 44 | 0 | I | 8.5 | | • | | 08/2002–09/2002 | Offerle et al. (2006) |
| Lo02s | **Lodz**, Poland ('Residential') | 19.456°E 51.809°N | Cfb | 6 | 10 | 78 | 13 | 0 | R | 7.7 | | • | | 08/2002 | Offerle et al. (2006) |
| Lo05n | **Lodz**, Poland ('Narutowicza') | 19.4809°E 51.7732°N | Cfb | 2 | 29 | - | - | - | R S O | 16 | | • | • | 06/2005–09/2010 | Fortuniak and Pawlak (2015) |
| Ma01 | **Marseille**, France ('CAA') | 5.379°E 43.290°N | Csa | 2 | 59 | 14 | 27 | 0 | R C | 15.6 | | • | • | 06/2001–07/2001 | Grimmond et al. (2004) |
| Mb03h | **Melbourne**, Australia ('high density') | 145.030°E 37.856°S | Cfb | 3 | 46 | 34 | 21 | 1 | R | | | • | | 12/2003–05/2004 | Coutts et al. (2007) |
| Mb03l | **Melbourne**, Australia ('low den.' or 'Surrey Hills') | 145.106°E 37.821°S | Cfb | 6 | 29 | 44 | 15 | 3 | R | 12 | • | • | | 03/2004–07/2004 | |
| Mb03m | **Melbourne**, Australia ('medium den.' or 'Preston') | 145.015°E 37.731°S | Cfb | 6 | 45 | 38 | 18 | 1 | R | 12 | • | • | | 08/2003–08/2004 | |
| Me93 | **Mexico City**, Mexico ('School of Mines') | 99.133°W 19.433°N | Cwb | 2 | 54 | 1 | 44 | 2 | R | 18.4 (±6.6) | | • | | 11/1993–12/1993 | Oke et al. (1999), Grimmond and Oke (2002) |
| Me98 | **Mexico City**, Mexico ('School No. 7') | 99.133°W 19.433°N | Cwb | 2 | - | <2 | - | - | - | - | | • | | 12/1998 | Tejeda-Martínez & Jauregui (2005) |
| Me03 | **Mexico City**, Mexico ('Iztapalapa') | 99.074°W 19.359°N | Cwb | 3 | 50 | 12 | 38 | 0 | R S O | 12 | • | | | 04/2003 | Velasco et al. (2005) |

Table A2.1 (cont.)

| Code | City, country site-name | Lon./lat. | Köppen climate zone | LCZ[a] | $\lambda_b$ (%) | $\lambda_v$ (%) | $\lambda_i$ (%) | $\lambda_{oth.}$ (%) | Land-use[b] | $z_H$ (m) | T | E | G | A | Measurement period | Selected references |
|---|---|---|---|---|---|---|---|---|---|---|---|---|---|---|---|---|
| Me06 | **Mexico City,** Mexico ('Escandon') | 99.176°W 19.404°N | Cwb | 2 | 57 | 6 | 37 | 0 | R C | 12 | | • | • | • | 03/2006, 02/2009–* | Velasco et al. (2011) |
| Mi95 | **Miami,** United States ('Dade county fairground') | 80.367°W 25.733°N | Am | 6 | 33 | 42 | 20 | 6 | R | 8.0 (±2.1) | | • | | | 05/1995–06/1995 | Grimmond and Oke (2002), Newton et al. (2007) |
| Mn05 | **Minneapolis,** United States ('Roseville', 'KUOM') | 93.188°W 44.998°N | Dfa | 6 | 10 | 77 | 22 | 2 | R | 6 | | • | • | • | 11/2005–06/2009 | Peters and McFadden (2012) |
| Mo05 | **Montreal,** Canada ('Rue Fabre') | 73.603°W 45.545°N | Dfb | 3 | 36 | 29 | 35 | 0 | R | 9.5 | | • | • | • | 03/2005–04/2005 | Lemonsu et al. (2008) |
| Mo07u | **Montreal,** Canada ('Rue des Ecores') | 73.592°W 45.547°N | Dfb | 3 | 27 | 29 | 44 | 0 | R | 8 | | • | • | • | 10/2007–10/2009 | Bergeron and Strachan (2011, 2012) |
| Mo07s | **Montreal,** Canada ('Roxboro') | 73.810°W 45.501°N | Dfb | 6 | 12 | 50 | 31 | 1 | R | 7 | | • | • | • | 11/2007–10/2009 | Bergeron and Strachan (2011, 2012) |
| Mu06 | **Münster,** Germany ('Kaserne') | 7.650°E 51.958°N | Cfb | 2 | 40 | 35 | 23 | 2 | R C | 15 | | • | • | • | 03/2011–06/2011, 08/2006 | Dahlkötter et al. (2010) |
| Mx01 | **Mexicali,** Mexico | 115.467°W 32.550°N | Bwh | - | - | - | - | - | - | - | | • | | | 03/2001 | Garcia-Cueto et al. (2003) |
| Na74 | **Nantes,** France ('CSTB') | - | Cfb | - | - | - | - | - | - | 10 | • | | | | 1974–1977 | Duchêne-Marullaz (1979), Roth (2000) |
| Ny05 | **New York,** United States ('Madison Square Garden') | 73.994°W 40.750°N | Cfa | 1 | - | - | - | - | R C O | 50–300 | • | | | | 03/2005 | Hanna et al. (2007) |

| | | | | | | | | | | | | | | |
|---|---|---|---|---|---|---|---|---|---|---|---|---|---|---|
| Ok03p | **Oklahoma City**, United States ('Park Avenue') | 97.515°W 35.469°N | Cfa | 1 | - | - | - | - | R C O | - | • | | | 06/2003–07/ 2003 | Nelson et al. (2007), Hanna et al. (2007) |
| Ou03 | **Ouagadougou**, Burkina Faso ('Residential') | 1.517°W 12.367°N | Bsh | - | ~30 | ~10 | < 10 | ~60 | R | N/A | | • | | 02/2003 | Offerle et al. (2005a) |
| Ob10s | **Oberhausen**, Germany ('Borbeck') | 6.910°E 51.494°N | Cfb | 9 | 16 | 53 | 26 | 5 | R | 6.2 | | • | | 08/2010–07/ 2011 | Goldbach and Kuttler (2013) |
| Ob10u | **Oberhausen**, Germany ('Falkensteinstrasse') | 6.866°E 51.476°N | Cfb | 2 | 48 | 17 | 35 | 0 | R C | 14 | | • | | 08/2010–07/ 2011 | Goldbach and Kuttler (2013) |
| Ph11 | **Phoenix**, United States ('West Phoenix') | 112.143°W 33.484°N | Bwh | 6 | 26 | 52 | 22 | 0 | R | 4.5 | | • | • | 12/2011–* | |
| Ro04 | **Rome**, Italy ('Collegio Romano') | 12.480°E 41.898°N | Csa | 2 | 60 | 5 | 35 | 0 | R C | 20 | | | • | 2004-2006 | |
| Sa92 | **Sapporo**, Japan ('Ishikari') | 141.244°E 43.136°N | Dfb | 3 | 25 | - | - | - | R | 7 | • | | | 07/1991, 11/ 1992 | Oikawa and Meng (1995) |
| Sc91u | **Sacramento**, United States ('Urban', Charles Peck Elementary School) | 121.300°W 38.390°N | Csa | 6 | 36 | 47 | 12 | 5 | R | 4.8 (±0.2) | | • | | 08/1991 | Grimmond et al (1993), Grimmond and Oke (1995), Grimmond and Oke (2002) |
| Sg94 | **Los Angeles**, United States ('San Gabriel') | 118.050°W 34.133°N | Csb | 6 | 29 | 37 | 31 | 0 | R | 4.7 | | • | | 07/1994 | Grimmond and Oke (1995), Grimmond and Oke (2002) |
| Sl76 (105) | **St. Louis**, United States ('Site 105') | ~90.204°W 38.600°N | Cfa | 8 | 25 | 15 | 60 | 0 | C I | 5.5 | • | • | | 07/1976–08/ 1976, 10/1976–11/ 1976 | Clarke et al. (1982), Roth (2000) |
| Sl76 (107) | **St. Louis**, United States ('Site 107') | ~90.243°W 38.668°N | Cfa | 3 | 25 | 60 | 15 | 0 | R | 9.7 | • | • | | 07/1976–08/ 1976, 10/1976–11/ 1976 | |

Table A2.1 (cont.)

| Code | City, country site-name | Lon./lat. | Köppen climate zone | LCZ[a] | $\lambda_b$ (%) | $\lambda_v$ (%) | $\lambda_i$ (%) | $\lambda_{oth.}$ (%) | Land-use[b] | $z_H$ (m) | T | E | G | A | Measurement period | Selected references |
|---|---|---|---|---|---|---|---|---|---|---|---|---|---|---|---|---|
| Sl76 (111) | **St. Louis,** United States ('Site 111') | ~90.287°W 38.581°N | Cfa | 6 | 15 | - | - | 0 | R I | 7.5 | • | | | | 07/1976–08/1976 | |
| Sl05 | **Salt Lake City,** United States ('Murray') | 111.916°W 40.648°N | Dfb | 6 | 17 | 49 | 16 | 18 | R | 4.4 | | • | • | • | 2005, 2007–* | Ramamurthy and Pardyjak (2011) |
| Sl06 | **Singapore,** Singapore ('Telok Kurau') | 103.911°E 1.314°N | Af | 3 | 39 | 12 | 49 | 0 | R | 9.1 | | • | • | • | 03/2006–03/2007, 09/2007–03/2008 | Velasco et al. (2013) |
| Sw11 | **Swindon,** United Kingdom ('BWY') | 1.798°W 51.585°N | Cfb | 6 | 19 | 40 | 40 | 1 | R | 5.5 | | • | • | • | 05/2011– | |
| Tk98 | **Tokyo,** Japan ('Setagaya') | 35.661°E 139.666°N | Cfa | 3 | 61 | 15 | 24 | 0 | R | 8.5 | • | | | | 10/1998 | Kanda et al. (2002) |
| Tk01 | **Tokyo,** Japan ('Kugahara') | 139.694°E 35.583°N | Cfa | 3 | 33 | 21 | 38 | 8 | R | 7.3 (±1.3) | • | • | • | | 2001–2007 | Moriwaki and Kanda (2004) |
| Tu90u | **Tucson,** United States ('Sam Hughes') | 110.933°W 32.117°N | Bsh | 6 | 23 | 18 | 42 | 17 | R | 5.2 (±0.8) | | • | | • | 06/1990 | Grimmond and Oke (1995), Grimmond and Oke (2002) |
| To04 | **Toulouse,** France ('Monoprix') | 1.446°E 43.604°N | Cfb | 2 | 54 | 8 | 38 | 0 | R C S O | 14.9 | • | • | • | • | 02/2004–02/2005 | Masson et al. (2008) |
| Up79g | **Uppsala,** Sweden ('Gränby') [c] | 17.650°E 59.866°N | Dfb | 2[c] | (c) | (c) | (c) | (c) | R O | 8 | • | | | | 1978–1979 | Högström et al. (1978, 1982), |

466

| Code | Location | Coordinates | LCZ | | | | | | Land-use | Height | | | Period | References |
|---|---|---|---|---|---|---|---|---|---|---|---|---|---|---|
| Up79u | **Uppsala**, Sweden ('Upplandia') | 17.633°E 59.850°N | Dfb | 5 | 51 | 0 | 32 | 17 | R C | 15 | • | | 1978–1979 | Alexandersson (1979), Roth (2000), Taesler (1980) |
| Va92i | **Vancouver**, Canada ('Light Industrial') | 123.106°W 49.267°N | Cfb | 8 | 51 | 5 | 44 | 0 | I | 5.8 (±0.1) | • | | 07/1992–08/1992 | Grimmond and Oke (1999) |
| Va92o Va08o | **Vancouver**, Canada ('Oakridge') | 123.133°W 49.231°N | Cfb | 6 | 25 | 55 | 20 | 0 | R | 5.8 | • | • | 07/2008–08/2008, 07/2009–08/2009 | Christen et al. (2009) |
| Va83s Va89s, Va92s, Va08s | **Vancouver**, Canada ('Sunset') | 123.078°W 49.226°N | Cfb | 6 | 29 | 34 | 35 | 2 | R C | 5.3m (±0.5) | • | • | Intermittent since 1978; 07/1989, 07/1992–09/1992, 2001–2002, 05/2008–* | Cleugh and Oke (1986); Roth and Oke (1993, 1995), Grimmond and Oke (2002), Christen et al. (2011) |
| Zu86 | **Zurich**, Switzerland ('Anwandstrasse') | 8.521°E 47.378°N | Cfb | 2 | - | - | - | - | R C | 18.3 | • | | 11/1986–05/1988 | Rotach (1993a, b), Rotach (1995) |

(a) Local Climate Zone (see Section 2.1.4).

(b) Land-use in the tower source area – R: residential, C: commercial, I: industrial, S: institutional, O: other.

(c) Tower on land discontinuity – cover fractions and height depend on wind direction.

Table A2.2 List of rural micrometeorological sites/programmes referred from within tables and figures in this book that have simultaneous measurements available in an adjacent city. The dots in the highlighted columns indicate whether the site has published results on wind profiles and atmospheric turbulence (T), terms of the urban energy balance (E), greenhouse-gas fluxes (G), or aerosol and air pollutant fluxes (A).

| Code | City, Country Site-Name | Lon. / Lat. | Köppen Climate Zone | LCZ [a] | T | E | G | A | Measurement period | Simultaneously operated urban sites [b] | Selected references |
|---|---|---|---|---|---|---|---|---|---|---|---|
| Ba02r-1 | **Basel**, Switzerland ('Grenzach') | 7.675°E 47.536°N | Cfb | D | | ● | | | C4/2002–07/2002 | Ba02u1, Ba02u2, Ba02u3, Ba02s | Christen and Vogt (2004) |
| Ba02r-2 | **Basel**, Switzerland ('Village Neuf') | 7.558°E 47.619°N | Cfb | D/F | ● | ● | | | 05/2002–07/2002 | Ba02u1, Ba02u2, Ba02u3, Ba02s | |
| Ba02r-3 | **Basel**, Switzerland ('Lange Erlen') | 7.649°E 47.592°N | Cfb | D | ● | ● | | | 1991–* | Ba02u1, Ba02u2, Ba02s | |
| Lo02r | **Lodz**, Poland ('Airport') | 19.400°E 51.723°N | Cfb | D | | ● | | | 08/2002 | Lo02c, Lo02s | Offerle et al. (2006) |
| Mb03r | **Melbourne**, Australia ('Lyndhurst') | 145.225°E 38.073°S | Cfb | D | | | | | 11/2003–05/2004 | Mb03l, Mb03m, Mb03h | Coutts et al. (2007) |
| Mo07r | **Montréal**, Canada ('Coteau-du-Lac') | 74.165°W 45.328°N | Dfb | D | | ● | ● | | 11/2007–03/2009 | Mo07u, Mo07s | Bergeron and Strachan (2012) |
| Sa91d | **Sacramento**, United States ('dry', scrub grassland) | 121.616°W 38.500°N | Csa | D | | ● | | | 08/1991 | Sa91u | Oke et al. (1998) |
| Sa91w | **Sacramento**, United States ('wet', sod farm) | 121.650°W 38.483°N | Csa | D | | ● | | | 08/1991 | Sa91u | |
| Tu90d | **Tucson**, United States ('dry', desert scrub) | 110.933°W 32.116°N | Bsh | C | | ● | | | 05/1990–06/1990 | Tu90u | |
| Va83r | **Vancouver**, Canada ('Airport') | 123.193°W 49.200°N | Cfb | C | | ● | | | 07/1983–09/1983 | Va83s | Cleugh and Oke (1986) |
| Va08r | **Vancouver**, Canada ('Westham Island') | 123.177°W 49.086°N | Cfb | C | | ● | ● | | 07/2007–09/2009 | Va08s, Va08o | Christen et al. (2009) |

[a] Local Climate Zone (see Section 2.1.4).
[b] See Table A2.1 for urban reference sites.

# A3 | Glossary and Acronyms

**ABL** – see **atmospheric boundary layer**.

**Absolute humidity** ($\rho_v$) – the ratio of the mass of water vapour to the volume occupied by the mixture. Same as **vapour density**. Not conservative with pressure and therefore not with expansion, so not useful if large altitude change is involved.

**Absorption** – process by which incident **radiant energy** is retained by a substance.

**Absorptivity** ($\varphi_\lambda$) – the relative fraction of the **irradiance** of a particular wavelength reaching a surface that undergoes **absorption**. Values for absorptivity range between 0 (no **absorption**) to 1 (**blackbody**). See also **reflectivity** and **transmissivity**.

**Accuracy** – the extent to which results of a calculation or the readings of an instrument approach the true values of the calculated or measured quantities. Compare with **precision**.

**Active surface** – the principal plane of climatic activity in a 3-D land surface such as an **urban canopy**. It is where the majority of **radiation** exchange, energy and mass transformation, precipitation **interception** and **drag** on airflow occur.

**Adaptation** – the process of adjustment to actual climate or expected climate change and its effects. In human systems, adaptation seeks to moderate or avoid harm or exploit beneficial opportunities.

**Advection** – primarily used to describe predominantly horizontal motion in the atmosphere due to the mean wind and associated transport by **turbulent fluxes** of heat, mass or **momentum**.

**Aerodynamic resistance** ($r_a$) – total **resistance** to **turbulent** transfer of entities between a level in the atmosphere and its source. It is a measure of the facility, or lack of it, for **eddies** to transport properties, it is related to the surface roughness and **dynamic stability**.

**Aerodynamic roughness length** – see **roughness length**.

**Aerosol** – a system in which solid or liquid particles are dispersed in air. Same as **particulate matter**.

**Air pollutant** – substance which, when present in the atmosphere, may harm human, animal, plant or microbial health, or damage infrastructure or ecosystems.

**Air pollution** – atmospheric condition where **air pollutants** are present in **concentrations** of concern to the health of humans or ecosystems or may damage infrastructure.

**Albedo** ($\alpha$) – the ratio of the **shortwave radiation** reflected by a surface (**reflectance**) to the shortwave radiation reaching that surface (**irradiance**). See also **reflectivity**.

**Anabatic wind** – upslope wind due to **buoyancy** and the **pressure gradient force** arising from stronger surface heating on slopes and ridges relative to a valley, typically develops during daytime. Opposite of a **katabatic wind**.

**Anemometer** – instrument to measure wind speed.

**Anisotropic** – physical properties have different values when measured in different directions. The opposite of **isotropic**.

**Anthropogenic** – caused by human activity.

**Anthropogenic heat flux** ($Q_F$) – heat directly released to the atmosphere as the result of human activities such as the **combustion** of fuels. Urban anthropogenic heat fluxes include those due to **urban metabolism** associated with space heating and cooling of buildings, vehicles, industrial processes and human and animal metabolism.

**Anthropogenic water vapour** ($F$) – water vapour formed chemically during **combustion** of fuels in air or fugitive release from boiling water and dehumidification inside buildings. A component of the urban **surface water balance**.

**Anticyclone** – large-scale atmospheric circulation system in which (in the Northern Hemisphere) the winds rotate clockwise. Synonymous with 'High' pressure system.

**Aspect ratio** – see **canyon aspect ratio**.

**Atmospheric boundary layer** (ABL) – the bottom layer of the **troposphere** that is in contact with the surface of Earth. It is often **turbulent** and capped by an **inversion** (statically **stable**) layer (the **entrainment zone**) of intermittent **turbulence**.

**Atmospheric window** – the relative gap in the **absorption** spectrum for atmospheric gases (between about 8 and 13 μm).

**Attenuation** – any process in which the **flux** density of a 'parallel beam' of energy decreases with increasing distance from the energy source. Also in which wind decreases with height in a canopy.

**Available energy** – portion of the **surface energy balance** available for turbulent **heat fluxes**.

**Barrier effect** – the diversion of airflow around the physical bulk of a city, rather like water around a rock in a stream.

**Base flow** – the part of stream discharge not due to direct **runoff** from **precipitation** or snow melt; it is usually sustained by **groundwater**.

**Blackbody** – (or full radiator) an hypothetical body which absorbs all the **radiation** striking it, i.e. allows no reflection or transmission.

**Blending height** ($z_r$) – height at which flow gradually changes from being influenced by individual **urban elements** in the **roughness sublayer** to being independent of horizontal position in the **inertial sublayer**.

**Boundary conditions** – conditions that need to be satisfied, e.g. at the edges of a **domain** in a **numerical model** or **physical model**.

**Boundary layer** – layer of air adjacent to a surface influenced by the properties of the boundary. See also **atmospheric boundary layer**.

**Boundary layer urban heat island** ($UHI_{UBL}$) – the relative warmth of air in the layer between the top of the **urban canopy layer** and the top of the **urban boundary layer**, and that at similar elevations in the **atmospheric boundary layer** of the surrounding rural region.

**Bowen ratio** ($\beta = Q_H/Q_E$) – ratio of the **sensible heat flux** density ($Q_H$) to latent heat flux density ($Q_E$).

**Brightness temperature** ($T_{0,B}$) – temperature a **blackbody** would possess in order to produce the **radiance** that is detected by a remote **sensor**.

**Buoyancy** – (or buoyant force) the upward force exerted upon a parcel of fluid by virtue of the density difference between itself and the surrounding fluid.

**Canyon** – see **urban canyon**.

**Canyon aspect ratio** ($\lambda_s$) – non-dimensional ratio ($\lambda_s = H/W$) of the height of the buildings ($H$) to the street width ($W$) in an **urban canyon**. $W$ is

defined by the horizontal distance between the principal urban elements along the canyon sides (e.g. houses, trees).

**Capping inversion** – statically **stable** layer at the top of the **atmospheric boundary layer**.

**Carbon dioxide** ($CO_2$) – colourless gas, molecular mass 44.01 g mol$^{-1}$. Most relevant anthropogenically enhanced **greenhouse gas** released in cities in the processes of **combustion** and **respiration**, and taken up in **photosynthesis**.

**Cavity zone** – zone immediately before and after an **urban element** (e.g. building, tree) where the wind and **turbulence** are significantly reduced.

**Chaotic flow** – flow régime over **neighbourhoods** characterized by tall and irregular **urban elements** (e.g. in **Local Climate Zone** 1), so flow cannot be generalized and since it depends on the specific configuration of the elements and atmospheric conditions.

**Canopy layer** – see **urban canopy layer**.

**Canopy layer urban heat island** ($UHI_{UCL}$) – the relative warmth of the air contained in the layer between the ground surface and roof level (the exterior **urban canopy layer**) and the corresponding height in the **rural** near-surface layer.

**Canopy resistance** ($r_c$) – the integrated leaf resistance for all leaves in a vegetated canopy. It regulates exchange between the leaves of a canopy and the canopy air volume and combines the stomatal resistance and the leaf's laminar boundary layer resistance $r_c$.

**Canyon trapping** – the increase in radiative absorptive 'efficiency' of an urban canyon of a given wavelength or band as **canyon aspect ratio** increases.

**CCN** – see **cloud condensation nuclei**.

**CDD** – see **cooling degree-day**.

**CFD** – see **computational fluid dynamics**.

**CG** – cloud-to-ground lightning.

**City-country wind system** – see **urban heat island circulation**.

**Climate-sensitive urban design** (CSUD) – tools that are used to improve climates in cities (**thermal comfort**, air quality, etc.) and to mitigate the undesirable climatic outcomes of urbanization.

**Climatope** – geographically coherent areas at the **local scale** that exhibit roughly the same climatic characteristics, delineated primarily by properties of the **topography** and the near-surface climate relevant to regional and urban planning and land use. Different from **Local Climate Zone**, which describes a universal scheme rather than one specific to an area and/or planning objective.

**Cloud** – a visible aggregate of fine liquid water droplets and or ice particles suspended in the air above the surface. Clouds form by **condensation** of water vapour on **cloud condensation nuclei** for water clouds or **ice nuclei** for ice crystal clouds.

**Cloud condensation nuclei** (CCN) – **hygroscopic aerosol** particles that act as the site for **condensation** of water vapour into **cloud** droplets.

**Combined sewers** – sewers that carry any type of 'waste' water for disposal. With low flow water is routed to a treatment facility designed to remove and dispose of solids. In large storms to avoid backing-up the sewer must discharge some untreated contents, including sanitary waste, to an adjacent water body.

**Combustion** – controlled burning (oxidation) of fuels in engines, industrial processes and for space heating, which results in **emissions** of primarily water vapour and **carbon dioxide**, but also **air pollutants** as byproducts.

**Complete aspect ratio** ($\lambda_c$) – ratio of the total three-dimensional surface area of all **urban elements** such as buildings and trees to the total plan area.

**Complete urban surface** – the true 3-D interface between all **facets** comprising a city and the atmosphere.

**Complete surface temperature** ($T_c$) – the area averaged **surface temperature** that considers the temperatures of all **facets** of the **complete urban surface.**

**Computational Fluid Dynamics** (CFD) – a **numerical modelling** approach that attempts to solve the complete set of nonlinear partial differential equations governing viscous fluid flow around objects (e.g. buildings). See also **large eddy simulation.**

**Concentration** ($\chi$) – mass of a substance, or number of a quantity, contained in unit volume.

**Condensation** – process by which vapour becomes a liquid.

**Condensation nuclei** – see **cloud condensation nuclei**.

**Conductance** – inverse of **resistance**.

**Conduction** – transfer of energy in a substance by means of molecular motions without any net external motion.

**Convection** – mass motions within a fluid resulting in transport and mixing of properties (e.g. energy and mass). Usually restricted to predominantly *vertical* motion in the atmosphere.

**Convergence** – opposite of **divergence**.

**Cool roof** – refers to a roof made of **fabrics** and **surface cover** that are designed to lower daytime **surface temperature** on roofs by increasing **reflectivity**. Most often cool roofs increase surface **albedo** by replacing dark-coloured roofs with light-coloured materials but it also includes materials that are highly reflective in the near infra-red. Compare to **green roof**.

**Cooling degree-days** (CDD) – a measure of the departure of the mean daily temperature above a given standard (typically a base of 25°C): one cooling degree-day is accumulated for each degree above the standard. It provides an estimate of the energy demand for cooling of buildings.

**Coriolis force** ($\vec{F}_{Co}$) – apparent force acting on objects moving in a frame of reference that is itself moving; in the atmosphere as a result of Earth's rotation. It deflects airflow to the right (left) of its intended path in the Northern (Southern) Hemisphere.

**Country breeze** – see **urban heat island circulation**.

**CSUD** – see **climate-sensitive urban design**

**Cyclone** – large-scale atmospheric circulation system in which (in the Northern Hemisphere) the winds rotate anti-clockwise. Synonymous with 'Low' pressure system.

**Deposition** – See **dry deposition** or **wet deposition**.

**Deposition velocity** ($v_d$) – in **dry deposition** the quotient of the flux of an entity to the surface and the **concentration** at a standard height.

**Detention storage** – water retained during a flood to be gradually released later, this may be in overflow ponds, wetlands or cisterns.

**Dewfall** – **condensation** of water from the lower atmosphere onto objects at or near the ground.

**Dewpoint (temperature)** ($T_d$) – temperature to which a parcel of air must be cooled (at constant pressure and constant water vapour content) in order for **saturation** to occur.

**Diffuse irradiance** ($D$) – **shortwave radiation** reaching Earth's surface after having been **scattered** from the **direct-beam** by molecules or other agents in the atmosphere. See also **direct-beam irradiance**.

**Diffusion** – exchange of properties and mass between regions in space by apparently random motions on a very small (usually molecular) scale.

**Diffusivity** – rate of **diffusion** of a property.

**Dimensional analysis** – a method of determining possible relationships between meteorological variables based on their dimensions. The results are often expressed using dimensionless groups. The relationships are determined empirically using field or laboratory results.

**Direct-beam irradiance** ($S$) – that portion of **shortwave radiation** received in a parallel beam 'directly' from the Sun. See also **diffuse irradiance**.

**Dissipation** – the conversion of the **turbulent kinetic energy** contained in **eddies** to **sensible heat** by work done against viscous stresses.

**Dispersion** – the spreading of atmospheric constituents such as **air pollutants**. Dispersion is produced by molecular **diffusion** and **turbulence**. Not the same as **advection** which is due to transport by the mean wind.

**Divergence** (div) – spreading apart of a **vector** field. Because of the predominance of horizontal motions in meteorology this usually refers to the two-dimensional horizontal **divergence** of the velocity field. Opposite of **convergence**.

**Domain** – the space (area or volume) or time represented or contained in a **numerical model**, **physical model** or data set.

**Dose** – amount of a substance (e.g. **air pollutant**) or influence (e.g. **radiation**) received by a subject. Magnitude of **exposure**.

**Downscaling** – a method that derives **local-** to **mesoscale** information from **macroscale** models or data analyses.

**Downwash** – the downward movement of **air pollutants** immediately in the lee of a smokestack, building or sharp break of slope caused by a **wake** and the **cavity** circulation.

**Drag** – retarding force that slows air flowing past stationary objects or surfaces. Can be separated into **skin drag** and **form drag**.

**Dry deposition** – process in which gaseous and **particulate matter** are transferred to a surface by turbulence and molecular **diffusion**. See also **wet deposition**.

**Dynamic stability** – the ratio between thermal production (or suppression) to mechanical production of **turbulence** at a given location and time. See difference to **static stability**.

**EC** – see **eddy covariance**.

**Ecological footprint** – the cumulative effect of human consumption and waste practices in terms of the ecologically productive area needed to sustain these activities. Not to be confused with the **footprint** of sensors.

**Eddy** – (1) by analogy with a molecule a 'glob' of fluid that has a certain life history of its own; (2) circulation in the lee of an **urban element** brought about by pressure irregularities.

**Eddy covariance** (EC) – a direct method to determine a vertical **turbulent flux**. Calculated as the average of the instantaneous product of vertical velocity and a **scalar** quantity such as a **concentration** (of heat, water vapour, **carbon dioxide** or **air pollutant**) or **momentum**.

**Ejection** – an event in a **turbulent** flow that transports a **momentum** deficit from a lower to a higher velocity region (typically upwards). Opposite to a **sweep**.

**Emission** – the process in which an object or surface actively releases **radiant energy**, heat or mass (e.g. **air pollutants**, water vapour).

**Emission factor** – relates the mass of an **air pollutant** released to the mass of fuel **combusted** (or alternatively to an activity such as vehicle distance driven).

**Emissivity** ($\varepsilon$) – ratio of the total **radiant energy** emitted per unit time per unit area of a surface at a specified wavelength and temperature to that of a **blackbody** under the same conditions.

**Emittance** – the radiant flux density a surface releases by **emission** of **photons** (in W m$^{-2}$).

**Energy balance** – see **surface energy balance**.

**Entrainment** – process of mixing of fluid across a density interface bounding a region of **turbulent** flow. Relatively quiescent fluid is engulfed by **eddies** able to cross the interface. Important transport process in the **entrainment zone**.

**Entrainment zone** (EZ) – the region at the top of the **mixed layer**, characterized by a strong **inversion**, where **turbulence** and **thermals** overshoot and **entrain** air from the **free atmosphere** down into the **atmospheric boundary layer**.

**Equivalent carbon dioxide emissions** ($CO_2$e) – for a given **emission** of a well-mixed **greenhouse gas** it expresses the comparable mass of **carbon dioxide** emission necessary to cause the same integrated **radiative forcing** on the global **troposphere** (over a given time horizon). Obtained by multiplying the mass emitted of a well-mixed greenhouse gas by its **global warming potential**.

**Eulerian** – looking at fluid motion past fixed points in space at different times.

**Evaporation** – (or vaporization) the process by which a liquid is transformed into a gas, in the atmosphere usually water changing to water vapour.

**Evapotranspiration** ($E$) – combined loss of water to the atmosphere by **evaporation** and **transpiration**.

**Excess resistance** ($r_b$) – the difference between the boundary layer **resistance** for **momentum** and other entities (heat, water vapour, etc.). The transfer efficiency of momentum is higher than for other entities because it includes bluff body effects, not just skin friction (see **drag**), therefore it has a lower resistance. This necessitates adding an extra resistance when considering the transfer of other entities.

**Exposure** – (1) description of openness of a site or instrument especially in relation to ability to obtain representative measurements free of extraneous influences; (2) amount of time a person is affected by a particular influence, for example, sunlight, air pollution, nuclear radiation.

**EZ** – see **entrainment zone**.

**FA** – see **free atmosphere**.

**Fabric** – the materials comprising **facets** and infrastructure in cities, for example, asphalt, concrete, stone, wood, soil, etc.

**Facet** – the smallest **urban unit**, usually considered to be made of homogeneous **fabric** and flat geometry (walls, roofs, leaves, road segments) that together form 3-D **urban elements** (trees, buildings).

**Fanning** – plume shape from a point source in a **stable** environment wherein **turbulent** mixing is suppressed, especially in the vertical. Plume contents remain concentrated with small vertical extent. Lateral changes of wind direction are not similarly constrained. If wind direction oscillates a thin 'fan'-shape is formed; with no change in wind direction plume is a pipe-like shape.

**Fetch** – distance from a measurement site or other place of interest, measured in the upwind direction.

**Field of view** (FOV) – the 3-D angle of view for a remote **sensor** that in combination with the distance to the target defines the area 'seen'.

**Flow separation** – A condition that occurs in a **turbulent** flow, as represented by the surface streamlines breaking away from the surface.

**Flux** – rate of flow of some quantity.

**Flux density** ($Q$) – the **flux** of any entity through unit surface area (e.g.: radiant flux density, **heat flux density**, mass flux density, molar flux density).

**Fog** – water droplets in suspension in air near the ground that reduces **visibility** to less than 1 km. Fog is a cloud on the ground whereas **clouds** have their base above the surface.

**Footprint** – the **source area** of a **sensor** that affects the **turbulent flux** or a signal mixed by **turbulence** measured above. It is an elliptical patch on the surface the location of which depends on the roughness of the surface, sensor height, wind direction, wind speed, and **dynamic stability**. Not to be confused with **ecological footprint**.

**Forcing** – any process which drives flows of heat, water vapour and **momentum** in a system. In the **atmospheric boundary layer** (ABL) the processes may be of external origin (e.g. solar energy, **precipitation**, **pressure gradient force**), or imposed by one layer upon another (e.g. **synoptic** flow upon the top of the ABL) or from the surface (e.g. **drag**, **heat flux densities**, **evaporation**, **air pollutant** emission).

**Form drag** – retarding force when wind is hitting an **urban element** protruding into the atmosphere. A consequence of flow separation and pressure differences around the elements. See also **skin drag**.

**FOV** – see **field of view**.

**Free atmosphere** (FA) – that portion of the **troposphere** above the **atmospheric boundary layer** in which the effects of friction are negligible.

**Friction velocity** ($u_*$) – a scaling velocity defined as the square root of the ratio of the **Reynolds stress** to the density of air. It is used to describe the effect of the **drag** force in the **surface layer**.

**Front** – interface or transition zone between two air masses of different density.

**Frontal aspect ratio** ($\lambda_f$) – ratio of the total area of urban elements (e.g. buildings and trees) projected in the direction of the mean wind to the total plan area.

**Fumigation** – process of sudden downward mixing of **air pollutants** from an overhead plume. This occurs when **convection** reaches up to a plume that was previously in **stable** air.

**GCC** – global climate change.

**General circulation model** (GCM) – a **numerical model** representing **macroscale** (global) processes in the atmosphere, ocean, cryosphere and land surface.

**GHG** – see **greenhouse gas**.

**Global warming potential** (GWP) – a factor that defines the **radiative forcing** per mass of a greenhouse gas relative to that of **carbon dioxide** when emitted into the **troposphere**.

**GPS** – Global Positioning System.

**Gradient wind** ($\vec{u}_g$) – wind resulting from a balance of the horizontal **pressure gradient force**, **Coriolis force**, and the centripetal acceleration due to motion in a curved path in a **cyclone** or **anticyclone**.

**Greenhouse effect** – the **longwave** radiative effect of **greenhouse gases** and **clouds** in the **troposphere** by absorbing outgoing longwave radiation from Earth and re-emitting it partially back to Earth's surface.

**Greenhouse gas** (GHG) – radiatively active gases present in trace amounts in the atmosphere that contribute to the **greenhouse effect**. Major GHGs are **carbon dioxide**, water vapour, ozone, methane, nitrous oxide, chlorofluorocarbons, etc.

**Green roof** – a roof with a vegetated **surface cover**. Extensive green roofs consist of a lightweight structure with a thin soil layer that supports short flowering plants. Intensive green roofs allow for deep rooted vegetation and need a deep (> 150 mm) growing substrate. Compare to **cool roof**.

**Groundwater** – subsurface water occupying the saturated zone below the level of the water table.

**GWP** – see **global warming potential**.

**Haze** – particles suspended in the air that reduce **visibility** by **scattering** light. Usually composed of smoke, dust and **photochemical smog**.

**HDD** – see **heating degree-day**.

**Heat capacity** ($C$) – amount of heat absorbed (or released) by unit volume of a system for a corresponding temperature rise (or fall) of 1 degree.

**Heat flux density** – rate of flow of **sensible heat** or **latent heat** (in Joules) per unit surface area ($J\ s^{-1}\ m^{-2} = W\ m^{-2}$).

**Heat island** – see **urban heat island**.

**Heat island magnitude** – the difference between the highest **urban** temperature and a representative temperature at the same elevation of the surrounding **rural** area, over a specified period.

**Heat storage** – the residual term in the **surface energy balance** wherein a layer or volume has a net gain (loss) of heat resulting in a temperature increase (decrease) depending on the **heat capacity** of the materials.

**Heating degree-day** (HDD) – a measure of the departure of the mean daily temperature below a

given standard (typically a base of 18°C): one heating degree-day is accumulated for each degree above the standard. It provides an estimate of the energy demand for heating of buildings.

**Heatwave** – a sustained period (typically several days to several weeks) that is abnormally and uncomfortably hot and usually humid caused by **macroscale** weather (e.g. persistent **anticyclones**).

**Homeostasis** – the ability of the human body to maintain a stable core temperature under varying environmental conditions using a range of thermoregulatory responses.

**Horizon screening** – obstruction created by objects (buildings, trees, hills) that block a portion of the horizontal horizon view of the sky hemisphere above a point on a surface.

**Hydrologic cycle** – the cycle in which **evaporation** on ocean and land surfaces injects water vapour into the **troposphere**, and via circulation and the formation of **clouds** this water vapour is transported, causing **precipitation** elsewhere, where it provides **runoff** on the land surface, causes **infiltration** into soils, discharges into streams and flows out into the oceans.

**Hydrologic unit** – see **urban hydrologic unit**.

**Hygrometer** – instrument that measures the water vapour content of the atmosphere.

**Hygroscopic** – having a marked ability to accelerate the **condensation** of water vapour.

**Hysteresis loop** – in micrometeorology rate-dependent hysteresis is common, especially in the **surface energy balance**. There is a dynamic lag between the phases of the input (radiative) and output (**conductive** and **turbulent**) fluxes. If input is plotted against output a 'loop' is formed. The effect disappears if the input changes are slower (e.g. annual).

**IBL** – see **internal boundary layer**.

**Ice nuclei** (IN) – particles that serve as nuclei leading to the formation of ice crystals in **clouds**.

**IAUC** – International Association for Urban Climate

**Impermeable** – a **surface cover** that does not permit the passage of fluids like water, i.e. it is sealed to **infiltration**.

**Inertial sublayer** (ISL) – the part of the **surface layer** above the **roughness sublayer** where **turbulence** dominated by **wind shear** creates a logarithmic velocity profile. In this layer **Monin-Obukhov similarity** can be used to describe profiles and

**turbulent fluxes** and the vertical variation of turbulent fluxes when the altitude is small ($< 5\%$).

**Infiltration** – the passage of water through the soil surface into the soil.

**Insulation** –the ability of fabric to resist the transfer of heat.

**Interception** – process whereby **precipitation** is trapped on vegetation and other elevated **urban elements** before reaching the ground.

**Internal boundary layer** (IBL) – a layer within the atmosphere bounded below by the surface and above by a discontinuity in properties. IBLs are initiated at a **leading edge** as a result of horizontal advection.

**Inversion** – a departure from the usual decrease or increase with height of an atmospheric property. Most commonly refers to a temperature inversion when temperatures increase, rather than lapse, with height, but can also be a moisture inversion.

**IPCC** – Intergovernmental Panel on Climate Change.

**Irradiance** – total radiant flux density received by a surface in a given wavelength band (in W m$^{-2}$). See also **reflectance**.

**ISL** – see **inertial sublayer**.

**Isolated roughness flow** – flow régime found over **neighbourhoods** with **urban elements** that are widely spaced, so that the flow is able to readjust to the ground surface in-between elements.

**Isotropic** – having the same physical properties in all directions. Processes subject to isotropic properties can include atmospheric **turbulence** and **radiation**. Its antonym is **anisotropic**.

**Jetting** – flow acceleration in constricted sections of an **urban canyon**.

**Kármán's constant** (*k*) – see **von Kármán's constant**.

**Katabatic wind** – wind blowing down an incline (slopes, valleys) due to the greater density of cold air, typically occurs during nighttime. Opposite of an **anabatic wind**.

**LBL** – see **laminar boundary layer**.

**Lagrangian** – viewing the path of an air parcel as it passes through space over time.

**Lake breeze** – see **sea breeze**

**Laminar boundary layer** (LBL) – the layer immediately next to a fixed boundary in which laminar flow prevails.

**Land breeze** – local wind circulation present during night due to thermal differences between the air over a warmer water body and the adjacent cooler land. Blows at low level from land to sea (or lake)

and in the opposite direction aloft (return flow). Opposite flow directions to those of a **sea breeze**.

**Lapse rate** – decrease of an atmospheric variable with height. Usually refers to temperature unless otherwise specified.

**Large Eddy Simulation** (LES) – a **3-D numerical model** of **turbulent** flow. Used to study flow over cities, in which larger **eddies** are resolved and the effects of smaller ones (subgrid-scale), which are more universal in nature, are parameterized.

**Latent heat** – heat released or absorbed by a system when it changes between solid, liquid and gas states. In contrast to **sensible heat**.

**Latent heat flux density** ($Q_E$) – the **heat flux density** of **latent heat** away or towards a surface, object or individual, usually in form of a **turbulent flux**. Used in contrast to **sensible heat flux density**.

**Leading edge** – the line of discontinuity between two surfaces with different surface properties (e.g. roughness, moisture). The edge initiates a new **internal boundary layer**.

**Leading edge effect** – abrupt changes to **fluxes** at a **leading edge** and subsequent adjustments to fluxes and **scalars** that occur downwind until a fully adjusted **boundary layer** is formed.

**LCZ** – see **local climate zone**.

**LES** – see **large eddy simulation**.

**Lidar** (Light Detection and Ranging) – an active **remote sensing** technique that uses laser beams. It can employ an airborne downward-directed scanning laser rangefinder to produce detailed and accurate surveys of the atmosphere or surface (e.g. **urban structure**), or a ground-based system sending laser beams upward and using the backscatter to infer **aerosols, clouds** and wind.

**Liquid water content** – mass of liquid water in a **cloud** in a specified volume of dry air (g m$^{-3}$) or mass of air (g kg$^{-1}$).

**Local Climate Zone** (LCZ) – a universal scheme to classify **urban** and **rural** areas at the scale of **neighbourhoods** based on typical controls (**surface cover, urban structure**, construction material, and human activity) that create spatially uniform thermal climates. Different from a **climatope** which describes a geographic area of uniform climate.

**Local scale** –includes atmospheric phenomena with horizontal dimensions from about 100 m to 50 km, with lifetimes of less than one day. Usually associated with topographic features, for example, land/sea, mountain/valley winds and **urban heat islands**.

**Longwave** – **radiation** emitted by objects at temperatures found in the Earth-Atmosphere system. Wavelengths are typically in the range 3–100 μm. Also referred to as infrared (IR).

**Looping** – plume form typical of highly **unstable** environments. Plume contents are transported up and down and sideways by **thermals** and disperse due to both thermal and mechanical **convection**.

**Macroclimate** – the climate imposed by **macroscale** processes such as synoptic systems and global circulation in absence of landscape or **microscale** effects (**orography**, surface cover etc.). Often categorized by the Koppen scheme into tropical (A); arid (B); mild temperate (C); continental temperate (D); and polar (E) macroclimates.

**Macroscale** – class of atmospheric phenomena with horizontal dimensions of hundreds to thousands of kilometres and a lifetime of weeks to decades. Includes continental to global features such as jet streams, weather systems, monsoons and global climate distributions.

**Mean radiant temperature** ($T_{MRT}$) – is the equivalent temperature of the environment to which a person is exposed that generates the same radiation gain to the body as **longwave** and net **shortwave** receipt from the natural environment.

**Mesoscale** – class of atmospheric phenomena with horizontal dimensions of about 20 to 200 km such as the **urban heat island**, **country breezes**, **sea breezes**, valley winds, squall lines and thunderstorms.

**Metabolic heat** – the heat generated by chemical reactions in the human body that convert food to energy. Its magnitude is closely associated with the level of human activity.

**Metadata** – information that increases knowledge of the content of data. For a climate station it means exact information regarding station number, geographic location, immediate surroundings, instruments, observation frequency, methods of processing and archiving the data. Literally means 'data about the data'.

**METROMEX** – Metropolitan Meteorological Experiment. A multi-institutional project carried out in and around St. Louis, United States mainly designed to study urban influences on clouds, precipitation and thunderstorms. Main field phases

from 1971–1976, analysis and modelling continued until early 1980s.

**Microclimate** – the climate caused by **microscale** processes and properties of a landscape.

**Microscale** – class of atmospheric phenomena with horizontal dimensions from millimetres to 1 km with lifetimes of minutes to hours; at the upper end overlap exists with **local scale**. Includes **surface layer** turbulence, flow around buildings and in **urban canyons**, and climates of **urban elements** (e.g. buildings), **neighbourhoods** and parks.

**Mitigation** – A human intervention to reduce the causing effect of a problem, for example, reduction of emissions of air pollutants (including **greenhouse gases**) that consequently reduce the hazard, **exposure** and/or **vulnerability** to global climate change or to severe air pollution, etc.

**Mixed layer** (ML) – layer of air usually capped by an **inversion** within which atmospheric properties are stirred and uniformly mixed by **turbulence**.

**Mixing ratio** – ratio of the mass of water vapour to the mass of dry air in a sample. Same as **specific humidity**. See also **molar mixing ratio** for **air pollutants**.

**ML** – see **mixed layer**.

**Molar mixing ratio** ($r$) – describes mole of an **air pollutant** or trace gas per mole air, typically expressed in parts per million (ppm, $10^{-6}$) or parts per billon (ppb, $10^{-9}$).

**Momentum** – property of a particle given by the product of its mass with its velocity.

**Momentum flux** – the vertical flux of horizontal **momentum**, equal to the force per unit area, or the stress. Can be achieved by viscosity (viscous shear stress) and/or turbulent motions (**Reynolds stress**).

**Monin-Obukhov similarity** (MOS) – a framework relating the vertical behaviour of non-dimensionalized properties (flow, turbulence) in the **inertial sublayer** as functions of a set of key parameters.

**Nadir** – point in the sphere surrounding an observer or **sensor** that lies directly below. Diametric opposite of **zenith**.

**Natural area** – a geographic location where human changes to ecosystems are not evident. Contrast with **urban** and **rural** areas.

**NBL** – see **nocturnal boundary layer**.

**Neighbourhood** – in the context of this book a fundamental **urban unit** at the **local-scale** consisting of many similar **urban blocks**, often characterized by a particular land use.

**Net allwave radiation** – the difference between the absorbed and emitted **radiation** (both **shortwave** and **longwave**). Commonly used to refer to the **surface radiation budget** and expressed as a radiant flux density.

**Net ecosystem productivity** (NEP) – the net accumulation of carbon in an ecosystem, calculated as **photosynthesis** minus **respiration**.

**Neutral** – An atmospheric state usually found with strong winds and little surface heating. Neutral **static stability applies** when vertical motion is neither enhanced nor suppressed due to **buoyancy**. Under neutral **dynamic stability**, mechanical production is the only significant source of **turbulence**. Contrast with **stable** and **unstable**.

**Nocturnal boundary layer** (NBL) – the cool **stable** layer of air that forms next to the ground at night.

**Numerical model** – a set of mathematical equations able to simulate atmospheric processes and links properties (e.g. air temperature) to processes (e.g. **sensible heat flux densities**) and permits quasi-controlled experiments. See also **physical model**.

**Oasis effect** – the enhanced rate of **evaporation** that occurs from surfaces that are wetter than their surroundings because of the continual delivery of warmer, drier air both laterally and from above (**subsidence**). Rates may be more than doubled.

**Orography** – The relief of a landscape including elevation and aspect, part of the **topography** of a place.

**PAH** – Polycyclic aromatic hydrocarbons, a class of **primary air pollutants** emitted during incomplete **combustion**.

**PAN** – Peroxyacetyl nitrate. A **secondary pollutant** important in **photochemical smog**.

**PAR** – see **photosynthetically active radiation**.

**Parameterization** – an approximation to nature. Usually involves replacing a fully physical equation with a relation that is an intuitively- or physically-reasonable approximation or surrogate.

**Particulate matter** (PM) – solid or liquid particles in the air. Same as **aerosol**.

**Permeable paving** – allows rainwater to **infiltrate** into the substrate rather than **runoff** the surface directly. Permeable paving systems include surfaces manufactured from porous material and built of blocks with interstitial spaces filled with porous materials. Permeable paving can serve many purposes, including lowering **surface temperature** and water control. They are an example of **water-sensitive urban design**.

**Perturbation** – any departure introduced into an assumed steady state of a system.

**PET** – see physiological equivalent temperature.

**pH** – a measure of the acidity or alkalinity of a solution: defined as the logarithm of the hydrogen-ion activity. For **cloud** droplets or **precipitation**, a pH of $> 7$ is alkaline, a pH of 7 is neutral, and $< 7$ is acid.

**Photochemical grid model** (PGM) – a **numerical model** that simulates dynamical and chemical processes in the **atmospheric boundary layer** using **emission** inventories of **primary pollutants** to forecast **photochemical smog**.

**Photochemical smog** – degraded air quality in the **urban boundary layer** due to **emissions** of nitrogen oxides and **volatile organic compounds**, primarily from the **combustion** exhaust of vehicles, that lead to the formation of **secondary pollutants** such as ozone.

**Photon** – the basic unit of electromagnetic **radiation**, and therefore of light. A discrete unit of energy.

**Photosynthesis** ($P$) – the biological process of carbohydrate formation from atmospheric **carbon dioxide** and water in the chlorophyll-containing tissues of living plants through the action of **photosynthetically active radiation**.

**Photosynthetically active radiation** (PAR) – **shortwave radiation** between wavelengths 0.4 and 0.7 μm (approximately same as visible radiation) that can be used by plants in the process of **photosynthesis**.

**Physical model** – a simplified surrogate of a real world system, typically using a scaled representation of the system that permits quasi-controlled experiments. In urban applications, **urban elements** are scaled and/or greatly simplified. The model is exposed to selected conditions in order to assess the climate effects. Also known as a hardware or scale model; see also **numerical model**.

**Physiological equivalent temperature** (PET) – describes the environment to which a person is exposed in terms of the air temperature that would be required in reference conditions as simulated by a physically based model (MEMI). These conditions correspond to a person located indoors, wearing office attire and engaged in light work.

**Plan area fraction** – fraction of the plan area of a particular **surface cover** type to the total plan area (viewed vertically). Typical examples include building ($\lambda_b$), vegetated ($\lambda_v$), impervious ground ($\lambda_i$) plan area fractions.

**PM** – see **particulate matter.**

**PM10** – **Concentration** of **particulate matter** less than 10 μm in diameter that can enter human lungs.

**PM2.5** – **Concentration** of **particulate matter** less than 2.5 μm in diameter that can enter the human blood stream.

**PMV** – see **predicted mean vote.**

**Pollutant** – see **air pollutant.**

**Potential temperature** ($\theta$) – temperature a parcel of dry air would have if brought adiabatically from its present altitude to a standard pressure of 100 kPa. An isothermal vertical **profile** of potential temperature is therefore statically **neutral**, a lapse is **unstable** and an **inversion** is **stable.**

**Precipitation** ($P$) – all liquid or solid water that originates in the atmosphere and falls to the Earth's surface.

**Precision** – quality of being exactly defined. A measurement with small random error is said to have high precision. Hence precision and accuracy are not the same.

**Predicted mean vote** (PMV) – is an assessment of the degree of thermal discomfort expressed in terms of a value from $-3$ (extreme cold) to $+3$ (extreme heat). These votes are predicted based on the physically based comfort model (Fanger) designed for indoor conditions in which clothing and activity levels are standardized.

**Pressure gradient force** ($\vec{F}_{pn}$) – the force imposed on air parcels due to **macroscale** pressure differences caused by **cyclones** and **anticyclones.**

**Primary pollutant** – **air pollutants** directly emitted by surface processes (**combustion, evaporation** etc.) into the air. In contrast to transformation that creates **secondary pollutants.**

**Profile** – graph of an atmospheric quantity versus a horizontal or vertical distance, or time scale.

**Radar** (Radio Detection and Ranging) – a **remote sensing** technique based on electromagnetic waves to detect distant objects in the atmosphere that scatter or reflect radio signals (e.g. water droplets, insect swarms, birds, balloons or metallic chaff).

**Radiance** – rate at which radiant energy in a set of directions confined to a unit solid angle is transferred across unit area of a surface projected onto this direction (W m$^{-2}$ sr$^{-1}$). Unlike **irradiance** it is a property solely of the **radiation** field not of the orientation of the surface.

**Radiant energy** – the energy of any type of electromagnetic **radiation.**

**Radiation** – process by which electromagnetic waves are propagated through free space by virtue of joint undulatory variations in the electric and magnetic fields.

**Radiative forcing** (RF) – change in **net radiation** (in W m$^{-2}$) at the tropopause following a **perturbation** of the climate system relative to its previous, undisturbed state. This radiative forcing metric is used to quantify the impact of increasing **greenhouse gas concentration** and land-cover changes on the global climate system.

**Radiometer** – any instrument that senses radiant fluxes (**shortwave, longwave** or allwave). Facing upwards instrument 'sees' down-welling fluxes, facing downwards it 'sees' **radiation** emitted and/or reflected from the surface. Instruments with two faces back-to-back record **net radiation** (short, long or allwave) across the plane of the faces.

**Rainout** – removal of **air pollutants**, by incorporation as **condensation nucleus**, into **cloud** droplets, which are later removed by rain or snow.

**RCP** – see **representative concentration pathways.**

**Reflectance** – the radiant flux density a surface returns by reflection (W m$^{-2}$). See also **irradiance.**

**Reflectivity** ($\omega_\lambda$) – the relative fraction of the **irradiance** of a particular wavelength reaching a surface that is reflected. See also **albedo.**

**Relative humidity** ($RH$) – ratio of the **vapour pressure** to the **saturation vapour pressure** at the same temperature. Usually reported as a percentage.

**Remote sensing** – obtaining information about the properties of an object without coming into physical contact with it.

**Representative concentration pathways** (RCP) – Trajectories that describe different possible changes of **greenhouse gas concentrations** in the **troposphere** in the twenty-first century.

**Residence time** – average length of time an **air pollutant** or a trace gas remains in the atmosphere or a system (cloud, chamber).

**Residual layer** (RL) – the middle part of the **nocturnal boundary layer**: above the **roughness sublayer** but below the **capping inversion.** It is characterized by weak or sporadic **turbulence** remaining from the previous day's **mixed layer.**

**Resilience** – the capacity of an urban (or any other) system to cope with hazards or disturbances, responding or reorganizing in ways that maintain essential functions while maintaining the capacity for **adaptation** and transformation.

**Resistance** – measure of the difficulty of transporting entities such as heat, water vapour and **momentum** in a system (air, water, soil, plant and animal tissue, building materials). Analogy between fluxes and electrons in an electrical circuit (Ohm's Law). Resistances are assigned to significant parts of a flux pathway, for example, soil, roots, leaf, **stoma, laminar boundary layer, surface layer** and **mixed layer**. Systems and layers have bulk resistance values, for example, canopy and aerodynamic resistances. Inverse of **conductance**.

**Respiration** ($R$) – metabolic processes of organisms that release **carbon dioxide** to the atmosphere.

**Return period** – the average time until the next occurrence of a defined event.

**Reynolds stress** – the forces (per unit area) imposed on the mean flow by **turbulent eddies** causing a turbulent **momentum flux**.

**RF** – see **radiative forcing**.

**Risk** – probability of reoccurrence (frequency) of a hazard at a place.

**RL** – see **residual layer**.

**Roughness length** ($z_0$) – (also aerodynamic roughness length) a measure of the roughness of a surface to airflow. The theoretical height above the **zero-plane displacement** at which the mean wind speed becomes zero when the logarithmic wind **profile** is extrapolated downward towards the surface. It can be approximated from the arrangement, spacing and height of the **urban elements**. Roughness lengths for heat ($z_{0H}$) and water vapour ($z_{0V}$) also exist that represent the height at which turbulent exchanges occur above a surface; these lengths are lower than that of $z_0$ representing a greater resistance to the transfer of heat and water vapour. See also **excess resistance**.

**Roughness sublayer** (RSL) – the air layer extending from the surface up to about two to five times the height of the **urban elements** (e.g. buildings, trees) that includes the **urban canopy layer**. Within the RSL flow is 3-D and governed by the size and shape of the elements. Surface scaling such as **Monin-Obukhov similarity** is not likely to apply.

**Runoff** ($R$) – water, derived from **precipitation**, that ultimately reaches stream channels as overland flow. Occurs when excess water from storms or snowmelt flows over the ground due to the soil being saturated or having low permeability (roofs and pavement).

**Rural** – an area where effects of human activity are present due to land management (e.g. agriculture, forestry), human impacts are detectable but not dominant. See also **urban** and **natural area**.

**Saturation** – (1) the condition in which **vapour pressure** is equal to the equilibrium **saturation vapour pressure** over a plane surface of pure liquid water and (2) the state of soil moisture when all available pore spaces are filled with water.

**Saturation vapour pressure** ($e^*$) – the **vapour pressure** at a given temperature at which the vapour in a sample of air is in equilibrium with a plane surface of pure water.

**Scalar** – any physical quantity that can be described by its magnitude alone. Temperature and water vapour are scalars. Contrast with **vector**.

**Scattering** – the process by which small particles, suspended in a medium of a different index of refraction, diffuse a portion of the incident **radiation** in all directions.

**Screen-level** – refers to the height of the instruments in a weather screen (approximately 1.5 m above ground).

**Sea breeze** (or **Lake breeze**) – local air circulation during day due to pressure differences between air over a cooler water body and adjacent warmer land. The breeze blows at lower levels from the sea (lake) to the land and in the opposite direction aloft (return flow). Opposite flow directions to those of a **land breeze**.

**Sea breeze front** – horizontal discontinuity in temperature and humidity that marks the leading edge of the intruding maritime (lake) air of the **sea breeze**. The front may also exhibit a shift in wind direction, pollution content and enhanced cumulus clouds.

**SEB** – see **surface energy balance**.

**Secondary pollutant** – **air pollutants** formed in the atmosphere as a result of chemical reactions. In contrast to a **primary pollutant**.

**Sensible heat** – that heat energy able to be sensed (e.g. with a thermometer). Used in contrast to **latent heat**. See also sensible heat flux density.

**Sensible heat flux density** ($Q_H$) – the **heat flux density** of **sensible heat** away or towards a surface, object or individual, usually in form of a **turbulent flux**. Used in contrast to **latent heat flux density**.

**Sensor** – the part of an instrument that converts an input signal into a quantity able to be measured by the instrument system.

**Shear zone** – line or narrow zone across which there is an abrupt change in horizontal wind component

**Shortwave** – **radiation** emitted by the Sun in the approximate waveband from 0.15 to 3 or 4 μm. Also referred to as solar radiation. Includes the ultra-violet (UV, 0.14–0.4 μm), visible and **photosynthetically active** (VIS and PAR, 0.4–0.7 μm) and the near infrared (NIR, 0.7–3 μm) radiation.

**Similarity** – see **Similitude**.

**Similitude** – a matching of **physical model** characteristics to that of the real world prototype through consideration of geometric, dynamic and thermal characteristics. Full similitude exists when fluxes of mass, momentum and energy are matched between the model and the real world prototype.

**Skimming flow** – flow régime found over **neighbourhoods** with **urban elements** (e.g. buildings) that are closely spaced, so the flow skims over the tops of the elements. Air in the cavities (e.g. **urban canyons**) between the elements is not well coupled with the flow above.

**Skin drag** – Retarding force when air is moving parallel to a smooth surface due to the friction. See also **form drag**.

**Sky view factor** ($\psi_{sky}$) – The ratio of the **radiation** received (or emitted) by a planar surface to the radiation emitted (or received) by the entire sky hemisphere.

**SL** – see **surface layer**.

**Smog** –degraded air quality at the urban to regional scale associated with a visually polluted **urban boundary layer** (see also **photochemical smog**).

**Sodar** (Sound Detection and Ranging) – **remote sensing** technique using sound waves to determine wind speed and wind direction in the **atmospheric boundary layer**.

**Solar altitude** ($\beta$) – vertical direction of the Sun above the horizon expressed in degrees.

**Solar azimuth** ($\Omega$) – horizontal direction of the Sun relative to a reference direction (usually true north) expressed in degrees.

**Solar zenith angle** ($Z$) – vertical direction of the Sun relative to the **zenith** expressed in degrees. The reciprocal of **solar altitude**.

**Source area** – the surface area 'seen' by a **sensor**. For a **radiometer** oriented with its receiving surface parallel to level ground the **radiation** source area is a fixed circular patch the dimensions of which depend on the sensor height and its field-of-view. For a sensor of properties transported from the source by **turbulent** motions the source area is also called a **footprint**. Turbulent source areas are elliptical and vary according to the wind direction, surface roughness, stability and instrument height. Both types of source area are complicated by the 3-D form of the urban surface.

**Specific heat** ($c$) – amount of heat absorbed (or released) by unit mass of a system for a corresponding temperature rise (or fall) of one degree.

**Specific humidity** ($q$) – ratio between the mass of water vapour and the mass of moist air; for most purposes it can be equated with **mixing ratio**.

**Stable** – an atmospheric situation in which vertical motions are suppressed and/or limited. Generally occurs when the surface is cooling (e.g. night). For **static stability**, the situation when any vertical motion is suppressed due to **buoyancy**. For **dynamic stability**, stable conditions occur when thermal production is negative and **turbulence** is consumed or completely inhibited. Contrast with **unstable**.

**Stability** – see **static stability** and **dynamic stability**.

**Static stability** – tendency of air parcels having been displaced vertically through the atmosphere. After forcible movement up or down, due to **buoyancy** a parcel has tendency to be **unstable**, **stable** or **neutral**. See also difference to **dynamic stability**.

**Stomata** – minute pores on the surface of leaves allowing them to exchange mass with the atmosphere (**carbon dioxide** in **photosynthesis** and water in **transpiration**).

**Storm sewer** – a pipe system to drain and transport excess rain and ground water from paved areas and roofs, often fed by street gutters and drains. Many systems drain storm water, untreated, into rivers or coastal waters (see **combined sewers**).

**Street canyon** – see **urban canyon**.

**Structure** – see **urban structure**.

**Sublimation** – transition of a substance directly from the solid to the vapour phase or vice versa.

**Subsidence** – slow sinking of air, usually associated with high pressure areas.

**Subsurface urban heat island** ($UHI_{sub}$) – relative warmth of the ground under a city, including the soils and subterranean built **fabric**, compared with that in the surrounding **rural** ground at similar depths. See also **urban heat island**.

**Suburban** – the predominantly residential areas around central cities consisting mainly of single and multi-family homes with scattered shopping, school and hospital services and some light industry. See also **urban** and **rural**.

**Supersaturation** – the condition when the **relative humidity** exceeds 100%, i.e. an air volume that contains more water vapour than is needed to saturate it with respect to a plane surface of pure liquid water or ice. This can occur if there is no water surface, or **cloud condensation nuclei** or other suitable surface upon which to condense.

**Surface cover** – type of materials covering a given **urban** or **rural** area, for example, vegetation, buildings, roads, water, etc. Often expressed as a **plan area fraction**.

**Surface energy balance** (SEB) – a statement of the conservation of energy applied to a given surface. The main components of the balance are the **net radiation**, **sensible heat flux density** and **latent heat flux density** between surface and atmosphere and the **conduction** of **sensible heat** into the substrate. Also includes net horizontal fluxes due to **advection** and in urban areas the **anthropogenic heat flux**. If it is a layer non-zero energy changes in the volume create a **heat storage** term.

**Surface layer** (SL) – the lowest layer of the **atmospheric boundary layer**, tens of metres in thickness, adjacent to the surface where mechanical (forced) generation of **turbulence** by **wind shear** exceeds buoyant (free) production by thermal instability.

**Surface radiation budget** – also called the **net radiation**. The net result of all streams of incoming and outgoing **radiation** at a surface. The incoming consists of **shortwave** radiation from the Sun and sky and that reflected from surrounding objects plus the **longwave** radiation from the sky and environment. The outgoing consists of the shortwave reflected from the surface due to its **albedo** and both the longwave emitted by the surface, due to its temperature and **emissivity**, and any incoming longwave reflected by the surface.

**Surface resistance** ($r_s$) – canopy average of all leaf stomata to the transfer of water and **carbon dioxide** acting in parallel. It is a measure of the physiological control exerted by all vegetative components of the surface.

**Surface temperature** ($T_0$) – the temperature at the interface between a surface and the air. The temperature of every surface is the result of a unique energy balance due to the combination of its radiative, conductive and turbulent fluxes. See also **brightness temperature**.

**Surface urban heat island** ($UHI_{Surf}$) – relative warmth of the surface-air interface of a city compared with the corresponding temperature of the surface-air interface of its **rural** surroundings. Ideally uses the **complete surface temperature** of the respective surfaces. See also **urban heat island**.

**Surface water balance** (SWB) – a statement of the conservation of mass for water applied to a portion of the landscape. It accounts for all water that passes through, or is stored within the volume of that portion of the landscape. In an urban landscape the inputs are **precipitation**, piped water and **anthropogenic water vapour** which are balanced by the outputs of **runoff**, **infiltration** and **evaporation**.

**Sweep** – an event in a **turbulent** flow that transports excess **momentum** from a higher to a lower velocity region (typically downwards). Opposite to an **ejection**.

**Synoptic** – refers to the use of meteorological data obtained simultaneously over a wide area to present a comprehensive and nearly instantaneous picture of the atmospheric state.

**TEB** – see **Town Energy Balance model**.

**Thermal** – A coherent parcel of warm air moving upward in the **atmospheric boundary layer** due to **buoyancy**. Rising current generated when the atmosphere is heated by the ground sufficiently to make it absolutely **unstable**.

**Thermal admittance** ($\mu$) – surface thermal property that governs the ease with which a body takes up or releases heat. It is the square root of the product of the **thermal conductivity** and **heat capacity**. Also called **thermal inertia**.

**Thermal anisotropy** – directional variation in the upwelling thermal **radiance** from a surface. Over urban areas it is created when a remote **sensor** views a biased selection of the **complete surface temperature** such that it varies with the position from which it is viewed.

**Thermal comfort** – environmental conditions where an individual feels no **thermal stress** and exhibits no sign of **thermal strain**. In these circumstances, the net heat gain/loss of the body will be close to zero.

**Thermal conductivity** ($k$) – physical property of a substance describing its ability to **conduct** heat by molecular motion.

**Thermal diffusivity** ($\kappa$) – the ratio of the **thermal conductivity** to the **heat capacity** of a substance. It determines the rate of heating due to a given temperature distribution in a given substance.

**Thermal inertia** – see **thermal admittance**.

**Thermal strain** – describes the effort required by the thermoregulatory functions, such as skin temperature and sweat rate, to offset the **thermal stress** to which it exposed.

**Thermal stress** – describes the net effect of the environmental conditions to which the body is exposed on thermoregulatory functions. Stress is generally at a minimum when an individual is comfortable, which depends on activity and clothing levels, in addition to the environmental conditions.

**Topography** – a detailed description of the natural and anthropogenic features of a landscape. Can include forests, rivers, roads, water bodies, **orography**, built-up features.

**Town Energy Balance model** (TEB) – A numerical **urban climate model**.

**Tracer** – a release of any chemical compound not found in a natural system that is conserved during **advection** and so can be used to track flow and **dispersion** in the atmosphere. Main requirement is that the **residence time** of the compound is substantially greater than the transport process under study.

**Trajectory** – the path an air parcel will follow over time.

**Transmissivity** ($\Psi_\lambda$) – the relative fraction of the **irradiance** of a particular wavelength reaching a layer of air that passes through without **absorption** or reflection. See also **reflectivity** and **absorptivity**.

**Transpiration** – process by which water in plants is transferred as water vapour to the air.

**Troposphere** – lowest 10–20 km of the Atmosphere, characterized by decreasing temperature with height, appreciable water vapour, vertical motion and weather.

**Turbulence** – the irregular fluctuations occurring in a turbulent flow, due to the superposition of **eddies** of various size and duration. Capable of transporting atmospheric properties by **convection** (heat, water vapour, **momentum**, etc.) at rates far in excess of molecular **diffusion**.

**Turbulence intensity** – measure of how important **turbulence** is relative to the mean wind, expressed as the ratio of the standard deviations of the $u$, $v$ and $w$ wind components to the mean wind speed.

**Turbulent** – state of fluid flow in which the instantaneous velocities exhibit irregular and apparently random fluctuations (**turbulence**) so that in practice only statistical properties can be recognized.

**Turbulent flux** – transport of a quantity by **eddy** motions; the covariance between a velocity component and any other property.

**Turbulent kinetic energy** (TKE) – the kinetic energy of the 3-D, random, **turbulent** fluctuations of the wind components in a flow per unit mass.

**UBL** – see **urban boundary layer**.

**UCL** – see **urban canopy layer**.

**UHI** – see **urban heat island**.

**UHIC** – see **urban heat island circulation**.

**Universal thermal climate index** (UTCI) – describes the environment to which a person is exposed in terms of an air temperature that would be required in reference conditions (an individual walking outdoors) as simulated by a physically based outdoor model (UTCI-Faisla).

**Unstable** – an atmospheric state characterized by strong vertical motions and **thermals**, generally occurring when the surface is being heated (e.g. daytime). The situation when vertical motion is enhanced and accelerated due to **buoyancy**. **Dynamic instability**, the condition when thermal production is significant source of **turbulence**. Contrast with **stable**.

**Urban** – the quasi-continuous built-up area of a human settlement including the highest building density that is usually found in the central area. See also **suburban** and **rural**.

**Urban block** – a fundamental **urban unit** at the **microscale** delineated by the principal road network. Urban blocks are comprised of a number of **urban elements** (buildings, trees, streets). Several urban blocks arranged in repetitive patterns form **neighbourhoods**.

**Urban boundary layer** (UBL) – **internal boundary layer** formed when an **atmospheric boundary layer** flows across a city. It is a **mesoscale** phenomenon the properties of which reflect the nature of the urban surface.

**Urban canopy** – the assemblage of **urban elements** (buildings, trees, streets) extending from the ground up to the mean height of the elements. Analogous to a vegetation canopy formed by a crop or forest.

**Urban canopy layer** (UCL) – the layer of air within the **urban canopy**. Its character is dominated by the **microscale** properties of the many surfaces present in the urban canopy. It consists of an exterior layer of air between the buildings and trees, and an interior portion consisting of the air inside the buildings.

**Urban canyon** (street canyon) – the characteristic geometric form created by a street with its three facets (the canyon floor that is usually a road, and the two sides that are usually the walls of the flanking buildings).

**Urban climate** – the climate of an area that is mainly affected by the presence of a city.

**Urban climate model** (UCM) – a physical or **numerical model** capable of simulating climatic conditions around buildings and in cities.

**Urban climatology** – the study of **urban climates**.

**Urban dome** –shape of the **urban boundary layer** (UBL) when regional winds are absent. The shape of the UBL is thinnest at the edge of the city and grows in depth towards the centre. Compare to **urban plume**.

**Urban ecosystem** – an open system formed by the biological population of organisms (vegetation, animals, people), the abiotic environment (atmosphere, hydrosphere, pedosphere, lithosphere) and the built environment of cities. It can be viewed from the perspective of a habitat for humans and other organisms, or in terms of flows of energy, water and chemical elements.

**Urban effects** – the modifications to the climate of an area due to the presence of a city.

**Urban element** (also more narrow 'roughness element') – the primary 3-D **urban unit**, the repetition of these creates **urban blocks** and forms the **urban canopy**. Urban elements can be buildings, trees, streets, etc. They are made up from smaller **urban units** referred to as **facets**.

**Urban fabric** – see **fabric**.

**Urban form** – the static, physical properties that shape a city, including the overall dimensions of a city,

and at finer scales its **urban structure, surface cover** and **fabric**.

**Urban function** – the underlying uses and processes that shape a city, including its land use, its economy, and the cycles of the **urban metabolism**, which operate over different time scales.

**Urban heat island** (UHI) – characteristic warmth of a city, often approximated by temperature differences between **urban** and **rural** areas. Separated into **substrate urban heat island, surface urban heat island, canopy layer urban heat island** and **boundary layer urban heat island**.

**Urban heat island circulation** (UHIC) – the **mesoscale** circulation system of thermal breezes generated primarily by a **boundary layer urban heat island** in weak or calm regional airflow. The UHIC resembles a **sea breeze** system however in the city low-level breezes converge towards the city centre resulting in uplift over the city, **divergence** aloft and **subsidence** over the surrounding **rural** or **natural area**. Unlike the **sea breeze**, it can occur by day or night. Also called 'city-country wind system'.

**Urban hydrologic unit** – are similar to **natural** catchment basins but because of they are imposed on the landscape they may possess multiple inlets and outlets (some natural, others anthropogenic). These units nest into a multi-level hierarchical drainage system to which an urban **surface water balance** can be applied, typically at the scale of a **neighbourhood** or an entire city.

**Urban metabolism** – the processes accompanying **urban functions** involving the intake of resources such as energy, water and raw materials and the release (**emissions**) of heat, vapour, **particulate matter, air pollutants, greenhouse gases** and other products.

**Urban moisture excess** (UME) – at night the spatial distribution of **vapour pressure** in the **urban canopy layer** often shows a maximum positive difference in the city centre. This pattern is similar to that of the **canopy layer urban heat island**.

**Urban plume** – at the upwind edge of a city a new **internal boundary layer (urban boundary layer)** forms. Similarly, when the urban air advects back across the downwind **rural** area it isolates a residual urban boundary layer aloft. The resulting layer of urban-modified air resembles the shape of a giant chimney plume, where the city is the chimney orifice.

**Urban structure** – the geometrical arrangement of **urban elements** (buildings, trees, etc.) in an **urban canopy**. The structure can be described by measures of the dimensions of the elements and the spaces between them (street widths, building spacing).

**Urban units** – the hierarchical scale sequence of physical features that when amalgamated resemble the **urban form**. The smallest units are **facets** (walls, roofs), several of which create **urban elements** (buildings, trees), which combine to form **urban blocks**, which together form **neighbourhoods** and many neighbourhoods represent a synthetic **city**.

**UTCI** – see **Universal Thermal Climate Index**.

**UV** – ultraviolet **radiation**. See **shortwave**.

**Vapour density** ($\rho_v$) – (same as **absolute humidity**) ratio of mass of water vapour to the volume occupied by the mixture.

**Vapour pressure** ($e$) – the partial pressure exerted by the water vapour molecules (or any other gaseous compound) of an air sample.

**Vasoconstriction** – is a **vasomotor control** of the human body that causes an decrease in the diameter of blood vessels close to the skin, which impedes the rate of blood flow from the heart and lowers skin temperature. Contrast to **vasodilation**.

**Vasodilation** – is a **vasomotor control** of the human body that causes an increase in the diameter of blood vessels close to the skin, which enhances the rate of blood flow from the heart and raises skin temperature. Contrast to **vasoconstriction**.

**Vasomotor control** – a thermoregulatory response that alters the flow of blood from the heart to the skin and thereby increases (**vasodilation**) or decreases (**vasoconstriction**) skin temperature in response to warming and cooling, respectively.

**Vector** – a quantity that has both magnitude and direction at each point in space. Contrast with **scalar**.

**Ventilation factor** ($V_f$) – the product of **mixed layer** height ($z_i$) and mean horizontal wind. It is used to summarize meteorological controls on **dispersion** of **air pollutants** in the **atmospheric boundary layer**.

**VIS** – visible **radiation**. See **shortwave**.

**View factor** ($\psi$) – a geometric ratio that expresses the fraction of the **radiation** output from one surface that is intercepted by another. It is a dimensionless number between zero and unity. See also **sky view factor**.

**Visibility** – (1) the greatest distance at which it is just possible to see and recognize objects with the unaided eye; (2) the clarity with which an object can be seen.

**Volatile organic compounds** (VOC) – are hydrocarbons which have a relatively high **vapour pressure** at ambient temperatures so they can occur in the vapour phase in the atmosphere. Easily vaporize from solvents, paints, industrial processes, released during fuel **combustion** and emitted biologically from plants.

**von Kármán's constant** ($k$) – A dimensionless universal constant (0.4) in the logarithmic wind profile equation for the **surface layer** that characterizes the dimensionless wind shear for statically **neutral** conditions.

**Vortex** – any circular, rotary or spiral flow. The motion of the fluid swirling rapidly around a centre.

**Vorticity** – a measure of local rotation in a flow.

**Vulnerability** – potential impact of hazards (or impact of a change in frequency of hazards).

**Wake** – zone in the lee of an obstacle (e.g. building) that extends for 10 to 25 times the height of the obstacle downstream, wherein the wind remains slowed and **turbulence** is increased.

**Wake interference flow** – the flow régime over **neighbourhoods** of **urban elements** that are moderately spaced, so the flow disturbed by one element cannot readjust before hitting the next element.

**Washout** – removal of **air pollutants** due to capture by falling **precipitation**.

**Water-sensitive urban design** (WSUD) – tools that are employed to manage water quality and quantity in urban areas. WSUD mimic the **hydrologic cycle** and hydrologic properties of **natural** water systems. These include **permeable paving**, **green roofs**, **detention storage** ponds, swales and floodplain management.

**Wet deposition** – removal of **air pollutants** by incorporation into cloud droplets, and also their capture by rain drops or snowflakes (see **rainout** and **washout**). Contrast to **dry deposition**.

**WHO** – World Health Organization

**Wind shear** – variation of direction of any component of the local wind **vector**.

**Wind tunnel** – a facility used by aerodynamicists and meteorologists to simulate the flow of air past test objects (e.g. buildings, trees, vehicles) under controlled conditions. Flow in the tunnel is generated by a fan. The test objects are scaled models of the full scale system of interest. Mean flow and **turbulence** can be measured at various points in the vicinity of the model.

**WMO** – World Meteorological Organization

**WSUD** – see **water-sensitive urban design**.

**Zenith** – point in the sphere surrounding an observer or a **sensor** that lies directly overhead. Diametric opposite of **nadir**.

**Zero-plane displacement** ($z_d$) – The average height of the momentum sink **urban canopy**. It is a theoretical height derived from the logarithmic wind profile. In **neutral** stability the logarithmic

wind **profile** becomes a straight line if the height axis is shifted up by the amount of the zero-plane displacement length.

**Zoning** – land use management by designating areas for particular uses. It may specify land cover and building characteristics associated with that use.

The entries in this Glossary are compiled from diverse sources and those created by the authors. By far the largest source is: American Meteorological Society, 2000: *Glossary of Meteorology*, 2nd edn., Boston, American Meteorological Society.

American Meteorological Society, 2016: *Glossary of Meteorology*, 2nd edn. Available online at http:// glossary.ametsoc.org/wiki/Main_Page.

# References

Ackerman, B., 1978: Regional kinematic fields. In: *Summary of METROMEX, Vol. 2: Causes of Precipitation Anomalies*, Ackerman, B. et al. (eds.), Bulletin 63, Illinois State Water Survey, Urbana, 165–209.

Ackerman, B., 1987: Climatology of Chicago area urban–rural differences in humidity. *Journal of Climate and Applied Meteorology*, **26**, 427–430.

Ackerman, B., S.A. Changnon, Jr., G. Dzurisin, et al. 1978: *Summary of METROMEX, Volume 2: Causes of Precipitation Anomalies*, Illinois State Water Survey, Urbana, Bulletin 63.

Adderley C., A. Christen, & J.A. Voogt, 2015: The effect of radiometer placement and view on inferred directional and hemispheric radiometric temperatures of an urban canopy. *Atmospheric Measurement Techniques*, **8**, 2699–2714.

Aida, M., 1982: Urban albedo as a function of the urban structure – a model experiment. *Boundary-Layer Meteorology*, **23**, 405–413.

Akan, O., & R.J. Houghtalen, 2003: *Urban Hydrology, Hydraulics, and Stormwater Quality*, Wiley, New York, 392 pp.

Akbari H., S. Davis, S. Dorsano, J. Huang, & S. Winnett, (eds.), 1992: *Cooling our Communities: A Guidebook on Tree Planting and Light-Colored Surfacing*. EPA.

Albrecht, F., 1933: Untersuchungen der vertikalen Luftzirkulation in der Großstadt. *Meteorologische Zeitschrift*, **50**, 93–98.

Alexander, P.J., & G. Mills, 2014: Local climate classification and Dublin's urban heat island. *Atmosphere*, **5**, 755–774.

Alexandersson, H., 1979: A statistical analysis of wind, wind profiles and gust ratios at Gränby, Uppsala. Report No. 55, Dept. Meteorology, Univ. Uppsala.

Allwine, K.J., 2004: Overview of Joint URBAN 2003-An atmospheric dispersion study in Oklahoma City. Eighth Symposium on Integrated Observing and Assimilation Systems for Atmosphere, Oceans, and Land Surface, and Symposium on Planning, Nowcasting, and Forecasting in the Urban Zone, AMS 84th Annual Meeting, Seattle, WA, January 10–16, Paper J7.1.

Allwine, K. J., J. H. Shinn, G. E. Streit, K. L. Clawson, & M. Brown, 2002: Overview of URBAN 2000: A multiscale field study of dispersion through an urban environment. *Bulletin of the American Meteorological Society*, **83**, 521–536.

Anandakumar, K., 1999: A study on the partition of net radiation into heat fluxes on a dry asphalt surface. *Atmospheric Environment*, **33**, 3911–3918.

Andres, R.J., T.A. Boden, & G. Marland, 2015: Annual fossil-fuel $CO_2$ emissions: Mass of emissions gridded by one degree latitude by one degree longitude. Carbon Dioxide Information Analysis Center, Oak Ridge National Laboratory, U.S. Department of Energy, Oak Ridge, TN.

Angell, J.K., Hoecker, W.H., Dickson, C.R., & Pack, D.H., 1973: Urban influence on a strong daytime airflow as determined from tetroon flights. *Journal of Applied Meteorology*, **12**, 924–936.

Arya, S.P., 2001: *Introduction to Micrometeorology*, 2nd edn., Academic Press, Orlando, 420 pp.

Arnfield, A.J., 1982: An approach to the estimation of the surface radiative properties and radiation budgets of cities. *Physical Geography*, **3**, 97–122.

Arnfield, A.J., 2003: Two decades of urban climate research: a review of turbulence, exchanges of energy and water, and the urban heat island. *International Journal of Climatology*, **23**, 1–26.

ASCE, 2004: Outdoor human comfort and its assessment. In: *ASCE State of the Art Report*, Irwin, P.A. (ed.), Prepared by a task group of the Aerodynamics Committee. American Society of Civil Engineers.

ASCE, 2011: *Urban Aerodynamics: Wind Engineering for Urban Planners and Designers, American Society of Civil Engineers*, Reston, VA, 76 pp.

Ashley, W.S., M.L. Bentley, & J.A. Stallins, 2012: Urban-induced thunderstorm modification in the Southeast United States. *Climatic Change*, **113**, 481–498.

ASHRAE, 2009: *ASHRAE Handbook: Fundamentals – IP Edition*. American Society of

Heating, Refrigerating and Air-Conditioning Engineers, Atlanta, GA.

Atkinson, W., 1912: *The Orientation of Buildings or Planning for Sunlight*, John Wiley & Sons.

Atwater, M.A., 1971: The radiation budget for polluted layers of the urban environment. *Journal of Applied Meteorology*, **10**, 205–214.

Baetens, R., B.P. Jelle, & A. Gustavsen, 2010: Phase change materials for building applications: a state-of-the-art review. *Energy and Buildings*, **42**, 1361–1368.

Bailey, W.G., T.R. Oke, & W.R. Rouse, 1997: *The Surface Climates of Canada*, McGill-Queen's Press, 400 pp.

Balchin, W.G.V., & N. Pye, 1947: A microclimatological investigation of Bath and the surrounding district. *Quarterly Journal of the Royal Meteorological Society*, **73**, 297–323.

Balling, R.C., R.S. Cerveny, & C.D. Idso, 2001: Does the urban $CO_2$ dome of Phoenix, Arizona contribute to its heat island? *Geophysical Research Letters*, **28**, 4599–4601.

Balogun, A.A., J.O. Adegoke, S. Vezhapparambu, M. Mauder, J.P. McFadden, & K. Gallo, 2009: Surface energy balance measurements above an exurban residential neighbourhood of Kansas City, Missouri. *Boundary-Layer Meteorology*, **133**, 299–321.

Beirle, S., K.F. Boersma, U. Platt, M.G. Lawrence, & T. Wagner, 2011: Megacity emissions and lifetimes of nitrogen oxides probed from Space. *Science*, **333**, 1737–1739.

Belcher, S.E., 2005: Mixing and transport in urban areas. *Philosophical Transactions of the Royal Society, Series A*, **363**, 2947–2968.

Bell, M. L., & D.L. Davis, 2001: Reassessment of the lethal London fog of 1952: Novel indicators of acute and chronic consequences of acute exposure to air pollution. *Environmental Health Perspectives*, **109**, 389–394.

Benson, L., K. Petersen, & J. Stein, 2007: Anasazi (pre-Columbian native-American) migrations during the middle-12th and late 13th centuries – were they drought induced? *Climatic Change*, **83**, 187–213.

Bergeron, O., & I.B. Strachan, 2011: $CO_2$ sources and sinks in urban and suburban areas of a northern mid-latitude city. *Atmospheric Environment*, **45**, 1564–1573.

Bergeron, O., & I.B. Strachan, 2012: Wintertime radiation and energy budget along an urbanization gradient in Montreal, Canada. *International Journal of Climatology*, **32**, 137–152.

Bertaud, A., 2003: Clearing the air in Atlanta: transit and smart growth or conventional economics? *Journal of Urban Economics*, **54**, 379–400.

Bitan, A., 1988: The methodology of applied climatology in planning and building. *Energy and Buildings* **11**, 1–10.

Bitan, A., & A.A. Rehamimoff, 1991: Bet She'an master plan – climatic rehabilitation of an ancient historic city. *Energy and Buildings*, **15**, 23–33.

Blake, D. R., & F.S. Rowland, 1995: Urban leakage of liquefied petroleum gas and its impact on air quality in Mexico City. *Science*, **269**, 953.

Bluyssen, P.M., E.D. Oliveira Fernandes, L. Groes, et al., 1996: European indoor air quality audit project in 56 office buildings. *Indoor Air*, **6**, 221–38.

Bodri, L., & V. Cermak, 2007: *Borehole Climatology: a New Method on How to Reconstruct Climate*, Elsevier, 1049 pp.

Bogner, J., R. Pipatti, S. Hashimoto, et al., 2008: Mitigation of global greenhouse gas emissions from waste: conclusions and strategies from the Intergovernmental Panel on Climate Change (IPCC) Fourth Assessment Report. Working Group III (Mitigation). *Waste Management & Research*, **26**, 11–32.

Bonan, G., 2002: *Ecological Climatology: Concepts and Applications*, Cambridge University Press, 563 pp.

Bornstein, R.D., 1968: Observations of the urban heat island effect in New York City. *Journal of Applied Meteorology*, **7**, 575–582.

Bornstein, R.D., 1975: The two-dimensional URBMET urban boundary layer model. *Journal of Applied Meteorology*, **14**, 1459–1477.

Bornstein, R.D., & Q. Lin, 2000: Urban heat islands and summertime convective thunderstorms in Atlanta: three case studies. *Atmospheric Environment*, **34**, 507–516.

Bornstein R.D., & W.T. Thompson, 1981: Effects of frictionally retarded sea breeze and synoptic frontal passages on sulfur dioxide concentrations in New York City. *Journal of Applied Meteorology*, **20**, 843–858.

Bosselmann P., E. Arens, K. Dunker, & R. Wright, 1995: Urban form and climate. *Journal of the American Planning Association*, **61**, 226–239.

Bottema, M.M., 1993. *Wind climate and urban geometry*. Doctoral dissertation, Technische Universiteit Eindhoven.

Bowler, D.E., L. Buyung-Ali, T.M. Knight, & A.S. Pullin, 2010: Urban greening to cool towns and cities: A systematic review of the empirical evidence. *Landscape and Urban Planning*, **97**, 147–155.

Braham, R.R. Jr., 1974: Cloud physics of urban weather modification – A preliminary report. *Bulletin of the American Meteorological Society*, **55**, 100–105.

Braham, R.R. Jr., 1981: Urban precipitation processes. In: *METROMEX: A Review and Summary*, Changnon, S.A. Jr. (ed.), Meteorological Monographs, **18**, 75–116.

Braham, R.R. Jr., & M.J. Dungey, 1978: A study of urban effects on radar first echoes. *Journal of Applied Meteorology*, **17**, 644–654.

Braham, R.R. Jr., & D. Wilson, 1978: Effects of St. Louis on convective cloud heights. *Journal of Applied Meteorology*, **17**, 587–592.

Braham, R.R. Jr., R.G. Semonin, A.H. Auer, S.A. Changnon Jr., & J.M. Hales, 1981: Summary of urban effects on clouds and rain. In: *METROMEX: A Review and Summary*, Changnon, S.A. Jr. (ed.), *Meteorological Monographs*, **18**, 141–152.

Bretz, S., H. Akbari, & A. Rosenfeld, 1998: Practical issues for using solar-reflective materials to mitigate urban heat islands. *Atmospheric Environment*, **32**, 95–101.

Brimblecombe, P., 2011: *The Big Smoke: A History of Air Pollution in London since Medieval Times*, 2nd edn., Routledge, London.

Brooks, C.E.P., 1952: Selective annotated bibliography on urban climate. *Meteorological Abstracts and Bibliography*, **3**, 734–773.

Brown, M. J., 2004: Urban dispersion-challenges for fast response modeling. Proceedings of the AMS 5th Symposium on Urban Environment, Vancouver, BC, Canada, 23–26 August 2004, Paper 5.1.

Brown, R.D., & T.J. Gillespie, 1995: *Microclimatic Landscape Design: Creating Thermal Comfort and Energy Efficiency*, John Wiley & Sons, 208 pp.

Bünzli, D., & H.P. Schmid, 1999: The influence of surface texture on regionally aggregated evaporation and energy partitioning, *Journal of the Atmospheric Sciences*, **55**, 961–972.

Campbell, G.S., & J.M. Norman, 1998: *An Introduction to Environmental Biophysics*, 2nd edn., Springer-Verlag, New York, 286 pp.

Carrió, G.G., & W.R. Cotton, 2011: Urban growth and aerosol effects on convection over Houston. Part II: Dependence of aerosol effects on instability. *Atmospheric Research*, **102**, 167–174.

Carrió, G.G., W.R. Cotton, & W.Y.Y. Cheng, 2010: Urban growth and aerosol effects on convection over Houston Part I: The August 2000 case. *Atmospheric Research*, **96**, 560–574.

Center for International Earth Science Information Network-CIESIN-Columbia University, 2015: Gridded Population of the World, Version 4 (GPWv4): Population Count. Palisades, NY: NASA Socioeconomic Data and Applications Center (SEDAC).

Čermák, J., J. Kučera, & N. Nadezhdina, 2004: Sap flow measurements with some thermodynamic methods, flow integration within trees and scaling up from sample trees to entire forest stands. *Trees*, **18**, 529–546.

Chandler, T.J., 1965: *The Climate of London*, Hutchinson, London, 292 pp.

Chandler, T.J., 1967: Absolute and relative humidities in towns. *Bulletin of the American Meteorological Society*, **48**, 394–99.

Chandler, T.J., 1970: Selected bibliography on urban climate, WMO Publ. No. 276, TP 155, World Meteorological Organization, Geneva.

Chandler, T.J., 1976: Urban Climatology and its Relevance to Urban Design, WMO Technical Note No. 149, WMO N. 438, World Meteorological Organization, Geneva, 61 pp.

Chandler, T.J., 1976: The climate of towns. In: *The Climate of the British Isles*, Chandler, T.J. and S. Gregory, (eds.), Longman, 307–329.

Chandler, T., & G. Fox, 1974: *3000 Years of Urban Growth*, Academic Press, London.

Changnon, S.A., 1976: Effects of urban areas and echo merging on radar echo behaviour. *Journal of Applied Meteorology*, **15**, 561–570.

Changnon, S.A., 1978: Urban effects on severe local storms at St. Louis. *Journal of Applied Meteorology*, **17**, 578–586.

Changnon, S.A., 1980: More on the La Porte anomaly: A review. *Bulletin of the American Meteorological Society*, **61**, 702–711.

Changnon, S.A., ed., 1981: METROMEX: A review and summary. *Meteorological Monographs*, **18**

(40), American Meteorological Society, Boston, 181pp.

Changnon, S.A., 1999: A rare long record of deep soil temperatures defines temporal temperature changes and an urban heat island. *Climatic Change*, **42**, 531–538.

Changnon, S.A., 2003: Urban modification of freezing-rain events. *Journal of Applied Meteorology*, **42**, 863–870.

Changnon, S.A., 2004: Urban effects on winter snowfall at Chicago and St. Louis. *Bulletin of the Illinois Geographical Society*, **46**, 3–14.

Changnon, S.A. Jr., F.A. Huff, P.T. Schickedanz, & J.L. Vogel, 1977: *Summary of METROMEX, Volume 1: Weather Anomalies and Impacts*: Illinois State Water Survey, *Urban Bulletin* **62**.

Changnon, S. A., R.G. Semonin, A.H. Auer Jr., R.R. Braham, & J. Hales, 1981: *METROMEX: A Review and Summary*. Monograph 18, American Meteorological Society, Boston, 81 pp.

Charpentier A., 2011: On the return period of the 2003 heat wave. *Climatic Change*, **109**, 245–260

Chow, W., & M. Roth, 2006: Temporal dynamics of the heat island of Singapore. *International Journal of Climatology*, **26**, 2243–60.

Christen, A., 2005: Atmospheric turbulence and surface energy exchange in urban environments, Stratus, **11**, Institut Meteorologie, Klimatologie und Fernerkundung, Univ. Basel, Basel, 140 pp.

Christen, A., 2014: Atmospheric measurement techniques to quantify greenhouse gas emissions from cities. *Urban Climate*, **10**, 241–260

Christen, A., & R. Vogt, 2004: Energy and radiation balance of a central European city. *International Journal of Climatology*, **24**, 1395–1421.

Christen, A., E. van Gorsel, & R. Vogt, 2007: Coherent structures in urban roughness sublayer turbulence. *International Journal of Climatology*, **27**, 1955–1968.

Christen, A., B. Crawford, N. Goodwin, et al., 2009: The EPiCC Vancouver Experiment-how do urban vegetation characteristics and garden irrigation control the local-scale energy balance? Proc. of the AMS 8th Symposium on the Urban Environment, Phoenix, AZ, January 11–15, Paper J9.1A

Christen, A., M.W. Rotach, R. Vogt, 2009: The budget of turbulent kinetic energy in the urban roughness sublayer. *Boundary-Layer Meteorology*, **131**, 193–222.

Christen, A., N.C. Coops, B.R. Crawford, et al., 2011: Validation of modeled carbon-dioxide emissions from an urban neighborhood with direct eddy-covariance measurements. *Atmospheric Environment*, **45**, 6057–6069.

Christen, A., F. Meier, & D. Scherer, 2012: High-frequency fluctuations of surface temperatures in an urban environment. *Theoretical Applied Climatology*, **108**, 301–24.

Christen, A., T.R. Oke, C.S.B. Grimmond, D.G. Steyn, & M. Roth, 2013: 35 years of urban climate research at the 'Vancouver-Sunset' flux tower. *FLUXNET Newsletter*, **35**, 29–39.

Churkina, G., 2008: Modeling the carbon cycle of urban systems. *Ecological Modelling*, **216**, 107–113.

Churkina, G., D.G. Brown, & G. Keoleian, 2010: Carbon stored in human settlements: the conterminous United States. *Global Change Biology*, **16**, 135–143.

Cionco, R.M., 1965: A mathematical model for air flow in a vegetative canopy. *Journal of Applied Meteorology*, **4**, 517–522.

Clancy, L., P. Goodman, H. Sinclair, & D.W. Dockery, 2002: Effect of air-pollution control on death rates in Dublin, Ireland: an intervention study. *Lancet*, **360**, 1210–14.

Clark, J.R., 1969: Thermal pollution and aquatic life. *Scientific American*, **220**, 19–27.

Clarke, J.F., 1969: The nocturnal urban boundary layer over Cincinnati, Ohio. *Monthly Weather Review*, **97**, 582–589.

Clarke, J.F., J.K.S. Ching, & J.M. Godowitch, 1982: *An Experimental Study of Turbulence in an Urban Environment*. Technical Report US EPA, Research Triangle Park, N.C., USA.

Cleugh, H.A., & C.S.B. Grimmond, 2012: Urban climates and global climate change. In: *The Future of the World's Climate*, Henderson-Sellers, A. and K. McGuffie, (eds.), Elsevier, Amsterdam, 47–76.

Cleugh, H.A., & T.R. Oke, 1986: Suburban-rural energy balance comparisons in summer for Vancouver. *Boundary-Layer Meteorology*, **36**, 351–369.

Coceal, O., T.G. Thomas, I.P. Castro, & S.E. Belcher, 2006: Mean flow and turbulence statistics over groups of urban-like cubical obstacles. *Boundary-Layer Meteorology*, **121**, 491–519.

Cook, D.G., D.P. Strachan, & I.M. Carey, 1998: Health effects of passive smoking. 9.

Parental smoking and spirometric indices in children. *Thorax*, **53**, 884–893.

Coquillat, S., M.-P. Boussaton, M. Buguet, D. Lambert, J.-F. Ribaud, & A. Berthelot, 2013: Lightning ground flash patterns over Paris area between 1992 and 2003: Influence of pollution? *Atmospheric Research*, **122**, 77–92.

Corsmeier, U., R. Behrendt, Ph. Drobinski, & Ch. Kottmeier, 2005: The mistral and its effect on air pollution transport and vertical mixing. *Atmospheric Research*, **74**, 275–302.

Counihan, J., 1975: Adiabatic atmospheric boundary layers: A review and analysis of data from the period 1880–1972. *Atmospheric Environment*, **9**, 871–905.

Coutts, A.M., J. Beringer, & N.J. Tapper, 2007: Impact of increasing urban density on local climate: Spatial and temporal variations in the surface energy balance in Melbourne, Australia. *Journal of Applied Meteorology and Climatology*, **46**, 477–493.

Coutts, A.M., N.J Tapper, J. Beringer, M. Loughnan, M. Demuzere, 2012: Watering our cities: the capacity for water sensitive urban design to support urban cooling and improve human thermal comfort in the Australian context. *Progress in Physical Geography*, **37**, 2–28.

Crawford, B.R., & A. Christen, 2014: Spatial source attribution of measured urban eddy covariance $CO_2$ fluxes. *Theoretical and Applied Climatology*, **119**, 733–755.

Crawford, B.R., C.S.B. Grimmond, & A. Christen, 2011: Five years of carbon dioxide fluxes measurements in a highly vegetated suburban area. *Atmospheric Environment*, **45**, 896–905.

Crevier, L.-P., & Y. Delage 2001: METRo: A new model for road-condition forecasting in Canada. *Journal of Applied Meteorology and Climatology*, **40**, 2026–2037.

Crutzen, P.J., 2004: New directions: the growing urban heat and pollution "island" effect—impact on chemistry and climate. *Atmospheric Environment*, **38**, 3539–3540.

Dabberdt, W.F., F.L. Ludwig, & W.B. Johnson, 1973: Validation and applications of an urban diffusion model for vehicular pollutants. *Atmospheric Environment*, **7**, 603–618.

Dahlkötter, F., F. Griessbaum, A. Schmidt, & O. Klemm, 2010: Direct measurement of $CO_2$ and particle emissions from an urban area. *Meteorologische Zeitschrift*, **19**, 565–575.

Daigo, M., & T. Nagao, 1972: *Urban Climatology*, Asakura-Shoten, Tokyo, 214 pp. [In Japanese]

Davenport, A.G., C.S.B. Grimmond, T.R. Oke, & J. Wieringa, 2000: Estimating the roughness of cities and scattered country. *12th Conference on Applied Climatology*, American Meteorological Society, Boston, 96–99.

Davidson, B., 1967: A summary of the New York urban air pollution dynamics research program. *Journal of the Air Pollution Control Association*, **17**, 154–158.

Davis, M.L., & J.E. Pearson, 1970: Modelling urban atmosphere temperature profiles. *Atmospheric Environment*, **4**, 277–288.

Demirel, Y., 2012: *Energy: Production, Conversion, Storage, Conservation, and Coupling*, Springer, 508 pp.

de Dear, R.J., E. Arens, Z. Hui, & M. Oguro, 1997: Convective and radiative heat transfer coefficients for individual human body segments. *International Journal of Biometeorology*, **40**, 141–156.

de Dear, R.J., & G.S. Brager, 2002: Thermal comfort in naturally ventilated buildings: revisions to ASHRAE Standard 55. *Energy and Buildings*, **34**, 549–561

DePaul, F.T. & C.M. Sheih, 1986: Measurements of wind velocities in a street canyon. *Atmospheric Environment*, **20**, 455–459.

Dirks, R.A., 1974: Urban atmosphere: warm dry envelope over St. Louis. *Journal of Geophysical Research*, **79**, 3473–3475.

Dixon, P.G., & T.L. Mote, 2003: Patterns and causes of Atlanta's urban heat island-initiated precipitation. *Journal of Applied Meteorology*, **42**, 1273–1284.

Djuricin, S., D.E. Pataki, & X. Xu, 2010: A comparison of tracer methods for quantifying $CO_2$ sources in an urban region. *Journal of Geophysical Research*, **115**, D11303.

Dou, J., Y. Wang, R. Bornstein, & S. Miao, 2015: Observed spatial characteristics of Beijing urban-climate impacts on summer thunderstorms. *Journal of Applied Meteorology*, **54**, 94–105.

Douglas, I., 1983: *The Urban Environment*, Edward Arnold, London, 229 pp.

Douglas, I., D. Goode, M. Houck, & W. Rusong (eds.), 2010: *The Routledge Handbook of Urban Ecology*, Routledge, London, 688 pp.

Draxler, R., 1986: Simulated and observed influence of the nocturnal urban heat island on the local wind field. *Journal of Applied Meteorology*, **25**, 1125–1133.

Duchêne-Marullaz, P., 1975: Full-scale measurements of atmospheric turbulence in a suburban area. *Proceedings 4th International Conference on Wind Effects on Buildings and Structures. Heathrow, England*, 23–31.

Duchêne-Marullaz, P., 1979: Effect of high roughness on the characteristics of turbulence in cases of strong winds. Preprint 5th International Conference on Wind Engineering, Fort Collins, paper II-8, 15 pp.

Duckworth, F.S., & J.S. Sandberg, 1954: The effect of cities upon horizontal and vertical temperature gradients. *Bulletin of the American Meteorological Society*, **35**, 198–207.

Dyer, A.J., & B.B. Hicks, 1970: Flux-gradient relationships in the constant flux layer. *Quarterly Journal of the Royal Meteorological Society*, **96**, 715–721.

Einarsson, E., & A.B. Lowe, 1955: *A Study of Horizontal Temperature Variations in the Winnipeg Area on Nights Favouring Radiational Cooling*. Meteorology Division, Department of Transport, CIR-2647, TEC 214, Toronto.

Emmanuel, R., H. Rosenlund, & E. Johansson, 2007: Urban shading – a design option for the tropics? A study in Colombo, Sri Lanka. *International Journal of Climatology*, **27**, 1995–2004.

EPA, 2004: *Report to Congress: Impacts and Control of CSOs and SSOs*, US Environmental Protection Agency, Washington, D.C., Document NO. EPA 833-R-04-001.

EPA, 2008: Reducing Urban Heat Islands: Compendium of Strategies. Available at www.epa .gov/sites/production/files/2014-06/documents/ basicscompendium.pdf

Epstein, Y., & D.S. Moran, 2006: Thermal comfort and the heat stress indices. *Industrial Health*, **44**, 388–398.

Erell, E., D. Pearlmutter, & T. Williamson, 2011: *Urban Microclimate: Designing the Spaces between Buildings*, Earthscan, London, 266 pp.

Estournel, C., R. Vehil, D. Guedalia, J. Fontan, & A. Druilhet, 1983: Observations and modeling of downward radiative fluxes (solar and infrared) in urban/rural areas. *Journal of Climate and Applied Meteorology*, **22**, 134–142.

Eugster, W.J., W.R. Rouse, R.A. Pielke, Sr., et al., 2000: Land-atmosphere energy exchange in Arctic tundra and boreal forest: available data and feedbacks to climate. *Global Change Biology*, **6**, (Suppl. 1), 84–115.

Feigenwinter, C., R. Vogt, & E. Parlow, 1999: Vertical structure of selected turbulence characteristics above an urban canopy. *Theoretical and Applied Climatology*, **62**, 51–63.

Feigenwinter, C., R. Vogt, A. Christen, 2012: Eddy covariance measurements over urban areas. In: *Eddy Covariance-A Practical Guide to Measurement and Data Analysis*, Aubinet M., T. Vesala, & D. Papale (eds.), Springer, 377–397.

Feingold, G., W. Cotton, U. Lohmann, & Z. Levin, 2009: Effects of pollution aerosol and biomass burning on clouds and precipitation: Numerical modeling studies. In: *Aerosol Pollution Impact on Precipitation*, Levin, Z., & W.R. Cotton (eds.), Springer-Science+Business Media B.V., 243–276.

Fenger, J., 1999: Urban air quality. *Atmospheric Environment*, **33**, 4877–4900.

Fenner, D., F. Meier, D. Scherer, & A. Polze, 2014: Spatial and temporal air temperature variability in Berlin, Germany, during the years 2001–2010. *Urban Climate*, **10**, 308–331.

Ferguson, G., & A.D. Woodbury, 2007: Urban heat island in the subsurface, *Geophysical Research Letters*, **34**, L23713.

Fezer, F., 1995: *Das Klima der Städte*, Justus Perthes Verlag Gotha.

Fiala, D., G. Havenith, P. Bröde, B. Kampmann, & G. Jendritzky, 2012: UTCI-Fiala multi-node model of human heat transfer and temperature regulation. *International Journal of Biometeorology*, **56**, 429–441.

Finn, D., K.L. Clawson, R.G. Carter, J.D. Rich, & K.J. Allwine, 2008: Plume dispersion anomalies in a nocturnal urban boundary layer in complex terrain. *Journal of Applied Meteorology and Climatology*, **47**, 2857–2878.

Flamant, C., P. Drobinski, M. Furger, et al., 2006: Föhn/cold-pool interactions in the Rhine valley during MAP IOP 15. *Quarterly Journal of the Royal Meteorological Society*, **132**, 3035–3058.

Folberth, G. A., S.T. Rumbold, W.J. Collins, & T.M. Butler, 2012: Global radiative forcing and megacities. *Urban Climate*, **1**, 4–19.

Font, A., C.S.B. Grimmond, S. Kotthaus, et al., 2015: Daytime $CO_2$ urban surface fluxes from airborne measurements, eddy-covariance observations and emissions inventory in Greater London. *Environmental Pollution*, **196**, 98–106.

Fortuniak, K., 2008: Numerical estimation of the effective albedo of an urban canyon. *Theoretical and Applied Climatology*, **91**, 245–258.

Fortuniak, K., Klysik, K., & J. Wibig, 2006: Urban-rural contrasts of meteorological parameters in Łødź, *Theoretical and Applied Climatology*, **84**, 91–101.

Fortuniak, K., & W. Pawlak, 2015: Selected spectral characteristics of turbulence over an urbanized area in the centre of Łødź, Poland. *Boundary-Layer Meteorology*, **154**, 137–156.

Freitas, E.D., C.M. Rozoff, W.R. Cotton, & P.L. Silva Dias, 2007: Interactions of an urban heat island and sea breeze circulations during winter over the metropolitan area of São Paulo, Brazil. *Boundary-Layer Meteorology*, **122**, 43–65.

Frey, C.M., 2010: *On the determination of the spatial energy balance of a megacity on the example of Cairo*, Egypt. PhD. dissertation, University of Basel, Switzerland.

Frey, C.M., & E. Parlow, 2012: Flux measurements in Cairo. Part 2: On the determination of the spatial radiation and energy balance using ASTER satellite data. *Remote Sensing*, **4**, 2635–2660.

Frey, C.M., G. Rigo, & E. Parlow, 2007: Urban radiation balance of two coastal cities in a hot and dry environment. *International Journal of Remote Sensing*, **28**, 2695–2712.

Fujibe, F., 2011: Urban warming in Japanese cities and its relation to climate change monitoring. *International Journal of Climatology*, **31**, 162–173.

Fujibe, F., & T. Asai, 1980: Some features of the surface wind system associated with the Tokyo heat island. *Journal of the Meteorological Society of Japan*, **58**, 149–152.

Gaffen, D., & R.D. Bornstein, 1988: Case study of urban interactions with a synoptic scale cold front. *Meteorology and Atmospheric Physics*, **38**, 185–194.

Gaffin, S.R., M. Imhoff, C. Rosenzweig, et al., 2012: Bright is the new black – multi-year performance of high-albedo roofs in an urban climate. *Environmental Research Letters*, **7**, 014029.

Ganeshan, M., R. Murtugudde, & M.L. Imhoff, 2013: A multi-city analysis of the UHI-influence on warm season rainfall. *Urban Climate* **6**, 1–23.

Garcia-Cueto, R., E. Jauregui, & A. Tejeda, 2003: Urban/rural energy balance observations in a desert city in northern Mexico. Proceedings Fifth International Conference on Urban Climate, ICUC-5, Lodz, Poland, 1 – 5 September 2003.

Garratt, J.R., 1992: *The Atmospheric Boundary Layer*, Cambridge University Press, Cambridge, 316 pp.

Garratt, J.R., & R.A. Brost, 1981: Radiative cooling effects within and above the nocturnal boundary layer. *Journal of the Atmospheric Sciences*, **38**, 2730–2746.

Garratt, J.R., & M. Segal, 1988: On the contribution of atmospheric moisture to dew formation. *Boundary-Layer Meteorology*, **45**, 209–236.

Gauthier, W. A., L.D. Petersen, L.D. Carey, & H.R. Christian Jr., 2006: Relationship between cloud-to-ground lightning and precipitation ice mass: A radar study over Houston. *Geophysical Research Letters*, **33**, L20803.

Gioli, B., P. Toscano, E. Lugato, et al., 2012: Methane and carbon dioxide fluxes and source partitioning in urban areas: The case study of Florence, Italy. *Environmental Pollution*, **164**, 125–131.

Giometto, M.G., A. Christen, C. Meneveau, J. Fang, M. Krafczyk, & M.B. Parlange, 2016: Spatial characteristics of roughness sublayer mean flow and turbulence over a realistic urban surface. *Boundary–Layer Meteorology*, doi:10.1007/s10546-016-0157-6.

Girardet, H., 1999: Sustainable cities: A contradiction in terms? In: *The Earthscan Reader in Sustainable Cities*, Satterthwaite D., (ed.), Routledge, London, 413–425.

Givoni, B., 1998: *Climate Considerations in Building and Urban Design*, Van Nostrand Reinhold, New York, 464 pp.

Goldbach, A., & W. Kuttler, 2013: Quantification of turbulent heat fluxes for adaptation strategies within urban planning. *International Journal of Climatology*, **33**, 143–159.

Golubiewski, N. E., 2006: Urbanization increases grassland carbon pools: Effects of landscaping in Colorado's front range. *Ecological Applications*, **16**, 555–571.

Grace, J., 1983: *Plant-Atmosphere Relationships*. Chapman and Hall, London, 92 pp.

Grimmond C.S.B, & A. Christen, 2012: Flux measurements in urban ecosystems. *FluxLetter-Newsletter of Fluxnet*, **5**(1), 1–7.

Grimmond, C.S.B., & T.R. Oke, 1986: Urban water balance: 2. Results from a suburb of Vancouver, British Columbia. *Water Resources Research*, **22**, 1404–1412.

Grimmond, C.S.B., & T.R. Oke, 1995: Comparison of heat fluxes from summertime observations in the suburbs of four North American cities. *Journal of Applied Meteorology*, **34**, 873–889.

Grimmond, C.S.B., & T.R. Oke, 1999a: Aerodynamic properties of urban areas derived from analysis of surface form. *Journal of Applied Meteorology*, **38**, 1262–1292.

Grimmond, C.S.B., & T.R. Oke, 1999b: Heat storage in urban areas: Local scale observations and evaluation of a simple model. *Journal of Applied Meteorology*, **38**, 922–940.

Grimmond, C.S.B. & T.R. Oke, 2002: Turbulent heat fluxes in urban areas: Observations and a local-scale urban meteorological parameterization scheme (LUMPS). *Journal of Applied Meteorology*, **41**, 792–810.

Grimmond, C.S.B., H.A. Cleugh, & T.R. Oke, 1991: An objective urban heat storage model and its comparison with other schemes. *Atmospheric Environment*, **25B**, 311–326.

Grimmond, C.S.B., T.R. Oke, & H.A. Cleugh, 1993: The role of "rural" in comparisons of observed suburban-rural flux differences. Proceedings of the Yokohama Symposium on 'Exchange processes at the land surface for a range of space and time scales', IAHS Publication, **212**, 1–10.

Grimmond, C.S.B., C. Souch, & M.D. Hubble, 1996: Influence of tree cover on summertime surface energy balance fluxes, San Gabriel Valley, Los Angeles. *Climate Research*, **6**, 45 – 57.

Grimmond, C.S.B., J.A. Salmond, T.R. Oke, B.D. Offerle, & A. Lemonsu, 2004: Flux and turbulence measurements at a densely built-up site in Marseille: Heat, mass (water and carbon dioxide), and momentum. *Journal of Geophysical Research*, **109**, D24101.

Grimmond, C.S.B., T. King, F. Cropley, D.J. Nowak, & C. Souch, 2002: Local-scale fluxes of carbon dioxide in urban environments: methodological challenges and results from Chicago. *Environmental Pollution*, **116**, S243–S254.

Grimmond, C.S.B., M. Blackett, M. Best, et al., 2010: The International Urban Energy Balance Models Comparison Project: First results from Phase 1.

*Journal of Applied Meteorology and Climatology*, **49**, 1268–1292.

Gromke, C., & B. Ruck, 2008: Aerodynamic modelling of trees for small-scale wind tunnel studies. *Forestry*, **81**, 243–258.

Gubareff, G.G., J.E. Janssen, & R.H. Torberg, 1960: *Thermal Radiation Properties Survey*, Honeywell Research Center, Minneapolis, MN.

Guenther, A.B., P.R. Zimmerman, P.C. Harley, R.K. Monson, & R. Fall, 1993: Isoprene and monoterpene emission rate variability: Model evaluations and sensitivity analyses. *Journal of Geophysical Research*, **98(D7)**, 12609–12617.

Gultepe, I., R. Tardif, S.C. Michaelides, et al., 2007: Fog research: A review of past achievements and future perspectives. *Pure and Applied Geophysics*, **165**, 1121–1159.

Guo, J., K.A. Rahn, & G. Zhuang, 2004: A mechanism for the increase of pollution elements in dust storms in Beijing. *Atmospheric Environment*, **38**, 855–862.

Guttikunda, S. K., G.R. Carmichael, G. Calori, C. Eck, & J.-H. Woo, 2003: The contribution of megacities to regional sulfur pollution in Asia. *Atmospheric Environment*, **37**, 11–22.

Haberlie, A.M., W.S. Ashley, & T.J. Pingel, 2015: The effect of urbanisation on the climatology of thunderstorm initiation. *Quarterly Journal of the Royal Meteorological Society*, **141**, 663–675.

Haeger-Eugensson, M., & B. Bolmer, 1999: Advection caused by the urban heat island circulation as a regulating factor on the nocturnal urban heat island. *International Journal of Climatology*, **19**, 975–988.

Hage, K.D., 1975: Urban-rural humidity differences. *Journal of Applied Meteorology*, **14**, 1277–83.

Halawaa, E., & J. van Hoof, 2012: The adaptive approach to thermal comfort: A critical overview. *Energy and Buildings*, **51**, 101–110.

Hall, D.J., R. Macdonald, S. Walker, I. Mavroidis, H. Higson, & R.F. Griffiths, 1997: *Visualisation Studies of Flows in Simulated Urban Arrays, BRE Client Report CR39/97*, Building Research Establishment, UK, 14 pp.

Hallegatte, S., N. Ranger, S. Bhattacharya, et al., 2010: Flood Risks, Climate Change Impacts and Adaptation Benefits in Mumbai: An Initial Assessment of Socio-Economic Consequences of Present and Climate Change Induced Flood Risks and of Possible Adaptation Options, OECD

Environment Working Papers, No. 27, OECD Publishing.

Han, J.-Y., & J.-J. Baik, 2008: A theoretical and numerical study of urban heat island–induced circulation and convection. *Journal of the Atmospheric Sciences*, **65**, 1859–1877.

Han, J.-Y., J.-J. Baik, & A.P. Khain, 2012: A numerical study of urban aerosol impacts on clouds and precipitation. *Journal of the Atmospheric Sciences*, **69**, 504–520.

Han, J.-Y., J.-J. Baik, & H. Lee, 2014: Urban impacts on precipitation. *Asia-Pacific Journal of Atmospheric Sciences*, **50**, 17–30.

Hann, J. von, 1885: Über den Temperaturunterscheid zwischen Stadt und Land. *Österreichische Gesellschaft für Meteorologie, Zeitschrift*, **20**, 457–462.

Hann, J. von, 1895: Die Unterschiede der meteorologischen Elemente in der Stadt Paris und ausserhalbe derselben. *Meteorologische Zeitschrift*, **12**, 37–38.

Hanna, S.R., J. White, & Y. Zhou, 2007: Observed winds, turbulence, and dispersion in built-up downtown areas of Oklahoma City and Manhattan. *Boundary-Layer Meteorology*, **125**, 441–468.

Harman, I.N., M.J. Best, & S.E. Belcher, 2004: Radiative exchange in an urban street canyon. *Boundary-Layer Meteorology*, **110**, 301–316.

Helbig, A., J. Baumüller, & M. Kerschgens, (eds.), 1999: *Stadtklima und Luftreinhaltung*, Springer, Berlin, 467 pp.

Hellmann, G., 1892: Resultate des Regenmessversuchsfeldes bei Berlin 1885–1891. *Meteorologische Zeitschrift*, **9**, 173–181.

Hendel, M., M. Colombert, Y. Diab, & L. Royon, 2015: An analysis of pavement heat flux to optimize the water efficiency of a pavement-watering method. *Applied Thermal Engineering*, **78**, 658–669.

Herz, R.K., 1988: Considering climatic factors for urban land use planning in the Sahelian zone. *Energy and Buildings*, **11**, 91–101.

Hildebrand, P.H., & B. Ackerman, 1984: Urban effects on the convective boundary layer. *Journal of the Atmospheric Sciences*, **41**, 76–91.

Hinkel, K.M., F.E. Nelson, A.E. Klene, & J.H. Bell, 2003: The urban heat island in winter at Barrow, Alaska. *International Journal of Climatology*, **23**, 1889–1905.

Hobbs, P.V., 2000: *Introduction to Atmospheric Chemistry*, Cambridge University Press, New York, 262 pp.

Högström, U., H. Bergström, H. Alexandersson, 1982: Turbulence characteristics in a near-neutrally stratified urban atmosphere. *Boundary-Layer Meteorology*, **23**, 449–472.

Högström, U., R. Taesler, S. Karlsson, L. Enger, & A.-S. Smedman-Högström, 1978: The Uppsala Urban Meteorology Project. *Boundary-Layer Meteorology*, **15**, 69–80.

Holmer, B., & I. Eliasson, 1999: Urban-rural vapour pressure differences and their role in the development of urban heat islands. *International Journal of Climatology*, **19**, 989–1009.

Holmer, B., S. Thorsson, & J. Lindén, 2013: Evening evapotranspirative cooling in relation to vegetation and urban geometry in the city of Ouagadougou, Burkina Faso. *International Journal of Climatology*, **33**, 3089–3105.

Höppe, P., 1999: The physiological equivalent temperature–a universal index for the biometeorological assessment of the thermal environment. *International Journal of Biometeorology*, **43**, 71–75.

Howard, L., 1818: *The Climate of London, Vol.1*, London.

Howard, L., 1820: *The Climate of London, Vol.2*, London.

Howard, L., 1833: *The Climate of London, Vol.1–3*, Harvey and Darton, London.

Huff, F.A., 1977: 1971–1975 Rainfall pattern comparisons. In: *Summary of METROMEX, Vol. 1, Weather Anomalies and Impacts*, Changnon, S.A., et al. (eds.), Illinois State Water Survey Bulletin, 62, Urbana IL, 13–29.

Huff, F.A., 1978: Radar analysis of urban effects on rainfall. In: *Summary of METROMEX, Vol. 2, Causes of Precipitation Anomalies*, Ackerman, B. et al. (eds.), Illinois State Water Survey Bulletin, 63, Urbana IL, 45–52.

Hughes, M., & A. Hall, 2010: Local and synoptic mechanisms causing Southern California's Santa Ana winds. *Climate Dynamics*, **34**, 847–857.

Hunt, J.C.R., C.J. Abell, J.A. Peterka, & H. Woo, 1978: Kinematical studies of the flows around free or surface-mounted obstacles: applying topology to flow visualization. *Journal of Fluid Mechanics*, **86**, Pt. 1, 179–200.

Hupfer, P., & F.-M. Chmielewski, (eds)., 1990: *Das Klima von Berlin*, Akademie-Verlag, Berlin, 288 pp. [In German]

Hussain, M., & B.E. Lee, 1980: An investigation of wind forces on three-dimensional roughness elements in a simulated atmospheric boundary layer flow: Part II Flow over large arrays of identical roughness elements and the effect of frontal and side aspect ratio variations, Report BS 56, Dept. Building Science, Univ. of Sheffield.

Idso, C., S. Idso, & R. Balling, 2001: An intensive two-week study of an urban $CO_2$ dome in Phoenix, Arizona, USA. *Atmospheric Environment*, **35**, 995–1000.

Imamura, G., 1949: Air-raids disastrous fire and urban temperature of Tokyo. *Kagaku*, **19**, 273–275.

Imhoff, M.L., P. Zhang, R.E. Wolfe, & L. Bounoua, 2010: Remote sensing of the urban heat island effect across biomes in the continental USA. *Remote Sensing of Environment*, **114**, 504–513.

IPCC, 2007: *Climate Change 2007: Synthesis Report. Contribution of Working Groups I, II and III to the Fourth Assessment Report of the Intergovernmental Panel on Climate Change*, Core Writing Team, Pachauri, R.K, & A. Reisinger, (eds.), IPCC, Geneva, Switzerland, 104 pp.

IPCC, 2013: Summary for Policymakers. *In: Climate Change 2013: The Physical Science Basis: Contribution of Working Group I to the Fifth Assessment Report of the Intergovernmental Panel on Climate Change*, Stocker, T.F., D. Qin, G.-K. Plattner, M. et al., (eds.), Cambridge University Press, Cambridge, United Kingdom and New York, NY, USA, 3–29.

IPCC, 2014: Summary for policymakers. In: *Climate Change 2014: Impacts, Adaptation, and Vulnerability. Part A: Global and Sectoral Aspects. Contribution of Working Group II to the Fifth Assessment Report of the Intergovernmental Panel on Climate Change*, Field, C.B., V.R. Barros, D.J. Dokken, et al. (eds.), Cambridge University Press, 1–32.

Jacovides, C.P., M.D. Steven, & D.N. Asimakopoulos, 2000: Solar spectral irradiance under clear skies around a major metropolitan area. *Journal of Applied Meteorology*, **39**, 917–930.

Järvi, L., C.S.B. Grimmond, & A. Christen, 2011: The Surface Urban Energy and Water Balance Scheme (SUEWS): Evaluation in Los Angeles and Vancouver. *Journal of Hydrology*, **411**, 219–237.

Jáuregui, E., 1973: The urban climate of Mexico City. *Erdkunde*, **27**, 298–307.

Jáuregui E., 1990/91: Effects of revegetation and new artificial water bodies on the climate of northeast Mexico City. *Energy and Buildings*, **15–16**, 447–455.

Jáuregui, E., 1990/91: Influence of a large urban park on temperature and convective precipitation in a tropical city. *Energy and Buildings*, **15–16**, 457–463.

Jáuregui, E., 1993: Bibliography of Urban Climate in Tropical/Subtropical areas, WCASP 25, WMO/TD No. 552, World Meteorological Organization, Geneva.

Jáuregui, E., 1996: Bibliography of Urban Climatology for the Period 1992 – 1995, WMO/TD No. 759, World Meteorological Organization, Geneva.

Jáuregu, E., 2000: *El Clima de la Ciudad de México*, Plaza y Valdez, México, D. F., 129 pp. [In Spanish]

Jáuregui, E., & A. Tejeda, 1997: Urban-rural humidity contrasts in Mexico City. *International Journal of Climatology*, **17**, 187–196.

Johansson, E., 2006: Influence of urban geometry on outdoor thermal comfort in a hot dry climate: A study in Fez, Morocco. *Building and Environment*, **41**, 1326–1338.

Johnson, G.T., & I.D. Watson, 1984: The determination of view-factors in urban canyons. *Journal of Climate and Applied Meteorology*, **23**, 329–335.

Johnson, W.B., F.L. Ludwig, W.F. Dabberdt, & R.J Allen, 1973: An urban diffusion simulation model for carbon monoxide. *Journal of the Air Pollution Control Association*, **23**, 490–498.

Kaimal, J.C., & J.J. Finnigan, 1994: *Atmospheric Boundary Layer Flows*, Oxford University Press, Oxford, 289 pp.

Kanda, M., 2012: *Urban Weather and Climate*, Meteorological Monograph, 224, Meteorological Society of Japan, 296 pp. [In Japanese]

Kanda, M., & T. Moriizumi, 2009: Momentum and heat transfer over urban-like surfaces. *Boundary-Layer Meteorology*, **131**, 385–401.

Kanda, M., R. Moriwaki, M. Roth, & T.R. Oke, 2002: Area-averaged sensible heat flux and a new method to determine zero-plane displacement length over an urban surface using scintillometry. *Boundary-Layer Meteorology*, **105**, 177–193.

Kanda, M., M. Kanega, T. Kawai, R. Moriwaki, & H. Sugawara, 2007: Roughness lengths for momentum and heat derived from outdoor urban scale models. *Journal of Applied Meteorology and Climatology*, **46**, 1067–1079.

Kandlikar, M., & G. Ramachandran, 2000: The causes and consequences of particulate air pollution in urban India: A synthesis of the science. *Annual Review of Energy and the Environment*, **25**, 629–684.

Kastner-Klein, P., R. Berkowicz, & R. Britter, 2004: The influence of street architecture on flow and dispersion in street canyon. *Meteorology and Atmospheric Physics*, **87**, 121–131.

Kataoka, K., F. Matsumoto, T. Ichinose, & M. Taniguchi, 2009: Urban warming trends in several large Asian cities over the last 100 years. *Science of the Total Environment*, **407**, 3112–3119.

Kawai, T., M. K. Ridwan, & M. Kanda, 2009: Evaluation of the simple urban energy balance model using selected data from 1-yr flux observations at two cities. *Journal of Applied Meteorology and Climatology*, **48**, 693–715.

Kawamura, T., 1964: Some consideration on the cause of city temperature at Kumagaya City. *Geographical Review of Japan*, **37**, 560–565

Kawamura, T., 1985: Recent changes of atmospheric environment in Tokyo and its surrounding area. *Geographical Review of Japan*, **58B**, 83–95.

Keatinge, W.R., G.C. Donaldson, E. Cordioli, et al. 2000: Heat related mortality in warm and cold regions of Europe: Observational study. *British Medical Journal*, **321**, 670–673.

Kellett, R., A. Christen, N.C. Coops, et al., 2013: A systems approach to carbon cycling and emissions modeling at an urban neighborhood scale. *Landscape and Urban Planning*, **110**, 48–58

Kelliher, F.M., R. Leuning, M.R. Raupach, & E.-D. Schultze, 1995: Maximum conductances for evaporation from global vegetation types, *Agricultural and Forest Meteorology*, **73**, 1–16.

Kennedy, C.A., A. Ramaswami, S. Carney, & S. Dhakai, 2010: Greenhouse gas emission baselines for global cities and metropolitan regions. In D. M. Freire, et al. (eds.) *Cities and Climate Change: Responding to an Urgent Agenda*. Hoornweg, The World Bank, 15–54.

Kennedy, C., J. Steinberger, B. Gasson, et al., 2009: Greenhouse gas emissions from global cities. *Environmental Science & Technology*, **43**, 7297–7302.

Kenworthy, J.R., 2006: The eco-city: ten key transport and planning dimensions for sustainable city development. *Environment and Urbanization*, **18**, 67–85.

Koenig, L.R., 1981: Anomalous snowfall caused by natural-draft cooling towers. *Atmospheric Environment*, **15**, 1117–1128.

Koerner, B., & J. Klopatek, 2002: Anthropogenic and natural $CO_2$ emission sources in an arid urban environment. *Environmental Pollution*, **116**, S45-S51.

Komp, M.J., & A.H. Auer Jr., 1978: Visibility reduction and accompanying aerosol evolution downwind of St. Louis. *Journal of Applied Meteorology*, **17**, 1357–1367.

Kondrat'yev, K. Ya., 1973: The Complete Atmospheric Energetics Experiment. *GARP Public. Series*, No.12, ICSU, World Meteorological Organization, Geneva.

Kordowski, K., & W. Kuttler 2010: Carbon dioxide fluxes over an urban park area. *Atmospheric Environment.* **44**, 2722–2730.

Kottek, M., J. Grieser, C. Beck, B. Rudolf, & F. Rubel, 2006: World map of the Köppen-Geiger climate classification updated. *Meterologische Zeitschrift*, **15**, 259–263.

Kotthaus, S., T.E.L. Smith, M.J. Wooster, & C.S.B. Grimmond, 2014: Derivation of an urban materials spectral library through emittance and reflectance spectroscopy. *ISPRS Journal of Photogrammetry and Remote Sensing*, **94**, 194–212.

Kratzer, A., 1937: *Das Stadtklima*, Friedrich Vieweg und Sohn, Braunschweig, 143 pp. [In German]

Kratzer, P.A., 1956: *The Climate of Cities (Das Stadtklima)*. American Meteorological Society.

Kratzer, A., 1956: *Das Stadtklima*, 2nd edn., Friedrich Vieweg und Sohn, Braunschweig, 184 pp. [In German]

Krayenhoff, E.S., & J.A.Voogt, 2007: A microscale three-dimensional urban energy balance model for studying surface temperatures. *Boundary-Layer Meteorology*, **123**, 433–461.

Krayenhoff, E.S., A. Christen, A. Martilli & T.R. Oke, 2014: A multi-layer radiation model for urban neighbourhoods with trees. *Boundary-Layer Meteorology*, **151**, 139–178.

Krayenhoff, E.S., J.-L. Santiago, A. Martilli, A. Christen, & T.R. Oke, 2015: Parametrization of drag and turbulence for urban neighbourhoods with trees. *Boundary-Layer Meteorology*, **156**, 157–189.

Kremser, V., 1886: Vortrag über das Klima von Berlin. Festschrift d.59 vers d. d. Naturf. und Ärtze. Berlin, 37–38.

Kremser, V., 1908: Der Einfluß der Großstädte auf der Luftfeuchtigkeit, *Meteorologische Zeitschrift*, **25**, 206–215.

Kremser, V., 1909: Ergebnisse vieljähriger Windregistrierungen in Berlin, *Vortrag über das Klima von Berlin, Meteorologische Zeitschrift*, **26**, 259–265.

Krishan, A., 1996: The habitat of two deserts in India: hot-dry desert of Jaisalmer (Rajasthan) and the cold-dry high altitude mountainous desert of Leh (Ladakh). *Energy and Buildings*, **23**, 217–229.

Krzysztof, B., G. Jendritzky, P. Bröde, et al., 2013: An introduction to the Universal Thermal Climate Index (UTCI). *Geographia Polonica*, **86**, 5–10.

Kuc, T., K. Rozanski, M. Zimnoch, J.M. Necki, & A Korus, 2003: Anthropogenic emissions of $CO_2$ and $CH_4$ in an urban environment. *Applied Energy*, **75**, 193–203.

Kusaka, H., K. Nawata, A. Suzuki-Parker, Y. Takane, & N. Furuhashi, 2014: Mechanism of precipitation increase with urbanization in Tokyo as revealed by ensemble climate simulations. *Journal of Applied Meteorology and Climatology*, **53**, 824–839.

Kuttler, W., J. Miethke, D. Dütemeyer, & A.-B. Barlag, 2015: *Das Klima von Essen*, Westarp Wissenschaften, 249 pp.

Kuttler, W., S. Weber, J. Schonnefeld, & A. Hesselschwerdt, 2007: Urban/rural atmospheric water vapour pressure differences and urban moisture excess in Krefeld, Germany. *International Journal of Climatology*, **27**, 2005–2015.

Kuttler, W., & A. Strassburger, 1999: Air quality measurements in urban green areas-a case study. *Atmospheric Environment*, **33**, 4101–4108.

Kwartler M., & R. Masters, 1984: Daylight as a zoning device for midtown. *Energy and Buildings*, **6**, 173–189.

Lagouarde, J-P., P. Moreau, M. Irvine, J-M. Bonnefond, J.A. Voogt, & F. Solliec, 2004: Airborne experimental measurements of the angular variations in surface temperature over urban areas: case study of Marseille (France). *Remote Sensing of Environment*, **93**, 443–462.

Lagouarde, J-P., A. Hénon, B. Kurz, et al., 2010: Modelling daytime thermal infrared directional anisotropy over Toulouse city centre, *Remote Sensing of Environment*, **114**, 87–105.

Lamptey, B., 2010: An analytical framework for estimating the urban effect on climate. *International Journal of Climatology*, **30**, 72–88.

Landsberg, H.E., 1956: The climate of towns. In: *Man's Role in Changing the Face of the Earth*, Thomas, W.L. (ed.), University of Chicago Press, Chicago, 584–606.

Landsberg H.E., 1962: City air – better or worse. In: Symposium: Air over Cities, US Public Health Service, Taft Sanitary Eng. Center, Cincinnati, Ohio, Tech. Rept. A62–5, 1–22

Landsberg, H.E., 1970: Man-made climatic changes, *Science*, **170**, 1265–1274.

Landsberg, H., 1973: The meteorologically utopian city. *Bulletin of the American Meteorological Society*, **54**, 86–89.

Landsberg, H.E., 1981: *The Urban Climate*, Academic Press, New York, 269 pp.

Lareau, N.P., E. Crosman, C.D. Whiteman, et al., 2013: The persistent cold-air pool study. *Bulletin of the American Meteorological Society*, **94**, 51–63.

Lauscher, F., 1934: Wärmeausstrahlung über Horizonteinengung. *Sitz. Akad. in Wien*, **143**, 503–519.

Lauscher, F., 1959: Witterung und Klima von Linz, *Wetter und Leben*, No. **6**, 235 pp.

Lavery, S., & B. Donovan, 2005: Flood risk management in the Thames Estuary looking ahead 100 years. *Philosophical Transactions of the Royal Society of London A: Mathematical, Physical and Engineering Sciences*, **363**, 1455–1474.

Lazar, R., & A. Podessor, 1999: An urban climate analysis of Graz and its significance for urban planning in the tributary valleys east of Graz (Austria). *Atmospheric Environment*, **33**, 4195–4209.

Lebassi, B., J. Gonzalez, D. Fabris, et al., 2009: Observed 1970–2005 cooling of summer daytime temperatures in coastal California. *Journal of Climate*, **22**, 3558–3573.

Lee, J.K., A. Christen, R. Ketler, & Z. Nesic, 2017: A mobile sensor network to map carbon dioxide emissions in urban environments. *Atmospheric Measurement Techniques*, **10**, 645–665, doi:10.5194/amt-10-645-2017.

Lee, T.F., 1987: Urban clear islands in Central Valley fog. *Monthly Weather Review*, **115**, 1794–96.

Lemonsu, A., & V. Masson, 2002: Simulation of summer urban breeze over Paris. *Boundary-Layer Meteorology*, **104**, 463–490.

Lemonsu, A., V. Masson, & E. Berthier, 2007: Improvement of the hydrological component of an urban-soil-vegetation-atmosphere transfer model. *Hydrological Processes*, **21**, 2100–2111.

Lemonsu, A., R. Kounkou-Arnaud, J. Desplat, J.L. Salagnac, & V. Masson, 2013: Evolution of the Parisian urban climate under a global changing climate. *Climatic Change*, **116**, 679–692.

Lemonsu, A., S. Bélair, J. Mailhot, et al., 2008: Overview and first results of the Montréal Urban Snow Experiment 2005. *Journal of Applied Meteorology and Climatology*, **47**, 59–75.

Lenzholzer, S., 2015: *Weather in the City: How Design Shapes the Urban Climate*. Published by nai010, Netherlands 216 pp.

Lerner, D.N., 2002: Identifying and quantifying urban recharge: a review. *Hydrogeology Journal*, **10**, 143–152.

Lettau, H.H., 1970: Problems of micrometeorological measurements (on degree of control in out-of-door experiments). In: *The Collection and Processing of Field Data*, Bradley, E.F. & O.T. Denmead, (eds.), Interscience, New York, 3–40.

Lewis, J.E., & T.N. Carlson, 1989: Spatial variations in regional surface energy exchange patterns for Montréal, Québec. *The Canadian Geographer*, **33**, 194–203.

Li, D., & E. Bou-Zeid, 2013: Synergistic interactions between urban heat islands and heat waves: the impact in cities is larger than the sum of its parts. *Journal of Applied Meteorology and Climatology*, **52**, 2051–2064.

Li, T., R.M. Horton, & P.L. Kinney, 2013: Projections of seasonal patterns in temperature-related deaths for Manhattan, New York. *Nature Climate Change*, **3**, 717–721.

Li, X.-X., C.-H. Liu, D.Y.C. Leung, & K.M. Lam, 2006: Recent progress in CFD modelling of wind field and pollutant transport in street canyons. *Atmospheric Environment*, **40**, 5640–5658.

Li, X., X. Xia, L. Wang, et al., 2015: The role of foehn in the formation of heavy air pollution events in Urumqi, China. *Journal of Geophysical Research: Atmospheres*, **120**, 5371–5384.

Li, Z., F. Niu, J. Fan, Y. Liu, D. Rosenfeld, & Y. Ding, 2011: Long-term impacts of aerosols on the vertical development of clouds and precipitation. *Nature Geoscience*, **4**, 888–894.

Lichtfouse, E., M. Lichtfouse, M. Kashgarian, & R. Bol, 2005: $^{14}C$ of grasses as an indicator of fossil fuel $CO_2$ pollution. *Environmental Chemistry Letters*, **3**, 78–81.

Liezke, B., & R. Vogt, 2013: Variability of $CO_2$ concentrations and fluxes in and above an urban street canyon. *Atmospheric Environment*, **74**, 60–72.

Lietzke, B., R. Vogt, C. Feigenwinter, & E. Parlow, 2015: On the controlling factors for the variability of carbon dioxide flux in a heterogeneous urban environment. *International Journal of Climatology*, **35**, 3921–3941.

Lindberg, F., 2007: Modelling the urban climate using a local governmental geo-database. *Meteorological Applications*, **14**, 263–273.

Lindqvist, S., 1968: Studies on the local climate in Lund and its environs. *Geografiska Annaler: Series A, Physical Geography*, **50**, 79–93.

Littlefair, P., 2011: *Site Layout Planning for Daylight and Sunlight*, BR209, 2nd Ed. BRE, 84 pp.

Littlefair, P., M. Santamouris, S. Alvarez, et al., 2000: *Environmental Site Layout Planning: Solar Access, Microclimate and Passive Cooling in Urban Areas*. Construction Research Communications Ltd., 151 pp.

Lokoshchenko, M.A., 2005: About meteorological observations at the observatory of Moscow State University and their / representativeness of. *Vestnik of the Moscow State University, Series 5 Geography*, **6**, 25–31. [in Russian]

Loose, T., & R.D. Bornstein, 1977: Observations of mesoscale effects on frontal movement through and urban area. *Monthly Weather Review*, **105**, 563–571.

Lowry, W.P., 1977: Empirical estimation of urban effects on climate: a problem analysis. *Journal of Applied Meteorology*, **16**, 129–135.

Lowry, W.P., 1988: *Atmospheric Ecology for Designers and Planners*, Peavine Publications, McMinnville, Oregon.

Lowry, W.P., 1998: Urban effects on precipitation amount. *Progress in Physical Geography*, **22**, 477–520.

Lowry, W.P., & P.P. Lowry, 2001: *Fundamentals of Biometeorology: 2. The Biological Environment*, Peavine Press, McMinnville, OR, 680 pp.

Macdonald, R.W., 2000: Modelling the mean velocity profile in the urban canopy layer. *Boundary-Layer Meteorology*, **97**, 25–45.

Malhi, Y., 1996: The behaviour of the roughness length for temperature over heterogeneous surfaces. *Quarterly Journal of the Royal Meteorological Society*, **122**, 1095–1125.

Malmstadt, J.C., J.B. Elsner, & T.H. Jagger, 2010: Risk of strong hurricane winds to Florida cities. *Journal of Applied Meteorology and Climatology*, **49**, 2121–2132.

Martilli, A., 2014: An idealized study of city structure, urban climate, energy consumption, and air quality. *Urban Climate*, **10**, 430–446.

Martilli, A., & J. Santiago, 2007: CFD simulation of airflow over a regular array of cubes. Part II: analysis of spatial average properties. *Boundary-Layer Meteorology*, **122**, 635–654.

Martilli, A., A. Clappier, & M.W. Rotach, 2002: An urban surface exchange parameterisation for mesoscale models. *Boundary-Layer Meteorology*, **104**, 261–304.

Martinelli, L., T.P. Lin, & A. Matzarakis, 2015: Assessment of the influence of daily shadings pattern on human thermal comfort and attendance in Rome during summer period. *Building and Environment*, **92**, 30–38.

Martinuzzi, R., & C. Tropea, 1993: The flow around surface-mounted, prismatic obstacles placed in a fully developed channel flow. *Journal of Fluids Engineering*, **115**, 85–92.

Masson, V., 2000: A physically-based scheme for the urban energy budget in atmospheric models. *Boundary-Layer Meteorology*, **94**, 357–397.

Masson, V., C.S.B. Grimmond, & T.R. Oke, 2002: Evaluation of the Town Energy Balance (TEB) scheme with direct measurements from dry districts in two cities. *Journal of Applied Meteorology*, **41**, 1011–1026.

Masson, V., L. Gomes, G. Pigeon, et al., 2008: The Canopy and Aerosol Particles Interactions in TOulouse Urban Layer (CAPITOUL) experiment. *Meteorology and Atmospheric Physics*, **102**, 135–157.

Masson, V., C. Marchadier, L. Adolphe, et al., 2014: Adapting cities to climate change: A systemic modelling approach. *Urban Climate*, **10**, 407–429.

Matese, A., B. Gioli, F. Vaccari, A. Zaldei, & F. Miglietta, 2009: Carbon dioxide emissions of the city center of Firenze, Italy: Measurement, evaluation, and source partitioning. *Journal of Applied Meteorology and Climatology*, **48**, 1940–1947.

Matzarakis, A., H. Mayer, & M.G. Iziomon, 1999: Applications of a universal thermal index: physiological equivalent temperature. *International Journal of Biometeorology*, **43**, 76–84.

Matzarakis, A., F. Rutz, & H. Mayer, 2010: Modelling radiation fluxes in simple and complex environments: basics of the RayMan model. *International Journal of Biometeorology*, **54**, 131–139.

Mayer, P.W., W.B. DeOreo, E.M. Opitz, et al., 1999: The Residential End-uses of Water, Published by the AWWA Research Foundation and American Water Works Association, 310 pp.

Maykut, G.A., 1985: An introduction to ice in the polar oceans. Report 2nd printing, APL-UW 8510, Dept. Atmos. Sci., Univ. Washington, Seattle, WA.

Mays, K.L., P.B. Shepson, B.H. Stirm, A. Karion, C. Sweeney, & K.R. Gurney, 2009: Aircraft-based measurements of the carbon footprint of Indianapolis. *Environmental Science & Technology*, **43**, 7816–7823.

McCarthy, M.P., M.J. Best, & R.A. Betts, 2010: Climate change in cities due to global warming and urban effects. *Geophysical Research Letters*, **37**, L09705.

McGee, T.G. 1991: The Emergence of Desakota Regions in Asia: Expanding a Hypothesis, p3–25 in Ginsburg, N., B. Koppel, & T.G. McGee, (eds.), 1991: *The Extended Metropolis: Settlement Transition in Asia*, University of Hawaii Press, Honolulu, 352 pp.

McGregor, G.R., S. Belcher, J. Hacker, et al., 2006: *London's Urban Heat Island*. A Report to the Greater London Authority, Centre for Environmental Assessment, Management and Policy, King's College London, 110 pp.

McNaughton, K.G., & P.G. Jarvis, 1983: Predicting effects of vegetation changes on transpiration and evapotranspiration. In: *Water Deficit and Plant Growth*, Vol. 7, Kozlowski, T.T. (ed.), Academic Press, 1–47.

McPherson, E.G., D. Nowak, G. Heisler, et al., 1997: Quantifying urban forest structure, function, and value: the Chicago Urban Forest Climate Project. *Urban Ecosystems*, **1**, 49–61.

Meinders, E.R., K. Hanjalic, & R.J. Martinuzzi, 1999: Experimental study of the local convection heat transfer from a wall-mounted cube in turbulent channel flow. *Journal of Heat Transfer*, **121**, 564-573.

Menberg, K., P. Bayer, K. Zosseder, S. Rumohr, & P. Blum, 2013: Subsurface urban heat islands in German cities. *Science of the Total Environment*, **442**, 123–133.

Meroney, R.N., 1982: Turbulent diffusion near buildings. In: *Engineering Meteorology*, Plate, E. (ed.), Elsevier Scientific Publishing Co., Amsterdam, 481–525.

Mestayer, P., P. Durand, P. Augustin, et al., 2005: The urban boundary-layer field campaign in Marseille (UBL/CLU-ESCOMPTE): set-up and first results. *Boundary-Layer Meteorology*, **114**, 315–365.

Meyn, S., & T.R. Oke, 2009: Heat fluxes through roofs and their relevance to estimates of urban heat storage. *Energy and Buildings*, **41**, 745–752.

Miles, J.C., 2001: Temporal variation of radon levels in houses and implications for radon measurement strategies. *Radiation Protection Dosimetry*, **93**, 369–376.

Mills, G., 1997a: An urban canopy-layer model. *Theoretical and Applied Climatology*, **57**, 229–244.

Mills, G., 1997b: Building density and interior building temperatures: A physical scale modeling experiment. *Physical Geography*, **18**, 195–214.

Mills, G., 2008: Luke Howard and the climate of London. *Weather*, **63**, 153–157.

Mills, G., H. Cleugh, R. Emmanuel, et al., 2010: Climate information for improved planning and management of mega cities (needs perspective). *Procedia Environmental Sciences*, **1**, 228–246.

Modelski, G., 2003: *World Cities,-3000 to 2000, FAROS 2000*, Washington, DC.

Mölders, N., & M.A. Olson, 2004: Impact of urban effects on precipitation in high latitudes. *Journal of Hydrometeorology*, **5**, 409–429.

Monteith, J.L., & M.H. Unsworth, 2008: *Principles of Environmental Physics*, 3rd edn., Academic Press, Burlington, Massachusetts, 418 pp.

Moriwaki, R., & M. Kanda, 2004: Seasonal and diurnal fluxes of radiation, heat, water vapor, and carbon dioxide over a suburban area. *Journal of Applied Meteorology and Climatology*, **43**, 1700–1710.

Moriwaki, R., & M. Kanda, 2005: Flux-gradient profiles for momentum and heat over an urban surface. *Theoretical and Applied Climatology*, **84**. 127–135.

Moriwaki, R., M. Kanda, H. Senoo, A. Hagishima, & T. Kinouchi, 2008: Anthropogenic water vapor emissions in Tokyo. *Water Resources Research*, **44** (11).

Mount, L.E., 1979: *Adaptation to Thermal Environment: Man and His Productive Animals*. Edward Arnold, London, 333 pp.

Moya, J., L. Phillips, L. Schuda, et al., 2011: *Exposure Factors Handbook: 2011 Edition*. US Environmental Protection Agency.

Müller, N., W. Kuttler, & A.B. Barlag, 2014: Counteracting urban climate change: adaptation measures and their effect on thermal comfort. *Theoretical and Applied Climatology*, **115**, 243–257.

Munn, R.E., 1966: *Descriptive Micrometeorology*, Academic Press, New York, 253 pp.

Munn, R.E., 1970: Airflow in urban areas. WMO Tech. Note 108, World Meteorological Organization, Geneva, 15–39.

Munn, R.E., 1973: Urban meteorology: some selected topics. *Bulletin of the American Meteorological Society*, **54**, 90–93.

Munoz, R.C., & A.A. Undurraga, 2010: Daytime mixed layer over the Santiago Basin: Description of two years of observations with lidar data. *Journal of Applied Meteorology and Climatology*, **49**, 1728–1741.

Murakami, S., 1982: Wind tunnel modeling applied to pedestrian comfort. In: *Wind Tunnel Modelling for Civil Engineering Applications*, Reinhold, T.A. (ed.), Cambridge University Press, 486–503.

Myrup, L.O., 1969: A numerical model of the urban heat island. *Journal of Applied Meteorology* **8**, 908–918.

Nakayoshi, M., R. Moriwaki, T. Kawai, & M. Kanda, 2009: Experimental study on rainfall interception over an outdoor urban-scale model. *Water Resources Research*, **45**, W04415.

Nakayoshi, M., M. Kanda, R. Shi, & R. de Dear, 2015: Outdoor thermal physiology along human pathways: a study using a wearable measurement system. *International Journal of Biometeorology*, **59**, 503–515.

Nazaroff, W.W., 1992: Radon transport from soil to air. *Reviews of Geophysics*, **30**, 137–160.

Nelson, M.A., E.R. Pardyjak, J.C. Klewicki, S.U. Pol, & M.J. Brown, 2007: Properties of the wind field within the Oklahoma City Park Avenue street canyon. Part I: Mean flow and turbulence statistics. *Journal of Applied Meteorology and Climatology*, **46**, 2038–2054.

Nemitz, E., K. Hargreaves, A. McDonald, J.R. Dorsey, & D. Fowler, 2002: Meteorological measurements of the urban heat budget and $CO_2$ emissions on a city scale. *Environmental Science & Technology*, **36**, 3139–3146.

Newman, P., & J. Kenworthy, 1989: *Cities and Automobile Dependence: An International Sourcebook*, Gower, U.K., 388 pp.

Newton, T.L., T.R. Oke, C.S.B. Grimmond, & M. Roth, 2007: The suburban energy balance in Miami, Florida. *Geografiska Annaler Series A-Physical Geography*, **89A**, 331–347.

Ng, E., 2009: Policies and technical guidelines for urban planning of high-density cities – air ventilation assessment (AVA) of Hong Kong. *Building and Environment*, **44**, 1478–1488.

Nicholls, N.J., S. Hanson, C. Herweijer, et al., 2008: Ranking Port Cities with High Exposure and Vulnerability to Climate Extremes: Exposure Estimates. Environment Working Papers No. 1, Organization for Economic Co-Operation and Development.

Nishimura N., T. Nomura, H. Iyota, & S. Kimoto, 1998: Novel water facilities for creation of comfortable urban micrometeorology. *Solar Energy*, **64**, 197–207.

Niyogi, D., & P. Schmid, 2014: The history and future of the La Porte anomaly. Stanley, A. Changnon Symposium, AMS 94th Annual Meeting, Paper TJ6.3. Available online at: https://ams.confex.com/ams/94Annual/webprogram/Paper235964.html

Niyogi, D., P. Pyle, M. Lei, et al., 2011: Urban modification of thunderstorms: An observational storm climatology and model case study for the Indianapolis urban region. *Journal of Applied Meteorology and Climatology*, **50**, 1129–1144.

Nordbo, A., L. Järvi, S. Haapanala, J. Moilanen, & T. Vesala, 2013: Intra-city variation in urban morphology and turbulence structure in Helsinki, Finland. *Boundary-Layer Meteorology*, **146**, 469–496.

Nowak, D.J., & D.E. Crane, 2002: Carbon storage and sequestration by urban trees in the USA. *Environmental Pollution*, **116**, 381–389.

Nowak, D.J., & J.F. Dwyer, J.F., 2007: Understanding the benefits and costs of urban forest ecosystems. In: *Urban and Community Forestry in the Northeast*, Kuser, J.E. (ed.), Springer Netherlands, 25–46.

Nunez, M. & T.R. Oke, 1977: The energy balance of an urban canyon, *Journal of Applied Meteorology*, **16**, 11–19.

OECD, 2012: *Compact City Policies: A Comparative Assessment*, OECD Green Growth Studies, OECD Publishing, Paris.

Offerle, B.D., P. Jonsson, I. Eliasson, & C.S.B. Grimmond, 2005a: Urban modification of the surface energy balance in the West African Sahel: Ouagadougou, Burkina Faso. *Journal of Climate*, **18**, 3983–3995.

Offerle, B., C.S.B. Grimmond, & K. Fortuniak, 2005b: Heat storage and anthropogenic heat flux in relation to the energy balance of a central European city centre. *International Journal of Climatology*, **25**, 1405–1419.

Offerle, B.D., C.S.B. Grimmond, K. Fortuniak, K. Kłysik, & T.R. Oke, 2005c: Temporal variations in heat fluxes over a central European city centre. *Theoretical and Applied Climatology*, **84**, 103–115.

Offerle, B.D., C.S.B. Grimmond, K. Fortuniak, & W. Pawlak, 2006: Intraurban differences of surface energy fluxes in a central European city. *Journal of Applied Meteorology*, **45**, 125–136.

Oikawa, S., & Y. Meng, 1995: Turbulence characteristics and organized motion in a suburban roughness sublayer. *Boundary-Layer Meteorology*, **74**, 289–312.

Oke, T.R., 1973: City size and the urban heat island. *Atmospheric Environment*, **7**, 769–779.

Oke, T.R., 1974: Review of urban climatology, 1968–1973. WMO Tech. Note 134, World Meteorological Organization, Geneva.

Oke, T.R., 1976: The distinction between canopy and boundary-layer urban heat islands. *Atmosphere*, **14**, 268–277.

Oke, T.R., 1979: Review of urban climatology, 1973–1976. WMO Tech. Note 169, World Meteorological Organization, Geneva.

Oke, T.R., 1981: Canyon geometry and the nocturnal urban heat island: comparison of scale model and field observations. *Journal of Climatology*, **1**, 237–254.

Oke, T.R., 1982: The energetic basis of the urban heat island. *Quarterly Journal Royal Meteorological Society*, **108**, 1–24.

Oke, T.R., 1983: Bibliography of urban climate, 1977–1980. WCP No. 45, World Meteorological Organization, Geneva.

Oke, T.R., 1984: Methods in urban climatology. In: *Applied Climatology. Zürcher Geographische Schriften*, **14**, 19–29.

Oke, T.R., 1987: *Boundary Layer Climates*, 2nd edn., Routledge, London, 435 pp.

Oke, T.R., 1988a: The urban energy balance. *Progress in Physical Geography*, **12**, 471–508.

Oke, T.R., 1988b: Street design and urban canopy layer climate. *Energy and Buildings*, **11**, 103–113.

Oke, T.R., 1989: The micrometeorology of the urban forest, *Philosophical Transactions of the Royal Society, Series B*, **324**, 335–351.

Oke, T.R., 1990: Bibliography of urban climate, 1981–1988. WCAP No. 15, World Meteorological Organization, Geneva.

Oke, T.R., 1995: The heat island of the urban boundary layer: characteristics, causes and effects. In: *Wind Climate in Cities*. Cermak, J.E., Davenport, A.G., Plate, E.J. and Viegas, D.X., (eds.), Kluwer Academic, 81–102.

Oke, T.R., 1997: Urban environments, In: *The Surface Climates of Canada*, Bailey, W.G., T.R. Oke, & W.R. Rouse (eds.), McGill-Queen's University Press, Montréal, 303–327.

Oke, T.R., 2004: Initial guidance to obtain representative meteorological observations at urban sites, Instruments and Methods of Observation Programme, *IOM Report* No. **81**, WMO/TD No. 1250, World Meteorological Organization, Geneva, 51 pp. http://www.wmo.int/web/www/IMOP/publications/IOM-81/IOM-81-UrbanMetObs.pdf

Oke, T.R., 2008: Urban observations. In: *Guide to Meteorological Instruments and Methods of Observation*, Part II-Observing Systems, Chapter 11, WMO-No.8, 7th Edition, World Meteorological Organization, Geneva, II-11-1 – II-11-25.

Oke, T.R., & C. East, 1971: The urban boundary layer in Montréal. *Boundary-Layer Meteorology*, **1**, 411–437.

Oke, T.R., & R.F. Fuggle, 1972: Comparison of urban/rural counter and net radiation at night. *Boundary-Layer Meteorology*, **2**, 290–308.

Oke, T.R., G.T. Johnson, D.G. Steyn, & I.D. Watson, 1991: Simulation of surface urban heat islands under 'ideal' conditions at night. Part 2: Diagnosis of causation. *Boundary-Layer Meteorology*, **56**, 339–358.

Oke, T.R., C.S.B. Grimmond, & R.A. Spronken-Smith, On the confounding role of rural wetness in assessing urban effects on climate. Proceedings of the AMS 2nd Symposium on the Urban Environment Symposium, Albuquerque, NM, November 2–6, 59–62.

Oke, T.R., R.A. Spronken-Smith, E. Jauregui, C.S.B. Grimmond, 1999: The energy balance of central Mexico City during the dry season. *Atmospheric Environment*, **33**, 3919–3930.

Oleson, K.W., G.B. Bonan, J. Feddema, & T. Jackson, 2011: An examination of urban heat island characteristics in a global climate model. *International Journal of Climatology*, **31**, 1848–1865.

Omoto, Y., K. Hamotani, & H–H. Um, 1994: Recent changes in trends of humidity of Japanese Cities, *Journal of Japan Society for Hydrology and Water Resources*, **7**, 106–113.

Parsons, K.C., 2003: *Human Thermal Environments: The Effects of Hot, Moderate and Cold Environments on Human Health, Comfort and Performance*, 2nd Edition. Taylor & Francis.

Pawlak, W., K. Fortuniak, & M. Siedlecki, 2011: Carbon dioxide flux in the centre of Łódź, Poland —analysis of a 2-year eddy covariance measurement data set. *International Journal of Climatology*, **31**, 232–243.

Peng, S., S. Piao, P. Ciais, et al., 2012: Surface urban heat island across 419 global big cities. *Environmental Science and Technology*, **46**, 696–703.

Pérez-Lombard, L., J. Ortiz, & C. Pout, 2008: A review on buildings energy consumption information. *Energy and Buildings*, **40**, 394–398.

Perryman, N., & P.G. Dixon, 2013: A radar analysis of urban snowfall modification in Minneapolis–St Paul. *Journal of Applied Meteorology and Climatology*, **52**, 1632–1644.

Peters, E.B. & J.P. McFadden, 2012: Continuous measurements of net $CO_2$ exchange by vegetation and soils in a suburban landscape. *Journal of Geophysical Research*, **117**, G03005.

Peterson, J.T., E.C. Flowers, & J.H. Rudisill, 1978: Urban-rural solar radiation and atmospheric turbidity measurements in the Los Angeles Basin. *Journal of Applied Meteorology*, **17**, 1595–1609.

Phillips, N.G., R. Ackley, E.R. Crosson, et al., 2013: Mapping urban pipeline leaks: Methane leaks across Boston. *Environmental Pollution*, **173**, 1–4.

Pigeon, G., D. Legain, P. Durand, & V. Masson, 2007: Anthropogenic heat release in an old European agglomeration (Toulouse, France). *International Journal of Climatology*, **27**, 1967–1981.

Pigeon, G., A. Lemonsu, C.S.B. Grimmond, P. Durand, O. Thouron, & V. Masson, 2007: Divergence of turbulent fluxes in the surface layer: case of a coastal city. *Boundary-Layer Meteorology*, **124**, 269–290.

Piringer M., & K. Baumann, 1999: Modifications of a valley wind system by an urban area— Experimental results. *Meteorology and Atmospheric Physics*, **71**, 117–125.

Plate, E. J., 1999: Methods of investigating urban wind fields—Physical models. *Atmospheric Environment*, **33**, 3981–3989.

Popiel, C.O., J. Wojtkowiak, & B. Biernacka, 2001: Measurements of temperature distribution in ground. *Experimental Thermal and Fluid Science*, **25**, 301–309.

Quéré, C.L. et al., 2013: The global carbon budget 1959–2011. *Earth System Science Data*, **5**, 165–185.

Raman, A., A.F. Arellano, & J.J. Brost, 2014: Revisiting haboobs in the southwestern United States: An observational case study of the 5 July 2011 Phoenix dust storm. *Atmospheric Environment*, **89**, 179–188.

Ramamurthy, P., & E.R. Pardyjak, 2011: Toward understanding the behavior of carbon dioxide and surface energy fluxes in the urbanized semi-arid Salt Lake Valley, Utah, USA. *Atmospheric Environment*, **45**, 73–84.

Ramamurthy, P., E.R. Pardyjak, & J.C. Klewicki, 2007: Observations of the effects of atmospheric stability on turbulence statistics deep within an urban street canyon. *Journal of Applied Meteorology and Climatology*, **46**, 2074–2085.

Rees, W., 1997: Urban ecosystems: the human dimension. *Urban Ecosystems*, **1**, 63–75.

Rees, W., & M. Wackernagel, 1996: Urban ecological footprints: Why cities cannot be sustainable – and why they are a key to sustainability. *Environmental Impact Assessment Review*, **16**, 223–248.

Renou, E., 1855: Instructions météorologiques. *Société Météorologique de France*, **3**, 73–160.

Renou, E., 1868: Differences de température entre la ville et la campagne. *Société Météorologique de France*, **16**, 83–97.

Richards, K., 2005: Urban and rural dewfall, surface moisture, and associated canopy-level air temperature and humidity measurements for Vancouver, *Canada, Boundary-Layer Meteorology*, **114**, 143–163.

Roberts, S.M., T.R. Oke, C.S.B. Grimmond, & J.A. Voogt, 2006: Comparison of four methods to estimate urban heat storage, *Journal of Applied Meteorology and Climatology*, **45**, 1766–1781.

Robins, A., H. Cheng, P. Hayden, & T. Lawton, 2004: Flow Visualization Studies—I. DAPPLE–EnFlo 04 Note, 8 pp. www.dapple.org.uk/downloads.html.

Rogers, J.J.W., & P.G. Feiss, 1998: *People and the Earth*, Cambridge University Press, Cambridge, U.K. 364 pp.

Romanov, P., 1999: Urban influence on cloud cover estimated from satellite data. *Atmospheric Environment*, **33**, 4163–4172.

Romero, H., M. Ihl, A. Rivera, P. Zalazar, & P. Azocar, 1999: Rapid urban growth, land-use changes and air pollution in Santiago, Chile. *Atmospheric Environment*, **33**, 4039–4047.

Rosenfeld, D., U. Lohmann, G.B. Raga, et al., 2008: Flood or drought: How do aerosols affect precipitation? *Science*, **321**, 1309–1313.

Rotach, M.W., 1993a: Turbulence close to a rough urban surface. 1. Reynolds stress. *Boundary-Layer Meteorology*, **65**, 1–28.

Rotach, M.W., 1993b: Turbulence close to a rough urban surface. 2. Variances and gradients. *Boundary-Layer Meteorology*, **66**, 75–92.

Rotach, M.W., 1995: Profiles of turbulence statistics in and above an urban street canyon. *Atmospheric Environment*, **13**, 1473–1486.

Rotach, M.W., 1999: On the influence of the urban roughness sublayer on turbulence and dispersion. *Atmospheric Environment*, **33**, 4001–4008.

Rotach, M.W., 2001: Simulation of urban-scale dispersion using a Lagrangian stochastic dispersion model. *Boundary-Layer Meteorology*, **99**, 379–410.

Rotach, M.W., R. Vogt, C. Bernhofer, et al., 2005: BUBBLE – an urban boundary layer meteorology project. *Theoretical & Applied Climatology*, **81**, 231–261.

Roth, M., 1993: Turbulent transfer relationships over an urban surface. II. Integral statistics. *The Quarterly Journal of the Royal Meteorological Society*, **119**, 1105–1120.

Roth, M., 2000: Review of atmospheric turbulence over cities. *Quarterly Journal of the Royal Meteorological Society*, **126**, 941–990.

Roth, M. 2007: Review of urban climate research in (sub)tropical regions. *International Journal of Climatology* **27**, 1859–1873.

Roth, M., & T.R. Oke, 1993: Turbulent transfer relationships over an urban surface. I. Spectral characteristics. *Quarterly Journal of the Royal Meteorological Society*, **119**, 1071–1104.

Roth, M., & T.R. Oke, 1995: Relative efficiencies of turbulent transfer of heat, mass, and momentum over a patchy urban surface. *Journal of the Atmospheric Sciences*, **52**, 1863–1874.

Roth, M., C. Jansson, & E. Velasco, 2017: Multi-year energy balance and carbon dioxide fluxes over a residential neighborhood in a tropical city. *International Journal of Climatology*, **37**, 2679–2698.

Rouse, W.R., & R.L. Bello, 1979: Short-wave radiation balance in an urban aerosol layer. *Atmosphere-Ocean*, **17**, 157–168.

Rouse, W.R., D. Noad, & J. McCutcheon, 1973: Radiation, temperature and atmospheric emissivities in a polluted urban atmosphere at Hamilton, Ontario. *Journal of Applied Meteorology*, **12**, 798–807.

Rouse, W.R., P.D. Blanken, N. Bussières, et al., 2008: An investigation of the thermal and energy balance of Great Slave and Great Bear Lakes. *Journal of Hydrometeorology*, **9**, 1318–1333.

Rozoff, C.M., W.R. Cotton, & J.O. Adegoke, 2003: Simulation of St. Louis, Missouri, land use impacts on thunderstorms. *Journal of Applied Meteorology*, **42**, 716–738.

Rubin, J. I., A.J. Kean, R.A. Harley, D.B. Millet, & A.H. Goldstein, 2006: Temperature dependence of volatile organic compound evaporative emissions from motor vehicles. *Journal of Geophysical Research*, **111**, D03305.

Runnalls, K.E., & T.R. Oke, 2000: Dynamics and controls of the near-surface heat island of Vancouver, B.C. *Physical Geography*, **21**, 283–304.

Sachweh, M., & P. Koepke, 1995: Radiation fog and urban climate. *Geophysical Research Letters*, **22**, 1073–1076.

Sachweh, M., & P. Koepke, 1997: Fog dynamics in an urbanized area. *Theoretical and Applied Climatology*, **58**, 87–93.

Sailor, D.J., 2011: A review of methods for estimating anthropogenic heat and moisture emissions in the urban environment. *International Journal of Climatology*, **31**, 189–199.

Sailor, D.J., & L. Lu, 2004: A top–down methodology for developing diurnal and seasonal anthropogenic heating profiles for urban areas. *Atmospheric Environment*, **38**, 2004, 2737–274.

Sakakibara, Y., & E. Matsui, 2005: Relation between heat island intensity and city size indices/canopy characteristics in settlements of Nagano Basin, Japan. *Geographical Review of Japan*, **78**, 812–824.

Santamouris, M., A. Synnefa, & T. Karlessi, 2011: Using advanced cool materials in the urban built environment to mitigate heat islands and improve thermal comfort conditions. *Solar Energy*, **85**, 3085–3102.

Santella, N., D.T. Ho, P. Schlosser, et al., 2012: Atmospheric variability and emissions of halogenated trace gases near New York City. *Atmospheric Environment*, **47**, 533–540.

Satterthwaite, D., 2008: Cities' contribution to global warming: notes on the allocation of greenhouse gas emissions. *Environment and Urbanization*, **20**, 539–549.

Savelyev, S.A., & P.A. Taylor, 2005: Internal boundary layers: I. Height formulae for neutral and diabatic flows. *Boundary-Layer Meteorology*, **115**, 1–25.

Savijärvi, H., & J. Liya, 2001: Local winds in a valley city. *Boundary-Layer Meteorology*, **100**, 310–319.

Scheffe, R.D., & R.E. Morris, 1993: A review of the development and application of the Urban Airshed Model. *Atmospheric Environment Part B-Urban Atmosphere*, **27**, 23–39.

Schmid, H.P., 1994. Source areas for scalars and scalar fluxes. *Boundary-Layer Meteorology*, **67**, 293–318.

Schmid, H.P., & D. Bünzli, 1995: The influence of surface texture on the effective roughness length. *Quarterly Journal of the Royal Meteorological Society*, **121**, 1–21.

Schmidt, W., 1927: Die Verteilung der Minimumtemperaturen in der Frostnacht des 12.5.1927 im Gemeindegebiet von Wien. *Fortschritte der Landwirtschaft.*, **2**, (H.21), 681–686.

Schneider, A., M.A. Friedl, & D. Potere, 2009: A new map of global urban extent from MODIS satellite data. *Environmental Research Letters*, **4**, 044003.

Scholz, M., & P. Grabowiecki, 2007: Review of permeable pavement systems. *Building and Environment*, **42**, 3830–3836.

Seinfeld, J.H., & S.N. Pandis, 2006: *Atmospheric Chemistry and Physics*, 2nd edn. John Wiley and Sons, 1232 pp.

Semonin, R.G., 1981a: Cloud characteristics. In: *METROMEX: A Review and Summary*, Changnon, S.A., (ed.), *Meteorological Monographs*, **18**, 63–74.

Sham, S., 1987: *Urbanization and the Atmospheric Environment in the Low Tropics*. Penerbit Universiti Kebangsaan Malaysia, Bangi, 606 pp.

Shanahan, P., 2009: Groundwater in the urban environment. In: *The Water Environment of Cities*, Baker, L.A. (ed.), Springer Science+Business Media, 29–48.

Sharples, S., 1984: Full-scale measurements of convective energy losses from exterior building surfaces. *Building and Environment*, **19**, 31–39.

Shashua-Bar, L., & M.E. Hoffman, 2000: Vegetation as a climatic component in the design of an urban street. An empirical model for predicting the cooling effect of urban green areas with trees. *Energy and Buildings*, **31**, 221–235.

Shashua-Bar, L., D. Pearlmutter, & E. Erell, 2009: The cooling efficiency of urban landscape strategies in a hot dry climate. *Landscape and Urban Planning*, **92**, 179–186.

Shea, D.M., & A.H. Auer Jr., 1978: Thermodynamic properties and aerosol patterns in the plume downwind of St Louis. *Journal of Applied Meteorology*, **17**, 689–698.

Shepherd, J.M., 2013: Impacts of urbanization on precipitation and storms: Physical insights and vulnerabilities. In: *Climate Vulnerability*, Pielke, R. (ed.), Academic Press, 109–125.

Shepherd, J.M., & T.L. Mote, 2011: Can cities create their own snowfall? What observations are required to find out? Posted Sept 6, 2011 to *Earthzine*. Available online at: www.earthzine.org/2011/09/06/can-cities-create-their-own-snowfall-what-observations-are-required-to-find-out/

Shi, C., M. Roth, H. Zhang, & Z. Li, 2008: Impacts of urbanization on long-term fog variation in Anhui Province, China. *Atmospheric Environment*, **42**, 8484–8492.

Shreffler, J.H., 1978: Detection of centripetal heat island circulations. *Boundary-Layer Meteorology*, **15**, 229–242.

Shuttleworth, W.J., 2012: *Terrestrial Hydrometeorology*. Wiley-Blackwell, 448 pp.

Shuzhen Z., & C. Chao, 1985: *An Introduction to Urban Climatology*. Press of East China Normal University, 324 pp. [in Chinese]

Simmons, I.G., 1995: *Earth, Air and Water: Resources and Environment in the Late 20th Century*. Edward Arnold, London U.K.

Sini, J., S. Anquetin, & P. Mestayer, 1996: Pollutant dispersion and thermal effects in urban street canyons. *Atmospheric Environment*, **30**, 2659–2677.

Small, C., 2004: Global population distribution and urban land use in geophysical parameter space. *Earth Interactions*, **8**, 1–18.

Smith, D.M., & S.J. Allen, 1998: Measurement of sap flow in plant stems. *Journal of Experimental Botany*, **47**, 1833–1844.

Sobrino, J.A., R. Oltra-Carrió, J.C. Jiménez-Muño, et al. 2012: Emissivity mapping over urban areas using a classification-based approach: Application to the Dual-use European Security IR Experiment (DESIREX). *International Journal of Applied Earth Observation and Geoinformation*, **18**, 141–147.

Song, T., & Y. Wang, 2012: Carbon dioxide fluxes from an urban area in Beijing. *Atmospheric Research*, **106**, 139–149.

Soulhac, L., V. Garbero, P. Salizzoni, P. Mejean, & R.J. Perkins, 2009: Flow and dispersion in street intersections. *Atmospheric Environment*, **43**, 2981–2996.

Sportisse, B., 2009: *Fundamentals in Air Pollution*. Springer Science & Business Media, Dordrecth, 299 pp.

Spronken-Smith, R.A., 1994: *Energetics and cooling in urban parks*. Ph.D. thesis, Department of Geography, University of British Columbia, 203 pp.

Spronken-Smith, R.A., 2002: Comparison of summer- and winter-time suburban energy fluxes in Christchurch, New Zealand. *International Journal of Climatology*, **22**, 979–992.

Spronken-Smith, R.A., & T.R. Oke, 1998: The thermal regime of urban parks in two cities with different summer climates. *International Journal of Remote Sensing*, **19**, 2085–2104.

Spronken-Smith, R.A., T.R. Oke, & W.P. Lowry, 2000: Advection and the surface energy balance of an irrigated urban park. *International Journal of Climatology*, **20**, 1033–1047.

Stallins, J.A., & M.L. Bentley, 2006: Urban lightning climatology and GIS: An analytical framework from the case study of Atlanta, Georgia. *Applied Geography*, **26**, 242–259.

Stallins, J.A., & L. S. Rose, 2008: Urban lightning: Current research, methods and the geographical perspective. *Geography Compass*, **2/3**, 620–639.

Stallins, J.A., J. Carpenter, M.L. Bentley, W.S. Ashley, & J.A. Mulholland, 2013: Weekend-weekday aerosols and geographic variability in cloud-to-ground lightning for the urban region of Atlanta, Georgia, USA. *Regional Environmental Change*, **13**, 137–151.

Stearns, C.R., & H.H. Lettau, 1963: Two wind-profile measurement experiments in airflow over the ice of Lake Mendota. University of Wisconsin, Department of Meteorology, Annual Report, Madison. Wis., 115–138.

Steeneveld, G.J., M.J.J. Wokke, C.D. Groot Zwaaftink, et al., 2010: Observations of the radiation divergence in the surface layer and its implication for its parameterization in numerical weather prediction models. *Journal of Geophysical Research*, **115**, D06107.

Steinhauser, F., O. Eckel, & F. Sauberer, 1959: Klima und Bioklima von Wien. *Wetter und Leben*, **7**, 1–136.

Stewart, I.D., 2011: A systematic review and scientific critique of methodology in modern heat island literature. *International Journal of Climatology*, **31**, 200–217.

Stewart, I.D., & T.R. Oke, 2012: Local Climate Zones for urban temperature studies. *Bulletin of the American Meteorological Society*, **93**, 1879–1900.

Stewart, I.D., T.R. Oke, & E.S. Krayenhoff, 2014: Evaluation of the 'local climate zone' scheme using temperature observations and model simulations. *International Journal of Climatology*, **34**, 1062–1080.

Stohl, A., C. Forster, S. Eckhardt, *et al.*, 2003: A backward modeling study of intercontinental pollution transport using aircraft measurements. *Journal of Geophysical Research*, **108**, *4370*, D12.

Stone, B., 2004. Paving over paradise: how land use regulations promote residential imperviousness. *Landscape and Urban Planning*, **69**, 101–113.

Strikas, O.M., & J.B. Elsner, 2013: Enhanced cloud-to-ground lightning frequency in the vicinity of coal plants and highways in Northern Georgia, USA. *Atmospheric Science Letters*, **14**, 243–248.

Stull, R.B., 2000: *Meteorology for Scientists and Engineers*, 2nd edn. Brooks Cole Thomson Learning, 528 pp.

Sugawara, H., T. Narita, & T. Mikami, 2001: Estimation of effective thermal property parameter on a heterogeneous urban surface. *Journal of the Meteorological Society of Japan*, **79**, 1169–1181.

Sukopp, H., 1998: Urban Ecology—Scientific and Practical Aspects. In: *Urban Ecology*, Breuste, J., H. Feldmann, O. Uhlmann, (eds.), Springer, Berlin. 3–16.

Summers, P.W., 1964: *An urban ventilation model applied to Montréal.* PhD Thesis, McGill University, Montréal.

Sundborg, A., 1950: Local climatological studies of the temperature conditions in an urban area. *Tellus*, **2**, 222–232.

Synnefa, A., M. Santamouris, & I. Livada, 2006: A study of the thermal performance of reflective coatings for the urban environment. *Solar Energy*, **80**, 968–981.

Szymanowski, M., 2005: Interactions between thermal advection in frontal zones and the urban heat island of Wrocław, Poland. *Theoretical and Applied Climatology*, **82**, 207–224.

Taesler, R., 1980: Studies of the development and thermal structure of the urban boundary layer in Uppsala, Part I, Experimental program, and Part II, Data analysis and results, Report No. 61, Meteorological Institute, Uppsala University, Uppsala, Parts I and II consist of 57 and 177 pp., respectively.

Takahashi, M., 1959: Relation between the air temperature distribution and the density of houses in small cities of Japan. *Geographical Review of Japan*, **32**, 305–313.

Takebabayasi, H., & M. Moriyama, 2009: Study of the urban heat island mitigation effect achieved by converting to grass-covered parking. *Solar Energy*, **83**, 1211–1223.

Taniguchi, M., T. Uemura, & K. Jago-on, 2007: Combined effects of urbanization and global warming on subsurface temperature in four Asian cities. *Vadose Zone Journal*, **6**, 591–596.

Tapper, N.J., 1990: Urban influence on boundary layer temperature and humidity: results from Christchurch, New Zealand. *Atmospheric Environment, Part B*, **24**, 19–27.

Tardif, R., & R.M. Rasmussen, 2007: Event-based climatology and typology of fog in the New York City region. *Journal of Applied Meteorology and Climatology*, **46**, 1141–1168.

Tejeda-Martínez, A., & E. Jáuregui, 2005: Surface energy-balance measurements in the Mexico City region: A review. *Atmósfera*, **18**, 1–23.

Terjung, W.H., 1974: Urban climatology: with reference to the inter-relationship between external weather and the microclimate in houses and buildings. In: *Progress in Biometeorology, Vol.1*, Pt.1, Ch.4, Sect.5, Tromp, S.W. (ed.), Swets and Zeitlinger, Amsterdam, 168–180.

Theurer, W., E.J. Plate, & K. Hoeschele, 1996: Semi-empirical models as a combination of wind tunnel and numerical dispersion modelling. *Atmospheric Environment*, **30**, 3583–3597.

Thompson, W.T., T. Holt, & J. Pullen, 2007: Investigation of a sea breeze front in an urban environment. *Quarterly Journal of the Royal Meteorological Society*, **133**, 579–594.

Thornes, J.E., & J. Shao, 1991: Spectral analysis and sensitivity tests for a numerical road surface temperature prediction model. *Meteorological Magazine*, **120**, 117–124.

Trusilova, K. & G. Churkina, 2008: The response of the terrestrial biosphere to urbanization: land cover conversion, climate, and urban pollution. *Biogeosciences*, **5**, 1505–1515.

Tseng, Y.-H., C. Meneveau, & M.B. Parlange, 2006: Modeling flow around bluff bodies and predicting urban dispersion using large eddy simulation. *Environmental Science and Technology*, **40**, 2653–2662.

Tulet, P., A. Maalej, V. Crassier, & R. Rosset, 1999: An episode of photooxidant plume pollution over the Paris region. *Atmospheric Environment*, **33**, 1651–1662.

Turco, R.P., 2002: *Earth Under Siege*, 2nd edn. Oxford University Press, 552 pp.

Turner, D., 1970: *Workbook of Atmospheric Dispersion Estimates*. Office of Air Programs Publication No. AP-26 (NTIS PB 191 482). US Environmental Protection Agency.

Tyson, P.D., M. Garstang, & G.D. Emmitt, 1973: The structure of heat islands. Occas. Paper No.12, Dept. Geog. Environ. Studies, Univ. Witwatersrand, Johannesburg.

Um, H.-H., K.-J. Ha, & S.-S. Lee, 2007: Evaluation of the urban effect of long-term relative humidity and the separation of temperature and water vapor effects. *International Journal of Climatology*, **27**, 1531–42.

UN, 2015: World Urbanization Prospects: The 2014 Revision. United Nations, Department of Economic and Social Affairs, Population Division, ST/ESA/SER.A/366.

Upmanis, H., Eliasson, I., & S. Lindqvist, 1998: The influence of green areas on nocturnal temperatures in a high latitude city (Göteborg, Sweden). *International Journal of Climatology*, **18**, 681–700.

Urban Task Force, 1999: *Toward an Urban Renaissance*. Taylor & Francis, London.

Valentini, R., D.D. Baldocchi, & J.D. Tenhunen, 1999: Ecological controls on land-surface atmosphere interactions. In: *Integrating Hydrology, Ecosystem Dynamics and Biogeochemistry in Complex Landscapes*. Tenhunan, J., & P. Kabat, (eds.), Dahlem Workshop Report, John Wiley, 117–145.

Van den Heever, S., & W.R. Cotton, 2007: Urban aerosol impacts on downwind convective storms. *Journal of Applied Meteorology and Climatology*, **46**, 828–850.

Van Hoof, J., 2008: Forty years of Fanger's model of thermal comfort: comfort for all? *Indoor Air*, **18**, 182–201.

Van Vuuren, D.P., J. Edmonds, M. Kainuma, et al., 2011: The representative concentration pathways: an overview. *Climatic Change*, **109**, 5–31.

Vardoulakis, S., N. Gonzalez-Flesca, & B.E.A. Fisher, 2002: Assessment of traffic-related air pollution in two street canyons in Paris: implications for exposure studies. *Solar Energy* **36**, 1025–1039.

Vardoulakis, S., B. Fisher, K. Pericleous, & N. Gonzalez-Flesca, 2003: Modelling air quality in street canyons: a review. *Atmospheric Environment*, **37**, 155–182.

Varma, R., & D.R. Varma, 2005: The Bhopal disaster of 1984. *Bulletin of Science, Technology & Society*, **25**, 37–45.

Vautard, R., P. Yiou, & G. van Oldenborgh, 2009: Decline of fog, mist and haze in Europe over the past 30 years. *Nature Geoscience Letters*, **2**, 115–119.

Velasco, E., S. Pressley, E. Allwine, H. Estberg, & B. Lamb, 2005: Measurements of $CO_2$ fluxes from the Mexico City urban landscape. *Atmospheric Environment*, **39**, 7433–7446.

Velasco, E., S. Pressley, R. Grivicke, et al., 2009: Eddy covariance flux measurements of pollutant gases in urban Mexico City. *Atmospheric Chemistry and Physics*, **9**, 7325–7342.

Velasco, E., S. Pressley, R. Grivicke, E. Allwine, L.T. Molina, & B. Lamb, 2011: Energy balance in urban Mexico City: Observation and parameterization during the MILAGRO/MCMA-2006 field campaign. *Theoretical and Applied Climatology*, **103**, 501–517.

Velasco, E., M. Roth, S.H. Tan, M. Quak, S.D.A. Nabarro, & L. Norford, 2013: The role of vegetation in the $CO_2$ flux from a tropical urban neighbourhood. *Atmospheric Chemistry and Physics*, **13**, 10185–10202.

Vesala, T., L. Jarvi, S. Launiainen, et al., 2008: Surface-atmosphere interactions over complex urban terrain in Helsinki, Finland. *Tellus*, **60B**, 188–199.

Vogt, R., & E. Parlow, 2011: Die städtische Wärmeinsel von Basel-tages-und jahreszeitliche Charakterisierung. *Regio Basiliensis*, **52**, 7–15.

Vogt, R., A. Christen, M.W. Rotach, M. Roth, & A.N.V. Satyanarayana, 2006: Temporal dynamics of $CO_2$ fluxes and profiles over a central European city. *Theoretical and Applied Climatology*, **84**, 117–126.

Voogt, J.A., 2002: Urban heat island. In: Causes and Consequences of Global Environmental Change, Douglas. I. (ed.), Vol. 3 in *Encyclopedia of Global Environmental Change,* Munn, R.E. (ed-in chief), John Wiley & Sons, 660–666.

Voogt, J.A., & T.R. Oke, 1997: Complete urban surface temperatures. *Journal of Applied Meteorology*, **36**, 1117–1132.

Voogt, J.A., & T.R. Oke, 2003: Thermal remote sensing of urban climates. *Remote Sensing of Environment*, **86**, 370–384.

Vukovich, F.M., & J.W. Dunn, 1978: A theoretical study of the St. Louis heat island: Some parameter variations. *Journal of Applied Meteorology*, **17**, 1585–1594.

Wallace, J.M., & P.V. Hobbs, 2006: *Atmospheric Science: An Introductory Survey*, 2nd Edition. Elsevier, 504 pp.

Walter, H., & H. Lieth, 1967: *Klimadiagram-Weltatlas*. VEB Gustav Fischer Verlag, Jena

Wanner, H., & P. Filliger, 1989: Orographic influence on urban climate. *Weather and Climate*, **9**, 22–28.

Weber, S., & K. Kordowski, 2010: Comparison of atmospheric turbulence characteristics and turbulent fluxes from two urban sites in Essen, Germany. *Theoretical and Applied Climatology*, **102**, 61–74.

Webster, N. Jr., 1799: A dissertation on the supposed change in the temperature of winter, *Memoirs Connecticut Acad. Arts & Sci.*, **1**, 1–68.

Weissert, L.F., J.A. Salmond, & L. Schwendenmann, 2014: A review of the current progress in quantifying the potential of urban forests to mitigate urban $CO_2$ emissions. *Urban Climate*, **8**, 100–125.

Welty, C., 2009: The urban water budget. In: *The Water Environment of Cities*, Baker, L.A. (ed.), Springer Science+Business Media, 17–28.

Wennberg, P.O., W. Mui, D. Wunch, et al. 2012: On the sources of methane to the Los Angeles atmosphere. *Environmental Science & Technology*, **46**, 9282–9289.

Westcott, N.E., 1995: Summertime cloud-to-ground lightning activity around major Midwestern urban areas. *Journal of Applied Meteorology*, **34**, 1633–1642.

Whitby, K.T., 1978: The physical characteristics of sulphur aerosols. *Atmospheric Environment*, **12**, 135–159.

White, J.M., F.D. Eaton, & A.H. Auer, Jr., 1978: The net radiation budget of the St. Louis Metropolitan area. *Journal of Applied Meteorology*, **17**, 593–599.

WHO, 2014: 7 million premature deaths annually linked to air pollution. *WHO International*, www.who.int/mediacentre/news/releases/2014/air-pollution/en/.

WHO, 2006: *WHO Air quality guidelines for particulate matter, ozone, nitrogen dioxide and sulfur dioxide. Global update 2005. Summary of risk assessment.* Publications of the World Health Organization. Geneva. 22pp.

WHO, 2009: *Global health risks: mortality and burden of disease attributable to selected major risks.* Publications of the World Health Organization. Geneva. 62pp.

Wieringa, J., 1993: Representative roughness parameters for homogeneous terrain. *Boundary-Layer Meteorology*, **63**, 323–363.

Wieringa, J., 1986: Roughness-dependent geographical interpolation of surface wind speed averages. *Quarterly Journal Royal Meteorological Society*, **112**, 867–889.

Wilson, K.B., D.D. Baldocchi, M. Aubinet, et al., 2002: Energy partitioning between latent and sensible heat flux during the warm season at FLUXNET sites, *Water Resources Research*, **38**, 1294, doi:10.1029/2001WR000989.

Wittich, K.-P., 1997: Some simple relationships between land-surface emissivity, greenness and plant cover fraction for use in satellite remote sensing. *International Journal of Biometeorology*, **41**, 58–65.

Wittwer, W.C., 1860: Grundzüge der Klimatologie von Bayern. In: *Landes-und Volkskunde d. Kgr. Bayern, Vol.1*, Munich.

WMO, 2008: *Guide to Meteorological Instruments and Methods of Observation*, Part II-Observing Systems, Chapter 11, WMO-No.8, 7th Edition, World Meteorological Organization, Geneva, II-11-1 – II-11–25.

Woods-Ballard, B., S. Wilson, H. Udale-Clark, et al., 2015: *The SUDS manual* (Vol. 753). London: CIRIA.

Wunch, D., P.O. Wennberg, G.C. Toon, G. Keppel-Aleks, & Y.G. Yavin, 2009: Emissions of greenhouse gases from a North American megacity. *Geophysical Research Letters*, **36**, L15810.

Wurman, J., C. Alexander, P. Robinson, & Y. Richardson, 2007: Low-level winds in tornadoes and potential catastrophic tornado impacts in urban areas. *Bulletin of the American Meteorological Society*, **88**, 31–46.

Yakhot, A, H. Liu, & N. Nikitin, 2006: Turbulent flow around a wall-mounted cube: A direct numerical simulation. *International Journal of Heat and Fluid Flow*, **27**, 994–1009.

Yamartino, R. J., & G. Wiegand, 1986: Development and evaluation of simple-models for the flow, turbulence and pollutant concentration fields within an urban street canyon. *Atmospheric Environment*, **20**, 2137–2156.

Yoshikado, H., 1990: Vertical structure of the sea breeze penetrating through a large urban complex. *Journal of Applied Meteorology*, **29**, 878–891.

Yoshikado, H., 1992: Numerical study of the daytime urban effect and its interaction with the sea breeze. *Journal of Applied Meteorology*, **31**, 1146–1164.

Yoshino, M.M., 1975: *Climate of a Small Place*, University of Tokyo Press, Tokyo, 549 pp.

Yoshino, M.M. & S. Yamashita, 1998: *Encyclopedia of Urban Environment*, Asakura Publishing Co., Tokyo, 435 pp. [In Japanese]

Yow, D.M., 2007: Urban heat islands: observations, impacts and adaptations. *Geography Compass*, **1**, 1227–1251.

Zhang, J., & K.R. Smith, 2003: Indoor air pollution: a global health concern. *British Medical Bulletin*, **68**, 209–225.

Zhu, K., P. Blum, G. Ferguson, K.-D. Balke, & P. Bayer, 2010: The geothermal potential of urban heat islands. *Environmental Research Letters*, **5**, 044002.

Zrudlo, L.R. 1988: A climatic approach to town planning in the arctic. *Energy and Buildings*, **11**, 41–63.

# Subject Index

*Italic* page numbers refer to glossary entries, **bold** page numbers to sections or chapters where this subject is discussed in detail.

absolute humidity 255, *469*
absorption 15, **124**, 146, 202, *469*
absorptive efficiency 136
absorptivity 18, **123**, 124, *469*
accumulation mode 146
accuracy 18, *469*
acid deposition 329
acid fog 265
active remote sensor 46
active surface **15**, 170, 389, *469*
adaptation **361**, *469*
adiabatic process 66
advection
    definition 37, *469*
    measurement 48
    of heat 159, 180, 204
    of momentum 81
    of pollutants 315
    of water vapour 243, 252, 257, 435
advection fog 255
aerodynamic method 179
aerodynamic resistance 176, **183**, 301, *469*
aerodynamic roughness 436
aerodynamic roughness length 202
aerosols
    air pollution 299, 361
    chemical composition 146
    cloud formation 270
    cloud microphysics 282
    condensation 256
    definition *469*
    emissions 291
    invigoration 289
    measurement 32
    radiation **125**, 205, 268
    urban boundary layer 145
air conditioning 245, 248, 427
air pollutants 45
    above roof level 92
    at intersections 91, 302
    definition 295, *469*
    dispersion 58, 85, 93, 231, **299**
    emissions 4, 25, 34, **298**
    fog 265
    in urban canyons 89
    indoor 304

    radiation 123, 147
    recirculation 117, 353
    removal 301
    residence time 302
    transformation 302
air pollution **294–331**, 346, 455
air pollution episode 317
air quality 446
air quality guidelines 315
air temperature 202
aircraft measurements 57
airflow **77–121**, 431
Aitken nuclei 125
albedo
    definition **127**, *469*
    of cities **140–145**
    of facets 185
    of human skin 390
    planetary 361
    urban design 424
    urban fabrics 128
aligned array 23
ammonia 330
anabatic wind 347, *469*
anemometer 46, 454, *469*
anisotropic *469*
anisotropy 50
anthropogenic 3, 25
anthropogenic climate change 302, **360–384**
anthropogenic heat flux
    definition 159, *469*
    energy balance **160–168**
    global climate 362
    macroclimates 340
    urban design 427
    urban heat island 199
anthropogenic water vapour **247**, 264, *469*
anticyclone 266, *469*
applied climatology 14
artificial light 149
asbestos 306
atmospheric boundary layer 30, *469*
atmospheric moisture **254–269**
atmospheric window 128, 203, 362, *470*
attenuation 97, 147, *470*
available energy 177, *470*

barrier effect 113, 118, 287, *470*
base flow 245, *470*
basin 343
benzene 308
Bergeron-Findeisen process 272
bifurcation 287
biodetention system 442
biodiesel 364
biogenic emissions 299
biological air pollutants 306
biosphere 2, **5**
bird's eye view 15
blackbody 128, 390, *470*
blending height **33**, 98, *470*
blossoming 223
body area 388
Bolz relation 220, 222
boundary conditions 61, 67, 71, *470*
boundary layer *470*
boundary layer urban heat island 199, **226**, *470*
Bowen ratio **158**, 182, 340, 362, *470*
Bowen ratio-energy balance approach 179
brightness temperature 128, 202, *470*
brown sites 12
Brunt-Väisälä frequency 344
building climatology 18
building design 427
building energy model 163, 172
buildings 132
buoyancy **81**, 105, 228, *470*

canopy layer urban heat island 199, **213**, 257, 341, *470*
canopy resistance 178, 180, *470*
canopy ventilation 423
canyon aspect ratio 25, 60, 215, 415, *470*
canyon trapping 136, *470*
canyon venting 90
capping inversion 30, *470*
carbon cycle 368
carbon dioxide *470*
    combustion 298
    emissions 438
    global emissions 12
    greenhouse gas 296, **362**, 363
    plant gas exchange 176
    radiation 128, 146
    respiration 307
    urban enhancement 370
carbon monoxide 296, 298, 305, 308, 315
carbon pool 374
carbon sequestration 438
catchment 241
cavity zone 84, *470*
central city 8

chaotic flow 34, 95, *470*
Chinook 346
climate assessment 410
climate change 302, 360–384
climate risks 448
climate-sensitive urban design **408–452**, 459, *470*
climatic elements 333
climatography 14
climatope **28**, 411, *470*
clothing 389, 393
clothing area factor 391
cloud **271–294**, *471*
cloud condensation nuclei
    cloud formation 271
    definition *471*
    fog 265
    humidity 255
    smog 317
    wet deposition 301
cloud cover 153, 220, 275
cloud heights 276
cloud microphysics 282
coal 267, 317, 364
coalescence 272
coarse particle mode 146
coarse particulates 126
coastal flooding 448
coastal wind systems 350
coefficient of heat transfer 392
cold front 278
cold-cloud processes 272
combined sewers 241, *471*
combustion 247, **298**, 305, 365, *471*
compact city 444
complete aspect ratio **21**, 60, 130, 170, *471*
complete surface temperature 206, *471*
complete urban surface 15, 208, *471*
computational fluid dynamics **71**, 310, *471*
concentration 45, 271, **299**, *471*
condensation
    cloud 255, **270–293**
    cloud formation 271
    definition *471*
    dewfall 255
    energetics 158, **256**
    water cycle 239
conditional sampling 41
conductance **172**, 301, *471*
conduction 157, 202, *471*
coning 314
conservation laws 66
convection 157, 202, 243, *471*
convective available potential energy 284
convective clouds 290

convective heat transfer coefficient 75
convective precipitation 277
convergence *471*
cool pavement 425, 451
cool roof 406, 419, 426, *471*
cool walls 427
cooling degree-days 223, *471*
cooling rates 223
Coriolis force 63, 78, 346, *471*
counter urbanization 10
cyclone 282, 355, *471*

decoupling factor 178
deposition 255, *471*
deposition velocity 301, 437, *471*
desakota 9
detention storage 236, 241, 412, *471*
dewfall 158, 255, **264**, *471*
dewpoint temperature 191, **256**, 259, *471*
diabatic process 66
diesel 364
diffuse irradiance **135**, 389, *471*
diffusion *471*
diffusivity 301, *471*
dimensional analysis 42, 60, 105, *472*
direct numerical simulation 71
direct-beam irradiance **126**, *472*
discharge curve 250
dispersion
    definition *472*
    of air pollutants 312, 457
    of chemicals 295
    urban design 415
displacement zone 83
dissipation 81, *472*
divergence *472*
domain 61, 68, *472*
dose 295, *472*
downdrafts 221
downscaling 379, *472*
downwash 85, *472*
drag *472*
drag coefficient 386
droplet size 288
drought 246
dry deposition **301**, 304, 437, *472*
dynamic similarity 60
dynamic stability **81**, 231, *472*

ecological footprint 10, *472*
ecosystem 2
eddy 79, *472*
eddy covariance 176, 179, 371, *472*
ejection 80, *472*

Ekman wind spiral 113
emission 6, 37, 254, *472*
emission factor 325, 363–364, *472*
emission inventory 323
emissivity
    definition 123, *472*
    of cities **140–145**
    of urban canyons 143
    urban fabrics 129
    urban heat island 202
    urban surface 18
emittance 123, *472*
empirical models 74
energy balance 156–196, *472*
energy balance models 71
energy balance residual approach 163, 171
energy conservation principle 124, 157
energy efficiency 161
entrainment 31, 184, 257, 264, *472*
entrainment zone 30, *472*
equivalent carbon dioxide emissions
        365, *472*
Eulerian 68, *472*
evaporation
    advective effects 182
    definition *473*
    energetics 158, **248**
    human body 392
    humidity 254
    sources 239
    urban design 423
    urban heat island 218
evaporative cooling 190
evaporative emissions 299, 304, 318
evapotranspiration *473*
excess resistance 177, *473*
exposure
    air pollutants 295
    air pollution 447
    definition *473*
    instrument 46–47
    to weather hazards 448

fabric **6**, 14, 20, 426, *473*
facet 18, 156, *473*
Fanger model 396
fanning 300, 314, *473*
fetch 46, 48, *473*
field observations 45, 459
field of view 47, 206, *473*
fine particulates 126
fixed stations 52
flooding **357**, 448
floor space ratio 21

flow separation 82, *473*
flow-following methods 58
fluvial flooding 448
flux *473*
flux density *473*
Foehn 346
fog
  air pollution 455
  definition 256, *473*
  formation **265**
  harvesting 267
  urban heat island 197
footprint *473*
forcing *473*
form drag 80, 176, *473*
formaldehyde 306, 318
fossil fuels 10
fountains 441
free atmosphere 30, 78, *473*
freezing rain 279
friction velocity 100, *473*
front 221, 355–356, *473*
frontal aspect ratio **22**, *473*
frontal fog 265
frost 223
Froude number 344
fugitive emissions 299
fumigation 230, 314, *473*
fumulus 270
function 14

garden irrigation 245
gardens 3, 187, 438
gasoline 364
Gaussian plume model 313
general circulation model 378, *473*
geometric similarity 60
geothermal heat flux 234
giant particles 125
global climate change 361
global land-cover change 361
global positioning system 56
global warming potential 365, *473*
gradient wind *474*
gravitational settling 301, 304
green roof 244, **426**, 442, *474*
green wall 427
greenhouse effect *474*
greenhouse gas 295, 361, **362**, 424, *474*
grey water 251
ground temperature 205
groundwater *474*
  drinking water 241
  formation 5, 239

pipe leakage 247
subsurface heat island 234
temperature 236
urban hydrology 252
water table 252
growing degree-days 223
gusts 401

haboob 332
hail 273, 279
halocarbons 146, 296, 363
Harmattan 412
hazards 381
haze 126, 268, **272**, 302, *474*
heat capacity 18, **168**, 202, *474*
heat flux density *474*
heat island magnitude **205**, *474*
heat sharing 170
heat storage change 159, **168–175**, 211, *474*
heat stress 223, 450
heat venting 192
heating degree-days 223, 323, 367, *474*
heatwave **358**, 385, 451, *474*
heliodon 61
hills 344
hinterland 10, 360
hoar frost 256
homeostasis 386, *474*
homogeneity 28
horizon screening 22, 137, 216, 429, *474*
horizontal averaging 38
human respiration 245, 368, 387, 392
humidex 400
humidity **254–269**, 455
hurricanes 355–356
hydraulic flow 345
hydraulic jump 344
hydrogen gas 364
hydrologic cycle 239, *474*
hydroperoxyl radical 302
hydrosphere 2, **5**
hydroxyl radical 302
hygrometer 46, 256, *474*
hygroscopic *474*
hysteresis loop 173, *474*

ice fog 265, 267
ice nuclei 255, 271, *474*
ideal gas law **66**
impervious fraction 449
impervious surface fraction 252
inadvertent climate modification 2
indoor air pollution 304
indoor climates 406

indoor comfort 406
indoor thermal comfort 396
Industrial Revolution 7, 148, 267, 454
inertial sublayer **33**, 52, **98**, 176, *474*
infiltration 5, 238, **251**, *475*
inflection point 96
inner region 30
instruments 46
insulation 211, 388–389, *475*
interception **243**, *475*
internal boundary layer
  airflow 108
  definition 34, *475*
  observations 48
  urban design 419
  urban heat island 227
International Association for Urban Climate 454,
  459
intersections 91, 302
intertropical convergence zone 333
inversion 158, 346, *475*
invigoration 289
irradiance 126, 339, *475*
irrigation 23, 245, 262
irrigation ban 246
isolated roughness flow **87**, *475*
isotopologues 368
isotropic 81, *475*

jetting 90, 433, *475*

katabatic flow 226, 411, 415
katabatic wind **347**, 349, *475*
Kirchhoff's Law 123, 128
Köhler curve 271
Köppen macroclimates 336

La Porte anomaly 274
Lagrangian 68, *475*
lake breeze 439, *479*
Lambertian 127
laminar boundary layer 176, 301, *475*
land breeze 351, 439, *475*
land-use management 419
lapse rate *475*
Large aerosol 125
large eddy simulation 71, 310, *475*
latent heat 202, *475*
latent heat flux density **157**, 339, 423, *475*
latent heat of vaporization 158, 248
lead 296
leading edge 49, *475*
leading edge effect 181, *475*
Leighton relationship 318

lidar 23, *475*
light pollution 127, 148
lightning 272, 279
liquefied petroleum gas 364
liquid water content *475*
lithosphere 2, **5**
Local Climate Zones
  definition **25**, 48, *475*
  urban heat island studies 205
  urban planning 423
local scale 198, *475*
local scaling 98
lofting 315
longwave radiation 123, **127**, 150, 202, *476*
looping 301, *476*
low emissivity glass 130
Lowry's framework **35**, 262, 332, 377
lysimeter 179

macroclimate **333**
macroscale *476*
managed system 5
maximum urban heat island 215
mean building height 21
mean radiant temperature 390, *476*
megalopolitan plume 31
MEMI model 401
mesoscale *476*
mesoscale models 72
metabolic heat 159, 387, *476*
metadata 54, 226, 377, *476*
methane
  emissions 299, 373
  greenhouse gas 296, 363
  photochemical smog 318
  radiation 128, 146
METROMEX 146, 275, 458, *476*
microorganisms 307
microscale 198, *476*
mini-lysimeter 179
Mistral 347
mitigation 361, *476*
mixed layer 43, 176, 228, 346, *476*
mixed layer depth 316
mixing ratio *476*
mobile measurements 55, 308
model domain 61
model validation 73
molar mixing ratio 299, *476*
mold 307
molecular diffusivity 176, 301
momentum *476*
momentum flux *476*
Monin-Obukhov similarity **105**, *476*

mountain wind **348**, 349

multiple reflection 134

nadir *476*

natural area 4, *476*

natural gas 364

natural hazards 381, 448

near infrared radiation 149

neighbourhood 20, 423, *476*

net allwave radiation

    definition **126**, *477*

    energy balance 157

    human body 390

    macroclimate 339

    urban–rural difference 151

net ecosystem productivity 375, *477*

neutral 81, *477*

nitric acid 302

nitric oxide 299, 302

nitrogen dioxide 295, 299, 302, 309, 325

nitrogen oxides 296, 305, 315, 318, 324, 437

nitrous oxide 128, 146, 296, 363, **364**

nocturnal boundary layer 30, 233, *477*

noise barrier 433, 448

nucleation mode 145

Numerical discretization 69

numerical modelling 42, **66**, *477*

oasis effect 181, *477*

Objective Hysteresis Model 173

Obukhov length 105

Ohm's Law 172

orography 198, 332, 412, *477*

outdoor physical models 63

outdoor thermal comfort 397

outer region 30

ozone

    air pollution 296, 298, 315

    air quality management 437

    greenhouse gas 364

    indoor air quality 305

    photochemical smog 318

    radiation 128

    ultraviolet absorption 149

    urban plume 347

parameterization 68, 74, 171, *477*

park 3, 20, 181, 403, 408, **438**

park cool island 215

particulate mass density 148

particulate matter

    air pollution 296, 299, 315

    definition *477*

    emissions 305, 308

    human exposure 305

    measurement 32

    plant damage 437

    removal 429

    smog 317

    transport 339

    visibility 148

passive remote sensor 47

pavement watering 441

paving 424

pedosphere 2, **5**

Penman-Monteith equation 178

percolation 239

permeability 18

permeable paving 425, 442, *477*

peroxyacetyl nitrate 320

perturbation *477*

phenology 192

photochemical grid model 322, *477*

photochemical smog 265, 318, 437, 448, *477*

photodissociation 302

photon 302, *477*

photosynthesis 363, **367**, 375, 435, *477*

photosynthetically active radiation *477*

phreatic layer 239

physical climatology 457

physical modelling **60**, *477*

physiological equivalent temperature 397, 401, *477*

pigmentation 390

pipe leakage 245, **247**, 252

piped water supply 244, 246

plan area fraction 20, 60, *478*

plan view 15

Planck constant 318

Planck's Law 123

pluvial flooding 448

PM10 299, *478*

PM2.5 299, *478*

podium 404

polycyclic aromatic hydrocarbons 296, 305, 308

ponds 5, 442

population distribution 333

postural balance 386

potential temperature 176, 230, *478*

power law 102

precipitation

    definition *478*

    macroclimates 335

    surface water balance 243

    urban effects **270–293**, 455

    washout 301

    water cycle 239

precision *478*

predicted mean vote 396–397, *478*

pressure gradient force 66, 77–78, *478*
primary pollutants 298, *478*
profile *478*
prototype 60
psychrometric constant 178
public transit 444

radar 23, 274, *478*
radiance 127, *478*
radiant energy *478*
radiation **122–155**, 157, 202, 431, *478*
radiation fog 265–266
radiation source area 49
radiation-mass interactions 124
radiative forcing 302, 361, *478*
radiative heat transfer coefficient 391
radiative properties 202
radioactivity 306
radiocarbon 368
radiometer *478*
radiosonde 57
radon 306
rain gardens 449
rain gauge 273
rainout 126, 301, *478*
random array 23
Rayleigh scattering 148
reflectance 127, *478*
reflection 15, 122
reflective coating 451
reflectivity 18, 124, *478*
relative humidity **255**, 271, *478*
remote sensing 46, **59**, 459, 478
renewable energy 10
representative concentration pathways 379, *478*
residence time 302, *478*
residual layer 30, 233, 320, *478*
resilience 381, *479*
resistance 172, *479*
respiration
    biogenic emissions 299
    by plants 375
    carbon dioxide 363, 367
    definition *479*
    in soils 375
    of humans 392
    urban enhancement 236
return period **355**, *479*
Reynolds decomposition 78
Reynolds number 60
Reynolds stress 80, 181, *479*
ridges 344
risk 381, *479*
rivers

    contamination 250
    surface water balance 241
    urban design 413, 439, 449
    water cycle 238
roads 186
roadside station 308
roofs 185, 265, 426
roughness length **102**, *479*
roughness sublayer **33**, 48, 176, 303, *479*
runoff 238, 240, **249**, 424, *479*
rural 4, *479*

salt particles 271
sanitary sewers 241
Santa Ana winds 347
sap flow measurement 179
saturation 255, *479*
saturation vapour pressure 255, *479*
scalar *479*
scale 18, 30
scattering 126, 146, 148, *479*
screen-level *479*
sea breeze 351, 439, *479*
sea breeze front 351, 353, *479*
secondary pollutants 298, *479*
sensible heat *479*
sensible heat flux density
    definition *480*
    energy balance **157**, 339
    of human body 391
    urban design 423
    urban heat island 200
sensor 45, *480*
settling velocity 125–126, 301
sewer leakage 247
shading 122, 416, 431, 451
shadow mask 132
shear zone 88, 96, 98, 186, *480*
shivering 387
shortwave 123, 202, *480*
shortwave irradiance **126**, 146, 333
shortwave reflectance 127
similitude 60, *480*
skimming flow **87**, 104, 232, 308, *480*
skin drag 80, 176, *480*
sky view factor
    definition **22**, *480*
    heat island magnitude 209, 213
    human bioclimate 399
    of an urban canyon 189
    radiation 122, 131
    urban design 417
    urban heat island 215
slope winds 348–349

smog 197, 265, 316, *480*

snow accumulation 244, 413

snow cover

   albedo 143

   energy balance 170, 193

   precipitation 286

   radiation 154

   urban heat island 218

snow drift 413

snow removal 143

snowfall 143, 243, 279

snowmelt 193

snowout 301

sodar *480*

sodium chloride 146

soil moisture 221, 251

soil temperature 205

soil water 251

solar access 122, 399, 413, 429

solar altitude 127, 418, *480*

solar azimuth 127, *480*

solar constant 124

solar envelope 416

solar radiation 123

solar zenith angle 127, 202, 334, *480*

source area 41, 46, 131, 163, *480*

space cooling 248

space heating 260

specific heat 18, **168**, *480*

specific humidity 255, 263, *480*

spectral reflectivity 390

sprinkling 245, 249

squall line 278

stability *480*

stable 81, *480*

staggered array 23

stagnation point 83

static stability 219, 230, *480*

steam fog 265

Stefan-Boltzmann Law 123

stomata *480*

storm sewer 238, 241, *480*

storm systems 355

stormwater management 427

streamlines 83

streams 238

streets 431

stress 80

sublimation 158, 286, *480*

subsidence 73, 284, 345, *481*

subsurface climates 34

subsurface urban heat island 198, 234, *481*

subsurface temperature 205, 234

suburban 9, 423, *481*

sulphur-based smog 317

sulphur dioxide

   air pollution 296, 339

   guidelines 315

   removal 437

   smog 267, 317

   transformation 295, 302

sulphur hexafluorid 59, 365

sulfuric acid 302, 317

sun path diagram 334

supercooled 272

supersaturation 255, 271, *481*

surface cover **6**, 15, 20, 239, *481*

surface energy balance 156–196

   definition *481*

   energetics 339

   link to water balance 239

   radiation fluxes 155

   surface temperature 200

surface layer 30, **32**, *481*

surface patchiness 180

surface radiation budget **123**, *481*

surface resistance 178, 301, *481*

surface temperature 68, 200, *481*

surface urban heat island 198, **206**, 341, 390, *481*

surface water availability 180

surface water balance 239, 241, *481*

surface wetness 251

suspension 305

sustainability 10–11, **444**

swale 442

sweating 392

sweep 80, *481*

swimming pools 245

synoptic weather 30, 353, *481*

tall buildings 93, 430

terrestrial radiation 123

tethered balloon 58

tetroon balloon 58

theodolite 57

thermal admittance

   definition *481*

   energy balance 185

   heat storage **169**

   rural 221

   urban heat island 202, 213

   urban surface 38

thermal anisotropy 171, 206, *481*

thermal comfort 256, 435, *482*

thermal conductivity 18, **168**, 202, *482*

thermal diffusivity 18, **168**, *482*

thermal indices 400

thermal inertia *482*

thermal infrared radiation 123
thermal mass scheme 171
thermal properties 168
thermal similarity 61
thermal strain 394, *482*
thermal stress 358, 394, 445, 451, *482*
thermals 81, 186, 228, *481*
thermometer 46, 454
thermoregulation 386
thermoregulatory effect 164
thunderstorm electrification 291
thunderstorm 221, 250, **272**, 278
tobacco smoke 305
topography **342**, 412, *482*
tornado 355
total suspended particulate 126
Town Energy Balance model
  climate projections 383
  definition *482*
  heat storage 172
  numerical modelling 71
  precipitation 284
  urban heat island circulation 115
toxic shock 250
tracer **58**, *482*
traffic produced turbulence **91**, 310
trajectory *482*
transmission 124
transmissivity 18, 124, *482*
transpiration 241, *482*
traverses 55, 213, 370
trees 433, 442
tropical cyclone 355
troposphere *482*
turbulence **78**, *482*
turbulence intensity *482*
turbulent flux *482*
turbulent heat fluxes **175–184**
turbulent intensity 79
turbulent kinetic energy 79, 300, *482*
turbulent source area 49
turbulent transport 158
typhoons 355

ultrafine particulate matter 125, 299
ultraviolet radiation 149, *484*
Universal Thermal Climate Index 401, *482*
unstable 81, *482*
upslope fog 265
upwind–downwind differences 40
urban 'cool island' 210
urban area 9
urban atmosphere 2, 6
urban biosphere 5, **5**

urban block **19**, 431, *482*
urban boundary layer
  air temperature 205
  definition 31, *482*
  humidity 262
  observations 46
  precipitation 270
  radiation 145
  urban heat island 199
urban canopy *483*
urban canopy layer
  air pollution 303
  definition **34**, *483*
  humidity 257
  precipitation distribution 243
  radiation 128
  urban heat island 199
urban canyon
  air pollution 308
  airflow 89
  definition 19, *483*
  energy balance 187
  radiation 134
  urban design 431, 433
  urban heat island 204
urban climate 35, *483*
urban climate models 66, **71**, 172, *483*
urban climatology xix, 14, *483*
urban density 417
urban design **408–452**
urban development 3
urban dome 30, 228, 302, *483*
urban ecology 2
urban ecosystem 2, *483*
urban effects 38, 332, 445, *483*
urban element 18, 34, 156, *483*
urban energy balance 341
urban fabric *483*
urban forest 438
urban form **6**, 14, 408, *483*
urban function **6**, 25, 408, *483*
urban heat island
  air quality 448
  characterization **197–237**
  definition *483*
  formation **197–237**
  global climate change 382
  heatwaves 450
  historical studies 457
  precipitation 270
  types 198
urban heat island circulation 114, 282, 320, 348, *483*
urban hydrologic unit 242, *483*
urban hydrology **238–253**, 459

urban hydrosphere **5**, 5
urban lithosphere 2, **5**
urban metabolism
  definition **3**, **23**, *483*
  emissions 6, 20, 323
  global climate change 360
  resource use 444
  surface properties 20
urban meteorology 14, 457
urban moisture excess **257**, 263, *483*
urban pedosphere **5**
urban planning **408–452**
urban plume
  air pollution 302, 325
  boundary layer heat island 228
  definition **37**, *483*
  urban design 419
urban sprawl 8
urban structure **6**, 15, 20, **21**, *484*
urban surface 15
urban units 18, *484*
urban water management 239
urbanization **7**, 361
urban–rural differences 39
UTCI-Faisla model 401

vadose layer 239
valley 343
valley wind **347**, 349
vapour density *484*
vapour pressure 255, *484*
vapour pressure deficit 255
vasoconstriction 388, *484*
vasodilation 387, *484*
vasomotor control 387, *484*
vector *484*
vegetation
  energetics 178
  energy partitioning 182
  seasonality 192
  urban biosphere 5
  urban design 423
  urban trees 433
ventilation 427, 448
ventilation factor 316, *484*
ventilation pathway 413, 419
venting 233
vertical soundings 57
view factor 131, *484*
viscous stress 81

visibility **147**, 266–267, *484*
visible radiation 149, *484*
volatile organic compounds
  air pollution 296, 308
  definition *484*
  evaporation 299
  indoor air quality 305
  photochemical smog 318
von Kármán's constant **101**, *484*
vortex 83, 308, *484*
vorticity *484*
vulnerability 381, 448, *484*

wake 82, *484*
wake interference flow **87**, *484*
wake zone 84
walls 187
Walter-Lieth climate diagrams 337
warm-cloud processes 271
washout 126, 302, *484*
water **238–253**, 439
water ban 246
water consumption 244
water features 441
water flume 62
water management 239
water quality 5, 250
water-sensitive urban design 241, 443, *484*
water spray 451
water storage 251
water table 252
water vapour deficit 183
weather hazards 448
weekday–weekend differences 40, 291
wet deposition *484*
wetlands 5, 442, 449
wind **77–121**, 455, 457
wind chill index 400
wind comfort 401
wind shear 80, *484*
wind shelter 413
wind tunnel 44, **62**, *484*
wind vector 78
windbreak 427
World Meteorological Organization 454

zenith *485*
zero-plane displacement 17, 99, *485*
zoning 447

# Geographical Index

Abidjan, Cote d'Ivoire
  coastal flooding 382
Adelaide, Australia
  energy consumption 11
Alexandria, Egypt
  coastal flooding 382
Amsterdam, Netherlands
  coastal flooding 382
  energy consumption 11
Athens, Greece
  greenhouse gases 366
  radiation 150
Atlanta, United States
  anthropogenic heat 165
  precipitation 284
  shadows 127
  urban density 11

Baghdad, Iraq
  urbanization 7
Baltimore, United States
  carbon dioxide fluxes 371
Bangkok, Thailand
  coastal flooding 382
  energy consumption 11
  greenhouse gas emissions 367
Barcelona, Spain
  greenhouse gases 366
  street design 430
  urban density 12
  urban design 422
Barrow, United States
  irradiance 334
  sun path diagram 334
  urban heat island 341
Basel, Switzerland
  aerodynamic resistance 183
  air temperature 211
  airflow 90
  atmospheric stability 231
  BUBBLE 52, 459
  canopy layer heat island 222
  carbon dioxide fluxes 371
  energy balance 160, 190
  heat storage 175
  radiation 154
  surface temperature 201, 211

Bath, United Kingdom
  climatography 456
Beijing, China
  air pollution 41, 339
  greenhouse gases 366
  urbanization 7
Berlin, Germany
  climatography 455
  energy consumption 11
  surface temperatures 203
  temperature record 378
Bhopal, India
  disaster 295
Boston, United States
  energy consumption 11
  methane emissions 370
  temperature record 378
Boulder, United States
  carbon storage 376
  wind 345
Brasilia, Brazil
  climate diagram 338
  macroclimate 337
Brisbane, Australia
  energy consumption 11
Brussels, Belgium
  energy consumption 11
  green roof 427

Cairo, Egypt
  air pollution 294
  albedo 142
  macroclimate 335
  street orientation 416
  turbulent fluxes 179
Calgary, Canada
  urban design 446
Chicago, United States
  airflow 59
  anthropogenic heat 165
  energy balance 160, 190
  energy consumption 11
  freezing rain 280
  green roof 426
  heat storage 175
  humidity 259
  precipitation 284

tornado 355
Chongqing, China
  fog 266
Christchurch, New Zealand
  boundary layer heat island 230
  humidity 259, 263
Cologne, Germany
  subsurface heat island 235
Columbus, United States
  radiation 154
Copenhagen, Denmark
  air pollution 309
  energy consumption 11

Daegu, South Korea
  dispersion 310
Dallas, United States
  urban design 449
Delhi, India
  air pollution 305
Denver, United States
  carbon storage 376
  energy consumption 11
Detroit, United States
  energy consumption 11
Dhaka, Bangladesh
  coastal flooding 382
Donora, United States
  air pollution disaster 457
Dubai, United Arab Emirates
  radiation 154
Dublin, Ireland
  air pollution 317
  coastal flooding 450
  greenspace design 408
  street design 432
  urban form 423
  urban design 446

Edmonton, Canada
  canopy layer heat island 218
  humidity 258
Essen, Germany
  carbon dioxide fluxes 372
  ozone 320
  smog 320

Fairbanks, United States
  anthropogenic heat 165
Fermont, Canada
  urban form 413
Fez, Morocco
  human comfort 402
Florence, Italy

methane emissions 373
  urban design 440
Freiburg, Germany
  green design 430

Gaborone, Botswana
  urban heat island 342
Gothenburg, Sweden
  humidity 259
  street design 434
Graz, Austria
  airflow 349
  katabatic flows 415
  urban planning 415
Guangzhou, China
  coastal flooding 382
  urbanization 7

Hai Phong, Vietnam
  coastal flooding 382
Hamilton, Canada
  radiation 151
Ho Chi Minh City, Vietnam
  coastal flooding 382
Hong Kong, China
  anthropogenic heat 165
  energy consumption 11
  extreme winds 355
  sea breeze 95
  street design 431–2
Houston, United States
  convective clouds 290
  energy consumption 11

Indianapolis, United States
  urban plume 374

Jaisalmer, India
  urban design 432

Karachi, Pakistan
  drinking water 244
Khartoum, Sudan
  climate diagram 338
  irradiance 334
  macroclimate 337
  sun path diagram 334
Kobe, Japan
  coastal flooding 382
Kolkata, India
  coastal flooding 382
Krakow, Poland
  greenhouse gases 368

Krefeld, Germany
  humidity 260
Kumagaya City, Japan
  climatography 456

La Porte, United States
  precipitation anomaly 274
Lanzhou, China
  air pollution 349
Leicester, United Kingdom
  humidity 257
Lima, Peru
  fog 266
Linz, Austria
  climatography 456
Lodz, Poland
  canopy layer heat island 221
  carbon dioxide fluxes 372
  climate diagram 338
  energy balance 340
  humidity 260
  macroclimate 338
London, United Kingdom 94
  air pollution 267, 307, 454
  air pollution disaster 457
  airflow 119
  canopy layer heat island 218–19
  climatography 456
  ecological footprint 2
  emission inventory 323
  energy consumption 11
  flooding 357
  greenhouse gases 366, 374
  groundwater 252
  historical climate 45
  humidity 259
  piped water 247
  smog 317
  thermal comfort 412
  urban design 440
  urbanization 7
Los Angeles, United States
  air pollution 353, 457
  airflow 353
  energy consumption 11
  greenhouse gases 366
  methane emissions 364
  radiation 150
  Santa Ana wind 347
  smog 318
Ludwigshafen, Germany
  wind tunnel 64

Madrid, Spain
  greenhouse gases 366
Malaga, Spain

urban design 446
Manchester, United States
  urbanization 2
Manhattan, United States
  airflow 113
  cold front 357
  daylight access 432
  heat related mortality 395
  sea breeze 352
Marrakech, Morocco
  urban form 122
Marseille, France
  ESCOMPTE 459
  heat storage 173
Melbourne, Australia
  carbon dioxide fluxes 372
  energy consumption 11
  green roof 426
Meuse Valley, Belgium
  air pollution disaster 457
Mexico City, Mexico
  canopy layer heat island 214
  dust 439
  energy balance 191
  groundwater 252
  human respiration 368
  humidity 261
  park climate 438
  urbanization 1
Miami, United States
  climate change 382
  coastal flooding 382
  extreme winds 356
  sea breeze 95
  urban design 422
Milan, Italy
  fog 266
  street design 434
Minneapolis–Saint Paul, United States
  carbon budget 285, 375
  snowfall 284
Mito, Japan
  temperature record 377
Monterrey, Mexico
  urban heat island 342
Montreal, Canada
  anthropogenic heat 165
  boundary layer heat island
    228, 230
  carbon dioxide fluxes 372
  climate diagram 338
  energy balance 193, 340
  EPiCC 459
  irradiance 334
  macroclimate 338
  radiation 151

snow 144
sun path diagram 334
Moscow, Russia
  cloud cover 275
  macroclimate 336
  thermal comfort 412
  wind 101
Mumbai, India
  climate change 382
  coastal flooding 382
  extreme rainfall 358
Munich, Germany
  climatography 455
  fog 267
  urban design 446

Nagoya, Japan
  energy consumption 11
New Delhi, India
  groundwater 252
  thermal comfort 412
New Orleans, United States
  coastal flooding 382
New York, United States
  air pollution 457
  airflow 113, 119
  boundary layer heat island 230
  coastal flooding 382
  cold front 357
  cool roofs 426
  daylight access 432
  drinking water 244
  energy consumption 11
  freezing rain 280
  greenhouse gases 365–6
  piped water 247
  sea breeze 352
  shadows 127
  Urban Air Pollution Dynamics
    Program 457
  urban heat island 455
  urban plume 325
Nimes, France
  urban design 440

Oberhausen, Germany 405
Ogaki City, Japan
  canopy layer heat island 214
  climatography 456
Oklahoma City, United States
  airflow 112–13
  Joint URBAN 2003 459
  tornado 355
  wind tunnel 44
Orlando, United States
  canopy layer heat island 222

Osaka, Japan
  coastal flooding 382
  park design 441
Oslo, Norway
  greenhouse gases 366
Ouagadougou, Burkina Faso
  humidity 261
  urban heat island 217

Panama City Beach, United States
  airflow 86
Paris, France
  air pollution 308, 454
  airflow 119
  carbon cycling 369
  climate change 383
  climatography 455
  energy consumption 11
  greenhouse gases 366
  pavement watering 441
  urban heat island circulation 114
  urban plume 325
Perth, Australia
  climate diagram 338
  energy consumption 11
  macroclimate 338
Phoenix, United States
  carbon dioxide 370
  energy consumption 11
  haboob 332
  soil respiration 368
Pietermaritzburg, South Africa
  airflow 119
Portland, United States
  dispersion 310

Reykjavik, Iceland
  anthropogenic heat 165
Rhone river valley 347
Richmond, Canada
  surface heat island 208
Riyadh, Saudi Arabia
  urban plume 325
Rocky Mountains 346
Rome, Italy
  human comfort 403
Rotterdam, Netherlands
  coastal flooding 382
  greenhouse gases 366
Rouyn, Canada
  anthropogenic water vapour 247

Sacramento, United States
  energy balance 182
  humidity 261
  urban heat island 451

Sacramento, United States (cont.)
  urban rural contrast 194
Salt Lake City, United States
  anthropogenic heat 165
  cold air pooling 346
  dispersion 349
  tracer release 58
San Francisco, United States
  climatography 456
  energy consumption 11
  fog 266
  wind comfort 416
Santiago, Chile
  cold air pooling 346
Sao Paulo, Brazil
  anthropogenic heat 165
  greenhouse gases 366
  sea breeze 352
  urban form 14
Sapporo, Japan
  thermal admittance 171
Seoul, South Korea
  humidity 262
  urban design 442
Shanghai, China
  climatography 458
  coastal flooding 382
Shenzhen, China
  coastal flooding 382
Singapore
  anthropogenic heat 165
  carbon dioxide fluxes 372
  climate diagram 338
  energy balance 340
  energy consumption 11
  irradiance 334
  macroclimate 337
  street design 434
  sun path diagram 334
  thermal comfort 412
  urban design 439
  urban form 156
Southern California
  sea breeze 353
St. Louis, United States
  aerosol 146
  airflow 111, 116
  boundary layer heat island 230, 233
  canopy layer heat island 218
  cloud cover 285
  hail 279
  heatwave 385
  humidity 263
  METROMEX 458
  urban heat island circulation 115

St. Petersburg, Russia
  temperature record 378
Stockholm, Sweden
  energy consumption 11
  temperature record 378
Stuttgart, Germany
  urban planning 411
Swindon, United Kingdom
  carbon dioxide fluxes 372
Sydney, Australia
  anthropogenic heat 164 5
  energy consumption 11

Tahoua, Republic of Niger
  climate-sensitive planning 412
Taipei, Taiwan
  extreme winds 355
Tel Aviv, Israel
  sea breeze 95
  urban design 437
Thames Barrier 357
Tianjin, China
  coastal flooding 382
Tokyo, Japan
  anthropogenic heat 166
  climate diagram 338
  cloud cover 275
  coastal flooding 382
  energy balance 160, 190, 340
  energy consumption 11
  greenhouse gases 366
  heat storage 175
  humidity 262
  macroclimate 338
  sea breeze 95, 351–2
  temperature record 377
  tornado 355
  urban heat island circulation 115
  urban river 238
  urbanization 7
  visibility 148
  wind 77
Toronto, Canada
  energy consumption 11
  temperature record 378
Toulouse, France
  anthropogenic heat 165
  CAPITOUL 459
  radiation 150
Trieste, Italy
  wind 345
Tucson, United States
  humidity 261
Turin, Italy
  fog 266

Ulaanbaatar, Mongolia
  climate diagram 338
  macroclimate 339
Uppsala, Sweden
  canopy layer heat island 214, 218, 226
  climatography 456
Urumqi, China
  airflow 350

Vancouver, Canada
  canopy layer heat island 217, 221
  carbon cycling 368
  carbon dioxide fluxes 372
  energy balance 160, 186, 189–90
  EPiCC 459
  fog 254
  garden irrigation 245
  heat storage 174
  land-cover change 252
  radiation 152
  street design 435
  surface heat island 206
  thermal admittance 221
  urban boundary layer 31

  urban tree canopy 436
  water balance 246, 249
Veracruz, Mexico
  urban heat island 342
Vienna, Austria
  climatography 456
  energy balance 186
  energy consumption 11

Washington DC, United States
  airflow 119
  energy consumption 11
  rain garden 450
Winnipeg, Canada
  canopy layer heat island 214
  subsurface heat island 236
Wroclaw, Poland
  urban heat island 356

Zaporozhye, Ukraine
  CAENEX 458
Zurich, Switzerland
  energy consumption 11

Printed in the United States
By Bookmasters